Descartes-Agonistes

STUDIES IN HISTORY
AND PHILOSOPHY OF SCIENCE

VOLUME 27

For further volumes:
http://www.springer.com/series/5671

John Schuster

Descartes-Agonistes

Physico-mathematics, Method & Corpuscular-Mechanism 1618-33

Springer

John Schuster
Campion College
Sydney, NSW, Australia

Unit for History and Philosophy of Science
University of Sydney
Sydney, NSW, Australia

ISSN 0929-6425
ISBN 978-94-007-4745-6 ISBN 978-94-007-4746-3 (eBook)
DOI 10.1007/978-94-007-4746-3
Springer Dordrecht Heidelberg New York London

Library of Congress Control Number: 2012950907

Printed on acid-free paper

Springer is part of Springer Science+Business Media (www.springer.com)

Preface

This work is a contribution to the historical literature both on Descartes and on the Scientific Revolution, particularly the crucial first half of the seventeenth century. It should not be confused with my earlier work on Descartes in my 1977 Princeton dissertation. Over the years I have seen that text cited a fair few times. I even recently encountered a colleague who reported having seen it referred to in print as some sort of 'underground classic'. But rather than have readers cite that work, I would today prefer that they instead contemplate it as a vestige of how some history of science theses were constructed in departments of history in the 1970s.

Between 1981 and 1999 I published almost no first order research on the natural philosophical career of Descartes. The exception was my suite of papers (one in collaboration with Evelleen Richards) on how to model grand doctrines of method (including Descartes') as 'mythic speech'. That research was stimulated by the post-Kuhnian sociology of scientific knowledge of that period. In those years, beyond the work on method, I was more concerned with the historiography of the Scientific Revolution; problems of Bachelardian/Kuhnian theorizing of the rise of experimental fields; and, with the demands of course and degree structure design in history and philosophy of science, under the compulsion of which I eventually wrote two open access, introductory textbooks, one of which was recently revised for translation and publication in Mandarin.

In the late 1990s, initially prompted by suggestions and invitations from Stephen Gaukroger, I returned to Cartesian studies in the history of science. I was involved in the editing of two thematic collections about Cartesian science and natural philosophy, one with Gaukroger and John Sutton, the other with Peter Anstey. More importantly, since 2000 I have published a series of papers concerning Descartes' optics, his hydrostatics (with Stephen Gaukroger), his vortex celestial mechanics, and (with Judit Brody) his previously little appreciated strategies of pro-Copernican systematisation in the *Principia philosophiae*. My 'late twentieth century' work on the structure, periodization and process of the Scientific Revolution has informed this work, along with another historiographical category that has been attracting attention in recent years. That is 'physico-mathematics', an actor's term in the later sixteenth and early seventeenth century, first brought to the notice of historians of science in an important way by the work of Peter Dear. Indeed, this category—decoded in the early work of Descartes and also elaborated as an historiographical theme—plays a central role in the present text, because this book is largely concerned with the trajectory of Descartes between 1618 and 1633: that is, from being a physico-mathematician who was seeking piecemeal corpuscular-mechanical grounding to his work, to becoming a corpuscular-mechanical systematiser with recognizable physico-mathematical conceptual stitches holding together large parts of the system. However, unlike my earlier approach to Descartes, in the present work that trajectory is now extended to include the *Principles of Philosophy*. There, based on my most recent research, in collaboration with Judit Brody, a novel and perhaps surprising interpretation is offered about Descartes' strategies of systematization and his enrolment of novel matters of fact—about sunspots, *novae* and variable stars—in the service of the very daring version of realist Copernicanism that text offers to the discerning reader, then and now.

In sum, then, my recent papers on Descartes, and my earlier work on his method, form the backbone of the present book, while those papers themselves draw upon my wider concerns with explaining the Scientific Revolution and with articulating and putting to work the category of physico-mathematics. Additionally, there are deployed in the present text some bits and pieces mined from my earliest work on Descartes. In almost all cases, however, that mining is not crude, nor are the extracts left unrefined. The historiographical categories and insights contained in my recent papers have determined how slices of my earliest work have been selected, shaped and placed in the present argument.

Accordingly, themes from my original dissertation which have not been topics of development in my intervening published work are revised extensively, if they reappear here. For example, in 1977 I had an overly optimistic view of the thesis concerning the 'crisis of the seventeenth century' and its relation to the rhythm of the Scientific Revolution (despite some terminological strictures against Popkin and Rabb mentioned at the time). I then passed through a long stage of deconstructivist scepticism about the entire notion, expressed only in conference and seminar papers, but not in print. However, I now return to it in modified form. It plays through my view of the 'inflection' (rather than rupture) in Descartes' career, identity structure and agenda, in 1629–1630, between the abandonment of the unfinished *Rules for the Direction of the Mind* and the emergence of the project of his first system of natural philosophy, *Le Monde*. Extending this approach in the present work, I endeavor to take stock of Descartes' self-understanding of his role and agenda at several key moments in his natural philosophical career; for example, when in 1618 he follows Isaac Beeckman into the alluring if under-determined realm of physico-mathematics; or when, soon after, he diverts along the ultimately delusional paths of *mathesis universalis* and universal method; or when, nearing completion of *Le Monde*, he momentarily oscillates between, on the one hand, still craving an '*a priori* science' and, on the other, bemoaning his lack of a 'complete natural history' of facts, necessary to ground his effort; or when he decides, in the *Principles of Philosophy*, vastly to outbid and out theorize even his younger self in the game of promoting a radical, infinite universe, realist Copernicanism.

All this further points to a characteristic and goal of the present book which I would hope all readers try to bear in mind as they proceed. This work, like almost all of my research, is deeply imbued with historiographical claims, insights and advice. The story, or at least my story, of the young Descartes' trajectory in natural philosophy, method and physico-mathematics could not have been told without the guidance, framing and explanatory fruits of my own career-long concern with historiographical problems and historical category formation in the history of science. I learned this style of problematising, and category formation and testing, from my initial mentors in the history of science, and early modern social and economic history—Tom Kuhn, Mike Mahoney, Ted Brown, Ted Rabb and Lawrence Stone. Some people discern in my historical style the overhang of a youth misspent in trying to become a physicist (whilst obsessively reading history of all kinds). It was this unpromising material on which these maestros of the Princeton History Department tried to work.

Those history of science colleagues who in the long time since then have been most important in influencing my work have been those who have in one way or another prompted my concerns along these lines, rather than, say, pointing out this or that Cartesian detail. Amongst these historiographical benefactors I would list Jerry Ravetz, Bob Westman, Floris Cohen, Wilbur Applebaum, Keith Hutchison, John Henry, Simon Schaffer, Richard Yeo, Peter Harrison and Peter Dear. But pride of place goes to Stephen Gaukroger, since the late 1970s my Cartesian sparring partner, occasional collaborator and constant exemplar of scholarly application and distinction.

My concerns with explanatory and interpretive resources and categories have always made me a consumer of cognate work in sociology of science, sociology of knowledge and other areas of history of science which might confer some heuristic guidance upon my own deliberations. These are patent in the present work. Over the years the most important contributions to my own 'concept formations' in these areas have come from the following (the last of whom I admit I never had the pleasure of meeting in person): Stephen Shapin (despite some tangential but overt differences over historical details), Barry Barnes, Trevor Pinch and Pierre Bourdieu. I also acknowledge the work of history of science/medicine/technology colleagues who have one way or another concretely affected or deflected my historiographical concerns: Ivan Crozier, David Mercer, Adam Lucas, Jan Golinski, Larissa Johnson Aldridge and Luciano Boschiero amongst former undergraduate or doctoral students of mine, as well as former or present colleagues, Evelleen Richards, Alan Chalmers, Barry Brundell, David Miller, Katherine Hill (Neal), Ofer Gal and Victor Boantza. When it comes to highly technical scientific issues, or tangled textual matters, rather than overtly historiographical questions, my debts are to such past and present history of science (or mathematics) masters as Alan Gabbey, Alan Shapiro, Jed Buchwald, Henk Bos, Eric Aiton and Bruce Eastwood.

For all the above reasons, the present book is not well categorized as history of philosophy. I was not professionally trained as an historian of philosophy, let alone as a philosopher as such. Of course, from the beginning I have benefitted from philosophers' research on Descartes, including even the oldest French Cartesian scholarship from the turn of the last century: from Boutroux to Brunschvicg; from Gilson to Gueroult, from Mouy to Marion. To these we can add the burgeoning ranks of Anglophone historians of early modern philosophy who have worked on Descartes. In my case the most notable help has come from the works of Dan Garber (supplemented in person in his case), Roger Arieu, Des Clarke, Peter Machamer, Gary Hatfield and Denis Sepper. But, the philosopher who most shaped my approach to Descartes, and indeed to parts of history of science in general, was the late Gerd Buchdahl, a model colleague in the distant past, and, though he would have denied it, a master of historiographically relevant conceptualization, *malgré lui*.

Nevertheless, the exception of Buchdahl rather proves the rule, for I do not, in general, find amongst the cohorts of professional historians of early modern philosophy treatment of the same sorts of properly historical questions—relevant to micro as well as macro/comparative studies—that one finds amongst at least some historians of science, especially those trained in, and concerned with, other areas of general history. I very often benefit from the technical insights and wrangling of historians of philosophy, but I cannot think of more than a small number of occasions when an historiographical insight or problem of some import to this project has been stimulated by such a practitioner. Nor do I expect that the kinds of categories explored and deployed in this work in the service of an historical (partial) biography in context will be of particular interest or relevance within the empire of history of philosophy. However, exceptions will be welcomed and engaged, as they always have been.

A simple glance at the contents of Chap. 2, my initial statement of key concepts and historiographical issues that will be deployed and articulated throughout the work, will show this. In compensation I expect that the sort of categories and models set up there will interest, or usefully incense, intellectual and social historians and historians of science concerned with how nature-knowledge disciplines develop and interact with each other and their wider contexts. In short, while I hope historians of philosophy read (at least parts of) this book, I am speaking as an historian—with wide and eclectic interests in that discipline beyond mere, say, history of physics or natural philosophy—to historians, in a language of questions, concepts and categories, many of my own making or revision, which remains unremittingly historical.[1]

In closing, I must acknowledge one philosopher whose work, more than any other, haunts my present effort—and my earlier one. That scholar is the redoubtable Norman Kemp Smith, whose *New Studies in the Philosophy of Descartes* (1952) gave me my first clues about how to deal historically with the young Descartes. I picked up that work in 1972 on the prompting of Tom Kuhn, who had written a positive review of it in *Isis* in 1955, when he himself was contemplating a scientific biography of Descartes. Many reading this Preface will know that Kemp Smith had

[1] Any historian of philosophy reading this might refer to my short review of Ted McGuire and Peter Machamer's recent brilliant rational reconstruction of the 'mature Descartes' in *Descartes' Changing Mind* (2009) to see how I envision the difference between their work and any instance of a suite of possible historical approaches, whilst granting the complete legitimacy of their terms of discourse. *Renaissance Quarterly* 63 (2010): 579–581.

two attempts at the study of Descartes. His *New Studies* post dated his original *Studies in the Cartesian Philosophy* by 51 years. Perhaps Descartes exercises a certain compulsion upon some of his scholars. After all, I am here presenting my second attempt at 'a Descartes for historians of science'—and students of history in general—following a gap of a mere 35 years. I say this not in any way to equate my efforts with those of Kemp Smith, but simply to acknowledge the effect of his second book on me at my most formative moment, and to register the fact that such Cartesian obsession, with which I might well be charged, is neither unprecedented, nor necessarily unfruitful.

Mount Keira John Schuster

A note on the use of this book: A glance at the table of contents of this volume and a reading of Sect. 1.4, 'Overview of Argument' will show that, despite its length, this book embodies a tightly knit, and highly iterative, argument. The key categories and frames of interpretation, as well as the central questions addressed, are treated intensively, rather than diffusely. Core questions and concepts recur during the course of the narrative with increasing articulation and contextualization. Two of the chapters, Chaps. 2 and 6, are largely devoted to exposition of interpretative concepts and frames required at those stages of the argument. There is a high level of internal cross referencing both within chapters and between sections and sub-sections of differing chapters. Use of the table of contents, which has over 160 entries, and the extensive internal cross referencing, provide the best reader's map to both the pattern of the argument and its underlying conceptual architecture. These resources also make perfectly clear where key figures other than Descartes enter the story. Additionally, many readers will be approaching this volume in its digital manifestation, rendering it easy to design one's own complete searches for topics, persons and categories.

All these facts conduced to the decision not to provide a standard index, a piece of apparatus that tends to enforce a particular and atomized picture of the contents of a book. That may be appropriate to factually exhaustive and circumstantially rich discourses. But, it does not suit works such as this, which attempts a detailed narrative *cum* explanation of Descartes' activities in a limited number of intellectual disciplines, the narrative/explanation being shaped by, and iteratively articulating, a number of historiographical concepts and frames of interpretation, some of which began life as contemporary actors' categories. Readers should note that internal cross references to specific sections of this book (rather than to entire chapters), always use the full numerical code for the section in question, as it appears in the Table of Contents: the first numeral always denotes the chapter, the second the section and the third, if needed, the sub-section. Thus, a cross reference to 'Sect. 8.2.1' refers the reader to Chapter 8, Section 2, Sub-Section 1. This protocol applies to cross references in the text and in footnotes, and even to cross references within a given chapter.

Contents

1 Introduction: Problems of Descartes and the Scientific Revolution 1
1.1 Prologue: The 'Young' and the 'Mature' Descartes, Natural Philosopher 1
1.2 Descartes and the Historians of Science 3
1.3 Key Pitfalls (and Opportunities) Facing Descartes' Biographers (Even Authors of Quite Truncated Biographies) 8
1.3.1 The Problem of Method and Its Texts: *Regulae* and *Discours* 8
1.3.2 The Problem of Descartes the Natural Philosopher, and of Natural Philosophy as a Wide and Dynamic Field of Discourse and Contention 10
1.3.3 Scientific Biography and the Historiography of Science 13
1.4 Overview of the Argument 19
References 26

2 Conceptual and Historiographical Foundations—Natural Philosophy, Mixed Mathematics, Physico-Mathematics, Method 31
2.1 Jesuit neo-Scholasticism for the *noblesse de robe* 31
2.2 In Search of Proper Categories and Angle of Attack 35
2.3 Constructing the Category of Natural Philosophy, Part 1—Natural Philosophizing as Culture and Process 37
2.4 Some Heuristic Help: Modeling Modern Sciences as Unique, Agonal Traditions in Process 44
2.5 Constructing the Category of Natural Philosophy, Part 2: The Dynamics and Rules of Contestation of Natural Philosophizing 49

2.5.1 Articulation on Subordinate Disciplines: Grammar and Specific Utterance ... 50
2.5.2 Find or Steal Discoveries, Novelties or Facts, Including Experimental Ones ... 53
2.5.3 Bend or Brake Aristotle's Rules About Mathematics and Natural Philosophy: The Gambit of 'Physico-Mathematics' ... 56
2.5.4 "Hot Spots" of Articulation Contest: Additional Causes and Effects of Heightened Turbulence in the Field of Natural Philosophizing ... 59
2.5.5 Modeling System Construction and Contestation—The 'Core', 'Vertical' and 'Horizontal' Dimensions of a Natural Philosophical System ... 62
2.5.6 The Mechanics of Responding to 'Outside' Challenges and Opportunities ... 65
2.6 The Special Status of the Problem of Method ... 70
2.7 Phases and Stages in the 'Scientific Revolution' Seen as an Unfolding Process in the Field of Natural Philosophizing, with Its Attendant Articulations to Other Domains ... 77
2.8 Looking Forward—What Kind of Natural Philosopher/Physico-Mathematician Was René Descartes? ... 88
References ... 93

3 'Recalled to Study'—Descartes, *Physico-Mathematicus* ... 99
3.1 Introduction ... 99
3.2 Beeckman: Mentor and Colleague in Physico-Mathematics and Natural Philosophy ... 104
3.2.1 Corpuscular-Mechanical Natural Philosophy and the Values of the Practical Arts ... 104
3.2.2 Beeckman's Causal Register, Principles of Mechanics and Version of Physico-Mathematics ... 108
3.3 Exemplary Physico-Mathematics: The Hydrostatics Manuscript of 1619 ... 112
3.3.1 Stevin, Archimedes and the Hydrostatic Paradox ... 113
3.3.2 The Hydrostatics Manuscript [1] The Micro-Corpuscular Reduction ... 114
3.3.3 The Hydrostatics Manuscript [2] The Force of Motion ... 121
3.4 What's the Agenda: Descartes' Radical Form of Physico-Mathematics ... 124
3.5 The Physico-Mathematics of Natural Fall ... 128
3.5.1 Introduction—The Study of Fall as [Abortive] Physico-Mathematics ... 128
3.5.2 Beeckman's Problem, and His Version of Descartes' Solution ... 130

3.5.3 Descartes' Solution—Triumphs and Pitfalls of a Physico-Mathematics of Fall 134
3.5.4 How and Why Descartes Hit a Pitfall 139
3.5.5 The Physico-Mathematics of Fall Stalls—Too Many Laws, Too Many Causes, No Measurements 145
3.6 A Physico-Mathematical Foray into Optics (1620) 153
3.7 Conclusion: Options, Pitfalls and Trajectories 163
References 165

4 Descartes *Opticien*: The Optical Triumph of the 1620s 167
4.1 Genealogical Detective Work—Hints, Clues and the Problematical Text of the *Dioptrique* 167
4.2 Cartesian Dynamics in *Le Monde* 169
4.3 Making Sense of the Proofs of the Laws of Reflection and Refraction in the *Dioptrique* 173
4.4 Descartes' Dynamical Premises: Demonstrative Efficacy and Empirical Weakness 178
4.5 Descartes' Route to the Law of Refraction 1619–1627 184
4.5.1 The Mydorge Letter of 1626/1627 184
4.5.2 Lens Theory and the Date of the Material in Mydorge's Letter 186
4.5.3 Traditional Geometrical Optics and the Discovery of the Cosecant Form of the Law 188
4.6 The Dynamical Premises for the Deduction of the Sine Law of Refraction: Their Pre-History and History 1618–1629 190
4.7 The Mechanical Theory of Light 1620–1628 194
4.7.1 Expository Strategy and Working Distinctions 194
4.7.2 Reprise—The Optical Fragment of 1620 195
4.7.3 Light as an Instantaneously Transmitted Mechanical Impulse 1626–1628 197
4.7.4 Light as Mechanical Impulse and the Explanation of the Law of Refraction 1626–1628—The Balance Beam Model 199
4.8 Full Circle: Cartesian Dynamics, Optics and the Tennis Ball Model 1628–1633 204
4.8.1 The Exemplar for Descartes' Laws of Dynamics in His Physico-Mathematical Optics 204
4.8.2 In a Spin Over Tennis Balls and *Boules* of Second Element: Cartesian Dynamics, Optics and the Problem of Color 209
4.9 Grist for the Method Mill: Method and Optics in Rule 8 of the *Regulae ad directionem ingenii* 215
4.10 Conclusion: Looking Forward—Mathematics and Method: 1618–1629 220
References 221

5 Analytical Mathematics, Universal Mathematics and Method: Descartes' Identity and Agenda Entering the 1620s ... 225
5.1 Introduction: The Struggle Over Mathematics, Universal Mathematics and Method ... 225
5.2 The Universal Mathematics of 1619: Rule 4 of the *Regulae* ... 230
5.3 Reading Rule 4: Method and Universal Mathematics ... 235
5.4 Straining at the Classical Bit: Descartes' Early Work in Analytical Mathematics ... 239
5.5 Genesis and Dating of Universal Mathematics ... 242
5.6 The Core of Descartes' Method Discourse in the Early *Regulae* ... 248
5.7 The Making of Cartesian Method–Talk, Winter 1619–1620 ... 257
5.8 Conclusion: Descartes' Unfolding Agendas and Identities 1618–1620 ... 260
References ... 262

6 Method and the Problem of the Historical Descartes ... 265
6.1 The Way Forward: Between Naïve Belief and Pure Debunking ... 265
6.2 The Cult of Method in Descartes Studies ... 266
6.3 Descartes' Method as Mythic Speech: Where 'Myth' Is Not a Colloquial Term of Abuse ... 270
6.4 The Failure of Adequate Redescription: An Example of Descartes Attempting to 'Methodologize' a Field of Inquiry ... 273
6.5 The Structural Levels and Underlying Metaphors in Descartes' [or Anybody's] Method Discourse ... 280
6.6 The First Two Structural 'Effects': Adequate Redescription' and 'Application' ... 283
6.7 The Third and Fourth 'Effects': The 'Unity' and 'Progress' of a Method Discourse ... 286
6.8 The Rhetorical Functions of Cartesian (and Other) Method Discourses ... 293
6.9 Rethinking Method and the Career of Descartes ... 297
6.9.1 The Original Inscription of Descartes' Method: Bricolage, Self-Deception and Self-Definition ... 297
6.9.2 The Failure of the *Regulae*, the Birth of the System and the Problem of the Cynical *Discours de la méthode* ... 299
References ... 303

7 Universal Mathematics *Interruptus*: The Program of the Later *Regulae* and Its Collapse 1626–1628 ... 307
7.1 Introduction—Toward the Renewed Project of *Mathesis Universalis* in the Later *Regulae* ... 307
7.2 Rule 8: The Emergence of the Project of the Later *Regulae* ... 311

7.3 Rule 12: From 'Most Splendid Example' to the Articulation of the Machinery of Universal Mathematics 314
7.4 Rules 14–18: The Machinery of Universal Mathematics 320
7.5 The Structure of Universal Mathematics in the Later *Rules* and Its Legitimatory Functions 328
7.6 The Instability of the Later *Regulae* and the Beginnings of the Origins of Cartesianism 334
7.6.1 Discursive Corpuscular–Mechanism in Tension with Genuine Mathematization and the Aims of the O-P-P Nexus 334
7.6.2 Ambushed by the Unexpected Manifestation of the Problematic of Modern (Cartesian) Epistemology 339
7.6.3 Analytical Mathematics Works and Is Useful, But Resists Mapping onto the Legitimatory O-P-P Nexus 343
7.7 Conclusion—The Project of the Later *Regulae* and the Inflection of Descartes' Agenda and Identity Toward Systematic Natural Philosophy and Metaphysical Grounding 345
References 347

8 Reinventing the Identity and Agenda: Descartes, Physico-Mathematical Philosopher of Nature 1629–1633 349
8.1 The Problem of Descartes' Career 'Inflection Point' and How to Approach It 349
8.2 Fundamental Intellectual Agendas and Projects 1629–1633 352
8.2.1 The Emergence of Cartesian Metaphysical Dualism 353
8.2.2 Some Voluntarist Theology and Its Strategic Uses 360
8.2.3 Plenist or Holistic Realism 373
8.3 Events and Interactions Partially Shaping the Motives for and Content of *Le Monde* 384
8.3.1 Abandonment of the Later *Regulae* Actually the Most Important 'Event' of All 384
8.3.2 The Chandoux Episode and Relations with Cardinal Bérulle: Method and/or Metaphysics and the Defeat of Scepticism 385
8.3.3 Challenge of Renewed Interaction with Beeckman 390
8.3.4 The Galileo Affair and Its Perceived Meanings 394
8.4 The Chronology of *Le Monde* 399
8.4.1 Introduction: Sliding and Tinkering Toward a System of Natural Philosophy 399
8.4.2 Optics—Physico-Mathematical, Mixed and Practical 400
8.4.3 'Tous les Phainomenes Sublunaires'—The Parhelia: Grasping Opportunities and Pregnant Problems 404

8.4.4 *'Tous les Phaenomenes de la Nature'—Le Monde Begins to Crystallize* ... 405
8.4.5 Reprise: In a Spin Over Light and Color ... 407
8.4.6 Descartes at Work, November 1629 to April 1630 ... 408
8.4.7 Spring to Autumn 1630—Writer's Melancholy, Visioning, Spinning and Dealing with Beeckman ... 410
8.4.8 Organizing the System and Himself—Investigations, Decisions, Hesitations ... 413
8.5 An Exercise in Counter-Factual History: If the *Regulae* Had Not Failed ... 416
8.6 Aspiration, Identity and Strategy at the Birth of *Le Monde*: Between Natural History and A Priori 'Science' ... 418
References ... 421

9 Reading *Le Monde* as Pedagogy and Fable ... 425
9.1 Introduction ... 425
9.2 The Ground Plan of *Le Monde* ... 426
9.3 The Common Sense of Corpuscular-Mechanism pour les honnêtes hommes—The Opening Chapters of *Le Monde* ... 428
9.4 Why the Fable of the Mechanistic World? ... 440
9.5 Working Out the Fable: Chapters 6–8 of *Le Monde* ... 442
9.5.1 Cosmogony, Matter–Extension and the Introduction of Motion and Its Laws ... 442
9.5.2 The Laws of Nature ... 444
9.5.3 Vortex Formation, Stability Principle and [Re-]Introduction of the Elements ... 446
References ... 450

10 'Waterworld': Descartes' Vortical Celestial Mechanics and Cosmological Optics in *Le Monde* ... 453
10.1 Introduction: Uncommon Vortices ... 453
10.2 Descartes' Vortical Celestial Mechanics in *Le Monde* ... 455
10.2.1 Charitable Hermeneutics—Principles and Aims ... 455
10.2.2 Size and Speed Distributions of Vortex Corpuscles and the Role of Central Stars ... 458
10.2.3 Locking and Extruding—Unpacking the Technical Core of the Vortex Celestial Mechanics ... 464
10.2.4 Genealogical and Systematic Dimensions of the Vortex Celestial Mechanics ... 469
10.3 Beeckman's Cosmic Balancing Acts—The Last Genealogical Step to the Vortex Mechanics ... 471
10.4 Descartes' Celestial Vortex Mechanics as a 'Science of Equilibrium' ... 475
10.5 Applying the Vortex Mechanics to Local Fall, the Moon and the Tides: An Exercise in Charitable Interpretation ... 478

10.5.1 Aims of This Section and the Strategy of Reading 478
10.5.2 Local Fall as Downward Extrusion (In Ultimate Search for an Orbit—Like the Moon's) 479
10.5.3 Vortex, Earth, Sea and Moon: Descartes' Theory of the Tides 484
10.5.4 Summing Up the Vortex Mechanics and Its Consequences on Charitable[1] Reading 486
10.6 Some Intricacies of the Theories of Local Fall and the Moon—Bearing the Imprint of the Genealogy of Physico-Mathematics 486
10.6.1 The Intricacies of the Theory of Local Fall and Weight 487
10.6.2 Some Intricacies of Descartes' Moon Theory 495
10.7 Cosmological Theory of Light in Relation to Celestial Mechanics and the Style of Cartesian Physico-Mathematics in Corpuscular-Mechanical Model 499
10.7.1 Reprising the Dynamics and Laws of Nature 499
10.7.2 The Cosmological Theory of Light as Tendency to Motion Transmitted Instantaneously Through the *Boules* of the Vortex 501
10.7.3 The Physico-Mathematical Style and Protocols and the Celestial Optics 507
10.7.4 The Physico-Mathematical Hydrostatics of 1619 as Precise Exemplar for the Celestial Optics 509
10.8 Matching of Evidence: The Appearance of Comets in Relation to the Celestial Mechanics, the Cosmological Optics and the Style of Cartesian Physico-Mathematics in Corpuscular-Mechanical Mode 515
10.9 Looking Forward: *Le Monde* as Natural Philosophical System and Gambit 521
References 522

11 *Le Monde* as a System of Natural Philosophy and Gambit in the Field 525
11.1 *Le Monde* as Competitive Gambit in the Natural Philosophical Field 525
11.2 *Le Monde* as a System: 'Core', 'Horizontal' and 'Vertical' Dimensions and 'System-Binding' Moves 527
11.2.1 Horizontal Analysis of the System in *Le Monde* 529
11.2.2 'System–Binding' Moves on the Horizontal Plane 532
11.2.3 The Vertical Dimension of Systematicity 533
11.3 Remediating Problems and Taking the Best Steps Forward in the *Principia* 538
References 540

12 **Cosmography, Realist Copernicanism and Systematising Strategy in the *Principia Philosophiae*** ... 543
12.1 More than Remediation: the *Principia* as a Triumph of Novel and Daring System-Binding ... 543
12.2 Cosmogony, Cosmology and Cosmography: Key Categories and Insights ... 544
12.3 Matter and Element Theory in *Le Monde* and the *Principia philosophiae* ... 549
12.4 Inter-Vortical Phenomena ... 557
12.5 Co-opting and Re-framing Gilbert's 'Cosmic' Magnetism ... 558
12.6 Claims About Sunspots from Galileo and Scheiner to Descartes ... 561
12.7 Gaining Strategic Leverage: Sunspots as Explananda and Explanans in the *Principia Philosophiae* ... 566
12.8 Claimed Matters of Fact About Novae and Variable Stars Before Descartes ... 568
12.9 Extending the Strategy: Seizing upon Novae and Variable Stars in the *Principia Philosophiae* ... 571
12.10 Raising the Cosmographical Stakes: Genealogy of the Earth and All Other Planets in All Other Systems ... 574
12.11 Radical Realist Copernicanism and the Grand Cosmographical Gambit ... 576
12.12 Conclusion: Cosmographical System and Strategy in the *Principia*, the Culmination of Descartes' Natural philosophical Trajectory ... 582
References ... 586

13 **Conclusion: The Young and the Mature Descartes *Agonistes*** ... 589
13.1 Coda: Descartes' 'Youthful' Struggles Reconsidered ... 589
13.2 Epilogue: The Mature, Public and Published Descartes *Agonistes* ... 593
References ... 600

Appendix 1: Descartes, Mydorge and Beeckman: The Evolution of Cartesian Lens Theory 1627–1637 ... 603
A.1 Introduction ... 603
A.2 Mydorge's Refractive Instrument: Cosecants, Not Sines ... 604
A.3 Mydorge's Synthetic Propositions 3 and 4 on Anaclastic Surfaces: An 'Antique' Version of the Sine Law ... 605
A.4 Relating Mydorge's Propositions 3 and 4 to Descartes' Analogues in the *Dioptrique*: From 'Antique' to 'Natural' Representation of the Sines, Thanks to Isaac Beeckman in October 1628 ... 607
A.5 Decoding Mydorge's Proposition 5: The Cosecant Form Leads to the 'Discovery' of the 'Antique' Sine Form Then Used Synthetically in Propositions 3 and 4 ... 610

A.6 A Reconstruction of Descartes and Mydorge's First Analysis of the Anaclastic Problem, with Cosecant Law of Refraction to Hand 613
A.7 The Kramer-Milhaud Thesis: Discovering the Law of Refraction by Analysis of the Anaclastic Problem 615
A.8 Conclusions 617
References 619

Appendix 2: Decoding Descartes' Vortex Celestial Mechanics in the Text of *Le Monde* 621

A.6 A Reconstruction of Descartes and [illegible]
First Analysis of the [illegible]
[illegible]

Appendix [illegible]
[illegible]

Chapter 1
Introduction: Problems of Descartes and the Scientific Revolution

1.1 Prologue: The 'Young' and the 'Mature' Descartes, Natural Philosopher

In this book, I attempt to reconstruct key aspects of the early career of Descartes from 1618 to 1633; that is, up through the point of his composing his first system of natural philosophy, *Le Monde*, in 1629–1633. I focus upon the overlapping and intertwined development of Descartes' projects in physico-mathematics, analytical mathematics, universal method, and, finally, systematic corpuscular-mechanical natural philosophy.[1] My concern is not simply with the conceptual and technical aspects of these projects; but, with Descartes' agendas within them, and his construction and presentation of his intellectual identity in relation to them. Hence, my subject matter is selective and ultimately limited in relation to the potential field of concerns in which intellectual historians and historians of science and philosophy might place Descartes, or even the young Descartes. Nevertheless, as explained below in Sect. 1.3.3, my focus on technical projects, agendas and identity well fits the scope and aim of scientific or intellectual biography.

On my analysis, Descartes' technical projects, agendas and senses of identity all shift over time, entangle and display great successes and deep failures. This motivates my choice of title, 'Descartes *Agonistes*': In all three dimensions—projects, agendas and identity concerns—the young Descartes struggles and contends, with himself and with real or virtual peers and competitors between 1618 and 1633, as he morphs from a mathematically competent, Jesuit–trained graduate in neo-Scholastic Aristotelianism to aspiring prophet of a firmly systematized corpuscular-mechanism,

[1] Physico-mathematics is defined in a preliminary way below, in Sect. 1.3.2, and more fully in Sect. 2.5.3; Descartes' early projects within the intended scope of this discipline are explored in Chap. 3. Also see below, Sect. 1.4 'Overview of Argument', for more on the way this category will recur, be studied and explored throughout this work.

J. Schuster, *Descartes-Agonistes: Physico-mathematics, Method & Corpuscular-Mechanism 1618-33*, Studies in History and Philosophy of Science 27,
DOI 10.1007/978-94-007-4746-3_1, © Springer Science+Business Media Dordrecht 2013

passing through stages of being a committed *physico-mathematicus*, advocate of a putative 'universal mathematics', and projector of a grand methodological dream.

I argue that Descartes' evolving program in physico-mathematics was the central, but not exclusive element in this complicated story, and thereby I indicate how the more usual tales of Descartes' development can and must be retold around this axis. Unlike my previous work, the present book establishes that early on Descartes was very far from being interested in constructing a systematic natural philosophy, and that his commitment to corpuscular mechanism, though real, was more peripheral than central to the evolving tangle of his projects and concerns.

Indeed, it was only through the maturation of his physico-mathematics, and the simultaneous collapse of alternative grand projects in method and universal mathematics, that he saw his way, finally, to becoming a systematic philosopher of nature. He then offers in *Le Monde* his first version of a corpuscular-mechanical system with, as he saw it, unusually strong grounding in physico-mathematical achievements and practices. Descartes' early career, leading to the composition of *Le Monde*, is explored in Chaps. 3, 4, 5, 6, 7 and 8. Chapters 9, 10, and 11 present what I suspect may be the most extensive conceptual and technical dissection ever undertaken of *Le Monde*.

But, despite what has just been proposed, why take *Le Monde* as such an important event in investigating the early career of Descartes? It is perfectly clear that there are many ways to define, and to inquire about, the intellectual projects of the 'young' Descartes, and thence, by contrast to define some works or achievements as constituting the 'mature' or late Descartes. For example, to name but two: Taking 'method' and its supposed applications or examples as central, one might view the aborted *Regulae ad directionem ingenii* as marking the end of his early development, and the publication of the *Discours and Essais* in 1637 to mark not only the public, but the mature Descartes; or, privileging the trajectory of his work in metaphysics, one might take the *Discours* as marking the end of a long apprenticeship in the field, which still left Descartes far short of the mature position[s] he finally staked out in the *Meditations*, *Objections and Replies* and portions of the *Principia* itself.

Now, the present work aims to be neither tendentious nor precious about this issue. The matter is quite straightforward, once the goals and interpretative framework of this book are stated. My focus, as already noted, is upon *Descartes as a philosopher of nature*. The main concern is with natural philosophy and those fields, such as optics and mechanics, as well as programs such as physico-mathematics, that bore on the practice of natural philosophy, and vice versa. Therefore, the early trajectory of Descartes, *his early career in this sense*, consists of those events, processes and achievements that led to writing *Le Monde*, his first systematic work in that field. This immediately dictates that whatever other scholars and studies may define as the mature Descartes, in this work *the mature Descartes is the mature natural philosopher*, whose definitive, natural philosophical system is his *Principia Philosophiae* (1644, 1647).

It is for this reason that the penultimate chapter of the book undertakes an extensive analysis of the *Principia* as a systematic work in natural philosophy.

This inquiry will reveal a number of novel and surprising conclusions, concerning the content and intent of the *Principia* as a daring gambit in natural philosophy and in realist Copernican cosmology and cosmography.[2] These findings will serve both to put into perspective our main target, Descartes' achievement and intentions in *Le Monde*, and to mark out a field of further inquiry into what exactly the *Principia* was intended to accomplish within the natural philosophical contest of Descartes' generation. Overall, about three-quarters of the book are devoted to the trajectory of the young Descartes, leading up to the composition of *Le Monde*, whilst the final quarter consists of the detailed dissection of *Le Monde*, and the comparison of it to the *Principia*, the latter treated in the rather unconventional manner just outlined.

Finally, readers should rest assured that although this work declines to define the early and mature Descartes in the alternative terms mooted three paragraphs above, the *Regulae* as well as the *Discours* and *Essais* are treated herein, insofar as required by our focus on Descartes' natural philosophical career. As the 'Overview of Argument' in Sect. 1.4 below shows, within the perspective adopted in this book, the *Regulae* are examined extensively as a nodal work, especially in regard to their content and the history of their composition between 1618 and 1628. In addition, although our approach dictates that we not focus upon the *Discours* and *Essais* as a publishing event or putative statement of program *per se*, several aspects of Descartes' 1637 public manifestation come under examination at appropriate points in our argument—most importantly his work on the law of refraction of light and theory of lenses in the *Dioptrique*; his purported method and tendentious autobiography in the *Discours*, and his corpuscular–mechanical explanation of color in the *Météores*.

1.2 Descartes and the Historians of Science

As recently as a generation ago, apprentice historians of science, especially Anglophones, faced a daunting challenge, linguistic and cultural, in coming to grips with Descartes the mathematician, natural philosopher and 'scientist'. They tended to encounter a Descartes already unpalatably familiar from undergraduate excursions in philosophy or the history of ideas: an eerily contextless and anonymous author of atomic, isolated, putatively timelessly relevant texts, the *Discourse on Method* and the *Meditations*, the Father of Modern Philosophy (epistemology) with whom their philosophy instructors were, apparently, in constant, but critical discussion.

With the help of equally frustrated but more experienced senior colleagues, one soon realized that the best sources for the study of Descartes' mathematics, science and their relation to his philosophical projects were old, rare and mostly in French,

---

[2] Cosmography is initially defined below in Sect. 1.4, 'Overview of Argument', regarding Chap. 12, and in Chap. 12, itself (Sect. 12.2).

bearing names such as Brunschvicg, Milhaud, Boutroux, Dreyfus-Le Foyer, and Klein.[3] As one who slogged through some of these tomes and others like them, I can say that the lack of consensus they displayed about Descartes' science, methodological and natural philosophical project(s) was on the whole challenging and exhilarating. At least here was a Descartes relevant to the history of science, a working mathematician, anatomist, student of geometrical optics, natural philosopher and methodologist. And if this Descartes did not have much of a context, either intellectual, institutional or political, at least he had dense and much debated projects and trajectories, imputed strategic plans and occasionally mooted failures, defeats and tactical retreats.

I suspect I am far from alone in possessing several large canvas-bound loose leaf notebooks of 1970s vintage, filled with the erudite thoughts of these earlier twentieth century masters, along with my own almost random counterpoint of hesitant query, contrast and even the odd idea. Luckily, I did not then appreciate the degree to which many of these works, and their mutual conflicts, were themselves the products of arcane Parisian politico-religious-cultural-pedagogical battles which would have required a second, very French and very privileged, life-time to unravel, if at all. Most Anglophone neophyte Cartesian scholars were, I think, similarly insulated from this insight, which might have extinguished all hope of penetrating, let alone contributing to this archive.

As for contemporary Anglophonic Cartesian studies, most remained firmly anachronistic. The number that gave access to even bits of Descartes' scientific concerns and projects would be counted on not more than one hand. Looking beyond Scott's (1952) even then dated but still occasionally useful work, one found inspiration chiefly in the brilliant but scattered work of people like Gabbey (1971), Sabra (1967), and Aiton (1972).[4] For an overall picture of Descartes' intellectual projects and transformations, it was not idiosyncratic to rely on the then generation old work of Norman Kemp Smith.[5] For Descartes' cultural, religious, and, if you like, political context and motivation, one of course resorted to Popkin.[6] As for the historiographical traditions descending from Bachelard, Koyré and Kuhn in which many were being trained, there was relatively little help to be found. These had shared one key premise: method does not really account for work at the research coalface or for trajectories of research. For Bachelard, much of Descartes technically is 'pre-science', the 'philosophy of the sponge' to be exact, and although in an odd way Bachelard was inviting us to consider what we might now term the field of natural philosophical discourse and struggle, the insight was, in my view, impossible at the time.[7] Koyré was more to the point, in a negative way, by stressing Descartes' slippages and mistakes, compared to

[3] Brunschvicg (1927), Milhaud (1921), Boutroux (1900), Dreyfus-Le Foyer (1937), and Klein (1968).

[4] Scott (1952), Gabbey (1971), Sabra (1967), and Aiton (1972).

[5] Smith (1952).

[6] Popkin (1964).

[7] Bachelard (1965) 79. Cf Schuster and Watchirs (1990) 7–11.

the ultimate achievement of scientificity in mechanics by Galileo.[8] As for Kuhn, early in his career he had called for serious work on Descartes in a series of reviews in *Isis* and even apparently had contemplated some himself. However, in retrospect, we can see how the socio-cognitive category of natural philosophizing eluded Kuhn, just as it had Koyré.[9]

Since then many of these limitations have been swept aside. First year philosophy does not seem to live or die on the *Meditations*. Many students may first meet Descartes in a history of science course, where the readings can take advantage of a renaissance of Anglophonic, serious, state of the art scholarship in the history of science and history of philosophy. This has depicted Descartes as a significant player in that historical passage of natural philosophical contention, physico-mathematical invention, and grand methodological posturing called the Scientific Revolution. He is a player who perhaps failed in terms of his own vision of his mature projects and aims, but whose interventions shifted the ground of debate in several key areas of natural philosophy, mathematics and the technical sciences.[10] In the wake of these developments we have also been able to benefit from full scale intellectual biographical studies anchored in history of science and/or history of philosophy, which exploit and extend for Anglophones these relatively recently expanded horizons.[11]

The present book similarly operates on the basis of these improvements in our understanding, and expansion of our archive of reliable history of science studies of Descartes. Like much of the newer literature, and in accord with my own apprentice studies in this direction, this volume also concentrates on Descartes' early career—the period residing between his early and problematical methodological visions, and the emergence of his mature philosophical works, the *Discourse on Method* (1637), *Meditations* (1641) and *Principles of Philosophy* (1644)—and hence with his natural philosophical and mathematical concerns in those years. It also has something of the form of an intellectual biography, albeit in deliberately truncated and narrowed form. This is because whilst there is little need at the moment for another full blown intellectual biography of even the young Descartes, there is room for a very detailed reconstruction of his concerns, precisely in the overlapping domains he himself would have denominated physico-mathematics, mixed mathematics, natural philosophy and method.

We have learned much in recent years on both a factual and historiographical level about what these terms, and the activities they denote, meant to Descartes in particular and to the overlapping communities of practitioners to which he belonged.

---

[8] Koyré (1939). The time period I am discussing pre-dates Mephem's valuable English translation of this classic.

[9] Kuhn (1953) and (1955). Cf. Schuster and Watchirs (1990) pp.11–13. Kuhn himself informed me at some stage in the early 1970s that he had intended in the early to mid 1950s to write a book about 'Descartes' science'.

[10] For example, the Anglophonic studies collected at intervals of half a generation in, respectively, Gaukroger (1980), Voss (1993), and Gaukroger et al. (2000).

[11] Shea (1991), Garber (1992), and Gaukroger (1995).

We have become attentive to what young natural philosophers were taught at university by their neo-Scholastic Aristotelian masters about the (limited) relations that were supposed to exist between the study of natural philosophy, the cognate discipline of mathematics, and the curiously hybrid 'mixed mathematical' disciplines taken to be subordinate to both. We have learned that the term physico-mathematics signaled a questioning of the Scholastic Aristotelian view of the mixed mathematical sciences as subordinate to natural philosophy, non explanatory and merely descriptive.[12] Somehow, the mixed mathematical disciplines would become intimately related to natural philosophical issues of matter and cause—they were to become, as I shall explain below, more 'physicalised', more closely intertwined with natural philosophizing, regardless of which species of natural philosophy one pursued. Finally, we have also realized that we have to be careful about such notions as the possibility of an efficacious and transferable general method in the sciences; and that one needs to look carefully at how systems of discourse, especially highly contested systems, such as those in natural philosophy, were pursued by active players. All this means that surviving evidence of the young Descartes' involvement in these activities can be looked at anew, and perhaps folded into a tentative diachronic reconstruction of his intertwined trajectories in these fields. This inquiry will facilitate better large scale studies of the mature Descartes, as well as improved

[12] My use of the terms mixed mathematical sciences and subordinate sciences follows that stated in Gaukroger and Schuster (2002) p.537, which introduced a discussion of the young Descartes' enterprise in physico-mathematical hydrostatics: 'The term "mixed mathematics" had been framed by Aristotle to refer to a group of disciplines intermediate between natural philosophy, which dealt with those things that change and exist independently of us, and mathematics, which deals with those things that do not change but have no existence independently of us, since numbers and geometrical figures have (contra Plato) an existence only in our minds. (Aristotle, *Metaphysics* Book E.) A physical account of something — such as why celestial bodies are spherical — is an explanation that works in terms of the fundamental principles of the subject matter of physics, that is, it captures the phenomena in terms of what is changing and has an independent existence, whereas a mathematical account of something — such as the relation between the surface area and the volume of a sphere — requires a wholly different kind of explanation, one that invokes principles commensurate with the kinds of things that mathematical entities are. (Aristotle, *Posterior Analytics*, 75a28-38; Cf. 76a23ff and *De caelo* 306a9-12.) In the *De caelo*, 297a9ff, for example, we are offered a *physical* proof of the sphericity of the earth, not a mathematical one, because we are dealing with the properties of a *physical* object. In short, distinct subject matters require distinct principles, and natural philosophy and mathematics are distinct subject matters. However, Aristotle also recognizes subordinate or mixed sciences, telling us in the *Posterior Analytics,* 75b14-16, that "the theorem of one science cannot be demonstrated by means of another science, except where these theorems are related as subordinate to superior: for example, as optical theorems to geometry, or harmonic theorems to arithmetic." Whereas physical optics — the investigation of the nature of light and its physical properties — falls straightforwardly under natural philosophy, for example, geometrical optics "investigates mathematical lines, but *qua* physical, not *qua* mathematical." (*Physics*, 194a10.) The question of the relation between mixed mathematics, on the one hand, and the "superior" disciplines of mathematics and natural philosophy, which did the real explanatory work on this conception, remained a vexed one throughout the Middle Ages and the Renaissance, but so long as the former remained marginal to the enterprise of natural philosophy the problems were not especially evident.'

understanding of Descartes as a symptomatic player in these fields at a critical phase in what we call the Scientific Revolution.

The resulting study is therefore limited in time, the biographical narrative stopping after Descartes' first extensive project in systematic natural philosophy, *Le Monde*, largely finished but left unpublished in 1633 (although, as noted, we will end by jumping forward to consider his mature natural philosophy in the *Principia philosophiae*). More importantly, this study is quite focused (some might say limited) in terms of the intellectual or, if you will, the psychological space it canvasses.[13] Not every documentable concern of Descartes in the years 1618–1633 will be examined, only those bearing on his agendas, products and arguable self-images or identities in relation to natural philosophy, physico- and mixed mathematics and method. The colorful but more often than not wild motivational claims of some of the recent popular work on Descartes will be eschewed. For example, whether or not Descartes was an active, or more laid back, agent or informant for the Jesuits in the Counter Reformation politics of the time is not terribly relevant to studying and reconstructing his expert practices and aims in fields of expert endeavor (which themselves might be interrelated of course). Similarly, even the best intellectual history and biography is prone to assert (or assume) billiard ball models of causation and motivation: Descartes engaged in this or that expert and technical aim or activity because of some particular event, episode or one-off encounter, as, for example, he set out to do metaphysics and conquer scepticism because in the late 1620s he happened to cross paths with Cardinal Bérulle. This study eschews all such simplistic notions of how large (and often quite speculatively asserted) contextual forces or commitments immediately shaped Descartes' (or anybody else's) dense, long-term, complicated and expert engagements in expert activities and disciplines. All of these have their own more narrow social, organizational and rhetorical/persuasive economies, which do need to be factored into biographical stories.[14]

I sympathize with the possibility that Descartes had and pursued aims of contemporary religio-political relevance. That is part of the reason why I shall advance a model of how the game of natural philosophizing was played that allows players to have articulated and expressed such outside aims and values in terms of moves and positioning inside the expert field. Did Descartes' natural philosophizing have a Counter Reformation salience and message? The answer is, 'yes'. Is that because he

[13] We shall shortly see that the term 'psychological' is misplaced or misleading. Whilst I intend continually to attempt to reconstruct Descartes' relevant cognitive and motivational structures, this is not an exercise in psychological diagnosis, therapy or causal explanation. My bearings come from phenomenological sociology and so deal with evidence–based hypothetical reconstructions of the contents of Descartes' and others' structures of knowledge, relevance and motivation. See below, Sect. 1.3.3.

[14] We shall put these principles to work throughout this volume. A particularly salient example of their application occurs in Chap. 8, where we consider the problem of the break or inflection in Descartes' career following the collapse of his project of the *Regulae* in 1628. Here the issue of avoiding simplistic event to event causation and unrealistic images of rupture and revolution will be to the fore, and an alternative mode of description and explanation will be employed.

was a Jesuit agent.[15] The answer is, 'we do not definitively know, and even if he was, there was more in his background and view of the world which would have invited such gambits inside his natural philosophizing'. Similarly, we can easily agree that Descartes' physico-mathematics and his later emerging mechanistic system of natural philosophy expressed values of utility, domination of nature and the import of material practices for high cultural systems of knowledge. But, was this because he had met one or two practical mathematicians, or because he belonged to a 'class' whose interests would be served by enunciating such views at an elite cultural level? Neither popular and dramatic story telling, nor old fashioned vulgar Marxist social structural imprinting will be employed here. Again, we shall work from the inside out, placing Descartes in the more immediate and expert field of contention, natural philosophizing, and then and only then we shall ask, 'With what cultural items and agendas would he personally have been acquainted and concerned to articulate to his own particular weavings of natural philosophical utterance?' Neither class affiliation at one end, nor one-off biographical episodes at the other end, are likely to throw much light on such moves and strategies inside what we shall term the field of natural philosophical contention.

In short, the project I here propose is entangled in two intersecting spaces of interpretative possibilities and problems: (1) The general problematic of scientific biography (or intellectual biography) as it presents itself today; and, (2) the particular constraints and contours of a 'scientific' biography of Descartes, especially as such a project handles issues of the relation of Descartes to the wider processes of the Scientific Revolution. In the following Section we turn to these two spaces, starting with the latter.

1.3 Key Pitfalls (and Opportunities) Facing Descartes' Biographers (Even Authors of Quite Truncated Biographies)

1.3.1 The Problem of Method and Its Texts: Regulae and Discours

Although the message has perhaps not yet spread as widely as would be desirable in Cartesian studies and intellectual history generally, we now have excellent grounds for accepting, on the basis of the work of some historians and sociologists of science, the general proposition that no doctrine of method, whether Descartes' or anybody else's, ever has guided and constituted the actualities of scientific

[15] Grayling (2005). Another option, perhaps more plausible in my view, is that speculatively advanced by Harold Cook (in his contribution to the symposium 'Wil de Echte Descartes Opstaan?' at the Descartes Centre for the History of the Sciences and the Humanities, University of Utrecht, 2 October 2008), that Descartes may have been throughout his mature career some sort of 'intelligencer' in the service of Richelieu.

practice—conceptual and material—in the literal ways that such methods proclaim for themselves. This raises immediate and catastrophic implications for some traditions of Cartesian studies. Just as, until relatively recently, some major interpretations of the Scientific Revolution have tended to be dominated by heroic tales of the discovery, perfection and application of the scientific method; so, much the same sort of 'cult' of method has prevailed in Cartesian studies, within which certain broad 'sects' can be isolated, as I have argued elsewhere and shall revisit in Chap. 6.[16] The solution to these difficulties and what will be applied throughout this volume, is a full-blooded 'atheism' concerning the existence of any unique, efficacious and transferable general method in the sciences, and a corresponding willingness to ask what method-talk is, and how it works if it isn't what it says it is.[17]

In the next chapter, dealing with historiographical and conceptual issues, we shall locate the original grounds of such full-blooded modern 'atheism' about method in the history of science in the writings of Koyré and Kuhn, building on those of Bachelard. The problem for a long time was that whilst the historiographies of Koyré and Kuhn effectively debunked method discourse as having no role whatsoever in the dynamics of the sciences, they left the matter with a merely negative conclusion. It took the work of Feyerabend, on the rhetorical and propaganda functions of Galileo's and Newton's methodological pronouncements, to begin to point toward the political functions which method discourse can have in the life of the sciences. This was despite the fact that method doctrines do not function literally in the ways they proclaim, being as Koyré and company insisted, impotent and fruitless in those respects. Feyerabend's initiative was then extended in a 'post-Kuhnian' literature within the history and sociology of science, which explored what method discourse does in the sciences, if it does not and cannot do what had traditionally been claimed for it. Broadly speaking, this work suggested that method discourses are often deployed as rhetorical weapons in those negotiations and struggles over the framing and evaluation of knowledge claims which go on at all levels of scientific activity, from the laboratory bench, through published texts, to disciplinary debate and its necessarily associated micro-politics of groups and institutions.

In the case of my prior work, and with reference to Descartes as an example, it has been possible to address directly the Bachelard-Koyré-Kuhn problematic of method: how to overcome mere debunking by analysing how it was the very discursive structures of method doctrines (Descartes' included) which guaranteed both their lack of efficacy *and* their creation of literary effects of that efficacy.[18] This model of method discourse as a kind of mythic speech, is, I contend, essential to a consistent development of post-Kuhnian method atheism. It is not adequate to decode the rhetorical and political deployments of method-talk without accounting for the genuineness of historical actors' belief in the efficacy, unity and transferability of scientific method. Even if we do not believe in method, we must not attribute false consciousness to

[16] Schuster (1984, 1986, 1993).

[17] See also, J.A. Schuster and R. Yeo (1986) pp. ix–xxxvii.; E. Richards and J.A. Schuster (1989), pp.697–720.

[18] Schuster (1986, 1993).

historical believers in method; rather we have to unravel the mechanisms, quite obvious textual-discursive mechanisms, that help to solidify their belief in method. Otherwise, it would seem that even the best equipped biography must opt for belief in method, a carping debunking or a slippage between the two stools. From this perspective, it follows that in so far as biographical writing about Descartes is a function of the larger historiographies of method and of science, it too requires reformation. I have argued that with such a model in hand of the discursive structure and dynamics of Descartes' method, it proves possible to understand both the appeal and the necessary lack of efficacy of that doctrine, and also to address some of the actual micro-political and rhetorical functions which Descartes' method discourse did in fact perform.[19] This material will be revisited and applied in the present volume.

In this exercise, some key points need to be grasped quite firmly: Descartes' natural philosophical, scientific and mathematical work does not emerge from a method and neither does his order of study and biographical trajectory; nor, finally is his mature, metaphysically legitimated system of nature methodologically 'deduced' from first principles. By the same token, and putting all nostalgia for the good old days of Cartesian studies aside, one cannot take seriously the autobiography in the *Discours* as anything less than a method articulated, post-facto, self-legitimating narrative. This is the perspective from which this volume will consider Descartes' first methodological pronouncements of 1619, and his (ultimately abortive) project of a detailed work on methodology, his *Regulae ad directionem ingenii*, written, and developed, in several segments between 1619 and 1628. The *Regulae* reside at the intersection of most of these interpretative insights and caveats, and so it constitutes a critically important obstacle and challenge to any 'method-smart' modern biography of Descartes. One must ask: What is the text of the *Regulae*; when was it written; what are its subject and aim; and what was the fate of this abandoned project, premising all the answers on the new view of method as discourse which structurally cannot accomplish what it structurally so convincingly says it can accomplish. In short, the sorting out of Descartes' method discourse, the reconstruction of its genesis and the identification of its discursive structure and dynamics, are all necessary conditions for our recovery of an historical rather than mythological Descartes. This theme will play an important part in our story.

1.3.2 The Problem of Descartes the Natural Philosopher, and of Natural Philosophy as a Wide and Dynamic Field of Discourse and Contention

Descartes was from beginning to end a natural philosopher; that is to say, a well educated and increasingly skilled member of what we shall term the culture or field of discourse of natural philosophizing. Trained by the Jesuits in their brand of

[19] Schuster (1980, 1986, 1993).

Counter–Reformation, neo-Scholastic Aristotelianism, the young Descartes moved under the tutelage of Isaac Beeckman in 1618–1619. With Beeckman, the young Descartes adopted a radical stance on the place of the mixed mathematical disciplines, such as geometrical optics and hydrostatics, within his preferred brand of natural philosophy, which he, like Beeckman, now took to be of a corpuscular-mechanical type, although, as we shall see, he then held this natural philosophy in a piecemeal manner. This program they termed 'physico-mathematics'. For reasons we shall explore in this volume, about a decade later he moved to design an explicit system of corpuscular-mechanical natural philosophy, partly resting upon and partly transcending his earlier physico-mathematical work. There are two sets of questions here. We must ask, 'What was the game of early modern natural philosophizing about; what were its structure, dynamics and rules of participation?' Additionally we must ask, 'What was the genealogy and what were the specific characteristics of Descartes' brand of natural philosophy; that is, his particular brand of corpuscular mechanism, and how over the period 1618–1633 did it relate to his endeavors in physico-mathematics?'

To take the latter question first, one must grasp the impetuses and aims of Descartes' early commitment to corpuscular–mechanism, as well as its initially, piecemeal, non–systematized character. This, in turn, provides a foundation for assessment of the post 1628 systematization of his natural philosophy and its relation to his attempt at metaphysical and theological grounding. So, one must look for the specificity of content, style and aim in his earliest natural philosophical initiatives, growing from his interaction with and apprenticeship under Isaac Beeckman. Clearly, considering the strictures on method stated above, there can be no question of a methodologically articulated fairy tale of the sources or structure of that corpuscular–mechanism. Nor, should it be simply assimilated to some latter day textbook definition of what mechanical philosophy was 'in general'. Descartes' early mechanism was quite particular and it remained so in explanatory style and aim, even in later systematic form. For example, it fundamentally involved, as its doctrine of causation, a (spatially and temporally) punctiform 'dynamics' of micro-corpuscles where instantaneously manifested/altered force of motion and directional modes of force of motion ('determinations') were the operative concepts.[20]

Most importantly, we shall learn that in his early years, roughly 1618–1629, Descartes was not at all interested in putting forward a system of natural philosophy.

---

[20] On this conception of Descartes' dynamics as the 'causal register' of his corpuscular mechanism, see Schuster (2000), Gaukroger and Schuster (2002), Schuster (2005) and below, Sects. 3.3.3, 4.2, 4.6, 4.7.4, 4.8.1, 8.2.2, and 9.5.2. Any doctrine of natural philosophy had at the very least to specify a theory of matter and a theory of causation. And, if it had aspirations to systematicity, there needed to be added a doctrine of cosmology— dealing with the cosmic structuring of matter and patterns of causality holding in situ—as well as a doctrine of method concerning how such doctrines could be derived and/or justified. However, as should be clear, not all utterances or claims of a natural philosophical nature had to be parts of a system, and whilst conflict of systems was more characteristic of the early and mid seventeenth century, a muting of systematic claims and overt conflicts about them characterized the natural philosophical field in the later stages of the century. See below, Sect. 2.7.

His commitment to an unsystematized corpuscular-mechanism was real. However, it remained secondary to his primary intellectual agenda and domain of practice, in physico-mathematics, which, as explored in more detail in Chaps. 2 and 3, denoted a commitment to radically revising the received Aristotelian interpretation of the mixed mathematical sciences as subordinate to natural philosophy, merely instrumental and non explanatory. The mixed mathematical disciplines were to become more 'physicalised', meaning they would become intimately related to *natural philosophical issues of matter and cause*.[21] Prior to his post–1628 turn to systematic natural philosophizing, Descartes' corpuscular-mechanism was used to provide the ultimate explanatory terms—via the eminently natural philosophical categories of matter and cause—for otherwise quite piecemeal exercises in this physico-mathematics. And, as we shall see, it was primarily in physico-mathematical exercises in hydrostatics and especially optics, that he applied corpuscular-mechanical explanations, and worked out some of the principles of his dynamics of corpuscles, later deployed more systematically in *Le Monde.*

Descartes' later more systematized and self-consciously grounded and legitimated project in corpuscular-mechanism, beginning with *Le Monde*, built upon and articulated this basis, following the collapse of the project of the *Regulae*. However, it is of course important to recognize that method did not dictate these moves or their content; and that the system of mechanism was legitimated and grounded metaphysically and theologically, not 'deduced' from metaphysics or theology. Moreover, in accord with an understanding of Descartes as a player in an already given field of contention about natural philosophy, many of his maneuvers in constructing and defending his mechanism in the latter portion of his life require explanation in terms of the micro-politics of persuasion and contention in the *agon* of natural philosophizing, *and hence should not be seen as the peculiar (brilliant or wrong headed) private gymnastics of a detached and aloof mind.*[22]

Now, while adequate intellectual biographies can be constructed bearing in the mind the above strictures about Descartes the natural philosopher, I would contend that something is still missing, as a condition of the very best of state of the art historical practice. Biographers also need to be aware of the context in which natural philosophizing was done; in other words, the sort of universe of discourse and contention in which men like Descartes constructed and 'sold' systems of natural philosophy. Early modern natural philosophers risk being marginalized in historical gazes which refuse to recognize the 'culture of natural philosophical discourse' in and through which they struggled. After all, we no longer 'do' systematic natural philosophy in their manner and probably have not since the late eighteenth century. But it can be argued that the sub-culture of natural philosophizing, in all its richness, diversity and contention was the very locus and main prize in the critical phase

[21] Gaukroger and Schuster (2002) and below, Sect. 2.5.3; Chap. 3 passim.

[22] We shall see early examples of his tactics in the construction of *Le Monde*, below in Chaps. 9, 10, and 11.

(in the early and mid–seventeenth century) of that process we term the Scientific Revolution.[23] 'Descartes savant' is actually Descartes the philosopher of nature, and so it is important to analyze his natural philosophical projects as such, within an understanding of the larger domain of natural philosophizing. What we have here, as in the case of method, is a question of new historiographical categories and questions: 'What is natural philosophical discourse in the early modern period; how was it in play and process in the period?' These issues are the key to the period if they are theorized and articulated fruitfully.[24] For this reason, a major part of the next chapter will be devoted to a categorical construction and historiographical house-cleaning about early modern natural philosophy, as a field of discourse, intellectual sub-culture and organizational network.

1.3.3 Scientific Biography and the Historiography of Science

Before we outline the argument of this volume, one final set of pitfalls and opportunities must be canvassed. These concern the very legitimacy and possibility of intellectual biography. I have suggested that new perspectives on method and natural philosophy strengthen the case for fruitful modern intellectual biographical study of Descartes. This claim, however, collides with the broader problematic of biography in the history of science. Scientific biography has been disparaged in some quarters, on theoretical grounds, as unnecessary or misleading, because it tends to court the history of ideas and because of its failure to address the group agon of knowledge making/breaking sub-communities.[25] Additionally, and with more justification, the

---

[23] On the dynamics and trajectory of natural philosophizing in the seventeenth century as context for our study of Descartes' projects and agendas, see below, Sects. 2.5 and 2.7, particularly the latter for the dissection of the key phases in the trajectory of natural philosophizing in the Scientific Revolution.

[24] Schuster 1990, 1995, 2002; Schuster and Watchirs 1990; Schuster and Taylor 1996.

[25] The problems with traditional history of ideas are well known. The critique began with the school of Quentin Skinner and John Dunn who debunked the traditional, if often tacit, assumption that ideas have causal power; that earlier ideas (texts, books, core concepts) can 'influence' later thinkers. Skinner and his colleagues insisted that intellectual historians not work in terms of the force or influence of ideas, but rather try to construe actors' intentions regarding the adoption, revision and deployment of intellectual resources contingently available to and known by them. The loci classici are John Dunn (1968) and Quentin Skinner (1969) .

Post Kuhnian sociology of scientific knowledge, especially of the early 'Edinburgh School' in the work of Barry Barnes and Steven Shapin, also insisted that articulation of concepts cannot occur via influence but rather through later actors' access to, and appropriation, reinterpretation and redeployment of, earlier intellectual or 'cultural' resources (Barnes 1982; Shapin,1992). They widened this assault so that it did not appear merely psychologistic, and focused it on the seemingly inhospitable terrain of history of science. The actors doing the appropriating, negotiating and redeploying of resources are not treated as isolated, albeit intending and judging, agents; rather an actor's (formulations of) tactics and goals are taken into account as a function of the sort of network, tradition or community into which he or she is trying to launch claims and have them taken

theoretical distaste for biography also signals a turn away from the forms of psychologism that characterized some earlier scientific biographies, and toward 'social' terms of explanation. The de facto alignment of advocates of contextualist and social constructivist explanations against biography (and especially over against biography as dealing with the 'inner contents' of actors) is, however, simplistic and unnecessary in my view, since biographical study may bring to light things and processes otherwise thought quite valuable from such contextualist and sociology of scientific knowledge perspectives. These include diachronic understandings of the forms of discursive and material practice that today's historians and sociologists of scientific knowledge desire. Biographical study can place a microscope on one actor's path through disciplinary struggles and negotiations, allowing for the real possibility of grounding contexts in the biographer's considered reconstructions or modelings of an actor's perceptions of relevant goals, costs and benefits. Moreover, biographical study can do this without resorting to simplistic forms of home-spun psychologizing.[26]

---

up in some form by others. If one wants to speak about an actor's intentions, one must locate him as a contender inside a contested field, where success can only be achieved by one's competitors taking on board and redeploying one's own earlier claims. In short, the precepts of Skinnerian history of ideas and the implications of post-Kuhnian sociology of science, all comport to the conclusions that it definitely is not a question of how the past of the tradition forces or 'influences' present moves; but, of how later players mobilize and deploy resources for their present moves.

Given all this, the modus operandi of earlier historians of ideas can be understood on this basis: Classical historians of ideas misunderstood the fact, which we now grasp, that actors continually adapt, interpret and redeploy available cultural and discursive resources of perceived relevance and interest. They took this as an indication of the existence of an order of causation in the realm of ideas alone. It is not surprising that a discourse on 'influences', progressive continuity and the filiation of ideas resulted. Moreover, because of their acknowledged professional skill, and vested interest, in dealing with texts, and systems and relations of ideas, rather than with the social organization and interactional dynamics of their production, dissemination and (re-)interpretation, earlier historians of ideas were inclined to believe that ideas, especially scientific ideas, have a special and autonomous cognitive status.

[26] Thus it would seem that the core concern of an intellectual or scientific biography is the subject's trajectory or course of engagement in the fields, traditions and disciplines in question in the study. Otherwise it would be hard to see how such a work could transcend the mere narrative of ordinary life events and contingencies. However, Don Howard (2008) has also forcefully pointed out recently that there is always therefore a danger of over dramatizing the biographical subject in scientific biography as the nodal hero of all the intersecting fields and forces. As he observes, it is an empirical question (and judgment) how important the biographical subject's intersection with, and intervention in, the wider fields really was. Now, it is clear that Descartes was a first-class player in his fields of activity, and that his work was later widely discussed and renegotiated—hence he is, in a non-Whiggish sense, a significant figure. It is also obvious that if we do not come to grips with the actual fields of play, nothing useful in scientific biography is likely to eventuate. Hence, our concern in the next chapter in setting out the structure and dynamics of the intellectual fields and disciplines—natural philosophizing and the mixed mathematical sciences—through which we will trace the young Descartes' trajectory. See Schuster (2009) for an analysis of a recent biography of Descartes which does not adequately theorise the fields in which the young Descartes worked.

The interpretative or phenomenological sociology that resided behind many of the first developments in the sociology of scientific knowledge can provide the historian with considerable heuristic insight in this regard, by suggesting that historical actors are best seen as appropriating and interpreting available cultural resources for the attainment of ongoingly renegotiable goals. The array of resources available to an actor is patterned over time by his social location, affiliations and experiences. Similarly, goals and interests are socially transmitted and enforced. An actor's trajectory through the domains, fields and networks which constitute his social environment is open to empirical investigation and theoretical articulation by the historian as to their structures, dynamics and interrelations. In such a model, as adumbrated, say, in the writings of Schutz and Luckmann,[27] actors are not internalistic or externalistic cultural dopes; they can reinterpret resources and renegotiate goals; and they can gloss and legitimate their actions in public, a process which can contribute to molding the very environments in which they move.[28]

This, it should be obvious, provides a general basis for studying individual actors and texts in biographical mode; but one should note the special salience that this approach has for focusing the need for historians to attempt to construct and depict both the social contexts of actors and their 'inner mental contents'. One may see this especially clearly by considering a strong and influential statement to the contrary from within the school of post-Kuhnian sociology of scientific knowledge, amongst which the most persuasive has probably been from Bruno Latour. In his classic earlier works, he completely rejected the invocation not only of contexts, but of actors' cognitive contents, and thus *a fortiori* rejected biographical study of the garden variety historical sort.[29] However, the case for biography as serious historiography informed by sociology of science perspectives can emerge directly from confronting Latour's denials of the historian's craft.[30] Over against Latour's repeated insistence that such large contextual structures as 'society', 'economy' or state' are not to be invoked in explaining the construction and stabilization of facts and artifacts, one can argue, on historians' terrain, that historians' considered models of larger contexts, and models of actors inner cognitive/interest/relevance structures are fundamental to the tasks of narrative and explanation. Historians, including biographers, attend to two related genres of tracing. First, historians must manufacture models of what I have called the 'inner contents' of actors, their cognitive and interest structures; their meaning and relevance structures. These structures are imputed

[27] A. Schutz (1970) and A. Schutz and T. Luckmann (1973).

[28] Macro-social forces need not be the major elements in any situation directly faced by an actor, and played by him, but neither are they ruled out *a priori*. Similarly, cognitive structures, discourses etc. are amongst the resources in play, they are not the strings of a puppet. One finds this sort of model in the writings of Schutz, Luckmann and others, and to some degree it appears to have informed some styles of post-Kuhnian sociology of scientific knowledge.

[29] Especially Latour (1987), which may be considered his systematic attempt to displace Kuhn at the pinnacle of theory in 'science studies'. But also see Latour (1986, 1988).

[30] Schuster (1991).

to actors at given moments in the process or interaction under study. Secondly, historians must also manufacture models of relevant aspects of context, proximate or distant. That is, those aspects of context taken as relevant to the shaping, constraining and empowering of the actors with their 'inner contents'. All this obviously involves layers of theory—tacit or explicit on the part of the historian—concerning the nature of contexts and of contextual relations to actors. This, needless to say, is the sort of thing one is doing in historically theorizing about, for example, the 'domain of method discourse' or 'the field of natural philosophical struggle', as I do in this volume.

So, contrary to the 'socio-technical networking' model of Latour of his co-workers, with its 'just-so' stories of rational (that is, contextless and contentless) combatants, actors come into encounters to negotiate or construct knowledge with historically reconstructable and imputable internal grids of cognition, interest and value. And those grids come from somewhere, do they not? Actors do not drop in as *dei ex machina* miraculously gifted with inner socio-cognitive contents (except in method stories or Latour stories, that is). So historians appeal to the existence of contexts, whether mediate or quite macro, which have shaped, constrained and variously empowered those actors and their inner contents.[31] These models of 'insides' and of contexts are historians' *sui generis* professional constructs. Historians mobilize both sorts of tracing in accounts of passages of historical action, and biographies are simply one genre of this procedure. As such, these accounts, including biographies, obviously are simultaneously descriptions and explanations. They are also one-off 'life stories', or narratives, of actions in and of the structures so constructed.[32]

Given all this, biographers, far from despairing, should rejoice in the good news that diachronic reconstructions, including biographies of individuals, arguably have the character of adequate historical explanations. This is especially true for intellectual biographies when the diachronic threads of relevant material and discursive practices are fed through the biography; for example, as in this volume, when the larger tidal rhythms of the field of natural philosophizing are themselves woven through a biography of an early modern natural philosopher. So, intellectual biography need not be a chronology of ideas thinking themselves. Rather, such biography is simply a form of contextual history brought to a focus in one actor's life trajectory

[31] Historians can and must make judgments about what context(s) and what models of such context(s) are to be mobilized in their accounts. For example, it is typical of method-centric accounts of history of science that they similarly conjure away such inner contents and outer contexts: they happily leave all the little rational actors to agree straightaway on the latest falsifying test (Popper), or the degenerating state of such and such a research program (Lakatos). For an early critique of Lakatosian method-talk from an equally early post-Kuhnian perspective, see Schuster (1979).

[32] The foregoing should perhaps be taken in spirit of Evans (1999), Chaps. 1 and 8. in his highly intelligent refutation of post-modernist declarations of the end of historical explanation and narrative. Interestingly, my point here is that the kind of history of science which can provide adequately post- post-modern explanation cum description makes use of some of the tools of the very phenomenological sociology, and early post–Kuhnian sociology of scientific knowledge that later inspired the irrational, anti-historical histrionics of the post-modern demolition squad.

in the form of a narrative.[33] In biography, the historian still has to model macro structures and sub-cultures and construct a narrative of how the actor's ongoing pursuit of projects played through or played into these, as well as into shifting social contexts which, as usual, also tended to condition resources and goals.

Along the way, the biographer–historian can exploit other subsidiary diachronic aspects of this sort of model. These include the fact that part of the contingent structure of resources and problems confronting an actor is his own product, the upshot of his own engagement in earlier situations and resulting investments. Moreover, there is the actors' propensity to struggle to establish accounts of the meanings of past actions and events. Actors do this in their own perceived interests and within culturally available forms of discourse (such as method-talk) and it helps them struggle to constitute the situations in which they find themselves (such as a contest for natural philosophical hegemony in the age of the Baroque.). Our study of Descartes in the following pages will see plenty of action of these latter types. These matters are now typically grasped within the category of 'identity', meaning not so much the ideas an actor possessed and advocated, but his role(s) and agenda(s) as he saw them, and as he presented them. Accordingly, we shall have quite a bit to say about Descartes' shifting and evolving construction and presentation of his intellectual identities, agendas and self-understandings in the years in question.

It is worth noting by way of further explicating these concerns that they were in part inspired by, and aim to articulate what Stephen Gaukroger has premised as the requisites of an intellectual biography at the beginning of his own more comprehensive and full spectrum study of Descartes. Gaukroger argues that an intellectual biography must do more than establish a sequence of intellectual pursuits, that it must try 'to establish a rationale for them both in terms of the subject's motivations and in terms of a specific cultural and intellectual context within which those motivations are shaped and bear fruit'. Of course he did not mean by this the simplistic reduction of the subject to his contexts. Rather, Gaukroger saw intellectual biography, in general and in the case of Descartes, as having three interacting axes, not to be collapsed together, the enterprise involving addressing all three in their mutual relations: They are (1) the relations between Descartes' personal development and the cultural environment in which he lived and worked; (2) the relations between Descartes' personal development and his intellectual development; and (3) the relations between his intellectual pursuits and the cultural and intellectual environment in which they were pursued.[34] This in turn entailed a very important heuristic or regulative point about procedures and aims that was woven into Gaukroger's project and arguably should be reflected in any study that wishes, within its own limits, to be similarly conceptually and historiographically sound: In so far as a distant figure

[33] First put this way in Schuster (1995), p.113. following oral expression by Fr. Barry Brundell and myself in a conference paper, 'The Making of Mechanism 1620–1640: Descartes, Gassendi and the So-called Crisis of the Seventeenth Century', *Annual Conference of the Australasian Assoc. for the History, Philosophy and Social Studies of Science*, May 1983.

[34] Gaukroger (1995), p.8.

such as Descartes is known to us only by his formal works; a few texts and fragments unpublished in his life time; and a finite set of correspondence, our access to his personal development is achieved primarily in terms of the self images he adopts or conveys, those literally spoken or enacted by him or entailed in his discourse and reported behavior.

I completely agree that we can and should produce reasoned, evidence–based heuristic pictures of Descartes' shifting senses of identity, agenda and self-understanding, much as Gaukroger succeeded in doing. However, there are some adjustments to be made due to the fact that our study here, compared to Gaukroger's, is a very much truncated intellectual biography, limited in space, time and scope of intellectual activities examined. In recompense, we aim here to achieve improved and deepened understandings of some matters, most often of a technical history of science nature, reaching beyond even Gaukroger's own considerable efforts.[35] For that reason, I suggest we make the following substitutions in Gaukroger's three rules: for 'personal development' in general, let us put 'reconstruction from time to time of Descartes' identity, sense of personal agenda and self-understanding as a practitioner of natural philosophy and related fields'; for 'intellectual development' across the full spectrum, let us substitute 'intellectual development in the field of natural philosophy and related fields'; and for 'intellectual pursuits', let us limit ourselves to natural philosophizing and related disciplines, as and when related. In this way, our adoption and scaling down of Gaukroger's principles and procedures is brought into affinity with the kind of biographical approach we just presented—following Schutz and Luckmann and the thrust of post-Kuhnian sociology of scientific knowledge—involving attempts to map at certain stages of life Descartes' lived 'structures of relevance' as chosen or 'sedimented' into him from those available in his culture.

In sum, I think it can safely be concluded that it is a mistake to dismiss biography as necessarily falling short of serious, theoretically articulated historical inquiry.[36] In particular, the sociology of scientific knowledge and contextual history of science, far from precluding biography, require it, can facilitate it, and in turn can be enriched by it.[37] In Chap. 2, I shall attempt to show in the discussion of method and of natural

[35] Efforts which I note here I have publically approved of in the past; and had the opportunity to collaborate upon with Gaukroger from time to time in the decade and a half since publication of his intellectual biography of Descartes. (Schuster 1995; Gaukroger and Schuster 2002; Gaukroger et al. 2000).

[36] One can criticize all one wants the targeting for study of 'geniuses of the Scientific Revolution'. The simple fact of the matter is that we shall not have a serious historiography of that process, broad or narrow, social or intellectual, until the geniuses are properly understood, in so far as that is possible. And one clear way forward, amongst others, is on the front of the kind of scientific/intellectual biography envisioned here.

[37] There is no pretence here of having noted, let alone solved, all the problems involved in taking biography seriously as just another species of both history and sociology of knowledge. But at least some breathing space has been gained for the style and aims of the present study. An example of yet another problem or puzzle for biographers, working in the style advocated above, would be this: Starting (and ending) points for biographical narrative must be a matter of convention and convenience.

philosophy how a small bit of that historian's modeling of categories used to depict context and actor's resources might proceed. We can therefore finally turn to an overview of what this 'truncated' intellectual or scientific biography will contain.

1.4 Overview of the Argument

Chapter 2 sets the stage for the fully biographical chapters to come, by considering a number of conceptual and historiographical issues which frame the entire project, and which, in my view, are also indispensable for approaching the larger problem of explaining and narrating the so-called Scientific Revolution. The matters dealt with have mostly been broached in parts of my earlier work, and are brought together here on a 'need to know' basis. That is, they are arranged not as a putative primer or textbook on historical technique for dealing with the Scientific Revolution and natural philosophical players within it, such as Descartes. Rather, they have been selected and arranged based on ways substantive parts of the later argument depend upon, and express them. Amongst the matters dealt with the most important are a model for how the increasingly competitive and turbulent culture of natural philosophizing worked in the era of Descartes, including the question of the place and import of the subordinate mixed mathematical sciences, and the meaning Descartes and others attached to the idea, and project of a physico-mathematics, rendering those mixed sciences more properly 'natural philosophical'.[38]

Chapter 3 deals with the early physico-mathematics of Descartes, which he pursued at first in conjunction with his mentor Isaac Beeckman, who also had conveyed to him his first inkling of corpuscular-mechanism as an approach to natural philosophy. I argue that in 1618–1619 Descartes' emphasis was less on his newly acquired knowledge of corpuscular-mechanical ontology than on a commitment,

They, especially starting points, may therefore have an old fashioned history of ideas feel to them, as the actor is adduced *ex machina* equipped with certain ideas; identified as participating in certain traditions; and posited as starting with such and such a structure of aims and relevances. Methodologically speaking, there seems to be no way to avoid this, which is only to say that biography, in which social history and interpretive sociology of knowledge are brought to a focus in one actor's life trajectory in the form of a narrative, has special problems. But, for that reason, it does not cease to belong to the categories of proper historical discourse.

[38] Other matters dealt with in Chap. 2, and deployed later in the book include: An explication of the generic rules of construction and contestation for natural philosophers and natural philosophies; a model for dealing with the problem of 'external or contextual' drivers of natural philosophical utterance; an heuristic model for talking about and assessing the nature and degree of systematicity of a natural philosophy; an outline of the main phases in the trajectory of natural philosophizing and its subordinate disciplines in the period of the so-called Scientific Revolution, so that Descartes' location and role can be better identified; and finally, on the basis of the foregoing, an initial sketch of 'what sort of philosopher of nature was the young René Descartes'? I am presently engaged in writing a larger explanatory/descriptive work on early modern natural philosophy and the subordinate sciences, based on the articulation of the model of the dynamics of the field of natural philosophizing, and the other matters, mooted in Chap. 2.

similarly inherited from Beeckman, to a program of physico-mathematics in and of his natural philosophical work. We see that early on he was paying more attention to being an aspiring physico-mathematician within the field of natural philosophy (wherein he was leaning toward a corpuscularian agenda), than he was to articulating and enunciating details of corpuscular structures and behaviors. In the physico-mathematical program, the traditional Aristotelian view of the mixed mathematical sciences as subordinate to natural philosophy and largely devoid of explanatory scope and power was to be radically challenged. The mixed mathematical disciplines were intended to become more integrally linked related to questions of matter and cause, in other words to questions of a natural philosophical type, which, in the case of Descartes and Beeckman, meant an unsystematized, but firmly held, corpuscular-mechanism.

The chapter deals with three case studies of Descartes' physico-mathematics: his manuscript on hydrostatics and the hydrostatic paradox; his well known work with Beeckman on the nature of accelerated fall; and a curious, widely overlooked but extremely important geometrical and physical optical fragment on refraction of light adapted and explicated from bits of the work of Kepler. Although the material on fall is better known, my emphasis is on the first and third cases. The hydrostatics manuscript turns out to be the key case, most revealing of the style and aims of Descartes' physico-mathematics articulated to, and through, an embryonic corpuscular-mechanism. Understanding his agenda on this basis allows us to understand the third fragment, which in turn is critically important to my examination in Chap. 4 of Descartes' later successes in physico-mathematical optics in the mid and late 1620s, including his discovery and attempted mechanistic explanation of the law of refraction of light. As to the material on accelerated fall, it takes on a different appearance than it has in the traditional literature, because we view it across the natural philosophical *cum* physico-mathematical preoccupations of Descartes and Beeckman.[39]

Chapter 4 reconstructs the genealogy of Descartes' discovery of the law of refraction, initial development of a theory of lenses, and first attempts to explain the law through a mechanistic theory of light in the years 1626–1928. These events of the mid to late 1620s constitute the greatest of Descartes' achievements in mixed- and physico-mathematics and were also of the utmost importance for his emergence, from the late 1620s, as a systematic corpuscular-mechanical natural philosopher.[40]

[39] For reasons discussed in Chap. 3, Descartes' *Compendium of Music*, also composed at this time, in close connection with the tutelage of Beeckman, will not figure in our considerations. The simple reason is that the work is not a serious exercise in physico-mathematics, but, with some tiny exceptions, remained firmly within the traditional conception of mixed mathematics. See Chap. 3 Note 8.

[40] The development of Descartes' lens theory is discussed in an addendum to Chap. 4, Appendix 1 'Descartes, Mydorge and Beeckman: the Development of Cartesian Lens theory 1627–1637'. The material is unremittingly technical, hence its relegation to an Appendix. But it is also crucial for confirming our reconstruction of the timing and technique of Descartes' discovery of the law of refraction, hence its inclusion.

He would use the discovery of the law of refraction as a putative example of his supposedly all conquering method. More importantly, the optical work led him to the mature formulation of the central concepts of his dynamics—the causal register of his emerging system of corpuscular-mechanism. That system was first embodied in the text, *Le Monde* (1629–1933), tellingly subtitled *' traité de la lumière'*, that is, a veritable treatise on light, in which the recently polished dynamics, itself a product of the optical work, ran a corpuscular-mechanical theory of light in its cosmological setting and a corpuscular-mechanical theory of 'celestial mechanics'. The optical triumph of the 1620s was both the climax of the early physico-mathematical agenda of the young Descartes, as well as the exemplar for important parts of his mature, systematic natural philosophical work to come. It has posed difficulties of understanding for scholars of Descartes; but, with the exception of his youthful fantasy of a universal method, is perhaps the most important part of his early work to reconstruct on the path to a plausible interpretation of his early career.

Taken together, our case studies in physico-mathematics in Chap. 3 and the story of the great physico-mathematical work in optics of the mid 1620s in Chap. 4, are indispensable to properly executing our reconstruction of Descartes' struggles, first, between 1618 and 1628, in physico-mathematics articulated to natural philosophizing, and later, after 1628 in systematic natural philosophy bearing the genealogical imprints of that earlier physico-mathematics. But there is much more to the story of Descartes' struggles in these years, and the complicated—indeed largely unintended way—he arrived at the decision to practice systematic natural philosophy.

Chapters 5, 6, 7, and 8 therefore address additional complexities and layers of technical endeavor, agenda and identity spanning the period as far back as 1618 and reaching forward to the composition of Descartes' first system of corpuscular-mechanical natural philosophy in *Le Monde* (1629–1633). This part of the story is complex in itself, since it deals with both solid mathematical work, and rising, concatenating, and increasingly unrealistic and unrealizable aspirations of a general methodological type. It is also shown that Descartes' tortured trajectory in mathematics and method intersected and articulated with the story of physico-mathematics and natural philosophizing told in Chaps. 3 and 4, which in fact cannot be fully understood on its own, but only when this second dimension of the young Descartes' struggles is brought to light. Chapters 5, 6, 7, and 8 therefore form a bridge between the story of Descartes as a young physico-mathematician and the later chapters on his first systematic natural philosophy, written from 1629 to 1633. They deal with his other projects of the years 1618–1629, which were meant to encapsulate, and vastly transcend 'mere' physico-mathematics. The failure in the late 1620s of this vision—in its final, highly crafted version in the *Regulae ad directionem ingenii*—invited, or drove, Descartes toward an explicit vocation in systematic natural philosophy, a program he had never before embraced.

Chapter 5 shows that since his early days with Beeckman, Descartes had pursued a set of projects related to physico-mathematics, but far outstripping even it in potential scope and invested hopes. From 1618, Descartes had pursued an analytical, problem-solving oriented agenda in mathematics, which in these respects resembled his physico-mathematics, or so he thought. Indeed, the parallels he perceived between his mathematical and physico-mathematical work triggered in 1619–1620 his dream of a unified analytical approach to all mathematically based disciplines—practical, pure and physico-mathematical—to which he appropriated the already circulating name 'universal mathematics'. Moreover, that overheated conception quickly gave way to the even more encompassing mirage of a universal method, which remained with him from 1619 to the late 1620s, when, after his optical breakthrough, he picked up universal mathematics and method again in detail. These compounding enlargements of his mathematical and physico-mathematical agendas are traced in this chapter. We learn that Descartes' analytical mathematics, and his dreams of universal mathematics and a universal method, involved their own complicated genealogy, which interacted in intended and unintended ways with his work in physico-mathematics and (piecemeal) natural philosophy. Between 1618 and 1629, Descartes, it therefore turns out, was not just struggling to work out a physico-mathematics with possible corpuscular-mechanical bearings. He was also a master analytical mathematician and dreamer of gigantic and seductive methodological fancies, all of which arguably affected his shifting and evolving self-understandings and agendas. Once we have appreciated these addition sedimentary layers of his projects and aspirations in the 1620s, we can more fully grasp the implications of the term *agonistes* in our title. Moreover, our rising sense of the turbulence of Descartes' intellectual journey in the 1620s is only heightened, once we take seriously the actual pitfalls and delusions he encountered regarding his project in method, and accompanying identity as prophet of its triumph. It was the collapse of his methodological program, and identity, that actually set the stage for his mid career emergence as a systematic natural philosopher.

But, to reach that stage we shall first need to pause, in Chap. 6, for something of a conceptual and historiographical interlude. We stop our narrative briefly to consider the problem of how exactly to handle the young Descartes' belief in his own method. As discussed briefly above and in more detail below in Chap. 2, modern scholarship in history, philosophy and sociology of science debunks the idea that there is, or can be, a universal, efficacious and transferable 'scientific method'. The work of scholars such as Koyré, Kuhn and Bachelard has left us sceptical of the idea of an efficacious general method; and Feyerabend, along with leading sociologists of scientific knowledge, have explored the rhetorical and legitimatory use of methodological discourse in the sciences. But, we still need a way of dealing with historical actors' belief in their own method claims and their tendency to define their intellectual agendas and identities in these terms. We cannot, in short, believe in Descartes' method, but neither can we merely debunk his own apparent belief in it. We need to understand how methodological doctrines create for believers their appearances of unity, efficacy, applicability and progress whilst remaining, for the very same reasons, structurally incapable of delivering what they promise. In Chap. 6, a model of methodological discourse is adduced for these purposes, allowing us, at long last, to deal more appropriately with the historical Descartes and his apparently genuine belief in his own method. The findings of this chapter are applicable to any and all general method doctrines, and to figures espousing claims to possession of any sort of universal method. So, as with our other conceptual and historiographical considerations in Chap. 2, they have implications for Scientific Revolution historiography and the history of science in general.

Chapter 7 then returns to the main line of our narrative, explaining how all the relevant tendencies and projects of the young Descartes *agonistes*—physico-mathematics, universal mathematics and universal method—came to a climax and inflection point in the late 1620s. Working partly in the shadow of Marin Mersenne and his cultural battle against both radical scepticism and radical (religiously heterodox) natural philosophies, Descartes launched out, trying to realize his earlier dream of a methodologically sound 'universal mathematics'. Riding on his physico-mathematical and more purely analytical mathematical results and the confidence they fed into his dream of method, he worked himself into an intellectual and vocational dead end. We learn that this project, inscribed in the latter portions of his unfinished *Rules for the Direction of the Mind*, did not blossom into a magisterial work of method and universal mathematics, but collapsed in 1628 under its own weight of self-generating problems and contradictions. From this point on, Descartes arguably did not believe in his method, although he continued to exploit it for purposes of public presentation and justification of his work. Additionally, Descartes now entered upon a process of rapid change of direction of his intellectual agenda, and correlatively, his self-understanding and identity.

Chapter 8 deals with how Descartes struggled to redefine his projects and his vocation, given the collapse of the project of the *Regulae* in 1628. We find that it was only at this point that he set out to become something we have not seen him intending to become at any previous moment, the author of a systematic, radical, pro-Copernican and corpuscular-mechanical, new philosophy of nature, embodied first in *Le Monde*, which we study in Chaps. 9, 10, and 11. Accordingly, Chap. 8 deals with the events, episodes and tendencies that have variously been taken to have caused his move to systematic mechanism and/or dualist metaphysics. In accord with the historiographical principles for dealing with the dynamics of intellectual traditions, which guide this study, we shall eschew explanations by way of complete ruptures of agenda and identity, as well as explanations which assign sufficient causality to particular events and episodes. Rather, we shall take a structured approach, superimposing consideration of particular developments and events upon examination of fundamental intellectual commitments and agendas, to produce a denser description/explanation of what was more a 'process of inflection' of Descartes' projects rather than a 'crisis and break', 'rupture', or, a simple continuity. Chapter 8 also includes a detailed reconstruction of the course of writing *Le Monde* 1629–1633.

In sum, then, Chaps. 5, 6, 7, and 8 demonstrate that much of the story of Descartes' *agonistes* is precisely the story of the intended and unintended entanglements of two trajectories: in physico-mathematical natural philosophy and in analytical mathematics, which latter was promoted to fantasy programs in universal mathematics and method. The entire process was marked by determined planning, unintended shifts and some spectacular insights, some decisively fruitful, some disastrously misleading, all in turn conditioned by the varied environments in which Descartes moved. It is the story of these struggles that will finally bring us, fully prepared, to become readers of *Le Monde*.

Chapter 9 explicates the opening sections of *Le Monde* and is particularly concerned with their non-Scholastic rhetoric and organization, given their having been written in the vernacular, in *honnête homme* style with the core elements of the system presented as a fable. This chapter serves to clear the ground for the rest of our reading of *Le Monde*, so that we can appreciate both the systematicity it displays, and the genealogy of some of its key concepts in Descartes' earlier physico-mathematical strivings and results. To that end, special attention is paid to Descartes' curious cosmogonical fable, his matter theory *cum* theory of elements and the initial delineation of the vortex theory. (A number of these topics in *Le Monde* are revisited in Chap. 12, which examines their alteration and articulation in the *Principia philosophiae* as part of the daring, new systematizing strategy of that mature natural philosophical text.)

Chapter 10 then brings together two lines of investigation about the natural philosophical structure and aims of *Le Monde*: First of all, it focuses on showing that Descartes' famous and to some degree notorious vortex celestial mechanics was a serious intellectual construct and hence also a serious gambit in the natural philosophical contest. Indeed, we learn that the vortex mechanics was Descartes' technical answer to the natural philosophical challenge posed by realist Copernicanism. Often simplistically glossed and dismissed, the vortex celestial mechanics is the veritable 'engine room' of the argument of *Le Monde*, and the lynch pin of the corpuscular-mechanistic system of *Le Monde*.[41] A charitable reading of the vortex mechanics is offered, which takes us a long way into the details of the system, with considerable coherence being displayed—from the explanations of stars and stellar vortices, through planetary orbits, the behavior of satellites and comets, as well as local fall and tidal phenomena on planets, whilst the same vortex mechanics, element theory and dynamics explain, in broad terms, the existence and behavior of light in the cosmological setting.

Secondly, Chap. 10 explores the ways in which the celestial mechanics at the heart of *Le Monde* is a hybrid entity. On the one hand, it is shown that the vortex celestial mechanics has a genealogy reaching back through the physico-mathematics studied in earlier chapters. But, on the other hand, we also learn that the vortex mechanics was clearly a piece of generic natural philosophical discourse, understandable as such by any member of the educated culture of natural philosophizing, and playing the central role in this new corpuscular-mechanical system of natural philosophy. The larger lesson is that *Le Monde* was simultaneously the climax of Descartes' trajectory in physico-mathematics and the first iteration of a systematic natural philosophizing, emergent from that carapace.

Chapter 11 concludes our detailed analysis of *Le Monde*, by building on the structural and conceptual dissection of the text executed in the previous two

[41] Because my detailed presentation of the celestial mechanical vortex model in Chap. 10 is synthetic and the result of a rather complex process of interpretation, I have added a second Appendix, 'Decoding Descartes' Vortex Celestial Mechanics in the Text of *Le Monde*', dealing with the blow by blow textual exegesis underpinning the synthetic presentation in the main body of my text.

chapters. It has two main aims: Firstly, *Le Monde* is examined as a competitive bid for supremacy in the natural philosophical field. This is done by viewing it in relation to key natural philosophical aspirations and strategies of similar contemporary actors, such as Kepler, who, like Descartes, were attempting to displace Aristotelianism, install some version of realist Copernicanism, and create alternative hegemonic natural philosophical syntheses. Descartes had become a committed systematizer in natural philosophy, but, in tune with the rising competitive temper of the time, he was daring and aggressive in the ways he constructed and presented his system. His building of *Le Monde* out from and in part upon physico-mathematics (as he understood it over time), shows this daring, as does his creative engagement with realist Copernicanism and his direct attack on the key 'hot spot' (as we term it in Sect. 2.5.4) in relations between mixed mathematics and natural philosophy—concerning the causes of celestial motions in Copernican astronomy.

Then, secondly, *Le Monde* is assessed in terms of its strengths and weaknesses as a *system* of natural philosophy, using the model of natural philosophical systematicity developed in Sect. 2.5.5, as part of the explication of conceptual and historiographical foundations for this study. Such an assessment is necessary for the following reason: Although *Le Monde* marked a node and climax in Descartes' career, it was obviously a particularly transient and occluded one, rather internal to Descartes' development, not a public marker. That is, despite our finding that *Le Monde* was indeed a bold and highly systematic discourse in natural philosophy, we will also learn that it was, by the standards of the later *Principia*, a prentice work. This is not simply due to the obvious fact of its being earlier than the *Principia*, but because of the more interesting finding that *Le Monde*, Descartes' first step toward systematic natural philosophizing, débouched from a complex and often highly successful trajectory in physico-mathematics. On the one hand, *Le Monde* was a systematic, and deeply interesting culmination of Descartes' fifteen years of struggle in physico-mathematics, mathematics, method and emergent, piecemeal corpuscular-mechanism. On the other hand, it was just the beginning of his work as an even more masterly systematizing philosopher of nature, who, in the end largely dropped his project of physico-mathematics, as well as the delusional dream of a universal method. The resulting examination of the systematicity of *Le Monde* leads to some striking individual examples of refinements and improvements in that regard, as displayed by the *Principia* to come.

This, in a sense, completes our reconstruction of the trajectory of the young Descartes, from physico-mathematician with some unsystematized corpuscular-mechanical leanings, in 1619, to systematic mechanistic natural philosopher, shaped in part by the course of his physico-mathematical endeavors, in 1633. However, one more step is required fully to round off our inquiry. Since, as we noted above, the *Principia* (rather than *Le Monde* or the *Discours and Essais*) obviously contains the mature statement of Descartes' system and strategies in natural philosophy, our study of how Descartes matured as a natural philosopher can only be properly closed through a detailed analysis of the *Principia* and comparison of it to *Le Monde* as we now understand it.

Chapter 12 therefore uncovers exactly how the *Principia* was designed to improve the systematic power, scope and consistency of Descartes' natural philosophy compared to what he initially presented in *Le Monde*. Following my own work, and previous, separate collaborative researches with both Jacqueline Biro and Judit Brody, a number of very novel and surprising answers emerge: I find that the center of gravity of Descartes' revised systematizing strategies in the *Principia* did not reside in his metaphysical grounding of the natural philosophy; or in his now elaborate teaching concerning the laws of motion and collision. Rather, I argue that Descartes' systematizing strategy focused mainly upon weaving ranges of novel matters of fact—concerning sunspots, novae and variable stars, and the structure and formation of all planets (including the Earth)—into explanatory and descriptive narratives with cosmic sweep and radical realist Copernican intent. These gambits were "*cosmographical*" (the natural philosophical relating of heavens and earth in contemporary usage), and they were characteristic of radical realist Copernican natural philosophers, who reasoned that in a realist Copernican system, findings about the Earth, now a heavenly body, could be used for cosmological purposes. This tactic began with Copernicus himself, and ran through the contributions of Bruno, Gilbert and Galileo, down to, as established here, their most radical form in Descartes himself, with the *Principia*, properly decoded, expressing a stunningly bold case for a realist, infinite universe Copernicanism. It is the vast system-binding cosmographical gambit of Descartes, entraining the use and reframing of key, available matters of fact—in turn leveraged into explanatory resources within the system—that best characterises the natural philosophical difference between *Le Monde* and the *Principia.* By this point, and on this reading, the Descartes, *physico-mathematicus* of corpuscular leanings of 1619, had, on the one hand evolved into an entirely discursive and highly systematic natural philosopher of traditional type, albeit, on the other hand, a natural philosopher possessed of dramatically new goals and strategies within the field.

Chapter 13 takes the form of both a *Coda*, rounding out and underlining the key themes of this study, and an *Epilogue*, surveying some salient points about the subsequent career of the mature Descartes, as well as his (somewhat surprising) relations to the next phase in the Scientific Revolution, both now arguably more properly understandable as a result of our studies in the previous 12 chapters.[42]

The two Appendices, as signaled above in Notes 40 and 41 then follow.

References

Works of Descartes and Their Abbreviations

AT = *Oeuvres de Descartes* (revised edition, 12 vols.), edited by C. Adam and P. Tannery (Paris, 1964–76). References are by volume number (in roman) and page number (in Arabic).

SG = *The World and Other Writings*, edited and translated by Stephen Gaukroger (Cambridge,1998).

[42] The character of the later seventeenth century phase of the Scientific Revolution, as well as that of the preceding critical or turbulent phase in which Descartes' work falls, are delineated as part of the historiographical modeling in, Sect. 2.7.

MM=*René Descartes, The Principles of Philosophy*, translated by V. R. Miller and R. P. Miller (Dordrecht, 1991).

MSM=*Rene Descartes, Le Monde, ou Traité de la lumière*, translated by Michael S. Mahoney (New York, 1979).

CSM(K)=*The Philosophical Writings Of Descartes*, 3 vols., translated by John Cottingham, Robert Stoothoff, and Dugald Murdoch, and (for vol. 3) Anthony Kenny, (Cambridge, 1988) References are by volume number (in roman) and page number (in Arabic).

HR=The Philosophical Works of Descartes, vol I translated by E.S. Haldane and G.R.T. Ross (Cambridge, 1968 [1st ed. 1911]).

Other

Aiton, E.J. 1972. *The vortex theory of planetary motion*. New York: Neale Watson Academic Publications.

Bachelard, G. 1965, 1st ed. 1938. *La formation de l'esprit scientifique*, 4th ed. Paris: Vrin.

Barnes, Barry. 1982. *T.S.Kuhn and social science*. London: Macmillan.

Boutroux, P. 1900. *L'imagination et les mathématiques selon Descartes*. Paris: Félix Alcan.

Brunschvicg, L. 1927. Mathématique et métaphysique chez Descartes. *Révue de Métaphysique et de Morale* 34: 277–324.

Dreyfus-Le Foyer, H. 1937. Les conceptions Medicale de Descartes. *Revue de Metaphysique et de Morale* 44: 237–283.

Dunn, John. 1968. The identity of the history of ideas. *Philosophy* 43: 85–104.

Evans, Richard. 1999. *In defense of history*. New York: Norton.

Gabbey, Alan. 1971 (1980). Force and inertia in the seventeenth century: Descartes and Newton. In *Descartes: Philosophy, mathematics and physics*, ed. S. Gaukroger, 230–320. Sussex: Harvester Press.

Garber, Daniel. 1992. *Descartes' metaphysical physics*. Chicago: University of Chicago Press.

Gaukroger, Stephen (ed.). 1980. *Descartes: Philosophy, mathematics and physics*. Sussex: Harvester Press.

Gaukroger, Stephen. 1995. *Descartes: An intellectual biography*. Oxford: OUP.

Gaukroger, S., and J.A. Schuster. 2002. The hydrostatic paradox and the origins of Cartesian dynamics. *Studies in History and Philosophy of Science* 33: 535–572.

Gaukroger, S.W., J.A. Schuster, and J. Sutton (eds.). 2000. *Descartes' natural philosophy*. London: Routledge.

Grayling, A.C. 2005. *Descartes: The life of René Descartes and its place in his times*. London: Free Press.

Howard, Don. 2008. Time for a moratorium? Isaacson, Einstein and the challenge of scientific biography. *Journal of Historical Biography* 3: 124–133.

Klein, Jacob. 1968. *Greek mathematical thought and the origin of algebra*. Cambridge, MA: MIT Press.

Koyré, A. 1939, Eng. Trans. 1978. Etudes Galiléennes. Paris: Hermann & Cie. English trans. Galileo Studies. Trans. J. Mepham. Hassocks, Sussex: Harvester.

Kuhn, Thomas S. 1953. Review of *The Scientific Work of René Descartes* by J.F. Scott and of *Descartes and the Modern Mind* by A.G.A.Balz. *Isis* 44: 285–287.

Kuhn, Thomas S. 1955. Review of *New studies in the philosophy of Descartes* by Norman Kemp Smith and *Descartes' philosophical writings*, ed. N.K Smith, and of *The method of Descartes*, by L.J. Beck, *Isis* 46: 377–380.

Latour, Bruno. 1986. Visualisation and cognition; thinking with eyes and hands. *Knowledge and Society: Studies in the Sociology of Culture Past and Present* 6: 1–40.

Latour, Bruno. 1987. *Science in action*. Milton Keynes: Open University Press.

Latour, Bruno. 1988. The Prince for machines as well as for machinations. In *Technology and social process*, ed. B. Elliott. Edinburgh: Edinburgh University Press.

Milhaud, Gaston. 1921. *Descartes savant.* Paris: Alcan.

Popkin, Richard. 1964. *The history of scepticism from Erasmus to Spinoza.* Berkeley: University of California.

Richards, E., and J.A. Schuster. 1989. The myth of feminine method: A challenge for gender studies and the social studies of science. *Social Studies of Science* 19: 697–720.

Sabra, A.I. 1967. *Theories of light from Descartes to Newton.* London: Oldbourne.

Schuster, John. 1979. Kuhn and Lakatos revisited. *British Journal for the History of Science* 12: 301–317.

Schuster, J.A. 1980. Descartes' Mathesis Universalis: 1619–28. In *Descartes: Philosophy, mathematics and physics*, ed. S. Gaukroger, 41–96. Sussex: Harvester.

Schuster, John. 1984. Methodologies as mythic structures: A preface to the future historiography of method. *Metascience: Review of the Australasian Association for the History, Philosophy and Social Studies of Science* 1–2: 15–36.

Schuster, J.A. 1986. Cartesian method as mythic speech: A diachronic and structural analysis. In *The politics and rhetoric of scientific method*, ed. J.A. Schuster and R.R. Yeo, 33–95. Dordrecht: Reidel.

Schuster, J.A. 1990. The scientific revolution. In *The companion to the history of modern science*, ed. R.C. Olby, G.N. Cantor, J.R.R. Christie, and M.J.S. Hodge, 217–242. London: Routledge.

Schuster, J.A. 1991. 'Bruno's [No history required] Tour of the past', *Combined research program in science & technology analysis. Working paper number 1.* University of Wollongong, Wollongong, ed. Mark Rix. University of Wollongong, Wollongong. Available through the National Library of Australia and on my web site: http://descartes-agonistes.com/

Schuster, J.A. 1993. Whatever should we do with Cartesian method: Reclaiming Descartes for the history of science. In *Essays on the philosophy and science of René Descartes*, ed. S. Voss, 195–223. Oxford: OUP.

Schuster, John. 1995. Descartes *Agonistes:* New tales of Cartesian mechanism. *Perspectives on Science* 3: 99–145.

Schuster, J.A. 2000. Descartes *Opticien*: The construction of the law of refraction and the manufacture of its physical rationales, 1618–29. In *Descartes' natural philosophy*, ed. S. Gaukroger, J.A. Schuster, and J. Sutton, 258–312. London: Routledge.

Schuster, J.A. 2002. 'L'Aristotelismo e le sue Alternative'. In *La Rivoluzione Scientifica*, ed. D. Garber, 337–357. Rome: Instituto della Enciclopedia Italiana.

Schuster, J.A. 2005. "Waterworld": Descartes' vortical celestial mechanics: A gambit in the natural philosophical contest of the early seventeenth century. In *The science of nature in the seventeenth century: Changing patterns of early modern natural philosophy*, ed. Peter Anstey and John Schuster, 35–79. Dordrecht: Springer.

Schuster, J.A. 2009. Descartes—philosopher of the scientific revolution; or natural philosopher in the scientific revolution. *Journal of Historical Biography* 5: 48–83.

Schuster, J.A., and Graeme Watchirs. 1990. Natural philosophy, experiment and discourse: Beyond the Kuhn/Bachelard problematic. In *Experimental inquiries: Historical, philosophical and social studies of experimentation in science*, ed. H.E. Le Grande, 1–47. Dordrecht: Kluwer.

Schuster, J.A., and Richard R. Yeo. 1986. Introduction. In *The politics and rhetoric of scientific method*, ed. J.A. Schuster and R.R. Yeo, ix–xxxvii. Dordrecht: Reidel.

Schuster, John, and Alan B.H. Taylor. 1996. Seized by the spirit of modern science. *Metascience* 9: 9–26.

Schutz, Alfred. 1970. *Reflections on the problem of relevance.* New Haven: Yale University Press.

Schutz, Alfred, and Thomas Luckmann. 1973. *The structures of the life-world.* Trans. R.M. Zaner and H.T. Engelhardt. London: Heinemann

Scott, J.F. 1952. *The scientific work of René Descartes, 1596–1650.* London: Taylor and Francis.

Shapin, Steven. 1992. Discipline and bounding: The history and sociology of science as seen through the externalism-internalism debate. *History of Science* 30: 333–369.

Shea, W. 1991. *The magic of motion and numbers: The scientific career of René Descartes*. Canton: Science History Publications.
Skinner, Quentin. 1969. Meaning and understanding in the history of ideas. *History and Theory* 8: 3–53.
Smith, Norman Kemp. 1952. *New studies in the philosophy of Descartes*. London: Macmillan.
Voss, Stephen (ed.). 1993. *Essays on the philosophy and science of René Descartes*. Oxford: OUP.

Chapter 2
Conceptual and Historiographical Foundations—Natural Philosophy, Mixed Mathematics, Physico-mathematics, Method

2.1 Jesuit neo-Scholasticism for the *noblesse de robe*

Born on 31 March 1596 in La Haye (now Descartes) in Touraine, Descartes was the product of a relatively recently risen family of the *noblesse de robe*. His father, Joachim, was a conseiller of the Parlement of Brittany in Rennes. Joachim's father, Pierre, had been a prominent physician, as had been his mother's father. Descartes' mother was Jeanne Brochard, daughter of René Brochard, lieutenant générale of the Presidial of Poitiers. Her mother was Jeanne Sain, whose father, had been a merchant from a family of merchants. Descartes' family styled themselves noble. René Brochard, Joachim and his father called themselves, écuyers, squires, and thus appropriated themselves to the lowest rank of the nobility. As early as 1547, Pierre Descartes had sought and obtained the all important exclusion from the taille. In early adulthood, René Descartes sported the title Sieur du Perron, after a small seigneurie which he was due to inherit. He was accustomed to bearing a sword, and served for a time as a gentleman volunteer in the army of Maurice of Nassau. Descartes' family thus offers a rather typical example of the extrusion of provincial upper professional and mercantile families into the ranks of the administrative nobility, from which position they proceeded to try to ape the proprietary and social habits of the landed nobility.

Although Descartes eventually trained in law, he did not follow the profession of his father and older brother, Pierre, who, in 1618, also became a conseiller in the Parlement of Brittany. Though they might aspire to the status and bearing of the landed nobility, and often acted out of the narrow self-interest of men holding venal offices and defending provincial liberties and nepotistic excesses, royal officials of the generation of Joachim Descartes also had a broader role to play in upholding the King's law and administration. If anything, this key role could only have appeared in sharp contrast to the background of administrative paralysis, political maneuvering and popular and religious discontent, and endemic civil war in evidence in the late sixteenth century and again (absent civil war as such) after the assassination of Henri IV in 1610. It is not drawing too long a bow to conjecture that for the post-adolescent

J. Schuster, *Descartes-Agonistes: Physico-mathematics, Method & Corpuscular-Mechanism 1618-33*, Studies in History and Philosophy of Science 27,
DOI 10.1007/978-94-007-4746-3_2, © Springer Science+Business Media Dordrecht 2013

Descartes, who in 1615 and 16 considered a career in the law, idealized hopes for good order and administration—to be asserted and actualized in the face of apparently endemic unrest and turbulence—weighed more upon his student mind than did those cynical, short range calculations of interests—in patronage, bureaucratic infighting, provincial politics and financial wrangling—that would have daily absorbed the mental energies of experienced *nobles de robe*, such as his father.

For young René Descartes had been well prepared both for those sorts of idealized insights and for possible careers in the law or the military by his years spent as a good and loyal student of the Jesuit order. Between 1606 and 1614, or possibly 1607 and 1615, he was educated at the Collège de la Flèche, which had only opened in January 1604, shortly after Henri IV had allowed the Jesuits back into France.[1] Henri intended La Flèche to be a training ground for an elite and militant nobility, loyally monarchist in politics and orthodox in the Catholic faith to which Henri himself had converted to end the Civil Wars and consolidate his claim on the thrown. The Jesuits of course were interested in training theologians and future Jesuits, but equally so in molding a devoted and active elite of lay gentlemen of affairs. They upheld the original inspiration of the Order that proper religion, and consequently proper morals and order, could be restored to a world threatened by heresy and chaos by men of action, well trained and prepared to deploy all the spiritual, intellectual and material means at their disposals.

The curriculum at La Flèche was of a neo-Scholastic type modified by important concessions to humanist reforms and content. The first five years were devoted to classical languages and literature, embodying also rhetoric, which Descartes fondly remembered in the *Discours de la méthode*. The heart of the course consisted in the three final years of 'philosophy', divided respectively into logic, physics and mathematics, and metaphysics and ethics. In accordance with the Ratio studiorum, the philosophy curriculum was Aristotelian in essence, and pride of place amongst commentators was given to St. Thomas Aquinas. The logic showed little influence

[1] The dates for Descartes' attendance at La Flèche are notoriously disputed, several brackets of nine years' study having been offered. In the view of most commentators, 1606–1614 seems to remain the most likely period, despite Rodis-Lewis having mounted considerable argument for the period 1607–1615 (Cf. Gouhier 1958, 158–19, but also 19–20 and Note 143; and Rodis-Lewis 1992, 25–7). The line of argument for 1607–1615 turns to a large degree on accepting Sirven's (1928, 41–46) original correction of the identity of Descartes' instructor for the 3 year philosophy course, showing it had not been the previously identified Père Fournet, but rather Descartes' later friend and correspondent, Père Noël, who had been in charge 1612–1613 through 1614–1615. (Cf. Gilson 1947, 479) Three of the four most recent Anglophone biographies of Descartes (Gaukroger 1995; Watson (2007) and Grayling 2005) support the period 1606–1614, while Desmond Clark follows Rodis-Lewis. Garber (1992) p. 5 agrees Noël was Descartes' philosophy instructor whilst not explicitly choosing between the two time frames. The outcome of this issue has little bearing on the key themes of this volume, unless, contingently, proof of one or the other time bracket also were to throw up evidence bearing on the way the dialectic of Descartes' concerns for physico-mathematics, natural philosophy and method are depicted in this volume. It obviously does have consequences for understanding Descartes' non-trivial later relations with Noël, renewed after the publication of the *Discours*.

of the sixteenth century innovations in the teaching of dialectic. It relied on Porphyry, the *Categories* and *Topics*, a small part of the *Prior Analytics* and the material on demonstration in the *Posterior Analytics*, framed in the sixteenth century introductory texts of Toletus and Fonseca. The teaching of physics, or natural philosophy, depended upon Aristotle's *Physica*, *de Caelo* and a small part of *de Generatione*, backed up by the modern Coimbrian Commentaries and others such as those of Toletus and Rubius. The third year metaphysics instruction featured parts of Aristotle's *De anima*, *De generatione* and *Metaphysics*, with commentaries by Suarez, the Coimbrians, Toletus and others, while the *Nichomachian Ethics* with the relevant Coimbra commentary formed the core of the ethics curriculum.[2]

None of this, of course, signaled that the Jesuits purveyed a closed or fossilized system of thought. Within limits, they prided themselves on their openness to new currents in natural philosophy and the subordinate, mixed mathematical sciences, as well as bits of the practical mathematical arts. It is well known, for example, that during Descartes' stay at La Flèche, the College celebrated Galileo's discovery of the 'Medician planets' as part of a commemoration of the death of Henri IV. More generally, the mathematical portion of second year philosophy curriculum largely consisted in the study of areas of mixed and practical mathematics, rather than remaining at the level of elementary geometry, arithmetic and astronomy. Such practical mathematical arts as geography, mechanics and military architecture were touched upon. It appears that Descartes was able to familiarize himself with the textbook of algebra by Father Clavius of the Jesuits' Roman College.[3] Descartes' philosophy and mathematics master, Jean François, was especially interested in the mathematical arts and in distinguishing their supposedly useful and legitimate application from the taint of suspicion of what he would categorize as magic.[4]

Their attention to mixed and practical mathematics spoke well for the Jesuits' awareness of the needs and changing aspirations of their clientele. The educated gentleman-officer was increasingly expected to command a knowledge of practical mathematical arts. This shift in emphasis in the training of the secular elite in the late sixteenth century is indicative of a temporary lowering of caste barriers to the acceptance of the mechanical arts, including practical mathematics, as elements of higher culture.[5] For those exposed to this sort of education, it could become a stimulus to the later pursuit by some, and acceptance by many, of approaches which tried to bring mixed and practical mathematics into closer contact with inquiry into matter and

---

[2] Sirven (1928), Chap. 1; Garber (1992) 5–9; Gaukroger (1995) gives the best and most fully contextualized account, 38–62, citing in turn the most important sources; see also Clarke (2006) 15–36.

[3] de Dainville (1954) 6-21, 1-9-123; Gaukroger (1995) 57–9.

[4] E. Gilson (1947) 120, 126–7, 129–30. On Clavius and the Jesuits' version of 'relating' mathematics to natural philosophy and its relation to Descartes' emerging project of physico-mathematics, after meeting Beeckman in late 1618, see below Sect. 2.5.3.

[5] For an early and forceful statement, cf. Ravetz (1975) 369. Such points about Descartes' generation may now be obvious to historians of science; they were not at the time, particularly in the Anglophonic world.

cause; that is, into the high cultural realm of natural philosophizing.[6] A great deal of the present volume will be taken up with how, why and with what consequences over time, the young Descartes pursued daring, sometimes brilliantly successful, sometimes deeply misleading and disappointing, initiatives in this very sphere where natural philosophizing and the existing mathematical fields were put into novel, even revolutionary relations. Let us therefore consider briefly, and in the spirit of an initial exposure, what might be made of Descartes' educational experience for interpreting his trajectory in natural philosophy and the subordinate mixed mathematical sciences over the years, down through the early 1630s.

First of all, Descartes' experience with practical and mixed mathematics was to be critically important for his later work, although, as we shall see, not in some vulgar sense of immediately or proximately suggesting the idea of a supposedly general method, or some sense of unification of mathematics and natural philosophy. Indeed, our main theme in this volume revolves around this problem, and so we shall need to be very careful about the categories mixed mathematics, natural philosophy, and 'physico-mathematics'. An initial delineation of these matters takes up the latter portions of this chapter. For the moment, we note only that as Descartes explained in the *Discours* (in ways needing to be taken with some grains of salt), in the several years after he left La Flèche, his exposure to the mathematical arts left as it were a fascinating if elusive intellectual and aspirational residue, which continued to intrigue him, even as his commitment to official neo-Scholastic Aristotelian natural philosophy began to wane:

> I was most keen on mathematics, because of its certainty and the incontrovertibility of its proofs; but I did not yet see its real use. Believing as I did that its only application was to the mechanical arts, I was astonished that nothing more exalted had been built on such sure and solid foundations....[7]

Even if this only vaguely reflects Descartes' attitudes at the time he left La Flèche, it is worth remembering and pondering as we go along in this volume. Impressed by the clarity and rigor of mathematics, including practical and mixed mathematics, he was also aware (1) that in Aristotelianism no proper explanations, those dealing with matter and cause, could be formulated in mathematical terms; and, (2) that the mixed mathematical disciplines were given a subordinate and non-explanatory role in the accepted neo-Scholastic map of the domain of natural philosophizing. As we shall see in the next chapter, four or five years later in the Netherlands Descartes was, with the initial help of Isaac Beeckman, to be seized by an awareness that it might be possible, under the label of a 'physico–mathematics', to render the mixed mathematical fields, such as mechanics, hydrostatics and optics more relevant to natural philosophizing so that matter and cause

[6] This curriculum mix also contributed in some cases, including presumably Descartes', to a growing awareness of new aspirations and values to be associated with natural philosophy, in particular an emphasis on operative, cumulative and correctable knowledge claims. Rossi (1970) 137–45 was perhaps the first to see the matter quite this way.

[7] AT VI, p.7; This translation is from Ian Maclean's scholarly and vigorous rendering: Maclean (2006) 9.

could be in part pursued by mathematical techniques. Soon after that, as we learn in Chap. 5, he was to inflate his aspirations to a 'universal mathematics', and then a universal method of mathematics-like tenor and procedure. And, we learn in later chapters, even after those over blown agendas failed in ways he recognized, leading him to return to the more traditional, and discursive, style of natural philosophizing, he remained focused on some important 'physico-mathematical' chromosomes in his own approach, accordingly considered by him superior to others, and at least wistfully still able to be termed 'mathematical'.

Similarly, his Jesuit education also had a number of important consequences for Descartes in the realm of natural philosophy. First, as has been demonstrated by a long line of brilliant studies, from Gilson, down through the contemporary work of Garber, Arieu, Gaukroger and Des Chene,[8] the neo-Scholastic training provided Descartes with a philosophical, indeed conceptual vocabulary which would inform much of his later natural philosophizing, regardless of how radically opposed his work became to the content and values of the dominant Aristotelianism. Moreover, in the neo-Thomist ambiance of the Jesuit school, his vision of natural philosophy took on the precise form of including a commitment to the mutual accommodation of natural philosophy and orthodox Catholic theology, according to which natural philosophy could not but serve as a necessary and efficacious propaedeutic to at least some of the elements of faith. Despite various temptations and exposures to the fideist and libertine currents of his day, Descartes was by the late 1620s to set himself against them, as well as natural philosophical novelties which intimated unsavory theological complications. We shall discern traces of this in his attempted bold synthesis in the later portions of his *Regulae* written in the late 1620s. When that daring gambit, deeply related to his youthful excursions in physico-mathematics and method failed, he further recommitted himself to the form or grammar of the Jesuits' articulation of natural philosophy and theology, if not to its content on the natural philosophy side, by redirecting his work toward an explicit system of natural philosophy with worked out metaphysical and theological underpinnings.

2.2 In Search of Proper Categories and Angle of Attack

To speculate, as we just have, about possible uses and consequences of Descartes' education for his early trajectory in natural philosophy and its related disciplines gets us only so far. It certainly prevents our slipping into simplistic tales about the meaning of his education: For example the conceit that Jesuit mathematical training provided the initial influence for Descartes' supposed later creation of a new, modern conception of mathematicized scientific thinking; or, that because his philosophical vocabulary was initially and deeply neo-Scholastic, all we are faced with is a case of subtle continuity and filiation of ideas between Descartes and his 'forerunners'.

[8] Gilson (1947), Garber (1992), Arieu (1999), Gaukroger (1995), Des Chene (1996).

No such stories will do, as we can already see. But how exactly do we pursue the more promising line of taking seriously the problem of Descartes' commitments, work, and shifting identities in natural philosophy and subordinate mixed mathematical disciplines?

The answer comes from reflecting on something about his education which is both deeper and yet more subtle and difficult to discern or properly factor into historical analysis than anything mentioned previously. It is tied up with this seemingly simple claim—*Descartes' education rendered him an adherent of natural philosophical culture, if not in the end a devotee of Aristotle and his modern renovators.* That is, as mooted in our opening Chapter, I want to attempt a maneuver which is new to Descartes studies and which should be applied across the board to all players in the so-called Scientific Revolution—I wish to focus on Descartes' emersion in a Europe wide 'culture of natural philosophy', a culture propounded to every young educated man in Europe via his neo-Scholastic Aristotelian education, whether Catholic or Protestant in allegiance. For, despite all the shifts Descartes' agendas and technical preoccupations were to undergo, we shall see that it was only in two brief, and unsuccessfully consummated moments, that he ever envisioned leaving behind and marginalizing the culture or, as we shall term it, the field of natural philosophizing. We turn, therefore, in the remainder of this chapter to exploring the categories of natural philosophizing and its related disciplines that we shall need to deploy later in our attempt to describe and explain the trajectory of the young Descartes *agonistes*.

What I mean by the culture or field of natural philosophy is not to be identified solely with the content of Scholastic Aristotelianism, because there were old and revived, as well as newly designed alternatives meant to displace it. Nor is this culture to be thought of as fossilized and static, just because it was mainly purveyed to post-adolescents in universities. Quite apart from the flux and contention within universities in the late sixteenth and early seventeenth century, the culture of natural philosophy was much more widely characterized at this time by conflict, turmoil and hence, for the keen eyed historian, displays a certain dynamic of change and process. Hence, I shall propose a model for dealing with natural philosophy as a wide, dynamic and contested intellectual culture; a 'cultural process' model, in other words, of what the game of natural philosophizing was about, its rules, its patterns of change, what a natural philosopher of radical or conservative leanings might concern himself with, and how he might conceive of his activities and his identity in the game.[9]

We shall pursue this aim in three stages: First, in the next section (2.3), we shall introduce broadly the idea of a large and contested and deeply institutionalized field or culture of natural philosophizing in which much of what we call 'the Scientific Revolution' took place. Then, in Sect. 2.4 we shall take something of a necessary detour, in order to gain some heuristic and analogical purchase on the agonal character of natural philosophy and its modes and rules of contestation. This we shall do by

[9] Note, my terminology, *'natural philosophizing'* invoking a 'doing' and 'contesting' by natural philosophers in a disciplinary domain.

looking at the agonistic knowledge making traditions par excellence which are, in complex ways, descendents of this very natural philosophical turbulence of the seventeenth century: That is, we shall examine the nature of the modern natural sciences as contested, discovery seeking, and highly rhetorically couched traditions and fields of contestation. Only then, with some valuable modeling under our belts, will we return in Sect. 2.5 to flesh out in several dimensions the details of the dynamics and rules of natural philosophizing as Descartes experienced them, knew them, and played with and upon them. With this major work done, the three final sections of the chapter will take advantage of our model to set up the remainder of the argument of this book: Sect. 2.6 will alert us of the considerable dangers posed to critical historiography by the fact that Descartes apparently believed, for a least part of his career, in the efficacy of his own general method. A way out of these dangers will be suggested, to be applied later in Chap. 7, when we actually meet Descartes the young and enthusiastic methodologist. Section 2.7 will sketch a periodization of the Scientific Revolution, between roughly 1500 and 1750 which follows from, and further articulates, our model of natural philosophizing. This will allow us to locate Descartes firmly in what we shall call the 'critical period' or 'period of civil war amongst natural philosophers' in the early to mid seventeenth century. Finally in Sect. 2.8 we shall ask, on the basis of our findings in this Chapter, 'What kind of natural philosopher was Descartes?', providing a preliminary answer which will set us on our concrete historical reconstruction from Chap. 3 onward. The reader should be aware that every single concept, category and historiographical idea in this chapter will be put to work somewhere in the rest of this book.

2.3 Constructing the Category of Natural Philosophy, Part 1—Natural Philosophizing as Culture and Process

It is often said that periodization is the indispensible armature of historical inquiry, a wise enough statement in itself; but, only half, or indeed less than half of the story. Historical understanding also requires conceptualization of the kinds of entities, structures and processes in play. Only this allows serious narration *and* explanation to be slung across a periodization armature. There is no point telling stories and giving explanations about entities and processes whose nature and dynamics have not been carefully thought out. These categories or models are heuristic and revisable, but they are ignored, or frozen, at the peril of one's life as a serious historian. So, to inquire about the 'Scientific Revolution', and the young Descartes' trajectory within it, we need categories both of periodization and of entities and processes, with the latter taking priority, starting with this Section.

In the early modern period, the central discipline for the study of nature was natural philosophy.[10] Hence, I employ the category 'natural philosophy' in strict

[10] Anstey and Schuster (2005).

preference to Science, Modern Science, new science, experimental science, etc. 'Natural philosophy' is THE appropriate historical category with which to think through our problems.[11] Natural philosophy is first of all an actor's term, but, if we metaphorically treat natural philosophy—and other important categories for the history of science—as an *iceberg*, actors' usages are merely the tip. We must also theorize the bottom of the iceberg, by modelling the structure and dynamics of the game of natural philosophizing, including points that did not or could not have been known to the players. In the intellectual construction of a model of the structure and dynamics of natural philosophizing (or of any sort of knowledge making/breaking field, tradition or sub-culture), one employs prior concepts of others and one's own, along with appeals to historical evidence and analogy. The resulting model or category then becomes an object of inquiry and tool of explanation and interpretation, meaning that the point of such models is both heuristic and explanatory: They can guide our understanding and inquiry about the past and about other accounts of the past; they can be woven into narratives and fed into explanations of players' decisions and actions; the presumed dynamics and structure of the category in question can similarly guide questions about larger processes and trends above and beyond particular actors' grasps; and finally, any such critical category is itself continually open to revision and articulation in the light of evaluation of cases to which it has been applied, and of outcomes of expert historiographical debates.[12]

Categories of analysis and of explanation/narrative are always in use in historical writing, whether or not one tries to pretend that they do not exist, or that one has particular access to divinely given eternal and nugget–like facts and/or to permanent conceptual truths of historiography. Not being conscious and critical of one's categories, and not being willing to construct and revise them, simply means that one's

[11] To place the evolution of natural philosophy, and in particular the shifting patterns of its relations to other enterprises and disciplines, at the center of one's conception of the Scientific Revolution is not novel, and more scholars are realizing the value of such a perspective, but neither is it obvious or agreed upon in the scholarly community. Many older discussions, and some contemporary ones, are marred by a tendency to lump the culture of natural philosophizing under an anachronistic label of 'science', thus obscuring the possibility of speaking convincingly about the internal texture and dynamics of the culture of natural philosophy and its patterns of change over the period. H. Floris Cohen's massive survey of Scientific Revolution historiography (Cohen 1994) illustrates that the term 'natural philosophy' has been endemically present in the literature, but not systematically theorized, often serving as a synonym for 'science' or (some of) the sciences. Recent attempts to delineate the category of natural philosophy and deploy it in Scientific Revolution historiography include Schuster (1990, 1995); Schuster and Watchirs (1990), Andrew Cunningham (1988, 1991), Cunningham and Williams (1993), Dear (1991, 2001a), Peter Harrison (2000, 2002, 2005), and John Henry (2002). Cohen (2010) makes explicit use of a version of the category of natural philosophy in his comparative and macro historical analysis of the rise of modern science in seventeenth century Europe.

[12] My understanding of the category 'natural philosophy' as constructed here is similar to that of Maravall (1973) when he speaks of an 'historical structure' (applied to the question of understanding 'Baroque culture'). See also Schuster (2012a)

The field, sub-culture or tradition of Natural Philosophy

Nature is a coherent unity, to be studied systematically under the dimensions of 1.matter 2.cosmos 3. causation 4. method

•Hegemonic Scholastic Aristotelianism[s]

Challenger Genera:

•*Neo-Platonic varieties, high tide 1580-1620*

•*Mechanistic varieties, high tide 1640-80*

•*Newtonianism[s] from 1690s*

Fig. 2.1 Natural philosophy—generic structure, competing genera

key categories are unexamined, merely trendy, and very likely to be shown to be seriously inadequate, even to one's own stated aims in narration-explanation. So, I intend here to produce a working heuristic model of early modern natural philosophy as a dynamic, elite sub-culture and field of contestation, by theorizing about its structure, dynamics and its process over time.[13]

When one 'Natural philosophized' one tried systematically to explain the nature of matter, the cosmological structuring of that matter, the principles of causation and the methodology for acquiring or justifying such natural knowledge[14] (Fig. 2.1). The dominant genus of natural philosophy was, of course, Aristotelianism in various neo-Scholastic species, but the term applied to alternatives of similar scope and aim; that is, to any particular species of the various competing genera: neo-Platonic,

[13] Thus, in this spirit, I would contend that Kuhn's original notion of 'normal science' was such an ideal typical model of the structure and dynamics of how 'a mature science' functions over time and may change and be affected by endogenous and exogenous forces. His model has been modified in important subsequent theoretical, case study and historical work by 'post-Kuhnian' sociologists and historical sociologists of scientific knowledge. In this chapter, for reasons that will become clear, we shall have cause to mobilize both Kuhn's own model and that suggested by his creative followers, respectively in Sects. 2.6 and 2.4.

[14] The common method training, allowing of course for the unending technical debates about method, its meaning, contents, scope etc., is of the upmost importance. This is not because knowledge was actually discovered and demonstrated by method. As noted earlier, modern sociology and philosophy of science has put paid to that notion, in the writings of Bachelard (1949), Kuhn (1970), Feyerabend (1975), Schuster and Yeo (1986a), Schuster (1986, 1993) and others. Rather method discourse provided universally understood packaging and rhetorical framing for claims of natural philosophical type, and by means of the tools of logic provided natural philosophical players, as subjective agents, the technical capability for reflexively criticizing, comparing, overthrowing and radically reworking the claims of others and of themselves.

Chemical, Magnetic, mechanistic or, later, Newtonian. Early modern natural philosophers learnt the rules—or template for—natural philosophizing at university whilst studying hegemonic neo-Scholastic Aristotelianism. Even alternative systems followed the rules of this game. All natural philosophers and natural philosophies constituted one sub-culture in dynamic process over time.

We should not simply equate 'natural philosophy' to Scholastic Aristotelianism. Nor should we accept the popular historiographical conceit that after about 1660 'natural philosophy' died and was replaced by an essentially different activity, Science.[15] As we shall see in Sect. 2.7, the 'Scientific Revolution, in its most turbulent or critical phase, in the early and mid Seventeenth century, was a set of transformations, a civil war, inside the seething, contested culture of natural philosophizing. That culture then continued to evolve under internal contestation, and external drivers, and variously elided and fragmented into more modern looking, science-like, disciplines and domains, plural, over a period of 150 years from 1650.[16]

That there was a European culture of natural philosophizing depended upon a High Medieval development of world historical import—the establishment of a European system of universities all teaching and arguing about variants of a Christianized Aristotelian corpus in logic and natural philosophy.[17] This fact continued and evolved right into the early and mid seventeenth century and beyond.[18]

[15] Schuster and Taylor (1996, 1997); Schuster (2002)

[16] Other contemporary knowledge systems, such as natural history and natural theology also need to be theorized in this manner and the entire set examined for their dynamics and articulations over time. For a cognate model of seventeenth century natural theology see Aldridge (2009).

[17] The distinguished historian of medieval science, David Lindberg, writes of the Christianized Scholastic Aristotelian undergraduate curriculum in the high medieval universities, 'For the first time in history, there was an educational effort of international scope, undertaken by scholars conscious of their intellectual and professional unity, offering standardized higher education to an entire generation of students.' Lindberg (1992) 212. He is pointing to the unique fact of the extensive European institutionalization of a religiously more or less acceptable version of one genus of ancient natural philosophy, Aristotelian. And we might add, generation upon generation of students was thus produced.

[18] We now know a lot more about neo-scholastic education at the turn of the seventeenth century, thanks to efforts of scholars like Ian Maclean (2007) and Dennis Des Cheyne (1996): especially about the tools and habituses of thought imbibed by years of study of the host of dense, printed neo-Scholastic texts of the late sixteenth and early seventeenth century. Picture the tens of thousands of educated men in each generation, who had been taught Aristotelian logic and related tools of thought as well as large swathes of natural philosophy derived from Aristotelian doctrine about cause, matter and how—methodologically—you get knowledge about them. That's the core of what all the players were on about—even the rebels wanted 'regime change' in natural philosophy not total destruction. This 'brute historical fact' of institutionalized acculturation of educated European men into one genus of natural philosophy is a continuing, necessary bass line, underscoring a process best understood within a sharpened and refined understanding of what the field or culture of natural philosophizing in a larger sense was all about. Hence, I hold that most of what we conceive of as the process and the products of the 'Scientific Revolution' took place within patterns of change, internal contestation and contextual shaping in this evolving field or culture of natural philosophizing. I have written several overviews of the Scientific Revolution in this style. Schuster (2002), also Schuster and Watchirs (1990) and Schuster (1990).

Although specific concepts constitutive of Scholastic Aristotelianism were displaced during the seventeenth century, this occurred inside the continuing, contested life of the larger 'field' or 'tradition' of natural philosophizing. We should not throw out the living baby of the 'culture' of natural philosophizing' with the bath water of large chunks of neo-Scholastic Aristotelianism.[19]

A Scholastic Aristotelian education taught that nature has a coherent, systematic unity; that nature not only can be studied by specific means, but that correspondingly systematic knowledge of it can be obtained. This template for the generic contents and aims of natural philosophizing applied to all jostling species of the genus Scholastic Aristotelianism and to all natural philosophical challengers. Additionally, Scholastic Aristotelianism, dominant institutionally and in the cultural experience of educated men, entrained an entire geography of knowledge: It framed the way in which other disciplines were conceived, and related to each other, and to natural philosophizing.

How natural philosophical claims were positioned in relation to other enterprises and concerns is particularly important. Some disciplines were considered superior to natural philosophy (such as theology); others cognate with it (such as mathematics); or subordinate to it. One may think of the subordinate disciplines as an *entourage* of more narrow traditions of science-like practice: (Fig. 2.2). These included the subordinate mixed mathematical sciences, such as geometrical astronomy, geometrical optics, mechanics, statics, and music theory, as well as the bio-medical domains, such as anatomy, medical theorizing and proto-physiology in the manner of Galen.[20] The members of this entourage changed and interrelated over time. In the seventeenth century, some were disputed; some were created; all changed; new or newly revamped entourage members evolved.[21] Still other fields,

[19] Hence, my category modeling can lead to the production of heuristic advice for historiographical practice. For example, whether one studies Descartes, as I do here, or the Royal Society, as I sometimes do elsewhere, *A site where natural philosophers natural philosophize is a natural philosophical site, not a non-natural philosophical site. Since the field is pan–European and cosmopolitan, a natural philosopher even alone in his study is in a natural philosophical site, and at the very least virtually in communication with some intended sub-set of the pan European natural philosophical audience. He is not a 'mind alone', opposed to some new form of communicatable and networked knowledge making/breaking in the new scientific organizations, which are also natural philosophical sites.* There might have been some new registers of natural philosophizing at the early Royal Society, as we mention below in Sect. 2.7 but no break or rupture had occurred in the ongoing dynamics of the culture. See Schuster and Taylor (1996) and Boschiero (2007) for parallel results for the Florentine Accademia del Cimento.

[20] Below, in Sect. 2.5, when we return to more detailed modeling of natural philosophizing, we will learn more about the relations that hold between a particular system of natural philosophy and its particular selection and weighting of subordinate disciplines; in other words what is meant in Fig. 2.2 by the indications (1) that subordinate disciplines can supply *support* for a system and even *shape* its content and direction, whilst (2) a given system *orders*, *prioritizes* and *imposes core concepts* upon its entourage of subordinate disciplines.

[21] Mechanics meant something different to Galileo and to Descartes, and both had left behind Stevin or Benedetti's notions of the domain. A mutant novelty, a discourse of 'celestial physics' emerged in Kepler and Descartes. See below Sects. 2.5.3 and 2.5.4, Schuster (2005), and above all Chap. 10 below.

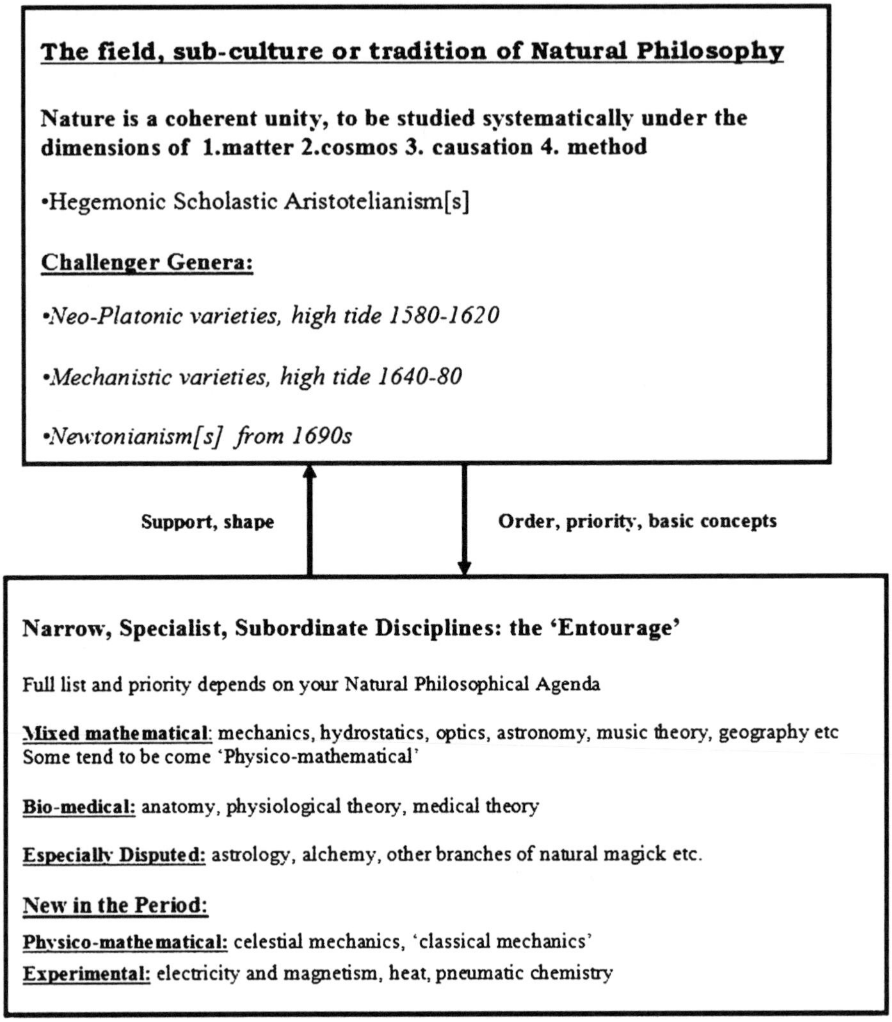

Fig. 2.2 Generic structure of natural philosophy and possible entourage of sub-ordinate fields. In a given system of natural philosophy: (1) the particular entourage of subordinate disciplines lends support to and can even shape the system; while (2) the system determines the selection of and priority amongst entourage members, and imposes core concepts deployed within them

disciplines or domains of concern were considered (by some players at least) as of some sort of relevance to natural philosophizing, as for example pedagogy or the practical arts, including practical mathematics. The positioning of natural philosophical claims in relation to other enterprises always involved two routine maneuvers: the drawing or enforcing of boundaries and the making or defending of linkages or articulations (including efforts to undermine others' attempts at bounding and linking).[22] This set–up created the 'objective field of possible moves' in which natural philosophers carried out their own specific systematizing and linking strategies—claiming new linkages or defending older ones—depending upon their respective proclivities, aims and skills, and hence producing for their own natural philosophies particular linkage profiles, with a selection and weighting of subordinate fields.[23]

This almost completes our initial examination of the category of natural philosophy. But before we take the necessary detour mentioned earlier, which will pave the way for an expansion of the model in Sect. 2.5, we need to deal with one aspect of my more complete model: its overall manner of dealing with natural philosophy as a *dynamic and evolving sub-culture*.[24] To explicate this notion, I invoke Marshall Sahlins' way of analyzing cultures as dynamic historical entities in terms of their mechanisms of change and adaptation over time to exogenous and endogenous challenges. Insisting on the need for an historical category of culture in anthropology, he argues that cultures display specificity of response to outside impingement, they are not simply imprinted upon or pushed around. The dynamics of response, over time, characterizes the culture.[25] Similarly, my overall model of natural

[22] Cf. Anstey and Schuster (2005). We shall refine the concept of boundary-work, including how we think about players' contestation about it, below in Sect. 2.5.6.

[23] This manner of conceptualizing a competitive creative 'field', or fields of modern science, of course derives originally from the seminal and suggestive work of Bourdieu (1971a, b, 1975) to be discussed in Sect. 2.4. It has some analogical applicability inside our model of natural philosophizing, which will be more apparent after our discussion of contestation in modern sciences in the next section, and when we return to the dynamics of natural philosophizing in Sect. 2.5.

[24] I put the matter this way because my full model of natural philosophizing has five major theoretical dimensions, of which this is only one. Limitations of space mean that even the more developed model to follow in Sect. 2.5 will not canvass these issues. For the record, the five theoretical dimensions of the model are: (i) natural philosophy as intellectual tradition in the manner of post-Kuhnian science dynamics with a dash of Quentin Skinner; (ii) as a competitive creative field in the manner of Bourdieu; (iii) as an evolving field of claims governed by rules of utterance, with apologies to the younger Foucault; (iv) as an historically dynamic sub-culture of the larger culture in the manner of Marshall Sahlins; (v) and as a network of institutions, in a much revised manner of Mertonian sociology as refracted through my work with Alan Taylor on the 'organization of the experimental life' at the early Royal Society. The full version of the model will be presented in my work in progress on 'The Fate of Natural Philosophy at the Dawn of Modern Science: A recasting of the plot of The Scientific Revolution.'

[25] Sahlins (1993) at pp. 25 ,15. '[Cultural orders] reveal their properties by the way they respond to diverse circumstances, organizing those circumstances in specific forms and in the event changing their forms in specific ways. Here, then, in a historical ethnography—an ethnography that extends, say, over a couple of centuries—here is a method for reconciling form and function in a logic of meaning, for discovering the relatively invariant and mutable dimensions of structures....the currently fashionable idea that there is nothing usefully called "a culture"—no such reified entity—since the limits of the supposed "cultures" are indeterminate and permeable...paradoxically... misreads a cultural power of inclusion as the inability to maintain a boundary. It is based on an underestimate of the scope and systematicity of cultures, which are always universal in compass and thereby able to subsume alien objects and persons in logically coherent relationships.' Shapin (1992) speaks of sciences as cultures in process in analogous ways.

philosophizing includes conceptualizing it as a sub-culture, tradition or field in dynamic process—defined over time by the resultant of its players' combats over claims, where some of those claims have to do with attempts to respond culturally to variously perceived, and represented, contextual structures and forces, threats and opportunities. These moves are not determined by a universal logic, may express considerable novelty, but remain specific to the (evolving) culture. This approach also allows one to deal with all types of contextual drivers or causes asserted by externalists in their explanations of the 'rise of science', as we shall see later in Sect. 2.5.6.[26] Indeed this dimension of the model is so important that I term the entire model of the field of natural philosophizing a *cultural process model.* But before we arrive there, or indeed arrive at any of the more advanced parts of the model of natural philosophy, we have to stop, detour and seek important heuristic help in our modeling from the one place in the modern academic landscape where there has been a sustained attempt to model the agonistic processes of knowledge construction and negotiation in expert communities; that is, in the realms of post-Kuhnian analysis of how mature, modern scientific traditions function. A look at these results will supply us with useful hints as to how to proceed to deal with that earlier cultural entity, natural philosophy, which, it turns out, first displayed during the heated period of the seventeenth century the sorts of agonistic dynamics that led, eventually, to its own dissolution and the formation of the earliest examples of those very modern expert traditions of scientific research which have been the object of post-Kuhnian scrutiny.

2.4 Some Heuristic Help: Modeling Modern Sciences as Unique, Agonal Traditions in Process

It is an obvious and trivial claim to assert some kind of link between the culture and dynamics of the modern sciences and the Scientific Revolution. But what if one could fruitfully link the culture and dynamics of the modern sciences to specific aspects of the process in the Scientific Revolution; that is, to some key elements of the structure and dynamics of natural philosophizing. This is precisely what I seek to do. We therefore need to acquaint ourselves with those traits of modern sciences which, I shall suggest, express competitive, cognitive and rhetorical accounting genes first implanted in Western seeking of natural knowledge during the most turbulent phases of the civil war in natural philosophizing of the early to mid seventeenth century. Reading our results backwards, we will then

[26] On internalism/externalism, Schuster (2000a), Shapin (1992). These ideas are applied to the problems of externalist explanations of the role of practical mathematics and mathematicians in the Scientific Revolution in J.A. Schuster, 'Consuming and Appropriating Practical Mathematics and the Mixed Mathematical Field, or Being 'Influenced' by Them: The Case of the Young Descartes', available on my website: http://descartes-agonistes.com.

have some hints about how further to model the structure and dynamics of natural philosophizing, especially in its turbulent phase which coincided with the adult life of René Descartes. To these ends, we have to examine an ideal typical model of the agonal dynamics of modern scientific disciplines, grounded, I shall argue, in reflection on state of the art findings in contemporary HPS and sociology of scientific knowledge (SSK).[27]

We need to grasp the historically unique and, on reflection, quite peculiar tradition dynamics of the modern sciences. Indeed, to think of the modern sciences as consisting in social, cognitive and rhetorical accounting *traditions* goes against three centuries of both popular and elite Enlightenment rhetoric emphasizing the anti-authority, anti-tradition essence of 'Science', and focusing attention upon heroic discoveries by isolated individuals, struggling with the sole help of 'scientific method', to extract from nature discoveries of significant fact and theory. The realization that each of the sciences is a tradition of theorizing, material practice, social organization and communication is one of the lasting achievements of critical history, philosophy and social studies of science of the mid to late twentieth century.[28] A related advance is the realization that the discourse concerning isolated, non-tradition bound and method-wielding heroic discovers is not an accurate representation of how the sciences work, but rather an *accounting rhetoric* used within the sciences, their pedagogy and public representations as part of the mechanisms of contestation and accounting for change.[29]

A corrected, or post-Kuhnian reading of Thomas Kuhn's model of science dynamics is a good entry point for the conclusions we need to canvass. Kuhn,

---

[27] The model presented below is ideal typical. It is not meant to capture the precise social and cognitive dynamics of any particular modern (that is, post early nineteenth century) scientific discipline. As an ideal model, it invites complexification on a case by case basis by considering variants, deviances and even emerging long term shifts affecting the sciences as a whole. One suspects that the sorts of ideal models arising from post-Kuhnian thinking in HPS and SSK are better attuned to what Ravetz (1971) called the classical academic science of the late nineteenth and early twentieth century, rather than the industrial/military science of the mid and late twentieth century or the emerging post-modern transdisciplinary sciences of today, which are more than ever deeply engrafted onto government and corporate funding drivers, and strongly tinged by deliberate plays for attention from the educated public and elite policy makers, thus diluting and shifting the classical point of reference in peer competition and approval, via expert communication networks and status systems.

[28] I put the matter this way, because where the writ of 'revolution and rupture' has not run in imagining 'Science', it has not been unusual to think of Science, or the sciences as old fashioned history of ideas traditions, consisting in so-called filiations of ideas, plays of 'forerunners' and 'final accomplishers' and the like. Cf above Chapter I, Note 25.

[29] For the literature on the politics and rhetoric of scientific method, see above, Chapter I, Notes 16 and 17. 'Method-talk', as I call it, is flexibly used by players inside science to account for achievements, failures and allocate credit. It is part of the self-identity of many practicing scientists and an important part of the public imaging of science and its constituent disciplines. We shall see that much of this also applies not only to Descartes' own illusions about method, but also to those of some of his loyal scholars.

properly understood, was fully committed to the idea that the sciences are many, not one Science, and that his theorizing was aimed at providing an ideal typical account of how any given mature science functions, the motor of tradition dynamics in any given science as it were. In simplistic readings of Thomas Kuhn's view of this motor or tradition dynamics, one has rigid—frozen—paradigms facilitating puzzle solving research, until dysfunction, crisis and revolution install a new puzzle solving paradigm, equally rigid. Against this, post-Kuhnians have explicated 'normal science' dynamics using micro-sociological tools.[30] In this approach the cultural resources in play in a tradition of research, and constituting that tradition at that moment, are constantly subject to re-negotiation and modification. Suppose a problem is solved by advocating a shift in some aspect of 'the paradigm', however so slight. This means the problem solution involves feed-back alterations to the paradigm—conceptual, instrumental, normative. Such alterations—if negotiated into place by the expert community[31]—carry over into subsequent rounds of problem-solving, where further alterations may be negotiated. Post-Kuhnian historians and sociologists of science call such negotiated alterations of the paradigm 'discoveries', especially when they involve the conceptual/theoretical 'objects of inquiry' in the discipline[32], rather than, say, its instrumental techniques and standards, or norms of adequate procedure and argument. So, normal scientific research involves as its core aim and *raison d'être*, 'discoveries', negotiated significant modifications of the paradigm, or prevailing disciplinary culture at any given moment in its life.[33] Modern sciences are expert research traditions, in which claims are constantly made about 'discoveries', which are contested, debated and negotiated.[34] The acceptance of such a claim (often in quite revised form) into the working resources of the tradition affects for the time being both the tradition make up and the nature and directions of immediately

[30] Ravetz (1971), Schuster (1979), Barnes (1982), M. Mulkay (1979), Latour and Woolgar (1979), Knorr-Cetina (1981), Collins (1985).

[31] Of course the form of the discovery claim negotiated into place, and accounted back to the presumed individual discoverer, can differ greatly from that originally published, let alone imagined, by the first inventor[s] of the claim.

[32] The expression '(intellectually constructed) objects of inquiry' is Ravetz's (1971) term of art in his own early and brilliant sophistication of Kuhn's original model of 'normal science'.

[33] This 'post-Kuhnifies' the partially separate development of the so-called attributional model of scientific discovery. (Brannigan 1980, 1981; Schaffer 1986); For a textbook level exposition of a case study of these issues of post-Kuhnian notions of discovery and 'revolution' see Schuster (1995a) Chaps. 4 and 5.

[34] Again standard HPS fare: a claim to a significant discovery is not just a claim to have found some atheoretical nugget of fact in the world (not possible in any case); but a claim, simultaneously, to introduce new or changed reports about external affairs linked to some modification/renegotiation of previously accepted conceptual framework. As Kuhn more or less said many years ago, a discovery claim is not just in the form 'that x is the case' but also 'what therefore in revised theory terms is at stake'. (Kuhn 1977) If nobody's previous or newly minted theories are at stake, the discovery claim is about everyday trivia stated in ordinary language: e.g., 'I found the pencil I lost yesterday'; rather than a really big and significant discovery claim like this one made by Lavoisier in the late eighteenth century: 'Gents, 'phlogiston' does not exist, but 'oxygen' and the 'weightless fluid of heat' (caloric) do.' See Schuster (1995a) Chaps. 4 and 5.

subsequent work. Hence, any *facts* or *theories* originating as significant discoveries of this type (and hence themselves the product of contestation and negotiation at their births) have non-trivial subsequent histories as play continues in the discipline.[35] (Remembering of course that any fact in such a situation is necessarily and always a heavily 'theory-loaded' fact.)

According to this sort of ideal model, modern scientific disciplines constitute unusual sorts of traditions in which tradition–modifying alterations are constantly sought, and fought over. The only people who can accredit, and subsequently make use of, a 'significant discovery' are the very peers and competitors of a claimant; and the highest status and rewards accrue to the members who are credited with the most significant of such 'discoveries'.[36] Note, also, that this sort of post-Kuhnian modeling throws into high relief the sorts of rhetorics and accounting resources that players (and other commentators, popular or expert historians of sciences for example) use in self-understanding their roles and moves, and in representing them to each other, and to wider publics. These rhetorics involve resources for story telling about universal scientific methods, heroic, isolated discoverers as well as about the bad influence of this or that sort of bias or prejudgment; but, they do not capture the actual dynamics of these traditions as we now understand them. Instead, they are part of the very weave of how the traditions function, since they are, amongst other things, actor's possessions.[37]

---

[35] There are other important implications, crucial in comprehending post-Kuhnianism, but less important for our present concerns: For example, we have learned to see paradigms, the working core of a research tradition at a given point in time, as fluid and constantly open to greater or less renegotiation, around 'significant discoveries'. Hence, 'revolutions' are merely relatively large renegotiations within continuing traditions, not battles between armies from incommensurable intellectual planets. *Normal practice within a tradition* should be seen as a process of social negotiation of change, continually shaped by the distribution of resources and power amongst the players.

[36] These traditions, by virtue of their own living dynamics, are not and cannot be frozen into shape over any significant period of time, and even the greatest authority is in principle, and in fact, subject to revision or even rejection as play unfolds. What is taught to initiates in advanced textbooks in one generation is typically radically different from what is taught in the next. However, this should not be read simplistically: At any given moment the cultural package in play in a scientific tradition is highly structured. Some tools, concepts, standards, and even instruments and protocols are more deeply sedimented into the main line of current research trajectories than others, and hence will be subject to revision and renegotiation in different ways, and on different time scales, than elements of the tradition that are more marginal and in play. Think of Newton's laws of motion. Basic to mechanics and celestial mechanics as they have been since the early eighteenth century, they have been changed in mathematical expression (Euler) and generalization of systematic presentation (Lagrange, Hamilton); and they have been restricted or changed in domain of application (advent of special relativity, and trajectory of subsequent developments). So, fundamentals, the equivalent of holy writ, there may be, but only 'for the time being and until further notice', as the early Edinburgh and Bath 'Schools' of SSK taught us always to remember.

[37] Let us put this another way: There are multiple scientific traditions (another point always stressed by Kuhn). We now know that not one of them has an essence—by rupture or accretion—only endless, competitive claim-making and claim breaking and a shifting consensual tool kit of theory, standards and hardware. This plays into new understandings about supposedly general, universally efficacious 'scientific methods': they cannot explain what they claim to explain, the essence of scientific practice in any and all sciences. Rather, they are attractive, indeed, I would argue *mythopoeically seductive*

Post-Kuhnian modeling of scientific tradition dynamics also depends upon and further stimulates inquiry into the organizational structures and dynamics of such fields and how these are intimately involved in their knowledge-making, knowledge-breaking dialectic. Pierre Bourdieu's approach is particularly striking, especially for our purposes here, because it again highlights the historical peculiarity of scientific traditions, and because it may even be applied, with care, to certain moments in the dynamics of the natural philosophical field.[38] Bourdieu places members of any creative tradition, including science, as players in a field, in a peculiar agonistic, mutually competitive, relation, involving an economy of material and symbolic resources, strategies, positions etc. Bourdieuian players seek a monopoly of the sorts of cognitive and social power at stake in their particular field: To try to survive in their field they have certain amounts of symbolic and material resources (or capital) which they can deploy, strategically, in attempts to secure more resources and more power over the determination of the social and cognitive stakes at risk in the field in the next rounds of play. For Bourdieu, unlike ethnomethodologists or discourse analysts, a 'system of objective relations' exists at any given moment among the positions already won and occupied in the field, via previous rounds of struggle. Bourdieu insists that the system of relations should not be reduced to or conflated with the micro-interactions and moment to moment strategies 'which it in fact determines'. There is more to the field than the ethnographically recordable discourse and posturing of the players.[39]

Let's recall our conclusion that HPS/SSK theorists have concentrated upon the idea of players' bids to alter significantly, but not catastrophically, the terms of the

accounting resources, deployed within these traditions by the members for self-understanding and for packaging of one's own claims, and attacking those of opponents (as we shall see in detail in Chap. 6, on the basis of our preparations for the deconstruction of Descartes' method in Sect. 1.3.1 and below Sect. 2.6.

[38] An application Bourdieu himself does not envision, indeed he writes as though Science were one generic field, rather than as though he were modeling, in a generic way, any given scientific field. His model of course is an ideal type to which empirical fields approximate. Bourdieu (1975, 1971a, b).

[39] Cf. the discussion above, at Note 23 and corresponding text. Bourdieu says the system is 'objective'. We need to interpret this claim into the terms of historian's craft: It means that the field exists as an analyst's model, an historian's model of the internal political economy of the field at a given moment. As in any model in historiography—for example my model of natural philosophy presented below, or the post-Kuhnian model of research dynamics in a scientific tradition—it is an intellectual construct, category, constellation of concepts, constructed using social theory, bits of other historical findings, and appeals to evidence about the field or discipline in question. It then functions, as Bourdieu suggests, as the ultimate object of study *and as* an explanatory resource for understanding particular plays and processes in the field. Such models of fields and traditions need to be constructed by historians as objects of inquiry/objects providing explanations. It is not a 'bad' thing for historians that fields or traditions exist in this manner—that is the condition of, and the nature of, our knowledge of them as historians. Nor does it mean as tendentious post-modernists proclaim, that no material reality past or present exists or can be referred to by these models (as first cousins to theories in natural sciences, they are as good or as bad about 'reality' as are the natural sciences) We shall apply these same methodological reflections to the model of natural philosophy as a tradition of practice/field of discourse.

objects of inquiry—thus making normal science dynamic and reducing the plausibility of classical Kuhnian revolutions. In a complementary manner, Bourdieu sees the strategic imperative of players, given their different positions and resources as follows: A player attempts to produce claims that are both do-able within the limits of his symbolic capital and most likely to prove significant and attractive to his competitors. It is these peers who accredit his work by taking it up and redeploying it in their own subsequent construction of bids: Just as the attractiveness and significance of the first player's bid to these same competitors depended upon his having made use of and redeployed their claims and those of their common predecessors. Whilst Ravetz, Barnes and others saw the dynamics of a scientific tradition in post-Kuhnian terms as a negotiated drift of concepts, standards and aims, Bourdieu focussed on the motivated, tactical play of differentially resourced and placed players. But what they are playing for—the production of non-trivial, new claims that will be taken up and used by peer-competitors—maps directly onto the post-Kuhnian conception of ongoing negotiation into place of 'discoveries', which shift the terms of practice in subsequent rounds of research. Bourdieu can thus be read as showing us how to deal with the internal political-economy of a field, the agon that drives on the play of negotiation of the conceptual fabric and tools, glossed and understood as 'making discoveries'. So a tradition of science gets a social and political 'inside' and a motor, and we may speak of 'agonistic traditions of scientific practice'.

In sum, we wind up with a post-Kuhnian/Bourdieuian model of modern sciences as peculiarly agonistic traditions, compulsively and continuously manufacturing and negotiating novel shifts of tradition practice, and awarding credit for these shifts using a rhetoric of individual methodologically based heroic discovery. We can now return to the main line of our argument, using the insights just gained to articulate further our model of natural philosophizing begun in Sect. 2.3.

2.5 Constructing the Category of Natural Philosophy, Part 2: The Dynamics and Rules of Contestation of Natural Philosophizing

Our focus now will be on the modes and types of contestation and competition amongst natural philosophers and the rules (revisable and negotiable of course) of such engagement. Our modeling will be thematic, unfolding in terms of a half dozen further dimensions of our model. But our modeling will also be recursive and cumulative, in that discussion of earlier dimensions will be applied and articulated by discussion of later ones, while the deferred discussion of certain dimensions is dictated by the need first to sediment into the model more basic points. Additionally, although the presentation of the model is generic and meant to be applicable, with modifications, at all stages and phases in the unfolding of the process of the Scientific Revolution, between the early sixteenth century and the mid eighteenth century, we shall keep an eye on the heightened contestation that marked the critical phase of the Scientific

Revolution—the period of civil war amongst natural philosophies—through which Descartes lived and worked.[40]

First we should note what amounts to an objective condition of the field:[41] that virtually any natural philosophical utterance, by any player, was ultimately referred back to a template initially learned through neo-Scholastic training in Aristotelian natural philosophy which taught the possibility of systematic, unified and true knowledge of nature (*Scientia*), expressed through systematically related doctrines of matter, cosmology, causation and method. Superimposed upon this in the critical period of the early to mid seventeenth century was the fact that Scholastic Aristotelianism, with its extensive and hegemonic institutional base, provided the target of strategies of displacement and alternative institutionalization, whilst competition amongst members of different broad genera of natural philosophizing—Aristotelian, neo-Platonic, Magnetic, qualitative atomistic, and finally mechanistic—also heated up, as has been long recognized.

2.5.1 Articulation on Subordinate Disciplines: Grammar and Specific Utterance

As mentioned in Sect. 2.3, every natural philosophy necessarily had a *profile of linkages or articulations* onto a selection and weighting of subordinate fields of inquiry. Natural philosophers had different interests and skills within the *entourage* of subordinate disciplines, and even different lists of what was within or without its boundaries. Onto these structural conditions in the grammar of natural philosophizing, the critical period of the early and mid seventeenth century superimposed increasing competition amongst natural philosophers to co-opt and shape entire members of the entourage of sub-disciplines and to practice them under the aegis of one's preferred natural philosophy. This phenomenon was an index of contestation and was objectively intensified by the fact that the subordinate fields had been more intensively cultivated during the later sixteenth century than previously, displaying more dense interrelations amongst themselves.[42]

---

[40] As noted earlier: the penultimate section of this chapter will outline the three main phases or stages in the Scientific Revolution correlative with this sort of modeling of natural philosophy and its dynamics. This will help us place Descartes in the critical or civil war phase, and also aid in our discussion of 'what kind of natural philosopher Descartes was' in the final Section of this chapter.

[41] The term 'objective' is used here in the sense of Bourdieu (see above Note 39), whereby we denote the (model-derived) organization and dynamics of a competitive field, existing above and beyond the immediate control, or even necessarily the understanding, of actors in the field, and not capable of being instantly or unilaterally modified by the actions of such players in their respective micro contexts. These notions may of course be related back to the iceberg metaphor offered earlier, at the commencement of Sect. 2.3 and the related historiographical observations laced into my argument.

[42] As is argued below in Sect. 2.7, in describing the main phases of the Scientific Revolution and Descartes' place in that temporal process.

As usual with our model, we must first ask what in general was involved the articulation of a natural philosophy to a subordinate field at the level of grammatical possibility, before examining particular, highly contested co-optations of such fields. Consider the situation of the mixed mathematical fields, under Aristotelianism, where they were considered to be intermediate between natural philosophy and mathematics. A natural philosophical account of something was an explanation in terms of matter and cause, but for Aristotle, mathematics could not provide that. The mixed mathematical sciences, such as optics, mechanics, astronomy or music theory, used mathematics not to provide explanations, but instrumentally to represent physical things and processes mathematically in ways that might be useful but certainly were not true to reality as defined by natural philosophical explanation stories of matter and cause. For example, for Aristotelians, the investigation of the physical nature of light would fall straightforwardly under natural philosophizing, an issue of invoking appropriate principles of matter and cause. In contrast, the mixed mathematical science of geometrical optics studied ray diagrams, in which geometrical lines represented rays of light, and phenomena such as the reflection and refraction of light were dealt with in a descriptive, mathematical manner. This might be useful, but it was, according to Aristotle, incapable of providing proper explanations, dealing with the physical nature, properties and causal behavior of light.[43]

The question of the relation between the mixed mathematical disciplines, on the one hand, and the 'superior' discipline of natural philosophy, which did the real explanatory work on this conception, was thus dominated by the entrenched, grammatically definitive, Scholastic viewpoint. However, as the competition amongst differing approaches to natural philosophy heated up in the early seventeenth century, many natural philosophers hostile to Aristotelianism proposed a more central explanatory role for mathematics in natural philosophy, and sophisticated Scholastic Aristotelians also began to loosen the Aristotelian marginalization of mathematics as non-explanatory. And it is here that we begin to see a competitive dynamic develop, out of attempts to bend or elude the template, the 'declaratory' rules of subordination of the mixed mathematical sciences to Aristotelian natural philosophy.[44]

Geometrical astronomy, the exemplary case of a mixed mathematical science, provides illuminating insights. The fine details and elaborate geometrical tools of Ptolemaic astronomy fell outside any plausible realistic interpretation, offered merely appearance-saving geometrical models and could not provide natural philosophical explanations in terms of matter and cause. However, a deeper grammatical

[43] Cf. Chap. 1 Note 12.

[44] I term the widely taught rule of subordination of mixed mathematics to natural philosophy 'declaratory' to denote that it was publicly proclaimed, but not necessarily binding or agreed to by relevant players, as we shall see in more detail in Sect. 2.5.3 below. The usage mirrors the distinction in U.S. cold war nuclear strategy, between publicly available and academically discussed declaratory doctrine, compared to then secret actual war plans within the military establishment. See Mark Rix (1997).

gaze clearly shows that the fundamental concepts of Ptolemaic astronomy were shaped by Aristotelian natural philosophy: the finite earth-centered cosmos, the distinction between the celestial and the terrestrial realms, the primacy of uniform circular motion. Hence, even in the relations of Aristotelian natural philosophy to Ptolemaic geometrical astronomy, there were some, albeit thin, linkages of a causal and matter theoretical nature that grounded the discipline and linked it to its 'parental' natural philosophy. When Copernican astronomy came to be hotly debated in the later sixteenth and early seventeenth centuries, it was not as an instrumental predictive device, but rather as a system with realistic claims about the cosmos, implying the need for a non-Aristotelian natural philosophy, able to explain its physical workings. However, the articulation of a natural philosophy to a mixed mathematical science could be much looser than the Copernican example implies. As just noted, under Aristotelianism geometrical optics consisted largely in geometric ray diagrams, their rules of construction and a set of canonical puzzles, such as the behavior of mirrors, the rule governing reflection, the explanation of the rainbow and other curious optical effects. Broadly speaking, virtually any natural philosophical theory of matter could have been used to provide an explanatory 'voice over' for this science, from Scholastic 'propagation of species', through the transport of atoms or the propagation of neo-Platonic immaterial substances, to the mechanists' passage of light corpuscles or propagation of mechanical pressures or tendencies to motion in a medium. Only later, during the critical phase of the Scientific Revolution, in the optical work of Johannes Kepler and René Descartes, do we begin to find attempts to bring into more intimate and organic relation new findings in geometrical optics and the matter/cause considerations of their respective natural philosophical programs. Both of these examples illustrate the ways in which articulation of a subordinate field to one's brand of natural philosophy involved acceptance or non-acceptance (or bending) of the template Aristotelian rules, and also entailed that the discipline in question should be conceptually flavored (in terms of matter and cause explanations) and pursued as part of, and in support of, that favored natural philosophy—a generic type of gambit which we shall discuss under the contemporary label 'physico-mathematics' in Sect. 2.5.3.[45]

But there was an even more radical way in which concern with a putatively subordinate field could be played as a gambit into the contest of systematic utterances, for, in a way, entire natural philosophies could be launched, or differentiated off from a broader genus, by borrowing their core conceptual and normative resources from a particularly privileged, more narrow discipline.[46] Indeed, and not

[45] This kind of move operated at an individual basis, but over time, such 'physico-mathematizing' moves could themselves aggregate and form patterns of largely unintended change in the subordinate disciplines in question, as we shall mention in Sect. 2.7 and Note 64 below.

[46] For example, what differentiated natural philosophies of a Chemical type from the wider set sharing neo-Platonic ontologies was the way they linked the more widely shared neo-Platonic ontology and commitment to the possibility of natural magic to their own particular concern with the content and value structure of chymical arts and practices, including especially the use of chymical knowledge in medicine. This, in effect, was the natural philosophical master stroke of

surprisingly, articulating one's natural philosophy onto a favored version of a favored discipline was a two edged sword, because natural philosophical opponents would then be stimulated to co-opt and 'sanitize' (of opposing natural philosophical valencies) the domain in question: Perhaps the most profound level at which this strategic battle was carried out was where entire disciplines and their value structures were at stake.[47]

2.5.2 *Find or Steal Discoveries, Novelties or Facts, Including Experimental Ones*

There was, especially in the critical period of the Scientific Revolution in the early to mid seventeenth century, a competitive production of novel experiments and facts, accompanied by scrambles to deflect, co-opt, steal or reinterpret others' claims, whether amongst nominal members of the same natural philosophical genus, or across such families.[48]

Paracelsus to whom the Chemical Philosophers of the critical period looked for inspiration. The varieties of mechanical philosophy battened upon and projected the supposed meanings and promise of mechanics, their construction upon metaphorical amplifications of the supposed content and meanings of various strands in the domain of mechanics being obvious, although space requirements mean we leave them tacit at this point. What were constructed were still natural philosophies, within the common field of natural philosophizing, but the Aristotelian limitations on the rules or terms of construction were being radically challenged and shifted. As we shall learn in Chap. 3, Beeckman's corpuscular mechanism was keyed to a reading and amplification of dynamical interpretations of mechanics, as in the pseudo Aristotelian *Mechanical Questions*. Descartes' corpuscular-mechanism, surprisingly was keyed in part to the purely static mechanics and hydrostatics of Stevin (and Archimedes) much overlaid as it developed with material from his own 'physical' optics. (Gaukroger and Schuster 2002; Schuster 2000, 2005)

[47] So, versions of the Chemical philosophy depended for both technical and value orientation on the notion of a spiritualized yet practically productive alchemy. In this energized and articulated spiritual form, alchemy powerfully expressed moral-psychological aspirations, a search for redemption through esoteric knowledge and successful practice. These powerful sentiments were partially shared, and certainly co-opted, in the programs of Bacon, Descartes and their later seventeenth century followers. For mechanists, the nature and 'control' of alchemy was therefore a particularly strategic issue. In Bacon, Descartes and their mechanist followers, the values and aims which Paracelsianism and later the Chemical philosophy invested in alchemy were co-opted, sanitized of radical political and religious resonances and made acceptable to intellectually progressive but socially conservative elites, a ready audience for the mechanical philosophy. Alchemy itself was de-spiritualized and reduced to applied mechanistic matter theory, whilst the search for personal justification and social benefit would now be achieved through proper method and well grounded results, rather than esoteric insight and wisdom.

[48] As we shall see later, in the succeeding phase in the late seventeenth century, the emphasis falls more on production of one's own novel facts and experimental outcomes. This contrast correlates with there having been more contestation about systems in the critical period, and more contestation within and about crystallizing more narrow domains of inquiry in the later seventeenth century.

Note first of all, that any given natural philosophy was capable of stimulating new developments—discoveries of fact, production of new instruments or experiments—conditioned and shaped by the natural philosophy in question. Aristotelianism itself could still provide deep conceptual orientations for narrow specialist pursuits, of which those of William Harvey, extending the deeply Aristotelian 'comparative anatomy' program of Hieronymus Fabricius at Padua, are only the best known.[49] Aristotelians continued to contend about experimental discoveries and instruments well into the middle of the seventeenth century. The novelties in Gilbert's work heavily conditioned by, and in turn affecting the shape of, his neo-Platonic natural philosophy are well known.[50] Similarly, the manner in which Kepler's, optical, astronomical and celestial mechanical discoveries were shaped by his version of a neo-Platonic philosophy of nature will be touched on below.[51]

The increasing imperative and willingness to pursue novelties and embed them within one's own natural philosophical agenda was a remarkable phenomenon, and one bound to alter the cozy world of Scholastic 'teaching and learning' (to use a modern policy epithet). But, this did not simply amount to a lust to fill cabinets of curiosities: *To be important in the history and dynamics of natural philosophizing, novelties had to be pursued and coveted within and for natural philosophical purposes.* Nor was competition to produce novelties and discoveries the whole story, because at this stage natural philosophers vigorously attempted to appropriate the discoveries and novelties of others, or to negate them. (Only later was genuine novelty of claim seemingly required.) Moreover, all this appropriation or negation was tactical; that is, if a discovery or claim was particularly significant in the architecture of a competing system that claim had to be appropriated, down played, reinterpreted or neutralized.

So, in a nice example, Harvey's ultra significant claims about the motion of the heart and blood became a target in an extended game of inter-systemic competitive football: Descartes was happy to appropriate Harvey's epochal, yet clearly Aristotelian based discovery of the circulation of the blood and motions of the heart, radically altering the latter (to the point of arguably contradicting it) to fit his mechanistic program in physiology.[52] Within his radical Chemical natural philosophy, Fludd endorsed the discovery of his friend Harvey, but invested

---

[49] Cunningham (1985).

[50] William Gilbert's *On the Magnet* (1600) is arguably the most influential and impressive new natural philosophical gambit of the turn of the seventeenth century. To call Gilbert 'the father of electrical or magnetic science' rather misses the point that his program involved a new natural philosophical agenda and content, on which see below, Note 55, as well as, Sect. 12.5, where we discuss Descartes' co-optation of Gilbert's work in the *Principia philosophiae*.

[51] Chemical natural philosophies were not bereft of new claims that were quite plausible to a wide range of contemporaries, as illustrated by Paracelsus's iatrochemical treatments and later by van Helmont's chemical novelties, such as the beginning of the construction of the concept of 'gas'. Pagel (1982), Hannaway (1975).

[52] Descartes, *Treatise on Man*, AT XI, 123–6, 167–70.

its meaning with mystical connotations in ways that only committed aficionados of his natural philosophy could appreciate. The tactical cross fire became even more entangled when Gassendi, another early mechanist of quite different stripe from Descartes, and in competition with him, bid to refute Fludd's interpretation of the meaning of the circulation, before going on to reaffirm, against Harvey (and Descartes) the Galenic pores in the septum of the heart on the basis of first hand witnessing of anatomical facts! For Gassendi, this *Galenic claim* vindicated the identity of venous and arterial blood, one of Harvey's marquee claims. Hence, with Harvey, Gassendi endorsed the 'anatomists' way' of first hand experience, yet also preserved a key conceptual claim of Galen, the 'physiology expert', that both Harvey and Descartes were determined to kill off, or at least fatally co-opt.[53]

What was at issue here was not merely staking a first discovery claim. The players were happy to co-opt, and reinterpret, each others' claims. Nor was symbolic capital assigned just to new matters of fact.[54] Novelty, discovery and dramatic observations of matters of fact were all in play, but often second hand, since borrowing and renegotiation were endemic, because, and this is crucial, the entire contest was about systematic natural philosophical advantage, not toting up of unique, novel discoveries. That pursuit would gain more privilege later in the century, but not during the critical period of natural philosophical turbulence.[55]

---

[53] On Fludd and Gassendi's maneuvers and negotiations: Debus (1977) 206–224, 253–279; Debus, (1970). On the anatomists' way, Wear (1983, 1990)

[54] For example, Gassendi's observational claim only confirms Galen, and is subservient to the larger natural philosophical contestation in which he is involved.

[55] Descartes' extended strategic encounter with Gilbert's work on magnetism, in his *Principles of Philosophy*, a case of massive co-optation of previously claimed, often dramatic novelties, illustrates all the above points. What was novel in Gilbert's experimentation was co-opted by Descartes, without the addition of a single new experiment. For Descartes the nub of the encounter lay elsewhere. Gilbert's natural philosophical exploitation of the magnet was dictated by his concern to establish a novel system of Magnetic natural philosophy of distinctly neo-Platonic flavor and embodying and supporting a modified Tychonic cosmology. This was the 'significance' of the magnet work that had to be appropriated, reframed, and tamed to the imperatives of Descartes' program. Gilbert's natural philosophizing of the magnet was too important and impressive a gambit in the natural philosophical field to be ignored by his natural philosophical competitors. So, Descartes efforts were directed at re-glossing Gilbert's experimental work in mechanistic terms, rather than at extending the number and type of magnetic experiments. Descartes devoted considerable attention to preserving and capturing the 'cosmic' significance of magnetism, the keynote of Gilbert's system. He replaced Gilbert's story of the cosmos making and binding role of the spiritual magnetic force with a mechanist's story of an equally cosmic magnetism which was now the purely mechanical effect of a species of corpuscle of particular, and peculiar, shape and size, moving in and through suitably configured aggregations of ordinary 'third matter'. We shall recur to some of these points when commenting on the 'system-binding' strategies of Descartes in Chap. 12, but there we will go beyond the above mentioned rather defensive tactics of Descartes to show how his co-optation of Gilbert was actually part of a hitherto little noticed, but vast, novel systematizing gambit which resides at the heart of the *Principia*.

2.5.3 *Bend or Brake Aristotle's Rules About Mathematics and Natural Philosophy: The Gambit of 'Physico-Mathematics'*

Historians have traditionally talked about a movement to 'mathematicize science', invoking Kepler, Galileo, Descartes and others in the early seventeenth century stages of a process leading to Newton. These developments are better understood as products of contestation and renegotiation in a particular corner of the natural philosophical field, involving challenges to the dominant Aristotelian template rules about how the mixed mathematical sciences should relate to natural philosophy. I address this problem making use of the category 'physico-mathematics', which, like natural philosophy, is both an actors' term from the time, and a category to be fleshed out for historiographical use. Exploring physico-mathematics throws more light on the dynamics of natural philosophical contestation in the critical phase, and illuminates what used to be denoted by the term 'mathematization'.[56]

Recall our sketch of the dominant Scholastic view of how the mixed mathematical disciplines related to the 'superior' discipline of natural philosophy, where we cited examples of attempts to articulate geometrical astronomy and optics much more closely to anti-Aristotelian natural philosophies, bringing the matter and cause dimensions of the natural philosophy into play inside the target discipline. This is what one means by players attempting to render the mixed mathematical disciplines more physico-mathematical. It is not the *mathematization* of natural philosophy, but the *physicalization* (tighter natural 'philosophication' as it were) of disciplines taught by Aristotelianism to be merely instrumental and non-explanatory.[57]

---

[56] The category of 'physico-mathematics' was first systematically explored as a thread in the process of the Scientific Revolution by Dear (1995). Gaukroger and Schuster (2002) first explored in detail what the category meant to the young Descartes in relation to his work on hydrostatics, with Beeckman (See Chap. 3 below).

[57] On the terminology of 'physicalization' of the mixed mathematical sciences, rather than mathematization of natural philosophy, the following genealogy should be noted: Gaukroger and Schuster (2002) 538, 545, 547 came close to saying this, as did Schuster (2002) 347. The conception has thus far been made clear in the following conference papers: J.A. Schuster, '*Descartes agonistes—The 'Real' Descartes Stands Up: How the agendas, identities, rebellions, successes, failures and delusions of 'youth' (1618–1633) generated the historians 'mature Descartes*', Invited Lecture for 'Nacht van Descartes', Descartes Centre for the History of the Sciences and the Humanities, University of Utrecht, and Studium Generale, University of Utrecht, October 2008; John Schuster, *'What was Seventeenth-Century Physico-Mathematics?'* for the session on 'Connecting Disciplines: Mathematics, Natural Philosophy and Reason in the Early Modern Era,' Sixth Joint US/UK/Canadian History of Science Societies Quadrennial Conference, Oxford University, July 2008; J.A. Schuster, *'From Natural Philosophy to Science(s): Transformations (Intended and Unintended), Not Ruptures, in Early Modern Knowledge Network—the Disputed Case of the Early Royal Society,'* First International Conference of ARC Network of Early European Researchers (NEER), University of Western Australia, July 2007; and J.A. Schuster, *'What was the Relation of Baroque Culture to the Trajectory of Early Modern Natural Philosophy,'* Second International Workshop of the Baroque Science Project, Unit for History and Philosophy of Science, University of Sydney, February 2008. The conception of physicalization of the mixed mathematical sciences is

Outcroppings of such 'physico-mathematical' initiatives began to appear in the sixteenth century, for example, regarding the natural philosophical status of mechanics.[58] The heightened natural philosophical contestation of the early seventeenth century intensified the proliferation, and competition amongst, physico-mathematical gambits, a number of which can be identified in the period. There were competing varieties of physico-mathematics: (1) As Peter Dear has found, some leading Jesuit mathematicians pursued what I would term a 'conservative' sort of physico-mathematical program within the confines of Scholastic natural philosophy and its institutions. They argued that the mixed mathematical fields should enjoy a status as 'separate but more or less equal' to natural philosophizing' thus in a way liberating the pursuit of the mixed sciences from Aristotelian constraints (but without fully cashing out their potential to deal with matter and cause).[59] (2) As noted, there were attempts reaching back into the sixteenth century to bring mechanics, particularly a dynamical approach to the simple machines into natural philosophy. This was a physico-mathematical program of long duration and complex internal structure, consisting in a series of attempts, from the early sixteenth century onward, to move one or another of the constituent texts or sub-disciplines grouped under the label 'mechanics'—such as the statics and hydrostatics of Archimedes, the so-called Medieval science of weights, the more diffuse science of machines, or the pseudo-Aristotelian *Mechanical Problems*—into closer contact with natural philosophizing. We find in this domain, whether the term physico-mathematics is deployed or not, varied, sustained and serious attempts to do the very opposite of the strategy of Dear's Jesuits, that is, divorce mechanics from natural philosophizing (it was already sufficiently divorced from natural philosophy in the declaratory Aristotelian view). Rather, the common denominator—whether expressed through classificatory arguments, rhetoric about values and aims, or downright technical gambits—was to modify natural philosophizing by bringing in mechanics, and to shift the valencies of mechanics by making it relevant to, even central to, natural philosophizing; that is, seeking explanations in terms of matter and cause. This is radical, rather than 'inventive yet conservative' 'physico-mathematicizing'. (3) There was Kepler's profound neo-Platonising of mixed mathematics and redirecting the thus physicalized disciplines back into natural philosophy, while also creating a new physico-mathematical field, celestial physics; (4) Beeckman's linking of an emergent corpuscular mechanism to dynamical interpretations of the simple machines, which we shall study in, Sect. 3.2.2; (5) Descartes' very radical attempts to ground a corpuscular-mechanism and determine the principles of its doctrine of causation (laws controlling force and determination of motion) through exploitation of hydrostatical and optical

discussed in John Schuster, 'Consuming and Appropriating Practical Mathematics and the Mixed Mathematical Fields, or Being 'Influenced' by Them: The Case of the Young Descartes 1619–1637', available on my website: *http://descartes-agonistes* and has been explicitly and categorically stated in print in Schuster (2012, 2012a).

[58] Hattab (2005), following Laird (1986), Rose and Drake (1971).

[59] Peter Dear (1995)

inquires of a physico-mathematical character.[60]; and finally (6) Galileo's rather more piecemeal physico-mathematical excursions, including his construction of a sui generis new kinematical science of motion.[61]

Physico-mathematicians hostile to Aristotelianism claimed that mathematics could play an explanatory role in natural philosophy, thus promoting the 'physicalization' of the mixed mathematical sciences, which, in turn required unprecedented, tight, articulations between their respective innovations in the mixed mathematical sciences and their respectively favored natural philosophies. Moving between mixed mathematics and novel natural philosophizing, the usual suspects—Galileo, Kepler, Descartes, Gilbert, Mersenne and Beeckman—variously produced their much more 'physico-mathematical' versions of the old fields, supportive of their natural philosophical agendas. Consider, for example, how the traditional mixed mathematical field of geometrical optics developed 'physico-mathematically' inside the natural philosophical turbulence in the early seventeenth century: In their optical work Kepler (1604) and Descartes (1637) each sought closer articulation between optical innovation, on the one hand, and natural philosophical explanations, on the other. That is, new natural philosophical theories of matter and cause were taken more intimately to control technical details in geometrical optics, and in turn, technical details in geometrical optics exerted pressure on the exact nature of those natural philosophical claims about matter and cause.[62] Under such pressures mixed mathematical geometrical optics began to evolve into a much more obviously 'physico-mathematical' discipline, in which innovating natural philosophers extracted natural philosophical capital out of optical work, whilst unintentionally there emerged at each turn a more dense, relatively more independent domain of physico-mathematical optics—*a disciplinary*

[60] Gaukroger and Schuster (2002), Schuster (2000, 2005). These will be our main topics in Chaps. 3 and 4 below.

[61] In regard to Galileo historiography, it is worth adding that he presents a difficult case, in that he was every bit as avid as other radical players to appropriate and make natural philosophical capital out of mixed mathematics, and quite technically expert at this tactic as well. But, because he did not pursue a systematic natural philosophy throughout his career, as opposed to trying to establish a realist Copernican cosmology and a strong anti-Aristotelian stance, his strategy and results look more modern to us than do the strivings of a Kepler or Descartes. But, if we think the issue through in contemporary categories, we can plausibly conclude that Galileo, like Kepler and Descartes, was specifically and pointedly breaking the declaratory Scholastic rules about subordination of mixed mathematics, and that his pro-realist Copernican cosmology campaign and anti-Aristotelian agenda amount to substantial gambits in the field of natural philosophizing, short of advocating a 'new system'.

[62] Kepler practiced geometrical optics under, and in the service of, a neo-Platonic natural philosophy and conception of light. He got brilliant results in the theory of the camera obscura, theory of vision, and, to some degree, the theory of refraction and the telescope. Descartes, as we shall see in later chapters, emulated Kepler's technical optical achievements but in competition with his neo-Platonic natural philosophical program, practiced geometrical optics under his version of a mechanical conception of light. He achieved a simple and workable version of the law of refraction and a general theory of lenses. Conversely, as we shall also see, essential details of Descartes' mechanistic system were shaped by his optical successes. Cf. also Schuster (2000, 2005).

area was crystallizing as a function of being batted around in the natural philosophical ruck.[63]

Finally, before we leave the issue of physico-mathematics, it will be useful to reflect on what it means to talk about players within the field of natural philosophy obeying, or bending 'rules'. This is motivated by our important finding that *attempts to found a physico-mathematics were parts of the larger picture of contestation and competition within the natural philosophical field, not steps beyond or away from it.*[64] Members of a culture may in the normal sense of politics try to take it over or marginalize opponents within it; they are not usually involved in the wholesale destruction of it. Hence, we should view the physico-mathematicians as indeed rebels, but not in the sense of intending the destruction of the culture of natural philosophizing, but rather as attempting to alter the rules under which the natural philosophical game subsequently would be played.[65] So, returning to the our main theme, in my terminology, by the first third of the seventeenth century, the given, template-derived rules about the status of the mixed mathematical sciences were the subject of an unprecedentedly vexed debate and a turbulent state of play.

2.5.4 "Hot Spots" of Articulation Contest: Additional Causes and Effects of Heightened Turbulence in the Field of Natural Philosophizing

Just as the overall intensity and 'spatial' extent of contestation increased in the critical phase of the Scientific Revolution in the early to mid seventeenth century, so also

---

[63] Hence, what was involved was the long term emergence of a more autonomous new field of 'physico-mathematical' optics, which, like other such fields, tended to become more autonomous of natural philosophizing per se, and develop embryonic tendencies toward disciplinary independence, as we shall see in a bit more detail in Sect. 2.7 below.)

[64] Of course at a macroscopic level, these sorts of individual and local gambits contributed to a pattern of change in the field of natural philosophy which involved consequences and outcomes unintended by any particular player or group of players, the most important of which was the long term tendency for specialist disciplines (emergent physico-mathematical ones, as well as new experimentally based ones) to crystallize off from natural philosophizing and for the latter to dissipate over time, as we shall discuss briefly in Sect. 2.7 below. Cf Schuster and Watchirs (1990), Schuster (2002).

[65] Innovation in natural philosophy, as in any particular more narrow scientific tradition, then or now, is not limited to significantly new claims about conceptual content, or technique. Innovation can also be pursued in regard to rules and values. We need to know how to calibrate and describe such radical gambits. Furthermore, we have observed (in the spirit of interpretive sociology) that Aristotle's rules about natural philosophy and the mixed mathematical sciences were actually 'declaratory' rules; formally invoked and usually obeyed in practice; but constantly open to renegotiation in practice and challenge at the level of formal principle. The letter of Aristotle's distinction was hard to practice and was violated in many instances such as astronomy, where natural philosophical and mixed mathematical commitments at least overlapped, and where the entire issue eventually became inflamed by the realist Copernican challenge—a mixed mathematical theory that claimed natural philosophical truth and demanded perforce the overthrow of Aristotelianism to achieve it. Similarly, it was open for non-Aristotelian rebels like Descartes and Beeckman to try to renegotiate the rule.

sites of particular inflammation of contestation appeared, which, unsurprisingly, I term 'hot spots' in the field of natural philosophizing. At a hot spot: (1) the inflammation was new, not having appeared in the late Medieval period or in the initial 'Renaissance' phase of the Scientific Revolution (1500–1590), as we shall term it below;[66] and (2) a dual process of change took place, involving, on the one hand, the target—the subordinate science, theory, instrument, novelty or discovery in question—and, on the other hand, the natural philosophies contending, as it were, to exploit the target in question. The target in a hot spot was often pushed along an unintended trajectory toward becoming a small domain of inquiry with relative independence from natural philosophy; whilst the future shape and success of natural philosophies struggling about the target was often at stake at the hot spot of contention. An example of such a hot spot involved the claims of Harvey discussed above. Not only were Harvey's claims contested, and revised, by natural philosophical combatants for natural philosophical ends, but, over the next two generations we see an unintended trajectory, as a domain of experimental physiological inquiry emerged at this site, leading on to later English experimental natural philosophers, so-called, investigating issues not only about 'cardiology', but about the functions of respiration, the blood, the lungs and the atmosphere.[67] In this case a new, relatively autonomous domain of inquiry started to crystallize only to suffer a foreshortened and ultimately abortive trajectory.

All the characteristic features of a hot spot are even more apparent in that most important, exemplary and historically consequential instance, where some astronomically concerned natural philosophers, and natural philosophically engaged astronomers wanted to articulate realist Copernicanism to natural philosophical claims, which in the nature of the case had to be non-Aristotelian. Recall that in the traditional alignment of Aristotelian natural philosophy and Ptolemaic astronomy, what I term the 'declaratory' position on the merely instrumental and non-explanatory status of astronomy was slightly but necessarily compromised in practice by small but noticeable articulations of Aristotelian natural philosophy onto Ptolemy's astronomy.[68] These high order conditions of Ptolemaic model building were not going to be altered within Aristotelianism. No matter how much Aristotelians debated the marginal elements of their system; and no matter what elaborations needed to be added to Ptolemaic planetary models to improve predictive utility, no natural philosophical 'field-altering' controversy would take place at this site. This was no hot spot, and never likely to become one under the prevailing hegemonic rules of the field of natural philosophizing.

---

[66] Hence, the salience of significant novelties and discoveries, immediately up for contestation in the field. As interesting novelties emerged across increasingly dynamic and interrelating subordinate fields, the struggles over them increased. Merely gazing at, or hording or collecting curious new facts may have been a popular pastime, but it was not central to the natural philosophical agon—contention about curiosities was!

[67] Frank (1980), Anstey (2000).

[68] The Ptolemaic cosmos was finite and spherical with a motionless, spherical earth more or less centrally located. Combinations of uniform circular motion prevailed and the celestial and the terrestrial realms were distinct.

Copernicus himself, with his realist claims for his astronomical theory, had been de facto attempting what we can now discern as a 'physico-mathematical' move: This theory of astronomy had natural philosophical implications contradicting the prevailing Aristotelianism, and in effect demanding a systematic replacement, although Copernicus offered nothing substantial along these lines. A hot spot developed in the natural philosophical field, between systematic natural philosophical theorizing and the formerly relatively tame sub-ordinate mixed mathematical science of geometrical astronomy, only when some later players took Copernican realism more seriously for their own reasons and for their own agendas.

Supporters of realist Copernicanism needed to adduce a framework of non-Aristotelian natural philosophy, a new theory of matter and cause, adequate to explaining the heliocentric cosmos. Implicitly or explicitly, they had to bid to radicalize the grammar of relation between mixed mathematics and natural philosophical explanation. The entire late sixteenth and early seventeenth century debate over realist Copernicanism (culminating in the embryonic emergence in Kepler's and Descartes' respective philosophies of nature of a discourse of 'celestial physics') was a phenomenon of competition at a now inflamed site within the natural philosophical field—no realist Copernicanism, no inflammation.[69] But why be a realist Copernican, unless you intend a quite radical overhaul of Aristotelian natural philosophy (and its rules) as such?[70] Furthermore, it was only in articulations of natural philosophy onto realist Copernicanism that the issue or possibility of a 'physico-mathematical' astronomy arose. The cutting edge here was the embryonic emergence of that field

[69] What is meant in Descartes' case by his having a discourse on celestial mechanics or physics will be fully discussed when we arrive at his vortex celestial mechanics in *Le Monde*, in Chap. 10. For the moment, it can be foreshadowed that in *Le Monde* Descartes had a complex articulation strategy spanning astronomy, optics and a new challenging utterance in natural philosophy. His vortex theory of celestial motion, which formed the core of the natural philosophy, was the engine room of a now 'infinite universe' realist Copernicanism, and also explained the higher registers of the theory of light, and hence, he hoped, articulated onto his dazzling physico-mathematical achievements in geometrical optics. See also Schuster (2005).

[70] The rhythm of this process is fascinating, and important. Copernicus, a realist himself, staked his claims about the natural philosophical truth of his mixed mathematical theory upon the truth value of the 'cosmic harmonies' his astronomical models for the motions of each of the planets displayed when considered together as a 'cosmological' package or assemblage. Copernicus himself was either too timid, or unprepared, to force the realist issue more deeply into natural philosophical issues of cosmic matter and cause—What were planets, including the earth, that they could so move, and what moved them? His own answers were famously lame, even in contemporary terms, rather poor attempts at twisting Aristotelian matter and cause discourse to finesse the natural philosophical problems of his system. Instead, it was Tycho who, toward the end of the century, kicked off the eventual crisis of natural philosophy/astronomy articulation by linking his favored version of quasi Copernican astronomy to significantly altered (Aristotelian) claims in natural philosophy. Gilbert weighed into the contest with arguably the most innovative and consequential natural philosophical vision of his generation. Then, in short order, Kepler subsumed his brand of Copernicanism within physico-mathematical explanations which in turn resided at the centre of his version of a neo-Platonic natural philosophy. The situation was similar with Descartes, for in *Le Monde* he staked the truth of his natural philosophy on the truth of his version of a physically explained Copernicanism. (Schuster 2005; Gaukroger 1995; and below, Chap. 10).

we retrospectively term celestial mechanics, and which Kepler, its first self-conscious advocate, called celestial physics. The relevant work of Kepler and Descartes tended toward a physicalization of certain astronomical questions.[71] The old mixed mathematical science of Ptolemaic astronomy was passing, not simply as a particular theory, but the very genus 'astronomy as mixed mathematics' was giving way to physico-mathematical problematics in astronomy and celestial mechanics.[72]

2.5.5 *Modeling System Construction and Contestation — The 'Core', 'Vertical' and 'Horizontal' Dimensions of a Natural Philosophical System*

It is now time to think through what we might mean by systematization and a systematizing élan in natural philosophy. This model of making systems and competing about them can help us understand the rules and dynamics of conflict; its typical modes; competition over co-optation of fields and novel discoveries and the development of hot spots. As such, it summarizes and interpretively solidifies many of our findings so far. To address this modeling problem, I have begun to develop the idea of systematicity of a natural philosophy as another 'iceberg' category. Like 'natural philosophy' itself, systematicity is an actor's category, in their hands to negotiate from instance to instance, but it has certain contours we can model, beyond what they might have enunciated, and which we can use as a regulative tool for describing and assessing player versus player moves, and long-run dynamics and trends, whilst also looking to refine the category by critical reflection on its application to concrete cases.[73]

---

[71] The natural philosophical strivings of Descartes and Kepler, which were pursued with special attention to the subsumption of astronomy, i.e., Copernican astronomy, variously interpreted, and to its problem of celestial causation, raised a number of crucial topics and opportunities for natural philosophical inquiry and construction, quite apart from what arose later and was taken on board as a result of the use of the telescope: What was the nature of the earth as a planet, what could be gathered about the earth, for example, about its structure, its magnetism (Gilbert), its tides (Galileo and Descartes), the nature of local fall, that would support its construal as a planet amongst planets and allow for the motions realist Copernicanism required of it; what caused the celestial motions; what physical role did the sun (and all stars in multiple planet system versions of Copernicanism) play in those motions; did the nature and behaviour of comets throw any light on these problems? We shall later see that both *Le Monde* (Chap. 11) and the *Principles* (Chap. 12) intentionally played upon these issues.

[72] An outcome occluded and hidden in the turbulence of early and mid seventeenth century natural philosophy, but quite clear in the wake of the reception of Newton's work two generations later.

[73] Additionally, by offering us a view of what systematizing was about, particularly in the heated critical phase of the Scientific Revolution, it gives us a set of interpretive measures by which to perceive and gauge the processes that set in after the critical phase. As we shall see in Sect. 2.7, one of those later seventeenth century trends was the muting of contestation over systems and the tendency for quasi-autonomous, more narrow successor fields of inquiry to emerge from natural philosophy, along fault lines forming amongst subordinate fields and domains previously pursued, in part at least, for systematizing ends, thus signaling the slow but inevitable dissolution of the field of natural philosophizing.

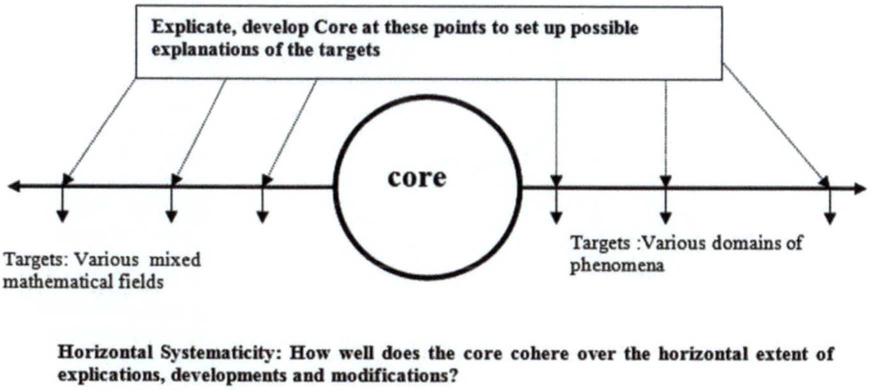

Fig. 2.3 Horizontal dimension of articulation of system

I articulate systematicity using the concepts of the explanatory 'core' of any natural philosophy; its 'horizontal' and 'vertical' articulations toward particular domains of explanation; and the idea of 'system-binding moves'. This model will be mobilized later in relation to *Le Monde* (Chap. 11) and the *Principles* (Chap. 12). First, by the *core*, we mean its central, enunciated doctrines of matter and cause and any especially significant explanatory cases within it.[74] By the *horizontal articulation* of the system (Fig. 2.3), we denote the explication/modification of the core in order to try to launch explanations of results, 'matters of fact', or 'solid findings' in various sub-disciplines and sub-domains of inquiry. Across the horizontal level, one asks how well these articulations of the core cohere over the spectrum of applications to differing domains.

By the *vertical articulation* of the system (Fig. 2.4), we mean how fully and coherently any and all of the various sub-disciplines (such as fields of mixed mathematics) or domains of inquiry (such as local motion and fall, or magnetism) are grasped and explained by the (articulated) core of the system, and what sort of program of further inquiry, if any, is possible in any given case. In this way, we explore, horizontally, the arguable coherence of *extension* of the core to cover various sub-domains, and, vertically, the arguable *depth* and *strength* of the core's explanatory grasp of those various domains.[75] *System binding moves* occur across

[74] For example, we shall see in Chap. 11 in the case of Descartes' system of corpuscular mechanism in *Le Monde*, we mean the matter/element theory, the dynamical principles and laws of motion and, as an exemplary explanatory case, his vortex celestial mechanics itself. Then in Chap. 12, we shall uncover a hitherto unnoticed vast system-binding strategy in the *Principles*, far outstripping Descartes' accomplishment in *Le Monde*.

[75] 'Arguable' means, of course, that there is no essentially correct or final answer. We can observe actors struggling over such judgments and as historians may sometimes have to evaluate for ourselves in the interest of explanation.

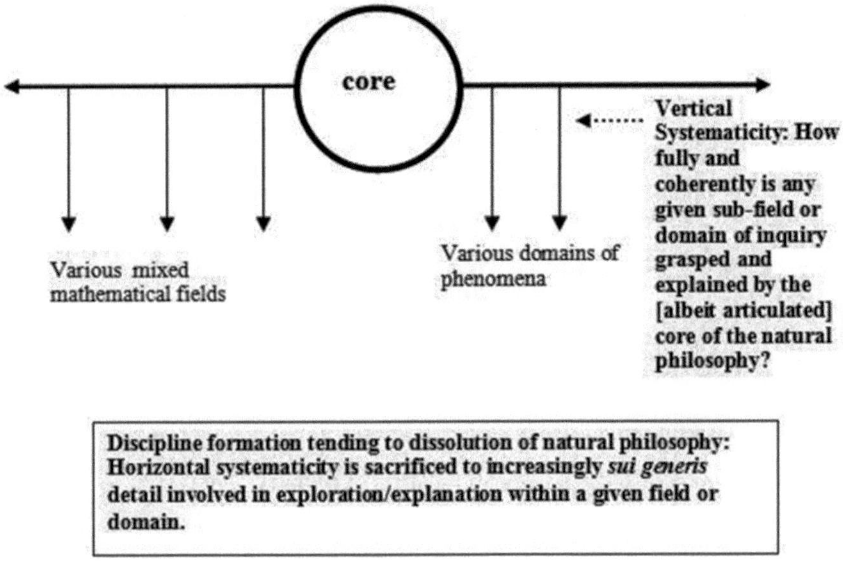

Fig. 2.4 Vertical dimension of articulation of system

the horizontal dimension and will tend to be missing where systematicity is not an important actor's aim or intention.[76]

Now, of course, at the actor's level the criteria for assessing the goodness of a natural philosophy and the modes of applying such criteria to cases, were themselves objects of negotiation, part of the weave of the contestation in natural philosophy itself. Hence, we would expect that each and every type and focus of contestation we have discussed thus far could be mobilized in such constructions of putative systematicity and contestations thereof. Nevertheless, what is proposed by my articulation of the idea of systematicity is not meant as the only, best or truest way of sizing up any system, and certainly not a set of criteria any actor embraced fully explicitly and exclusively. Rather, it is a self-consciously designed analytical tool for dissecting systems of the time, in the interest of building better accounts of the process of natural philosophizing, a tool reflecting to some degree some of the goals and standards the actors arguably used.

Finally, two immediate heuristic benefits of this model need to be mentioned: First, analyzing a natural philosophy in the vertical dimension—where subordinate fields or novel results or non-natural philosophical claims might be attempted to be co-opted—can show us what was and was not happening in the dynamic heart of a system, where reduction and co-optation of subordinate domains was expected, and often rhetorically claimed, but where, in the nature of the beasts under examination, smooth success is hard to find, never consensually granted by everyone, and very

[76] For example, when we study *Le Monde* in Chaps. 10 and 11, we shall see that Descartes makes a number of elegant and clever moves that arguably bind the system together and lend extra theoretical credibility to some of his claims. Things that look rather ad hoc from one angle, look highly systematic, almost inevitable, if we tease out the system binding logic with a perspective informed by a category of systematicity

much dependent upon the eye of the beholder. On the player's level, this was part of the field of possible contestations that made the game so inviting and difficult.

Secondly, the model gives us some heuristic insight concerning the long run tendency, from the mid seventeenth century, for natural philosophizing to be pursued less in terms of explicit systems, and more as fragments of 'experimental natural philosophy', eliding slowly toward the crystallization of more specialized domains of inquiry, which began to look more like separate disciplines. Analysis of natural philosophical work by vertical and horizontal articulation suggests a rule of thumb: In a given natural philosophy, to the extent that vertical articulations within subordinate domains are dictated by and co-opted toward the strengthening of horizontal systemic considerations, that natural philosophy is, and indeed is intended to be, a system. To the extent that investigations within subordinate domains take on a life of their own—meaning amongst other things that horizontal systematic articulation is neglected, rather *ad hoc*, or merely rhetorically asserted—that natural philosophy is tending toward the genus 'experimental natural philosophy', which appears more frequently, indeed endemically, in the later phases of the Scientific Revolution. Moreover, to the extent that various natural philosophies tended to treat specific sub domains in the latter way, relatively autonomous of horizontal articulation concerns, those sub domains took on *sui generis*, quasi disciplinary characters, and over time floated more free of any particular natural philosopher's systematizing ambitions.[77]

2.5.6 The Mechanics of Responding to 'Outside' Challenges and Opportunities

The next dimension of our model concerns how we may more seriously treat the problems colloquially spoken of in terms of 'contextual or macro forces' somehow 'shaping or influencing' the content, values, agendas and directions of natural philosophizing. This problem has a long pedigree in the so-called internalist/externalist debate about the historiography of science, as well a special relevance to our topic, because of the equally long discussion in wider historical circles about a general, sceptical, or cultural crisis in the second and third generations of the seventeenth century.[78] We will not be directly concerned in this study with the larger reaches of the latter problem, although we shall touch upon it when we meet Descartes expanding his intellectual horizons, and the presumed cultural import of his method, in Paris in the mid and late 1620s, in Chaps. 7 and 8. Nevertheless, the issue of how we should think through the possible causal role[s] of larger socio-political, religious and cultural structures and events in terms of our model of natural philosophizing is an important one, with implications for studying the Scientific Revolution and any players within it.[79] Fortunately, the trajectory we have thus far followed in constructing

[77] The assertions in the last four sentences of this paragraph will be canvassed in a bit more detail below in Sect. 2.7, where we overview the key phases and stages in the process of the Scientific Revolution.

[78] On internalism/externalism, see the literature cited Note 26 above.

[79] This part of the model and its historiographical applications, including to the problem of the crisis of the seventeenth century, will therefore play a large role in my study in progress of 'The Fate of Natural Philosophy at the Dawn of Modern Science: A recasting of the plot of The Scientific Revolution'.

historical categories and structures of interpretation, leads rather straightforwardly toward a solution of this problem, indeed a solution which simply explicates and further applies the approach followed thus far.

After two generations of development of methodological criticism from both the school of Quentin Skinner and the school of post-Kuhnian sociology of scientific knowledge, we cannot appeal to the 'influence' of ideas upon other ideas, nor can we revisit vulgar Marxism and its cognates, wherein social and economic structures imprint corresponding constellations of ideas upon leading thinkers, who just happen for these purposes to be cultural dopes.[80] A promising avenue does, however, arise directly from within our model of natural philosophizing. Indeed, the way to deal with 'contextual drivers, shapers or causes' of 'thought' is built into our model of a dynamic agonistic field or tradition, in which competing players deploy resources, and follow (or attempt to revise) rules of engagement, in order to construct claims whose value, longevity or otherwise, is entirely in the hands of their peers and successors in the evolving field. The modeling here follows directly from Sahlins' conception of the historicity of cultural dynamics, discussed above in Sect. 2.3, by extending the idea of natural philosophical players competing over articulations of their preferred natural philosophy onto subordinate fields. Quite macro entities—social structure, economic forces, political structures and forces—can be brought into the explanatory machinery, but not in the form of causing, imprinting or influencing the ideas of actors.[81] Rather, natural philosophers responded to challenges and forces and decided to bring them into play in the form of revised claims, skills, material practices and values in the field. To do that, the 'things' being brought in had to be represented *to and by them (not us!)* in appropriate form. In my model's terminology, a player had to *articulate* his natural philosophical claims upon some available representation of the 'contextual' or 'external' things of

[80] See J.A. Schuster, 'Consuming and Appropriating Practical Mathematics and the Mixed Mathematical Field, or Being 'Influenced' by Them: The Case of the Young Descartes', cited above note 26. Nor do we want to follow normal intellectual history practice, as evidenced in this particular area by, for example, Popkin (1964) with his hypostatized, growing then resolved 'sceptical crisis'. The technique is to give thick enough, untheorized descriptions so that a de facto and largely tacit explanation emerges something along the psychologistic lines of 'great thinkers somehow get it into their heads to address the great challenges hanging about in the cultural atmosphere, and hence their intellectual output somehow reflects or is shaped by them'. (We shall encounter this sort of difficulty below in Chap. 8 in reconstructing Descartes' path of inscription of *Le Monde*.) On the problems of the older style socio-economic 'imprinting' Marxist historiography of science, see Schuster (2000a); on transcending the older history of earlier ideas or thinkers influencing later ideas or thinkers, see above, Chap. 1, Note 25.

[81] Note also that nobody is being naively essentialist about these macro entities. Historians' representations of them are also categorical constructions, woven out of relevant evidence, previously accepted claims, metaphors, and arguments. As we said in Chap. 1, 'historians must also manufacture models of relevant aspects of context, proximate or distant'. However, as also previously stressed, this no more means that these constructs completely lack real reference than that the theoretically couched objects of natural scientific inquiry do so as well. There was a French economy in the seventeenth century, and a French state. We know them through the evidence based, conjectural and revisable models we make of them, exactly as the case with 'natural philosophy'. We judge and revise those models in the light of expert debate concerning the explanations and narratives we offer using them.

relevance and concern to him, his allies and opponents—just as players variously *articulated* their own natural philosophical views onto a selection and weighting of the subordinate disciplines.[82] The players do the acting; they are not forced, imprinted, influenced or caused to do anything by large scale contextual features, let alone such features as later historians model them. Rather, from a natural philosophical player's perspective, available and appropriately thinkable/writable representations of things about contextual structures and features were intentionally mobilized, used, reshaped and deployed strategically in natural philosophical claims.[83]

Hence, in modeling terms, we are now talking about the boundaries of the field of natural philosophizing, or more properly, the shifting ways in which players accounted, acted upon, and competed over, what they took to be the boundaries of the field at any given moment.[84] According to this model, there were no fixed, essential boundaries of the field of natural philosophizing; no permanent and always honored account of what was inside natural philosophy and what was outside: what was relevant to natural philosophical utterance and what was not.[85] Rather, (1) the utterances

---

[82] The term articulation is used here in extension of our use of it in Sects. 2.3 and 2.5 above, partly as inspired by Sahlins' model of cultural process and, as some readers will note, in partial emulation of the young Foucault (1972). My thinking about this began a number of years ago in conjunction with suggestions from, and collaboration with, Dr Ivan Crozier, formerly of the Science Studies Unit, University of Edinburgh, now in the Department of History, University of Sydney.

[83] Our own models of the relevant macro structures and processes can therefore enter into our overall explanation, but not as drivers or printers of natural philosophical ideas. Rather, we use our knowledge or modeling of contextual structure and process to deepen our understanding of a given natural philosophical gambit. Such a gambit will initially be explained by appealing to the actor's decisions to mobilize into natural philosophical utterance bits and pieces of his representations of the kinds of things we denominate as larger contextual features. We can then extend our understanding by locating the actor's representations of those features, in a realm unknown to him in our form, but known to us through considered, evidence based, rational model building, that is, by framing our description of the situation with our models of the contextual features in play. Descartes did not think about natural philosophy the way he did because he was influenced by the rise of the *noblesse de robe*. But, there is much about his own cognitive make up and self understandings that arguably was sedimented through his experience in, and reflection about, the lives and training of many of his relatives and himself. We see him mobilizing bits and pieces of these available representations into his discourse in, and about, natural philosophy, for example in his autobiography in the *Discourse on Method*. Similarly, Descartes built the values of utility and progress in domination of nature into his natural philosophizing. He was not forced to do this by the rise of the commercial capitalist economy or centralization of the state, nor did these macro processes imprint the ideas in his head. Rather he himself imbibed rhetoric and literature by others already representing things about the changing commercial and political situation of the time. That is how he thought about such things, and when he wanted to bring such wider considerations into natural philosophizing, he did not wait to be driven or impressed, rather he decided to mobilize certain representations for certain agendas and types of claims in natural philosophy.

[84] To this end, I have also benefited from post-Kuhnian sociology of scientific knowledge scholars' concept of 'boundary work' in disciplines or professions (Gieryn 1983), but, as some readers will sense, my conceptions of boundary maintenance and work upon field or disciplinary boundaries are wider, more historical and tempered by a much modified 'Foucault' passed through the filter of Bourdieuian sociology of agonistic fields.

[85] Hence, shifting views in this regard can be seen as involving tacit or explicit 'rules' for natural philosophizing.

of dominant figures and groups tended to create, and recreate, a 'leading or hegemonic picture' of those boundaries and how to articulate natural philosophy onto them, whilst (2) articulation upon boundaries was an essential part of the competitive dynamics of the field. The university neo-Scholastic Aristotelians' possession of dominant institutions was crucial; but, competitors challenged the way dominant players articulated utterances to boundaries in order to define the field. In general, the dominant utterances in the field carried a particular *selection, weighting and thematization* of articulations on boundaries.[86] Challengers could reorder these selections, weightings and contents, and also modify existing articulations, or bring in new ones.

For example, one might say that in the university teaching of Aristotelianism, a virtual articulation was present to whatever version of orthodox religion dominated that particular polity and university. However, the traditional exclusion of discussion of theology in the undergraduate course meant that this articulation was tacit, not thematized in the body of undergraduate natural philosophical teaching. In effect, a rule existed about not explicitly articulating natural philosophy to theology from the natural philosophers' side of the fence. But, competing utterances from non- and anti-Aristotelian challengers could mobilize explicit and deeply developed articulations onto religion. To bring in religion in an explicit way involved devising new utterances, new articulations in depth and degree of thematisation in accord with favored religious and theological commitments, claims and agendas.[87]

[86] My emphasis on selection, weighting and content of boundary articulations seems to me an important conceptual point, requiring more 'articulation' on my part. I can say, however, that I believe it pushes beyond the customary 'boundaries' in how sociology of scientific knowledge work on boundary management has been conceived and applied in case studies.

[87] This is what we mean by challenging the choice, depth and weighting of an articulation. Similar points attach to politics, or more particularly to issues about the nature and role of 'the state', and the contemporary tortured issues of sovereignty, church governance *vis à vis* the state, and issues of civil order and legitimate rebellion (all of which could count as elements in a larger 'crisis' perceived and responded to by some natural philosophical players). Most Aristotelian teachers of natural philosophy in the university environment would have *left largely unsaid within natural philosophy* its linkages to the local political status quo, and to the institutional arrangements that supported the very existence of that particular university and its natural philosophical functions. A Bacon or Hobbes, however, articulated natural philosophical utterance in part upon such particular evaluations of these political issues. But this is not to say that politics or political doctrines or agendas 'influenced' the natural philosophical utterances of Bacon or Hobbes. Rather, it is to say in the first instance that within the field of natural philosophy they saw fit to mobilize and deploy such articulations in an effort to win the natural philosophical agon, and through it, partially to support their properly political aims, now recursively expressed, amongst other ways, through natural philosophy. So we do not deny their aims and aspirations in the actual domain of politics—but, we must demarcate and understand before we associate and explain. Hobbes would have liked to have won in politics as well as natural philosophy, and his possession of a natural philosophy well articulated to a particular view of the state, and the causes and cures of civil wars, was in his view a weapon in the real political field, as his novel articulation upon politics was in his view a weapon and argument in his favor in the natural philosophical field. To conflate the two fields of play or link them by 'influence', contextual imprinting' or an intimate psychology of motive may paint a pretty picture of Hobbes, but it will probably ruin our ability to do the history of either natural philosophy or politics (or their precise modes of interrelation in the actions and discourse of such interestingly innovative figures).

This, for example, would provide a more precise meaning (and tool kit for study) to a the well worn formula that in the second and third generations of the seventeenth century *'some natural philosophers responded to a perceived 'crisis'—cultural, religious or sceptical— with cultural moves inside natural philosophy'*.[88] In general, contending players, with differing agendas and perspectives, were always in the process of making out the boundaries and relations of the field (from their perspectives and agendas) by articulating utterances in the field upon (their selection and weighting of) boundary structures and discourses. If outside entities and forces seemed to some to be particularly threatening and challenging (if, hence, a crisis was in progress), the variety, intensity and scope of competing articulations would rise, and it did!

Our suggestions for handling the contextual shaping or driving of natural philosophy will be applied below in our periodization of the Scientific Revolution in Sect. 2.7, especially when we look the critical or 'civil war' in natural philosophy stage. For the moment, in concluding this initial and exploratory construction of the category of natural philosophy that we have attempted in this section, we can say that a prima facie case has been established for giving serious consideration to the category of 'natural philosophy' in Scientific Revolution historiography: The issue is not whether we should entertain such a category, but rather how to design it conceptually and set it to work in explanation and narrative. The key dimensions of the model in its present stage may be enumerated as follows: (i) the field or culture of natural philosophizing encompassed more than merely its hegemonic Aristotelian variants; (ii) families or genres of natural philosophy differed with each other; yet (iii) they obeyed common rules about the production and content of natural philosophical claims; (iv) the entire field was marked by competitive struggle,

---

To recur to the parallel ruminations of Marshall Sahlins (1993) on the need for an historical category of culture: This is analogous to his critique of post-modernist views of indigenous cultures as simply the decrepit or sad results of a steamrollering impact or imprinting by Western imperialism. He argues that such pessimistic sentimentality systematically neglects the specificity of response to Western impingement from an indigenous culture, and the fact that even the history of imperialism must take note of the dispersion and effects of such culture specific responses over time. Similarly 'politics' or 'social factors' impinging upon natural philosophy and philosophers did not denature, or collapse the latter. Rather, politics were played by some natural philosophers, as part of doing natural philosophy and often as part of their engagement with politics. Correlatively, natural philosophy as a (sub-)culture needs to be studied historically, with close attention to contestations within it, including responses to, and articulations upon, 'contextual factors'—large and small, structural or ephemeral.

[88] Similarly, it can be argued that the practical arts and their practitioners did not influence natural philosophers, but rather that certain natural philosophers articulated their natural philosophical utterances in part upon resources from and about the domain of practical arts. I apply similar arguments to the more specific issue of the relation between practical mathematics and mathematicians and the 'Scientific Revolution' in J.A. Schuster, 'Consuming and Appropriating Practical Mathematics and the Mixed Mathematical Field, or Being 'Influenced' by Them: The Case of the Young Descartes', cited above note 26. Cf. also Note 107 below, on the suggestive findings of Paolo Rossi which can also now be interpreted along these lines.

which (v) in turn was linked to and shaped the development of the sub-ordinate sciences; (vi) natural philosophical claims or utterances were variously linked to agendas and beliefs about other neighboring fields of discourse in theology, politics, pedagogy and the practical arts; and finally (vii) the dynamics of this field, the tissue of unfolding intended and untended consequences of the various plays and gambits, largely constituted the sequence of developments identified as 'the Scientific Revolution'. This did not occur in any simple or linear sense. The natural philosophical game evolved significantly and eventually natural philosophy dissolved as a cultural field and institution—hence our concern in Sect. 2.7 below with stages and phases in the period of the Scientific Revolution. But, before we reach that point we need to clarify and remove one of the most important pitfalls, and illusions, standing in the way of a proper understanding of Descartes in the context of his natural philosophical and physico-mathematical concerns: the problem of what to do with his claims about, and apparent belief in, his 'method'.

2.6 The Special Status of the Problem of Method

If there is one concept, and agenda, shared by Descartes (or at least the young Descartes) and his contemporaries—as well as by many today, including scholars of these matters—which could derail the entire thrust of the present study it is this: belief in the in principle or in fact existence of a unique, universal, transferable and efficacious general method of discovery and/or justification for rational disciplines, outside of the realm of faith. If there is such a method, there is no need for any of the historiographical criticism or categorical construction we have undertaken. The history of science becomes a simple tale of heroic figures who first struggled against obstacles and opponents to piece together, and then employ, this method.

Indeed, until relatively recently interpretations of the Scientific Revolution tended to be dominated by heroic tales of the discovery, perfection and application of the scientific method. Descartes, Bacon, Galileo, Harvey, Huygens and Newton were singularly successful in persuading posterity, historians of science included, that they contributed to the invention of a single, transferable and efficacious scientific method. The earliest systematic studies of the history and philosophy of science, the writings of d'Alembert, Priestley, Whewell and Comte, attempted to distil from the historical progress of science a sense of that method, so that its further perfection and wider application could insure the future growth of the sciences.[89] In the early twentieth century, pioneer professional historians of science, such as George Sarton and Charles Singer, saw the elucidation of the scientific method as one of the chief functions of the study of the history of science.[90] Subsequently, a

[89] Cf. Priestley (1767) v–vi, Whewell (1837) 5, Whewell (1980) 3–4.

[90] Singer (1917–21) vi, Sarton (1921–22) 25, Sarton (1924) 26.

thriving sub-discipline of the history of science concerned itself with the history of methodological ideas in (supposed) relation to the larger course of the history of science,[91] and later, Karl Popper, Imre Lakatos and their followers sought to revive the link between theorizing about the purported scientific method and re-writing a 'method-centric' history of science.[92]

In Chap. 1, we have already foreshadowed the nature of the historiographical challenge we face in this area, by noting the way in which modern work in history and philosophy of science has, for the attentive at least, rendered inoperative belief in such a general, transferable and efficacious method, so that scholars of Descartes and of the Scientific Revolution generally, need to be aware of this conceptual ground note to any and all narratives and explanations they may attempt. We now need to pay more sustained and minute attention to this issue, in order, firstly, to make certain that our categorical and historiographical housecleaning to this point is not undermined by sloppy thinking, forgetfulness and backsliding, and secondly, because later in Chap. 6 we shall have to delve even further into this question, and seek not the reasons why such grand methods do not work (to be explored here) but why and how any rational actor, such as Descartes for example, could come to believe that an efficacious general method could exist—how, in short, we can sympathize with the genuine belief that actors have in doctrines which we are certain cannot accomplish what they believed them able to accomplish.

As we noted briefly in the preceding chapter, it has become increasingly clear to some historians and sociologists of science that the traditional belief in the existence of a single, transferable, efficacious scientific method is highly dubious. The work of Alexandre Koyré, Gaston Bachelard and Thomas S. Kuhn especially pointed in this direction, but only in the last 30 years or so have their insights been followed up in attempts to revise the 'believer's' historiography of method. Although Koyré—the doyen of post World War II French and Anglo-American internalist history of science—firmly believed in scientific progress, he did not consider it the product of applying a general scientific method. Rather, for Koyré, progress depended upon the adoption of appropriate metaphysical presuppositions and the pursuit of science within them. His classic example was Galileo's mechanics, which, he argued, owed nothing to any methodological achievement, but issued from Galileo's brilliance in working and arguing his case within the framework of a loosely 'Platonic', mathematical metaphysics. Similarly, Aristotelian physics had not failed for lack of a method, but largely because it had had the wrong conceptual presuppositions, ones too close to untutored commonsense about motion.[93] The point for Koyré was that a general, transferable method is neither necessary nor sufficient for the pursuit of science. 'No science has ever started with a treatise on method and progressed by the application of such an abstractly derived method,' Koyré intoned, commenting

[91] For example, Crombie (1953), Randall (1961).

[92] Popper (1959), Lakatos (1978).

[93] Koyré (1939, 1978, 1956, 1969).

on the *Discours*, and at least some historians of science have tended, correctly, to agree.[94]

Bachelard's early work slightly pre-dated that of Koyré, and seems to have been subtly refracted in the thinking of both Koyré and Kuhn. In this process, Bachelard's scepticism about method was not brought to the fore, and even with the wider dissemination of his writings over the past thirty years, the implications of his work for undermining the cult of method have not been sufficiently articulated. However, those implications are quite clear in the core of his work. For Bachelard, each field of science consists in a set of interlinked, mathematicized concepts which interact dialectically with the instrumentalities through which the concepts are objectified and materialized.[95] To paraphrase Bachelard, the meaning of a concept must include the technical conditions of its material realization.[96] When a science is created, an artificial technical realm comes into being, in which phenomena are literally manufactured under the joint guidance of the system of mathematicized concepts and the instruments and experimental hardware in which those concepts have been realized. In an ironic jibe at positivist dogma, Bachelard termed any such realm of theoretically dominated artificial experience a '*phénoméno-technique*', thus signifying that the phenomena of science are not discovered but made, not natural but artificial, being created and commanded in the light of theory and theory-loaded instruments. In Bachelard's view, therefore, each science is unique and self-contained; each has its own specific system of concepts and related instrumental armory. No single, transferable, general scientific method can explain the genesis of any science or its contents and dynamics.

Kuhn, too, can hardly be said to have focussed upon the demystification of method in his theoretical or historical writings. But, just as we have seen with Koyré and Bachelard, there is in Kuhn a clear denial of the role traditionally ascribed to method, and that denial relates directly to the major premises of his position. In effect, Kuhn's approach vastly strengthened Koyré's assertion that grand set-piece doctrines of method are irrelevant to the practice of the sciences. The key point resided not in Kuhn's conception of 'scientific revolutions', but rather was implicit in his view of routine, 'normal', 'puzzle-solving' research within a 'paradigm'. Here we shall delve into that concept, looking at the ways it underscores Kuhn's scepticism about general methods. Before we do that, however, we must note that here we are recurring, at least at first, to a more primitive conception of paradigms than we managed to achieve earlier in Sect. 2.4, where we discussed the nature of discovery-seeking, knowledge-making and knowledge-breaking expert traditions in the modern natural sciences. Recall that in Sect. 2.4, we discussed Kuhn's notion of a paradigm and of a normal tradition of puzzle solving research based on it. There we were interested in how post-Kuhnian research in sociology and history of science has broken down Kuhn's stark, black and white dichotomy of normal versus revolu-

---

94 Koyré (1956).

95 Bachelard (1975a, b, 1949); Lecourt (1975) 40–47, 60–70.

96 Bachelard (1975a) 61. Cf. Gaukroger (1976) 212–23.

tionary types of research. This leads us toward the post-Kuhnian conception of expert scientific research traditions as dynamic, constantly re-negotiated and focussed on claims to, and debates about, 'discoveries'. The acceptance of such discoveries into the working resources of the tradition, affects both the tradition (the paradigm) and the nature and directions of subsequent work. Here, in deconstructing the possibility of efficacious general methods, we begin by referring more directly to Kuhn's own, rather static, view of a paradigm, although, as will be shown, nothing about the argument we shall build on this notion is undermined by then re–introducing the post-Kuhnian concept of dynamic and ever changing discovery-seeking 'normal' traditions of research.

So, taking the notion in its original form as set forth by Kuhn himself, recall that a Kuhnian paradigm is that entire discipline-specific culture which at a given time governs cognition, action and evaluation within a given mature tradition of scientific inquiry. For Kuhn, a paradigm consists first of all in a 'metaphysics', a set of deep conceptual presuppositions, which need not be of Koyré's Platonic type. A paradigm also contains the central concepts and law sketches of the field, and all the instrumental hardware and experimental procedures considered relevant to the posing and solving of problems within the paradigm. Kuhn stresses the theory-loading, or, more precisely, the paradigm-loading of the instruments and procedures. Standards and norms for the adequate use of instruments and procedures are also part of the paradigm, being inherent in the theoretical and craft training necessary to become proficient in paradigm-based research. One learns these and other parts of the paradigm through a course of practice on piecemeal, already solved problems—'paradigms' in the narrow sense (later designated 'exemplars'), bearing some relation to Bachelard's *phénoméno-techniques*. There is also a negotiable pecking order of unsolved problems and their correspondingly negotiable degrees of 'significance' or 'anomalousness', which forms a resource for selecting, shaping and evaluating courses of research and their results.[97]

Assuming that such paradigms, or anything like them, guide normal research in the various traditions of research in the sciences, it then becomes highly unlikely that some single method guides the history of the sciences, individually or collectively. The elements making up a particular paradigm, and hence making possible, for the time being, a particular tradition of research, are unique to that field and are a sufficient basis for its practice. Moreover, if each field has such a unique and self-contained conceptual fabric and associated mode of practice, then it is irrelevant to our understanding of its cognitive dynamics to re-describe, gloss or otherwise 'account' for them by the use of heroic tales of method.[98] This point also holds for

[97] Presumably none of this surprises readers of Kuhn (1970), especially the 'Postscript'; Kuhn (1977), Chap. 13; Ravetz (1971) 71–240, Barnes (1982), Bachelard (1975a, b).

[98] The parallel to Bachelard's conception is particular strong at this point. Bachelard saw the various 'philosophical' glosses on scientific practice, such as instrumentalism, empiricism, rationalism and conventionalism as each, in a specific manner, missing the 'point' of how real sciences are constituted by *phénoméno-techniques*. Bachelard (1949) 4–5, Bachelard (1975b) 61.

Concepts	**Metaphysics**	**Tools**
Standards	**Aims**	**Exemplars**

Fig. 2.5 The Kuhnian disciplinary matrix of elements in a paradigm

all the traditions of scientific research existing at any moment: Each has its own particular paradigm, and whilst neighbouring or cognate fields might share certain paradigm elements in common, there is no reason to assume, as methodological accounts must, that there is some identity or long term convergence among paradigms.

The radical anti-methodism to be extracted from Kuhn's position can be illustrated using a pair of figures. First, in Fig. 2.5, we illustrate the general point that any given field of science has at any given moment its own paradigm, its own versions of the generic elements displayed in the matrix: (1) basic concepts and law sketches; (2) metaphysics; (3) tools and instrumentalities (including the theories and standards thereof); (4) standards of relevance and of adequacy for the selection of problems and for the formulation and evaluation of knowledge claims; (5) disciplinary goals of any internally or externally generated sort; (6) concrete achievements, exemplars, instantiating laws, concepts and standards.

Then, at any given moment, the domain of the sciences may then be represented as in Fig. 2.6, where we have n *sui generis* fields, each with its own particular constellation of matrix elements, constituting, for the time being, its own paradigm. Field 1 has its own sui generis matrix of concepts (C), metaphysical presuppositions (M), theory-loaded tools and instruments (T), standards (S), aims and goals (A), and exemplars (E); so does Field 2 and every other field down to Field n. The sciences; that is research traditions, are thus many, not one. True, neighboring and cognate fields may share certain elements in common; concepts in one field may be taken up (under translation) as tools in another; or, groups of fields may have emerged under the aegis of a common metaphysical umbrella. But none of this argues the identity or even the long term convergence among paradigms.

In Kuhnian terms, each field or tradition of research has, at each moment, its own 'method', inextricable from the contents and dynamics of its paradigm. But, to speak of some putatively common, transferable, efficacious, universal scientific method or epistemology—Baconian, Cartesian, Newtonian, Popperian—is merely to float above the lived, thought and practiced life of each of the sciences, and fallaciously to substitute an externally prompted discourse for the dense and varied cultures of the several paradigms. There are, in short, no unified and literally

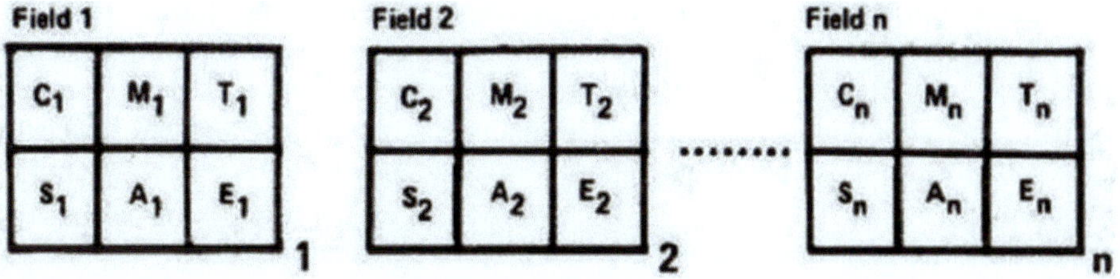

Fig. 2.6 A set of n coexistent, *sui generis* paradigms or n conceptually and materially *sui generis* research traditions at a given moment of time

applicable methods. No method discourse corresponds to, or maps onto, any given domain of scientific practice, let alone a number of such domains.[99]

So much, then, for a deconstruction of the possibility of there being some universal, transferable, efficacious general method of the sciences, working from a crude Kuhnian model of paradigms and the Kuhnian principle that 'the sciences are many, not one'. We now need to increase the stakes by noting an important further claim: *It makes no difference to the foregoing argument that whilst Kuhn saw his paradigms as rather static, until they entered an anomaly induced crisis leading perhaps to a radical and incommensurable shift to a new (rigid) paradigm, we post-Kuhnians have learned to see paradigms as fluid and constantly in principle open to greater or less renegotiation, around 'significant discoveries'.* The dynamism of a set of Kuhnian paradigms in no way suggests that they are any less *sui generis* and unrelated as (ongoingly negotiated) research sub-cultures. Additionally, and importantly, the two perspectives—crude Kuhnian and post-Kuhnian—actually collapse into one if we append to discussion of Kuhn's paradigms the rider 'at any given moment', (which the reader will notice we have in fact done throughout the above argument). In that case, the anti-method argument from Kuhn's conception of multiple, co-existent, *sui generis* static paradigms becomes indistinguishable from the same argument made in terms of post-Kuhnian multiple, co-existent, *sui generis* and dynamic paradigms.

Moreover, the post-Kuhnian work in the sociology of scientific knowledge is not merely consistent with the above argument, but in fact considerably deepens it. Recall that post-Kuhnianism unfreezes Kuhn's metaphor of routine 'puzzle solving', and suggests that even in normal research there is a constant, subtle revision

[99] In more colloquial terms, I instruct undergraduates on these points with the following axioms and conclusions: (1) There are scores of scientific disciplines and sub-disciplines. Each one has its own unique research 'coalface'. (2) Workers at each different coalface use theories, assumptions and techniques specific to that discipline or sub-discipline. (3) Even experimental techniques and instruments are shaped or loaded by theories. (4) So, each coalface is constituted by a collection of theories, assumptions and techniques unique to that coalface. (5) Each coalface has its own 'method' of going on with research. (6) The idea of a unique, single, transferable simple method for each and every coalface, past, present and future is highly implausible.

and negotiation of the elements in the paradigm.[100] Therefore, normal research always involves bids to make small, but significant, alterations in the prevailing disciplinary objects of inquiry. Such bids exert feedback effects on some of the elements of the paradigm—conceptual, instrumental, evaluative—if they are successful. So, in post-Kuhnian perspective, normal science may be 'puzzle solving', but it is a peculiar version of that activity, because the pieces, the rules of assembly, and the ultimate 'picture' keep changing as the players play and negotiate.[101] Hence, to recur to the anti-method argument that disciplinary 'method' is inextricable from a particular paradigm, now the situation worsens, because that disciplinary 'method' is also in flux, inextricable from the socio-cognitive dynamics of the field. No general doctrine of method can command or describe the static picture derivable from a popular reading of Kuhn himself, let alone this new, dynamic, post-Kuhnian picture of how a research tradition functions.

Indeed, the post-Kuhnian case against method did not stop here, but extends further to the issue of the social and political organization of normal fields and communities. If a research tradition is not in the grip of a total and immobilizing consensus (until the next 'revolution'), and if 'significant' research always involves a negotiated outcome (revolving around some sort of 'discovery') which alters the next rounds of disciplinary play, then a normal field must have a social and political life sufficient for the carrying out these knowledge-making and knowledge-breaking maneuvers, and for keeping them, most of the time, within the (actor) accounted realm of the 'non-revolutionary' (hence acceptable and 'non-cranky'). Accordingly, as we have seen in Sect. 2.4, discussing Barnes, Bourdieu and post-Kuhnian SSK generally, attention shifts to the micro-politics of scientific specialty groups to see how they manage, negotiate, refine, accept and reject bids to modify the paradigm, i.e. bids to have accomplished 'significant' results and discoveries. So, again to recur to the anti-method argument, in this view the 'method' of a discipline is not simply identified with its own particular paradigm—itself in flux over time—but further with the political and social structure and dynamics of the specialist community. The construction of scientific knowledge cannot be explained apart from the social processes in and through which that activity takes place.[102] No invocation of a general method can explain the manufacture and transformation of knowledge by paradigm-bearing and paradigm-negotiating communities, including the historically contingent socio-political structures of those communities. Method discourse abstracts from and floats above the proper cognitive and social complexity of scientific fields, and so it misses everything that now appears to be of importance in understanding the discovery–producing and hence tradition–altering dynamics of the sciences.

---

[100] Ravetz (1971), Mulkay (1979), Latour and Woolgar (1979), Knorr-Cetina (1981), Collins (1985), and above all Barnes (1982).

[101] For early 'derivations' of this position from the writings of Kuhn see Ravetz (1971) and Schuster (1979).

[102] These points were first brought out and displayed in full-scale contextualist studies in the history of science at the time of the emergence of these sorts of post-Kuhnian perspectives, for example, Rudwick (1985), Shapin and Schaffer (1985), Desmond (1982).

So much, then, for a formal deconstruction of the likely existence of universal, transferable and efficacious general methods, wherein we have, as it were, elaborated Koyré and Kuhn's debunking of general methods, by articulating it with reference to post-Kuhnian understandings of tradition dynamics in the sciences. We shall next meet the issue of grand methods in Chaps. 5, 6, and 7 in the context of Descartes' own methodological dreams and projects. There we will be in a position to move beyond the pure debunking of method. Rather, we will be able to understand in a more sympathetic, and 'anthropological' manner, how it could be that a Descartes (or anybody else) could genuinely believe in the efficacy of the kind of general method doctrine which we firmly know cannot work in the manner it claims. As noted in Chap. 1, this approach is necessary to any serious biographical understanding of the issue of method in Descartes' life. Mere debunking will not do; but, neither will it suffice to avoid the issue, or worst of all, to concede, if only tacitly, that there may after all have been something to his grand methodological claims. Our more sympathetic and biography-enhancing approach will be made possible through the construction of a model for how method discourses systematically mislead their inventors and believers—a model based in part on the very articulation of Kuhn's position we have outlined in the present section. For the moment, however, we must conclude our conceptual and historiographical exercise, by first sketching a periodization of the Scientific Revolution which follows from and articulates our model of natural philosophizing, before then locating Descartes in terms of our modeling and periodization by asking, 'What kind of natural philosopher was Descartes?' The preliminary answer to that question will bring us to the commencement of our biographical study proper.

2.7 Phases and Stages in the 'Scientific Revolution' Seen as an Unfolding Process in the Field of Natural Philosophizing, with Its Attendant Articulations to Other Domains

I offer here a periodization regarding the flow and dynamics of natural philosophizing, with its variously associated superior, cognate and subordinate disciplines and domains of concern. It marks out the central plot of the period called the Scientific Revolution.[103] The periodization categories are: (1) The Scientific Renaissance 1500–1600; (2) The Critical Period (or Period of Civil War in Natural Philosophy) 1590–1660; (3) The Period of Relative **C**onsensus, **M**uting of Systemic Conflict, New Institutionalization, End of *Scientia* and Incipient **F**ragmentation of the Field 1660–1720 (which will be abbreviated as CMF period below.)

The **Scientific Renaissance** displays in the realms of the subordinate sciences of the 'entourage' (Fig. 2.2 above), as well as that of natural philosophy, many of the

[103] For more details, and somewhat varying emphases, see Schuster (1990, 2002) and Schuster and Watchirs (1990).

scholarly aims and practices which already characterized the treatment of classical literature, history and languages in earlier stages of the Renaissance. The established humanist practices of textual recovery, editing, translation, commentary and printing increasingly focused on the scientific, mathematical and natural philosophical heritage of classical antiquity. These developments came late in the Renaissance considered as a larger historical epoch, but they mark the first stage and essential pre-condition for the further process of the Scientific Revolution.

There was a marked increase in the recovery, reconstruction and extension of the existing subordinate entourage sciences, the timing of which differed from field to field.[104] This took place amid the catalyzing influence of the pedagogical and philosophical assault on Scholastic philosophy; the reassertion of Platonizing modes of thought which helped revalue mathematics as the key to knowledge; and the more general trend toward recasting the ideal of knowledge in the image of practice, use and progress, rather than contemplation, commentary and conservation.

In natural philosophy, a wide and confusing array of non- or anti-Aristotelian approaches was made available through the recovery or improvement, assimilation and publication of alternatives. Outside of the universities, in princely courts, print house and workshops of master artisans, anywhere the practice of a subordinate science or practical art fell outside the preview of Aristotelianism, the practitioner could be set at odds with School philosophy and reach for rhetorical tools against it. Yet, throughout the sixteenth century 'orthodox' Scholastic Aristotelianism was officially entrenched as central to the education of all men with any serious concerns in natural philosophy. Indeed, from the late sixteenth century Scholastic Aristotelianism enjoyed renewed vigor in the rapidly rigidifying curricula of institutionalized forms of Protestantism and a militant, post-Tridentine Catholic Church. Hence the sixteenth century produced no crisis of natural philosophy.

In many ways, Descartes' Jesuit neo-Scholastic education at *La Flèche* embodied the results of these patterns of change. A critical juncture in the history of natural philosophizing and its subordinate disciplines was about to eventuate, and the young Descartes was both a product of these processes and an influential player in them.

The **Critical Period (or Phase of 'Civil War in Natural Philosophy')** of the Scientific Revolution (roughly 1590–1650) interests us most at the moment, as it

[104] In mathematical astronomy, the Renaissance phase is discernible from the late fifteenth century, whilst in mathematics and geometrical optics the pace of the Renaissance phase only accelerates in the later sixteenth century. In astronomy Copernicus could enter into the highly technical tradition of planetary astronomy basing himself on the prior labors of Regiomontanus and Peurbach, the late fifteenth century renovators of the field, who themselves had tried to appropriate and perfect the tradition as it had emerged from the later Middle Ages. In geometry the process of assimilation and purification is even easier to discern, for the century saw not only improved texts and commentaries on Euclid's *Elements*, but the recovery, translation and edition of the texts of higher Greek mathematics, of Apollonius, Archimedes and Pappus, not to mention Diophantus, who was critically important for the typically Renaissance development of the emergence of the mathematical art of algebra as a subject of theoretical import and structure. Anatomy and medical theory followed more closely upon astronomy, the program of editing and publishing the complete body of Galen's works culminating in the 1520s and 1530s. In each case, there was an initial stage of recovery, improvement, and, if necessary, translation of texts.

embraces the life-time of Descartes and the intellectual, and institutional, context in which were played out those features of his career under study here. The critical period was characterized by a conjuncture unique in the history of pursuit of natural knowledge, whether in classical antiquity, medieval Islam or Renaissance Europe: On the one hand, there was an unprecedented burst of conceptual transformation in the subordinate entourage sciences—optics, mechanics and astronomy, as well as anatomy/physiology—and in the crucial discipline cognate to natural philosophy, mathematics. The achievements of a Kepler, Galileo, Descartes or Harvey exemplify a wider pattern of accelerating transformation and mutual interaction among the subordinate entourage sciences in the two generations after about 1590. In this period, 'Renaissance' lines of development culminated, whilst the major figures of these generations worked out previously unexplored and unexpected orientations in the entourage sciences, thereby radically transforming them.[105] On the other hand, in natural philosophy the tendencies corrosive of Aristotelianism—the challenge of Paracelsianism, of Hermetically or alchemically tinged neo-Platonism, of calls for the re-evaluation of practical knowledge, of anti-Aristotelian rhetoric tied to the practice of the mathematical arts or classical mathematical sciences—all took on a greater urgency. There was a heightened, often desperate competition amongst systematic natural philosophies (some tied to utopian and irenic programs of religious, social and intellectual reform) which, later in this phase, eventually issued in the construction and initial successful dissemination of the mechanical philosophy.[106]

[105] The great stature, and frequently Whiggish interpretation of these men, and the Janus-like quality of their work, stems from their engagement with the classical subordinate sciences, mixed mathematical and 'bio-medical', at just the moment when characteristic lines of sixteenth century work were pushed to their apparent limits, and intended or unintended steps through these limits unexpectedly opened radically altered conditions and possibilities of investigation. H. Floris Cohen (2010) in his multi-phased interpretation of the process called the Scientific Revolution, also emphasizes the first generation of the seventeenth century as the crucial moment (consisting of three overlapping transformations, in realist mathematical science; corpuscular-mechanism; and a 'Baconian' style of experimentation) which took European natural philosophy and sciences for the first time beyond any previous revival of classical sources, such as had occurred in medieval Islam, high medieval Europe or even the European Renaissance itself, without in any of these previous cases reaching such a transformative watershed. A short version of part of this argument appeared in Cohen (2005). See my essay review of Cohen [DOI 10.1007/s11016-012-9645-6] in *Metascience* (2012), focused upon his conceptualization of this phase of the Scientific Revolution.

[106] About the earlier historiography of this heightened contestation, the following can be said in summary fashion: It has been obvious since Lenoble's (1943) work that in what we are calling the period of 'civil war in natural philosophy', families of natural philosophies competed in respect of the values, aspirations and religious resonances they endorsed and condemned. The classic work of Rattansi (1963, 1964) and Easlea (1980) took up this topos. However, we really begin to see the contestation in play when we contemplate the inter and intra family competition, arising from the fact that natural philosophy had that entourage of subordinate, more narrow traditions of science-like practice, including the mixed mathematical sciences, and the 'bio-medical' domains, such as anatomy, medical theorizing, and proto physiology in the manner of Galen. Hence, all the competition and contestation in the critical period was more serious than even the traditional literature suggests, since it is obvious that the competing families of natural philosophies actually consisted of quite individual systems, and that the situation was actually more like every man for himself.

Within and between both sets of developments the Renaissance themes of the re-evaluation of practical knowledge and the desire for domination of nature continued to be echoed. Now, however, they sounded more urgently and in a new key, marked by the crystallization of the hitherto disparate debate on the plane of 'high' natural philosophical utterance, as figures such as Bacon and Descartes systematically assimilated them to natural philosophical discourse.[107]

Out of this proliferation and climactic struggle amongst competing systems and their advocates there emerged varieties of the mechanical philosophy, which were designed and sold by a handful of innovators in an effort to finesse and resolve (in their own favor) the natural philosophical conflict of the age. By the mid-seventeenth century the cultural dominance of Aristotelianism collapsed (although it continued

---

The existence of clear genera did not prevent, and indeed it undoubtedly enflamed, a tendency, even for natural philosophers of similar genealogical stripe—neo-Platonic, proto or emerging mechanist, 'magnetic', or chemical—to compete with each other as well: Kepler vs. Fludd; Descartes vs. Gassendi vs. Hobbes; Libavius and other latter day Paracelsians vs. the heritage of Paracelsus himself.

[107] This sense of appropriation by natural philosophers of pre-existing discourse and rhetoric of the practical arts was the great insight of Paolo Rossi (1970) which becomes all the move obvious when one superimposes our model of natural philosophy onto his interpretation. Rossi's book, after all, is one of the great works on the Scientific Revolution. A naive summary runs like this: In the sixteenth century lots of books were written extolling the value of practical knowledge and the status of men of practice. These books issued from pedagogues, master artisans, courtiers, physicians, surgeons and others. Later these same revaluations and images become central in Bacon, Descartes and Hobbes. Most readers take this as an improvement on the vulgar Marxist notion of imprinting by structures upon actors. But how exactly should we understand Rossi? Do the new values float into Bacon's and Descartes minds, do they 'influence' these thinkers in some way: Is this a parallelism of ideas, or some contextual imprinting of them? Is this in the end a history of ideas, or some sort of mitigated Marxist account, or what? It is hard to answer, unless one has a model of the structure and dynamics of the field of natural philosophizing. Using it, we can interpret Rossi as having described a diffuse sixteenth century field of non-natural philosophical discourse on the practical arts. That discourse was itself articulated upon structural changes in sixteenth century Europe: changes in state and economy, to be modeled in state of the art social and economic history. Utterances in that discourse, that is, representations of the practical arts and their values, were later co-opted and redeployed, by Bacon and Descartes, into debates inside the natural philosophical field, as part of their respective strategies for advancing their overall claims in the natural philosophical agon. They were now articulating upon 'the practical arts' in this mediated sense. We need no implausible direct constitution of Science or natural philosophy by technical demands of a changing economy and state structure, à la Hessen (1931) or Zilsel (1942a, b). Similarly, we do not need a history of ideas notion of the 'influence' of this literature upon Descartes and Bacon. They were not being influenced by something in the society or economy that others were missing; nor were they 'reflecting' the interests of some particular group or class magically imprinted upon them. They were simply re-working, and projecting in the natural philosophical field, already available discursively embodied representations and revaluations of the meaning of the practical arts. In the first instance, the explanation of their behavior arises from their positions, tactics, resources and goals in the field of natural philosophy. Recalling our appeal to Sahlins' call to historicize the understanding of 'culture' in anthropology, we, like Sahlins, would see these natural philosophical 'natives' adapting to big, hard changes and forces by culturally specific moves; moves that are not determined by a universal logic, and may even possess novelty, but which are specific to the (evolving) culture. Finally, if all this reminds us of the discussion in Sect. 2.5.6 of 'the mechanics of responding to outside challenges', that is because precisely the same model and strategy of explanation are in play here.

supreme in most universities for another generation). The mechanical philosophy, in its several species, became the dominant genus. Hence my image of a 'civil war in natural philosophy', with multiple regime change: from Aristotelianism to mechanism, which had averted a threatened neo-Platonic take over. These struggles over natural philosophical systems on the one hand shaped struggles within the entourage of subordinate fields, and also, on the other hand, involved concerted attempts to articulate favored versions of natural philosophy to particular representations of matters political, theological, pedagogical or related to the content or values of the practical arts.

This is precisely where the description of change in natural philosophy and the entourage of subordinate sciences needs to be linked to the context of heightening political, religious and intellectual turmoil, denominated as the 'general crisis of the seventeenth century'. In terms of our model of natural philosophy we can now say the following about the troubled and turbulent age of the civil war within natural philosophizing: A genuine sub-culture of natural philosophy existed, in which systems of nature had significant and contested articulations to religious, political and social discourses. The equally really existing contextual problems and tensions, sometimes labeled a 'general crisis' of the seventeenth century, were interpreted by players through the filter of natural philosophizing, thus suggesting that the problems of the age had some of their basis in natural philosophical contention and dissensus. This raised the stakes in finding and enforcing the 'true' philosophy of nature, since natural philosophy was arguably part of these problems and part of their solutions. Hence, in the generation of civil war within natural philosophy, the proliferation of desperate and daring initiatives in neo-Platonic, alchemical, magical and Hermetically tinged natural philosophy, which in turn, elicited from some few individuals the equally sweeping, desperate as well as sudden invention of corpuscular-mechanism.[108]

[108] All of the major innovators in natural philosophy, whether or not part of the eventually triumphant mechanist party, should be viewed as actors responding to the context of religio-political-cultural 'crisis' of their generation. The careers of all the major figures in natural philosophy display certain similar strategies and aims, shaped by the needs of innovating in natural philosophy, because natural philosophy itself was thus placed in the turbulent culture of the age. They all aimed to fill a perceived void of natural philosophical authority, and they all overtly rejected Scholastic Aristotelianism, whilst remaining to varying degrees dependent upon its vocabulary and conceptual resources (hence giving endless work to historians of the continuity of ideas). Additionally, they all resonated, on the plane of natural philosophical discourse, some positive interpretation of the sixteenth century revaluation of the practical arts; and they all drew models and exemplars from the accrued catalogue of achievements in the practical arts and subordinate sciences of that century, although the choice and weighting of privileged items did vary greatly. In addition, most of the innovators stressed proper method and pedagogy as a salient feature of a new natural philosophy, as being necessary for establishing its truth and facilitating its dissemination and triumph. Their strivings grew in all cases from a sensitivity to the apparently irreconcilable divisions within the politics, religion (and natural philosophy) of the age. They also shared the perception that Aristotelianism could neither deal with those divisions, nor grasp or stimulate the proliferation of novelties in the practical arts and subordinate sciences. Beyond all this there was the suspicion, characteristic of the self-understanding of natural philosophers, that natural philosophical dissension was itself a conditioning cause of the larger political and religious conflicts, which, accordingly, could be wholly or partially cured by the installation of a true philosophy.

Educated men with natural philosophical interests recognized an imperative to find, and install, the 'proper' system of natural philosophy, because it was widely believed that the 'correct' program for natural knowledge would *ipso facto* provide much needed support for 'correct' religion, as well as a set of directives or for the improvement of both the moral and practical aspects of life. This powered and shaped the proliferation of alternative programs to Aristotelianism, and the eventual emergence of mechanism out of the competitive turbulence thus created. The stakes—political, moral and religious—inside the natural philosophical field were high. That there was no consensus on correct religion casts a poignant light on this struggle and explains its intensity as well as, to some degree, its ultimate lack of closure: There did not even emerge an agreed mechanistic system within the broad mechanistic consensus and, of course, the adherents of mechanism, Protestant and Catholic, remained unreconciled. No wonder René Descartes, as a radical and bold player in the natural philosophical contest of the age, in our view deserves the epithet, Descartes *agonistes*.

The founders of mechanism, such as Descartes, hoped to resolve the conflict of natural philosophies in a way which was to them cognitively progressive, but religiously and politically conservative. They exploited and co-opted recent achievements in the classical sciences, including the realist Copernican initiative, and amplified the premium placed upon mathematics and operative knowledge by sections of Renaissance opinion, whilst they avoided the perceived religious, political and moral pitfalls of the alchemical, Paracelsian, Hermetic and eclectic, 'qualitative' atomistic systems. Accordingly, the selection and molding of discursive resources to form the mechanistic systems was a nice and dangerous task. It involved endorsing some values and aims characteristic of the magical-alchemical systems, whilst explicitly opposing them as such. Mathematics was construed in terms of the sober geometry typical of the practical mathematical arts, to avoid any hint of neo-Platonic mathematical fancies; and yet, as in neo-Platonism—as opposed to Scholastic Aristotelianism—mathematics was to be the very language of nature. Experience was identified with experiment, itself rhetorically modeled upon the dissection and reassembly of machines, so as to marginalize alchemical and Paracelsian accounts of experience as an affect-laden, spiritually sanctioned and uplifting intuition of otherwise hidden relations and correspondences; and yet, as in natural magic and Paracelsianism—as opposed to Scholastic Aristotelianism—operative command over nature was sought through an active experiential engagement with nature. The mechanical philosophy was also constructed to embody an arguably orthodox 'voluntarist' vision of God's relation to nature and to mankind, so as to avoid collapsing the divine into nature and/or elevating man to the level of a 'magus', a status unacceptable to mainstream orthodox Catholic and Protestant thought alike. Accordingly, mechanism was neither the finest fruit of detached, rational 'modern' thought finally asserting itself to end 'the confusion', nor was it simply or directly, the reflection of some long rising merchant, administrative or craftsman-technologist groups, who for some contingent reason invented mechanism between 1630 and 1650.

Since we will be attending closely to what we can reconstruct about Descartes' shifting structures of agenda, self-understanding and identity, it is worth closing this description of the critical period with a look at how one might characterize these structures more generally amongst the anti-Aristotelian players in the critical period.[109] Note, first of all, that trying to run rings around Scholastic institutions and thinkers is a proclivity of the critical phase, indeed a characteristic of Baroque culture in general, although, to be sure, not a new pastime.[110] However, unlike Renaissance humanism, early and mid seventeenth century natural philosophizing displayed specific forms of anti-Aristotelianism focused on *strategies of displacement of hegemonic Aristotelianism within a continuing and contested game of natural philosophy*. Many contenders desired system change within the culture of natural philosophizing—a bold, determined change of regime— not the destruction of the game as such. These are the players Stephen Toulmin picked out in *Cosmopolis–The Hidden Agenda of Modernity* as the anti-Renaissance, self-proclaimed heroes of intellectual and cultural salvation.[111] The established rules of Renaissance humanism would have to go, as well as the taken for granted institutional hegemony of neo-Scholastic Aristotelianism. However, one did not have to be the mature Descartes, Hobbes or Bacon to be involved in this rather Baroque-looking penchant for rule breaking and bending. The vogue of seeking out novelty and discovery, not in the first instance a feature of the Scholastic culture of commentary and disputation,[112] meant that natural philosophies and natural philosophers were under pressure to change as the entire field came to be more contested and turbulent. A host of *de facto* or actually declared neo-Scholastic rules of the game can be cited as under threat from aggressive and individualistic (hence, if you like, Baroque looking) players. Neo-Scholasticism taught 'don't change the mixed mathematical sciences and their relation to natural philosophy'. But, some bold innovators tried to do so, and in doing so they created, fomented and explored the new domain of physico-mathematics, as we have seen. Neo-Scholasticism, in its deeply institutionalized customs of pedagogy and content, said, 'do not explicitly articulate natural philosophical claims on religious/political challenges, agendas and debates'. But some bold innovators tried to do so. Neo-Scholasticism also held *de facto*, but strongly, by means of its customs of pedagogy and content, 'do not bring in "inappropriate"

[109] Some of the points in this and the next paragraph were stimulated by participation in some of the Workshops and Seminars of the Baroque Science Project, headed by Ofer Gal at the Unit for History and Philosophy of Science, University of Sydney. Working papers in this area by me may be found at the Project website: www.usyd.edu.au/baroquescience/ The final result is Schuster (2012a).

[110] Clark (1992)

[111] Toulmin (1990)

[112] Which of course is not to say that no seeking of novelty and curiosities went on in Scholastic circles, teaching and textbooks, only that it was not the leading edge of these phenomena, rather the reluctant follower. Gascoigne (1990), Reif (1969), Schmitt (1973), Dibon (1954).

values, aims or players, particularly anything related to practical arts, material practice, instruments, and images and rhetoric concerning the status and value of same'. Although many bold innovators did so.

In the Critical Period, natural philosophical rules and norms, explicit or implicit, and practices, well entrenched and firmly reproduced from academic generation to generation, were all under threat of reformation, deformation or outright rejection. The self image, self-understanding, and correlated public posturing of the rebels and challengers was one of isolated, heroic, honor seeking, black and white decisive decision-making and action-taking. Interestingly, students of the cultural manifestations of this period, such as Carl Friedrich, stress that the Baroque was about rule bending and rule breaking, as well as about especially self-regarding and anguished matters of identity and honor. We easily discern these sorts of 'Baroque personalities' in the political and military figures of the age—Richelieu, Wallenstein, Gustavus Adophus, Maurice of Nassau, and Olivares—who displayed these cultural identity garments and proclivities at the same time that they forged new or revised forms and concretions of power (and of legitimations of power).[113] The natural philosophical players for the biggest stakes in the Critical Period seem similarly to display these traits.[114] To contest for systemic hegemony meant that one was a lone combatant against the rest, including the massed ranks, and deeply entrenched network of bastions of neo-Scholasticism. It would be an heroic effort, and one perhaps poignantly overlaid with intimations of tragic failure. We cannot know the nice biographical cum psychological channels through which the favored identity garments and protocols came to be lived and expressed. It is, however, clear that the situation in natural philosophizing seemed to many to demand such self-understandings, and public imagings, and that it was further enflamed by the presence of such personalities.

As noted earlier, the stakes in natural philosophizing were now very high, at least in some players' minds. And for such players, there were now numerous avenues open through which to pursue and express the traits of rule bending and honor seeking whilst natural philosophizing: *Is natural philosophy to become mathematical, that is more physico-mathematical?* In what sense, who gets the credit? What is the role and identity of the natural philosopher in that sense? *Is good and true natural philosophy to be decided more in terms of co-opting and explaining novel discoveries?* In which realms, by what techniques? What is the role and identity of the natural philosopher in this sense? *Can natural philosophy articulate to political philosophy, medicine, theology or not, and on whose terms?* What then is the role and identity of the natural philosopher? *Is natural philosophy meant to produce useful results?* Which ones? How? What then is the role and identity of the natural philosopher?

[113] Friedrich (1962) 41–46 and *passim.*

[114] With the exception of the gentle, genial (and resigned to unending crisis?) Gassendi, a man for that reason well recognized by historians as interestingly generationally displaced (too late for the scientific renaissance, too early for the age of consensus, muting and fragmentation). I thank my former University of New South Wales colleague, Dr. Barry Brundell for enlightening discussions on this and related points. See Brundell (1987).

Since all these channels were potentially open, and various gambits available within them, the overall goal of replacing Aristotelianism by producing the really best and truest natural philosophy got supercharged, at least for bold rebels, even if not for most Scholastics. René Descartes, with his particular concerns in natural philosophy, its rules and its subordinated disciplines, as well as his struggles over identity, agenda and self-understanding, was perhaps the exemplary figure of this moment in the cultural process of natural philosophy.

Turning now to the third and final stage in the Scientific Revolution, we find the **'CMF' Period (1660–1720)**, or in its full title, 'The Period of Relative **C**onsensus, **M**uting of Systemic Conflict, New Institutionalization, End of *Scientia* and Incipient **F**ragmentation of the Field'. Descartes, of course, did not live to see and grapple with the post-1650 developments of the CMF phase. However, as is well recognized, his work in natural philosophy and the subordinate sciences was hugely 'influential' (appropriated, used, renegotiated) in the CMF period, despite the less than overwhelming success of his own intended system.[115] Additionally, and a bit surprisingly, as we shall see in our concluding Chap. 13, Descartes even displayed a few of the tendencies which were to become more apparent in the two generations after his death.

The CMF Period was marked by the dissemination and widespread acceptance of mainly loosely held varieties of the mechanical philosophy, and by the endemic melding of these variants to Baconian rhetoric of method and experiment; the muting of contestation over systems (at least in public, especially in the new 'scientific' institutions); and the tendency for quasi-autonomous, more narrow successor fields of inquiry to emerge from natural philosophy, as natural philosophy itself began to undergo a slow century and a half process of final dissolution.[116] That is, the formerly more coherent—if internally contested—domain of natural philosophizing began to fragment into and *débouche* onto a suite of successor, more narrow and modern science-like domains. These included the emergent master science, classical mechanics, as well as evolved versions of the old mixed mathematical fields, now crystallized as more experimental and physico-mathematical; and a host of emergent new fields which solidified further in the eighteenth century.[117] Over the course of the next century, natural philosophy faded and died, and modern sciences emerged, along

---

[115] Dear (2001a, b), Schuster (2000b, 2002), Clarke (1989).

[116] Another, related ironic upshot of the 'civil war in natural philosophizing' was that natural philosophizing as a whole—the entire field of all these plays and turbulence—became, from the mid-seventeenth century, more autonomous of other cultural forms such as theology, as well as other branches of philosophy, whilst, at the same time beginning to undergo the process of fragmentation and dissolution just mentioned. (Schuster 2002)

[117] As to Newton, I hold that we misunderstand the rhythm of the development of early modern science by focusing too intently upon Newtonian celestial mechanics and physics. It is arguable that given the state of the natural philosophical field, including the subordinate sciences, the consensually held experimental form of corpuscular-mechanism, and its attendant sciences in their institutional, rhetorical and technical garb of the CMF stage, might have proceeded qualitatively rather undeterred for some considerable time had Newton not contingently intervened. Our periodization and plot—focusing on the trials of natural philosophy—should take this into account, seeing the process in terms of three phases or moments, punctuated, contingently by Newton, rather than aiming for him, or finding some clear closure in him. See Schuster and Watchirs (1990),

with an increasing armory of philosophical and other meta-scientific rhetoric and ideology constitutive of the onset of a wider 'modern scientific culture'.[118]

Additionally, and perhaps most tellingly as a function of these changes, the long held ideal of *Scientia*—systematic, unified and certain knowledge of nature—was destroyed. Aristotelianism had promised and promoted *Scientia*, and at the height of the process of the Scientific Revolution in the critical phase of the early and mid seventeenth century, other types of systematized natural philosophies challenged neo-Scholastic Aristotelianism, and some of these, notably Cartesianism, also seemed to promise *Scientia* on their own terms. But, in the CMF period, as natural philosophizing began its long process of fragmentation into a number of diverse and narrow special domains or disciplines of natural inquiry, which begin to look like sciences in our modern sense, the ideal of a unitary, systematic edifice of scientific knowledge was effectively rendered null and void.[119]

Natural philosophers also found themselves doing some of their natural philosophizing within the confines of new institutions, where they played the institution's organizational patterns in ways advantageous to them in institutional *and* natural philosophical terms. These institutions were additional nodes in the Europe wide field of natural philosophizing, not the exclusive ones, and, they were not the incubators of an essentially new, unified Modern Science, replacing a natural philosophizing supposedly barred from their precincts. This is because 'Modern Science' is not a unitary, 'Scientia' like entity, but rather a rhetorical and ideological label for the much messier, historically evolving and multiplying suite of expert scientific traditions and institutions—the very fact that, as we have seen, formed the ground note for Kuhn's model of science dynamics, and his anti-methodism.

Having noted the beginning of this process of dissolution of natural philosophizing into more modern looking, more narrow, highly internally competitive and 'discovery-seeking' disciplines, we may now reverse the direction of argument used above in Sects. 2.3 and 2.4. There we took some guidance in constructing our model of natural philosophizing from post-Kuhnian models of the agonistic dynamics of subsequently emerging more narrow, expert and modern scientific traditions of

---

Schuster (1990, 2002). Material in this and the preceding note implicitly touch upon the problem of how to think through the eighteenth century fragmentation and dissolution of natural philosophy into successor experimental and physico-mathematical sciences. Kuhn and Bachelard initially, if problematically, theorized this issue, later addressed and revised in Schuster and Watchirs (1990); Schuster (2002) as well as Schuster and Taylor (1996, 1997). See also Chap. 11 Note 11.

[118] By the late eighteenth century, all these tendencies contributed to the dissolution of the 500 year long European culture of systematic natural philosophizing and the emergence in its wake of that more typically nineteenth century institutional, professional and disciplinary ecology of the sciences which we might actually call 'modern'. Thus giving us the well known historiographical problem of the so-called 'Second Scientific Revolution', which, of course, was not a revolution at all. But, that is another story in the macro history of the natural sciences.

[119] The best treatment of these larger processes, with an emphasis on the intertwining of intellectual and social history is Gaukroger (2006), whose central motif may perhaps be captured by the notion that late in the Scientific Revolution, *Scientia* had definitely died, but the processes leading to the emergence of our modern, more socially and institutionally encompassing 'scientific culture' had begun to germinate.

research. It is now possible to suggest that the key cultural 'genes' found at the heart of modern, agonistic, discovery-seeking scientific disciplines were inherited from the culture of natural philosophizing as it passed through its critical and CMF phases in the seventeenth century Scientific Revolution.

The slow but powerful processes toward fragmentation of natural philosophy into successor domains and disciplines, unleashed originally during the critical phase, and clearly in play in the following CMF phase, carried the élan of continuous competition and contestation from the earlier period right into the structure and dynamics of the successor fields. Competing over systems disappeared, as did mere co–opting and copy–catting of others' discoveries. Transcribed into the successor fields were the peculiar tradition dynamics according to which a scientific tradition exists through, and for the purpose of, producing accredited novelty, a trait first expressed, in confused and desperate form, during the heated contestation of the critical phase. After all, modern sciences are by historical standards very odd beasts. They are continuously reproduced expert traditions whose very dynamics, and *raison d'être* in rhetoric and in practical activity, consists in the unremitting, competitive and concerted struggle to construct, and have implanted into the tradition, *significantly tradition-altering achievements*, which are proffered on a contested basis, and only have effect after being revised and negotiated into place by peer competitors of the initial proponents. Both the actual, messy, competitive and political 'mangle of practice' inside scientific traditions,[120] and the channels of crisp method rhetoric through which they are understood and accounted for, seem to bear just legible hallmarks that say—*"forged by somewhat rebellious master practitioners in the white heat of the early to mid seventeenth century natural philosophical crisis and initially polished by the experimental and physico-mathematical natural philosophers of the late seventeenth and early eighteenth century."*[121]

---

[120] The term derives of course from Andrew Pickering's (1995) brilliant and illuminating study of knowledge construction in modern physics.

[121] These conclusions also involve important insights about the history and deployment of 'method-talk' in the CMF stage of the Scientific Revolution. As the processes we have ascribed to the CMF period continued, actors' legitimatory and packaging rhetorics (typically rhetorics of method, as I have argued in previous publications on this issue) evolved to meet the needs of players with these new sorts of aims and agendas. For example, even before being further popularized by Newton, a method–discourse concerning 'speculative' vs. 'experimental' (natural) philosophy flourished in late seventeenth century England and was deployed, mainly by self-styled advocates of the latter, against real or imagined adversaries of the former stripe (Anstey 2005). All mid to late seventeenth century users of this rhetoric were inside the field of natural philosophizing—they had not really escaped to some other space. And, although those favoring the 'experimentalist' side of the rhetoric might have proclaimed the death and overcoming of natural philosophy (and fooled some subsequent historians), it was in fact a way of positioning themselves and their work in a field still inhabited not only by themselves, but by others, including a few players and texts of overtly theoretical, systematic and contentious natures. Once we understand that, we see that the ongoing secular process toward fragmentation of natural philosophy, and crystallization of more narrow and more modern looking successor fields, makes no difference to the argument, as some domains became more autonomous, *sui generis* and discipline–like, they still enjoyed the genetic endowment of this rhetoric of experimental method.

In sum, to take an overview of the changes in natural philosophy in the early modern era we have just sketched, I think we can say that coming out of the late medieval and into the early sixteenth century, natural philosophizing was largely, indeed almost entirely, a university based, in-house game of competing versions of Aristotelianisms. A subsequent 'scientific renaissance' stage followed in the sixteenth century, especially heating up in its last two decades, during which the game was opened to a widening range of available sources, eclectic takes on recovered alternatives, and increasingly bold and religiously implicated moves and claims in the field. The mechanistic gambits of the critical period of the early to mid seventeenth century can then be seen as direct responses to the already disturbed and turbulent state of the field. But, even that turbulence and bold stake-claiming did not and could not last forever. The conflict of systems in the early to mid seventeenth century was what first attracted the attention of historians of science to the problem of natural philosophy as an actor's and historiographical category—starting with Lenoble's great work in the 1940s down through Rattansi in the 1960s and Ravetz in the 1970s, to scholars of my own generation. But it also diverted us from looking at natural philosophy as a field, institution and tradition with a longer and wider life than indicated by the nodes of vicious confrontation in 'the age of the Baroque'. (Just as the contemporary phase of vicious religio-political, civil and international warfare was only a phase in the longer process of crystallization of some states—and of the state system itself—and failure of others).

2.8 Looking Forward—What Kind of Natural Philosopher/ Physico-Mathematician Was René Descartes?

Let us draw this chapter long excursion into historiographical and conceptual foundations to a close by refocusing on our subject, the young René Descartes. We know that he was trained as, and would eventually practice as, a philosopher of nature. But, what sort of natural philosopher was he; which subordinate fields was he

---

As a result, an interesting macro picture of the evolution of methodological accounting rhetoric emerges, which can now embrace the picture of the Scientific Revolution explored here. The matter might be envisioned as follows: The history of method discourse tracked and reflected the shifting dynamics and contents of natural philosophizing, and its fate—a long process running from the sixteenth century dominance of neo-Scholastic discussions of method, through the methodological prophets of 'the Baroque' such as Bacon and Descartes, with then some new threads of method–discourse being forged and deployed as the mid and later seventeenth century history of natural philosophizing unfolded. Later, with deepening fragmentation of the field and emergence of descendant fields, virtually the only dimension of natural philosophizing (of the original four—matter, cause, cosmology and method) that survived into the nineteenth century with its rationale and practice little changed was the dimension of 'method'. It became the last vanishing ghost of the living field of natural philosophizing. All the issues and implications involved in this Note are being pursued as part of my present project on: 'The Fate of Natural Philosophy at the Dawn of Modern Science: A recasting of the plot of The Scientific Revolution.'

concerned with, and how did those concerns both reflect and shape his larger natural philosophical agendas and outcomes; and what stages and phases did his natural philosophical and related activities display, in relation to what we now know about the field of natural philosophizing and its larger processes of change in the seventeenth century? The simple answer, based on common knowledge and the little we have said so far, is of course that he would become a corpuscular-mechanist natural philosopher, a devotee of a radical 'physico-mathematicising' approach to the mixed mathematical sciences, a realist Copernican of an extreme type and a dedicated proponent of the replacement of hegemonic Scholastic Aristotelianism by his own brand of natural philosophy. We also know he would pursue these activities during a period of heightened turbulence and contention in the field of natural philosophizing, and that many of his traits as a philosopher of nature would seem to set him, and many of his contemporaries, apart from the style and concerns of natural philosophizing in the later seventeenth century phase of the Scientific Revolution, with its muting of public theoretical controversy; consensual but increasingly piecemeal corpuscular-mechanism; new organizational forms; and creeping processes of death of *Scientia*, fragmentation of natural philosophy, and embryonic crystallization of narrow successor fields.

However, such generalities, whilst true and certainly useful for more cursory depictions of Descartes, will not disclose the kinds of details and dynamics we are after, nor suggest how to arrive at them. For that we need to concentrate on two related sets of considerations: first of all, an understanding of the main domains and fields he pursued, most notably natural philosophy as we have modeled it here; and, secondly the caveats and hints about pursuing Descartes' intellectual biography set out in Chap. 1, especially the imperative to reconstruct, at various moments in Descartes' trajectory, his own likely understanding of his agenda, his intellectual identity and his key concepts and skills. By proceeding this way, and bearing in mind our modified version of Gaukroger's suggested three interacting heuristic axes of intellectual biographical study, also offered in Chap. 1, we can better discourse biographically about issues of development, change, continuity and rupture. This will tend to keep us from neglecting changes in Descartes' doctrines and states of self–understanding and agenda; thus in turn helping us avoid the homogenization or rendering monolithic and monocausal of elements of his thought; and preventing misconceptions about what he was trying to achieve and why he employed the means he did.

Excellent, and innovative in a way, that his Jesuit education into neo-Scholastic Aristotelianism had been, there was nothing in it that would have prompted or armed the young Descartes in the direction of corpuscular-mechanism or any radical form of physico-mathematics.[122] It had, however, made him one of the natural philosophically literate. Tacitly and explicitly he had imbibed the aims and grammar of natural

[122] Notwithstanding the acquaintance he would have made with the very conservative form of physico-mathematics advanced by some of the Jesuit mathematicians. Cf. above, text accompanying Note 59, point [1] concerned with Peter Dear's important findings on the topic.

philosophizing, in a well taught neo–Scholastic Aristotelian form. He had benefitted from that short historical moment when sectors of the Catholic gentlemanly elite were actually trained to some degree, at university level, in some practical mathematics, and so also acquainted at least in passing with the rudiments of the idea that the utility of some forms of mathematics was to be valued, and to be pursued on that basis, legitimately, by the gentlemanly officer or functionary. But the brush with natural philosophy of the potential future officer, lawyer or royal administrator could easily have remained frozen at that level, never to be enlivened or explored again. After his legal studies, Descartes was travelling, not unlike thousands of other young educated gentlemen of the time, seeking experience and adventure, when there occurred the absolutely critical event of his entire natural philosophical career. This was his meeting with, and initial mentoring by Isaac Beeckman. The Dutchman, eight years his senior, is of supreme import to Descartes, philosopher of nature. Beeckman was one of only two people from whom anyone could have imbibed a corpuscular-mechanical perspective at the time (the other being the brilliant but unpublished Englishman, Thomas Harriot).[123]

Corpuscular-mechanism, as Beeckman and later Descartes and others practiced it, was not simply the adoption of some sort of ancient atomist matter theory. There were plenty of advocates of such styles of what historians now call 'qualitative atomism'. What set off corpuscular-mechanism as a unique genus of natural philosophy was the addition to atomism of a commitment to devise a 'mechanics' or 'science of motion' embodying laws governing the motion and exchanges of motion in the world of micro corpuscles. This would be the causal dimension of such natural philosophies. Qualitative atomists had no such imperative, the causal registers of these natural philosophies being filled out from traditional notions of spiritual or immaterial forces, attractions, repulsions, antipathies and sympathies. The search for a 'mechanics' to 'run' the world of micro particles was one sense in which the traditional mixed mathematical science of mechanics was being articulated and renegotiated in a physico-mathematical direction. It is important to note that the conventional topos according to which Descartes' mechanism comes from a, quite real, fascination with automata and mechanical contrivances, is misleading at this point. Simply to revel in automata, or read engineers and students of more traditional mechanics, like Simon de Caus, would not make you a corpuscular-mechanist: there were many devotees of mechanics and mechanical contrivances who were not corpuscular-mechanical natural philosophers.

However, Descartes was not at this point in 1618–1619 a systematic philosopher of nature, corpuscular-mechanist or otherwise. Neither he nor his mentor displayed the slightest interest in weaving a systematized version of their corpuscular-mechanism for private or public consumption. Rather, something else was on their minds, and it was tied up with how and why they held to corpuscular-mechanism as a preferred stance in natural philosophizing. That something else was what they explicitly and

[123] Gaukroger (1995) Chap. 3; Gaukroger and Schuster (2002). To see Harriot working out his own mechanics of corpuscles on analogy to the behaviour of light, see Smith (2008).

proudly called physico-mathematics, viewed in much the way we have generically described it earlier in Sect. 2.5.3. Beeckman and Descartes were committed radicals, challenging the declaratory Aristotelian rules about the nature and status of the mixed mathematical sciences. They worked on several projects which we shall examine in considerable detail, all devoted to moving bits and pieces of some of the mixed mathematical fields into direct, organic connection with natural philosophy, meaning for them a corpuscular-mechanical natural philosophy which would supply, in a piecemeal way, the explanatory framework of matter and cause involved in resolving each separate problem. So, we can and will picture the young, rather radical Descartes in late 1618 and early to mid 1619 as follows: In natural philosophy, he held a marginal and radical species, corpuscular-mechanism, but he held it loosely, not pursuing a system or advocating the program as such. It was subordinated to the real program, in physico-mathematics. His identity, perhaps his dream, was that of a corpuscular-mechanical *physico-mathematicus*, not a systematic corpuscular mechanist.

This, by the way, will show the value of attending not only to natural philosophizing, its structure and dynamics, but also simultaneously to how players address given but negotiable rules of natural philosophizing, and how, additionally, they position themselves as players. Simply to say Descartes becomes interested in corpuscular-mechanism is not enough; nor is it enough to recognize that this was one species of a number of increasingly competing programs in a turbulent field; nor is it enough to recognize that the rules and grammar of natural philosophizing were conveyed through university teaching of Aristotelianism; nor that *qua* rules they were in principle contestable and renegotiable. No, one also has to see that holding a natural philosophy might or might not involve commitment to systematization and that what a player imagined about his agenda and identity in the field will help explain much of how he dealt with and worked within the landscape just described.

Given all that, it would be a relatively easy problem with which to deal, if Descartes had simply remained this kind of Beeckmanian, piecemeal corpuscular-mechanical *physico-mathematicus*, and if correspondingly, this identity and agenda of his could have been smoothly actualized. Neither of these conditions were fulfilled. We shall see that Descartes' early physico-mathematics, which was both a vast agenda and possible intellectual identity, remained in practical terms a scene of very mixed results. So, even early on there was a tension between program and identity, on the one hand, and results, on the other. Moreover, this kind of tension was to escalate for the young Descartes in these very early years, because his trajectory did not remain fixed on physico-mathematical corpuscular mechanism. It became even more complex and tortured, and for long periods between 1618 and 1628 it even effectively took him out beyond, and as he would see it, above natural philosophizing, to a pair of related and extremely radical, indeed delusionary, imperialistic knowledge programs within which corpuscular-mechanical physico-mathematics would be reduced to only a small fiefdom.

We shall learn that in 1618–1619, Descartes envisioned in quick succession two breathtaking projects reaching beyond physico-mathematics: universal mathematics and universal method. First, he imagined his universal mathematics as meant to encapsulate and transcend 'mere' physico-mathematics by unifying it with the techniques and protocols he was also working on, in a piecemeal fashion, in analytic mathematics. Then, in a peak of excitement later in 1619 (around the time of his famous dreams), he envisioned his universal method which was meant to absorb universal mathematics itself and move on much further to be applicable and efficacious in all rational disciplines. We shall have much to say about these projects, their relations, and the reasons for their inevitable sterility and failure. Additionally, as suggested by Sect. 2.6 above, we shall also find out why projects such as Descartes' method offer such a convincing and attractive appearance to their inventors, and sympathetic audiences. Importantly, each of these programs offered Descartes resources for imagining his agenda and identity. With universal mathematics and method in mind, as he seems to have held them during the 1620s, he was in fact proposing to pass beyond natural philosophizing, by first limiting natural philosophy to physico-mathematics of his own style (a vast culling, or reforming of the field) and second, by subordinating even that type of natural philosophy to a subsidiary role in the larger proposed field of universal mathematics, to be 'worked' by the method.

This version of the young Descartes illustrates that he was aiming to become the leader of, and model for, a kind of intellectual program in which natural philosophizing would become 'mathematical', in the sense of physico-mathematical and be subsumed with all other mathematically based disciplines into universal mathematics. Indeed, all rational disciplines, including universal mathematics, would come under the sway of common procedures of a universal method, itself devised by extension and articulation of properly mathematical procedures. However, as we shall see, the young René Descartes eventually discovered the impossibility of his dreams of universal mathematics and method, when he tried to work them out in full in an expanded version of his *Rules for the Direction of the Mind* between 1626 and 1628.

We shall examine in considerable detail how and why that happened, and when we reach that point, we shall also see something completely explicable and unexceptional happen. With the collapse of his methodological dream, agenda, and identity, that is, with the failure of his project of the *Rules for the direction of the mind* in 1628, Descartes retreated to more familiar and densely populated terrain. He quickly evolved or inflected into being, by agenda and identity, a systematic philosopher of nature, as he developed his project for *Le Monde*, his first system of natural philosophy, a bold corpuscular-mechanical and realist Copernican vision which bore at its core some of the conceptual DNA of his earlier physico-mathematics. His more pure analytical mathematical work was separated off from natural philosophizing, in accord with the traditional Aristotelian position on natural philosophy and mathematics being cognate fields. The substitute for his previous grandiose programs became the attempt to ground and legitimate both natural philosophy and mathematics, by an overarching dualist metaphysics. Thus was the embryonic form of mature Cartesianism hatched. We shall see just how far Descartes was intending to design *Le Monde* as a coherent system of natural philosophy; how *Le Monde* bore some conceptual DNA from his physico-mathematical results and aspirations; and what his strategies of co-optation, emulation and replacement involved *vis à vis* his competitors such as Beeckman and Kepler. In all of this, our modelling of the systematicity of a natural philosophy in Sect. 2.5.5 will be of much use.

Such, in brief, will be our answer to the questions: What kind of natural philosopher was the young Descartes? How did he practice? What at various points did he envision as his agenda and identity? What tactics did he follow? How did he play with and upon the common rules and grammar of the field? And, where did he arrive by the time he had written his first system of natural philosophy? The detailed answers will be worked out in the next nine chapters by reconstructing his natural philosophical, physico-mathematical and methodological trajectories in their concrete and evolving circumstances.[124] As indicated, we must begin with the most important stage in the entire story, Descartes' second natural philosophical education under Isaac Beeckman, and his entry into the aspirational realm of physico-mathematics.

References

Works of Descartes and Their Abbreviations

AT = *Oeuvres de Descartes* (revised edition, 12 vols.), edited by C. Adam and P. Tannery (Paris, 1964–76). References are by volume number (in roman) and page number (in Arabic).

SG = *The World and Other Writings*, edited and translated by Stephen Gaukroger (Cambridge,1998).

MM = *René Descartes, The Principles of Philosophy*, translated by V. R. Miller and R. P. Miller (Dordrecht, 1991)

MSM = *Rene Descartes, Le Monde, ou Traité de la lumière*, translated by Michael S. Mahoney (New York, 1979).

CSM(K) = *The Philosophical Writings Of Descartes*, 3 vols., translated by John Cottingham, Robert Stoothoff, and Dugald Murdoch, and (for vol. 3) Anthony Kenny, (Cambridge, 1988) References are by volume number (in roman) and page number (in arabic).

HR = The Philosophical Works of Descartes, vol I translated by E.S. Haldane and G.R.T. Ross (Cambridge, 1968 [1st ed. 1911])

Other

Aldridge, Larissa. 2009. 'Kaleidoscopic natural theology: The dynamics of natural theological discourse in seventeenth and early eighteenth century-England,' Unpublished Ph.D. dissertation. University of New South Wales.

[124] Observations on Descartes' later career, and his relations with the emergent tendencies of the succeeding CMF period, will be reserved for our concluding 'Coda and Epilogue' in Chap. 13.

Anstey, Peter. 2000. Descartes' cardiology and its reception in english physiology. In *Descartes' natural philosophy*, ed. S. Gaukroger, J.A. Schuster, and J. Sutton, 420–444. London: Routledge.
Anstey, Peter. 2005. Experimental versus speculative natural philosophy. In *The science of nature in the seventeenth century: Changing patterns of early modern natural philosophy*, ed. Peter Anstey and John Schuster, 215–242. Dordrecht: Springer.
Anstey, Peter, and John Schuster. 2005. Introduction. In *The science of nature in the seventeenth century: Patterns of change in early modern natural philosophy*, ed. P. Anstey and J.A. Schuster, 1–7. Dordrecht: Springer.
Arieu, R. 1999. *Descartes and the late scholastics*. Ithaca: Cornell Universiety Press.
Bachelard, Gaston. 1949. *Le rationalisme appliqué*. Paris: Presses Universitaires de France.
Bachelard, G. 1975a. *Le Nouvel Esprit Scientifique*, 13th ed. Paris: Presses Universitaires de France.
Bachelard, G. 1975b. *La Formation de l'Esprit Scientifique*, 9th ed. Paris: Vrin.
Barnes, Barry. 1982. *T.S.Kuhn and social science*. London: MacMillan.
Boschiero, Luciano. 2007. *Experiment and natural philosophy in seventeenth century Tuscany: The history of the Accademia del Cimento*. Dordrecht: Springer.
Bourdieu, Pierre. 1971. Intellectual field and creative project. In *Knowledge and control: New directions for the sociology of education*, ed. M.F.D. Young, 161–188. London: Collier-MacMillan.
Bourdieu, Pierre. 1971a. Systems of education and systems of thought. In *Knowledge and control: New directions for the sociology of education*, ed. M.F.D. Young, 189–207. London: Collier-MacMillan.
Bourdieu, Pierre. 1975b. The specificity of the scientific field and the social conditions of the progress of reason. *Social Science Information* 14: 19–47.
Brannigan, Augustine. 1980. Naturalistic and sociological models of the problem of scientific discovery. *The British Journal of Sociology* 31: 559–573.
Brannigan, Augustine. 1981. *The social basis of scientific discoveries*. Cambridge: Cambridge University Press.
Brundell, Barry. 1987. *Pierre Gassendi*. Dordrecht: Reidel.
Clark, William. 1992. The scientific revolution in the German nations. In *The scientific revolution in national context*, ed. Roy Porter and Mikulas Teich, 90–114. Cambridge: CUP.
Clarke, Desmond. 1989. *Occult powers and hypotheses: Cartesian natural philosophy under Louis XIV*. Oxford: Clarendon.
Clarke, Desmond. 2006. *Descartes, a biography*. Cambridge: CUP.
Cohen, H.Floris. 1994. *The scientific revolution: A historiographical inquiry*. Chicago: University of Chicago Press.
Cohen, H.Floris. 2005. The onset of the scientific revolution: Three near–simultaneous transformations. In *The science of nature in the seventeenth century: Changing patterns of early modern natural philosophy*, ed. Peter Anstey and John Schuster, 1–33. Dordrecht: Springer.
Cohen, H.Floris. 2010. *How modern science came into the world: Four civilizations, One 17th–century breakthrough*. Amsterdam: Amsterdam University Press.
Collins, Harry. 1985. *Changing order*. London: Sage.
Crombie, A.C. 1953. *Robert Grosseteste and the origins of experimental science, 1100–1700*. Oxford: Clarendon.
Cunningham, A. 1985. Fabricius and the "Aristotle project" in anatomical teaching and research at Padua. In *The medical renaissance of the sixteenth century*, ed. A. Wear et al., 195–222. Cambridge: CUP.
Cunningham, Andrew. 1988. Getting the game right: Some plain words on the identity and invention of science. *Studies in History and Philosophy of Science* 19: 365–389.
Cunningham, Andrew. 1991. How the *Principia* got its name; or, taking natural philosophy seriously. *History of Science* 24: 377–392.
Cunningham, Andrew, and Perry Williams. 1993. De-centring the 'big picture': *The Origins of Modern Science* and the modern origins of science. *British Journal for the History of Science* 26: 407–432.
de Dainville, F. 1954. Enseignement des mathématiques dans les Collèges Jesuites de France au XVIIe et XVIIIe siècles. *Révue d'Histoire des Sciences* 7: 6–21.

Dear, Peter. 1991. The church and the new philosophy. In *Science, culture and popular belief in renaissance Europe*, ed. S. Pumfrey, P.L. Rossi, and M. Slawinski, 119–139. Manchester: Manchester University Press.

Dear, Peter. 1995. *Discipline and experience: The mathematical way in the scientific revolution*. Chicago: Chicago University Press.

Dear, Peter. 2001a. *Revolutionizing the sciences: European knowledge and its ambitions, 1500–1700*. Princeton: Princeton University Press.

Dear, Peter. 2001b. Religion, science and natural philosophy: Thoughts on Cunningham's thesis. *Studies in History and Philosophy of Science* 32: 377–386.

Debus, A.G. 1970. Harvey and Fludd: The irrational factor in the rational science of the seventeenth century. *Journal of the History of Biology* 3: 81–105.

Debus, A.G. 1977. *The chemical philosophy*, vol. I. New York: Dover.

Des Chene, D. 1996. *Physiologia: Natural philosophy in late Aristotelian and Cartesian thought*. Ithaca: Cornell University Press.

Desmond, A. 1982. *Archetypes and ancestors: Paleontology in Victorian London, 1850–1875*. London: Blond & Briggs.

Dibon, Paul. 1954. *La Philosophie neérlandaise au siècle d'or.* Vol I, *L'enseignement philosophique dans les universités a l'époque précartesienne*, 1575–1650. Paris: Elsevier.

Easlea, Brian. 1980. *Witch-hunting, magic and the new philosophy: An introduction to the debates of the scientific revolution 1450–1750*. Sussex: Harvester Press.

Feyerabend, P.K. 1975. *Against method*. London: New Left Books.

Foucault, Michel. 1972. *The archaeology of knowledge*. Trans. Alan Sheridan. London: Tavistock.

Frank, Robert G. 1980. *Harvey and the english physiologists: Scientific ideas and social interaction*. Berkeley: University of California Press.

Friedrich, Carl. 1962. *The age of the baroque: 1610–1660.* (1st ed 1952). New York: Harper & Row.

Garber, Daniel. 1992. *Descartes' metaphysical physics*. Chicago: University of Chicago Press.

Gascoigne, John. 1990. A reappraisal of the role of universities in the scientific revolution. In *Reappraisals of the scientific revolution*, ed. D.C. Lindberg and R.S. Westman, 207–260. Cambridge: Cambridge University Press.

Gaukroger, S. 1976. Bachelard and the problem of epistemological analysis. *Studies in History and Philosophy of Science* 7: 189–244.

Gaukroger, Stephen. 1995. *Descartes: An intellectual biography*. Oxford: OUP.

Gaukroger, Stephen. 2006. *The emergence of a scientific culture*. Oxford: Clarendon.

Gaukroger, S., and J.A. Schuster. 2002. The hydrostatic paradox and the origins of Cartesian dynamics. *Studies in History and Philosophy of Science* 33: 535–572.

Gieryn, Thomas. 1983. Boundary-work and the demarcation of science from non-science: Strains and interests in professional ideologies of scientists. *American Sociological Review* 48: 781–795.

Gilson, E. (ed.). 1947. *René Descartes, Discours de la Méthode: Texte et Commentaire*. Paris: Vrin.

Gouhier, H. 1958. *Les Premières Pensées de Descartes*. Paris: Vrin.

Grayling, A.C. 2005. *Descartes: The life of René Descartes and its place in his times*. London: Free Press.

Hannaway, O. 1975. *The chemists and the word: The didactic origins of chemistry*. Baltimore: Johns Hopkins University Press.

Harrison, Peter. 2000. The influence of Cartesian cosmology in England. In *Descartes' natural philosophy*, ed. S. Gaukroger, J.A. Schuster, and J. Sutton, 168–192. London: Routledge.

Harrison, Peter. 2002. Voluntarism and early modern science. *History of Science* 40: 63–89.

Harrison, Peter. 2005. Physico-theology and the mixed sciences: The role of theology in early modern natural philosophy. In *Early modern natural philosophy: Patterns of change in the culture of natural philosophy in the seventeenth century*, ed. Peter Anstey and J.A. Schuster, 165–184. Dordrecht: Springer.

Hattab, Helen. 2005. From mechanics to mechanism: The *Quaestiones Mechanicae* and Descartes' physics. In *The science of nature in the seventeenth century: Changing patterns of early modern natural philosophy*, ed. Peter Anstey and John Schuster, 99–129. Dordrecht: Springer.

Henry, John. 2002. *The scientific revolution and the origins of modern science*, 2nd ed. MacMillan: London.
Hessen, Boris. 1931. The social and economic roots of Newton's "Principia". In *Science at the crossroads, papers presented to the international congress of the history of science and technology held in London from June 29th to July 3rd, 1931, by the delegates of the USSR*, 149–212. London: Kniga (England) Ltd.
Knorr-Cetina, Karin. 1981. *The manufacture of knowledge: An essay on the constructivist and conventional character of knowledge and cognition*. Oxford: Pergamon.
Koyré, A. 1939, Eng. Trans. 1978. *Etudes Galiléenes*. Paris: Hermann & Cie. English trans. *Galileo Studies*. Trans. J. Mepham. Hassocks, Sussex: Harvester.
Koyré, Alexandre. 1956. The origins of modern science. *Diogenes* 16: 1–22.
Koyré, A. 1969. *Metaphysics and measurement: Essays in scientific revolution*. London: Chapman & Hall.
Kuhn, Thomas S. 1970. *The structure of scientific revolutions*, 2nd ed. Chicago: University of Chicago Press.
Kuhn, Thomas S. 1977. *The essential tension: Selected studies in scientific tradition and change*. Chicago: University of Chicago Press.
Laird, W.R. 1986. The scope of renaissance mechanics. *Osiris* 2: 43–68.
Lakatos, I. 1978. Falsification and the methodology of scientific research programmes. In *Imre Lakatos: Philosophical papers*, vol. I, ed. J. Worrall and G. Currie, 8–101. Cambridge: C.U.P.
Latour, Bruno, and Steve Woolgar. 1979. *Laboratory life, the social construction of scientific facts*. London: Sage.
Lecourt, D. 1975. *Marxism and epistemology: Bachelard, Canguilhem, Foucault*. Trans. B. Brewster. London: New Left Books.
Lenoble, Robert. 1943. *Mersenne ou la naissance du mécanisme*. Paris: J.Vrin.
Lindberg, David. 1992. *The beginnings of Western science*. Chicago: University of Chicago Press.
Maclean, I. Ed. and Trans. 2006. *René Descartes: A discourse on method*. Oxford: OUP.
Maclean, I. 2007. *Logic, signs and nature in the renaissance*. Cambridge: CUP.
Maravall, José Antonio. 1973. *The culture of the baroque: Analysis of a historical structure*. English Trans, 1986. Minneapolis: University of Minnesota Press.
Mulkay, Michael. 1979. *Science and the sociology of knowledge*. London: George Allen & Unwin.
Pagel, W. 1982. *Joan Baptista Van Helmont*. Cambridge: CUP.
Pickering, Andrew. 1995. *The mangle of practice: Time, agency and science*. Chicago: University of Chicago Press.
Popkin, Richard. 1964. *The history of scepticism from Erasmus to Spinoza*. Berkeley: University of California.
Popper, K.R. 1959. *The logic of scientific discovery*. London: Hutchinson.
Priestley, J. 1767. *The history and present state of electricity with original experiments*. London: J. Dodsley et al.
Randall, J.H. 1961. *The school of Padua and the emergence of modern science*. Padua: Editrice Antenore.
Rattansi, P.M. 1963. Paracelsus and the Puritan revolution. *Ambix* 11: 24–34.
Rattansi, P.M. 1964. The Helmontian-Galenist controversy in seventeenth century England. *Ambix* 12: 1–23.
Ravetz, J.R. 1971. *Scientific knowledge and its social problems*. Oxford: OUP.
Ravetz, J.R. 1975. 'Science, history of', *Encyclopedia Britannica*. 16: 366–372
Rief, Patricia. 1969. The textbook tradition in natural philosophy, 1600–1650. *Journal of the History of Ideas* 30: 17–32.
Rix, Mark. 1997. Discipline and *threatened* punishment: The theory of nuclear deterrence and the discipline of strategic studies, 1946–1960'. Unpublished Ph.D. dissertation, University of Wollongong, Department of Science and Technology Studies.

Rodis-Lewis, G. 1992. Descartes' life and the development of his philosophy. In *The Cambridge companion to Descartes*, ed. J. Cottingham, 21–57. Cambridge: CUP.

Rose, Paul Lawrence, and Stillman Drake. 1971. The Pseudo-Aristotelian *Questions of mechanics* in renaissance culture. *Studies in the Renaissance* 18: 65–104.

Rossi, Paolo. 1970. *Philosophy, technology and the arts in the early modern era*. New York: Harper and Row.

Rudwick, M.J.S. 1985. *The great Devonian Controversy: The shaping of scientific knowledge among gentlemanly specialists*. Chicago: University of Chicago Press.

Sahlins, Marshall. 1993. Goodbye to *tristes tropes:* Ethnography in the context of modern world history. *The Journal of Modern History* 65: 1–25.

Sarton, G. 1921-22. Introduction to the history and philosophy of science. *Isis* 4: 23–31.

Sarton, G. 1924. The new humanism. *Isis* 6: 9–34.

Schaffer, Simon. 1986. Scientific discoveries and the end of natural philosophy. *Social Studies of Science* 16: 387–420.

Schmitt, Charles. 1973. Towards a reassessment of renaissance Aristotelianism. *History of Science* 11: 159–193.

Schuster, John. 1979. Kuhn and Lakatos revisited. *British Journal for the History of Science* 12: 301–317.

Schuster, J.A. 1986. Cartesian method as mythic speech: A diachronic and structural analysis. In *The politics and rhetoric of scientific method*, ed. J.A. Schuster and R.R. Yeo, 33–95. Dordrecht: D. Reidel.

Schuster, J.A. 1990. The scientific revolution. In *The companion to the history of modern science*, ed. R.C. Olby, G.N. Cantor, J.R.R. Christie, and M.J.S. Hodge, 217–242. London: Routledge.

Schuster, J.A. 1993. Whatever should we do with Cartesian method: Reclaiming Descartes for the history of science. In *Essays on the philosophy and science of René Descartes*, ed. S. Voss, 195–223. Oxford: OUP.

Schuster, John. 1995. Descartes *Agonistes:* New tales of Cartesian mechanism. *Perspectives on Science* 3: 99–145.

Schuster, J.A. 1995a. *An introduction to the history and social studies of science* Open Learning Australia. This work may be found on my website http://descartes-agonistes.com/.

Schuster, J.A. 2000. Descartes *Opticien*: The construction of the law of refraction and the manufacture of its physical rationales, 1618–1629. In *Descartes' natural philosophy*, ed. S. Gaukroger, J.A. Schuster, and J. Sutton, 258–312. London: Routledge.

Schuster, John. 2000a. Internalist and externalist historiographies of the scientific revolution. In *Encyclopedia of the scientific revolution*, ed. W. Applebaum. New York: Garland Publishing.

Schuster, J.A. 2002. L'Aristotelismo e le sue Alternative'. In *La Rivoluzione Scientifica*, ed. D. Garber, 337–357. Rome: Instituto della Enciclopedia Italiana.

Schuster, J.A. 2005. "Waterworld": Descartes' vortical celestial mechanics: A Gambit in the natural philosophical contest of the early seventeenth century. In *The science of nature in the seventeenth century: Changing patterns of early modern natural philosophy*, ed. Peter Anstey and John Schuster, 35–79. Dordrecht: Springer.

Schuster, J.A. 2012. 'Physico-mathematics and the search for causes in Descartes' Optics—1619–1637', *Synthèse* 185:467–499. [published online Dec. 2011 DOI 10.1007/s11229-011-9979-4].

Schuster, J.A. 2012a. What was the relation of Baroque culture to the trajectory of Early Modern Natural Philosophy? In O.Gal and R. Chen-Morris eds. *Science in the Age of the Baroque, Archives internationales d'histoire des ideés* 209: 1–35.

Schuster, J.A., and R.R. Yeo (eds.). 1986a. *The politics and rhetoric of scientific method*. Dordrecht: D. Reidel.

Schuster, J.A., and Graeme Watchirs. 1990. Natural philosophy, experiment and discourse: Beyond the Kuhn/Bachelard problematic. In *Experimental inquiries: historical, philosophical and social studies of experimentation in science*, ed. H.E. Le Grande, 1–47. Dordrecht: Kluwer.

Schuster, John, and Alan B.H. Taylor. 1996. Seized by the spirit of modern science. *Metascience* 9: 9–26.

Schuster, John, and Alan B.H. Taylor. 1997. Blind trust: The gentlemanly origins of experimental science. *Social Studies of Science* 27: 503–536.

Shapin, Steven. 1992. Discipline and bounding: The history and sociology of science as seen through the externalism-internalism debate. *History of Science* 30: 333–369.

Shapin, S., and S. Schaffer. 1985. *Leviathan and the air-pump: Hobbes, Boyle and the experimental life*. Princeton: Princeton University Press.

Singer, C. (ed.). 1917–21. *Studies in the history and method of science*. Oxford: Clarendon Press.

Sirven, J. 1928. *Les années d'apprentissage de Descartes*. Albi: Imprimerie Coopérative du Sud-Ouest.

Smith, Russell. 2008. Optical reflection and mechanical rebound: The shift from analogy to axiomatization in the seventeenth century, Part 1. *British Journal for the History of Science* 41: 1–18.

Toulmin, Stephen. 1990. *Cosmopolis, the hidden agenda of modernity*. New York: The Free Press.

Watson, Richard. 2007. *Cogito, Ergo Sum, the life of René Descartes*. Boston: Godine.

Wear, Andrew. 1983. William Harvey and the "way of the anatomists". *History of Science* 53: 223–249.

Wear, A. 1990. The heart and the blood from Vesalius to Harvey. In *The companion to the history of modern science*, ed. R.C. Olby, G.N. Cantor, J.R.R. Christie, and M.J.S. Hodge, 571–574. London: Routledge.

Whewell, W. 1837. *History of the inductive sciences*, vol. I. London: John W. Parker.

Whewell, W. 1980. *The philosophy of the inductive sciences*. London: J. W. Parker.

Zilsel, Edgar. 1942a. The sociological roots of science. *The American Journal of Sociology* 47: 544–562.

Zilsel, Edgar. 1942b. The genesis of the concept of physical law. *Philosophical Review* 51: 245–279.

Chapter 3
'Recalled to Study'—Descartes, *Physico-Mathematicus*

3.1 Introduction

When Descartes received his baccalaureate degree and licentiate in civil and canon law from Poitiers, on November 9 and 10, 1616, he faced the problem of the choice of a career appropriate to a man of his social background and education. He might have followed his father and brother in the legal and administrative career of an aspiring *noble de robe*. Certainly everything in his immediate family background and education conduced to this choice. Alternatively, since he had been raised on the Jesuits' diet of neo-Scholastic natural philosophy, ethics, mixed mathematics and perhaps some military architecture, and was a minor a landowner in his own right, and, moreover, belonged to an age when aspiring French merchants and magistrates yearned to emulate the status and habits of the feudal nobility, he could have tested his fortune as a gentleman soldier. Apparently, after some hesitation, or perhaps just a period of deliberation, Descartes passed into the Low Countries sometime in 1618 and enrolled as a gentleman-volunteer in the army of Prince Maurice of Nassau, Stadholder of six of the seven United Provinces. It was by no means unusual at that time for young French Catholic gentlemen to pursue military careers in the service of the Protestant, and hence intensely anti-Habsburg, Dutch Republic. These were the closing years of the tense Dutch-Spanish truce, due to expire in 1621, and of mounting political and religious tension in the Holy Roman Empire, which would erupt into the thirty Years War within a matter of months. The moderately pro-Protestant and stridently anti-Hapsburg policy of France had perhaps been more clearly defined whilst Henri IV had lived, and, it would be revived and extended by Richelieu under the pressure of widening war as the 1620s unfolded. Still, even in the confused years of the minority of Louis XIII, service with the Dutch was a clear and acceptable option for a young French gentleman.

Descartes later claimed to have chosen a military career in order to begin to allay early sceptical doubts by studying 'the great book of the world', and he even implied that his experiences were meant to serve as a sort of propaedeutic to his natural

J. Schuster, *Descartes-Agonistes: Physico-mathematics, Method & Corpuscular-Mechanism 1618-33*, Studies in History and Philosophy of Science 27,
DOI 10.1007/978-94-007-4746-3_3,

philosophical career. His remarks, really rhetoric in the service of his method-based myth of his own life, served up in 1637 in the *Discourse on Method*, may have tickled a romantic nerve in generations of uncritical admirers. But, given Descartes' background and education, there was really nothing very romantic, or (methodologically) prophetic in his move to the United Provinces. There is no evidence for supposing that in 1618 he was in search of practical experience and worldly knowledge such as could assuage a seriously articulated scepticism, or in some way prepare him for the career of the future author of the *Discourse*. One is tempted rather to conclude that Descartes was a bit disillusioned with the lack of guidance his education had offered in the choice of a career, and that perhaps he had had enough of bookish learning, for the time being, or perhaps even for good. But even if it could be shown that Descartes harbored such commonly observed attitudes, little light would be cast on the subsequent direction of his career as a physico-mathematical natural philosopher and aspiring prophet of method.

By contrast, what is known about his stay in the United Provinces is of the utmost importance for interpreting his career as a natural philosopher of a so-called physico-mathematical type and as a self proclaimed prophet of method. In November 1618, while garrisoned in Breda, the 22 year old Descartes had the good fortune to make the acquaintance of Isaac Beeckman, a 30 year old Dutch scholar, who had recently taken his medical degree at Caen in Normandy.[1] This was to be the beginning of a decisive period in the development of Descartes' views about natural philosophy. For two months at the end of 1618, Descartes and Beeckman worked together, speculating upon and resolving various problems in natural philosophy, mechanics, theory of music, hydrostatics and mathematics. After Beeckman's departure for Middelburg, early in 1619, the two men continued to correspond at least until Descartes set off on his travels in Germany and the east in late April.[2] In one of those letters Descartes confessed that Beeckman had 'recalled' him to 'erudition' and led him back to 'serious occupations'.[3] If anything this was an understatement, for Descartes' work and study with Beeckman set the tone of his career as a natural philosopher. In effect, Descartes served a second natural philosophical apprenticeship with Beeckman. This fortified him with a new vision

[1] Beeckman was born in Middelburg on 10 December 1588. He was first intended for the reformed ministry and studied theology at Leiden between 1607 and 1610. There he also came in contact with Rudolph Snel, the Ramist practical mathematician and pedagogue. This connection is of potentially great significance for the interpretation of Beeckman's career, for Snel offers a prime example of the tendency of late sixteenth century Ramism to concern itself with problems of the practice and pedagogy of the mechanical arts and applied mathematics. See Hooykaas (1981) and biographical note by C. de Waard in Mersenne (1932–88) ii. 217; Mahoney (1981); Ong (1958), 305; Vollgraff (1913). The most important work on Beeckman of the last generation without doubt is Klaas van Berkel (1983). The author informs me that the long awaited English translation of this work is presently being prepared by Maartin Ultee for the Johns Hopkins University Press. (Personal communication 15 September 2009.)

[2] The last extant letter from this period dates from 29/4/1619, AT. X. p. 164.

[3] Descartes to Beeckman, 23/4/1619, AT. X. pp. 162–3.

of the aims and content of a physico-mathematical style of natural philosophy and displaced the Scholastic vision purveyed at La Flèche.[4] Beeckman also apparently stimulated Descartes' return to the study of mathematics.[5] But here Beeckman's influence was less decisive, for Descartes' earliest recorded work in mathematics already shows deep conceptual concerns which did not form an important part of Beeckman's intellectual armory.[6]

Beeckman famously was, along with Thomas Harriot, one of the very first genuine corpuscular-mechanical natural philosophers in Europe and hence a unique resource for Descartes' own orientation toward that species of natural philosophy. But, we misunderstand the situation and the nature of their intellectual relations, if we assume [1] that Descartes made a crisp, total, systematic, explicit and immediate conversion to corpuscular-mechanism; and [2] that corpuscular-mechanical ontology was the only dimension to their cultural exchange. As to the former point, Descartes did not become a clear and explicit advocate of corpuscular-mechanism in 1619–1620. Indeed, he only undertook a systematic elaboration of corpuscular mechanism after 1628. But, it is safe to say he had from Beeckman a leaning toward corpuscular-mechanism, which peeps out from some of his early natural philosophical work, and even that some of his early talk of force, attraction, penetration and

[4] With a few notable and important exceptions—Gaukroger (1995), van Berkel (1983), Shea (1991), and Garber (1992)—who all published subsequent to my (1977). Cartesian scholars have tended to minimize the import of Descartes' friendship with Beeckman. The literature has understandably focused attention on the metaphysical and epistemological aspects of Descartes' thought, and to the extent that it has dealt with Descartes' natural philosophy at all, it has usually stressed the novelty of his enterprise. The lack of appreciation of the similarities between the natural philosophical enterprises of Descartes and Beeckman has perhaps been reinforced by an implicit bias toward accepting Descartes' account of their relations. That account derives in large measure form Descartes' correspondence concerning a dispute with Beeckman which broke out in the early 1630s when Descartes was writing *Le Monde*. Descartes' complaint rested on Beeckman's remarks to Marin Mersenne to the effect that the had been Descartes' 'master' for ten years; that he had taught Descartes whatever he knew about music; and that he had invented many natural philosophical ideas and recorded them in his *Journal* long before Descartes decided to put similar ones into print. Additionally, as Descartes had learned, Beeckman was also on the verge of preparing his disparate natural philosophical manuscripts for publication (see below notes 11 and 12 and corresponding texts). Beeckman's claims are undoubtedly exaggerated, and the novelty of Descartes' natural philosophical vision, emerging in *Le Monde*, cannot be denied. But one should not ignore the very real, if somewhat elusive nature of Beeckman's influence on Descartes' career in natural philosophy, and hence one should not dismiss Beeckman's dismay at the prospect of Descartes publishing a system of natural philosophy. Descartes' debt to Beeckman was quite complex, not only on the basis of their early interaction, but also in the light of Descartes' dealings with him in the late 1620s, after an absence of 10 years. Descartes was to emulate some of the natural philosophical concerns of Beeckman and part of his style of explanation; but, he also found inadequacies in Beeckman's work and developed his own ideas partly in response to them, as we shall see in Sect. 10.3.

[5] Descartes alludes to the study of 'mechanics' and 'geometry' in the correspondence with Beeckman: 26/3/1619, AT, X. p. 159 l.13; and 23/4/1619, AT. X. p. 162 l.15.

[6] On Descartes' early mathematical work, interest in analytical procedures and a general science thereof, see below Chaps. 5, 6, and 7.

generation was, like Beeckman's endemic use of similar terms, meant to gloss implied, unarticulated and rather vague underlying commitments to corpuscularly mediated natural processes.[7] We are going to see several reasons why Descartes' natural philosophical talk remained so allusive and non-committal. Perhaps the most important reason of all was that his emphasis was less on his newly acquired knowledge of corpuscular-mechanical ontology than on a commitment, similarly inherited from Beeckman, to a program of physico-mathematics in and of his natural philosophical work. We shall see that early on he was paying more attention to being an aspiring physico-mathematician within the field of natural philosophy (wherein he was leaning toward a corpuscularian agenda), than he was to articulating and enunciating details of corpuscular structures and behaviors.

As we saw in Chap. 2, the term physico-mathematics, used historiographically to characterize players regardless of whether or not they adopted the term, denoted a commitment to radically revising the conventional Scholastic Aristotelian view of the mixed mathematical sciences as subordinate to natural philosophy, non explanatory and merely descriptive. The mixed mathematical disciplines were somehow to become more intimately related to natural philosophical issues of matter and cause—they were to become more 'physicalised', more closely intertwined with or integrated into natural philosophizing, regardless of which specific genre of natural philosophy the budding physico-mathematician endorsed. As we shall see, Beeckman's approach involved a physico-mathematical agenda in, or for, his favored brand of corpuscular-mechanical natural philosophy. One dimension of this admittedly vague program involved a dynamical approach to mechanics and the simple machines—in the well known sixteenth century style of the (pseudo-Aristotelian) *Mechanica*—as a template for drawing up rules of corpuscular impact behavior for the natural philosophy. Under Beeckman's tutelage, Descartes also identified himself as a physico-mathematician, and it is his physico-mathematical work, in several telling cases, that we examine in this chapter.

However, before we become immersed in Descartes' early physico-mathematics, we should, as it were, write ourselves a memorandum not to lose sight completely of the more purely mathematical work he also initiated in the period 1618–1620, to which we shall return in Chap. 5. Our promissory note should read as follows: In this early period Descartes pursued an analytical, problem-solving oriented agenda in mathematics, to which, in his view, his physico-mathematics bore a striking resemblance, in that it, too, was piecemeal, problem oriented, and aimed in part to physico-mathematically 'analyze' findings in mixed mathematics 'back to' their natural philosophical causes. Indeed, we shall later learn that the parallels he perceived between his mathematical and mixed mathematical work triggered in 1619–1620 his dream of a unified analytical approach to all mathematically based disciplines—practical, pure and physico-mathematical—to which he appropriated the already circulating term 'universal mathematics'. Moreover, that overheated

[7] See examples of this below, and in Beeckman's case later in 1620s in the context of celestial mechanical speculations in Sect. 10.3.

conception quickly gave way to the even more encompassing mirage of a universal method. We shall trace these compounding enlargements of his mathematical and physico-mathematical agenda in Chap. 5. For the moment we should note first of all their common origin with the 'physico-mathematisation' of corpuscular-mechanism in the early years 1618–1620. And secondly, we should flag the fact that much of our story of Descartes *agonistes* is precisely the story of the intended and unintended entanglements of these two trajectories—in physico-mathematical natural philosophy, and in analytical mathematics, promoted to fantasy programs in universal mathematics and method—marked as they were by determined planning, unintended shifts and some spectacular insights, some decisively fruitful, some disastrously misleading, all in turn conditioned by the varied environments in which Descartes moved.

Returning, then, to the present chapter on Descartes' physico-mathematical genre of natural philosophy in its embryonic forms, we shall examine three case studies of this work: his manuscript on hydrostatics and the hydrostatic paradox; his well known work with Beeckman on the nature of accelerated fall; and a curious, widely overlooked but extremely important geometrical and physical optical fragment on refraction of light adapted and explicated from bits of the work of Kepler. The first two cases derive from a document itself entitled 'Physico-Mathematica' which dates from the end of 1618 or beginning of 1619.[8] The third fragment may be dated around 1620. Although the material on fall is better known, our emphasis will be on the first and third cases. The hydrostatics manuscript will turn out to be the key case, most revealing of the style and aims of Descartes' physico-mathematics articulated

[8] As is the case with all the early writings, no exact date can be assigned to the hydrostatics manuscript. Some internal evidence suggests that Descartes composed it shortly before Beeckman left Breda at the beginning of 1619: see AT x. 69 l.15 and 74 l.23 which seem to imply that Beeckman and Descartes had recently discussed these problems in person. Adam and Tannery note that the 'Physico-Mathematica' were misplaced in Beeckman's *Journal*, having been transcribed along with the *Compendium Musicae* between two entries for 20 April 1620 (AT. x. 26–7). By that time Descartes himself was in Germany and no longer in contact with Beeckman. If, as seems to be the case, the 'Physico-Mathematica' were composed around the same time as the *Compendium of Music* which was a New Year's gift to Beeckman, then it again seems very likely that the hydrostatics manuscript dates from late 1618 or early 1619. The *Compendium* is not treated in this chapter on Descartes as a physico-mathematician for the simple reason that this early work of Descartes shows hardly any traits of physico-mathematics, staying almost entirely within the realm of traditional mixed mathematics. Zarlino's views on consonance are followed, but derived as much as possible from geometrical considerations. There is a brief early passage, inserted according to a suggestion by Beeckman, dealing with the physical vibrations actually made by a string, but it does not affect the tenor of the bulk of the piece. At no point do physico-mathematical protocols of the sort we will unpack here make an appearance. On the content and tenor of the *Compendium* the key work is by Floris Cohen (1984). See also the discussion in Gaukroger (1995), pp. 74–80, who points out that at this early stage Descartes is oblivious to recent developments by Benedetti, Vincenzo Galileo and Beeckman himself, who all recognised difficulties with Zarlino's arithmetical treatment of consonance and thus turned their attention to conceptualizing consonance, and its problems, as due to the coincidence of sound vibrations (variously physically explicated). Descartes keeps the *Compendium* within the realm of mixed mathematics, rather than opening up this potentially physico-mathematical domain.

to, and through, an embryonic corpuscular-mechanism. Understanding his agenda on its basis will allow us to understand the third fragment, which in turn will be critically important to our examination in Chap. 4 of Descartes' later successes in physico-mathematical optics in the mid and late 1620s, including his discovery and attempted mechanistic explanation of the law of refraction of light. As to the material on accelerated fall, it will take on a different appearance than it has in the traditional literature, because we shall view it across the natural philosophical cum physico-mathematical preoccupations of Descartes and Beeckman. To them it was not simply a search for a descriptive law of accelerated fall, but rather an exercise in the physico-mathematics of fall, meaning that an exact descriptive law was sought in order to reveal, or confirm, natural philosophical insights about the causes of the phenomenon. Our two *physico-mathematici* were stymied in this project by two interrelated problems: they could achieve no conviction either about the correct descriptive law, or about the nature of the causes at work. Hence, this approach helps explain what their work did and did not accomplish, and why it is not simply a case of trying, but failing, to 'be Galileo'. In contrast, regarding the physico-mathematics of hydrostatics, Descartes judged that he had a solid achievement and way forward; whilst in physico-mathematical optics, although his initial foray was inconclusive, he had an agenda, and indeed a style of practice, that would lead less than a decade later to the greatest of his physico-mathematical feats, precisely in the realm of optics and the refraction of light.

So, our cases are indispensable to properly commencing our reconstruction of Descartes' struggles, first, between 1618 and 1628, in physico-mathematics articulated to natural philosophizing, and later, after 1628 in systematic natural philosophy bearing the genealogical imprints of that earlier physico-mathematics. But, the precondition to understanding the cases is to comprehend Descartes' mentor, Beeckman, as an anti-Aristotelian, corpuscular-mechanical natural philosopher, interested in some sort of radical take on the relations of mixed mathematics and natural philosophy, which he pursued under aims and values redolent of the rise of elite estimation of the status of the practical and mechanical arts.

3.2 Beeckman: Mentor and Colleague in Physico-Mathematics and Natural Philosophy[9]

3.2.1 Corpuscular-Mechanical Natural Philosophy and the Values of the Practical Arts

Beeckman was one of the very first individuals in Europe to pursue consistently the idea of a micro-mechanical approach to natural philosophy. He conceived of a redescription of all natural phenomena in terms of the shape, size, configuration and

[9] Material in this section broadly follows Gaukroger and Schuster (2002).

motion of corpuscles. He insisted that what we have termed in Chap. 2 the 'causal register' of this account, that is, the principles of all natural change, had to be derived from the transdiction of the presumed mechanical principles of macro-phenomena, in particular the behavior of the simple machines. As discussed in Chap. 2, this was, of course, an exemplary 'physico-mathematical' move, because it promoted findings in the practical and mixed mathematical field of mechanics to the level of natural philosophy, indeed, to the very core of the causal register of that natural philosophy. Beeckman offered on a first-hand basis an approach to natural philosophy which was not available to Descartes from any other contemporary source.[10]

For most of his natural philosophical career Beeckman was no systematizer. His natural philosophical inquiries have a disorganized, almost random character, bespeaking more the humanist commonplace book than the Ramist attention to methodical textuality he surely learned from the elder Snel. However, at the end of the 1620s he edited his notes on mechanics and cosmology into the form of a reasonably systematic account with a view to publication.[11] Descartes, who was beginning to put together the material for *Le Monde* at this time and was evidently disconcerted to learn that Beeckman had a similar project in mind, directed a barrage of abuse against Beeckman, calling into question his abilities and his originality. As a result Beeckman abandoned plans for the book.[12] His *Journal* is filled with questions ranging from embryology to celestial mechanics and from logic to applied mathematics, all addressed in short entries, rarely as much as a page in length. Beeckman prided himself on the spontaneous character of his inscriptions, which he thought offered a more genuine insight into the questions posed than any prearranged program of scholarship.[13] In fact, he may have had a point, because his random speculations did focus his attention on troublesome details of applying micro-mechanical principles to specific questions, without the baggage of textual systematization and metaphysical or theological legitimation. This makes the *Journal* a unique source of insight into the values, aspirations and presuppositions constitutive of the emerging corpuscular mechanical genre of natural philosophy. As we shall see, Descartes' hydrostatics manuscript, emerging in this natural philosophical milieu, arguably displays in the case of Descartes a similar 'naïve', non-systematic, stage in the early formulation of a corpuscular mechanical approach.

Beeckman's views on natural-philosophical explanation seem to stem from his unexamined faith in the truth and relevance of the theory and practice of the mechanical

---

[10] Of course, in his own natural philosophizing Descartes would eventually employ a very different notion of just what the principles of mechanics are which provide the causal dimension of his mechanical philosophy. In addition, unlike Beeckman, Descartes would later be drawn into serious concern about the metaphysical grounding of his natural philosophy and the epistemic status of his claims. Nevertheless, from Beeckman came the inspiration for a new species of natural philosophy, as well as a considerable portion of its content.

[11] It was edited by his brother Abraham after Beeckman's death and appeared as Beeckman (1644).

[12] For the details of this episode, see van Berkel (2000).

[13] Beeckman (1939–53) ii. 99.

arts, and practical mathematics, as he had learnt them working with his father laying water conduits, and reading the works of Stevin and the Snels. In the *Journal*, with its hundreds of pages of natural-philosophical speculations, interleaved with practical questions drawn from the mechanical arts, one can detect the merger of natural philosophy with the re-evaluation of the aims and limits of knowledge which had emerged in discussions of technology and the practical arts in the later sixteenth century.[14] He consistently held Aristotelian and neo-Platonic notions of immaterial causes and agencies to be 'unintelligible' and hence useless in natural philosophy,[15] and he frequently insisted that natural philosophy speak in terms of imaginable things and processes, rather than entities of the pure understanding.[16] No doubt Beeckman conceived himself to be attacking traditional modes of philosophical discourse in the name of common sense; but, his 'common sense' was precisely the educated, and to that degree sophisticated, common sense of the theory, practice, and ideology of the mechanical arts and practical mathematics. No simple mechanic would appeal to teleological processes, occult virtues, or immaterial causes to account for the functioning of a simple mechanical device, although devotees of natural magic or Platonising trends in natural philosophy might. Explanations in the mechanical arts rested on the appeal to a clear picture of the structure and interaction of the constitutive parts of the apparatus. As simple mechanical and hydrodynamical devices showed, only motion or pressure can produce the rearrangement of parts and hence produce work, and for theoretical purposes, the causes of motions and pressures are other motions and pressures.

What Beeckman was demanding in natural philosophy was the application of the criteria of meaningful communication between mechanical artisans—the appeal to a picturable or imaginable structure of parts whose motions are controlled within a putative theory of mechanics. His central contention was that there is no point in talking about effects if you cannot imagine in this way how they are produced. The exemplar of imaginatively controlled efficacy resides in the mechanical arts, where one can command nature at a macroscopic level. Hence, it was characteristic of Beeckman's translation of the imperatives of the mechanical arts into the terms of natural philosophy, that he was not overly concerned with metaphysical objections to his doctrines. Transdiction from the macroscopic to microscopic realms did not pose epistemological difficulties for Beeckman, as it would later for Descartes and other more 'scholastified' mechanists, bearing the discursive weight of their Aristotelian educations. The only constraint he placed on transdiction was the eminently

[14] See Paolo Rossi (1970), 1–62, and our historiographical observations above Sect. 2.7.

[15] Beeckman (1939–53) i. 25.

[16] Beeckman to Mersenne, 1 October 1629, Mersenne (1932–88) ii, 283, 'nihil enim in philosophia admitto quam quod imaginationi velut sensile representatur.' Cf. the demands that Descartes was to place on mathematics and 'mathematical' natural philosophy in the latter portions of the *Regulae*, written in the late 1620s, as well as his insistence on the 'figurate' representation of problems to be solved, both in mathematics and in optics and natural philosophy generally, on which in general see Sepper (2000) and which we will see illustrated in the early physico-mathematical work below.

'mechanical' one of questioning whether the widely differing surface to volume ratios of macroscopic bodies and corpuscles would entail any differences in their mechanical behavior in various systems.[17] Nevertheless, Beeckman did not pursue such a natural philosophy because he had read Stevin, studied with Rudolph Snel, and made an early career in the mechanical arts. Rather, it was Beeckman's education and pedagogical vocation, and his objectively correct image of himself as a man of learning and polite interests which instilled in him the cultural value of the pursuit of natural philosophy.[18] In accordance with our 'cultural process model' of natural philosophy in Chap. 2, the point is that what helped make him unique within the culture of natural philosophy was the way his desire to be a natural philosopher was refracted by his early experience in the theory and practice of the mechanical arts. The *Journal* testifies to his private goal of reforming natural philosophy in the name of values of mechanical intelligibility and utility.

Beeckman held a fundamentally atomistic view of nature. His atoms possess only the geometrical-mechanical properties of size, shape, and impenetrability (being absolutely hard, incompressible and non-elastic). Motion is conceived as a simple state of bodies, rather than an end-directed process which they undergo. Moreover, the possession of motion is not mediated by any metaphysical conception of an internal moving force, *impetus*, or virtue. All other qualities, including the four elemental qualities of Aristotle, arise from the diverse ways in which various atomic structures constituting bodies impinge upon our sense organs.[19] Indeed, Beeckman devoted much of his speculation about matter to devising a four element theory within the assumptions of his atomic doctrine.[20] Beside allowing cooptation of traditional modes of explanation still very much alive in Aristotelianism and Galenic medicine, Beeckman's element theory allowed him, in certain contexts,

---

[17] Beeckman (1939–53) ii. 77–8. Similarly, Aristotelian 'philosophical' arguments against the existence of the void carried less weight against his atomism than the transdiction of the 'metaphysical' objection that perfectly hard atoms lacking pores cannot undergo rebound (*ibid.* p. 100). He was obviously disturbed by his inability to conceive of a convincing macroscopic model for hard body rebound. Mechanical common sense seemed to indicate atoms do not exist.

[18] Prior to 1616 Beeckman had spent a few years in the trade of candle making and also followed his father's craft of laying water conduits, especially for breweries. Many of the notes in his *Journal* reveal that Beeckman saw connections between practical questions raised in relation to his craft activities and the teachings he had received from the elder Snel, as well as the writings of Willebrord Snel and Simon Stevin. Beeckman, however, did not plan on remaining a practitioner of the mechanical arts, albeit a highly educated and philosophically literate one. In 1618 he took an M.D. degree at Caen. From November 1619 he was Conrector of the Latin School at Utrecht, and in December 1620 he moved to Rotterdam, where his brother was Rector of the Latin School. Beeckman gave lessons and became Conrector in 1624. He also founded a 'collegium mechanicum', or society for craftsmen and scholars interested in natural philosophical questions with technical import. In 1627 he became Rector of the Latin School at Dordrecht, a position he held until his death in 1637.

[19] Beeckman (1939–53) ii. 86.

[20] Ibid, pp. 86, 96; cf. Beeckman (1939–53) iii. 138, 'Ignis minimum non est atomus sed homogeneum ex atomis compositum.'

to de-emphasize atoms as explanatory elements. This was important, because he was impressed by arguments showing the impossibility of rebound after collision of perfectly hard atoms, and because he had difficulty reconciling atomic theory with the phenomena of elasticity.[21] Accordingly, he built his traditional elements out of congeries of atoms and manipulated the elements as functional units of explanation,[22] without, however, explaining what structural features the congeries had that enabled them to possess the required property of elasticity that their constituent parts lacked.

3.2.2 *Beeckman's Causal Register, Principles of Mechanics and Version of Physico-Mathematics*

Unlike previous advocates of atomism, and prior to any of the great mechanists of the later seventeenth century, Beeckman sought to explain the behavior of his atoms by applying to them a causal discourse modeled on the principles of mechanics; that is, the science of the simple machines, articulated in a dynamical way, in the tradition of the pseudo-Aristotelian *Mechanica* or *Mechanical Questions*, rather than in an Archimedean mathematically rigorous statical manner.[23] It is here that very precise bearings emerge for Beeckman's own understanding of the meaning of

---

[21] Beeckman (1939–53) ii. 100–1.

[22] Beeckman (1939–53) iii. 31 Beeckman's theory of light provides a good example: He held light to be corporeal and to consist in the finest particles of elemental heat or fire. Because light can be reflected and refracted (to Beeckman refraction was a form of internal reflection), it cannot consist in isolated atoms; therefore, light, heat and fire had to be conceived as second order homogenous composites made up of numerous atoms and void space.

[23] See Gaukroger and Schuster (2002), p. 545. For example, the basic principle behind the *Mechanica's* treatment of the lever (set out in a number of passages in Aristotle) 'holds that the same force will move two bodies of different weights, but it will move the heavier body more slowly, so that the velocities of the two bodies are inversely proportional to their weights. When these weights are suspended from the ends of a lever, we have two forces acting in contrary directions, and each body moves in an arc with a force proportional to its weight times the length of the arm from which it suspended. The one with the greater product will descend in a circular arc, but if the products are equal, they will remain in equilibrium.' In contrast, the purely statical and mathematical approach of Archimedes 'makes statics a mathematical discipline independent of any general theory of motion, whilst that of the *Mechanica* makes statics simply a limiting case of a general dynamical theory of motion, a theory which is resolutely physical. In other words, the *Mechanica* account comes as part of a package which is driven by Aristotelian dynamics, above all by the principle of the proportionality of weight and velocity. This did not stop a number of mathematicians, such as Benedetti, Tartaglia, and Galileo, from trying to revise the package, hoping they could salvage the dynamical interpretation of the beam balance and simple machines while jettisoning the natural philosophy that lay behind it, but the pivotal role this natural philosophy had played meant that such a revision could never be successful, as we shall see below when we consider Galileo's attempt to realize this program. The Archimedean account, by contrast, comes without any dynamical, or more broadly speaking physical, commitments: put more strongly, it comes without any physical content.'

'physico-mathematics', and hence for the resources his physico-mathematics could offer to Descartes, who adapted and translated them into a different, and we shall see, even more radically anti-Aristotelian form.

By 1613 or 1614 Beeckman formulated a concept of inertia holding for both rectilinear and curved motions.[24] He insisted that motion, once imparted to a body, is maintained at the same speed, unless destroyed by external resistances. In the absence of external constraints there is no reason why the state of motion of the body should alter:

> Everything once moved never comes to rest unless due to an external impediment. Moreover, the weaker the impediment, the longer the moving body moves.... A stone thrown in a vacuum is perpetually moved; but the air hinders it by always striking it anew and thus acts to diminish its motion. Indeed, what the philosophers say, that a force is impressed in the stone, seems without reason. For who can conceive in his mind what that force would be, or how it would continue to move the stone, or in what part of the stone it would find its seat? But someone can easily conceive in his mind that motion in a vacuum never comes to rest, because no cause changing the motion is present; for nothing is changed without some cause of change.[25]

Combining his principle of inertia with his atomic ontology, Beeckman concluded that the only possible mode of external constraint or resistance that can be exerted on an inertially moving body is corpuscular impact. Conversely, only corpuscular collision and transfer of motion can account for the initiation of motion of resting bodies which have resisted the passage of inertially moving bodies. Ultimately, therefore, only the transfer of motion can account for change in the position, arrangement and disposition of atoms, and hence furnish the principle of all natural change.[26]

Beeckman eschewed metaphysical elaboration of concepts of internal moving forces or *impetus* as the cause of the continuation of inertial motion.[27] His attitude seems to have been that the idea of motion is sufficiently well understood, and that it is motion *per se*, the state of traversing space in time, which is imparted to bodies

---

On the influence of the *Mechanica* in the sixteenth and seventeenth centuries, see Duhem (1905–6), Rose and Drake (1971), and Laird (1986). The *Mechanica*, which is probably the work of Strato or Theophrastus, was traditionally attributed to Aristotle, an attribution which Duhem and Carteron (1923) follow. The work is Aristotelian in tenor, but has the peculiar feature that whereas Aristotelian natural philosophy confines itself to natural processes, for it is these that follow from the nature of things, the subject matter of the *Mechanica*, as is explained in the opening sentence of the work is 'those phenomena that are produced by art despite nature, for the benefit of mankind.'

[24] Beeckman (1939–53) I. 24–5. I have employed the typescript translation by the late Michael S. Mahoney, Princeton University.

[25] Ibid.

[26] Beeckman even tried to explain the centrifugal tendency of bodies moving in circular motion in resisting media as the result of the combination of circular inertia and differential resistance of the medium on different parts of the body. Beeckman (1939–53) i. 253.

[27] Ibid, 25.

at the beginning of their movements. All this again points to the hard-headed 'common sense' of macro-mechanics which controls Beeckman's conceptualizations. His 'mechanics' of atoms—the causal register of his natural philosophy—was constructed within the limits of a mechanical artisan's belief in the priority of explanations appealing to the motions, resistances, and displacements of parts, and requiring no further verbal explication.

Central to such a mechanics was the problem of furnishing rules of collision, specifying the outcomes of exchanges of motion on the atomic level.[28] Since Beeckman's atoms are perfectly hard, he formulated rules applicable to what we would term perfect inelastic collisions. He measured the quantity of motion of corpuscles by taking the product of their quantity of matter and their speed. Significantly, Beeckman linked his measure of motion to a dynamic interpretation of the behavior of the balance beam. He evaluated the effective force of a body on a balance beam by taking the product of its weight and the speed of its real or potential displacement, measured by the arc length swept out in unit times during real or imaginable motions of the beam—the classic *Mechanica*-based procedure, but with his own dynamical gloss. Beeckman was able to build up a set of rules of impact, by combining certain intuitively symmetrical cases of collisions with the dictates of the inertial principle and an implicit concept of the conservation of the directional quantity of motion in a system. His treatment of symmetrical cases of collision and his notion of the conservation of motion owed their form and their putative legitimacy to the model of the balance beam, interpreted in a dynamic rather than static fashion.[29] In this very 'physico-mathematical' way, he wanted to transfer findings from a particular—*Mechanica* centered—interpretation of mechanics to the causal register of his natural philosophical discourse. Indeed, Beeckman's commitment to a dynamical interpretation of the principles of the simple machines and his belief in a

---

[28] See Appendix I in Mersenne (1932–88) ii. 632–44, which includes de Waard's notes.

[29] Beeckman's rules fall into two broad categories: (1) cases in which one body is actually at rest prior to collision, and (2) cases which are notionally reduced to category (1). The concept of inertia and the stipulation that only external impacts can change the state of motion of a body provide the keys to interpreting instances of the first category. The resting body is a cause of the change of speed of the impacting body and it brings about this effect by absorbing some of the quantity of motion of the moving body. Beeckman invokes an implicit principle of the directional conservation of quantity of motion to control the actual transfer of motion. In each case the two bodies are conceived to move off together after collision at a speed calculated by distributing the quantity of motion of the impinging body over the combined quantities of matter of the two bodies. For example, in the simplest case, in which one body strikes an identical body at rest, '...each body will be moved twice as slowly as the first body was moved...since the same impetus must sustain twice as much matter as before, they must proceed twice as slowly.' And he adds, analogizing the situation to the mechanics of the simple machines, '...it is observed in all machines that a double weight raised by the same force which previously raised a single weight, ascends twice as slowly.' (Beeckman 1939–53, i. 265–6) Instances of the second category of collision are assessed in relation to the fundamental case of collision of equal speeds in opposite directions (*ibid*, 266). Being perfectly hard and hence lacking the capacity to deform and rebound, the two atoms annul each other's motion, leaving no efficacious residue to be redistributed to cause subsequent motion.

correspondence between these principles and the rules of corpuscular collision run right through his work.[30] Beeckman consistently demanded a dynamical approach to statics, the theory of simple machines and mechanics in general, including hydrostatics. This dynamical approach inheres in a set of rules or principles about the motion or tendencies to motion of bodies, which may also be read into the behavior of fundamental corpuscles and atoms, to provide the causal register for our explanatory discourse about them.

To state this more generally, one can say that the style of Beeckman's natural philosophy demanded that macroscopic phenomena be explained through reduction to corpuscular-mechanical models. The *Journal* offers hundreds of examples of this sort of enterprise. In many cases merely qualitative reports of phenomena are so reduced; but in other cases one was dealing with quantitative representations of phenomena already achieved in the practical or mixed mathematical disciplines, as is the case in Beeckman's reading of laws of collision out of exemplary findings in

---

This symmetrical case, which was also generalized to cases of equal and opposite quantities of motion arising from unequal bodies moving with compensating reciprocally proportional speeds, derives from a dynamical interpretation of the equilibrium conditions of the simple machines. Instances in which the quantities of motion of the bodies are not equal are handled by annulling as much motion of the larger and/or faster moving body as the smaller and/or slower body possesses (Beeckman 1939–53, i. 266.) This in effect reduces the smaller and/or slower body to rest. The outcome of the collision is then calculated by distributing the remaining unannulled motion of the larger and/or swifter body over the combined quantities of matter of the two bodies (ibid). It is obvious that Beeckman viewed this case through a two-fold reference to the simple machines; for the first extracts as much motion as can conduce to the equilibrium condition for symmetrical cases, and then he invokes the principle cited just above in this note to determine the final outcome.

[30] Beeckman (1939–53) iii. 133–4. Consider for example his commentary in 1629 upon a remark made by Mersenne in his *Traité de l'Harmonie universelle* (1627) to the effect that, 'Vitesse ou tardivité du mouvement cause de tout ce qui se fait par bilances.' As Beeckman's entry shows, he fundamentally agreed with this dynamicist interpretation of the principles of the simple machines: 'The reason for this fact can be rendered very easily by those things which I wrote a little before concerning motion. For it follows from them that a sphere twice as heavy [as another sphere], that is, having twice as much matter, but moving twice as slowly [as the other sphere], will be stopped after colliding with it, that is, both spheres will be at rest. For I specified that mass and motion compensate for one another [*se reciprocari*]. The same thing must also be concluded concerning the balance.' Despite some confusions Beeckman introduced in the explication of this point, his central contention is clear enough: even macroscopic equilibrium is a consequence of the laws of motion and impact, because it can be explained through a dynamical interpretation of countervailing motions on the model of the laws of collision. He closes with a clear statement of this point: 'One should not doubt how an account is given here of the theory of equilibrium [*in isorhopicis*] by means of motion. For even if there is no motion when bodies hang in equilibrium, motions would however take place immediately if an external force, a weight, etc. were to displace these weights from equilibrium. Moreover, all bodies that return to their own places as soon as they are moved from them never change their places of their own accord. Thus stones never ascend spontaneously and in the absence of an external force. Bodies which are at rest in our vicinity never spontaneously move.... The cause of equilibrium therefore can be motion, even if the bodies in equilibrium are not moved. For the cause of equilibrium is past and future motion. During the present, to be sure, the body is at rest because past and future motions occasion rest.'

the (dynamical) interpretation of the simple machines. Descartes' own venture in physico-mathematics began, as we shall now see, when Beeckman questioned him about Stevin's 'paradoxical' hydrostatical findings. Beeckman's queries arguably sat squarely within his own practice of physico-mathematics as we have just described it. Beeckman wanted to see what his new friend and fellow 'physico-mathematicus' could do about reducing Stevin's work to corpuscular mechanical terms, thereby fundamentally explaining it. This was what Beeckman and Descartes were envisioning when in 1618 they congratulated themselves on being among the very few '*physico-mathematici*' in Europe. What they meant was that only they unified the mathematical study of nature with the search for true corpuscular-mechanical causes.[31] Hence, it is to this, the first and most important of our case studies in physico-mathematics that we now turn.

3.3 Exemplary Physico-Mathematics: The Hydrostatics Manuscript of 1619[32]

In 1586 Simon Stevin, the great Dutch engineer, algebraicist, maestro of the mechanical arts and practical mathematics, at the height of his considerable powers, did something that still incites interest, and admiration, amongst those literate in mechanics and curious about its history—he produced in strict, deductive Archimedean fashion, a proof for what is in effect a special case of the hydrostatic paradox. In late 1618 or early 1619 Descartes and Beeckman—who, as a student of the elder Snel, was also an intellectual descendant of Stevin—tried to improve upon Stevin's work. That is, he attempted to provide a deep natural philosophical explanation for Stevin's result. Descartes' treatment of the hydrostatic paradox is given in a report from Beeckman, which Stephen Gaukroger and I have termed, 'the hydrostatic manuscript'.[33] Beeckman set this task to Descartes as an exercise in their style of 'physico-mathematics'. Moderns literate in physics find nothing admirable, or even comprehensible in the young mechanist's machinations—and neither do

[31] AT x. 52. 'Physico-mathematici paucissimi'. In this regard Beeckman was to note in 1628 that his own work was deeper than that of Bacon on the one hand and Stevin on the other just for this very reason. Beeckman (1939–53) iii. 51–2, 'Crediderim enim Verulamium [Francis Bacon] in mathesi cum physica conjugenda non satis exercitatum fuisse; Simon Stevin vero meo judico nimis addictus fuit mathematicae ac rarius physicam ei adjunxit.'

[32] Material in this section broadly follows Gaukroger and Schuster (2002).

[33] The text, *Aquae comprimentis in vase ratio reddita à D. Des Cartes* which derives from Beeckman's diary, appears at AT x., 67–74, as the first part of the *Physico-Mathematica*. See also the related manuscript in the *Cogitationes Privatae*, AT x. 228, introduced with, 'Petijt e Stevino Isaacus Midlleburgensis quomodo aqua in funda vasis b…'.The expression 'hydrostatics manuscript' appears in Schuster (1977, 1980, 2005), Gaukroger (1995), and Gaukroger and Schuster (2002).

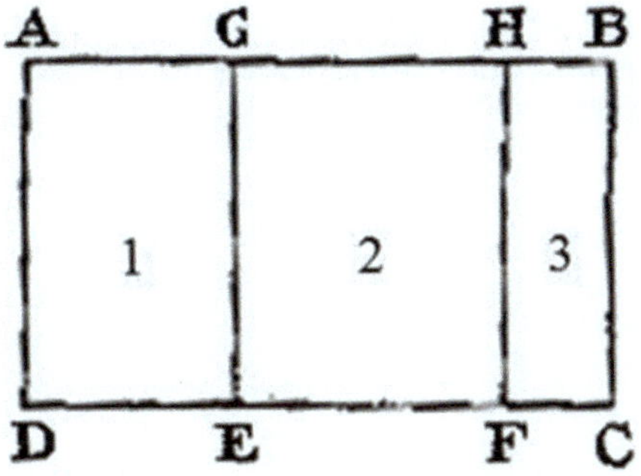

Fig. 3.1 Stevin, *Elements of hydrostatics* (1586) in Stevin (1955–66) I p. 415

Descartes scholars or more contextually oriented historians of science. But, we shall see this strange fragment as a valuable window into Descartes' early physico-mathematical agenda.

3.3.1 Stevin, Archimedes and the Hydrostatic Paradox

In his *Elements of Hydrostatics* 1586, Stevin demonstrated that a fluid can exert a total pressure on the bottom of its container that is many times greater than its weight. In particular, he showed that a fluid filling two vessels of equal base area and height exerts the same pressure on the base, irrespective of the shape of the vessel and hence, paradoxically, independently of the amount of (weight of) fluid contained in the vessel. Stevin's line of argument in establishing the hydrostatic paradox proceeds entirely on the macroscopic level of gross weights and volumes. The mathematical character of his proof depends upon his insistence on the maintenance of a condition of static equilibrium, understood in terms of the fundamentals of Archimedes' hydrostatics.

Stevin (Fig. 3.1) proves that the weight of a fluid upon a horizontal bottom of its container is equal to the weight of the fluid contained in a volume given by the area of the bottom and the height of the fluid measured by a normal from the bottom to the upper surface.[34] He employs a *reductio ad absurdum* argument applied to the gross statical properties of fluids: ABCD is a container filled with water. GE and HF are normals dropped from the surface AB to the bottom DC, notionally dividing the water into three portions, 1, 2 and 3. Stevin has to prove that on the bottom EF there rests a weight equal to the gravity of the water of the prism 2. If there rests on the bottom EF more weight than that of the water 2, this will have to be due to the water beside it, that is water 1 and 3. But then, there will also rest on the bottom DE more weight than that of the water 1; and on the bottom FC also more weight than that of the water 3; and consequently on the entire bottom DC there will rest more weight

[34] Stevin (1586); reprinted and translated in Stevin (1955–66), i. 415.

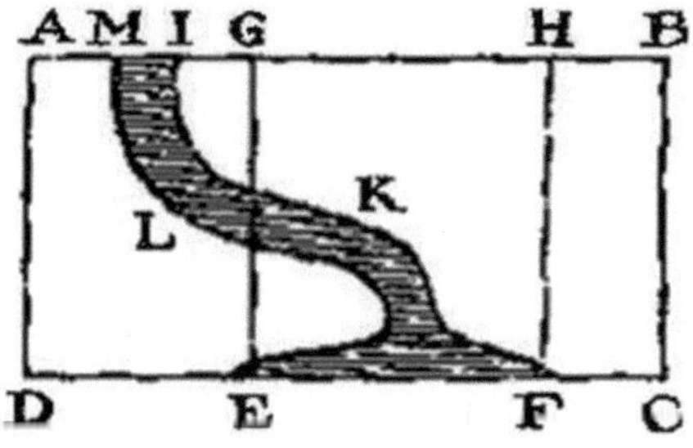

Fig. 3.2 Stevin, *Elements of hydrostatics* (1586) in Stevin (1955–66) I p. 417

than that of the whole water ABCD, which would be absurd. The same argument applies to the case of a weight of water less than 2 weighing upon bottom EF.[35]

Stevin then ingeniously argued that various portions of the water can be notionally solidified, or replaced by a solid of the same density as water. This permits the construction of irregularly shaped volumes of water, to which, paradoxically, the theorem can still be applied. Take, for example, Stevin's Corollary II (Fig. 3.2)[36]:

He shows therefore that on bottom EF there rests a weight equal to that of a volume of water whose bottom is EF and whose height is GE. Stevin goes on to apply these findings to oblique bottoms and thus to the sides of containing vessels.

3.3.2 The Hydrostatics Manuscript [1] The Micro-Corpuscular Reduction

The hydrostatics manuscript is concerned with four problems of hydrostatics, of which we shall only need to discuss one in detail. Descartes takes as given the following conditions (Fig. 3.3):

> … let there be four vases of equal width at the base, of the same weight when empty and of the same height. Let A be filled with only as much water as B can contain, and let the remaining three be filled as much as possible.

[35] *Ibid*, i. 415. 'We have to prove that on the bottom EF there rests a weight equal to the gravity of the water of the prism GHFE. If there rests on the bottom EF more weight than that of the water GHFE, this will have to be due to the water beside it. Let this, if it were possible, be due to the water AGED and HBCF. But this being assumed, there will also rest on the bottom DE, owing to the water GHFE, because the reason is the same, more weight than that of the water AGED; and on the bottom FC also more weight than that of the water HBCF; and consequently on the entire bottom DC there will rest more weight than that of the whole water ABCD, which (in view of ABCD being a corporeal rectangle) would be absurd. In the same way it can also be shown that on the bottom EF there does not rest less than the water GHFE. Therefore, on it there necessarily rests a weight equal to the gravity of the water of the prism GHFE.'

[36] *Ibid*, i. 417. 'Let there again be put in the water ABCD a solid body, or several solid bodies of equal specific gravity to the water. I take this to be done in such a way that the only water left is that enclosed by IKFELM. This being so, these bodies do not weight or lighten the base EF any more than the water first did. Therefore we still say, according to the proposition, that against the bottom of EF there rests a weight equal to the gravity of the water having the same volume as the prism whose base is EF and whose height is the vertical GE, from the plane AB through the water's upper surface MI to the base EF.'

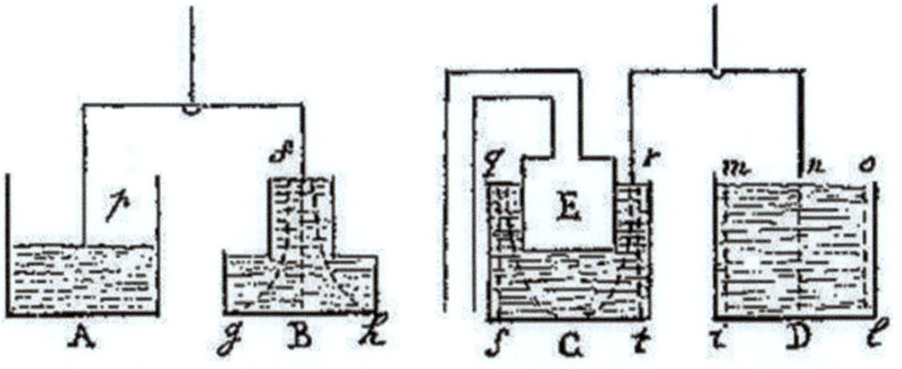

Fig. 3.3 Descartes, Aquae comprimentis in vase ratio reddita à D. Des Cartes, AT X, p. 69

It is proposed to show that,

> the water in vase B will weigh equally upon the base of the vase as does the water in D upon its base, and consequently each will weigh more heavily upon their bases than the water in A upon its base, and equally as much as the water in C upon its base.[37]

The first statement contains the key problem, because it refers to Stevin's important findings about the hydrostatic paradox. Beeckman, with his theoretical and practical concerns in hydrostatics and his corpuscular-mechanical aims in natural philosophy, was probably curious about how Descartes would explain this fundamental but strange result.

While Stevin's approach is geometrical, Descartes' analysis and explanation are by contrast based on an attempt to reduce the phenomenon to micro-mechanical terms. The hydrostatics manuscript implies the judgment that Stevin's macro-geometrical arguments and results can only be truly explained in terms of the corpuscular-mechanics of fluids. Thus, it reflects the opinion of Descartes and Beeckman, cited earlier, that they were among the very few true 'physico-mathematici' in Europe. It also foreshadows Beeckman's later claim, also cited earlier, that Stevin was too given to mathematics and insufficiently concerned with the physical causes residing behind his mathematical representations of macroscopic bodies and phenomena.[38] While the judgments of the two physico-mathematici corresponded, we shall see

[37] AT x. 68–9. This is the second of the four puzzles posed in the text, the others are: '(First), the vase A along with the water it contains will weigh as much as vase B with the water it contains. … Third, vase D and its water together weigh neither more nor less than C and its water together, into which embolus E has been fixed. Fourth, vase C and its water together will weigh more than B and its water. Yesterday I was deceived on this point.' (Descartes' latter point is not to be confused with his proof in the text that the water in vase B and vase C will weigh equally upon their respective bases—another case exemplifying the hydrostatic paradox and argued in a manner similar to the case we are treating in detail.)

[38] Beeckman (1939–53) iii. 51–2.

that in the end the detailed understandings of physico-mathematics on which these judgments were reached differed rather significantly.

Descartes' intention of reducing the problem to micro-mechanical terms emerges in the opening four paragraphs of the report. He complains that a full account would require a good deal of explanation of the 'foundations' of his 'mechanics'.[39] (It will soon become apparent that had this eventuated, it would have been intended as some sort of system of concepts concerning the causation of corpuscular behavior—a dynamics, as we are terming it, for the corpuscular-mechanical realm, meant to ground the sorts of arguments employed in the remainder of the manuscript.) In any case, Descartes asks us to accept a series of assumptions. First, he claims that of the various ways in which bodies may 'weigh-down' [*gravitare*], only two need be discussed: the weight of water on the bottom of a vessel which contains it, and the weight of the entire vessel and the water it contains.[40] Descartes' later discussion shows that by the weight of the water on the bottom of the vessel he does not intend the gross weight of the quantity of water measured by weighing the filled vessel and subtracting the weight of the container itself. He means instead the total force of the water on the bottom arising from the sum of the pressures exerted by the water on each unit area of the bottom. Secondly, the term 'to weigh down' is explicated as 'the force of motion by which a body is impelled in the first instant of its motion'. Descartes insists that this force of motion is not the same as the force of motion which 'bears the body downward' during the actual course of its fall.[41] Finally, one must, Descartes contends, attend to both the 'speed' and the 'quantity of the body', since both factors contribute to the measure of the 'weight' or force of motion exerted in the first instant of fall. He explains that,

> if one atom of water about to descend would be twice as fast as two other atoms, the one atom alone will weigh as much as the other two together.[42]

These three suppositions mark the first appearance of some fundamental notions of Cartesian natural philosophy, at least in forms from which later, mature versions are clearly descended. Weight or heaviness reduces to the mechanical force exerted by a particle in its tendency to motion of descent. Moreover, as will be more clear later in the text, the 'weight' of a body is now affected by the mechanical constraints and conditions of its surroundings. Far from being an essential and invariable quality of bodies, weight is now a derivative mechanical quality, jointly determined by the size of the body and its tendency to motion, as conditioned by the given configuration of neighboring bodies. To be sure, Descartes does not offer anything

[39] 'In order to set out fully my reasoning concerning the questions which have been proposed, I would first have to explain a great deal concerning the foundations of my Mechanics; but, since time does not permit this, I shall try to explain the matter briefly.' AT X p. 67.

[40] AT X. 68.

[41] AT x. 68. In the *Cogitationes Privatae* (AT X. 228) the inclination to motion is described as being evaluated 'in ultimo instanti ante motum'.

[42] AT x. 68.

like his mature version of these ideas. For example, there is little hint of the systematic composition and resolution of tendencies to motion which he would employ later, especially in his mechanistic optics. Nor does Descartes generalize from 'weight' considered as a tendency to motion to the decomposition of real motion into momentary states of tendency to motion. This move, which will be the key element in his mechanistic optics and general system of dynamics (i.e. what in 1619 in its embryonic form he is calling his 'mechanics'), only emerged in the 1620s.[43] On the other hand, Descartes does develop and articulate the concept of tendency to motion to a certain extent, as he struggles to apply it consistently throughout the manuscript. We should also note that Descartes' measure of the force of motion shows the imprint of Beeckman's ideas and their common source in a dynamical interpretation of the simple machines and ultimately therefore in the *Mechanica* tradition. As Milhaud long ago observed, this is apparent both in Descartes' concern with the first instant of descent, which is, so to speak, permanently maintained in dynamical equilibrium, and, in the evaluation of force as the product of quantity of matter and potential or nascent velocity.[44]

Descartes next solves the problem of accounting for the hydrostatic paradox. But, where Stevin offered an argument from macroscopic conditions of equilibrium, Descartes manufactures a curious exercise in ad hoc micro-mechanical reductionism. He proposes to demonstrate the statement by showing that the force on each 'point' or part of the bottoms of the basins B and D is equal, so that the total force is equal over the two equal areas.[45] He does this by claiming that each 'point' on the bottom of B is, as it were, serviced by a unique line of 'tendency to motion' propagated by contact pressure from a point (particle) on the surface of the water through the intervening particles. (See Fig. 3.3.)

> For example, let there be determined in one base the points g, B, h; in the base of the other, i, D, l. I contend that all these points are pressed by an equal force, because they are each pressed by imaginable lines of water of the same length; that is, from the top part of the vase [water level] to the bottom. For line fg is not to be reckoned longer than fB or [any] other line. It does not press point g in respect to the parts by which it is curved and longer, but only in respect to those parts by which it tends downward, in which respect it is equal to all the others.[46]

At least the latter portion of this passage is initially plausible. Assuming the points on the bottoms are indeed served by unique lines of tendency transmitted from points on the surface; then, in so far as we are only concerned with the tendency to descend, we may compare the lines of tendency in respect to their vertical

---

[43] On Descartes' optics and its connection to his mature dynamical conceptions, see Schuster (2000), and below Sect. 3.6 and Chap. 4.

[44] Milhaud (1921), 34.

[45] Descartes consistently fails to distinguish between 'points' and finite parts. But he does tend to assimilate 'points' to the finite spaces occupied by atoms or corpuscles. Throughout we shall assume that Descartes intended his points to be finite and did not want his 'proofs' to succumb to the paradoxes of the infinitesimal.

[46] AT x. 70.

'components'—a nice Stevinite touch, it may be noted. However, the procedure of mapping the lines of tendency is quite curious. Descartes can perhaps be taken to imply that when the upper and lower surfaces of the water are similar, equal and posed one directly above the other, then unique normal lines of tendency will be mapped from each point on the surface to a corresponding point directly below on the bottom. But, when these conditions do not hold; that is, when the upper surface of the water differs from the lower in respect to size and/or shape, or when it is not directly posed above the bottom, then some other unstated rules of mapping come into play. It would seem that in the present case, the area of the surface at f in the basin B is precisely one-third that of the bottom, so that each point or part on f must be taken to service three points or parts of the bottom. The problem, of course, is that no explicit criteria or rules for mapping are, or can be, given. Descartes makes no attempt to justify the three-fold mapping from f. He merely slips it into the discussion as an 'example' and then proceeds to argue that *given the mapping*, f can indeed provide a three-fold force to g, B and h.

In fact, the demonstration continues solely as a justification of the three-fold efficacy of f, rather than as a general demonstration of the problem, such as we might expect:

> It must be demonstrated, however, that point f alone presses g, B, h with a force equal to that by which m, n, o press the other three i, D, l. This is done by means of this syllogism. Heavy bodies press with an equal force all neighboring bodies, by the removal of which the heavy body would be allowed to occupy a lower position with equal ease. But, if the three points g, B, h could be expelled, point f alone would occupy a lower position with as equal a facility as would the three points m, n, o, if the three other points i, D, l were expelled. Therefore, point f alone presses the three points simultaneously with a force equal to that by which the three discrete points press the other three i, D, l. Therefore, the force by which point f alone presses the lower [points] is equal to the force of the points m, n, o taken together.[47]

Let us note the structure of this argument, for it is of some significance in understanding important aspects of Descartes' later natural-philosophical views. The demonstration depends on taking the mapping as given and then imagining g, B and h to be removed or the spaces below them opened. Descartes then asks whether it is not obvious that f would descend with equal ease toward each one of the three points and that it thus exerts a tendency to descend upon each one of them. In addition, it is implied that in working out the hypothetical case of descent, Descartes imagines away the rest of the fluid, *qua* fluid. That is, it is in effect hypothetically solidified, so that its behavior does not complicate the postulated mechanical relations between f and g, B and h. There is thus a three-fold displacement away from what one might consider the original terms of the problem: Descartes assumes an ad hoc mapping; invokes a hypothetical voiding and consequent motion; and, finally, implicitly solidifies parts of the fluid not involved in the first two steps. The proof of this 'example' is then taken as a general demonstration, without any indication as to how the procedure is to be generalized to all the points or parts in the surfaces.

---

[47] AT X. 70–1.

This idiosyncratic mode of argument has a much greater significance than might appear at first sight, because Descartes will make precisely the same moves in important areas of his later natural philosophy. In fact, the three-fold technique by which Descartes evades the problem forms, as it were, a fairly consistent motif or style of explanation in his later work. Much of his aerostatics and cosmological mechanistic optics will employ these sorts of arguments.[48] The key to this later style of natural philosophical argument will reside in his propensity for explanations based on the attribution of tendencies to motion to corpuscles in various states of rest, motion, and spatial relation. Often, tendencies to motion will be represented by geometrical lines which in turn are analyzed in order to yield the required explanation. But, Descartes will never make explicit the rules guiding the attribution of 'lines of tendency'; just as, in the present problem, he baldly presents his mapping of tendencies, yet cannot justify or even rationalize his choice. In his later work he will typically try to establish the mechanical efficacy of the lines of tendency chosen for the problem, granted their existence and precise configuration in the first place. As we have just seen, this *post facto* justification proceeds by means of the hypothetical voiding of the region toward which the relevant particles are said to tend. An analysis of the resulting hypothetical motion is used to buttress the claim for the efficacy of the particular configuration of lines in question. All this in turn renders comprehensible the *de facto* 'solidification' of parts of the medium, which he employs in this problem and will use again in his theory of light in its cosmological setting. The solidification is the conceptual corollary of mapping lines of tendency between specially chosen 'privileged surfaces', and those privileged surfaces will also reappear in the cosmological theory of light in *Le Monde*.[49]

What, then, should we make of the young Descartes' performance so far? For Stevin's formally rigorous and conclusive geometrical demonstration, Descartes substitutes a very different kind of account. Descartes did not and could not have denied the rigor of Stevin's account. If he was conceding rigor to Stevin's analysis, what was Descartes seeking to accomplish? The answer is that he was seeking

---

[48] On the aerostatics see Sect. 8.2.3.3 below. On the cosmological version of mechanistic optics in *Le Monde*, see Sect. 10.7 below. By 'cosmological' mechanistic optics, I mean the physical theory of light as a mechanical tendency to motion caused by the corpuscular-mechanical character of the sun and other stars, as well as their vortices, and governed by certain rules of dynamics.

[49] This mode of explanation haunts so much of Descartes' physical thought that one could venture to suggest that it goes a long way toward accounting for the curiously tendentious and idiosyncratic character of much of his later natural philosophical discourse. On the one hand, we can say, Whiggishly, that, after all, this style of explanation really consists in a connected sequence of ad hoc manipulations. The manipulations masquerade as clarifications, while in fact they condition a progressive loss of contact with the original aims of the problem. They close in on themselves, forming a superficially tidy universe of discourse increasingly irrelevant to the problem at hand and insulated from any fruitful return to new empirical information. On the other hand, and taking Descartes' part as it were, we shall argue in Chap. 10, concerning *Le Monde* that the persistence of these protocols and motifs shows the manner in which Descartes' first system of natural philosophy still bore the traces of (and was partially constituted by) his previous engagements with 'physico-mathematics'.

proper explanation, meaning explanation in terms of natural philosophy or physics. For all its oddness to us, this little exercise seemed to Descartes to bespeak the possibility of some new sort of agenda in the mixed mathematical science of hydrostatics and between it and a corpuscular or atomistic natural philosophy, or more generally between the mixed mathematical sciences, plural, and a new kind of natural philosophical discourse—atomist and mechanist. We see hints of this in his allusion to a 'Mechanics' which he intends to write. This, as we have seen, certainly would not have been a rehash of Archimedes or Stevin, but a compendium of the concepts and protocols he was developing for reducing macroscopic findings to talk about particles, their tendencies to motion and the geometrical representation of same. As noted, his later practice will tend to confirm this trajectory: he would elaborate dynamical concepts as a causal register for corpuscular-mechanical natural philosophizing, and he would invoke odd protocols of representation and argument descended from the section of the hydrostatics manuscript we have thus far surveyed.

The above points allow us to note in passing two further features of the hydrostatics manuscript, which just peak into view in the passages just surveyed, but which will have a long and important trajectory in his work, deeply intertwined with the evolution of his concerns with 'physico-mathematicized' natural philosophy. The first feature concerns a leaning toward, or a preference for, a problem solving, analytical, rather than deductivist approach to mathematical matters, including the mathematics that should be at play within natural philosophy. We can conjecture that this helped lead him initially to see, or reinforced an already rising preference for seeing, mathematics in terms of problem-solving analysis as opposed to demonstrative techniques.[50] The second feature involves the strategic importance Descartes places upon the set of geometrical lines and conditions he imposes upon the basic diagram of the filled vases. It is these lines of tendency to motion, linking particular points in privileged surfaces—along with correlative 'solidified' volumes—that set the terms for his discursive argument concerning underlying matter and cause. We shall see several further cases of this sort of 'figuring up' problems in physico-mathematics, meaning the way given diagrams, representing either common phenomena, or hard geometrical results in the mixed mathematical sciences, are further prepared for 'causal analysis' by the imposition of geometrical representations of key points, surfaces, lines of tendencies and 'frozen volumes'.[51]

---

[50] See Gaukroger (1995) 172–81, and Sepper (1996), 157–208. Indeed, Descartes would over time drive an agenda favoring analysis over synthesis in all mathematical pursuits: his view being that a geometrical demonstration does not reveal to us how a mathematical result is generated. Algebraic proofs, by contrast, have a transparency which reveals the path by which the conclusion is produced. The problem-solving, analytical approach to physico-mathematics in the manuscript hints at this later maturing agenda. In Chap. 5 we shall examine his early work in mathematics along these lines between 1619 and about 1625 and see how his pure mathematical work developed, and intersected with his physico-mathematical natural philosophizing, by means of his unifying dreams of first, a so-called 'universal mathematics', and then, his universal method.

[51] My use of the term 'figuring up', here and throughout, to denote Descartes' idiosyncratic protocols for problem preparation in physico-mathematics was suggested to me by reflecting on Sepper's seminal work on Descartes' early use of imagination in 'figuring things out' and 'figurate solution of problems' (Sepper 2000).

Given all this, what we next have to do is look at further embryonic yet profoundly promising developments in the remainder of the hydrostatics manuscript, concerning emergent 'principles of dynamics' or a causal register for his natural philosophizing.

3.3.3 The Hydrostatics Manuscript [2] The Force of Motion

A different set of conceptual problems was raised by a part of Descartes' argument which we have not yet examined. Descartes did resolve these problems with a degree of success. In so doing, he was able to begin to clarify some of the central ideas used later in his mature system of dynamics; that is, what in 1619 he terms his mechanics—the causal register in his nascent mechanistic natural philosophy. The difficulty involves an ambiguity or tension in the formulation of the concept of 'tendency to motion'. Descartes had two concepts through which to express the 'tendency to motion' of a body. On the one hand, in his second 'assumption' he spoke of the 'force of motion by which [a body] is impelled in the first instant of motion'. Here 'force of motion' is used in a manner similar to that in which it will later be employed in Descartes' mechanistic optics or in *Le Monde*. It bears the connotation of an efficacy or force characterizing the body during an instant (specifically the first instant) of its motion. By contrast, when Descartes specified the measure of 'tendency to motion' in his third 'assumption', he introduced the notion of speed:

> in that first imaginable instant of motion, we must take note also of the imaginable beginning of the speed by which the parts of the heavy body descend.[52]

Hence it turns out that one dimension of the instantaneous efficacy or 'force of motion' is constituted by the speed of the body. Conceptual tensions begin to appear at this point; because, in order to assimilate speed to instantaneous force, Descartes tries to introduce the notion of an 'imaginable beginning of speed'. This phrase deflects the kinematic connotation of speed over a finite interval of space or time toward an idea of instantaneous speed. However, the maneuver leads to a degree of ambiguity when Descartes later tries to evaluate real instantaneous tendencies (i.e. forces of motion) by reference to a set of hypothetical but 'kinematic' speeds. The kinematic connotation then reasserts itself, and Descartes is left saying that the body has a tendency to a triple speed when, in fact, it can attain only 'one' speed in case of a finite translation being actualized.

We see this issue played out in Descartes' explanation of the three-fold force of motion of the point f (Fig. 3.3). He first evaluates the total tendency to motion of f

[52] AT x. 68.

by attributing to it three units of instantaneous speed, arising from the three paths of descent caused by hypothetically voiding g, B and h:

> ... let all the lower points g, B, h and i, D, l be imagined to be opened at the same instant by the force of gravity of the superposed bodies. Certainly it will have to be conceived that in the same instant point f alone will move three times more quickly than each of the points m, n, o. For in that instant three places will have to be filled by the former [f], while only one place will have to be occupied by each of the points, m, n, o.[53]

Then he translates the result into a total force of motion, as we have already seen:

> Therefore, the force by which point f alone presses the lower [points] is equal to the force of the points m,n,o taken together.[54]

Descartes' argument can be rendered as follows: Point f will descend along all lines fg, fB and fh with the same 'natural' speed of descent. Since all three lines materialize at once, f must have three units of speed at once. But three speeds implies a three-fold force of motion and hence f can have as much 'weight' as m, n and o put together. The term 'speed' can mediate between the consequences of the three cases of hypothetical voiding and the reckoning of the total force of motion, because it signifies both the finite but hypothetical translations and a dimension of the measure of instantaneous force of motion.

Descartes quickly realized that the multiple speeds calculated for the hypothetical voiding are difficult to reconcile with the intuitively plausible idea that a body should be able to actualize its instantaneous force of motion as a commensurable real speed of descent. He saw that the dual role of 'speed' was to blame, for it allows one to slide easily between tendencies expressed as 'speeds'; and actualized tendencies measured by 'speeds'. Viewed in terms of the triple voiding, f has a three-fold instantaneous speed at the first moment of descent. But, if any real translation were to occur, it would obviously occur in one direction and at one speed only. We might say that f cannot really fall in three directions at once; or that its triple 'potential' speed can only be realized as a single unit of 'actual' speed. As Descartes put it,

> ... an objection can be offered, which in my opinion is not to be disregarded, and the solution of which will confirm the foregoing. All bodies of equal magnitude and weight, if they should be borne downwards, have some certain equal mode of speed, which they do not exceed unless they are impelled by some extraneous force. Thus it is wrongly assumed above that point f is inclined to move three times more quickly than any one of the points m, n, o, since it cannot be said to be impelled by any external force.[55]

To his credit, Descartes perceived that the difficulty is a conceptual one requiring a more precise notion of the relation between 'tendency to motion' and 'motion', as well as the avoidance of loose talk about multiple instantaneous speeds:

> I respond in this way to the objection. The antecedent is quite true; however, it is erroneously deduced from it that the point f is not able to incline to a triple velocity. For there are

[53] AT x. 70–1.

[54] AT x. 71.

[55] AT x. 71.

> two different considerations in relation to weight which must be distinguished: inclination to motion and motion itself. For bodies which tend downwards are not inclined to move to the lower place with this or that speed, but rather they are inclined to move there as quickly as possible. Whence it happens that point f is able to have a triple inclination, since there are three points through which it is able to descend. The points m, n, o each have a unitary inclination, since there is only one point through which each can move respectively.[56]

Through a conceptual reshuffling, Descartes is prepared to accept both horns of the dilemma. He grants that only one real speed can possibly be actualized and he still insists on the triple inclination. However, he is now expressing a modified understanding of inclination. It is now obvious that multiple inclinations are not and need not be translatable into multiple real motions. Clarification is achieved by insisting on a consistent dualism between 'motion' and 'tendency to motion', or 'speed' and 'inclination to speed': Descartes' phrase is *ad triplicem celeritatem propendere*. The real translation—motion or velocity—of a body cannot be evaluated in terms of the manifold tendencies to motion it may possess at any moment, owing to the mechanical conditions in which it is placed. Conversely, the fact that only one real translation can be attributed to a body does not alter the truth of mechanics that bodies, such as f, can press down on several bodies at once in several different directions.

The most striking thing about the passages just discussed is that they show Descartes in the very act of reformulating some of the concepts of his dynamics of corpuscles, his 'mechanics', as he struggles to solve the problem at hand. Descartes' mechanistic optics, as it developed in the later 1620s, and his general system of dynamics in *Le Monde*, are based on the configuration of concepts which begins to emerge in these passages. Cartesian mechanistic optics and natural philosophy will mainly depend on the analysis of instantaneous tendencies to motion, rather than finite translations. Indeed, Descartes dissolves real translation into a series of inclinations to motion exercised in consecutive instants of time at consecutive points in space. Moreover, many of Descartes' explanations will require the consideration of multiple tendencies to motion which a body may possess at any given instant, depending on its mechanical circumstances. In such cases, Descartes will be careful to employ the terms 'tendency to motion' or instantaneous 'force of motion', rather than 'motion' or 'speed', so that he may avoid the consequence that the real speed of a body varies with the number of different tendencies to motion one attributes to it at any given instant. In short, Descartes will insist that instantaneous tendency to motion can be resolved into various configurations of its 'components', but that real motion cannot be so analyzed, lest different sums result for the total quantity of motion of the body, the system to which it belongs, or the cosmos as a whole.

After responding to the objection that he has confused real motion with tendency to motion, Descartes adds that he described lines fg, fB, mi etc. 'not because' he

[56] AT x. 72.

wanted 'a mathematical line of water to descend, but rather for the easier comprehension of the demonstration'. He then closes the paragraph by remarking,

> For, since I speak here about things which are new and my own work, much must of necessity be supposed, unless they are to be explained in a complete treatise; therefore I judge that it is sufficient that I demonstrate that which I have undertaken.[57]

This treatise was presumably to deal with the 'mechanics' mentioned at the beginning of the manuscript. Hence, it would have contained the principles for an attempted justification of the hydrostatic argument. Such a treatise on 'mechanics' is also mentioned twice in the early correspondence between Descartes and Beeckman in the spring of 1619.[58] As noted above, it may be conjectured that Descartes' planned treatise of mechanics would have had to have been quite different from the classical model of treatises in statics or hydrostatics, such as those of Archimedes or Stevin. Unlike the latter thinkers, Descartes was not primarily interested in a macro-geometrical mechanics in which mathematical rigor was achieved by arguing through cases of static equilibrium. In order to legitimate the approach taken in the hydrostatics manuscript, which was, to a first approximation, a special exercise in Beeckman's kind of micro-mechanism, Descartes' treatise would have had to have dealt with the mechanics of corpuscles.[59] This could have included a micro-mechanics of moving particles concerned with the laws of collision, as already pursued by Beeckman; and, in addition, as our study of the entire manuscript has now made clear, a mechanics of force of motion and tendencies to motion, including a discussion of the representation of tendencies through geometrical lines—a style of mechanics more typically Cartesian, as evidenced in the manuscript and throughout his subsequent work.[60] This entire undertaking, in its embryonic and somewhat disjoint state in 1619, represented what Descartes then termed 'physico-mathematics'.

3.4 What's the Agenda: Descartes' Radical Form of Physico-Mathematics

There has been a tendency among those few commentators who mention Descartes' hydrostatical exercises to assimilate his treatment to that of Stevin. Milhaud, for example, maintains that Descartes proceeds geometrically, starting with definitions

---

[57] AT x. 72.

[58] AT x. 159 l.11-12; and 162 l.15.

[59] I say to a first approximation, because whilst superficially this seems to comport with the physico-mathematics of Beeckman, we shall soon see that it is, in underlying terms, much more radical—and intentionally so.

[60] The representation of corpuscular tendencies to motion by means of geometrical lines is a symptom of the more radical intentions of Descartes' species of physico-mathematics and also a partial indicator of its links to his aspirations for an analytical (rather than demonstrative) approach to mathematics, including mixed mathematical disciplines, which he intends to render more 'physical' or organically articulated to natural philosophizing.

or postulates and demonstrating results from these in a syllogistic way.[61] Rodis-Lewis also mentions his syllogistic path, noting his 'remarkable formal rigor'.[62] But actually what is remarkable is the absence of formal rigor, except for the one syllogism he presents, as noted above. Descartes substitutes, for Stevin's formally rigorous and conclusive geometrical demonstration, a very different kind of account which is, by the standards of Stevin's Archimedean statics, exploratory and inconclusive. Descartes, the talented and skilled mathematician, would not have denied the rigor of Stevin's account. So, we may ask—if Descartes was not concerned with Stevin's work *vis à vis* its rigor in the Euclidean or Archimedean sense, what was at stake—what was his agenda? We can hone in on this question by recalling what we now know about our actors' categories of natural philosophy, mixed mathematics and practical mathematics.

Stevin's explanation falls within the domain of mixed mathematics and Stevin, as usual, is eyeing off a range of follow-on practical applications.[63] The account Descartes substitutes for it falls within the domain of natural philosophy. The concern is to identify what causes material bodies to behave in the way they do. The geometrical account does not provide an *explanation* of the phenomenon, because it does not identify what causes the phenomenon. Fluids are physical entities made up, on Descartes' account, of microscopic corpuscles, the behavior of which determines the macroscopic behavior of the fluid. We need to understand the physical behavior of the constituent corpuscles, if we are to understand the behavior of the fluid, because this is what is causally responsible for its behavior. As we have seen, he speaks in terms of microscopic corpuscles whose movements or tendencies to movement are understood in terms of an emergent, but still largely tacit, theory of forces and tendencies, a causal discourse which he identifies as part of that 'Mechanics' upon which he claimed he was working.

This fully accords with the traditional view of the scope and aims of natural philosophy outlined in Chap. 2. Physical explanation involves the identification of what causes material bodies to behave in particular ways. This was understood to be the case whether, as in Aristotelianism, natural processes were explained primarily on the basis of causes identified with the nature or essence of the substance in question, or, as in neo-Platonic natural philosophies, brute matter was worked upon from the outside by various types of non-material causal agents. Theorizing about matter and an associated 'causal register' was traditionally taken as constitutive of natural philosophy, whatever disputes there might have been amongst Platonists, Aristotelians, Stoics, and atomists. And it was such a conception, reflected through Aristotle's

[61] Milhaud (1921), 34–7.

[62] Rodis-Lewis (1971) vol. 1, 30–1.

[63] As was the case with many master practitioners of the practical mathematical disciplines, Stevin envisions the applications of these results to more properly practical ends; that is, a key mathematical result will command a wide domain of application in a number of practical fields . Cf. Bennett (1998).

categorization of the mixed mathematical sciences as subordinate to given, previously established explanatory physical principles of matter and cause, that had effectively marginalized, or at least rendered problematic, mathematical approaches to natural phenomena within natural philosophy.

Now we more fully see what was at stake. This work on hydrostatical problems implied a radically non-Aristotelian vision of the relation of the mixed mathematical sciences to this emergent form of corpuscular-mechanical natural philosophizing. Descartes' aim seems to have been to shift hydrostatics from the realm of practical or mixed mathematics unambiguously into the realm of natural philosophy. This he tries to achieve by redescribing, in terms of his matter theory and embryonic concepts of dynamics, what it is that causes the pressure exerted by a fluid on the floor of the vessel containing it. He redescribes what causes the pressure in terms of the cumulative dynamical behavior of postulated microscopic corpuscles making up the fluid.

In terming this work physico-mathematics, Beeckman and Descartes were signaling a break with the traditional, Aristotelian modes of connecting, or not connecting, the mixed mathematical sciences with natural philosophy. Like their older contemporaries Kepler and Galileo, the two young mechanists were trying to renegotiate the standing of the mixed mathematical sciences in relation to natural philosophy, having rejected both the matter-theoretical and causal content of Aristotelianism and its grammar of subordination of mathematical sciences.

Descartes had learned from Beeckman that when you explain a machine by its parts and their motions, you simultaneously deal with it mechanically and in terms of its matter and the properties of that matter. In the hydrostatics manuscript, we see Descartes reducing Stevin's macro-analysis in descriptive geometry to the underlying 'machinery'—the material parts, their arrangements and motions, or rather their forces and tendencies to motion. The idea of 'underlying machinery' takes Descartes from mechanics as a general science of machines, which falls within practical and mixed mathematics, to mechanics, or dynamics, as a general causal account of underlying corpuscular machinery, that is, of matter and motion. We are going to see this agenda in play in many of the key moments in his work at the interface of mixed mathematics and mechanistic natural philosophy—that interface being the domain of what he calls physico-mathematics. In Sect. 3.6 below we shall learn that in 1620, he attempted precisely the same move in unpacking what he took to be a great insight of Kepler, who had suggested that light moves with more force in denser optical media and 'hence' is bent toward the normal in moving from a less to a more dense medium. Moreover, we shall find in the next chapter that the principal step in Descartes' constitution of a physico-mathematical optics—which in turn was to have an exemplary role in his mature natural philosophizing—occurred directly after his discovery of the law of refraction in 1626/1627 in a simple geometrical form (as a law of cosecants): He literally read out of his key geometrical diagram the principles of a micro-mechanical theory of light, which would then subsume the new macro-geometrical law that had prompted them in the first place.[64]

[64] For details see below Chap. 4 and Schuster (2000).

In short, after his initial interaction with Beeckman, Descartes almost always interpreted the search for causes in natural philosophy as the search for real corpuscular models worked according to principles of a mechanics, indeed a dynamics, specifying the causal principles at work in the microscopic realm. Those like Galileo, who theorized at the level of macroscopic geometrical regularities, would be accused of 'building without foundation', in much the same way that Beeckman identified 'physico-mathematics' with a proper balance of the mathematical and the physical (natural philosophical), using Stevin and Bacon respectively as examples of those who cleaved too much to the erroneous extremes of this continuum.[65]

However, granting all this, I am not saying that Descartes was slavishly following Beeckman—not in 1618 and certainly not later. If we look closely, we can see that even when he was pursuing his first physico-mechanical researches with Beeckman in 1618, Descartes' approach to this agenda was already much more radical than Beeckman's. What Descartes asserts in the hydrostatics manuscript does not map directly onto Beeckman's detailed conceptualization. Right from the start, he proceeds not via a dynamical interpretation of the *Mechanica* account of the lever, as we have seen Beeckman was doing, but rather via *Stevin's statically based neo-Archimedean account*, of all things, which he fleshes out in terms of the micro-corpuscularian model he learned from Beeckman, albeit with the details significantly revised. Neither in 1618 nor ultimately did he accept Beeckman's formulation of the principles of mechanics, or causal register of corpuscular mechanism. By the early 1630s and quite possibly even earlier, Descartes had invented a full system of dynamics, applied to corpuscles, as the causal dimension of his natural philosophical discourse. It was based on concepts owing little, if anything, to the teachings of Beeckman. Instead, it was largely grounded in his struggles over issues in geometrical and physical optics as they grew out of the work of 1618–1620.[66] We should see Descartes as consequentially much more radical than Beeckman in his interpretation for physico-mathematics and his agenda for its articulation.

But, what would Beeckman have thought at this stage.? Well, despite these rather profound differences, we can conjecture that Beeckman would have been pleased with Descartes' hydrostatic manuscript as a token of 'their' physico-mathematics. To Beeckman it would have seemed obvious Descartes was reducing Stevin's macro-analysis in descriptive geometry to the underlying 'machinery'—the material parts, their arrangements and motions, or, the kind of matter involved and its properties. At a general level, and as we said above, 'to a first approximation', this search for underlying machinery and its dynamic principles would have seemed similar to Beeckman's own unification of atomism and the *Mechanica* tradition. But Descartes was already potentially on a significantly different flight path, as his subsequent trajectory, especially his later results in optics, would show with absolute clarity. Nevertheless, we should remember that these differences can only be

---

[65] AT ii. 385.

[66] See below Sect. 4.6 and Schuster (2000).

assessed from within a perspective which recognizes the 'influence' of Beeckman. That is, we should note the role of Descartes' active adoption and modification of concrete and programmatic bits of Beeckman's work in the original formation of his own view of micro-mechanical natural philosophy, and its relation to the practical and mixed mathematical sciences, in particular as evidenced in the work on hydrostatics.

Next, therefore, we need to turn to the second of our early examples of physico-mathematics, the much better known work on naturally falling bodies. We can now approach the surviving materials in a different way than previous commentators, because we can frame our reading through what we have begun to know about Beeckman's stimulus to, and Descartes' embryonic agenda within, physico mathematics.

3.5 The Physico-Mathematics of Natural Fall

3.5.1 Introduction—The Study of Fall as [Abortive] Physico-Mathematics

Beeckman and Descartes' work on fall is contained in Beeckman's *Journal*, the second essay in the *Physico-Mathematica* of Descartes and further fragments in Descartes' *Cogitationes Privatae*.[67] This work has attracted considerable attention from Descartes scholars and historians of science—indeed much more attention than the hydrostatics manuscript, the early mathematical work (which we survey in Chap. 5) or the important fragment on optics from circa 1620, which below forms our third and final case study of the early physico-mathematics.[68] Attention is usually paid to the material on fall because it parallels that of Galileo on the law of falling bodies but, interestingly, involves several 'errors' and pitfalls related in turn to supposed reasons why Beeckman and Descartes failed to become Galileo; failed, that is, further to pursue the law, correct and confirm its form and publish the results. Alexandre Koyré famously analyzed Descartes and Beeckman's work in this manner,

[67] Beeckman's *Journal* (Beeckman 1939–53) contains Beeckman's statement of the problem, his remarks on the mathematical arguments of Descartes and his own further comments. *Journal* f105v-106r, cited in AT X 58–61. The *Journal* also contains a set of two short essays by Descartes which have been published under the title '*Physico–mathematica*', AT X 75–78. The first essay, as we have seen, concerns the hydrostatics. The second essay contains Descartes' version of his contribution to the discussion of accelerated fall. Finally, in the early fragments of Descartes, published in the Adam-Tannery edition under the title *Cogitationes Privatae*, one entry directly concerns the matters discussed with Beeckman about fall and several others on the related theme of the mathematical representation of motions. AT X 219–222.

[68] Duhem (1906–13) vol. III. 566ff, 399–405, 481ff. A. Koyré (1939) pt ii 28–39, pt iii 167–171. Hanson (1958) 43–49.

in his historiographically epochal, and iconic, *Études Galiléennes*, in order all the better to pave the way for his account of the modern science founding achievement of Galileo. Borrowing heavily from Koyré, Norwood Russell Hansen then presented the case in his path breaking, anti-inductivist philosophy of science tract, *Patterns of Discovery*, to illustrate the role of theoretical commitments in the process of discovery, with special attention in this case to the way errors and pitfalls could divert discovery processes from a straight, true and fruitful path.[69]

In this section we will have occasion to review and criticize Koyré and Hansen, but that is not our main concern, for the following reason: This section treats the material on fall in a new way, explicitly as a case study in the sort of physico-mathematical agenda that we have seen Beeckman and Descartes following. We shall not primarily see Beeckman and Descartes as 'failed Galileos' or as aspiring Hansonian discoverers, who happened to be interestingly conceptually confused. Rather, we shall interpret their work as a set of initiatives regarding 'the physico-mathematics' of natural fall. This means we take it that they were concerned not merely with the exact form of the law of fall (assuming, as we shall see, that they could agree that such a thing really existed!), but also with the physico-mathematical 'treatment' of such a law, a movement of analysis from the law back to its natural philosophical causes. There would be little point in working back, physico-mathematically, to causes, unless one were sure that there existed an exact descriptive law of mixed mathematical type, and that it had been found. Much about their work, its 'errors', pitfalls, and even its hitherto little noticed outcroppings of lightness, playfulness and speculation, will thus be explained. In addition we shall be able to set the case of fall alongside that of the hydrostatic paradox, as initiatives in physico-mathematics. In turn this will pave the way for our treatment of the 1620 optics fragment in the same way, allowing us to see it for the first time in its profoundly physico-mathematical character and, eventually, revealing its rich and complex relation to Descartes' great physico-mathematical achievements in optics during the late 1620s.[70]

Given all this, it is best to state from the outset, in general terms at least, what 'the problem' is with Beeckman and Descartes' work on fall, if it is not a failure to 'be Galileo'. We shall see that the problem facing our physico-mathematicians was that they could make no headway either on an agreed, geometrically expressed law of fall, or on the structure of (equally mathematical representable) causes that would explain it. Here is why: In physico-mathematics one ideally wants a crisp clean geometrical result at the 'empirical' level in the relevant mixed mathematical discipline, so that the natural philosophical causes of that result can be discerned by reading

[69] See also Jullien and Charrak (2002) 19–20, 89–96, 100–104, 107–112.

[70] My treatment of the material on fall differs considerably from that in Schuster (1977), which had been taken up and improved by Gaukroger (1995) 80–84. The chief difference is owing to my emphasis here on a new and improved understanding of the early physico-mathematical aspirations of Descartes. This frames the entire presentation and much of its content, although many technical details remain the same.

out from (or into) the diagrams in question. In the case of hydrostatics, this is obvious, as we have just seen. Descartes begins with diagrams of cases in which he knows that Stevin has demonstrated that the hydrostatic paradox occurs—these diagrams (cases) are ripe for what we have called his 'figuring up' of the problem, his reading in of further lines, supposedly bespeaking the underlying causes. In the case of his ultimately successful optical work of the later 1620s, Descartes will do the same thing; that is, take a geometrical representation of an arguably well established mixed mathematical result—a law of refraction—and read back from certain of the parameters of its geometrical representation to knowledge of the underlying causes of the phenomenon. A looser, more exploratory type of inquiry was also possible, as Descartes' 1620 optical work will illustrate in our next case study. Here, although he had no firm, convincing result about the law of refraction he was seeking, he had certain speculative suggestions from Kepler as to the geometrical form, and causes of the law, as well as his own initial guesses as to the nature of the causes, based on an improvement and articulation of Kepler's speculations. Descartes seemed to be implying that the 'gap', as we may term it, between a geometrically represented law-to-be-specified, on the one hand, and geometrically representable causes-to-be-specified, on the other hand, might be closed a bit, toward ultimate solution, by some play with both ends of the problem. We are about to see that the problem with fall, as it turned out for Beeckman and Descartes, was that this gap was just too wide, with both end points offering a confusing array of possibilities, some not easily to be discriminated one from the other, even on the basis of further evidence, assuming it could be obtained in some way. Leaving aside the facts that Beeckman and Descartes had some subsidiary differences of approach, and that the 'errors' attributed to Descartes by Koyré and Hanson are less damaging than they thought, the fundamental conclusion we shall reach is that the physico-mathematical inquiry into naturally accelerated fall petered out because the results at both ends of the gap were inconclusive and not likely to be improved, and Beeckman and Descartes knew it. For Descartes at least, optics was to prove a much richer field for uncovering not just natural philosophical causes, but core elements in his dynamics, that is, in the very causal register of his natural philosophy.

3.5.2 Beeckman's Problem, and His Version of Descartes' Solution

Beeckman entitled his first entry on fall 'Why the speed of a body falling in a vacuum always increases'.[71] However, the accompanying text shows that Beeckman did not so much seek a natural philosophical explanation of accelerated fall as

[71] AT X. p. 58.

assume one, whilst he concentrated more on the concepts to be employed in describing the macroscopic aspects of accelerated motion. He wrote,

> When there is a vacuum between the body and the earth, the body moves downwards, towards the centre of the earth, in the following way: in the first moment it covers as much space as possible as a result of the tractive force of the earth. In the second it keeps up this motion, to which a new motion is added due to the tractive force, so that in this second moment it covers a double space. In the third moment the double space is maintained and to it is added, by the earth's tractive force, a third, so that in one moment it covers a space three times the first.[72]

Beeckman asserts the existence of a terrestrial 'tractive' force acting repeatedly from 'moment' to 'moment'. Characteristically, Beeckman later explicates this tractive force as actually arising because of corpuscular collisions.[73] According to Beeckman's inertial principle, each time this 'terrestrial force' (corpuscular collision!) acts, it impresses upon the falling body a new degree of motion, which will be conserved in all subsequent 'moments'. Thus, the total motion of the body increases at each 'moment' and hence the space traversed during each subsequent 'moment' also increases. For the time being, this is all Beeckman has to say about the causation of the phenomenon. His problems reside elsewhere, particularly in his use of the term 'moment' of time. He, and we, need to work out what he is intending, and what requires further clarification.

The remainder of Beeckman's discussion clearly shows that he intends by 'moments' uniform intervals of time, which will be mathematically reduced to a continuous series of instants. But, despite the fact the central connotation of 'moment' is 'interval', it does not follow that Beeckman conceives the force to act continuously during each 'moment'. Of course, the literal sense of such expressions as '*Secundo [momento]...superadditur motus novus tractionis*' is that the force acts anew at each moment and continuously during the moment. Nevertheless, a close reading of the text shows that Beeckman is fundamentally concerned with the way in which reiterated applications of the 'tractive' force at the initial instant of each 'moment' give rise to increments of space traversed *during* consecutive 'moments'. Hence Beeckman manipulates the 'moments' as notional dividers which space out the reiterated instantaneous application of the force over time. We shall see that the rest of his discussion derives from precisely this conceptual orientation. Beeckman is not at all concerned with the idea that a force acting continuously over a 'moment' would itself give rise to a series of space increments which would have to be analyzed before any sums of space increments were taken over several 'moments' taken together. We are about to see that his argument moves in the opposite direction

---

[72] AT X p. 58, Koyré (1978) 80. Moventur res deorsum ad centrum terrae, vacuo intermedio spatio existente, hoc pacto: Primo momento, tantum spacium conficit, quantum per terrae tractionem fieri potest. Secundo, in hoc motu perseverando superadditur motus novas tractionis, ita ut duplex spacium secundo momento peragretur. Tertio momento, duplex spacium perseverat, cui superadditur ex tractione terrae tertium, ut uno momento triplum spacii primi peragretur. The translation has been slightly modified.

[73] AT X p. 61 Beeckman often speaks in this kind of shorthand for actually intended corpuscular-mechanical explanations.

entirely. Descartes will show him how to continuously reduce the intervals until he has a mathematical expression for the space increments arising from continuous instantaneous reapplications of the tractive force. In this connection his admission that the 'tractive' force is caused by corpuscular collisions lends weight to the contention that all along he is thinking in teams of instantaneous increments of motion imparted at the beginning of each interval.

Granting this interpretation of Beeckman's 'moments', we can return to the text in order to uncover the precise nature of the problem of describing fall which Beeckman posed for himself. In the text cited above, Beeckman has a tendency to translate immediately into spatial terms the amount of motion the body possesses at each moment. In the uniform intervals (or moments) between applications of the force, he implicitly assumes that the space traversed is proportional to the amount of motion possessed by the body, and that the amount of motion itself depends on the motion conserved from the previous moments and the unit increment of motion impressed by the application of the force at the commencement of the present moment. By the end of the passage he is most interested in the proportion between the spaces traversed and the number of moments accrued, being particularly focused upon the series of whole numbers which expresses the spaces traversed during successive moments. In fact, discussion of the space-time relation displaces any further consideration of the causal principles upon which the entire argument is based. It turns out that, for the time being at least, Beeckman is not interested in further natural philosophical inquiry into 'why the speed of a body falling in a vacuum always increases'. Rather, he is setting up the mathematical problem of correlating the unit spaces traversed with unit 'moments' elapsed, given that the space increments arise directly from increments of motion imparted in consecutive 'moments'.

This is important, because it was this manner of stating the problem of fall which dictated the form of the question Beeckman posed to Descartes. Beeckman asked Descartes to calculate how far a body in accelerated free fall would move from rest in one hour, given the distance it traversed from rest in two hours.[74] Both Descartes and Beeckman understood this question to entail the problem of first selecting a time frame for the spacing of the 'moments' to which the series of space increments are to be applied. Beeckman's initial conceptual framework provided only an abstract schema, according to which the distances traversed from rest during any arbitrary consecutive units of time are as the series of whole numbers. In any putatively real case of fall one has to demonstrate or suppose a particular magnitude for each of the 'moments' so that the space series may be summed, over the entire time of descent. I suggest that the question posed to Descartes reflects Beeckman's insight into the problem of summing spaces over definite intervals of time; although, to be sure, one cannot know what particular aspects of the problem were of immediate interest to Beeckman. For example, we cannot know whether he was initially puzzled about the consequences of making different assumptions about the magnitude of the 'moments', or, whether he had already seen the problem in terms of trying to reduce the 'moments' down to instants of time. In any case, it is plausible that he posed the question to

[74] AT X p. 60.

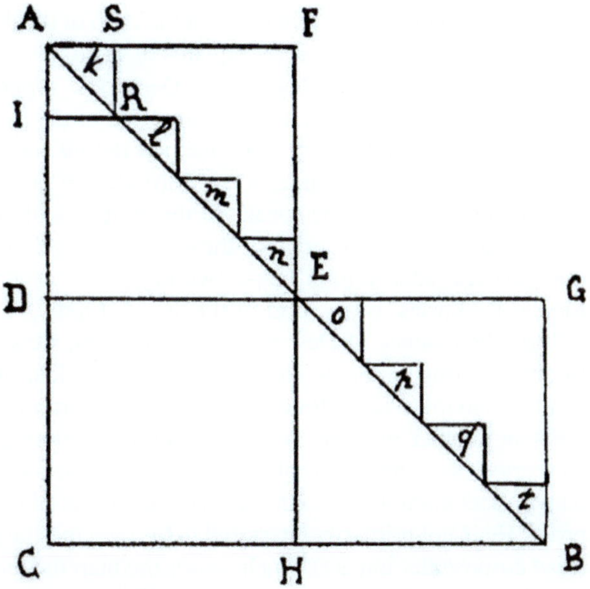

Fig. 3.4 AT X, p. 59

Descartes as the converse of 'given the distance fallen in one hour, find the distance fallen in two hours' so that Descartes would not be tempted to rely on the following simple idea: if the 'moments' are of an hour's duration, one unit of distance is traveled in the first hour, and two units in the second, so the ratio of the distance traveled in the two hours to the distance traveled in the first hour would be 3:1.

Beeckman next reports on Descartes' solution to the problem (Fig. 3.4), 'Haec ita demonstravit Mr. Peron'… .

> If the moments are not divided up, the space covered by a falling body in one hour will be ADE. The space covered in two hours will be double the proportion of the times, i.e., will be ADE to ACB, which is double the proportion of AD to AC. Let the moment of space that the body covers in falling for one hour be of some magnitude, e.g., ADEF. In two hours it will cover three similar moments i.e., AFEGBHCD. But AFED contains ADE and AFE. And AFEGBHCD contains ACB with AFE and EGB, i.e., with the double of AFE. Thus, if the moment is AIRS the proportion of the spaces will be ADE with *klmn* to ACB with *klmnopqt, i.e.,* once again, the double of *klmn*. But *klmn* is much smaller than AFE. Since, therefore, the proportion of space covered to space covered is composed of the proportion of one triangle with another triangle, with equal [magnitudes], added to these terms, and since these equal additions become ever smaller as the moments of space become smaller, it follows that these additions become null quantities when the moment has become a null quantity. Now, such is the moment of space through which the body falls. It remains, therefore, that the space through which the body falls in one hour is related to the space through which it falls in two hours as the triangle ADE to the triangle ACB…[75]

[75] AT X pp. 59–60 Koyré (1978) 80–81. Cum autem momenta haec sint individua, habebit spacium per quod res una hora cadit, ADE. Spatium per quod duabus horis cadit, duplicat proportionem temporis, id est ADE ad ACB, quae est duplicata: proportio AD ad AC. Sit enim momentum spatii per quod res una hora cadit alicuius magnitudinis videlicet ADEF. Duabus hours perficiet talia tria momenta, scilicet AFEGBHCD, Sed AFED constat ex ADE cum AFE; atque AFEGBHCD constat ex ACB cum AFE & EGB, id est cum duplo AFE.

Beeckman goes on to reiterate that the ratio of the distances fallen in one and two hours from rest is 1:4, or as the squares of the times.

For the moment let us leave aside Descartes' supposed contribution to the solution and concentrate instead on the structure of the proof as Beeckman recorded it. Beeckman labeled the entry 'the time of a falling body computed'.[76] But, in fact, he was computing the distances to be attributed to portions of the total time of fall, Throughout the proof, he speaks of the elements of area of the figure as *momenta spatii*, and he adds up series of these moments relative to intervals of time as the intervals are reduced to instants. Two key points need to be underscored before we look at Descartes' own report of this proof. Firstly, if the present report accurately reflects the essence of Descartes' own analysis, we see that Descartes here has no trouble relating *momenta spatii* (or their equivalent in his own terminology as we shall see) to intervals of time. Descartes will shortly be seen making an error on this point in another passage, so it is important to notice that Beeckman's version of Descartes' work has no problem in this respect. Secondly, as has already been hinted above, Beeckman tends to defocalize the natural philosophical substructure of this proof. He is not primarily interested in his own inertial principle or the tractive force (read corpuscular impact) which causes the increments of motion. Nor does he mention the direct relationship between the distance traversed in an interval of time and the degree of motion possessed by the body in that interval. Descartes, as we shall now see, presents the same mathematical argument, including the correct relating of distance to time, but his natural philosophizing—his dynamics that is—is more elaborate, for he relates intervals of time to increments of force [*vis*] which cause increasing *momenta motūs*, which, by implication, can be summed to indicate space traversed.

3.5.3 *Descartes' Solution—Triumphs and Pitfalls of a Physico-Mathematics of Fall*

Turning now to Descartes' version of the solution, in the *Physico-mathematica*, we see that the opening of his exposition, dealing with the mathematical movement

---

Sic, si momentum sit AIRS, erit proportio spatii ad spatium, ut ADE cum klmn, ad ACB cum klmnopqt, id est etiam duplum klmn. At klmn est multo minus quam AFE. Cum igitur proportio spatii peragrati ad spatium peragratum constet ex proportione trianguli ad triangulum, adjectis utrique termino aequalibus, cumque haec aequalia adjecta semper eo minors fiant, quo momenta spatii minora sunt; sequitur haec adjecta nullius quantitatis fore, quando momentum nullius quantitatis statuitur. Tale autem momentum est spatii per quod res cadit. Restat igitur spatium per quod res cadit una hora, se habere ad spatium per quod cadit duabus horis, ut triangulum ADE ad triangulum ACB.

[76] AT X p. 58.

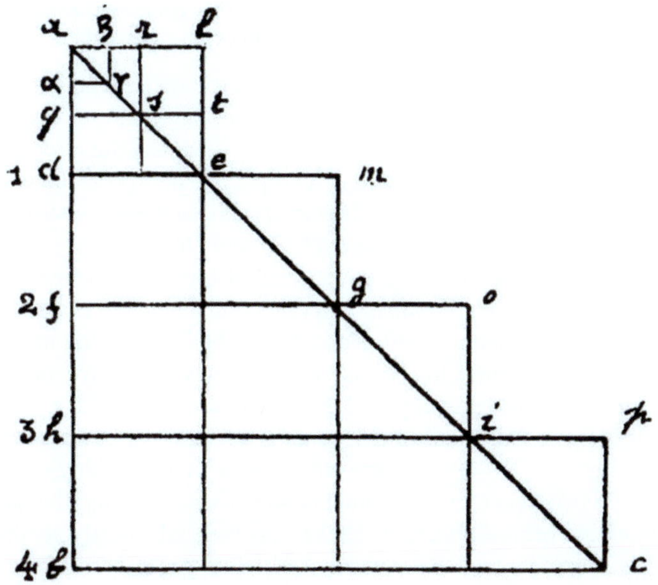

Fig. 3.5 AT X, p. 76

from finite to instantaneous time intervals, corresponds very well with Beeckman's report (Fig. 3.5).

> In the proposed problem, in which it is imagined that at each instant a new force is added to that with which the heavy body moves downwards, I say that this force increases in the same manner as do the transverse lines *de*, *fg*, *hi*, and the infinite other transverse lines that can be imagined between them. To demonstrate this I take as the first minimum or point of motion, caused by the first attractive force of the earth that can be imagined, the square *alde*. For the second minimum of motion we have the double, namely *dmgf*: the force in the first minimum persists and a new, equal force is added to it. Thus in the third minimum of motion there will be three forces, namely those of the first, second and third time minima, and so on. This number is triangular, as I will perhaps explain more fully elsewhere, and it appears here to represent the figure of the triangle *abc*. But, you will say, there are parts which protrude, *ale*, *emg*, *goi* etc., which are outside the figure of the triangle. Therefore, the figure of the triangle cannot represent this progression. But I reply that these protuberant parts come from the fact that we have given extension to the minima which must be imagined as indivisible and as containing no parts. This is demonstrated as follows. I divide the minimum *ad* into two equal parts at *q*; then *arsq* will be the [first] minimum of motion, and *qted* the second minimum of motion, in which there will be two minima of force. Similarly we divide *df*, *fh*, etc. Then we have the protuberant parts *ars*, *ste*, etc. Clearly they are smaller than the protuberant part *ale*.
>
> Furthermore, if I take a smaller minimum such as $a\alpha$, then the protuberant parts will be yet smaller, such as $a\beta\gamma$, etc. If, finally, I take as this minimum the true minimum, i.e., the

> point, then these protuberant parts will be nothing, for they could not be the whole point clearly, but only a half of the minimum *alde*, and a half of a point is nothing.[77]

This mathematical argument is clearly the source of Beeckman's analysis. In effect Descartes had shown Beeckman how to sum increments of areas applied to intervals along line *ab*, as the areas are reduced to lines and the intervals to points. It has been shown that Beeckman could offer a precise interpretation of the formalism in terms of the relation between the time intervals *and the* space series. Descartes' physical interpretation of the formalism does not seem to differ very much from that of Beeckman. When mentioning the increments of force, and hence of motion, *he sees that they are related to intervals of time*. It is true that whereas Beeckman terms the areas in question *momenta spatii*, Descartes insists on terming them *minima* or *punctii motūs*. But, it seems likely that Beeckman would have understood and agreed with this, for he too presupposed that the spaces traversed in equal intervals of time are directly proportional to the quantity of motion possessed by the body during those intervals. However, these differences in terminology are of course symptomatic of a difference in conceptual perspectives regarding the natural philosophical explication of motion and its causes. As we shall shortly see, these differences, plus the peculiar way Beeckman posed this problem to Descartes, were enough to produce an interesting mistake on the latter's part, which, in turn, has been the point of departure for accounts of this entire episode. We are going to see that Descartes' 'error' is less serious than has been made out; that explanations for this mistake have themselves been largely erroneous; and that perseverating on this aspect of the work has diverted attention from what I shall argue were the central physico-mathematical concerns and speculations that exercised Beeckman and Descartes, and perhaps even led to their giving up work on this issue.

[77] AT X pp. 75–7, Koyré (1978) 82–83 (translation slightly modified). In proposita quaestione, ubi imaginatur singulis temporibus novam addi vim qua corpus grave tendat deorsum, dico vim illam eodem pacto augeri, quo augentur Iineae transversae de, fg, hi. & aliae infinitae transversae, quae inter illas possum imaginari. Quod ut demonstrem, assumam pro primo minimo vel puncto motus, quod causatur a prime quae imaginari potest attractiva vi terrae, quadratum alde. Pro secundo minimo motus, habebimus duplum, nempe dmgf: pergit enim ea vis quae erat in primo minimo, & alia nova accedit illi aequalis. Item in tertio minimo motus, erunt 3 vires; nempe primi, secundi & tertii minimi temporis etc. Hic autem numerus est triangularis, ut alias forte fusius explicabo, & apparet hunc figuram triangularem abc representare. Immo, inquies, sunt partes protuberantes ale, emg, goi, etc., quae extra trianguli figuram exeunt. Ergo figura triangulari illa progressio non debet explicari. Sed respondeo illas partes protuberantes oriori ex eo quod latitudinem dederimus minimis, quae indivisibilia debent imaiginare & nullis partibus constantia. Quod ita demonstratur. Dividam illud minimum ad in duo aequalia in q; iamque arsq est primum minimum motus, et gted secundum minimum motus, in quo erunt duo minima virium. Eodem pacto dividamus df, fh, etc., Tune habebimus partes protuberantes ars, ste, etc., Minores sunt parte protuberante ale, ut patet. Rursum, si pro minimo assumam minorem, ut aα, partes protuberantes erunt adhuc minores, ut aβδ, etc., Quod si denique pro illo minimo assumam verum minimum, nempe punctum, tum illae partes protuberantes nullae erunt, quia non possunt esse totem punctum, ut patet, sed tantum media pars minimi alde; atqui puncti media pars nulla est.

Fig. 3.6 AT X, p. 77

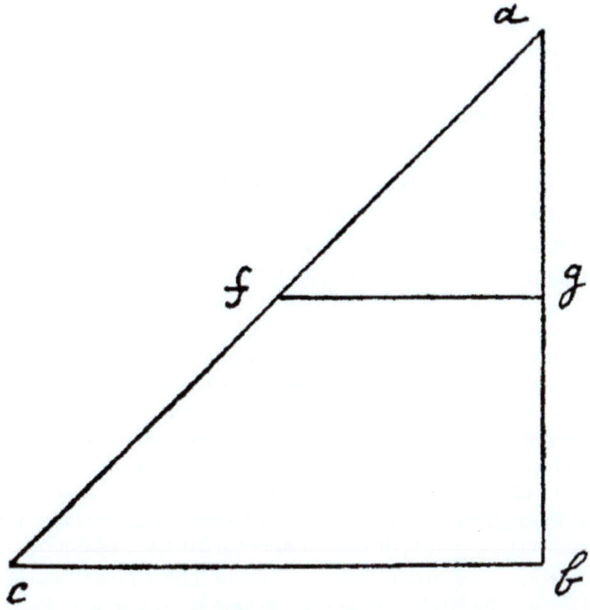

So, continuing the exposition in the *Physico-Mathematica*, we find Descartes purporting to apply his just announced geometrical formalism to Beeckman's problem with the help of yet another figure.

> From which it clearly follows that if we imagine, for example, a stone which is attracted by the earth, in a vacuum, from *a* to *b*, by a force which always remains equal to the first, persisting, force, then the first motion at *a* will be to the last at *b* as the point *a* is to the line *bc*. The part *gb*, which is half, will be covered by the stone three times as fast as the other half ag, because it will be drawn by the earth with three times the force. The space *fgbc* is three times the space *afg*, as is easily proved. And one can say this of the other parts proportionately. [78]

Beeckman's question and solution thus slipped from view. Somehow Descartes managed to transform the problem into the following form: given a completed motion along *distance ab* uniformly accelerated from rest at *a*, to find the ratio of the times taken to traverse *ag* and *gb*. Descartes still employed the *minima motūs*, but he apparently applied them to *ab* taken as the trajectory of the body, rather than as a time. In addition, he reintroduced time through the device of interpreting the areas *afg* and *fqbc* as sums of *minima motūs*, or 'total motions', which are inversely proportional to the times taken to traverse the distances to which they refer.

From the standpoint of subsequently emergent classical mechanics Descartes has committed some egregious errors, while his friend Beeckman has unerringly

---

[78] AT X p. 77. Koyré (1978) p.83 …si imaginetur, verbi gratia lapis ex a ad b trahi a terra in vacuo per vim quae aequaliter ab illa semper fluat, priori remanante, motum primum in a se habere ad ultimum qui est in b, ut punctum a se habet ad lineam bc; mediam vero partem gb triplo celerius pertransiri a lapide, quam alia media pars ag, quia triplo majori vi a terra trahitur: spatium enim fgbc triplum est spatii afg, ut facile probatur; & sic proportione dicendum de caeteris partibus.

pursued the correct answer. The need to explain Descartes' poor performance has driven commentators, such as Koyré and Hanson, to denigrate Descartes' supposedly excessively mathematical approach and his inability to grasp fundamental physical principles. According to Koyré, Descartes' difficulty lay in the fact he was too much a mathematician, too given to hasty geometrization of the terms of the problem, and that therefore he did not understand Beeckman's insights into the fundamental notions of what was, after all, to become classical mechanics. Koyré's ultimate claim is that this entire episode is just another case of Descartes' tragic scientific flaw, his tendency to '*géométriser à outrance*'[79].

[79] Koyré (1939) pt 2, pp. 32–33, 37; (1978 pp.83–84). 'C'est lorsqu'il essaie de traduire les résultats de son integration (of minima motūs) en terms d'espace que, emporté par l'élan de la representation imaginative et de sa tendence à la géométrisation à outrance, il tombe dans l'erreur'. Cf Hanson (1958) 45–46, 'The point of the problem of free fall eludes Descartes'. We shall see, and indeed already have seen in regard to the hydrostatics manuscript, that Descartes' views on causation within natural philosophy were marked not by a *géométrisation à outrance*, but if anything, by a 'dynamicization' *à outrance*'—a concern with imputing forces and tendencies to bodies at particular instants in their motions or tendencies to motion.

Koyré's indictment, however, runs to further particulars. In his view, the specific mistake of Descartes the mathematician was to have failed to exploit Beeckman's 'intellectual conquest', the principle of conservation of motion (1939 pt 2, 36; 1978, 83). To Koyré, Beeckman's notion that the conservation of uniform motion does not require a cause or explanation was clearly in the line of development of classical mechanics. By reintroducing the metaphysical concept of an internal moving force, Descartes fell back into the 'impetus physics' of the fourteenth century (1939 pt 2, 36; 1978, 83). It is correct to point out the contrast between Descartes' view of mechanics and that of Beeckman. In his later work Descartes would further develop the idea that the inertial motion of a body is caused by the continued action of an internal force of motion. In fact, much of his natural philosophy and mechanistic optics will be built around the analysis of the magnitude and components of directional magnitude of the force of motion possessed by a body at each moment of its motion. By contrast, Beeckman always seems to have entertained a 'modern' concept of motion, just because he did not mention impressed or internal moving forces. Nevertheless, Koyré's view can be shown to have been doubly misguided. In the first place, as Koyré himself showed, and subsequent research confirmed, the inertial concepts of both Descartes and Newton had significant residues of notions of impetus-like internal moving forces. Beeckman may have had a modern textbook notion of inertia, but the modern view itself emerged from the tradition of mechanics in which Descartes and then Newton forged the concept with strong dynamical overtones. Therefore, it is of little conceptual or historical significance to credit the 'progressive' nature of Beeckman's ideas over those of Descartes. Secondly, and more pertinently, it is erroneous to imply, as did both Koyré and Hanson, that Descartes' so-called impetus physics was responsible for his mistakes. (Koyré 1939 pt 2, 36; 1978, 83; Hanson 1958, 48). We are about to see that Descartes' concept of a conserved internal moving force mediates between the reiterated applications of the tractive force and the consequent motion actually produced and conserved. It is a conceptual elaboration, explicating the problem of the cause of the continued motion of the body. It need not have posed any mathematical difficulties. Beeckman's 'correct' demonstration of the time-space relation can be rewritten, substituting for *momenta spatii* more involved phrases relating increments of impressed internal moving force to *minima motūs* and thence to *momenta spatii*. Nor is this surprising, since the impetus theorists of the Parisian School following from Oresme were the first to derive the triangular representation of 'uniformly difform motions' in general (Clagett 1961, 331–418). Finally, it is amusing to note that if we reflect Koyré's views back onto our reading of the hydrostatics manuscript, we can speculate that had Descartes in fact been more of a pure geometer, and less of a physico-mathematician, he probably would have left Stevin's findings in the rather more rigorous form in which he had found them!

Let us therefore look at what was actually going on in the Beeckman/Descartes interchange. What natural philosophical and specifically dynamical principles did they entertain; did Descartes misunderstand his own or Beeckman's principles; and why, in fact, did he slip up in the latter stages of the solution of the problem, if his difficulty was not some vocational and epistemological blindness about how to geometrize natural philosophical issues? After all, Beeckman and Descartes were not trying to practice a Galilean mechanics they knew nothing about, and which did not yet exist in public, and which Descartes rejected when it was eventually published. They were trying to practice physico-mathematics, and that, perforce, could only be done in the piecemeal, problem-oriented way we have been examining. The game was ongoingly to devise and revise natural philosophical concepts—hopefully corpuscular-mechanical ones—in attempts to apply them to well formed and well grounded results in the mixed mathematical disciplines. It is quite possible, therefore, that there were more immediate, indeed contingent, reasons for Descartes' slip up.

3.5.4 How and Why Descartes Hit a Pitfall

In fact Descartes does not seem to have misunderstood Beeckman's principles for describing motion and its causes. A more equitable judgment might be that they shared a common conceptual approach but that each emphasized different elements in that structure at the expense of others. Beeckman's central insight was that in accelerated motion the spaces traversed in consecutive 'moments' of reapplication of the 'tractive force' (i.e. corpuscular impacts) are as the series of whole numbers. His thought moved smoothly from the initial dynamically interesting idea that, given his inertial principle, reiterated application of force (corpuscular impacts) produces increments of motion, to the central ideas that motions produce proportional distances in unit times, and hence that the proportionality of distance and time suffices for the description of the phenomenon of accelerated fall. Beeckman's particular view of his inertial principle was both symptomatic and constitutive of his position. As we know, his principle of inertia did not involve any notion of a conserved internal moving force which 'produces' the conserved motion. In the absence of external constraints, motion, once imparted, is conserved *qua* motion. No further natural philosophical qualms about the cause of motion disturbed Beeckman. The very wording of his inertial principle tended to prevent him from formulating any difficulties about the body's 'force of motion' or its conservation. Similarly, since the inertial principle directed his attention to 'motion *per se*', it was all the easier for him to attend consistently to the space relations arising from various degrees of intention of motion.

Descartes saw much the same pattern of underlying causal natural philosophical concepts, but he paid closer attention to articulating a different sector of it. His language, both here and in the hydrostatics manuscript, shows that he was more interested in the cause of motion than in the space-time relations derivable from a knowledge of those causes. Beeckman attended to the space series arising from the concatenation of conserved and impressed motion. Descartes focused on the relationship between impressed motive force and consequent motion. In his discussion of the problem, he insisted on employing and correlating the terms '*vis*' and '*minimum motūs*'. At each moment (of time!) it was a new increment of *force* which was added to the body, and the force caused an added *minimum motūs*. Likewise, it was the force which was conserved from instant (of time!) to instant, and the conserved force acted anew at each instant to cause another *minimum motūs*. As Descartes wrote in the opening section of his problem solution, in the second instant (of time) there will be twice the motion, for '*pergit ea vis quae erat in primo minimo, et alia nova accedit illi aequalis*'. Additionally, Descartes clearly did understand the time dependency of the force. When, opening his discussion, he wrote of the addition of new force to the body, he clearly stated that the force is added '*singulis temporibus*'.[80]

It is therefore probably quite fair to conclude that Descartes understood perfectly well the contention that reiterated acts of the tractive force result in the addition of increments of motion, which are then conserved during subsequent intervals of time. This presupposes his understanding the spirit of Beeckman's principle of inertia. The issue for Descartes—and this is typical of his dynamical thinking all the rest of his life—was that he did not agree to the letter of that principle. His own criteria of natural philosophical intelligibility demanded that an extra link be added to the chain of concepts. He insisted that each application of the tractive force, or each corpuscular impact, or whatever, impressed upon the body an increment of an internal moving force which was henceforth conserved. The internal moving force in turn 'causes' the increment in 'motion'.

All this amounted to a considerable shift in focus, and it did lead, as we have seen, to different results. But, it does not in itself constitute a misunderstanding of Beeckman's principle, and is at least compatible with Beeckman's manner of 'mechanics discourse'. There is no reason to assume that Descartes could not in principle have concluded with Beeckman that the spaces traversed by a body in equal times are as the force of motion, and hence as the 'motion' it possesses during those times. And, as just noted, Descartes opened his discussion with a clear recognition that the increments of force (or motion) are to be added relative to intervals of time expressed along a linear co-ordinate. If Descartes slipped when he addressed the question of distance traveled in one hour, given the distance traveled in two hours, it was not because his principles were incompatible with getting the right answer, or because he had some cognitive bias, as a mathematician, infecting his ability to think through physical questions. Rather, it would seem that his difficulties stemmed from a particular conjunction *in this case* between Beeckman's manner of posing the problem and his already emerging ideas about the dynamics of motion, which led to his tendency to speak of 'minima of motion' rather than 'moments of space'.

[80] In the hydrostatics manuscript, where of course no actual translation takes place, only instantaneously exerted tendencies to motion, the instants obviously are instants of time not of space or distance.

At first sight it might appear a trivial matter whether one seeks the distance fallen from rest in one hour, given the distance fallen in two hours; or, the distance fallen in two hours, given the distance fallen in one hour.[81] If one commands the basic functional relationship between distance and the square of the time in subsequent classical mechanics—as taught in algebraic form, for example, in one's high school physics textbook—the two problems are structurally identical. The point is that Beeckman and Descartes did not command a nice functionally expressed kinematics—nor were they seeking one as such. Instead, they were struggling, in the name of physico-mathematics, to find some workable match between the geometrical description of the phenomenon and some acceptable understanding of the underlying causal framework. We are going to see that it is plausible to speculate that they harbored real doubts about the form of the law, as well as the likely underlying structure of causes. For them the problems need not have been trivial 'inverses' of one another. As for Descartes, the fact that he started with a somewhat different articulation of dynamics—involving consideration of internalized causal forces, expressed as instantaneously exerted 'forces of motion', rather than with Beeckman's seemingly more parsimonious principle of inertia—did not help matters, given Beeckman's posing the question of space/time relations in 'inverse' fashion.

Descartes' initial demonstration for Beeckman of the 'triangular' character of uniformly accelerated motion is quite cogent in natural philosophical terms, as far as it goes. We have seen that for Descartes the solution of Beeckman's problem required the establishment of a mathematical relationship between [summed] *minima motūs* and time: The *minima motūs* arise from instantaneous unit increments of moving *vis*, and their conservation throughout the rest of the motion; and in turn the *minima motūs*, summed instantaneously over time, produce the 'triangular' accelerated motion. He shows a firm grasp, physically and mathematically, of how to relate *minima motūs* to time. Beeckman had no trouble adapting the proof, articulating it instead around *momenta spatii* (arising from instantaneous reiteration of moving force and the conservation, via the principle of inertia, of speed acquired to that point).[82] It would seem very uncharitable to assume that just because Descartes, in his version of the proof, does not explicitly translate summed *minima motūs* into distance, that he did not, or could not, comprehend that as the needed end point of the proof, since that was what the question was about. Beeckman easily inserts the needed category. So, we may ask, What did Descartes think the summed *minima motūs* expressed over his explicitly used time coordinate, if not distance? Yet Descartes, left to his own devices with the initial proof, went on to produce his 'mistaken' final solution to the problem (and indeed we shall shortly see him repeating

[81] In this spirit Koyré termed Beeckman's question the inverse of the latter one. Koyré (1939) part II, 28; (1978) 79.

[82] Remember that Beeckman himself is very clear from the first statement of the problem that reiteration of a moving force is imparting increments of motion, expressed in the condensed locution, italicized here: 'Secundo, in hoc motu pervererando *supperadditur motus novus tractionis*, ita ut duplex spacium secundo momento peragretur.' At one level the differences of conception and expression between the two physico-mathematicians were quite small.

that pitfall in another very interesting case). So we, like Koyré and Hanson, must ask why. But, rather than positing some supposedly life-long, mathematician's cognitive bias, infecting every corner of his physical reasoning, we shall find local and contingent reasons, more perhaps in the nature of a 'craftsman's pitfalls', when working near the limits of his previously achieved competence. In this connection, we shall see that Beeckman's manner of stating the problem did help Descartes step into a pitfall with his own attempted completion of the solution.

Let us put ourselves in Descartes' place in regard to his discussion and figure relating to the final solution of the problem. Beeckman's question requires consideration of a completed motion. Descartes was asked to compute the distance traversed in the first hour of the motion, given that traversed in the entire motion of two hours. The question explicitly requires that one initially stipulate the absolute distance covered in the full two hours. That is, as would be customary, in order to gain a geometrical foothold on the problem—even if one were working on the 'back of' the proverbial 'envelope'—one is tempted to draw a line or surface representative of the total distance traversed. Beeckman, of course, did not do this. He was presented with Descartes' finished diagram for the first part of the problem, and he had a sharp awareness of the space/time relations it embodied (as arguably did Descartes in regard to *that* diagram as suggested above). Consequently, Beeckman could see that the ratio of the distances is given directly by the areas associated with the time intervals. Descartes' first diagram presented Beeckman with what he wanted to see: the summing of *momenta spatii* over the two hours; and he immediately concluded the correct ratio of times and distances. Now, the evidence—in the form of Descartes' second diagram—suggests that Descartes started again to work from scratch on the specific question, 'Find the distance traversed in one hour, given the distance traversed in two hours'. He did this, as mathematicians are wont, by drawing a new diagram, embodying the 'givens' of the problem. Having cleverly established the basic 'triangular nature' of the accelerated fall (when the causes act instant to instant) via his first analysis and diagram, he now very reasonably decided that he had better start the second part of the problem by again signifying both the total distance and the total time of fall. He had no Galileo or modern physics tutor to peer over his shoulder and kindly suggest he not thereby run the risk of conflating the two givens. Instead, the unarticulated category of a 'completed motion' helped him to conflate (or more charitably, attempt elegantly to express) time and distance as one line. This is clear from his statement of the problem to be solved, in the *Cogitationes Privatae* and the *Physico-mathematica*, and the figures appended to them.[83]

In both cases line *ab* (Fig. 3.6) is intended as a representation of a motion completed in space. Descartes next introduces the already cleverly established triangle of *minima motūs*. Thinking in his accustomed dynamical terms concerning the correlated *minima motūs* and *vis*, he was perhaps hampered from turning the argument in the direction of the space relations of [summed] *minima motūs*, or the instantaneously exerted and conserved *vires* that cause them. (As noted, he had not

[83] AT X pp. 77, 219.

bothered, in the first part of his analysis, explicitly to move to distances as the results of summed *minima motūs*.) Acting consistently with his first analysis, he then concluded, correctly in his terms, that the areas of the figure reveal the 'total force' or 'total motion', whilst continuing still to suppress their previously implied signification as spaces as well.[84] It was then that he succumbed to the crucial pitfall: he immediately assumed that the total force or total motion—represented by areas in his figure—is inversely proportional to the elapsed time. That (mis–) step was facilitated and indeed shaped by the possibility of interpreting *ab* as also a distance, because, as is obvious, the intuitive and here inviting idea of an inverse proportionality between time and 'total motion' or 'total force of motion' depends on the prior postulation of a constant reference distance. Having, as usual, thought himself into the problem in terms of forces and motions, Descartes could glance at his nicely stipulated datum, *ab*, and slip into thinking that his total motions were being referred to a fixed distance. Had Descartes, like Beeckman, eschewed an articulated dynamics of causal forces and expressed *minima motūs*, and discoursed directly in terms of *minima spatii*, he might well have remembered that *ab* is also a stipulated, given time, and gone on to take the summed *minima spatii* to signify distances traversed. But having taken the areas as summed, instantaneously acting forces or resulting *minima motūs*, he was less likely to reason the long way around, as it were, that '*ab* is a time, so that the areas really are distances'. After all, in a sense, he already 'knew' when he started the second part of the solution, that *minima motūs* denote, amongst other things, distances, for this was the point that Beeckman made explicit on the implicit basis of Descartes' first analysis. Descartes himself continued to leave that conclusion implicit, even repressed as it were, as he again deployed, with increasing confidence, his own conceptual tools, the dynamical concepts of instantaneously exerted forces and correspondingly expressed *minima motūs*, which he knew how to 'integrate' over given 'lines', in this case the easily 'double purpose', given line *ab*.

One might say that Descartes was not geometrizing *à outrance* as Koyré maintained, but rather, quite in the manner of his emerging physico-mathematics, he was 'dynamicizing' reference diagrams, *à outrance*! We are presented not with evidence of some deep, dire, essential cognitive failure on Descartes' part, due to his 'being a mathematician' not a physicist. Rather, following a Kuhnian or Ravetzian understanding of research practices as craftsperson-like activities, we should say that Beeckman and Descartes were not seeking to found Galilean kinematics before the fact, but were straining, albeit in a piecemeal, problem oriented manner, to found and articulate physico-mathematics.[85] Descartes' proclivities, commitments,

---

[84] A signification that Beeckman, of course, made quite explicit!

[85] In addition to Thomas Kuhn's well known and seriously intended metaphor of expert, problem-oriented, scientific research as craftsman's work, see the profound development of that conceit by J.R. Ravetz (1971) and the convincing articulation of the notion in very many examples of the subsequent literature on 'sociology of scientific knowledge', particularly in the works of Karen Knorr-Cetina (1981), Trevor Pinch (1985) and Andrew Pickering (1995). I deploy the idea of pitfalls looming at the research coal-face, and of initial recognition of them, followed, one hopes, by gradual, crafty finding one's way around them, in the spirit of Ravetz's deep and still useful discussion.

and just achieved success with part of the problem, combined with the manner of statement of the task to push him into a set of pitfalls and missteps. This was despite the clear fact that in the opening stage of this same project he had, in physical and mathematical terms, provided just the solution that Beeckman could easily take up and deploy, because for that purpose, and in that context, Beeckman had an advantage: Beeckman's dynamics, his causal account of motion in natural philosophy, focused on inertia and conservation of motion per se, whereas Descartes deployed a dynamics of applied and internalized moving forces, and resulting *minima motūs*. Descartes' preferred style of 'tinkering' may have caused hitches and pitfalls in this case, but in other cases he could, and would, presumably have seen considerable success, able to be credited by his physico-mathematical research associate.[86] The hydrostatics manuscript embodies an example of this—the rigorous but 'superficial' (that is, merely mixed mathematical) work of Stevin is physico-mathematically co-opted and natural philosophically explained. In the next chapter, we shall see that Descartes' physicalisation of mixed mathematical optics, on the basis of the discovery of the law of refraction, would appear to him, and to Beeckman to whom he soon reported it, a very successful piece of physico-mathematics—no slipping and sliding on pitfalls in that case. Like any 'phénoméno-technical' practice, this physico-mathematics had its robust and extendable achievements and its lame or abortive initiatives as well. To make mistakes and encounter pitfalls in a living and developing craft is one thing; to be supposedly doomed to error and failure by some innate or acquired cognitive characteristic is another. The latter seems to be the stuff of fairy tales of scientific 'heroes or villains'—the heroes 'doomed to success' by happily possessing the inverse cognitive capacities. The former is the modus vivendi of competent people pursuing and articulating a tradition of cognitive and material practice.

3.5.5 *The Physico-Mathematics of Fall Stalls—Too Many Laws, Too Many Causes, No Measurements*

Most accounts of Beeckman and Descartes' texts on fall tend to concentrate upon Descartes' 'mistake'. However, if we take seriously the embryonic project of physico-mathematics, the material not only looks quite different—as we have seen so far—but, additionally, more of it comes into the frame of interpretation. This applies especially to some fragments in the Beeckman–Descartes exchange which are often overlooked. They confirm the senses in which this was an exercise in

[86] Recent luminaries of 'science studies' have described partially overlapping dimensions of the sort of craftsman-like grappling with scientific research problems alluded to here: Knorr-Cetina (1981) discussed 'tinkering toward success'; Latour (1987) borrowed and attractively articulated Levi-Strauss's conceit of 'bricolage', whilst Pickering (1995) describes 'the mangle of practice'. It is surely better to look to these authorities for heuristic guidance in understanding Descartes' problem-solving styles and struggles, than to spin out fantasy tales of 'methodological control', or, with Koyré, tales of ingrained, congenital epistemological blockages.

physico-mathematics, and indeed will help us to form some conjectures about why it was an unconsummated physico-mathematical project. They also provide more evidence against the Koyré/Hanson reading of this material and their reasons for Descartes' 'failure'.

The main entry in Beeckman's *Journal* containing the report of Descartes' proof ends with a set of 'physico-mathematical' reflections on the proportionality of space and time in free fall. Beeckman may not have been able to provide for himself the mathematical argument he attributes to Descartes. Yet, he was fully aware of the inner structure of the argument, and how it might accord with potentially ascertainable facts about local free fall. Consider his initial, extremely interesting, reflection:

> If the minimum moment of space has a finite magnitude, there will be an arithmetical progression. But one will not be able to know on the basis of one instance of fall, how far the body will fall in each hour; rather two instances would be needed in order that we might determine the quantity of the first moment.[87]

If the *momenta spatii* are not referred to instants of time, then their summation over unit times will be represented by the sum of an arithmetical progression, as we have seen earlier. This would also mean that the cause of the motion, the terrestrial traction/corpuscular collisions, would not be virtually continuous, but would arrive at (repeated, exact) finite intervals. Beeckman was clearly showing that he understood the physical question of causation turns on the size of the time intervals of action. Accordingly, he also concluded that no single measure of the distance and time of fall would suffice to establish the time dependency of the arithmetical series of spaces—in principle two measurements would have to be taken.

We may explicate Beeckman's thought as follows, allowing of course for the total impossibility of the sorts of measurements he discusses. Consider his case of distances travelled over one hour and two hours of fall respectively. Take a case where the ratio of the distances fallen does not 'arguably' result in the ratio of 1:4, to be expected if the causal 'terrestrial tractions' (corpuscular collisions) are effectively continuous. If the ratio of measured distances instead fall out 'reasonably' close to some other ratio of whole numbers, Beeckman is saying that this would indicate the causal force acts at intervals and not continuously. In this case the arithmetical progression of spaces could be fitted to the time intervals, thus giving the period between successive increments of motion and, hence, the 'unit interval' of causal action could be determined. This is what Beeckman means when he writes that two instances are needed to determine the 'quantity of the first moment'. For example, if the distance fallen from rest could be measured after one hour and then again after two hours, and if the ratio of distances were 'reasonably' judged to be 1:3, then we would know that the 'quantity of the first moment', the unit interval of causal action is one hour. If the measured ratio were reasonably judged to be, say, 6:21, then the unit interval of causal action would be 20 minutes, for the space series at 20 minutes intervals

[87] AT X p. 61 Si vero momentum minimum spatii sit alicuius quantitatis, erit arithmetica progressio. Nec poterit sciri ex uno casu, quantum singulis horis perficiat; set opus erit duobus casibus, ut inde sciamus quantitatem primi momenti.

would be 1+2+3 (=6) in the first hour, and 4+5+6 (=15) in the second and 6:21 overall first hour to total time. Beeckman goes on to observe that as the time intervals are reduced, the ratio of the distances fallen in one hour and two hours more closely approaches 1:4. At the natural philosophical—causal—level of analysis, with the 'traction' by 'corporeal spirits', Beeckman suggests that the ratio will not sensibly differ from 1:4; because, although the spirits act through distinct impacts, there are so many of them and they act so quickly that the [finite] intervals virtually vanish.[88]

Taken as a whole this latter portion of Beeckman's discussion shows a subtle understanding of the mathematical and natural philosophical implications of Descartes' formalism—this is physico-mathematics after all. Beeckman seems to think that the causal structure behind natural fall is indeed a continuously acting force, yielding, as Descartes has shown him, a mixed mathematical law of distances being proportional to the square of the time of fall from rest. But, Beeckman is saying that empirical investigation could in principle check this, and if it were found instead that the moving force acts discontinuously but at regular intervals, those intervals, and hence the causal structure, might be able to be determined. Beeckman is showing a healthy physico-mathematical concern with the underlying causal structure **and** the resulting mixed mathematical form of the law (if there is one).

Beeckman's remarks also seem to raise three troubling realizations for the two budding physico-mathematicians: [1] they have no way of performing any such measurements; [2] *a fortiori*, they cannot be sure of the mathematical form of the law of fall, or [3] about the causal structure behind it. In short, Beeckman's remarks reveal that this inquiry into fall may be physico-mathematics, but that it consists mainly of conjecture and speculative play with possibilities. To see that these concerns are more than our own modern projections, we need to look at a little noticed portion of this exchange, where we shall see that Descartes was in his own way, and contra Koyré and Hanson, also sensitive to the physico-mathematical aims and challenges of this project, and in particular seemed also to be playing on the wide open speculative nature of the issues under [2] and [3] above. Although we cannot with certainty reconstruct the give and take of their contributions, there are some fragments from Descartes which make it seem as though at this point he, as it were, stepped in to articulate further just the problems that Beeckman's remarks had highlighted. As we are about to see, Descartes put forward further speculations about the causation of fall and hence the form of the law—speculations that serve, amongst other things, to reinforce the conclusion that neither the law nor its causal framework are likely to be determined, and perhaps that a unique and simple law does not exist in this regard, rendering physico-mathematics of this domain pointless. In effect Descartes says to Beeckman, '*You have just argued very well, and your results are not a little troubling. Indeed, the situation is even more complex, because, Isaac, a completely different causal structure may be in play, and acting either in a continuous or discontinuous manner—what then is the mixed mathematical law of fall, and how shall we make any physico-mathematical progress in this inquiry?*'

[88] Ibid.

To the short summary of the problem of fall in the *Cogitationes Privatae*, Descartes appended a curious speculation about the geometrical representation of another possible sort of accelerated motion:

> The question could be posed differently: suppose the force of attraction of the earth remains equal to that which exists in the first instant, and a new force is produced (during each subsequent instant) while the pre-existing force remains. In this case the question will be solved by a pyramid.[89]

Koyré and Hanson both took this remark as further evidence of Descartes' mathematical hauteur, his lack of concern for the physical problem of fall and his geometer's delight in posing and solving yet another problem about possible relations of time and space. For Koyré the postulation of an attractive force increasing with time is a flight of mathematical fancy.[90] According to Hanson, 'Descartes never asks about the physical possibility of this hypothesis of growing force. It is a case of geometry….' He adds that in this case Descartes has 'the velocities (sic) increase in a cube-like way'.[91]

In fact, however, Descartes was not off on some merely mathematical wild goose chase. The notion of a cause of fall increasing in intensity over time is not implausible. As budding corpuscular-mechanical natural philosophers, neither Beeckman nor Descartes had any basis for preferring a speculative explanation of a constant cause of fall, as opposed to one that increased over time. Perhaps, for example, the flow of 'corporeal spirits' becomes more dense near the earth so that a gross body

---

[89] AT X p. 219 Aliter autem proponi potest haec quaestio, ita ut semper vis attractiva terrae aequalis sit illi quae primo momento fuit: nova producitur, priori remanante. Tunc quaestio solvetur in pyramide.

[90] Koyré (1939) pt 2, p. 32 'Comment un tel accroisement de la force attractive serait-t-it possible? Descartes ne se le demande pas. En fait ce n'est pas en physicien, c'est en mathematicien pur, en pur geométre, qu'il voit le problem.' This surely will not do, however, because there was no criterion of contemporary relevance to Beeckman and Descartes permitting a distinction between Beeckman's 'physics' and Descartes 'geometry'. Neither man had any firm basis for asserting the physical reality of any particular law of fall—as we are in the process of learning. Nor was the speculative corpuscular-mechanical explanation of one law any less plausible than that of another. Descartes' law of increasing force could be 'explained' just as well as Beeckman's law of uniform periodic impulse.

[91] Hanson (1958) 45–6. As we shall see, the 'cubic relation' that holds here—provided the force acts continuously, from instant to instant, a physical matter about which we have seen Beeckman and Descartes might have doubts— is that the distance travelled (or as Descartes would say, the sum of instantaneously exerted 'minima motus') will be as the cube of the time of fall. Hanson's text reads in full: '(Descartes) proposes another possible case, one in which the attractive force grows from moment to moment. In the second moment of its fall a body is attracted with twice the force of the first moment, in the third moment with a triple force. In this solution the velocities increase in a cube-like way and not as squares. Descartes never asks about the physical possibility of this hypothesis of growing force. It is a case of geometry—one more mathematical possibility.' Apparently, behind the façade of discussing natural philosophy, Descartes was really playing a mathematician's game of altering variables and solving new mathematical puzzles. What Hanson should have said, of course, is that if Beeckman's speculations were physics—that is physico-mathematics!—so were those of Descartes.

encounters an increasing *flux* of them (corpuscular impacts per unit of exposed surface area per unit time).[92] Nor is Descartes simply saying 'try a cubic relation rather than a square one', because the cubic relation, like the previous square one, only holds if the cause of fall acts continuously not at intervals—Beeckman has just been holding forth on this very matter.

Descartes, like Beeckman, has in mind issues of causation **and** geometrical expression of a law of fall. Additionally, Descartes was indeed engaging in speculation, but, again, in a very physico-mathematically relevant way, and in a mathematical idiom deriving from classical geometry, rather than a yet to be forged analytic geometry. All this can be demonstrated by Descartes' detailed exposition of 'pyramidal motion' in the *Physico-Mathematica.* This precious text makes very interesting reading, provided we allow for the fact that, once again, Descartes repeats his previous error in conflating the time coordinate for a distance coordinate, and adds another trivial slip of the quill as well. Descartes writes, again referring to Fig. 3.6 above.

> This problem can be solved in another, more difficult way. Let us imagine the stone remaining at point *a*, the space between *a* and *b* being a vacuum. And that for the first time, for example, today at nine o'clock, God creates at *b* a force which attracts the stone, and that at successive moments he creates ever new attractive forces, equal to that created at the first moment; and that combined with the previously created forces these pull the stone ever more powerfully, and even more powerfully given that in a vacuum a thing once set in motion moves for ever; and suppose that the stone, which was at *a*, arrives at *b* at ten o'clock. If we ask how long it takes to cover the first half of the path, i.e. *ag* and how long the remainder, I reply that the stone descends through the line *ag* in ⅛ (sic) of an hour and through the line *gb* in ⅞ of an hour (sic). Thus we must make a pyramid on a triangular base and of height *ab*, and divide the whole pyramid in some way by horizontally equidistant transverse lines. The stone will pass through the lower parts of the line *ab* as much faster as these parts are contained in larger sections of the pyramid.[93]

If we wish to understand this new version of the problem and its solution, we need first to look at the force law that Descartes is proposing: At each instant a new force is added (created) which is subsequently conserved and therefore produces a new increment of motion in each subsequent moment. Whereas Beeckman's force law gave rise to a series of motions or spaces in consecutive moments as 1, 2, 3, 4…, Descartes alternative causal regime will give rise to a series in consecutive moments as 1, 3, 6, 10, 15, 21 (Fig. 3.7).

---

[92] Koyré forgets that in Newtonian physics the acceleration of locally falling bodies also increases if only in a small manner.

[93] AT X pp. 77–78. Koyré (1978) 85 Aliter vero potest haec quaestio proponi difficilius, hoc pacto. Imaginetur lapis in puncto *a* manere, spatium inter *a* & *b* vacuu; iamque primum, verbi gratia, hodie hora nona Deus creet in b vim attractivam lapidis et singulis postea momentis novam et novam vim creet , quae aequalis sit illi quam primo momento creavit; quae iuncta cum vi ante creata fortius lapidem trahat & fortius iterum, quia in vacuo quod semel motum est semper movetur; tandemque lapis, qui erat in *a*, perveniat ad *b* hodie hora decima. Si petatur quanto tempore primam mediam partem spatii confecerit, nempe *ag*, & quanto reliquam: respondeo lapidem descendisse per lineam *ag* tempore 1/8 horae; per spatium autem *gb* 7/8 horae. Tunc enim debet fieri pyramis supra basim trangularem, cuius altitudo sit ab, quae quocunque pacto dividatur una cum tota pyramide per lineas transversas aeque distantes ab horizonte. Tanto celerius lapis inferiores partes lineae ab percurret, quanto majoribus insunt totium pyramidis sectionibus.

Units of Time	Units of motion due to [a] Reapplication of prior force Or initiation of a new force	[b] Conservation of units of motion obtaining in prior unit of time	Sum
Beeckman's Law of Fall#			
First	1	0	1
Second	1	1	2
Third	1	2	3
Fourth	1	3	4
Fifth	1	4	5
Sixth	1	5	6

 # In each unit of time the magnitude of the tractive force is constant

Units of Time	[a]	[b]	Sum
*Descartes' Pyramidal Law**			
First	1	0	1
Second	1+1	1	3
Third	2+1	3	6
Fourth	3+1	6	10
Fifth	4+1	10	15
Sixth	5+1	15	21

*** In each unit of time the magnitude of the tractive force increases by a unit.**

Fig. 3.7 Descartes' and Beeckman's laws of fall compared

Now, any seventeenth century mathematician would recognize this series as the Pythagorean 'triangular' numbers, that is, numbers arising from summing the units arrayed to form increasingly larger equilateral triangles as the next figure illustrates (Fig. 3.8).

Descartes reasoned that the pattern of increase of motion over time is as ever larger Pythagorean triangles; for in the first 'moment, the motion will be as 1; in the second moment, as 3 (rather than Beeckman's 2, because of the addition of a wholly new force in that moment); in the third moment as 6 (3 units of motion conserved from the second moment, and now 3 measures of force acting) etc. Implicitly following the argument of his previous proof, he moves to the limit as the moments are reduced to instants, resulting in his realization that the total motion would be represented not by a triangle as before, but now by a triangular pyramid, whose height represents the time, and whose triangular horizontal sections would represent

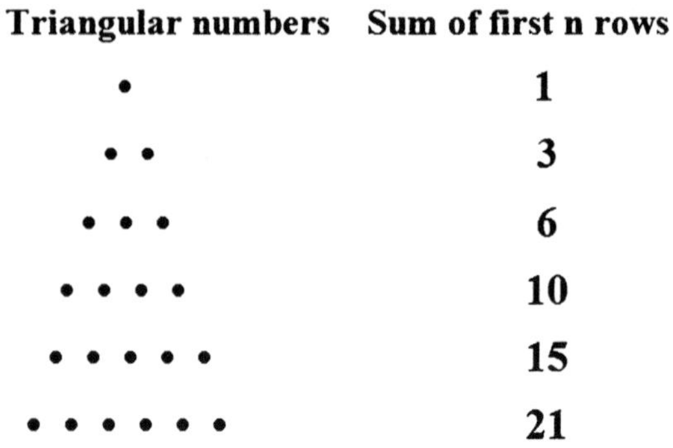

Fig. 3.8 Triangular numbers

successively larger instantaneous '*minima motūs*'.[94] At this point he again suffered the pitfall evidenced in his earlier proof. Instead of continuing to construe the height of the pyramid as the time, he slips into taking it as the distance travelled. He performs his summation of *minima motūs* over the first half and second half of this distance, and then takes the sums as inversely proportional to the time [sic] of travel over the referred distances. Since, according to Euclid, in a pyramid the volumes of the upper and lower halves are respectively 1/8 and 7/8 of the entire volume, Descartes uses these figures for the respective summed *minima motūs* and then somehow manages to attribute the faster motion to the first part of the trajectory, arriving at an inverted version of his own (pitfall marred) answer. (Although his verbal rendering in the last sentence corrects the former small slip.) If this new sort of additive causal structure acted continuously, the ratio of distances traversed over given time would indeed by as the cubes of the times. But is that all that Descartes was intending to explore and express? I think we should conclude that is not the case, if we take into account Beeckman's remarks on measurements, and the general tenor of this physico-mathematical project.

[94] Descartes' argument thus moves entirely within the confines set by the procedure of establishing an arithmetical series expressive of a force law (or causal regime) and then conceiving of a representative figure by intuitively reducing the 'moments' of application of the force to instants. He wrote in the *Cogitationes Privatae* (AT X p. 220 l.5-9) 'Ut autem huius scientiae fundamenta jaciam, motus ubique aequalis linea representabitur, vel superficie rectangula, vel parallelogrammo, vel parallelpipedo; quod augetur ab una causa, triangulo; a duabus, pyramide, ut supra; a tribus, aliis figuris.' This might at first glance seem reminiscent of the treatment of 'latitude of forms' stemming from Oresme and involving a looser kind of inquiry involving classification of types of motion mapped by reference to types of figural representations. Taking the entire exchange into account, however, it would seem that what Descartes envisions is just what we have been describing, a physico-mathematical inquiry into the modes of representation of various possible causal regimes covering natural fall. There were many possibilities, as no agreed, exact mixed mathematical law of fall had eventuated, and many causal regimes could be imagined, and geometrically represented. No closure of the physico-mathematical inquiry was reached, and it petered out in ramifying possibilities.

Descartes' text can be read, within the project of physico-mathematics, as a kind of extended response to, and articulation of, Beeckman's ruminations about causal structure, continuous or discontinuous causal action, and possible forms of the law of fall. Beeckman, assuming a cause of constant magnitude, has raised the issue of measurement to determine the 'time unit of causal action' in case the law is not arguably found to obey a simple relation between distances fallen and squares of time elapsed. That is, in case the cause of constant magnitude acts discontinuously. Beeckman in a sense 'plays' with the issue of what the time intervals of causal action are. Descartes, in his text, is playing not so much on the issue of discontinuous action of the force but the very structure of the force. Regardless of whether it acts continuously or discontinuously, it might not be constant, but rather grow linearly with time. In that case, if it acts instant to instant a 'pyramidal law' will result, yielding ratios of distances fallen as cubes of the respective times of fall. But, what if we imagine this increasing force to act discontinuously, as Beeckman had imagined the constant force to do? Here we go beyond the texts of Descartes, but one can imagine that this line of inquiry was also on Descartes' mind (in response to Beeckman). Let's explore it speculatively for a moment.

Note first of all that if Descartes' increasing cause of fall acts discontinuously, it will in principle require Beeckman's sort of 'two measurements' to determine the 'quantity of the first moment'. Now, not only would such measurements be mere pipe dreams, as in Beeckman's case, but in addition, the new, regularly increasing cause would be even harder to pin down to its unit interval structure, or to discriminate from the law arising from a (discontinuously acting) force of constant magnitude. To see why consider the series of motions or spaces in the first eight consecutive moments given by this law: 1, 3, 6, 10, 15, 21, 28, 36 compared to Beeckman's sequence of whole numbers, 1, 2, 3, 4, 5, 6, 7, 8, in the case of discontinuous action of his force of constant magnitude. Assuming one could 'catch' the unfolding moments near the beginning of the sequence, the numbers would not yield simple (to judge) ratios. For Beeckman's law the ratio of distances if one caught the 3rd and 4th moments would be 6:10, in Descartes law 10:20; or in the 4th and 5th moments 10:15 and 20:35. Assume such measurements could be made, could one discriminate between the law of constant force and the law of constantly increasing force? That applies to both forces acting in a non-continuous manner.

But, the problem runs deeper, because, in empirical terms, a continuously acting force of constant magnitude, yielding the distances as a time squared law, would be hard to tell apart from the discontinuous action of the continuously increasing force. Consider a measurement lucky enough to capture the first three moments of causal action according to the discontinuous version of Descartes' law and compare them to the ratios yielded over those times according to the continuously acting force version of Beeckman's law: The distance ratios would be 1st to 2nd moment: 1:4 (Beeckman) and 1:4 (Descartes)! A worrying identity! Or, taking the ratio of 1st to 3rd moment, the ratios of distances would be 1:9 (Beeckman) and 1:10 (Descartes); or 2nd to 3rd moment, the ratios of distances would be 4:9 (Beeckman) and 4:10 (Descartes)—hardly to be discriminated one from another using the fantasy measuring protocols that one might imagine. Backing up a step to more reasonable speculations, one can say, 'Surely actual measurements will be capturing exceedingly high numbers of 'intervals' even if they are finite, thus we might expect Beeckman's law to approximate to a law of squares and Descartes' to a law of cubes'. That is fine, but again it points out that Descartes and Beeckman, after this exchange, were faced with the problem that there might be various sorts of causal regimes accounting for natural fall; and hence various descriptive, mixed mathematical laws that might be found, if only such empirical work could be done. Even without allowing for the fact that they have not directly addressed their own idealization of the cases, 'motion in a void', they display no interest in, or commitment to, the idea of measurements, let alone the ability to carry any out.

In conclusion, three sets of reflections may be offered on this case study. First of all, as we foreshadowed at the start of this section, the physico-mathematical inquiry into fall petered out into play and speculation. There are too many possible and plausible regimes of natural philosophical causation, in continuously and discontinuously acting modes. There are too many resulting descriptive laws, laws that might well be impossible to determine one from another, even if measurement were possible. This is not a domain in which mixed mathematical practice might yield up a nice, given, simple, 'true' law to be open to natural philosophical explanation; nor is it one where a very narrow choice of possible causes is available, leading to a promising and unique geometrical regularity about which 'measurements' can be made. This is not hydrostatics, where as we have seen, Stevin's stunning and paradoxical results led Descartes, at least, to think he had made (quite radical) physico-mathematical capital; nor is it optics, where Descartes would eventually achieve profound physico-mathematical results.

Secondly, this outcome undoubtedly helped condition Descartes' cool and sceptical response to Galileo's kinematics when it appeared eighteen years later. As early as 1619 Descartes could have begun to form the opinion that the highly idealized study of fall, in search of some sort of descriptive, mixed mathematical law, was of no natural philosophical, that is, physico-mathematical import. Of course, we know that the search for and discovery of a law of falling bodies would be one of the key exemplars in the crystallization of classical mechanics during the course of the seventeenth century. But the study of falling bodies (meaning an attempted physico-mathematics of falling bodies) would play no role in Descartes' formulation of the causal register, the dynamics, that would sit at the heart of his later system of corpuscular mechanical natural philosophy. That dynamics also made a contribution to classical mechanics, but as we shall see, it would be derived physico-mathematically from important work in optics.

Finally, as was also signaled above, we have seen that Beeckman and Descartes' work on fall, viewed as physico-mathematics, begins to look rather different from how it has traditionally been interpreted. They were not simply striving for a 'Galileo–like' kinematic, or mixed mathematical law of fall. As physico-mathematicians, they were also looking for the causal explanation of such a law. This does not mean that the problem of mathematically describing fall was not important to Beeckman and Descartes. As physico-mathematicians they certainly wanted to find the descriptive law, if it existed. But simply to find and state such a law would have been to work, like Stevin, 'superficially', and without insight into causal issues, natural philosophical issues of matter and cause. They did not fail to find candidate laws, nor did they fail to find speculative candidate causes. The problem was that neither side of the equation could be well determined, so that some gain might be made toward determining the other. The physico-mathematics of fall, riven with little errors and pitfalls as it was, ended up looking like a poorly defined, or unsolvable problem from mixed mathematics. It was from mixed mathematical optics that Descartes would extract more physico-mathematical capital, hence, for our third and final case study, we turn now to his initial, halting, steps in rendering optics a physico-mathematical discipline.

3.6 A Physico-Mathematical Foray into Optics (1620)[95]

We turn now to the third case study of Descartes' early physico-mathematics. It deals with a fragment on optics and theory of light found in the 'Cogitationes Privatae' and datable from about 1620. It reads in part,

> Because light can only be produced in matter, where there is more matter there it is more easily generated; therefore, it more easily penetrates a denser medium than a rarer one. Whence, it happens that refraction occurs in the rarer medium from the perpendicular, in the denser medium toward the perpendicular.[96]

Close analysis of this fragment shows that Descartes was studying Kepler's optical masterpiece, the *Ad Vitellionem paralipomena* (1604) and that Descartes' text is a physico-mathematical 'reading' of a set of texts and figures in Kepler's work. Descartes was reading Kepler the way he had read Stevin: Seeking grist for the physico-mathematical mill, he attempted to elicit a physical theory of light, and perhaps the law of refraction, from a set of compelling geometrical diagrams and texts for refraction presented by Kepler. The 1620 optics fragment is little studied, apart from A.I Sabra's interesting speculation that it contains premises adequate for Descartes to have deduced from them his sine law of refraction of light, first published seventeen years later in the *Dioptrique* of 1637.[97]

The most important claims in Descartes' fragment are (1) that the 'penetration' of light varies positively with the density of the medium; and (2) that consequently light is refracted toward the normal in the denser medium, and away from the normal in the rarer one. It is essential to realize that in the traditional optical literature

---

[95] See Schuster (2000) 279–85, 287–89.

[96] AT X pp. 242–3: 'Lux quia non nisi in materia potuit generari, ubi plus est materiae, ibi facilius generatur, caeteris paribus; ergo facilius penetrat per medium densius quam per rarius. Unde fit ut refractio fiat in hoc a perpendiculari, in alio ad perpendicularem.'

[97] Reasons to reject Sabra's speculation (Sabra 1967) will emerge below in this section (see Notes 106, 111 and accompanying texts) and in Chap. 4 where Descartes' actual path to the law of refraction, discovered in 1626/1627, will be examined. See also Schuster (2000) 277–285.

there is no precedent for this sort of sketch physical theory of refraction. Earlier major authorities on optics, such as Alhazan, Witelo, Roger Bacon and Peckham, as well as contemporary ones such as Snel, had maintained in one fashion or another that media resist the passage of light in proportion to their densities, and that the path of motion normal to the refracting surface is the easiest or one of least resistance. From these premises opticians contrived to conclude that a ray obliquely entering a denser medium, and hence meeting increased resistance at the interface, must be refracted in a path lying closer to the easiest, normal path; and that a ray obliquely entering a rarer medium, and hence meeting decreased resistance at the interface, must be refracted into a path lying farther from the easiest, normal path.[98] Various explications were offered in attempting to link these conclusions to the premises. What one might term Kepler's 'official' qualitative theory of refraction, published in Chap. 1 of *Ad Vitellionem*, differed considerably from that of the Medieval and Renaissance perspectivists; but even he retained the stress on the denser medium weakening the incident light.[99]

It is quite obvious that Descartes' sketch theory of refraction rejects the central elements of the Medieval and official Keplerian theories of refraction. For example, refraction toward the normal in denser media in no way depends upon a weakening or obstructing of the incident light; quite the contrary, refraction toward the normal is said to depend directly upon the greater 'penetration' or 'generation' of light in denser media. *A fortiori*, there is no role for a compensating bending toward the easier, normal path, as in the Medieval theories. Nor does Descartes envision that a weakened parallel component causes the bending toward the normal, as in Kepler's official theory. So, Descartes certainly did not obtain his 1620 theory of refraction by reworking those of his predecessors. The conceptual resources upon which he was drawing are likely to have resided, if at all, in less obvious corners of the traditional optical literature. As suggested above, there is strong evidence that Descartes was reflecting upon certain parts of Kepler's work on refraction in *Ad Vitellionem*. This line of investigation was initially prompted by the concluding

[98] Lindberg (1968). On Snel's adherence to this type of conceptualization see Vollgraff (1913) 622–3.

[99] Kepler held that light is an immaterial emanation propagated spherically in an instant from each point of a luminous object. Refraction, he maintained, is a surface phenomenon, occurring at the interface between media. The movement of the expanding surface of light is affected by the surface of the refracting medium, because, according to Kepler, like affects like, hence surface can only affect surface, and the surface of the refracting medium 'partakes' in the density of the medium. He analyzed the effect of the refracting surface upon the incident light, by decomposing its motion into components normal and parallel to the surface. The surface of a denser medium weakens the parallel component of the motion of the incident light, bending the light toward the normal; a rarer refracting medium facilitates or gives way more easily to the parallel component of the motion of the incident light, deflecting it away from the normal. (The normal component of the motion of light is also affected at the surface by the density of the refracting medium, weakening or facilitating its passage, but not contributing to the change of direction). *Ad Vitellionem Paralipomena*, Chap. 1 Prop. 12, 13, 14, 20, in Kepler (1938ff) vol. II, 21–3, 26–7. I have termed this Kepler's official theory of refraction, because it is not his only articulated discussion of the causes of refraction (and their geometrical representation) offered in *Ad Vitellionem*, as we are now about to see.

portion of the 1620 fragment, not cited earlier, which discusses image places in the context of Kepler's new theory of vision, first published in *Ad Vitellionem.*[100] Examining the portions of *Ad Vitellionem* which deal with refraction, whilst bearing in mind Descartes' 'physico-mathematical' interests, brought to light two sets of passages which do seem to have provided the starting point for his curious 1620 theory of refraction.

The first and most important passage occurs in Chapter IV of *Ad Vitellionem*, where Kepler attempts to discover a simple law of refraction, by means of an analysis of its putative physical causes. Kepler asserts that there are two fundamental physical factors which any adequate theory of refraction must take into account: the inclination of the incident rays, and the densities of the media. (These points are consistent with his 'official' theory of refraction, described above.) He offers a geometrical construction representing these factors (Fig. 3.9).

Take AG incident upon a basin of water. The density of water is said to be twice that of air. Kepler lowers the bottom of the basin DE to LK so that the new basin contains 'as much matter in the rarer form of air as the old basin contained in the doubly dense form of water'. Kepler then extends AG to I and drops a normal from I to LK. Connecting M and G gives the refracted ray GM. Its construction involves the obliquity of incidence and densities of media.[101] Although Kepler then goes on to reject this construction on empirical grounds,[102] the question is, did this text speak to René Descartes, the 'physico-mathematician' and budding optician, and what did it say?

The first thing to notice is that Descartes' fragment and Kepler's text resemble one another in precisely those respects in which they are anomalous with regard to the traditional theories of refraction. Kepler's construction, like the Medieval theories and his own official theory, stresses the role of the greater density in bending rays towards the normal. But, in his figure Kepler directly represents the greater density (by lowering bottom DE) and he then utilizes that representation in an unmediated fashion to construct the refraction of the ray toward the normal. It is strongly implied that greater density is a direct cause of bending toward the normal. Kepler does not argue, as had the Medieval perspectivists, from greater density of the medium, to more resistance to the passage of light, and thence to a compensating

[100] Descartes' familiarity with Kepler new theory of vision and image formation has important implications for our reconstruction, in Chap. 4, of his later discovery of the law of refraction. Some time ago Dr Albrecht Heeffer, University of Gent, explored the Kepler/Descartes relation regarding these passages in the context of reconstructing Descartes' discovery and explanation of the law of refraction. During the course of an interesting and erudite discussion, '*The logic of disguise: Descartes' discovery of the sine law*', Dr Heefer did not cite my work (1977) and (2000; 2005) This was apparently a seminar or working paper at the University of Gent, History of Science Institute. I had the opportunity to confer with Dr Heeffer whilst he visited the HPS Unit, University of Sydney, March 2011, during which he kindly directed me to his published version of the original text, Heeffer (2006), which does cite my (1977) extensively.

[101] *Ad Vitellionem Paralipomena* Kepler (1938ff), vol. II, 81–5.

[102] Kepler (1938ff) 86.

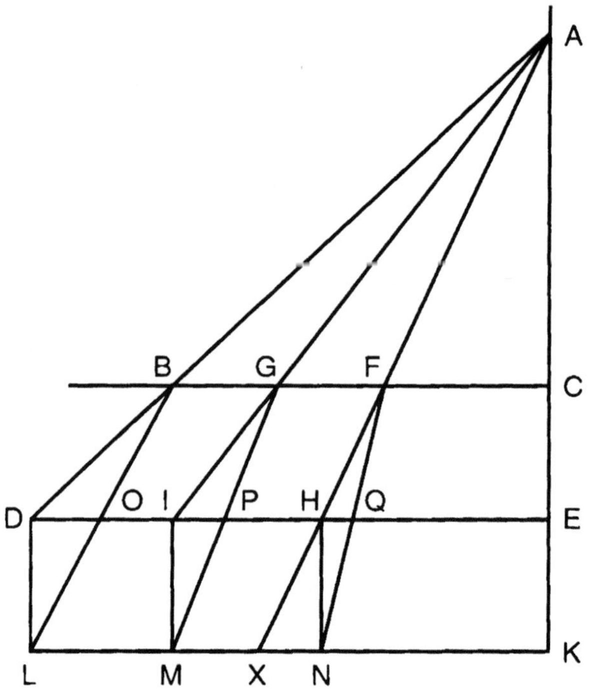

Fig. 3.9 Kepler's diagram representing the possible role in refraction of light of density of refracting medium and obliquity of incident ray. Kepler, *Ad Vitelionem Paralipomena* (1604) in (Kepler 1938) II, p. 85

bending toward the 'easier' normal path. Nor does he argue, as he had in his official theory, from greater density of the medium, to weakening of the parallel component of the motion of the light, and thence to bending toward the normal. *Descartes' fragment is peculiar in precisely this same respect.* There is no mention of a weakening of the light or of any of its components, nor of a compensating bending toward the normal. Instead, greater density is connected with greater 'generation'/'penetration', which apparently directly causes refraction toward the normal. Descartes' fragment would therefore appear to be based in some way upon Kepler's text and construction.

It is not difficult to see why Descartes, the aspiring 'physico-mathematician', would have been attracted to the non-traditional approach manifested in Kepler's text. Kepler was trying to penetrate beyond the mere phenomenon of refraction and to identify its physical causes. He wanted to represent geometrically the action of these causes and build the representations into a method of generating, by geometrical construction, the paths of refracted rays. If successful in empirical terms, this would be tantamount to possessing the sought for law of refraction of light. Descartes had already attempted to identify and geometrically represent the true causes of the paradoxical statical behavior of fluids, 'superficially' (if rigorously) mathematicized by Stevin. He probably saw Kepler's construction as a promising step toward the desired *physico-mathematization* of the problem of refraction, by both obtaining the descriptive, geometrical law of refraction and identifying its physical causes.[103]

This may explain Descartes' source and his motivation, but it does not yet elucidate the precise wording of his fragment. Here one has to be careful in teasing out the relationship between Descartes' fragment and Kepler's passage; for the fragment is not a simple verbal transcription of Kepler's construction technique (and verbal gloss), but rather an elaboration and explication of them. As we have seen, the two texts share the same anomalous posture *vis-à-vis* traditional theories of refraction. But within that broad similarity there resides an important residual difference. Kepler's construction technique does not focus upon, or work with, the parallel and normal components of the motion of the incident light or light ray. He directly represents the causally efficacious greater density of the lower medium and postulates a construction technique which uses that representation of density, and the obliquity of the incident ray, to manufacture a ray path bent toward the normal. The greater the obliquity of incidence and the farther the bottom DE has been lowered, the greater the resultant refraction toward the normal. In contrast, Descartes' fragment introduces the concept of 'generation'/'penetration' of light, which varies with density. It is the increased or decreased 'penetration' (itself the product of greater or lesser density) which causes refraction toward or away from the normal. Descartes, unlike Kepler, wishes to characterize the properties of the light or light ray itself and to insert the characterization between the talk of 'density' and of 'refraction' toward or away from the normal.

Why should Descartes have been led to view the Kepler diagram in these terms; why mention 'penetration'/'generation' at all; why not just say that greater or lesser density causes refraction toward or away from the normal? The answer would seem to be that Descartes, in interpreting Kepler's passage, was reintroducing quite customary questions about the comportment of the parallel and normal components of the motion of the incident light, or of the ray that represents it. Kepler, in other contexts in which he deals with refraction (and reflection), typically considers the comportment of these components, even though he does not always deduce changes in direction of light by (re-)composing altered components of its motion.[104] Descartes' contention that the 'penetration' of light varies with the density of the medium makes sense as a reading of Kepler's text, provided one takes Descartes to be thinking in terms of the comportment of the parallel and

[103] Schuster (1977) 336–9 and Schuster (2000) 281–285. Cf. also the problem solving techniques attributed to the young Descartes above in our analysis of the hydrostatics manuscript and the more general argument on this important issue by Sepper (2000).

[104] For example, Kepler's official theory of refraction (Note 99 above) dealt with the parallel and normal components of the motion of the light, asserting that both are weakened at the interface, whilst attributing the refraction to the alteration in the parallel component alone. In the traditional optical literature it was also thoroughly commonplace to attend to the comportment of the normal and parallel components of the motion of light when discussing its refraction and reflection.

normal components of the motion of the incident light or of the incident ray. When approached in this way, Kepler's diagram and construction technique would be taken as saying that the denser medium has the effect of increasing one or both of these components, hence causing refraction toward the normal. Only a little reflection is required to see that this in turn boils down to the claim that the normal component of the motion of the incident light increases upon entering a denser medium, while the parallel component can remain constant, increase in appropriate proportion, or even decrease.

The literal text of Descartes' optical fragment is therefore to be explained as follows. Descartes was pursuing the central idea of Kepler's passage, the direct causal role of greater or lesser density in bending light to or from the normal. But, Descartes translated that physico-mathematical insight into the customary mode of discourse about the parallel and normal components of the motion of light or of light rays, and so produced his proposition about 'penetration' varying with density. Hence, when Descartes writes of the 'penetration'/'generation' of light being directly related to the density of the medium, he is envisioning the behavior of the normal components of incident light rays. The magnitude of these components (the 'penetration') varies with the density of the medium. Increase in the normal component (with conservation or appropriate alteration in the parallel component) will bend the refracted ray toward the normal; decrease in the normal component (with conservation or appropriate alteration in the parallel component) will bend the ray away from the normal.[105] This also explains the entailment between the first and second sentences of the fragment, claimed by Descartes and first discerned by Sabra in his interesting analysis of part of this fragment.[106] Descartes can say that greater or lesser 'penetration' causes refraction toward or away from the normal, because he identifies greater/lesser 'penetration' with increase/decrease in the normal component, which can be represented in ray diagrams and used in the construction of refractions toward/away from the normal. Needless to say, given the argument of this chapter, Descartes' strategy here is pure physico-mathematics. He is eager to read a possible geometrical construction and representation of refraction back to a knowledge of its (mathematically representable) causes, and vice versa. Kepler's figure and construction may not capture the law of refraction—Kepler admits as much—but some such

[105] Again, our interpretation should be compared to Sepper's (2000) interesting thesis about how the young Descartes solved problems via strategies of figural representation. Here Descartes uses the routine representation of the components of rays to represent and articulate Kepler's interesting physical hypothesis. He was, in the language we developed above to describe his physico-mathematical practices, 'figuring up' the problem, by imposing upon Kepler's inviting conjecture and diagram the customary component analysis of rays.

[106] See above Note 97. Sabra (1967) drew attention to Descartes signalling the entailment, but incorrectly interpreted Descartes' premise concerning 'greater penetration in denser media' as applying independently of angle of incidence—thus allowing Sabra to deduce the sine law of refraction from the 'text'. We have seen that Descartes' premise applied to the normal component of the force or motion of the incident ray. Had Descartes carried out the resulting deduction, he would have arrived a tangent law of refraction.

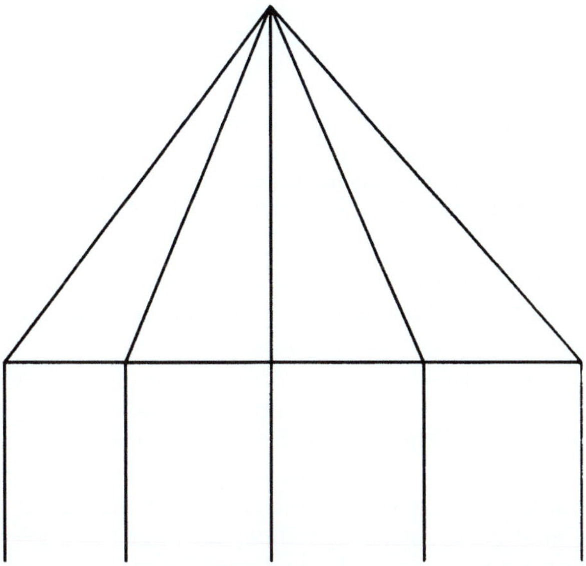

Fig. 3.10 Refraction by the most dense medium possible, after Kepler, *Ad Vitellionem Paralipomena* (Kepler, 1938) II. pp. 89–90, 107

inquiry of a similar type might. Descartes is working through what Kepler had done, in order to articulate and correct it.[107]

Our reading of Descartes' fragment—both in its detailed content and as a specimen of physico-mathematical procedure—can be confirmed by looking at a second set of passages in *Ad Vitellionem* which conditioned his thinking about the 'physico-mathematics' of refraction. Let us return to Descartes' fragment, and consider the sentence following on from the extract quoted above at the beginning of this section. Descartes continues,

> Moreover the greatest refraction of all should be in the densest medium of all....[108]

As it happens, in *Ad Vitellionem* Kepler twice considers the notion of 'the most dense medium possible', pointing out on both occasions that any ray entering such a medium will be refracted into the normal direction (Fig. 3.10).

---

[107] This tactic curiously foreshadows a similar process of adaption, criticism and modification which Beeckman and then Descartes would adopt toward Kepler's celestial mechanics in the late 1620s, a process that had its outcome in Descartes' vortex theory of planetary motion, a very seriously worked out theory indeed, as we shall learn in Chap. 10 Cf. Schuster (2005).

[108] AT x. 242–3. '...omnium autem maxima refractio esset densissimum, a quo iterum exiens radius egrederetur per eundem angulum.' In his analysis of the fragment, Sabra did not cite or discuss this remark; yet, it is of vital importance in understanding Descartes as a 'physico-mathematical' reader and-interpreter of *Ad Vitellionem*. See Sect. 4.6.

> In the most dense medium of all refractions are performed toward the perpendiculars themselves, and are equal in respect of (all) inclinations.[109]

And,

> ...if you should ponder what ought to occur in the most dense medium (or medium of infinite density), you would comprehend from the analogy of other media that, if there could be such a medium, it is necessary that all rays falling from one point onto the surface would be fully refracted, that is, after refraction they would coincide with the perpendicular itself.[110]

The context of these remarks is Kepler's official theory of refraction. The infinite density of the refracting medium destroys the parallel component of the motion of the light, leaving it only its normal component.

When Descartes echoes these passages in his fragment, the context is not Kepler's official theory of refraction, but rather the first two sentences of his own 1620 text, as we have learned to read them. Clearly, Descartes intended to present the case of the 'most dense medium' as a limiting case of the general proposition that 'penetration varies with density and causes refraction to or from the normal'. That is, when a ray enters the most dense medium possible, the normal component is infinitely (or as Descartes probably would have had it, indefinitely) increased and the ray bent into the normal, regardless of whether the parallel component suffers a finite increase, decrease or merely stays the same. If Descartes drew his limiting case from Kepler, this lends extra weight to the claim that the first two sentences of the fragment constitute a 'physico-mathematical' reading of the other passage in *Ad Vitellionem.* In sum, Descartes connected two lines of speculation present in *Ad Vitellionem* but not explicitly linked by Kepler: (1) The geometrical representation of the claim that 'the greater the density, the greater the refraction toward the normal'. And, (2) the claim that infinitely dense media would refract all incident rays into the normal. It was Descartes, not Kepler, who first related (2) to (1), using (2) to illustrate the limiting case of his own explicated version of (1), which related change in density to change in 'penetration' (normal component) to change in direction.

So, the optical fragment is a piece of highly interesting (and for the young Descartes increasingly typical!) physico-mathematics But what about the actual law of refraction? What did this physico-mathematical inquiry produce for Descartes? Well, we have now found that in the 1620 fragment Descartes embraced an assumption which would have hindered his deducing a sine law of refraction. He held that in two media the normal components of the force of light are in a constant ratio. Had he then assumed that the parallel components are constant, and attempted a

[109] Kepler (1938ff) II 107. 'In medio densissimo omnes refractiones fiunt ad ipsas perpendiculares suntque aequales inclinationibus.'

[110] *loc. cit.* pp.89–90. '...si perpendas, quid fieri debeat, medio existente plane densissimo (seu infinitae densitatis), deprehendes ex analogia mediorum caeterorum, oportere, si quod esset, omnes omnino radios ab uno puncto in superficiem huiusmodi illapsos, refringi plenarie, hoc est, coincidere post refractionem cum ipsos perpendicularis.'

physico-mathematical deduction of the law of refraction, he would have been led to a law of tangents.[111] We shall see in the next chapter that the obstacle posed by the 1620 optical fragment is critically important in reconstructing Descartes' path to the sine law of refraction in the mid to late 1620s.

Finally, we need to ask what sort of natural philosophical commitments—about matter and cause—were articulated to the physico-mathematical approach in the fragment. Descartes' physico-mathematics was meant, on the one hand, to 'physico–mathematicize' the mixed mathematical sciences—render them organic parts of natural philosophy, not subordinate, merely descriptive hangers on—and, on the other hand, it was supposed to produce from the terrain of 'to-be supplanted' mixed mathematical sciences, conclusions of natural philosophical relevance and import. To be sure, Descartes' optical fragment of 1620 makes no direct reference to a corpuscular-mechanical ontology. Indeed it appears to take a quasi-Aristotelian view of the nature of light, with Descartes writing of the 'generation' of light. (Although, if taken literally, this would imply light to be a substance rather the actualization of a potential property of the medium, as Aristotle held.) The generally Keplerian context of the fragment might suggest an underlying ontology of light as immaterial emanation. Yet, Descartes' apparent concern with quantifying the variation of 'penetration' (normal component) with density might also bespeak an unarticulated theory of light as mechanical impulse or tendency to motion. For example, in the hydrostatics manuscript of 1619, Descartes had, as we have seen, already explained gross weight as the product of summed corpuscular tendencies to (downward) motion, and he had analyzed the 'weight-producing' normal components of those tendencies.[112]

However, teasing deep and specific natural philosophical commitments out of the optical fragment of 1620 may be slightly beside the point. Descartes seems less interested in specific natural philosophical claims, or precise matter and cause discourse about light, than with generally explaining refraction in physico–mathematical terms, by relating density to 'generation/penetration' (magnitude of normal component), and expressing the relation geometrically. In so far as Descartes sought to explain

[111] Had Descartes assumed that the parallel component varies either directly or inversely with the density, he would have again deduced 'tangent laws' with slightly differing indices of refraction. There seems no way to proceed directly from the assumptions of 1620 to the sine law of refraction, unless one is prepared to introduce Newtonian complications about the variation in components as functions of the angle of incidence, a way of conceiving the problem foreign to Descartes in 1620, 1626, as well as 1637. Sabra, of course, assumed that penetration varied with density regardless of the angle of incidence, an assumption that does indeed yield the sine law, when conjoined with the assumption that the parallel component of the motion, force or penetration of the incident ray is unaffected by refraction. Sabra's error consisted in his construal of the first premise: Descartes was envisioning that the normal component of penetration varied with density. These matters are discussed in more detail below in Chap. 4.

[112] And in the case of the study of the physico-mathematics of fall, Beeckman and Descartes had both seemingly spoken a surface language of attractions and forces, arguably covering corpuscular-mechanical commitments about the causes of fall.

refraction by mathematicizing the density-penetration relation (which could have various specific natural philosophical explications), he was comporting himself as a *physico-mathematicus*. The question of how (or even whether) a corpuscular-mechanical ontology (or any other ontology) applied was pushed to the periphery, as was any unequivocal commitment about the physical nature of light. So, the optical fragment is every bit as physico-mathematical as the hydrostatics manuscript, but it eschews a definite commitment to a specific natural philosophical approach, speaking a generic matter/cause language of generation/penetration and density.

Two tentative reasons may be advanced as to why Descartes was so coy about specific natural philosophical claims in the optics fragment. First of all, as in the case of the study of fall, Descartes did not have in hand a firmly established descriptive geometrical law governing the phenomenon. So, reading back physico-mathematically from an established law to definite natural philosophical causes was not on the cards. The fragment is exploratory and preliminary, but certainly novel and pregnant with later work in optics. A second possible reason for his relative natural philosophical reticence may reside in Descartes having compared Kepler's approach to refraction with Beeckman's corpuscular speculations about the phenomenon, leading him to conclude that specific natural philosophical commitments would be premature. To explain refraction Beeckman explicitly employed his corpuscular-mechanical natural philosophy and a theory of light as the translation of light corpuscles. The macroscopic refraction of light results from a complex series of collisions between light corpuscles and the constituent particles of the refracting medium.[113] The explanation was qualitative and discursive, incapable of physico-mathematical treatment, and, if we may judge by Descartes' eager appropriation of Kepler's texts, was thought by Descartes as unlikely to lead to the discovery of the law governing refraction. Encountering Kepler's physico-mathematical approach to refraction, Descartes may well have faced a choice: either to pursue Beeckmanian qualitative corpuscular-mechanical speculations about light and refraction, or, to follow Kepler's 'obviously physico-mathematical' attempt to identify and mathematicize the causes of refraction as a step toward the discovery of the law. In the latter case a corpuscular-mechanical explanation need not have been rejected in principle, but merely deferred until such time as the law of refraction might be discovered (and indeed this is the pattern our analysis has suggested thus far.) When, in 1626–1628, Descartes did move forward in optics, we shall see that he first established a descriptive law of refraction by means of rather traditional geometrical optical techniques and then submitted the result to physico-mathematical treatment in an attempt to extract natural philosophical causes of a generally mechanistic type. But, he did not indulge in detailed corpuscular mechanical stories about light or its refraction, until he became engaged after 1628 in composing *Le Monde* and the earliest versions of the *Dioptrique*.[114]

---

[113] e.g. Beeckman (1939–53) III, 27–28.

[114] See Chap. 4, and Schuster (2000) 272–295.

Descartes' physico-mathematical encounter with Kepler's optics, recorded in the 1620 fragment, therefore probably affected his views about a physico–mathematical optics in two ways. First, it discredited, for the time being, detailed corpuscular-mechanical stories about light, media, sources and the micro-mechanics of refraction, because these eluded and obstructed attempts at mathematization. Second, at the level of physical theories of light—even quite unarticulated ones—it exerted pressure away from explicit kinematic models and toward models involving no passage of any material entity. Beeckman's kinematic fantasies were avoided, but there were permitted models of light as mechanical impulse or as tendency to motion, or indeed as Keplerian immaterial substance, or even as Aristotelian actualization of a potential property of the medium.[115]

In the end, viewing the optical fragment in the context of what we have seen in the other two case studies in this chapter, we may say that the fragment offers slim, but crucial evidence that Descartes was interested in a physico-mathematical agenda in optics, and that matters were necessarily fluid and inconclusive at the level of technical accomplishment—progress toward finding the law of refraction—and in relation to specific natural philosophical aims and valencies. In the next chapter we will be able go much further, to reconstruct the subsequent trajectory of his work in optics in the 1620s, including the next steps in the attempted physico-mathematical transformation of this traditional mixed discipline—his work on the law of refraction and its mechanistic explanations or rationales, and, his attempt to inscribe a methodological 'just-so' story of how all this work was accomplished.

3.7 Conclusion: Options, Pitfalls and Trajectories

This chapter arguably has made it clear, on the basis of the available evidence, that the hydrostatics manuscript was the key text and event in Descartes' early physico-mathematical experience. The hydrostatics manuscript exemplified the Beeckman/Descartes physico-mathematical agenda in profound ways: It showed how Descartes thought one might move a mixed mathematical science right into corpuscular-mechanical natural philosophy, and indeed much more. It held out the promise of arriving at principles of a dynamics, a causal register, for corpuscular-mechanical natural philosophy, and it seemed to forge a kind of protocol for the posing and solving of problems of physicalization of mixed mathematical findings and puzzles. The 1620 optics fragment looks as though it could have moved along similar lines, and we shall soon see that hint realized in Descartes' optical work of the later 1620s.

We have also learned that the extensive materials on accelerated fall are properly seen as exercises toward a physico-mathematisation of the domain—exercises which ran out into inconclusive results, due to multiplication of reasonable options

---

115 For Descartes' similar reaction to Beeckman's celestial mechanical speculations see Schuster (2005) and Sect. 10.3 below.

for both the law and its putative causes. Beeckman and Descartes were not Galileo, but not because of some failure of their 'scientific' or mathematical merit, but because, after some examination, accelerated natural fall did not seem to them to lend itself to solid physico-mathematical results in natural philosophy. To pretend that a kinematic law of fall did provide a royal road to natural philosophical insight—which is what Descartes thought Galileo was purporting in his great work of 1638—was to 'build without foundations', as he laconically commented at the time. Accelerated free fall was not the portal to wisdom about a physico-mathematicized corpuscular mechanism. It was to turn out that neither was hydrostatics to play that role. Rather it was optics, to which we now turn.

But, a note of caution and preparation must be registered first. The story of his next moves in physico-mathematical optics is not intended as some, albeit new fangled, linear, Whiggish tale of Descartes. His situation in 1620 was complicated and presented many options for further work. There was, first of all, as flagged earlier, the parallel work in analytical mathematics, which by late 1620 had also given birth to his overheated projects, first of 'universal mathematics' and then universal method. In physico-mathematics, optics would indeed play out fruitfully later; but, in 1620 it would not have been so clear to the budding *physico-mathematicus* what was the most promising exemplar and agenda. Moreover, Descartes was apparently not even involved in mathematical and natural philosophical matters over the next few years. When he moved back into Parisian circles in the mid 1620s, he effected his physico-mathematical breakthrough in optics, but his concerns (now under Mersenne's influence) were also unexpectedly extended to the political and religious implications of scepticism and rising radical philosophies of nature. This too would shape what he elected to select and develop out of his earlier repertoire of mathematical and natural philosophical concerns. He decided to deploy his new mechanistic optics (and a theory of perception) in the service of an attempt to articulate what his supposedly all conquering universal mathematics actually involved—all this, moreover, supposedly illustrating the power of his underlying universal method. This would answer to Mersenne's, and his own, cultural concerns. However, the resulting project, the bulk of the latter portion of the (unfinished) *Rules for the direction of the mind*, collapsed in 1628/1629. This in turn forced a detour into what we now see as the emergence in of Descartes, the systematic corpuscular natural philosopher, with *Le Monde* (1629–1633), a text which in a different way extends and articulates the impulses (and results) of his earlier physico-mathematics. In short, the future adventures of the brilliant and difficult disciple of Beeckmanian physico-mathematicized corpuscular mechanism, which we shall trace in the remaining chapters, were by no means inscribed in the early work we have reviewed in this chapter. That work, properly understood, is just the initial staging post for a surpassingly complex struggle played out over the next decade and a half, before anything like the public Descartes emerged into the wider projects and conflicts of his later life. With these caveats, then, we can turn first to the purest of his physico-mathematical feats, in optics, although even here the details of the trajectory were complex, surprising, and, for various reasons, occluded by Descartes from common view, when he finally came to publish them in his *Dioptrique* of 1637.

References

Works of Descartes and Their Abbreviations

AT = *Oeuvres de Descartes* (revised edition, 12 vols.), edited by C. Adam and P. Tannery (Paris, 1964–76). References are by volume number (in roman) and page number (in Arabic).

SG = *The World and Other Writings*, edited and translated by Stephen Gaukroger (Cambridge, 1998).

MM = *René Descartes, The Principles of Philosophy*, translated by V. R. Miller and R. P. Miller, (Dordrecht, 1991)

MSM = *Rene Descartes, Le Monde, ou Traité de la lumière*, translated by Michael S. Mahoney (New York, 1979).

CSM(K) = *The Philosophical Writings of Descartes*, 3 vols., translated by John Cottingham, Robert Stoothoff, and Dugald Murdoch, and (for vol. 3) Anthony Kenny, (Cambridge, 1988) References are by volume number (in roman) and page number (in arabic).

HR = The Philosophical Works of Descartes, vol I translated by E.S. Haldane and G.R.T. Ross (Cambridge, 1968 [1st ed. 1911]).

Other

Beeckman, Isaac. 1644. *Mathematicao-physicarum meditationum, quaestionum, solutionum centuria*. Utrecht: Petrus Daniel Sloot.

Beeckman, I. 1939–53. *Journal tenu par Isaac Beeckman de 1604 à 1634*, 4 vols, ed. C. de Waard. The Hague: Nijhoff.

Bennett, Jim. 1998. Practical geometry and operative knowledge. *Configurations* 6(2): 195–222.

van Berkel, Klass. 1983. *Isaac Beeckman (1588–1637) En De Mechanisering van Het Wereldbeeld*. Amsterdam: Rodophi.

van Berkel, Klaas. 2000. Descartes' Debt to Beeckman'. In *Descartes' natural philosophy*, ed. S. Gaukroger, J.A. Schuster, and J. Sutton, 46–59. London: Routledge.

Carteron, Henri. 1923. *La Notion de force dans la système d'Aristote*. Paris: Vrin.

Clagett, Marshall. 1961. *The science of mechanics in the middle ages*. Madison: Wisconsin University Press.

Cohen, H.Floris. 1984. *Quantifying music: The science of music at the first stage of the scientific revolution*. Dordrecht: Reidel.

Duhem, Pierre. 1905–6. *Les origines de la statique*, 2 vols. Paris: A. Herman.

Duhem, Pierre. 1906–13. *Etudes sur Leonard de Vinci*, 3 vols. Paris: A. Hermann.

Garber, Daniel. 1992. *Descartes' metaphysical physics*. Chicago: University of Chicago Press.

Gaukroger, S. 1995. *Descartes: An intellectual biography*. Oxford: OUP.

Gaukroger, S., and J.A. Schuster. 2002. The hydrostatic paradox and the origins of Cartesian dynamics. *Studies in History and Philosophy of Science* 33: 535–572.

Hanson, Norwood Russell. 1958. *Patterns of discovery*. Cambridge: CUP.

Heeffer, A. 2006. The logic of disguise: Descartes' discovery of the law of refraction. *Historia Scientiarum* 16: 144–165.

Hooykass, R. 1981. Isaac Beeckman. In *Dictionary of scientific biography*, vol. 1, ed. C.C. Gillespie, 566–568. New York: Scribners.

Jullien, Vincent, and André Charrak. 2002. *Ce que dit Descartes touchant la chute des graves: De 1618 à 1646, étude d'un indicateur de la philosophie naturelle cartésienne* Presses Universitaires du Septentrion.

Kepler, Johannes. 1938ff. *Gesammelte Werke*, ed. M. Caspar. Munich: Beck.

Knorr-Cetina, Karin. 1981. *The manufacture of knowledge: An essay on the constructivist and conventional character of knowledge and cognition*. Oxford: Pergamon.
Koyré, A. 1939, Eng. Trans. 1978. *Etudes Galiléennes*. Paris: Hermann & Cie. English trans. *Galileo Studies*. Trans. J. Mepham. Hassocks, Sussex: Harvester.
Laird, W.R. 1986. The scope of renaissance mechanics. *Osiris* 2: 43–68.
Latour, Bruno. 1987. *Science in Action*. Milton Keynes: Open University Press.
Lindberg, David. 1968. The cause of refraction in medieval optics. *British Journal for the History of Science* 4: 23–38.
Mahoney, M.S. 1981. Petrus Ramus. In *Dictionary of scientific biography*, vol. 11, ed. C.C. Gillespie, 286–290. New York: Scribners.
Mersenne, Marin. 1627. *Traité de l'Harmonie universelle*, 3 vols. Paris: Sebastien Cramoisy.
Mersenne, M. 1932–88. *Correspondence du P. Marin Mersenne*, 17 vols, eds. C. de Waard, R. Pintard, B. Rochot, and A. Baelieu. Paris, Centre National de la Recherche Scientifique.
Milhaud, Gaston. 1921. *Descartes savant*. Paris: Alcan.
Ong, Walter S.J. 1958. *Ramus: Method and the decay of dialogue. From the art of discourse to the art of reason*. Cambridge, MA: Harvard University Press.
Pickering, Andrew. 1995. *The mangle of practice: Time, agency and science*. Chicago: University of Chicago Press.
Pinch, Trevor. 1985. Towards an analysis of scientific observation: The externality and evidential significance of observational reports in physics. *Social Studies of Science* 15: 3–36.
Ravetz, J.R. 1971. *Scientific Knowledge and its Social Problems*. Oxford: OUP.
Rodis-Lewis, Geneviève. 1971. *L'Oeuvre de Descartes*, 2 vols. Paris: Presses Universitaires de France.
Rose, Paul Lawrence, and Stillman Drake. 1971. The Pseudo-Aristotelian *Questions of Mechanics* in renaissance culture. *Studies in the Renaissance* 18: 65–104.
Rossi, Paolo. 1970. *Philosophy, technology and the arts in the early modern era*. New York: Harper and Row.
Sabra, A.I. 1967. *Theories of light from Descartes to Newton*. London: Oldbourne.
Schuster, J.A. 1977. *Descartes and the scientific revolution 1618–34: An interpretation*, 2 vols, Unpublished Ph.D. dissertation, Princeton University.
Schuster, J.A. 1980. Descartes' Mathesis Universalis: 1619–28. In *Descartes: Philosophy, mathematics and physics*, ed. S. Gaukroger, 41–96. Sussex: Harvester.
Schuster, J.A. 2000. Descartes *Opticien*: The construction of the law of refraction and the manufacture of its physical rationales, 1618-29. In *Descartes' natural philosophy*, ed. S. Gaukroger, J.A. Schuster, and J. Sutton, 258–312. London: Routledge.
Schuster, J.A. 2005. Waterworld": Descartes' vortical celestial mechanics: A gambit in the natural philosophical contest of the early seventeenth century. In *The science of nature in the seventeenth century: Changing patterns of early modern natural philosophy*, ed. Peter Anstey and John Schuster, 35–79. Dordrecht: Springer.
Sepper, Dennis. 1996. *Descartes's imagination*. Berkeley: University of California Press.
Sepper, Dennis. 2000. Figuring things out: Figurate problem-solving in the early Descartes. In *Descartes' natural philosophy*, ed. S. Gaukroger, J.A. Schuster, and J. Sutton, 228–248. London: Routledge.
Shea, W. 1991. *The magic of motion and numbers: The scientific career of René Descartes*. Canton: Science History Publications.
Stevin, Simon. 1586. *De Beghinselen des Waterwichts*. Leiden, Inde druckerye van Christoffel Platin, by Françoyn van Raphelinghen.
Stevin, Simon. 1955–66. *The principal works of Simon Stevin*, 5 vols, eds. Ernst Cronie et al. Amsterdam: Swets & Zeitlinger.
Vollgraff, J.A. 1913. Pierre de la Ramée (1515–1572) et Willebrord Snel van Royen (1580–1626). *Janus* 18: 595–625.

Chapter 4
Descartes *Opticien*: The Optical Triumph of the 1620s

4.1 Genealogical Detective Work— Hints, Clues and the Problematical Text of the *Dioptrique*

This chapter reconstructs the genealogy of Descartes' discovery of the law of refraction, initial development of a theory of lenses, and first attempts to explain the law through a mechanistic theory of light. These events of the mid to late 1620s constitute the greatest of Descartes' achievements in mixed-and physico-mathematics and were also of the upmost importance for his emergence, from the late 1620s, as a systematic corpuscular-mechanical natural philosopher. He would use the discovery of the law of refraction as a putative example of his supposedly all conquering method. More importantly, the optical work led him to the mature formulation of the central concepts of his dynamics—the causal register of his emerging system of corpuscular-mechanism. That system was first embodied in the text, *Le Monde* (1629–1633), tellingly subtitled 'traité de la lumière', in which the recently polished dynamics, itself a product of the optical work, ran a corpuscular-mechanical theory of light in its cosmological setting.[1] The optical triumph of the 1620s is, from one point of view, the culmination of the physico-mathematical agenda of the young Descartes, whilst viewed prospectively, it is the exemplary basis and resource for large swathes of his mature, systematic natural philosophical work. Nothing could be more important to understand about the early career of Descartes, and nothing, with the exception of his fantasy of method, has proven so difficult and allusive to reconstruct.

The materials for this reconstruction are few and scattered, and this sort of reconstruction—especially one grounded in the realization that Descartes was a physico-mathematician leaning toward corpuscular mechanism—has not previously been attempted. For reasons that will become quite clear as we proceed, this inquiry takes

[1] The centrality of light and its action in the system of corpuscular-mechanical natural philosophy, as a set of phenomena and as an exemplar of action and explanation, will be discussed in Chap. 10.

J. Schuster, *Descartes-Agonistes: Physico-mathematics, Method & Corpuscular-Mechanism 1618-33*, Studies in History and Philosophy of Science 27,
DOI 10.1007/978-94-007-4746-3_4,

the form of a detective story. We have to start from work published much later—the *Dioptrique*, published in 1637 as one of the three '*Essais*' supporting the *Discours de la Méthode*—working back through scattered earlier hints and clues to uncover the genealogy of the discovery of the law, its application to lenses and attempted mechanistic explanations. We do have a slight head start, because we already know something about Descartes' physico-mathematics and embryonic corpuscular-mechanism, and the place that the optical fragment of 1620 holds in that enterprise. We know, for example, of his early interest in a physico-mathematized optics. We also know that at that point he did not have the law of refraction, and was working within a set of assumptions likely to hinder, rather than facilitate, its discovery. Finally, we have surmised—in, Sect. 3.6—that *in optics,* under the stimulus of Kepler, he was probably sceptical of kinematic-corpuscular models, and leaning toward instantaneous transmission (of an action or power). However, very little evidence survives between that document and the *Dioptrique* of 1637, and that evidence consists in fragments and scraps which can only facilitate a reconstruction of his discovery path, once we have found out how to decode what is going on in the *Dioptrique* itself. Therefore, that is our first problem, because the *Dioptrique* is by no means a straightforward text.

The *Dioptrique* certainly does not reveal on its surface the trajectory of Descartes' mixed and physico-mathematical optical struggles. Indeed, it has traditionally raised its own problems, puzzles and even accusations. For example, Descartes deduces the laws of reflection and refraction from a model involving the motion of some very curious tennis balls. Descartes' contemporaries tended not see any cogency in this model, nor did they grasp the theory of motion (actually his dynamics) upon which it is based.[2] These problems only further focused the question of how Descartes arrived at the law of refraction, if not through his dubious deduction. Later, suspicions were raised about whether Descartes had simply plagiarized the law from Willebrord Snel. If not, where had it come from?[3] To these contemporary and traditional problems, we may add a few more: how did the lens theory develop over time; how does this complex of optical work relate to the program of a physico-mathematics 'leaning toward corpuscular-mechanism', and indeed to his grandiose ideas of method? The *Dioptrique* offers no obvious answers to these problems, otherwise they would have been discerned long ago, and so its value to our detective story might seem rather dubious. Nevertheless, some of the puzzles of the *Dioptrique* can be resolved, and the resulting answers will in turn direct us to how to use the surviving hints and clues to construct the genealogy of Descartes, physico-mathematical '*opticien*'.

Our order of play in decoding the *Dioptrique* runs as follows: First I will show that the tennis ball model for reflection and refraction links quite coherently to Descartes' impulse theory of light *through* his dynamics of micro-corpuscles. We

---

[2] Fermat (1891–1922) t. II. 108–9, 117–24, 485–9; Mouy (1934) 55, Milhaud (1921) 110.

[3] It has long been well established that it is quite unlikely Descartes stole the law from Snel, as some contemporaries maintained. See Kramer (1882) and Korteweg (1896) pp.489–501.

have seen that dynamics mooted in his earliest physico-mathematical work, and it would be first worked out in some detail for *Le Monde* between 1629 and 1633. Nevertheless, we will also learn that the tennis ball model, even when given its proper dynamical basis, still poses a number of problems, which indeed were acknowledged at the time, by Descartes and his contemporary critics. Thus, we shall find that it is the very strengths *and* the weaknesses of the tennis ball model that provide us with further clues and tools for our main aim, the reconstruction of how the law of refraction was discovered. Given that reconstruction, we shall finally be able to explore some of the complicated relations between Descartes' geometrical optics and his attempts at mechanistic explanation in the 1620s, and we shall even be able to return to the *Dioptrique* in order to unpack the reasons for some of its allusive and misleading surface appearance. In sum, what we are after is a reconstruction of how, after 1620, Descartes' mixed- and physico-mathematical optics developed, down to the discovery of the law of refraction in 1626/1627; and how, after that discovery, he increasingly committed himself to mechanistic explanations of the law, instigating in effect a *physico-mathematical optics of firmly mechanistic tenor*, which, in turn, became exemplary for his emerging form of systematic corpuscular mechanism.

4.2 Cartesian Dynamics in *Le Monde*

This section examines the earliest articulated version of Descartes' dynamics, as offered in *Le Monde*. This will set the stage for our analysis of the tennis ball proofs in the *Dioptrique*. These will be fully explicated and consistently reduced to Descartes' actual mechanical theory of light, by means of an understanding of this dynamics. The rudiments of this dynamics of instantaneously exerted forces and determinations dates back to Descartes' earliest work in the hydrostatics manuscript of 1619, as we saw in the previous chapter. It was first fully articulated in *Le Monde*. As we shall learn in the course of this chapter, there was no straightforward evolution of clear conceptual possibilities between the embryonic dynamics of 1619 and the elaborated version in *Le Monde*. The intervening optical work, its triumphs and difficulties, embodies much of the tortuous path, from the first stirrings of a 'causal register' for corpuscular-mechanism in 1619, to the relatively mature Cartesian dynamics of *Le Monde*.

Descartes' elaborated dynamics of micro-particles in *Le Monde* had nothing to do with the mathematical treatment of velocities, accelerations, masses and forces. Rather, it was concerned with accounting for the motion, collision and tendency to motion of corpuscles. Descartes held that bodies in motion, or even merely tending to motion, can be characterized from moment to moment by the possession of two sorts of dynamical quantity: First, there is the absolute quantity of the 'force of motion'; secondly, there are the directional modes of that quantity of force; the directional components along which the force or parts of the force act. These directional modes of the quantity of force of motion, Descartes termed actions, tendencies, or most often determinations.[4] Descartes explains natural change mainly by instantaneously occurring corpuscular collisions. At the moment of a corpuscular impact, the God of the Cartesians instantaneously adjusts the quantities of force of motion and the determinations that will characterize the corpuscles concerned in the instant after the impact. God does this by following certain laws and rules of impact he has framed and 'ordinarily' follows. He, God that is, considers the force and determination relations of the two bodies just prior to impact, and upon impact God instantaneously rearranges those forces and determinations in accordance with the rules He has laid down. The laws and rules of impact are Divinely ordained prescriptions, stating what God will do about redistributing the dynamical quantities, given the conditions of the impact.[5]

Consider Descartes' first 'rule of nature' in *Le Monde*, which reads as follows:

> Each part of matter always continues to exist in the same state as long as other bodies do not constrain it to change that state. If it has a certain size, it will never become smaller, unless other bodies divide it... if a body has stopped in a given place, it will never leave that place unless others force it out; and if it has once commenced to move, it will continue along with the same force, until other bodies stop or retard it.[6]

We may take this to assert the conservation of the motion (or rest) of a body in the absence of external constraints. Closer inspection reveals a telling point. Descartes slips into speaking of the 'force of motion'. This is the quantity which is conserved. This is the force of motion we have been talking about. Descartes uses the term in relation to his Voluntarist understanding of ontology: God must continually support (or re-create) bodies and their attributes from moment to moment. This implies that in the final analysis a body in phenomenal translation, in motion, is really being recreated or continually supported at successive spatial points during successive temporal instants. In addition, and this is the key point, in each of those instants of re–creation, it is characterized by the Divine injection of a certain quantity of 'force of motion'. We should view the instantaneously conserved 'force of motion' as a kind of quantity of efficacy (the phenomenal mirror of the instantaneously injected Divine action).

The third law of motion in *Le Monde* specifies the direction in which the Divinely conserved *quantity* of force of motion is to act.[7] The force of motion is directed

---

[4] The understanding of determination used here develops work of Sabra (1967) 118–121, Gabbey (1980), Mahoney (1973), Gaukroger (1995), Knudsen and Pedersen (1968), Prendergast (1975), and McLaughlin (2000).

[5] It should also be noted that *Le Monde* itself contains a reference to the text of the *Dioptrique*, attributing the distinction between force of motion and directional force of motion to that text. AT X. 9. cf Alquié (1963) t. 1, 321 note 2. The importance of the priority of optics in the elaboration of the dynamics will emerge clearly from our reconstruction.

[6] AT xi. 38; SG 25–6; MSM 61.

[7] AT xi. 43–44: SG 29, 'I shall add as a third rule that, when a body is moving, even if its motion most often takes place along a curved line and, as we said above, it can never make any movement that is not in some way circular, nevertheless each of its parts individually tends always to continue moving along a straight line. And so the action of these parts, that is the inclination they have to move, is different from their motion (...leur action, c'est à dire l'inclination qu'elles ont à se mouvoir, est different de leur mouvement).' And,

along the tangent to the path of motion at the point under consideration. We have to be careful here. The third law does *not* say that merely a direction is conserved. Rather, it asserts that a *quantity* of force of motion is annexed to a privileged direction. That is, the law specifies a *directional quantity of force of motion*. It says that in the absence of external constraint, this directional quantity of force of motion would be conserved by God from instant to instant. This directional quantity of force of motion is, of course, that '*determination*' discussed above.[8] Let us call the directional quantity of force of motion directed along the tangent to the path of motion at a given instant the '*principal determination*' of a moving body. Following Descartes, one can decompose that directional quantity into components, also called determinations. In any given case, mechanical conditions and the spatial relations of bodies dictate which components of the principal determination come into play. We are going to see that in the demonstrations of the optical laws, the reflecting or refracting surfaces effectively dictate which components of the principal determination of a moving tennis ball come into play in the collision. The only other thing we have to remember is that determination, like force of motion, is a dynamical property predicated of moving bodies (or of bodies tending to motion), from instant to instant. Just as force of motion is injected by God from instant to instant, so is determination, which according to the third law, is only the directional magnitude of that force and the components into which it may be resolved. As God maintains or alters from moment to moment the absolute quantity of force of motion; so he also maintains or alters instantaneously the directional manifestations of that force—what Descartes calls the determinations.

Let us consider Descartes' chief example in *Le Monde* of the use of these concepts (Fig. 4.1). Consider a stone rotated in a sling. Descartes analyses the dynamical condition of the stone at the precise instant that it passes point A. By the first and third laws of motion, the force of the motion of the stone is directed along the

---

'This rule rests on the same foundation as the other two, and depends solely on God's conserving everything by a continuous action, and consequently on His conserving it not as it may have been some time earlier, but precisely as it is at the very instant He conserves it. So, of all motions, only motion in a straight line is entirely simple and has a nature which may be grasped wholly in an instant. For in order to conceive of such motion it is enough to think that a body is in the process of moving in a certain direction (en action pour se mouvoir ver un certain coté), and that this is the case at each determinable instant during the time it is moving.' (pp.29–30)

[8] In the passages discussing the third law, cited above, Descartes defines 'action' as 'l'inclination à se mouvoir'. He then says that God conserves the body at each instant 'en action pour se mouvoir ver un certain coté'. This would seem to mean that at each instant God conserves both a unique direction of motion and a quantity of 'action' or force of motion. In other words, the first law certifies God's instantaneous conservation of the absolute quantity of tendency to motion, the 'force of motion'. The third law specifies that as a matter of fact in conserving 'force of motion' or 'action', God always does this in an associated unique direction. The first law asserts what today one would call the scalar aspect of motion, the third law its necessarily conjoined vector manifestation. Just because he recognizes that some rectilinear direction is in fact always annexed to a quantity of force of motion at each instant, Descartes often slips into abbreviating 'directional force of motion' by the terms 'action', 'tendency to motion' or 'inclination to motion', all now seen in context as synonyms for 'determination'.

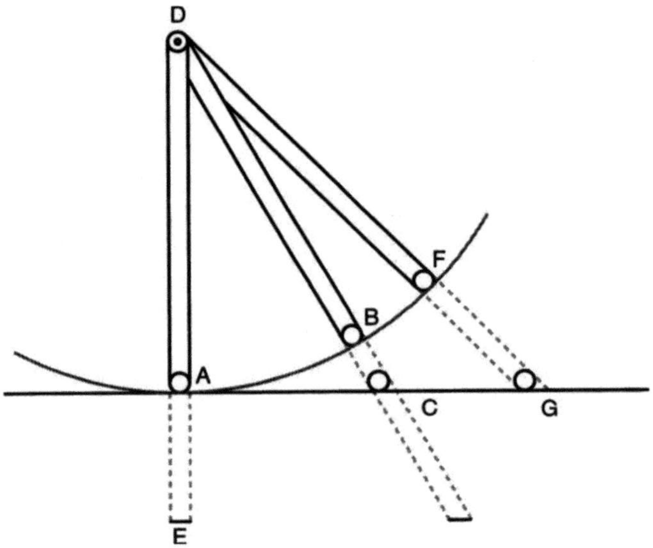

Fig. 4.1 Descartes' dynamics of the sling in *Le Monde* (Modified from AT XI, p.46)

tangent, that is along AG. If the stone were released and no other hindrances affected its trajectory, it would move along ACG at a uniform speed reflective of the conservation of its quantity of force of motion.[9] However, the sling constrains the privileged, principal determination of the stone and deflects its motion along the circle AF.

Descartes considers that the principal determination along AC can be divided into two components: one is a 'circular' determination along ABF; the other a centrifugal determination along AE. For present purposes, let us ignore the curious circular tendency. To discuss it would lead us further than we need to go into Descartes' manner of treating circular motion.[10] What Descartes is trying to do is decompose the principal determination into two components: one along AE completely opposed and hindered by the sling—so no actual centrifugal translation can occur—only a tendency to centrifugal motion; the other along the circle, which is as he says, 'that part of the tendency along AC which the sling does not hinder'.[11] Hence it manifests itself as actual translation. The choice of components of determination is dictated by the particular configuration of mechanical constraints on the system.

[9] *Le Monde*, AT xi. 45–6, 85. SG 30, 54–55; MSM 73–75, 147–151.

[10] *Le Monde* , AT xi. 85. For the sake of Whiggish edification it can be noted that had Descartes dealt with the centripetal constraint on the ball, offered by the sling, instead of the 'circular' tendency (which violates the first law in any case), he might have moved closer to Newton's subsequent analysis of circular motion. For an analysis of Newton's success and Descartes' pitfalls in dealing with circular motion, as a function of their respective theories of dynamics see Smith (2008a)

[11] *Le Monde*, AT xi. 85.

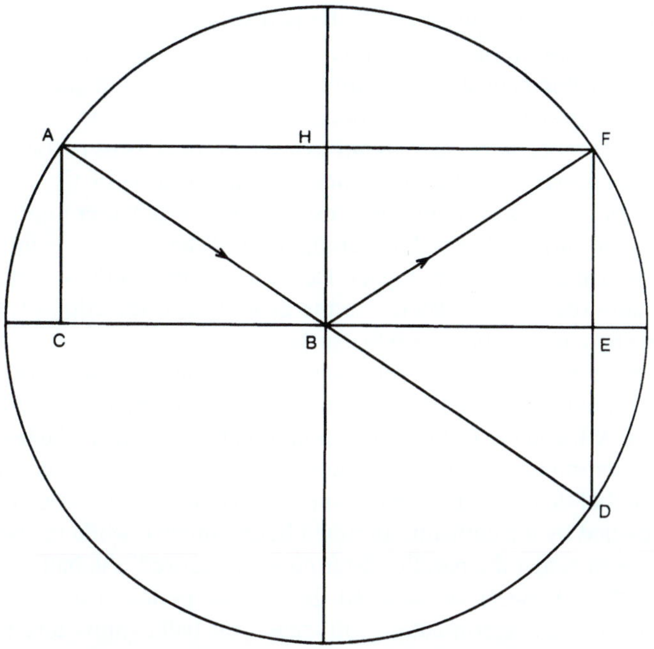

Fig. 4.2 Descartes' figure for reflection of light (Tennis Ball) in *Dioptrique*, AT VI, p.91

Now, leaving aside Descartes' theory of elements and his cosmology, his *basic* theory of light in *Le Monde* is that light is a tendency to motion, an impulse, propagated instantaneously through continuous optical media. So, in the dynamical language of *Le Monde,* light is or has a determination, a directional quantity of force of motion. Note that light, as a tendency to motion, can have a greater or lesser quantity of force—we can have weak light impulses or strong ones—but the speed of propagation in any case is instantaneous. This distinction between the force of light and its instantaneous speed of propagation is about to become very important, having been neglected for the better part of four centuries.

4.3 Making Sense of the Proofs of the Laws of Reflection and Refraction in the *Dioptrique*

We may now turn to the laws of reflection and refraction, as they are demonstrated using the tennis ball model in the *Dioptrique* of 1637. First the case of reflection (Fig. 4.2) Descartes takes a tennis ball struck by a racket along AB towards surface CBE. We neglect the weight of the tennis ball, its volume, as well as air resistance.[12]

[12] AT vi. 94.

The reflecting surface is considered to be perfectly flat and perfectly hard: upon impact it does not absorb any of the force of motion of the ball. The tennis ball is now virtually a mathematical point in motion; it bears a certain quantity of force of motion, divisible into directional components, or determinations.[13] The demonstration of the law of reflection is carried out as a geometrical locus problem. Descartes places two conditions upon the dynamical characterization of the ball. First, the total quantity of its force of motion is conserved before and after impact—no force can be lost to the surface. Second, the component of the force of motion parallel to the surface is unaffected by the impact. Descartes expresses these conditions geometrically, and uses them to determine the quantity and direction of the force of motion of the ball after impact with the surface.[14]

For the first condition, the conservation of the quantity of force of motion, we draw a circle of radius AB about B. Assume that prior to impact the ball took time *t* to travel along AB. Having lost no force of motion to the surface, the ball will, in an equal time *t* after impact, be located somewhere on the circle. The second condition is that the parallel determination, the component of force of motion along the surface, is unaffected by the collision. In time *t* before impact, while the ball traversed AB, Descartes says that the parallel determination 'caused' the ball to traverse the horizontal distance between AC and HB. In an equal period of time *t* after impact, the unchanged parallel determination will 'cause' the ball to move an equal distance toward the right.[15] We represent this by drawing FED so that the distance between FED and HB equals that between HB and AC. At time *t* after impact the tennis ball must lie somewhere on this line FED *and* it must also lie on the circle; that is it must be at F or D. The surface is impenetrable, so at time *t* after impact the ball must be at F. Geometrical considerations immediately show that the angle of incidence is equal to the angle of reflection.[16] This proof never takes into consideration the behavior of the component of force of motion perpendicular to the surface, the normal determination as we shall term it.[17]

I now propose to do something Descartes refused to do in the *Dioptrique*, even though it is perfectly feasible and follows easily in his overall physico-mathematical and natural philosophical perspective at the time. I shall translate the tennis ball proof into the terms of Descartes' theory of light, using his dynamics, taking both the theory of light and the dynamics from *Le Monde*. This is not difficult to do, because the tennis ball has already been stripped of all properties except location, force of motion and its determinations. It is already virtually a mechanical impulse, and that is all a ray of light is in Descartes' theory. *So we can assert the same things*

[13] On this interpretation of 'determination' in the *Dioptrique* see Sabra (1967) 118–21.

[14] AT vi. 95–6.

[15] AT vi. 95.

[16] AT vi. 96.

[17] Cf. Sabra (1967) 85, 110, Mahoney (1973) 379–80, and Westfall (1971) 65–6, were amongst the first scholars to appreciate this point. Previous students of Descartes' optics, such as Mach, Ronchi, Scott and Boyer, did not, as cited by Sabra (1967) 110.

about the tennis ball at the instant of impact as we would assert about a ray of light at the instant it meets a perfectly hard reflecting surface.[18] Consider in Fig. 4.2 a light ray, AB, a line of tendency to motion, or determination, impacting the surface CBE at B. The surface is perfectly hard, therefore the magnitude or intensity of the impulse is conserved. The parallel component of the impulse is unaffected by the collision.

The proof is again a locus problem. After impact, what are the orientation and magnitude of the force of the light impulse? The same two conditions apply. (1) unchanging total quantity of force of the ray; (2) conservation of the parallel component of the force of the ray. Represent (1) by a circle about A. Represent (2) by appropriately spacing FED parallel to HB and AC. Combining our conditions gives BF as the representation of the unchanged magnitude of the force of the ray and its new orientation. We have taken the diagram for the tennis ball model and re-interpreted it as a diagram about forces and determinations. This is obvious, provided (1) you attend to the very instant of impact; and (2) you take the circle and lines to represent the quantity and determination of the force of motion of the ball, as they are instantaneously rearranged in the impact. Descartes' vocabulary of 'forces', 'tendencies' and determinations is already reading the diagram that way, and later correspondence supports this, as we shall soon see. In this reading, the conceptual distance between the tennis ball model and the impulse theory of light virtually disappears.[19]

Let us now turn to the tennis ball model for the refraction of light (Fig. 4.3). Again consider a tennis ball struck along AB toward surface CBE. In this case the surface is a vanishingly thin cloth. The weight, shape and bulk of the ball are again neglected. It is taken to move without air resistance in empty geometrical space on either side of the cloth. In breaking through the cloth, the ball loses a certain fraction of its total quantity of force of motion, say one half. This fractional loss is independent of the angle of approach.[20] Again, two conditions are applied to the motion of the ball. First, the new quantity of force of motion (one half the initial amount) is conserved during motion below the sheet. Second, the parallel component of the force of motion, the parallel determination, is unaffected by the encounter with the cloth. Descartes takes the breaking through the cloth as an analogue to a surface collision, in which the parallel component is unaffected. We draw a circle about point B. Assume the ball took time t to traverse AB prior to impact. After impact it has lost one half of its force of motion, and hence one half of its speed. It therefore must take $2\,t$ to traverse a distance equal to AB. It arrives somewhere on the circle after $2\,t$.[21]

---

[18] This crucial point was first noted by Mahoney (1973) 378–9 in the course of his path breaking reinterpretation of Descartes' optical proofs in terms of relations amongst quantities and directional quantities of forces.

[19] See below note 25, and the argument in Sect. 4.4 below.

[20] AT vi. 97.

[21] AT vi. 97–8.

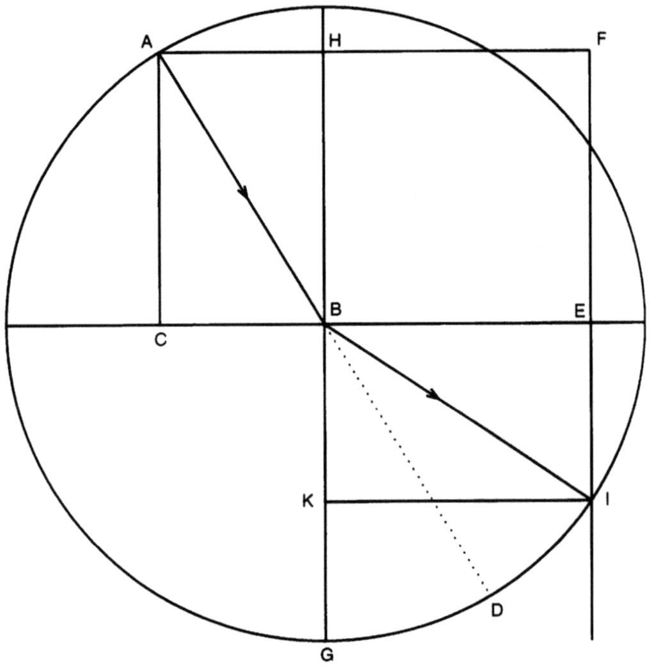

Fig. 4.3 Descartes' figure for refraction of light (Tennis Ball) in *Dioptrique*, AT VI, p. 96

Now, prior to impact the parallel determination 'caused' the body to move towards the right between lines AC and HBG.[22] But, after impact, the ball is taking *2 t* to move to the circumference of the circle, so its unchanged parallel determination has twice as much time in which to act to 'cause' the ball to move toward the right. Therefore set FEI parallel to HBG and AC, but make the distance between FEI and HBG twice as great as that between HBG and AC. At time *2 t* after impact the ball will be on the circle and on line FEI; that is, at point I, their intersection point below the cloth. The sine of the angle of incidence, AH, is to the sine of the angle of refraction, IK, as one is to two; that is as the force of motion in lower medium is to the force of motion in upper medium—which ratio is constant for all angles of incidence.[23]

Next, as we did in the case of reflection, let us sketch a proof of the law of refraction in the case of a light ray and Descartes' dynamics (Fig. 4.4). This will prove most instructive and consequential for our inquiry into how Descartes first constructed the law and how he subsequently came to design his dynamical rationale of it.[24] Consider a ray incident upon refracting surface CBE. Let length AB represent the magnitude of the force of the light impulse. The *orientation* and *length* of AB

[22] AT vi. 97.

[23] AT vi. 97–8. Descartes later supplies arguments concerning the mechanical structure of optical media to explain why light bends toward the normal when passing into a denser medium. AT vi. 103.

[24] Mahoney (1973) 379, was the first to suggest how the tennis ball model could be referred back to an imputed Cartesian dynamics in order to explicate Descartes' proof.

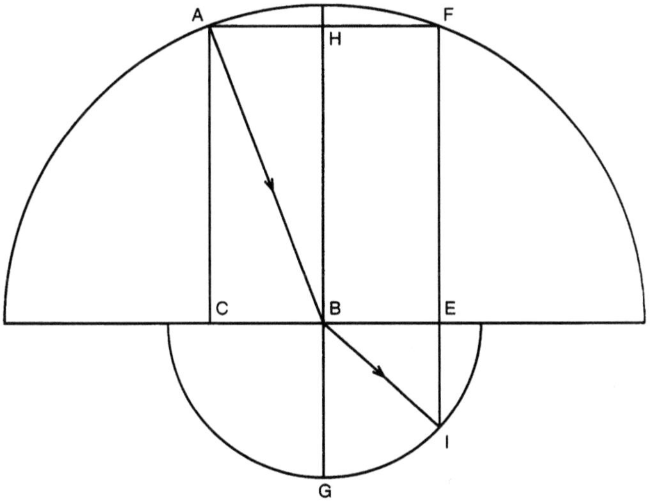

Fig. 4.4 Refraction of light using Descartes' dynamics and real theory of light

represent the principal determination of the ray. The force of the ray is diminished by half in crossing the surface, so we must draw a semi-circle below the surface about B with a radius equal to one half of AB; that is condition one. We also know that the parallel determination of the force of the ray is unchanged in crossing the surface; that is, condition two. The distance between AC and HBG represents that parallel determination. Therefore, we must set out line FEI parallel to the two former lines and with the distance between FEI and HBG equal to that between HBG and AC. Again the intersection of the lower semi-circle and line FEI gives the new *orientation* and *magnitude* of the force of the ray of light, BI and the law of sines (actually a law of cosecants) follows.

The case of the light ray (Fig. 4.4) requires manipulation of two unequal semi-circles. These directly represent the ratio of the force of light in the two media. In the tennis ball case (Fig. 4.3) we went from ratio of forces to ratio of speeds and hence differential times to cross *equal* circles. But in both cases, at bottom, we are attributing the same type of force and determination relations to the ball, and to the light ray, at the instant of impact.[25]

---

[25] It is noteworthy that Descartes himself thought about his tennis ball model proof in precisely the manner we have just used to render it in terms of his dynamics and apply it to light rays. He later wrote to Mydorge for Fermat to explain the manipulation of the speeds (forces of motion) and determinations in the tennis ball proof: (To Mydorge for Fermat, 1 Mar. 1638, AT ii. 20): 'The (principal) determination is forced to change in various ways, in accordance with the requirement that it accommodate itself to the speed (force of motion). And the force of my demonstration consists in the fact that I infer what the (principal refracted) determination must be, on the basis that it cannot be otherwise than I explain in order to correspond to the speed, or rather the force which comes into play at B.' Here Descartes views his proof in dynamical terms, as a deduction of the new refracted principal determination induced at the instant of impact with the surface, rather than in kinematical terms, as a deduction of the position of the tennis ball at a certain time after impact with the surface.

It is sometimes said Descartes fell into a contradiction, because in his theory of light, rays move instantaneously through any medium, whilst in the tennis ball model we must deal with a ratio of finite speeds. But, taking into account Descartes' dynamics and theory of light, we can now see that he had no problem: one must distinguish the *speed* of propagation of a light ray, which is instantaneous, from the *magnitude* of its force of propagation, which can take any finite positive value. The *speed* of Descartes' tennis ball corresponds not to the speed of propagation of light but to the intensity of the force of its propagation.

4.4 Descartes' Dynamical Premises: Demonstrative Efficacy and Empirical Weakness

Our analysis thus far goes some way toward vindicating the plausibility and coherence of Descartes' attempted demonstrations. Having decided in 1633 not to publish his first system of natural philosophy, *Le Monde*, Descartes offered the public in 1637 the *Discours de la méthode* and its three supporting *Essais*. The *Dioptrique* therefore appeared without the full backing of Descartes' principles of dynamics and real theory of light. Yet, we have now seen that the proofs were set up in such a way that their dependence upon the dynamics, and pertinence to the real theory of light, lurked between the lines, and hence could have been brought into the open in case of the eventual revelation of the full system. We have simply tried to read the proofs across a prior knowledge of the relevant contents of *Le Monde*. The dynamics of light which we can read out of *Le Monde* make good sense of the core aspects of the optical proofs. Using Descartes' dynamical principles, we can relate the tennis ball model back to the real theory of light, and hence vindicate Descartes of the traditional charge that the variable speed of the tennis ball bears no analogy within the real theory of light. We have also seen that recent interpreters are correct to interpret 'determination' as a coherent dynamical concept, denoting the directional magnitude of the instantaneously exerted force of motion. There are, however, definite limitations to this procedure of interpretive vindication. Even in our interpretation many problems surround Descartes' presentation, and the analysis of these problems is going to provide some signposts, both for the reconstruction of Descartes' route to the law of refraction and about its manner of 'demonstration'.

The difficulties with Descartes' theory of refraction arise from the very core of his presentation, from the two principal dynamical premises used in deducing the law of refraction. One may formulate his premises as follows:

1. For any two optical media, the quantity of the force of light in the upper 'incident' medium bears to the quantity of the force of light in the lower 'refracting' medium a constant ratio, characteristic of the two media and independent of the path of propagation, or

$$\frac{|Fi|}{|Fr|} = \text{const}$$

where |Fi| is the quantity of force of light in the upper medium and |Fr| the quantity of the force of light in the lower medium.

2. The component of the determination of the force of light parallel to the refracting surface is unaffected by the refraction of the ray, or

$$|\mathrm{Fi}|\sin \mathrm{i} = |\mathrm{Fr}|\sin \mathrm{r}$$

Combining (1) and (2), we obtain, following Descartes[26]

$$\frac{\sin \mathrm{i}}{\sin \mathrm{r}} = \frac{|\mathrm{Fr}|}{|\mathrm{Fi}|} = \frac{1}{\mathrm{const}}$$

We have seen that these premises can be grounded in Descartes' dynamics; that they mesh with his real theory of light as an instantaneously transmitted mechanical impulse; and that they allow a plausible deduction of the law of refraction in an idealized case, in which a vanishingly thin sheet, separating two void spaces, refracts an incident tennis ball, which, for all practical purposes, has been reduced to a point localization of an instantaneously exerted quantity and directional magnitude of force. But, although the premises work well in this limited and idealized context, as soon as one considers more complex and less idealized cases, they begin to reveal certain problems of empirical plausibility and logical consistency. Not only were these problems in principle capable of being detected by Descartes and other knowledgeable contemporaries, but in fact, as we shall now see, they were.

To put the matter in a nutshell, when one considers real space-filling media, Descartes' first dynamical assumption—path independent ratio of the force of light—seems to entail that optical media are isotropic, whilst the second dynamical assumption—conservation of the parallel determination—seems to entail that they are not. We are about to see that Descartes was aware of some of the difficulties consequent upon so construing the premises, and that he tried both to finesse and ignore them, whilst holding firm to the premises themselves. His determined investment in premises which permit derivation of the law of refraction, yet which are so empirically questionable in themselves, can provide us with clues about how and when the law was originally discovered and why the premises were devised. We shall first look at these difficulties in an abstract and slightly 'Whiggish' fashion,

[26] This derivation merely reworks Sabra's well known analysis of Descartes' demonstration. (Sabra 1967, 97–100, 105–6, 116.) The only difference is that here we deal with quantities of *forces* and their directional components (determinations), rather than with quantities of *speed* and their directional components, as Sabra did. The reason is that we have insisted upon the centrality of the former concepts for Descartes, and we have argued that Descartes could reduce phenomenal speeds to instantaneously exerted quantities of force of motion, so that speeds and tendencies to motion could be treated under the same conceptual and geometrical framework. We shall return to Sabra's analysis below in Sect. 4.6, concentrating on his contentions about the timing of Descartes' discovery and its possible relation to the optical fragment of 1620, discussed in, Sect. 3.6.

and then show how they manifested themselves in Descartes' articulation of his theory of refraction.[27]

At first sight Descartes' assumption (1) would seem to entail that optical media are isotropic, for the force ratio depends only upon the nature of the media and is independent of the incident and refracted paths of the tennis ball or light ray. The most superficial examination of assumption (2), however, shows that this must be an oversimplification. Assumption (2) maintains the conservation of the parallel component of the principal determination before and after refraction, and hence it entails that in refraction all dynamical changes affecting the ball or the ray in fact come about through variation in the normal component of the incident principal determination. Of course Descartes' proofs assign no quantitative or geometrically constructive role to the comportment of the normal component: the locus problems are solved using only the absolute quantities of force and the parallel components of the determination (laid off by lines normal to the refracting surface). Clearly, then, assumption (2) entails that Descartes' implied sense of 'isotropic' must differ from ours. His 'isotropic' media effect changes in the normal components of the determination of the incident ray which are complicated functions of the angle of incidence, while they leave the parallel component untouched.

Assumption (2), which raises difficulties for the isotropic character of optical media suggested by assumption (1), also generates some empirical implausibilities when considered on its own. While one can perhaps intuitively grasp how a vanishingly thin sheet might affect only the normal component of the incident determination, is this really plausible in the case of real space filling media? In such media, collision with the surface may well affect only the normal determination; but, what about the ball's or ray's subsequent penetration of a finite thickness of the medium? Would not the ball or ray now encounter altered conditions of motion (or of tendency to motion) in the direction parallel to the surface? If (1) really entails that media are isotropic in some sense, then the parallel component must be affected in precisely the same way as the normal component. So, depending upon how one views Descartes' implied notion of isotropic media, his assumptions are either contradictory or simply wildly implausible in an empirical sense: either (1) entails our notion of isotropic media while (2) denies it; or (1) entails path dependent variations in the normal component which are then most implausibly denied to the parallel component by (2) in the case of space filling media.

Returning to the *Dioptrique*, one finds that Descartes began to encounter difficulties reflective of these deeper problems, as soon as he moved beyond the case of the thin sheet separating two void spaces. When he turns to space filling media, Descartes harks back to Fig. 4.3 in which he now takes CBE to be the upper surface of a volume of water. He argues that if the tennis ball loses, as before, one–half of its force of motion in encountering the surface, then the derivation of the new

[27] We take it that in the spirit of Bachelard's epistemological and historiographical conception of *récurrence*, such analytical Whiggism is not at all a thing to be avoided. Cf. Gaukroger (1976) 229–34.

refracted principal determination will also follow as before and the ball will be refracted toward I.

> ...first of all, it is certain that the surface of the water must deflect it toward there in the same way as did the cloth, seeing that it takes from the ball the same amount of its force, and that it is opposed to it in the same direction.[28]

So, as one expects, refraction is still held to be an interface phenomenon, the new principal determination being set at the instant the ball encounters the surface, by the alteration of the quantity of force of motion, conjoined with the conservation of the incident parallel determination. It makes no difference, Descartes next argues, that the ball, after refraction, passes through a real, dense volume filling medium, for the medium is isotropic in the sense that *it offers the same resistance to the passage of the ball, regardless of the angle of path 'set' by the refraction at the interface.*

> Then, as for the rest of the body of water that fills all the space between B and I, although it may resist the ball more or less than did the air that we assumed to be there before, this is not to say that because of this it must deflect it more or less: for it can open in order to permit it passage, just as easily in one direction as in another, at least if we always assume, as we do, that neither the heaviness or lightness of this ball, nor its bulk, nor its shape, nor any other such foreign causes changes its course.[29]

Descartes apparently expects readers to accept that by appealing to the isotropic character of the medium, he can thus separate the setting of the refracted determination, at the moment of encountering the interface, from any mechanical effect the ball might undergo in passing through a finite thickness of the medium.

Descartes' strategy here seems to be to preserve at all costs the locus construction in Fig. 4.3, centering on the circle AHF and the lines AC, HB and FE, the representations of his two central assumptions. He fails to explain why the parallel component should be conserved during the passage of a finite thickness of the medium, and simply tries to persuade us that since media are isotropic in the Cartesian sense, whatever determination is set at the interface will be preserved within the medium. It was quite feasible for a contemporary reader to question Descartes' implied concept of isotropic media as both ad hoc and empirically implausible. In 1640 Père Bourdin explicitly questioned why the ball, in entering the water, is not retarded in moving from left to right, just as it is retarded in moving from high to low. Descartes' less than edifying response was that he had already dealt with this problem in the *Dioptrique* when he considered refraction through a thin sheet (sic):

> ...in order to show that it does not occur in the depth of the water, but only on its surface; and... that it is necessary to consider only the determination of the ball (*ver quel côte se détermine la bâle*) upon entering the water, because afterwards, whatever resistance the water exerts upon it will not change its determination.[30]

---

[28] AT vi. 98.

[29] AT vi. 98–9.

[30] to Mersenne, for Bourdin, 3 December 1640 AT iii. 250. Bourdin (1595–1653) an almost exact contemporary of Descartes, was a Jesuit, lecturing, since 1635, on natural philosophy and mathematics at the College of Clermont. He had most likely attended La Flèche during the time Descartes had been there, and had taught there from 1618 (Clarke 2006, 194).

This adds virtually nothing to the argument in the *Dioptrique*, and it in no way justifies Descartes' premises or answers Bourdin's penetrating query. What is at issue is, How can it possibly be, given Descartes' premises, that refraction does in fact only occur at the interface? Descartes' answer amounts to the claim that since, in fact, refraction occurs only at the interface, his premises explaining refraction must surely be adequate to that fact. And, indeed they are, if only one conceptually separates consideration of the causes of refraction at the interface from the effect upon the ray of the isotropic character of any finite thickness of the medium. Hence, we are forced to the following conclusion: The cash value of these maneuverings can only have been the staunch defense of the premises as such, and of the construction and demonstration which they ground.[31]

The difficulties posed by the two premises emerge more subtly when Descartes deals in the *Dioptrique* with the case of refraction toward the normal. In the tennis ball model the racket is taken to strike the ball again at the moment of incidence, thus increasing its speed, or quantity of force of motion, in a given ratio to the incident speed.[32] Commentators have often noted the sheer ad hocness of this strategy, as well as the even more damaging point that, in the real theory of light, there is virtually no analogue for this providentially adjusted stroke of the racket. But, it is less the ad hocness of the argument which interests us here than the deeper conceptual embarrassments of which it is merely a symptom. Note that according to Descartes' theory, the second stroke of the racket must act in the normal direction, for there can be no alteration in the parallel component of the determination. *This means that depending upon the angle of incidence, the racket acts in the normal direction to increase the normal component in such a manner that, as a consequence, the overall absolute quantity of force of motion is increased in just the prescribed ratio.* Descartes could hardly have failed to realize this, since it is an immediate consequence of the explicitly stated portion of his theory. However, he astutely avoided a clear indication that the racket must act in the normal direction (much less that its normal action is a function of the angle of incidence).

> But let us make yet another assumption here, and consider that the ball, having been first of all impelled from A toward B, is impelled again, once it is at point B, by the racket CBE which augments the force of its movement by for instance one-third, so that afterwards it can make as much headway in two moments as it previously made in three. This will have the same effect as if the ball were to meet, at point B, a body of such a nature that it could pass through the surface CBE one-third again more easily than through the air.[33]

Descartes' form of words is designed so as not to reveal to the reader the deeper consequences of the theory. His concern was well justified, because these consequences attach as well to the previous case of refraction away from the normal.

[31] Descartes is tacitly appealing on the empirical level to an indubitable fact: when dealing with a pair of homogenous media, refraction is an interface phenomenon. His dynamical premises are consistent with this fact, but they cannot be consistently articulated so as to allow the deduction of this fact, and this fact only.

[32] AT vi. 99–100.

[33] *ibid.*

Although the metaphor of penetrating a thin sheet tends to hide the relevant dynamical considerations, it remains the case that the *loss* of force of motion in a fixed ratio to the incident force of motion can only be accomplished on Descartes' premises through a path dependent decrease in the normal component of the incident determination. Descartes was and remained unwilling to bring these consequences into the open, for they threatened the plausibility of his central assumptions, and their presumed ties to his larger views on dynamics and the real theory of light. By what Cartesian mechanical means, after all, is such a path dependent variation in normal component to be effected, in the case of the decrease *or* increase of the incident force of motion? And if such a path dependent variation in normal component must occur, why then, to resume the earlier critique, does this not also occur in the parallel direction in the case of penetration of a finite thickness of the 'isotropic' refracting medium?

In sum, Descartes' two dynamical premises permitted a plausible deduction of the law of refraction, but they generated what seemed to some of his readers, and arguably to Descartes himself, to be crippling difficulties. His theory deals poorly with volume filling media, with refractions toward the normal, and more generally with the question of how it happens that the alteration in the normal determination is variable, depending upon the angle of incidence. Indeed, virtually the only strength of Descartes' central assumptions resides in their pleasing ability to rationalize the geometrical steps in his construction of the path of a refracted ray or ball. Descartes was willing to try to ride out likely accusations that the premises are empirically implausible, dynamically ad hoc, and in some interpretations, logically inconsistent, because the premises provided elegant and more or less convincing rationalizations for the geometrical moves in his demonstration. All this suggests that Descartes did not obtain his premises through a deep inquiry into the conceptual and empirical requirements of a mechanical theory of the propagation and refraction of light. It seems more plausible to associate the premises closely with the very geometry of the diagrams in which Descartes depicts and constructs the paths of refracted rays—as we have seen him doing here in the *Dioptrique*, once we understand the underlying dynamical rational of his proofs. The issue then turns on whether the premises are post-facto glosses of geometrical constructions arrived at in some other way; or, whether the diagrams themselves were invented to illustrate previously held dynamical principles concerning the behavior of light. In the following sections it will be suggested that the former hypothesis is the more likely. In particular it will be argued [1] that although Descartes held a number of unsystematized and abortive ideas about the mechanics of light as early as 1620, he discovered the law of refraction independently of any mechanical assumptions and through a process entirely within the bounds of a traditional mixed mathematics approach to optics; and, [2] that it was the geometrical diagrams expressing his newly found law which suggested to him a physico–mathematical insight into the precise form and content of his two dynamical premises and their mode of relation in explaining refraction.

In other words, having discovered the law of refraction at the level of a descriptive geometrical result in mixed mathematics, he worked back, in the style of physico-mathematics, reading underlying natural philosophical causes out of features of the diagram geometrically expressing the new found law. This is what in 1619 he had done in hydrostatics and had tried unsuccessfully to do with a law of falling bodies. It was also what he had abortively explored in the physico-mathematics of refraction in 1620. The discovery of the law of refraction and extraction of its dynamical rationale was to be his greatest moment as a *physico-mathematicus*. To understand it, we must first reconstruct how the descriptive law was found within the practices of traditional geometrical optics; and then how Descartes moved to extract causal insights from that discovery. It is this reconstruction that will ultimately explain the puzzle of why Descartes was so focused on keeping the premises, despite their vulnerability and dubiousness upon articulation: why, in short, he defended the premises at all costs, granting them special status. It was not just because they allowed 'deduction' of the law of refraction. It was also because, physico-mathematically, they had come from the well grounded mixed mathematical law! That was their ultimate warrant and import. Descartes was no fool, and he knew well what his physico-mathematics was supposed to produce, where it had seemed to work and where not. This is why seemingly obvious objections to his finished product were waved aside—after much trouble he had cracked an acclaimed classical problem, in a physico-mathematical way. He was justified in this procedure, whatever sniping was done by superficial critics, uninitiated into the ideal, and practice, of physico-mathematics.

4.5 Descartes' Route to the Law of Refraction 1619–1627

In this section we turn to the discovery of the law of refraction. As indicated above, our unearthing of the dynamical framework of the optical proofs will ultimately aid our detective work.

4.5.1 The Mydorge Letter of 1626/1627

Thomas Harriot discovered the law in exact form around 1598 and Willebrord Snel, who died in 1626, discovered it sometime after 1620.[34] Descartes, working with Claude Mydorge, discovered it in 1626/1627. The chief document supporting this conclusion is a letter from Mydorge to Mersenne.[35] It is well known to students of

[34] Lohne (1963, 1959), Vollgraff (1913, 1936), deWaard (1935–6). Here and throughout, 'exact form' of the law of refraction means not allowing for chromatic dispersion, they are working with and articulating assumption that all light rays are refracted in exactly same manner at a given interface. For the sequel, see Dijksterhuis (2004).

[35] Mersenne (1932–88) I 404–415.

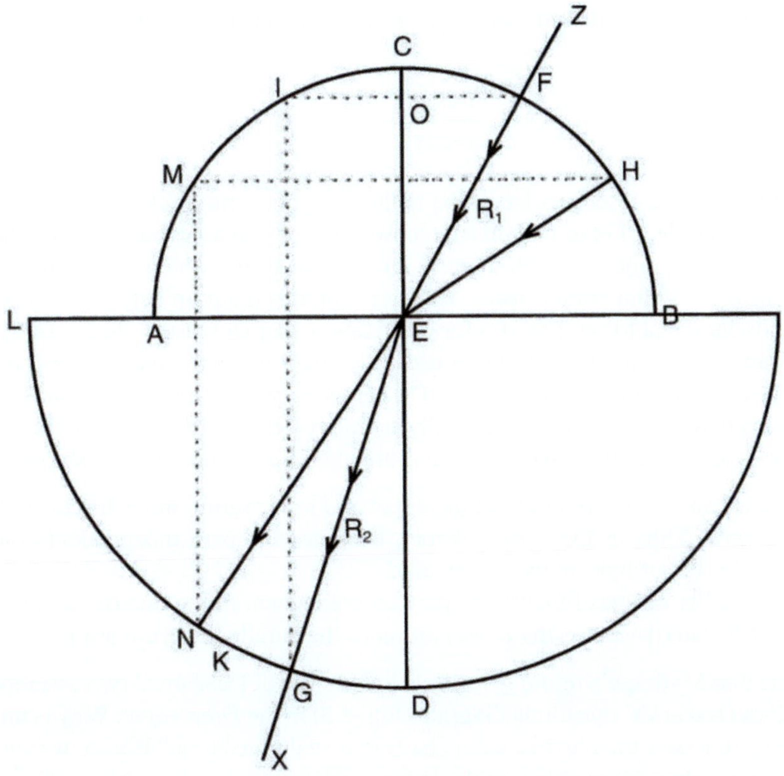

Fig. 4.5 Mydorge's refraction prediction device, Mersenne (1938–88) I, p.405

seventeenth century optics, but I suggest that it has not yet been properly understood. That depends upon its dating, and the dating depends upon its content.

Mydorge's first claim is that if he is 'Given the inclination and refraction of any one ray at the surface of any refracting medium' he can 'find the refraction of any other ray incident on the same surface.'[36] This is Mydorge's procedure: (Fig. 4.5) Ray ZE is refracted at surface AEB, along EX. Draw a semi-circle above AEB cutting the ray at F. Draw FI parallel to the surface. From I, where FI intersects the semi-circle, drop IG perpendicular to the surface until it cuts the refracted ray at G. Then with radius EG draw another semi circle about E, this time below the surface. This figure now permits the construction of the refracted path of any other incident ray, say HE. Draw HM parallel to the surface cutting the upper semi-circle at M. Drop MN normal to the surface until it meets the lower semi-circle. Connect E and N, then EN is the refracted ray.[37]

[36] *loc cit.* p.404.

[37] *loc cit.* p.405.

Mydorge observes that the law is given here as a law of cosecants. That is, taking the first ray

$$\frac{\text{cosec i}}{\text{cosec r}} = \frac{\text{R1 / OF}}{\text{R2 / OI}}$$

since OF = OI, the cosecants are as the radius of upper semi-circle is to the radius of lower semi-circle.[38] Let us call this the 'cosecants' or 'unequal radii' form of the law of refraction, compared to Descartes' *Dioptrique* form, which we shall call the 'sine' form or 'equal radius' form. We have seen this diagram before—it is mathematically identical to our Fig. 4.4 for refraction, using Descartes' theory of light as an instantaneous impulse. Mydorge uses two conditions to calculate the refracted ray. They are the same conditions that Descartes uses in his theory of light. The difference is that Mydorge states them only as rules of geometrical construction, while Descartes also gives them a dynamical rationale. The two conditions of course are:

1. the constant ratio of the radii of the upper and lower semi-circles for all angles of incidence. This, in Descartes' theory, becomes the path independent constant ratio of force of light in the two media.
2. The equality of lines FO, OI, the parallel component of the line representing the ray. This later becomes the conservation of the parallel determination of the ray.

Note that Mydorge's figure gives a clearer picture of Descartes' two assumptions than does Descartes' one circle diagram (Fig. 4.3) in the *Dioptrique*. Why is this so? And why did Descartes invoke tennis balls in actual translation? Before we can find out, we must date the *material* in the letter.

4.5.2 Lens Theory and the Date of the Material in Mydorge's Letter

Descartes' earliest recorded statement of the *sine* law of refraction dates from a report to Isaac Beeckman in October 1628.[39] Descartes consistently identified 1626/1627 as the crucial period for his optical studies.[40] He collaborated with Mydorge in that period, and Mydorge credited Descartes with the discovery of the law.[41] De Waard dated this letter from 1626, but that was merely a conjecture based

---

[38] loc. cit. p.406

[39] AT x. pp.336ff; also Beeckman (1939–1953) fol. 333v ff.

[40] Descartes repeatedly mentioned that during this period he recruited Mydorge and the master artisan Ferrier in an attempt to confirm the law and construct a plano-hyperbolic lens. Eg. Descartes to Golius, 2 February 1632, AT i. 239; Descartes to C. Huygens, December 1635, AT i. 335–6.

[41] In addition to the material cited in previous note, see Descartes to Ferrier, 8 October 1629, AT i. 32; 13 November 1629, AT i. 53ff; Ferrier to Descartes, 26 October 1629, AT i. 38 ff. In the mid 1620s Mydorge annotated Leurechon's *Récréations mathématiques*, a popular work dealing with mathematical tricks and fancies of a natural magical character. Leurechon's work was first published anonymously in 1624 and reprinted several times thereafter with additional notes,

on this collateral evidence.[42] Costabel, Shea and others date the letter from 1631 at the earliest.[43] But, evidence in the letter concerning the presentation of the law and the development of lens theory, *strongly* suggests this *material* is from 1626/1627, and is contemporary with the initial construction of the law and first articulation of lens theory.

After presenting the cosecant form of the law, Mydorge outlines a theory of lenses clearly antecedent to the theory of lenses offered in the *Dioptrique*. The key difference is that Mydorge does not initially use the sine law in constructing lens theory. Rather, starting with the cosecant form of the law, he only strikes a sine formulation in the course of his opening analysis of the anaclastic problem: it is a simple matter of adding a few lines.[44] He does not seem to know the sine form before that constructive maneuver. *Then* he deploys the sine form in the following synthetic demonstrations.[45]

Moreover, Descartes own synthetic lens theory demonstrations in the *Dioptrique* differ from those of Mydorge in another historically revealing way. Mydorge had set up the sines of the angles of incidence and refraction by reference to a semi-circle on one side of the interface.[46] In the *Dioptrique*, as we have seen, Descartes directly relates the sines to their respective rays.[47] Isaac Beeckman seems to have been the author of Descartes' more 'natural' representation of the sines. In October 1628 Descartes asked Beeckman to prove the refractive properties Descartes claimed for the hyperbola. Beeckman's proof is geometrically identical to Descartes' figure in the *Dioptrique* and was 'approved' by Descartes.[48] At the same time Descartes showed Beeckman an elegant proof for the ellipse case.[49] However, Descartes did not use that proof in the *Dioptrique*, probably because the sines of incidence and refraction are not related to their respective rays in the obvious way Beeckman achieved for the plano-hyperbolic case.

---

including those by Mydorge. I have consulted (Jacques Ozanam) *Les Récréations Mathématiques... Premierement revu par D. Henrion depuis par M. Mydorge* (Rouen 1669). Mydorge notes concerning the nature of refraction 'Ce noble sujet de refractions dont la nature n'est point esté cogneue n'y aux anciens, n'y aux modernes Philosophes et Mathematiciens iusque à present, doit maintenant l'honneur de sa découverte à un brave Gentilhomme de nos amis, autant admirable en scavoir et subilité d'esprit.' p.157.

[42] DeWaard admits that the copy he examined dated from 1631 at the earliest, Mersenne (1932–88) I. 404.

[43] Shea (1991) 243 note 38.

[44] Mersenne (1932–88) I. 411–413. The anaclastic problem is to define the refracting surface that will focus all parallel incident rays to one point.

[45] Mersenne (1932–88) I. 408–11. The textual and mathematical claims made in this and the next paragraph are documented in Appendix 1, 'Descartes, Mydorge and Beeckman—The Evolution of Cartesian Lens Theory 1627–1637'. The sceptical reader should examine this Appendix immediately after finishing the present Sect. 4.5.2.

[46] Mersenne (1932–1988) I 408–9.

[47] cf. Figure 4.3 above.

[48] AT x. 341–2; Beeckman (1939–1953) fol. 338r.

[49] ibid.

I conclude that in the *Dioptrique* Descartes used Beeckman's more 'natural' representation of the sines in both cases, ellipse and hyperbola, thus rejecting his own elegant ellipse proof and Mydorge's early 'one sided' representation of the sines. The Mydorge letter therefore contains Mydorge and Descartes' *earliest lens theory*, and arguably *their first form of the law*, the cosecant form. The *material in the letter*, if not the artifact itself, pre-dates October 1628, certainly predates composition of the *Dioptrique* and very plausibly is as early as 1626/1627—but not earlier as we shall soon see. So, this dating points to the cosecant form of the law as the first form Mydorge and Descartes possessed. And this, it transpires, is the key to reconstructing how they obtained it, because the other independent discovers first obtained it in the same *unequal radius form*.

4.5.3 Traditional Geometrical Optics and the Discovery of the Cosecant Form of the Law

To reconstruct how Descartes found the law, let us first follow Johannes Lohne's important analysis of how Thomas Harriot discovered the law, because, as we shall see, Mydorge's letter provides evidence for an identical path of discovery.

One obvious phenomenological expression of the behavior of refracted rays is the displacement of images of objects viewed under refracting media. Traditional geometrical optics had a rule for constructing the image locations of such sources. Lohne supposed that Harriot attempted to discover a general relation between the incident and refracted rays, using the image rule; and that the *cosecant* form of the law resulted from this strategy of research. The traditional image placement rule ran as follows (Fig. 4.6): AB is a refracting interface; a normal is dropped to AB at O, the point of incidence. E is a point source emitting ray EO, refracted at O to the eye at F. Experience teaches that E will not appear at E. Where does it *seem* to appear? The rule says that it will appear at I, which is the intersection point between the refracted ray FO drawn back into the first medium, and EG, which is the normal to the surface from E.[50]

Harriot used this rule in conjunction with observations made with a disk refractometer half immersed in water. Taking source points at 10° intervals around the lower circumference of the disk, he observed the corresponding angles of refraction. He then constructed the image places for the source points, by applying the image rule. With the source points located around the circumference of the disk, he found the calculated image places lie roughly on a smaller, concentric circle. If you suspect the plot is really a circle, a little trigonometric analysis gives you the cosecant form of the law. Harriot's key diagram (Fig. 4.7) is indistinguishable from Mydorge's diagram.[51]

---

[50] This principle appears in Alhazen, Pecham, Witello, Roger Bacon and Maurolico; cf Robert Smith, *A Compleat System of Optics* (Cambridge, 1738) para 212, cited in Turbayne (1959) 467.

[51] Lohne (1959) pp.116–7, (1963) 160. Gerd Buchdahl (1972) 284 provides a particularly clear statement of the methodological role played by the image principle in Harriot's discovery of the law.

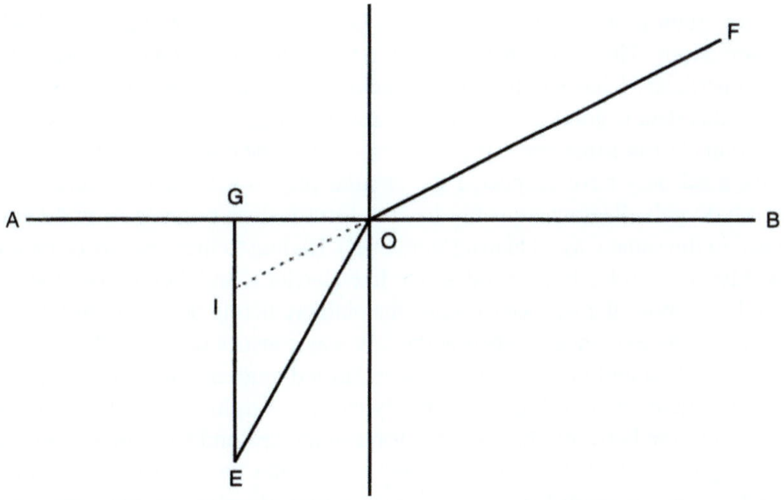

Fig. 4.6 The traditional image locating rule

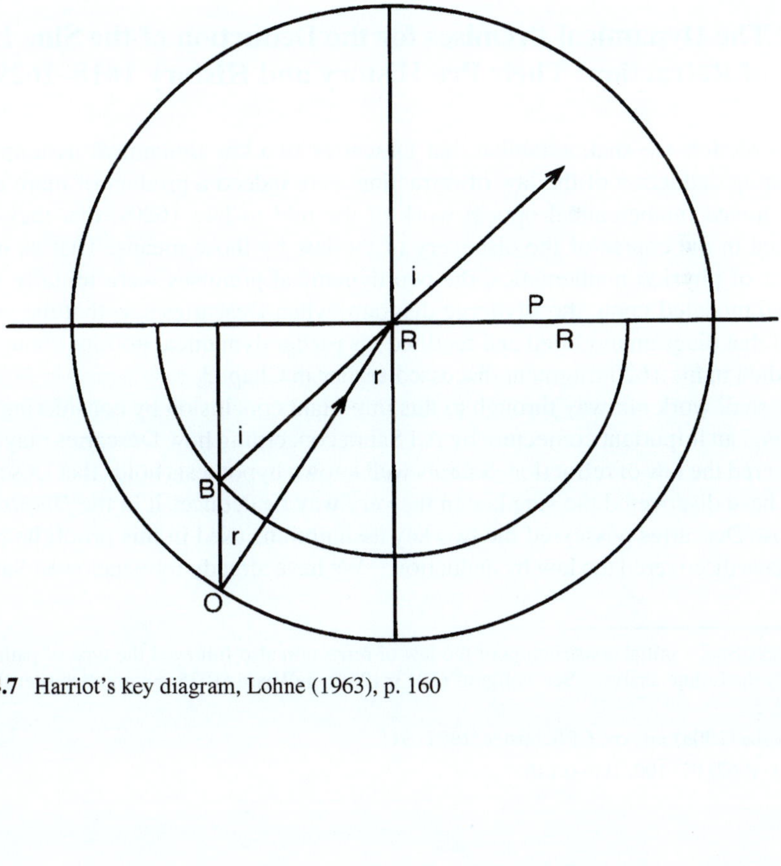

Fig. 4.7 Harriot's key diagram, Lohne (1963), p. 160

It is important to note that Mydorge and Descartes need not even have made *any* such observations. They could have used Witelo's rather cooked data for water/air and glass/air interfaces. I have followed calculations originally provided by Bossha and found that this data is good enough to give a strong suggestion of a semi-circular plot when used in Harriot's manner.[52] Trigonometricians of the power of Snel, Mydorge and Descartes need only have suspected the circular plot to seize upon it and explore it further. Mydorge's diagram arguably has the form it does because he and Descartes proceeded in the same way as Harriot (and Snel), leading to the same cosecant form of the law. Mydorge probably took a diagram like Harriot's, and then flipped the smaller semi-circle up above the surface to create the path predicting device in his letter.

In sum, strong evidence exists that the law was constructed by traditional optical means, using data and concepts familiar to skilled students of geometrical optics. This account involves nothing about the dynamics of light or of tennis balls. What then is the relation between the cosecant form of the law and Descartes' two dynamical assumptions? Did Descartes perhaps have the two assumptions prior to 1626/1627? And, if he did, is it still possible that, despite our reconstruction, he arrived at the law by deducing the cosecant form from the assumptions? Whilst that is, of course, logically possible, it is not supported by the existing evidence, as we shall learn in detail in the following section.

4.6 The Dynamical Premises for the Deduction of the Sine Law of Refraction: Their Pre-History and History 1618–1629

In this section we shall establish that Descartes two key dynamical assumptions permitting deduction of the law of refraction were indeed a product of more traditional mixed mathematical optical work of the mid to late 1620s; that they only emerged in the course of the discovery of the law by those means. That is, in the manner of physico-mathematics, the two dynamical premises were initially 'seen in', and modeled upon, the Mydorge diagram, when Descartes saw that the geometry of that diagram modified and rectified his earlier dynamical notions about light embodied in his 1620 fragment discussed earlier in Chap. 3.

We shall work our way through to this important conclusion by considering, and rejecting, an important conjecture by A.I Sabra concerning how Descartes may have discovered the law of refraction. Sabra's well known hypothesis holds that Descartes could have discovered the sine law in the very way he deduces it in the *Dioptrique*. Suppose Descartes possessed the two key assumptions used in this proof; he could then have discovered the law by deduction.[53] We have already foreshadowed Sabra's

Willebrord Snel's initial construction of the law of refraction also followed the type of path indicated by the Lohne analysis. See Vollgraff (1913, 1936), deWaard (1935–6), and Schuster (1977) pp. 313–5.

[52] J. Bossha (1908) xii–xiv. Cf Schuster (1977) 311.

[53] Sabra (1967) 97–100, 105–6,116.

argument in Sect. 4.4.[54] The first assumption is that the ratio of the force of light in two media is a constant for all angles of incidence

$$\frac{|Fi|}{|Fr|} = \text{const} \tag{1}$$

The second assumption is that the component of the force of light parallel to the refracting surface is unchanged by refraction.

$$|Fi| \sin i = |Fr| \sin r \tag{2}$$

Combining (1) and (2) we get the sine law.

$$\frac{\sin i}{\sin r} = \frac{|Fr|}{|Fi|} = \frac{1}{\text{const}}$$

The essential question is, did Descartes have the two assumptions before the Mydorge letter? Sabra made use of Descartes' optical fragment from 1620, which we analyzed in detail in Chap. 3. He claimed that the fragment implies possession of both assumptions. This means Descartes could have deduced the law any time from about 1620. Sabra cites the 1620s fragment only in part, as follows:

> Because light can only be produced in matter, where there is more matter there it is more easily generated; therefore, it more easily penetrates a denser medium than a rarer one. Whence, it happens that refraction occurs in the rarer medium from the perpendicular, in the denser medium toward the perpendicular.[55]

For Sabra the first sentence *is* assumption 1: the force of light is as the density of the media—independently of path. Sabra then notes the sentence: 'whence refraction occurs toward the normal in the denser medium, and away from the normal in the rarer medium'. He asks, how can Descartes say *that* unless he also has the second assumption? And, of course, given the two assumptions, Descartes could have deduced the law of refraction.[56] Sabra is thinking of a diagram very much like the Mydorge diagram (Fig. 4.5), which, of course, neatly represents these assumptions.

Let us recall what our study, in Chap. 3, of Descartes' optical fragment of 1620 revealed: Descartes did possess in 1620 some intriguing views about the dynamics of light, but these conceptions could not have directed him to the sine or cosecant law. Rather, they constituted an obstacle to his ever finding it. Our analysis of the entire fragment, and its likely contexts in Descartes' exploration of parts of Kepler's

---

[54] cf Note 26 above. Sabra, of course, spoke in terms of the 'speed' of light in the two media. The reader should note both here and in Sect. 4.4, we correct Sabra, speaking of the ratio of the 'force of light' in the two media. (For Descartes, the speed of the propagation of light being instantaneous, but with variable, finite, degrees of 'force', as explained above.)

[55] AT x. 242–3

[56] Sabra (1967) 106, 111.

optics, makes it clear that Sabra is mistaken: for, as we found, Descartes' first sentence does not contain or entail assumption (1). Rather, Descartes was assuming that the normal component of the force of light is increased in a denser medium. In other words in 1620 he held:

$$\frac{Fi \perp}{Fr \perp} = \text{const}$$

Rather than,

$$\frac{|Fi|}{|Fr|} = \text{const}$$

So, as was mentioned in Chap. 3, in 1620 Descartes embraced an assumption which would have hindered his ever deducing the sine law. Holding that in two media the normal components of the force of light are in a constant ratio, had he then assumed that the parallel components are constant, he would have gotten a law of tangents.[57]

How then did Descartes ever devise his two assumptions—and in particular why did he ever decide that the constant force ratio applies to media in a path independent manner? All the evidence examined thus far suggests that a likely answer is this: Descartes only formulated his two dynamical assumptions *after* he had constructed the law in cosecant form, using traditional means—issuing in the Mydorge diagram. The Mydorge diagram—the cosecant form—gives you the two assumptions *if you are looking to read them out of the diagram.* And in 1626 Descartes, physico-mathematician, was very interested to read out of his ray diagram some mechanical theory explaining that diagram. He did to the Mydorge diagram exactly what he earlier did to diagrams in Stevin and Kepler. That is, he took a geometrical picture of a macroscopic phenomenon, garnered by mixed mathematical procedures, and read out of it the underlying dynamical causes. Viewed through physico-mathematical spectacles, the Mydorge diagram was the locus where the two dynamical assumptions were forged and coordinated. In short, the two dynamical premises were modeled upon, or 'seen in', the Mydorge diagram, with Descartes realizing that the

[57] As we commented in Chap. 3, Notes 106 and 111 on this point: Had Descartes assumed that the parallel component varies either directly or inversely with the density, he would have again deduced 'tangent laws' with slightly differing indices of refraction. There seems no way to proceed directly from the assumptions of 1620 to the sine law of refraction, unless one is prepared to introduce Newtonian complications about the variation in components as functions of the angle of incidence, a way of conceiving the problem foreign to Descartes in 1620, 1626, as well as 1637. Sabra, of course, assumed that penetration varied with density regardless of the angle of incidence, an assumption that does indeed yield the sine law when conjoined with the assumption that the parallel component of the motion, force or penetration of the incident ray is unaffected by refraction. Sabra's error consisted in his construal of the first premise: Descartes was envisioning that the normal component of penetration varied with density.

geometry of that diagram clarified and modified his earlier, inefficacious, dynamical notions about refraction.

This reconstruction thus helps us understand why, after 1627, Descartes moved to a dynamical rationale for the law; and why that rationale took the form it did. Having been thwarted in his early attempt, in 1620, to arrive at the law of refraction by physico-mathematical analysis of its purported physical causes, Descartes would have seized upon the newly discovered, and arguably correct, cosecant form of the law of refraction. Acting in accord with the procedures of his physico-mathematics, he decoded the Mydorge diagram—representing the correct mixed mathematical form of the law—as a message concerning the causes of refraction. This account also helps us deal with the problem of why Descartes embraced such problematical dynamical premises for explaining refraction. Why, as we noted earlier, he used dynamical premises which simultaneously entail that optical media are, and are not, isotropic. The most likely answer is that having formulated (or 'seen') the premises by inspecting the geometry of the already discovered cosecant form of the law of refraction, he accepted and defended these premises because of their supreme value in grounding a deductive physical rationale for the law.[58] We are now going to see, in Sect. 4.7, how in the late 1620s and early 1630s the two key dynamical premises were variously articulated by Descartes, for the purpose of producing 'demonstrations' of the law. That is, as a physico-mathematician, he first 'read' his newly discovered mixed mathematical geometrical rule for refraction back to its presumed dynamical causes, understood as rather generically stated premises. He was now going to explore further mechanistic articulations of those premises. This was to be physico-mathematics in the grand style, at least as Descartes conceived it, based on a mixed mathematical breakthrough, and aiming at the elucidation of its causes and explanation through mechanistic models of light, if not yet at a fully articulated corpuscular-mechanical theory of light.

[58] The discerning reader will note a difficulty in this reconstruction. It has been argued that Descartes and Mydorge (as well as Snel) used the traditional image finding rule in their path of research leading to the law of refraction. But, unlike Harriot, the three later discoverers presumably were well aware of Kepler's new theory of vision, which cast grave doubt on the use of the traditional rule. Descartes, after all, was working on a mechanistic version of Kepler's theory of vision around the same time he and Mydorge discovered the law, and his 1620 optical fragment already indicates familiarity with Kepler's new work on vision. This fascinating issue cannot be addressed in full here. Suffice it to say that the problem is more Descartes' than our own. That is, there is evidence that Descartes suppressed discussion of his actual path of discovery for several reasons, one of which was the embarrassing point that his work depended upon an optical principle he could no longer accept. For example, his odd methodological story about how the law might be discovered, offered in rule 8 of the *Regulae ad directionem ingenii*, seems intended to occlude this fact, and to mythologize several of his other theoretical quandaries, under a cloak of persuasive, but necessarily vacuous 'method talk'. See below Sect. 4.9 and Chap. 6 where the issue of the efficacy of Descartes' method is discussed. These matters are also discussed Schuster (1993).

4.7 The Mechanical Theory of Light 1620–1628

4.7.1 Expository Strategy and Working Distinctions

Thus far, as promised, our investigation has allowed us to use the *Dioptrique* and the dynamics of *Le Monde* as a basis for decoding Descartes' actual intentions in his published optical proofs. This in turn prompted us to recognize the import of the Mydorge letter and the path of discovery of the law of refraction it revealed, and the way that Descartes and Mydorge's procedure resembled the purely geometrical optical techniques of the other independent discoverers of the law. We were then able to suggest a reconstruction how Descartes developed his two dynamical premises physico-mathematically out of his newly discovered 'mixed mathematical' law of refraction. Whilst doing this we have bracketed the question of what he actually took to be the nature of light in the crucial period of 1626–1628, before he launched into the composition of *Le Monde* and the *Dioptrique.* The reconstruction of the emergence of the two dynamical premises presupposed only what we already knew: (1) that since 1619 Descartes had thought of himself as a 'physico-mathematician'; (2) that, as we discovered in Chap. 3, he was, as a natural philosopher leaning toward corpuscular-mechanism of the Beeckmanian variety, but had done little to explore it, and certainly nothing to systematize it; whilst (3) his optical fragment of 1620 did not speak a mechanistic dialect, but may to some extent have implied one. Beyond that, the discussion was intentionally non-committal about details. Descartes was said to have realized that the parallel component of the force or motion of the incident light is conserved before and after refraction, and that the quantity of the force or motion of the light varies with the density of the medium and is path independent. Problems of exposition necessitated this strategy, because the evidence relating to Descartes' mechanistic theory of light in the period 1626–1628, which we are about to survey, can only be decoded on the basis of a prima facie account of how and when the law of refraction was discovered. So, in this section, we examine Descartes' commitment to a mechanistic theory of light between 1620 and 1628 with the goal of confirming and deepening the findings of Sect. 4.6.

When investigating Descartes' commitments to corpuscular–mechanism, and to a mechanistic optics, certain working categories need to be kept in mind. It is useful to distinguish between (1) fundamental ontological convictions in general, and (2) theories about the nature of light in particular. Furthermore, when considering (1) or (2), one needs to distinguish between (a) relatively articulated or systematized commitments or theories, and (b) relatively unarticulated commitments or theories. Combining these possibilities, one obtains a set of four broad analytic categories

(1a) A systematic corpuscular-mechanical ontology: such as is found in Descartes' two systematic treatises on the philosophy of nature, *Le Monde* (1629–1633) and *Principia philosophiae* (1644). This involves an elaboration of the corpuscular-mechanical structure of matter, leading on to a theory of 'elements', a theory of the 'cosmological' structuring of matter, and an explicit doctrine concerning the laws of motion, collision and tendency to motion, or what we have termed Cartesian dynamics.

(1b) An unarticulated corpuscular-mechanical ontology: such as is found in Beeckman's *Journal*, or in Descartes' work prior to his commencement of *Le Monde*. This involves a general belief in corpuscular-mechanism and piecemeal appeals to it in formulating particular explanations, without a sustained attempt to organize or mediate between these particular applications. Certain consistencies might run through these applications and, to that extent, one might speak of an 'element theory', 'cosmology' or 'dynamics' implied in them; but, in general, the more that the theme of systematization emerges, and claims to control the applications, the more articulated and systematized the ontology can be judged to be.

(2a) An articulated corpuscular-mechanical theory of light, such as is found in the explanations of light in *Le Monde* or *Principia philosophiae*. In the broadest sense this would therefore involve the attempt to explain the true nature of light, as part of the sort of system envisioned in (1a), in which the theory of light is articulated to the matter theory, cosmological setting and controlling principles of motion and dynamics.[59]

(2b) An unarticulated mechanical theory of light: such as we shall find in Descartes' optical work in 1626–1628. This would involve a loose commitment to the mechanistic nature of light, based on piecemeal and unsystematized appeals to mechanistic causes, and to 'causal principles' which have not quite taken the form of a systematized dynamics. This can involve a background belief in the corpuscular-mechanical character of matter and light.

One needs also to note that two broad options were open to Descartes in constructing a theory of light, whether under (2a) or (2b). Light could be taken to consist in the translation of pieces of matter, or, in mechanical impulses or tendencies to motion transmitted through media. Finally, under both (2a) and (2b), a theory of light could be elucidated, or applied, by means of explicit mechanical analogies. So, by the early 1630s Descartes had to hand his tennis ball model, which, as we have seen, was really offered under the tacit aegis of his (2a). Similarly we shall see that in the late 1620s he employed a balance beam model for the refraction of light, which was meant to clarify the version of (2b) which he then held.

4.7.2 Reprise—The Optical Fragment of 1620

Our starting point is the optical fragment of 1620, the third of our case studies of Descartes' early physico-mathematics in Chap. 3. The optical fragment of 1620, we recall, variously hints at a quasi-Aristotelian, or even a Keplerian, physical theory of light. Nevertheless, while the fragment makes no direct reference to a corpuscular-mechanical ontology, Descartes' apparent concern with quantifying the variation of

[59] One can also imagine slightly lesser degrees of articulation, involving, for example, merely a corpuscular-mechanical explanation of optical sources and media, but lacking cosmological articulation, and possibly lacking a highly articulated theory of dynamics.

'penetration' (normal component) with density might also indicate he held an unarticulated theory of light as mechanical impulse or tendency to motion. We saw, however, that mining deep ontological commitments out of the optical fragment of 1620 is beside the point. Descartes was more interested in explaining refraction by relating density to 'generation/penetration' (magnitude of normal component), and expressing the relation geometrically. Seeking to explain refraction by mathematicizing the density-penetration relation (which could have various specific natural philosophical explications), Descartes was comporting himself as a *physico-mathematicus*. But, the question of how a corpuscular-mechanical ontology (or any other ontology) might work into such a physico–mathematical inquiry was postponed, along with any firm commitment about the physical nature of light.

Additionally, we saw that Descartes' physico-mathematical encounter with Kepler's optics probably affected his views about ontology in two ways. First, it marginalized, for the time being, corpuscular-mechanical explanations of light, media, sources and refraction, because these did not seem to lend themselves to mathematization. Second, even at the level of unarticulated theories of light, his encounter with Kepler's optics devalued explicitly kinematic models, and raised the perceived value of models involving no passage of any material entity. Beeckman's kinematic models were avoided, whilst still potentially allowing for models of light as mechanical impulse or as tendency to motion, or indeed as Keplerian immaterial substance, or even as Aristotelian actualization of a potential property of the medium.[60] However, the search for a Beeckman-like a corpuscular-mechanical explanation of light need not have been rejected in principle, but merely deferred, until such time as the law of refraction might be discovered. In sum, the optical fragment offered evidence that Descartes was definitely interested in a physico-mathematical agenda in optics, and that matters were fluid and inconclusive at the level of technical accomplishment—progress toward finding the law of refraction—and in relation to specific natural philosophical aims and valencies.

This brings us back to the period of 1626–1628, and to Descartes' moves in physico-mathematics to attain a natural philosophical rationale for the newly discovered law of refraction. We shall now see that by 1626–1628 he was firmly convinced of an *unarticulated* theory of light (2b) as instantaneously transmitted mechanical impulse or tendency to motion. However, it was only in 1629/1630, when he began to compose *Le Monde*, that Descartes attempted to devise an *articulated* corpuscular-mechanical theory of light (2a) within his emerging system of mechanical natural philosophy (1a). Likewise, it was apparently at this same time that he designed the tennis ball model for use in the *Dioptrique*. The latter was his only foray into the 'corpuscular'-kinematic modeling of refraction, and its use is quite circumscribed. On the one hand, the tennis ball model is only a model for the corpuscular-mechanical theory of light as tendency to motion, and, on the other hand, the model itself is

[60] For Descartes' similar reaction to Beeckman's celestial mechanical speculations see Schuster (2005) 70–2 and below, Sect. 10.3.

essentially premised on the principles of his dynamics of instantaneously exerted forces and determinations, as we have seen.[61]

4.7.3 *Light as an Instantaneously Transmitted Mechanical Impulse 1626–1628*

Whatever the ambiguities of the 1620 fragment on the issue of the nature of light, one can be reasonably certain that by 1626 Descartes had opted for an unarticulated theory of light as mechanical impulse or tendency to motion, transmitted instantaneously through corpuscular media, although the microstructures of those media were not as yet a matter of concern, for the very reasons we have just canvassed. The main evidence on this point comes from parts of Descartes' *Regulae ad directionem ingenii* which he wrote in Paris between 1626 and 1628, after the discovery of the law of refraction, as well as from discussions he held with Beeckman in 1628. The former are discussed in this Section, the latter in Sect. 4.7.4.

We shall be examining the *Regulae* and Descartes' abortive dream of a universal method in great detail in Chaps. 5 and 6. For the present we are only interested in flagging some points which will be proven in those chapters, but which must now be posited as part of our inquiry into the evolution of Descartes' mechanistic theorizing about light. The key point for the moment is the following simple and unalloyed fact: *An unarticulated theory of light as an instantaneously transmitted mechanical impulse plays a central role in and between the lines of the latter portion of the text of the* Regulae *written in Paris between 1626 and 1628.* The *Regulae*, it will be shown, really consist in three main textual strata, written at different times between 1619 and 1628 with rather different aims in view.[62] The first stratum, consisting in a portion of Rule 4, is the remnant of a treatise which Descartes planned to compose in mid 1619 on the subject of 'universal mathematics'. Descartes conceived of this 'discipline' in mid 1619, viewing it as some sort of synthesis of his physico-mathematical project and his more purely mathematical researches into the generalization of analytical procedures, applied to classes of geometrical and algebraic problems. Later in 1619, this early project of universal mathematics was itself superseded by, and encysted within, the main lines of his method, the dream of a general analytical machinery suitable for all rational disciplines, mathematical or not. Accordingly, we shall see that Descartes' constructed his doctrine of method in the winter of 1619–1620, the results being recorded in the second stratum in the *Regulae*, rules 1–3, part of 4, and 5–11, excluding some material in rule 8.[63]

---

[61] On the larger functions and uses of the tennis ball model and Descartes' difficulties with it, see below Sect. 4.8.2.

[62] Below Chaps. 5 and 7 and Schuster (1980).

[63] Below Sect. 4.9 and Chap. 5, as well as Schuster (1986, 1993).

The third and final stratum of the *Regulae*, the one that interests us here, arose when Descartes arrived back in Paris in 1625 with his apparently effective method in hand. Seeking to emulate, and outdo, his friend Marin Mersenne, Descartes revived his *Regulae* project. He returned to the universal mathematics of 1619, which he now attempted to construct in detail, by expanding and extending his 1619/1620 text on method, that is, roughly rules 1–11 of the *Regulae*. This version of universal mathematics would appear to grow out of, and articulate, the doctrine of method. Interestingly, the shift from the second to third stratum in the *Regulae* can be located inside the present text of rule 8 and contains priceless evidence as to dating. The text in question contains a peculiar little methodological tale about how the anaclastic problem—to find the refracting surface that will focus all parallel incident rays to one point—might be solved on the basis of the prior discovery of the law of refraction. We shall look at this method story below in Sect. 4.9. What is important here is that the story, depending upon prior possession of the law of refraction, reinforces the dating of this third stratum of the *Regulae* after 1626/1627 (and prior to 1629 when the whole project collapsed). So, we have a dating just after the discovery of the law of refraction. And, finally, to get to the point about an unarticulated mechanistic theory of light, we find in this third stratum of the *Regulae*, that in order to underwrite his universal mathematics, Descartes, in rules 12–14, outlines a mechanistic theory of nervous function and perception.[64] In turn, a mechanistic theory of light as instantaneously transmitted impulse underpins this enterprise, which includes, prominently, a 'mechanization' of Kepler's new theory of vision.[65] So, whatever else we might know about Descartes' views on light immediately following

[64] Schuster (1980) 59–64, and, Sect. 4.7.3: In rule 12 Descartes claims that the external senses 'perceive in virtue of passivity alone, just in the way that wax receives an impression /figuram/ from a seal.' He intends no mere analogy: just as the wax is impressed with the image of the seal, 'the exterior figure of the sentient body is really modified by the object'. All of our sensations, whether of light, color, odor, savor, sound or touch, are ultimately caused by the mechanical disturbance of the external sense organs. From the sense organs the impressed 'figures' are transmitted instantaneously to the common sense via the nerves, by means of the passing of a pattern of mechanical disturbance. 'No real entity travels from one organ to the other', just as the motions of the tip of a pen are instantaneously communicated to its other end, for 'who could suppose that the parts of the human body have less interconnection than those of the pen'. Patterns in the common sense can then be imprinted in the imagination, either to be stored in memory for the future 'attention' of the *vis cognoscens*, or to be immediately attended to in sense perception. AT x. 412–4

[65] Schuster (1980) 61–2 and, Sect. 4.7.3 below: Although Descartes focuses upon the mechanical causation of sensation and perception, it is clear that a mechanical theory of light underpins the entire discussion. Whatever the essential nature of external objects may be, Descartes implies, they act upon the perceiving subject in a mechanical manner. In the case of visual perception, therefore, light (or the optical media through which it acts) mechanically impresses the 'figures'. Presumably light is an instantaneously transmitted mechanical impulse: Descartes' mention of instantaneous mechanical nervous action, and his analogy of it to the instantaneous transmission of motion from one end of a pen to the other, suggest that light is considered to act in the same fashion. Note also that although the pen analogy is applied to nervous action (see previous Note), it is similar to the analogy of the blind man's staff, used later in Partie 1 the *Dioptrique* to illustrate the instantaneous mechanical transmission of light. AT VI 85–6.

the discovery of the law of refraction, we can be virtually certain he was willing to write about, and rely upon, a mechanistic theory of light, vision, and nervous function. We shall now see him working directly and intriguingly on mechanistic models for the action of light, and its law of refraction, at precisely the same time.

4.7.4 Light as Mechanical Impulse and the Explanation of the Law of Refraction 1626–1628—The Balance Beam Model

The theory of light as an instantaneously transmitted mechanical impulse, unarticulated as it was in 1626–1628, would still have been sufficient to provide the conceptual framework for Descartes' physico-mathematical reading of the Mydorge diagram, as discussed in Sect. 4.6. Descartes, *physico-mathematicus*, operating with an unarticulated theory of light as mechanical impulse, could have read the Mydorge diagram as bespeaking the true physical premises necessary for the demonstration of the law of refraction, premises which corrected and reformed the ideas about density and penetration (normal component) evident in the 1620 fragment: (1) A light impulse, or ray, has a force, strength, or perhaps (retaining the language of 1620) a 'penetration', which varies when the impulse passes from one medium to another. For a given pair of media the ratio of these forces or 'penetrations' is constant and independent of the angle of incidence; (2) The force or 'penetration' of an impulse or ray may also be considered directionally, in the usual terms of components parallel and normal to the refracting surface. The force or penetration of the ray or impulse acting parallel to the surface must be unaffected by the refraction. This, of course, is a 'rational reconstruction' of how Descartes might have interpreted the Mydorge diagram, using a theory of light as mechanical impulse in the interests of designing a 'physico-mathematical' explanation of the new law. This rational reconstruction fills up the interpretive and evidential void left at the common terminus of our several lines of textual and contextual reconstruction. There is, however, a very remarkable piece of evidence, dating from 1628, which we are now finally in a position to examine, and which shows Descartes striving to elucidate how the theory of light as mechanical impulse could be used in the demonstration of the law of refraction. Although it does not record Descartes' initial 'physico-mathematical' reading of the Mydorge diagram, it is arguably a product of research and reflection which followed very closely upon that event.

In the autumn of 1628 Descartes paid a short visit to the United Provinces prior to his settling there permanently early the next year. On 4 October he met with his old friend Isaac Beeckman for the first time since early 1619. He sketched for Beeckman some of his discoveries of the previous nine years, including the work on lens theory (cf. Sect. 4.5.2 and Appendix 1). This was prefaced by a statement of the (sine) law of refraction, which Beeckman recorded in a short memorandum, illustrated by Fig. 4.8, in which for rays *aeg* and *cef*: $(ab/kg) = (cd/if)$. There immediately

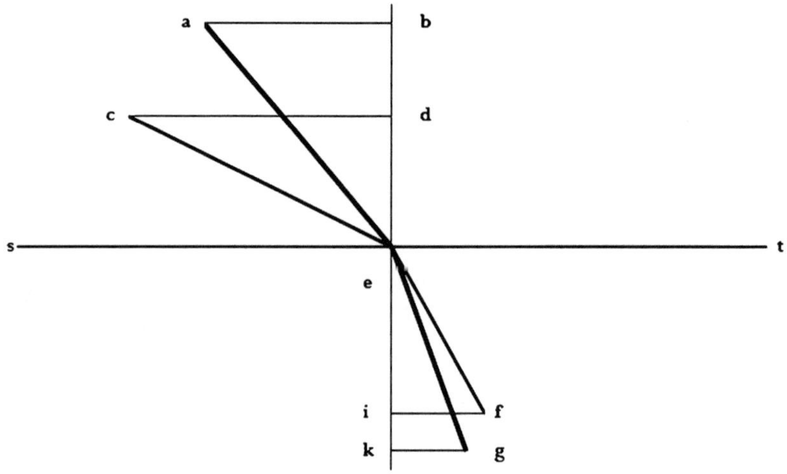

Fig. 4.8 Beeckman's 1628 illustration of discussion of the sine law of refraction

follows Beeckman's description of an analogy through which Descartes sought to explain the law to him,

> (Descartes) considers water to be under *st* and the rays to be *aeg*, *cef*. They seem to undergo the same (change) as the arms of an equal arm balance, on the ends of which are fixed weights, of which that in water is lighter and raises the arm.[66]

This passage certainly is cryptic; even so patient a Cartesian scholar as Gaston Milhaud was moved to dismiss the analogy as 'bizarre'.[67] But, Descartes' conception can be reconstructed, provided one is willing to grant that Beeckman, in an understandable way, garbled or mistook part of the sense of Descartes' exposition.

Let us take Descartes to be suggesting that the behavior of the incident and refracted rays of light is analogous to the behavior of an equal arm balance, the arms of which must be bent, or refracted, at the fulcrum to maintain equilibrium under varying conditions of loading (Figs. 4.9 and 4.10). The constant ratio of the force of light in a given pair of media is likened to the constant ratio of the 'effective' weights of identical bodies immersed in a pair of fluids differing in specific gravity. In Figs. 4.9 and 4.10 we have a balance whose equal arms can be pivoted about the fulcrum and fixed at the settings required to maintain equilibrium under differing conditions of 'effective' weight. The arms are loaded with two identical bodies of specific gravity SGb. The specific gravity of the upper medium, SGu, and the specific gravity of the lower medium, SGl, are each less than SGb, so the weights 'weigh down' from both ends of the balance. In Fig. 4.9, SGu > SGl and in Fig. 4.10,

[66] AT x 336; Beeckman (1939–53) fol. 333v.

[67] Milhaud (1921) 110.

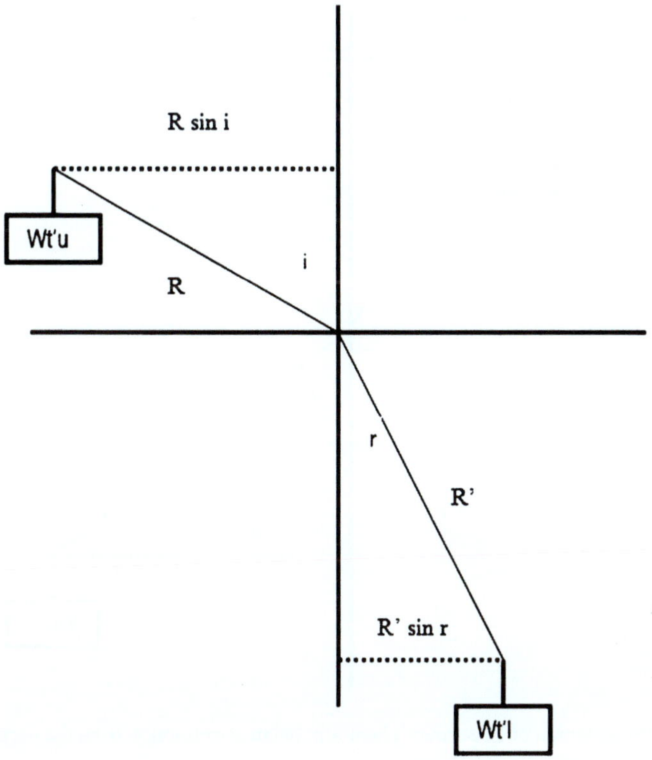

Fig. 4.9 Reconstruction of Beeckman's bent arm balance: refraction toward the normal

SGu<SGl. Then, in Fig. 4.9, the effective weight of body in the upper medium, Wt′u bears to the effective weight of the body in the lower medium, Wt′l, the ratio

$$\frac{\text{Wt}'\text{u}}{\text{Wt}'\text{l}} = \frac{\text{SGb} - \text{SGu}}{\text{SGb} - \text{SGl}} = \text{const.} \leq 1$$

And, in Fig. 4.10 the corresponding ratio is:

$$\frac{\text{Wt}'\text{u}}{\text{Wt}'\text{l}} = \frac{\text{SGb} - \text{SGu}}{\text{SGb} - \text{SGl}} = \text{const.} \geq 1$$

In either case at equilibrium,

$$(\text{Wt}'\text{u})(\text{R} \sin \text{i}) = (\text{Wt}'\text{l})(\text{R}'\text{sinr})$$

where r sin i and r′ sin r are the effective lever arms

$$\text{but, R} = \text{R}'$$

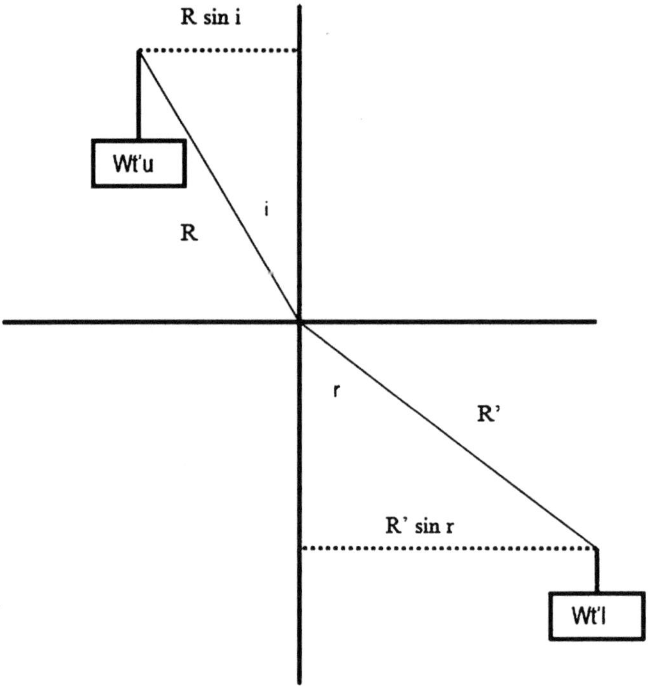

Fig. 4.10 Reconstruction of Beeckman's bent arm balance: refraction from the normal

therefore,

$$\frac{\sin i}{\sin r} = \frac{Wt'l}{Wt'u} = \text{const.}$$

Thus, if equilibrium is to be maintained, in Fig. 4.9 the right arm must be dropped toward the normal.

And, in Fig. 4.10 the right arm must be removed away from the normal. For a given pair of media, the 'refraction' of the right arm will always be given by the last equation, a veritable 'law of sines' telling us how to adjust the right arm at the fulcrum, for a given setting of the left arm, in order to maintain the condition of equilibrium.

Returning to the entry in Beeckman's *Journal*, we see that his diagram (Fig. 4.8) indicates refraction toward the normal in water, but that his discussion specifies that the weight on the right rises due to the buoyancy of the water being greater than that of the air. The inconsistency can be explained by Beeckman having garbled Descartes' explanation. Figure 4.10 illustrates what Descartes intended in the case of a real balance with weights immersed in air and water.[68] But, as we also know

[68] It would also illustrate the case of a 'tennis' or cannon ball whose motion is refracted away from the normal in water, as discussed later in the *Dioptrique* (AT vi. 97–8). Beeckman and Descartes might perhaps also have discussed this phenomenon in 1628.

from the *Dioptrique*, when Descartes switched from tennis balls to light rays, he had to argue that the force of light is greater in water than in air, in order to explain its refraction toward the normal in water. Accordingly, to apply the balance analogy to the case of light, Descartes must have claimed that the lower medium is rarer than the upper one, so that the effective weight of the body in the lower medium (analogous to the force of the refracted light) is greater than the effective weight of the body in the upper medium (analogous to the force of light in the upper medium), hence that Wt′l > Wt′u. This makes no sense if one still has in mind a real balance, with one arm plunged into a real vat of water. To make the balance germane to the behavior of light passing from air into water, one must abstract from the concrete situation and invoke different media with the appropriate ratio of densities. Beeckman may have become confused in the shift from the concrete case of a balance beam with weights in air and water, to the abstract case where the balance illustrates *by analogy* the force changes light undergoes in different media. In any case, Beeckman must have garbled the sense of his discussion with Descartes, for he cannot have both his figure and his text.

On this reading, Descartes was offering to Beeckman a particularly fine model for his two recently devised dynamical premises, as conceived against the background of the unarticulataed (category 2b) theory of light as instantaneous mechanical impulse (for example, as used in the later portions of the *Regulae*):

1. The path independent ratio of the force of light in the two media is modeled by the ratio of 'effective' weights, which depend on the ratio of the densities of the media.[69] The 'effective' weights, moreover, are beautifully 'path independent'. The weights hang down perpendicularly from the ends of the arms, regardless of the direction in which the left arm, the 'arm of incidence' if you will, has been set, and regardless of the direction then assumed by the right arm, the 'arm of refraction', in order to maintain equilibrium.
2. The conservation of the parallel component of the force of the light is modeled by the condition of equilibrium, which requires the equality of statical moments about the fulcrum.

One should also note that if, as seems likely, Descartes was thinking of his premises against the background of a theory of light as instantaneous impulse or tendency to motion, then the model is particularly apt for two further reasons. Firstly, weight may be interpreted as a tendency to motion (as Descartes did indeed conceive of it as early as 1619 in the hydrostatics manuscript), and hence as a kind of impulse reiterated from moment to moment; and, secondly, weight, like a tendency to motion or a light impulse, can be conceived to have a certain gross magnitude (measured by weighing), as well as specifiable components of 'directional magnitude'.[70]

[69] The only problem with Descartes' analogy of course is that greater force (effective weight) depends upon placement in a rarer medium and vice versa, thus implying a disanalogy between specific gravity and refractive 'density' of an optical medium

[70] As Stevin, the stimulus for the hydrostatic manuscript of 1619, had taught with his near approach to the parallelogram of forces, mainly applied to the non-vertical components of weight. Stevin (1955–66) Vol. 1. 183–5.

4.8 Full Circle: Cartesian Dynamics, Optics and the Tennis Ball Model 1628–1633

Our detective inquiry has now travelled almost full circle. It began with an analysis of Descartes' systematic dynamics in *Le Monde*, which was then used to unpack the tennis ball model and optical proofs in the *Dioptrique*. The reinterpretation of the *Dioptrique* was an important, yet secondary goal: The strategic aim was to take some bearings which could orient our reconstruction of Descartes' route to the law of refraction and of his physico-mathematical struggle to explain it in mechanistic terms. The analysis of the *Dioptrique* uncovered Descartes' two dynamical premises and the hidden radius form of the law of refraction to which they are best adapted. These findings provided questions and points of reference around which the reconstruction was developed. We can now reverse the process, using the reconstruction of the course of Descartes' optical researches in order to throw new interpretive light on two issues very significant in understanding the nature and direction of Descartes' post 1628 natural philosophical endeavors. They are (1) the origin and nature of his mature dynamics; and, (2) the reasons for the design and use of his tennis ball model. In regard to (1) we shall start by recalling that the originally discovered cosecant form of the law of refraction provided the basis for the two dynamical assumptions later used to explain it. We shall see that those assumptions, physico-mathematically 'read' out of the Mydorge diagram, in turn provided the exemplar for two of the three laws or 'rules' of nature in Descartes' systematic corpuscular-mechanical natural philosophy of *Le Monde*, thereby constituting a very large part of the core of his mature dynamics. In relation to (2) we shall discover that far from being central to Descartes' physical optical research, the tennis ball model was really a rather contingent element, explicable by the circumstances and needs which shaped the writing of the text of the *Dioptrique*, and consequently that it does not reflect the trajectory of Descartes' earlier optical researches and is likely to mislead us about them.

4.8.1 The Exemplar for Descartes' Laws of Dynamics in His Physico-mathematical Optics

As we have seen in Sect. 4.2, Descartes' mature dynamics treats bodies in motion or tending to motion in terms of two instantaneously acting dynamical properties: the absolute quantity of a force of motion, and its directional manifestations, expressed in law 1 and law 3 in *Le Monde*. We may now suggest that these principles derived from a further generalization of his original reading of the Mydorge diagram. Descartes first read the diagram for some basic principles of physico-mathematical optics, assumptions about the quantity and directional quantity of the force of light. But what about the laws of nature which he had to construct after 1629, when he began to write *Le Monde*? How better to base the laws of nature than to use as an exemplar the dynamical principles revealed by successful physico-mathematical research in optics: Light, after all is just an impulse, so its behavior clearly reveals the basic dynamics of forces and determinations. Descartes would have had every reason to be confident that his optical exemplar was well chosen and correctly analyzed, and so he would have had every reason to think that his dynamics of force and determination could be premised upon his having cracked the code of the physico-mathematics of refraction.

Whilst this interpretation is easy to state in brief terms, given our detective work so far, there is one important obstacle to its acceptance. This involves the quite reasonable, indeed necessary, consideration that it may have been the pressures and requirements of constructing, at long last, a systematic corpuscular mechanical philosophy that largely or entirely shaped the form and content of the laws of nature (dynamics) that lay at its center. We shall have to take these issues very seriously, showing just how far they reach, and how in the end both sets of drivers—requirements of systematicity and consistency, and the need to exploit the best and most exemplary physico-mathematical results as keys to nature—drove Descartes' inscription of his first and third laws of nature. Let us look therefore at the composition of *Le Monde* and some of the systematic needs that may in part have shaped the laws of nature enunciated therein.[71]

It was only in 1629 that Descartes began to construct a system of corpuscular-mechanical natural philosophy, supported by a systematic dualist metaphysics. At no time during his previous career as a physico-mathematician had he shown the slightest interest in systematizing the corpuscular-mechanism into which Beeckman had initiated him. Yet, by 1630 he had the main lines of his metaphysics, and by 1633 he was prepared to publish *Le Monde*, his first systematic treatise on natural philosophy. This reorientation in Descartes' projects had as its immediate cause the unexpected failure and collapse of the project of the third stratum of the *Regulae*, which he had begun three years before in Paris under the inspiration of Marin Mersenne. Recognizing by late 1628 that the *Regulae* suffered from fatal mathematical, epistemological and ontological problems, Descartes abandoned the text in mid stream, moved to the United Provinces, and spent his first six months there sketching the basics of his dualist metaphysics. This, he hoped, would resolve, or finesse, the fatal problems with the *Regulae*, whilst preserving its larger goals, influenced by Mersenne; that is, to defeat 'unorthodox' philosophies of nature, while avoiding (now defeating) scepticism.

By mid 1629 in the midst of this work, Descartes began to be drawn into the composition of a system of corpuscular-mechanical natural philosophy. At first he was attracted by the prospect of applying his optical results to explain the rainbow, as well as the unusual parhelia which had appeared at Rome the previous spring. By November 1629 he envisioned an entire system of corpuscular mechanism. Hence, by that time he would have had to contemplate the articulation of a theory of light

[71] Material in the next two paragraphs is explained in detail in Chaps. 7 and 8 below.

beyond the state reached in the *Regulae*. The execution of his plans, the designing and inscribing of the details of the system, took three years. Central to that systematic design were the laws of nature for his new mechanistic 'world', laws which embody his first expression of his mature ideas about dynamics—the causal register of his natural philosophy.

Let us recall our discussion in Sect. 4.2 of *Le Monde*'s first and third rules of nature. The first rule in *Le Monde* asserts that, in the absence of external constraints, God conserves from moment to moment a body's state of rest or motion, or more properly its force of rest or motion. The third rule of nature specifies that at each moment the conserved force of motion of a moving body (or of a body merely tending to motion) is directed along the tangent to its trajectory at the point under consideration. Taken together, these laws occupy some of the same explanatory space in a corpuscular-mechanical natural philosophy that had been occupied in the embryonic mechanics of Beeckman by his inertial principle. So, part of the explanation of their existence and shape has to do with Descartes explicitly refusing simply to adopt Beeckman's principle, and his decision to replace and reformulate the required concepts.

As we learned in, Sect. 3.2.2, as early as 1613/1614 Beeckman had enunciated an inertial principle in the following form:

> Everything once moved never comes to rest unless due to an external impediment. Moreover, the weaker the impediment, the longer the moving body moves. For, if something is thrown upwards and at the same time is moved circularly, it will not sensibly come to rest before its return to earth; and if it nevertheless were to come to rest, it would not do so due to a uniform impediment, but due to a non-uniform impediment, because one part after another of the air turn touches the thing moved.

And,

> ...a stone thrown in a vacuum is perpetually moved; but the air hinders it by striking it anew and thus acts to diminish its motion. Indeed, what the philosophers say, that a force is impressed in the stone seems without reason. For who can conceive what that force would be, or how it would maintain the stone in motion, or in what part of the stone it would find its seat?[72]

Beeckman's statements can be used as a benchmark to gauge the character and peculiarities of Descartes' dynamical conceptions. For example, Descartes has not one law, but two closely related ones. These deal not with the spatio-temporal translation of bodies, but with instantaneously manifested forces of motion. Correlatively, Descartes' mechanics deals with motions only in terms of their analysis instant by instant in terms of the momentarily exerted force of motion and its determinations. Where Beeckman stoutly refused to discourse about such internalized moving forces to explain the conservation of motion, Descartes asserts their existence, analyses their absolute and directional quantities, and refers their existence, causal efficacy and rules of 'behavior' to God's moment to moment rule–bound oversight of nature.

[72] Beeckman (1939–53) i. 24–5.

So, without referring to the trajectory of his work in optics, one can conclude that Descartes' laws of dynamics differ considerably from the analogous concepts in the mechanics of Beeckman. There is, moreover, good textual evidence to suggest that Descartes formulated these two laws of nature in this fashion under the pressure of conceptual constraints exercised by his emerging system of natural philosophy, in particular by his Voluntarist conception of God's relation to Nature, and by his plenist cosmology and insistence on impact and pressure as the sole intelligible modes of natural activity. The dead mechanical world of corpuscles depends for its existence upon God's moment to moment exercise of his freely willed conserving concourse. The continued existence of bodies and their properties is radically dependent upon the instant to instant reiteration of this divine action. It would seem to follow that the laws of natural change, which can only be laws concerning the conservation or alteration of bodies' motion and rest, must deal with divinely regulated instantaneous conservation or alteration of the forces of motion and rest. Moreover, in Descartes' plenum universe all real translations entail the displacement of matter around a closed path. No finite inertial rectilinear translations can occur. Presumably, it was therefore necessary in Descartes' third law to reformulate Beeckman's concept of rectilinear translation as an instantaneously exerted tendency to rectilinear motion. Similarly, Beeckman's approach raised the problem of the loss of motion through the presumably inelastic impact of perfectly hard ultimate principles. Descartes' Voluntarism, his conception of God's relation to nature, would not allow this running down of the world machine. He needed a law of divine conservation of the total quantity of (the force of) motion in the universe. But, this could not deal with the directional manifestations of motion or force of motion, what Descartes was to term 'determination'. It probably appeared to Descartes (as it later did to some of his readers) that determination could not be conserved. This was because the (scalar) sum of it present in any system naturally appeared to depend upon how a grid of components was applied to what we have termed the 'principal' determination, denoted by the third rule of nature. If rectilinear inertia could only be preserved in the form of the third law, because of systematic conceptual constraints, so, similarly cosmic conservation of (the force) motion could only be formulated in essentially scalar terms, in the manner of the first law.

There is additional evidence for Descartes having devised his two laws under the pressure of structural constraints involved in building his system of natural philosophy. Shortly after starting to compose *Le Monde*, Descartes reminded himself to employ Beeckman's inertial principle. In the margin of one of his letters to Mersenne he noted, 'We must remember to add that which has once been set in motion, will move forever in a vacuum, and I shall try to demonstrate this in my treatise.'[73] It would therefore seem likely that Descartes replaced Beeckman's inertial principle with his two new laws during the course of composing *Le Monde* and for the sorts of systematic reasons just discussed.

[73] To Mersenne December 18 1629, AT I 90

Important and enlightening as these structural explanations of Descartes' two laws may be, they do not fully explain their genesis. It was one thing for the perceived needs of the emerging system to exert pressure upon Descartes' formulation of the laws of nature; it was another thing for him to find a model or exemplar around which to elaborate his response to those pressures. I would argue that Descartes' prior physico-mathematical optical work provided him with the exemplary basis from which to elaborate a satisfactory response to those pressures.

There is, after all, an extremely close analogy between Descartes' first and third laws in *Le Monde* and his two unarticulated dynamical premises for the demonstration of the law of refraction. The first law establishes the principle of conservation of force of motion, regardless of the direction of the motion or its analysis into components. Hence it appears to transcribe and generalize the assumption that the force of light is related to the nature of the medium and is independent of the path of propagation. The third law specifies the (principal) tangential determination of the force of motion, and, by implication, licenses its analysis into components. It appears to transcribe and generalize the assumption that in the derivation of the refracted ray path, the determination of the light can be manipulated independently of the absolute quantity of the force of the light. The two rules together ground a general dynamical approach to mechanical explanation paralleling that implied in the particular instance of the optical proofs. The outcome of a collision, that is, the new quantities and principal determinations of the forces of motion of the bodies in question, is deduced by applying 'rules' of interaction to the quantities and directional quantities of the forces of motion obtaining just prior to the instant of collision.

Do these analogies bespeak a genetic relation? Let us recall that the reconstruction of the discovery of the law of refraction, the text of the later *Regulae* and Descartes' balance analogy of 1628 all suggest that by that date he thought he could demonstrate and explain the law of refraction by using his two dynamical premises, which he had read out of the Mydorge diagram, and which he maintained in the context of an unarticulated theory of light as mechanical impulse. But, when he began to compose *Le Monde* there arose the need not only to articulate his theory of light, a process leading eventually to his theory of elements, celestial mechanics and cosmological optics; but also to create the fundamental laws of this new 'nature', principles of natural change for the corpuscular world. The emerging shape of his plenist physics and Voluntarist theology showed that he could not simply appropriate Beeckman's formulation of the law of inertia. He needed an analogous principle dealing with tendencies to motion (a concept familiar to him since 1619), and so designed as to permit the conservation of force of motion in all possible circumstances of corpuscular collision. The optical premises were at hand and they precisely fit these needs. They embodied assertions about the comportment of instantaneously exerted tendencies to motion, in which quantity and directional quantity of force were distinguished, so that one could apply them in practice to the derivation of refracted paths without, it could now be seen, endangering a principle of conservation of force.

Optics practiced in the manner of physico-mathematics, moreover, was the obvious place for Descartes to look for exemplars for his desired cosmic laws. He believed light to be merely a mechanical impulse. Therefore the behavior of light would reveal the dynamics of impulses or tendencies to motion clearly and in paradigmatic ways obscured by systems in which real translation takes place. What is more, in the plenist and Voluntarist perspective he had now adopted, the dynamics of impulses was conceptually and ontologically prior to a dynamics of real translations and cycles of displacement. He possessed both the law of refraction and, thanks to his physico-mathematical 'analysis', the underlying dynamical principles needed to explain it. Thus, the refraction of light could be viewed as an exemplary case of mechanical interaction, involving the law-like instantaneous alteration of the force and determination of an impulse. How better, then, to determine the fundamental laws of nature than to work on the basis of the dynamical assumptions which were already informing his physico-mathematical understanding of the law of refraction? *In a word, Descartes' exemplar for the constitution of some of the cosmic laws of dynamics was his physico-mathematical understanding of the physical behavior of light, embodied in the law of refraction. From the unarticulated dynamical premises he elaborated the first and third laws of nature, working under the constraint, and through the medium of, his Voluntarist theology and plenist ontology—both sets of drivers acted and acted in a conjoint manner.* As Galileo's mechanics had its exemplars in certain ways of rendering problematical and then explaining pendulum motion and descent along inclined planes, so Cartesian dynamics had its exemplar in Descartes' physico-mathematical rationale for the law of refraction of light.

4.8.2 In a Spin Over Tennis Balls and Boules of Second Element: Cartesian Dynamics, Optics and the Problem of Color

Anyone the least familiar with the *Dioptrique* and who has followed the argument thus far will no doubt be wondering why Descartes chose to employ the tennis ball model in the first public exposition of his optics. In Sect. 3 we saw that the tennis ball demonstrations of the laws of optics make sense only when supplemented by a knowledge of Descartes' dynamics, which contemporary readers could only have gained from the suppressed *Le Monde*. We were able to recognize cryptic hints about Descartes' dynamics between the lines of the *Dioptrique* only after familiarizing ourselves with the relevant portions of *Le Monde*. What is more, we have discovered that kinematic tennis ball type models of light probably played no role in the long gestation of Descartes' physico-mathematical optics from the 1620 fragment down to the *Regulae* and bent arm balance beam analogy of the late 1620s. If our reconstructions are accepted, they seem to entail that Descartes committed a rhetorical miscalculation in the *Dioptrique*, when he suddenly elected to use a kinematic model for light and almost completely neglected to provide it with an adequate and explicit dynamical rationale which could link it to his real theory of light as a mechanical impulse.

The canons of historical interpretation suggest that perhaps there is something wrong with our reconstruction, if it entails such an unflattering picture of Descartes' capacities. In this section I want to avoid this conclusion by showing why Descartes himself probably believed that the tennis ball model could do an adequate job in the *Dioptrique*, despite certain gross limitations of which he was arguably aware. The answer resides in the demands of Descartes' theory of color, which figures prominently later in the *Dioptrique* and *Météores*. That requires the real spatial translation of balls or corpuscles, so that spin/speed ratios can account for colors; yet, you cannot have a ratio of a tendency to spin to a tendency to move. We are about to see that this problem partially explains Descartes' characteristic reticence about color theory at the level of his fully articulated mechanistic theory of light in its systematic setting. Using tennis balls at least allowed Descartes to finesse the problem in his 1637 texts. The tennis ball model could bear the weight of the color theory, and if one did not ask too many questions, it might seem to comport with the idea of the corpuscular basis of light in the behavior of his *boules* of second element. Unfortunately, his color theory and the mechanistic theory of light as tendency to motion transmitted through those *boules* did not cohere. Descartes, I suggest, knew this and struggled with the tensions it generated.

The first step toward grasping Descartes' rationale for the tennis ball model is to understand its wider range of functions in the *Dioptrique* and in the optical portions of the *Météores*. Thus far we have only discussed its use in the demonstration of the optical laws in the second discourse of the *Dioptrique*. In the *Météores* Descartes employed the model in a mechanistic explanation of the causes of the sensations of colors. Descartes was particularly interested in the production of spectral colors when a thin beam of light is refracted through a prism. The explanation of this phenomenon then served as the basis for the explanation of the colors of the rainbow and parhelia. These were among the first problems he addressed in 1629, when he began the work which eventually was embodied in *Le Monde*, the *Dioptrique* and the *Météores*.[74] One must appreciate the importance Descartes would have attached to a general solution to the problem of the (apparent) production of colors through the reflection and refraction of light.

According to Descartes, the tennis balls, whose rectilinear translation models the transmission of light, may also have spin imparted to them when they collide with 'reflecting' or 'refracting' surfaces. In certain situations the spin imparted to the balls is 'nearly equal to their motion in a straight line', and no colors result. But, in other situations, what we may term the ratio of 'spin to speed (of translation)' will be increased or decreased relative to the 'normal' ratio. Such non-normal spin to speed ratios are taken to explain the triggering of sensations of colors, red in the former case, 'blue or violet' in the latter.[75]

[74] To Mersenne, 8 October 1629, AT i. 23. His work at this time is discussed in more detail below, Sections 8.4.3, 8.4.4 and 8.4.5.

[75] *Météores* , AT vi. 331–32

Descartes lays the basis for this approach early in the *Dioptrique*, in the third of a series of analogies or 'comparisons' through which he proposes to explain and illustrate those properties of light relevant to the understanding of the *Dioptrique* and *Météores*, without having to enter upon the details of his 'philosophy' (element theory, dynamics and real theory of light at the corpuscular-mechanical level). The first two analogies explain properties of light travelling through uniform optical media.[76] To explain the phenomena which occur when light encounters a second medium, Descartes introduces the tennis ball model, to which he then adds the spin/speed articulation. He describes how one may impart spin to a tennis ball by grazing or 'cutting' it obliquely with a racket, and he points out how the same thing can happen when a ball bounces obliquely off uneven surfaces. Analogously, colors are produced when rays encounter uneven reflecting surfaces. And, as smooth regular surfaces do not graze the ball, so smooth regular reflecting surfaces do not endow the reflected light with the property of causing the sensation of colors.[77]

Later, in the *Météores*, the explanation of the generation of spectral colors through prismatic refraction, which is fundamental to the explanation of the rainbow and parhelia, proceeds on the basis set down at the beginning of the *Dioptrique*. Dropping all reference to macroscopic tennis balls, Descartes boldly descends to the micro level, to those '*petites boules d'une matiere fort subtile*', whose 'action or movement' constitutes the true nature of light, as, he says, was 'described' in the *Dioptrique*.[78] The *boules*, passing (or tending to pass)[79] through the pores of 'terrestrial bodies', can also acquire spin in certain circumstances. When such *boules* pass obliquely out of the glass prism into the air, their paths are, of course, refracted, and, entering a medium which alters their force of motion, they all acquire a uniform spin in the same direction 'equal to' their rectilinear motion. In this case no colors are produced. But if what we might term the 'beam' of *boules* is narrowed,

---

[76] First, he uses the analogy of the blind man's staff to illustrate the instantaneous propagation of light without the passage of any material (or immaterial) entity. The analogy clearly derives from the pen analogy used earlier in the *Regulae*. As the blind man receives from the far end of his staff only instantaneously conveyed tendencies or resistances to motion, so light rays are only lines of tendency to motion propagated instantaneously through the contiguous particles of optical media. (AT vi. 84–6) The second analogy deals with the rectilinear propagation of light rays, their propagation in infinitely many directions from a luminous point, and their ability to cross without impeding each other. Descartes' model is a vat filled with half crushed grapes and new wine. The analogy is carried out by manipulating putative lines of tendency-to-descend running from wine particles on the surface of the vat to hypothetically voided points on its bottom, a procedure clearly borrowed from the hydrostatics manuscript of 1619. (AT vi. 86–8). On a closely related set of observations, regarding Descartes' theory of light in its cosmic setting in *Le Monde*, see, Sect. 10.7.4 below.

[77] Although he will later deal with the production of colors through refraction of light, Descartes introduces the 'spin/speed' articulation of the tennis ball model in the case of reflection (AT vi. 90–1), because it is much more easily grasped in common sense terms, and because, he has not yet even shown how the simple tennis ball model can be applied to the law of reflection and then extended to the law of refraction.

[78] *Météores*, AT vi. 331.

[79] loc. cit. p. 332.

by blocking off with a shade all but a small area of exit on the refracting surface of the prism, then the *boules* in and near one side of the beam will have their spin/speed ratios increased above their normal amount, whilst those in or near the other side of the beam will have theirs lowered. In the former case the sensation of the color red will be produced in observers; in the latter case 'blue or violet'. The alteration of the spin/speed ratios necessarily follows from the fact that the *boules* at the edges of the beam must graze *boules* at rest, nestled amid the grosser particles of the shade (and of the air proper). Given their previously acquired uniform speed and sense of spin, the *boules* at one edge have their spin increased and those at the other edge have theirs decreased, and these respective effects also propagate inward from the edges of the beam to some distance, through the contact and interaction among the *boules* making up the beam.[80]

From Descartes' perspective the tennis ball model therefore works rather elegantly within the texts of the *Dioptrique* and *Météores*: In unarticulated form (that is without talking about 'spin') the model facilitates the deduction of the laws of reflection and refraction; then a simple articulation allows Descartes to explain the production of colors in these same processes. (In addition, the articulated model at least held out the promise of a general explanation of color phenomena, through the study of the reflection and absorption of light by the varied surfaces of colored bodies.) However, this elegance is achieved in Descartes' texts at some considerable cost, which is chargeable to his views about the real nature of light, and hence to the coherence of the system of natural philosophy he had just created. Descartes, we shall see, was well aware of this liability.

Unfortunately for Descartes, the model for the production of colors works only on condition that the balls, whether tennis balls or *boules* of 'subtle matter', undergo real rectilinear translation, and not merely a 'tendency to motion' or 'action'. 'Grazing' or 'cutting' imparts a real spin, and can do so in the systems of interest to Descartes only as the balls pass by the grazing or cutting surfaces.[81] In such cases there can be no question of merely a 'tendency to rectilinear motion', which might bear some ratio to a spin; or, even worse, to a 'tendency to spin'.[82] There simply is no coherent and convincing analogy in the real theory of light for the spin of the tennis ball or *boules*, or for their mode of acquisition of spin. The articulated tennis ball model therefore cannot be translated into the terms Descartes' real theory of light as an instantaneously propagated mechanical impulse. In this it differs from the unarticulated tennis ball

---

[80] *loc. cit.* pp.331–4. This piece of explanation in turn is fundamental to Descartes' groundbreaking work on the rainbow. The best modern explication of Descartes' research on this classic problem is Buchwald (2008), which also brilliantly demonstrates how within this work Descartes achieved the only instance in his corpus where a corpuscular-mechanical model is applied and further articulated with relation to novel experiments which have quantitative implications.

[81] At times Descartes speaks of a part of the speed of translation of a ball being converted into spin. (eg. AT vi. 90) He was no doubt thinking of everyday macroscopic analogies, such as a tennis ball appearing to lose some its incident speed upon acquiring a spin after bouncing obliquely on the ground.

[82] Descartes uses this infelicitous locution at AT vi. 333.

model used in the proofs of the optical laws. There the model and the real theory map onto each other, provided one attends to the crucial instant of impact with the reflecting or refracting surface, and concentrates upon the instantaneous, rule-bound alteration of the force and/or determination which occurs at that moment.[83]

Nevertheless, this difficulty need not have worried Descartes all that much in so far as he was concerned with the internal coherence and presentation of *Dioptrique* and *Météores*. Since the full details of his real theory of light and of his dynamics were not on display, because of his decision to abandon publication of *Le Monde* consequent upon the condemnation of Galileo, the tennis ball model could be deployed in these texts without appearing to violate the tenets of his real theory. The very absence of the full details allowed Descartes to write in the *Météores* of the translation of the *boules*, a violation of his real theory of light, but a neat and consistent sequel to the (superficially) kinematical optical proofs.

Looking more deeply into this, one realizes that at the level of the published texts the coherence of Descartes' presentation really turned on the dual character of the proofs of the optical laws: On the one hand, the tennis ball optical proofs were based on his dynamics and drew their cogency from the way they modeled instantaneous alterations of force and/or determination. Of course, their true character was only partly inscribed in the text, and for the most part had to be sought between the lines. The dynamical underpinnings were hinted at, and could be mobilized if questions arose, as occurred in the subsequent debates concerning the proofs, for example in Descartes' remarks cited above at Note 25. On the other hand, the optical proofs were presented in an overtly, if superficially, kinematical fashion. As such, they motivated and paved the way for the spin/speed articulation which would explain colors. This therefore marks our return full circle to the optical proofs in the *Diotprique* with which our detective work began. We now understand their implicit dynamical basis and their mode of overt presentation!

In Sect. 4.3 we in effect cast doubt upon Descartes' conceptual and literary skills when we discovered how little of the real dynamical rationale for the optical proofs is present in the *Dioptrique*. Now, however, we can appreciate that Descartes was cleverly adapting to the facts that *Le Monde* had been suppressed and that the *Dioptrique* and *Météores* would therefore appear without any extended discussion of dynamics or the real theory of light as an instantaneously propagated mechanical impulse. What from one perspective seems to have been a miscalculation in Descartes' presentation appears from this new perspective as a quite reasonable strategy of argument, adopted after he had decided that could not then publish *Le Monde* and the system it contained.

This interpretation assumes that Descartes was aware of the difficulty of identifying the spin/speed model with his real theory of light, and that he made his strategic decisions on that basis. Evidence on this score can be gleaned from both of Descartes'

[83] For, as we have established above, at the moment of impact, the tennis ball (reduced to a weightless, frictionless point) behaves exactly the way a light impulse would—indeed dynamically speaking the two are identical—and the superficially kinematical aspects of the model 'momentarily' drop from view.

treatises of systematic natural philosophy, *Le Monde* and *Principia philosophiae*, as well as from the *Météores* itself. Descartes never discussed the spin/speed explanation in detail in his systematic treatises: In *Le Monde* Descartes refuses to mention color as an essential phenomenon of light and so relieves himself of the onus of having to explain color in a manner inconsistent with the rest of his discussion.[84] His behavior, I contend, was quite intentional. Later, in *Principia philosophiae* he still avoided explicit discussion of the spin/speed explanation. With one exception, all questions about the causes of color were dealt with by referring the reader to the *Dioptrique* and *Météores*.[85]

But to say that Descartes was aware of this problem is not to suggest that it always haunted him with equal vigor. The intensity of the problem would have varied from context to context and from time to time. When, in 1644, Descartes finally published a system of mechanical natural philosophy, the problem would have loomed large and caused his evasions. But earlier, in the mid 1630s, when he was committed to suppressing *Le Monde* and only publishing the *Discours de la méthode* and its three *Essais*, he could well have been satisfied with the heuristic and organizational role played by the tennis ball model within the combined texts of the *Dioptrique* and the *Météores*.[86]

[84] When presenting his real theory of light in Chap. 14 of *Le Monde*, he lists 12 properties of light and explains them as arising from tendencies to motion transmitted through the spherical *boules* of his 'second element'. Color is not mentioned explicitly as one of these properties; but, it is implicitly contained in the last two properties, described in terms of capacity of the 'force' of a light ray to be increased or decreased 'by the diverse dispositions or qualities of the matter that receives them'. Descartes' 'explanation' of these properties makes no mention of color and seems intended more to elaborate the explanation of the tenth property, refraction. As for refraction and reflection themselves, Descartes passes up the opportunity to introduce the tennis ball model (or moving *boules*), and simply refers the reader to the *Dioptrique*. (AT x. 97-103)

[85] The exception occurs in an obscure corner of the final part of the French version of the treatise (*Prin*cipia IV 131, AT IXB. 270; MM 241), where Descartes explains the properties of colored glass. Leaving aside this limited and late passage, which is Descartes' and/or Picot's afterthought, we see that Descartes steadfastly refused to introduce the spin/speed model into his systematic work. And the likely reason for this is that the model cannot be made to agree with his real theory of light as a tendency to motion. Further evidence of Descartes' awareness of the problem, and its intractability, may be found in the *Météores*. In the passages discussed above (Note 82 above), Descartes twice writes of the *boules* 'tendency' to move and 'tendency' to spin. Evidently he was caught between the content and the grammar of his real theory, on the one hand, and the mechanical rationale of his spin/speed model, on the other. At this point of tension his discourse falters and wavers, despite the fact that here in the published text of 1637 he could (for the foreseeable future) have gotten away with the pretence that light consists in the translation (and spin) of *boules*.

[86] The little we know about the course of composition of the *Dioptrique* tends to confirm this picture of a Descartes reluctantly satisfied, for the time being, with the tennis ball model in the publications of 1637. The *Dioptrique* is first mentioned in a letter to Mersenne of 25 November 1630 (AT i. 179), over a year after the problems of parhelia and the rainbow had first stimulated his work on a system of corpuscular-mechanical natural philosophy. Descartes writes that he wishes to insert into the *Dioptrique* an explanation of 'the nature of light and colors', a task which has held him up for six months. This will virtually turn the *Dioptrique* into a 'system of physics', an 'abridgment

4.9 Grist for the Method Mill: Method and Optics in Rule 8 of the *Regulae ad directionem ingenii*

Descartes' work in physico-mathematical optics was so important, impressive and rich that he used and exploited it across a range of natural philosophical projects and initiatives. We have looked at Descartes' appropriation of his dynamical rationalization of the law of refraction in his attempt to frame general laws of nature in his first systematic natural philosophical treatise, *Le Monde*. We turn now to Descartes' attempt in 1626–1628 to weave a methodological tale of discovery around his experience in mixed– and physico–mathematical optics over the previous several years. Properly deciphered, Descartes' tale bears witness to some of the complexities, quandaries and pitfalls of his optical work, as revealed by our reconstruction. This episode will also prepare us for the next two chapters, where we turn to the complex trajectory of Descartes' work, and his aspirations, in analytical mathematics and methodology over the same period 1619–1628 which we have analyzed in respect of physico-mathematics in this and the previous chapter.

In rule 8 of the *Regulae* Descartes describes, in a carefully chosen subjunctive mood, how the law of refraction, the anaclastic curve, and the physical explanation

---

of *Le Monde*', and so acquit him of his promise to Mersenne, made in April 1630, to finish the system within three years. He adds that if the reception of the *Dioptrique* shows he can persuade people of the truth, then he will proceed to complete his treatise on metaphysics begun earlier in 1629.

Two main difficulties seem to have been haunting Descartes. First, the explanation of the nature of color had proven a most difficult proposition. One suspects this was not only due to the intricacies of his articulated tennis ball model, but also because of the dawning realization that it bore no convincing analogy in the real theory of the 'nature of light'. Second, Descartes was clearly still undecided about how much material from his emerging system of corpuscular-mechanism should or could appear in the *Dioptrique*. In the letter he toys with the idea of *adding* a section on the true nature of light and color, and thus implying that he already possessed some version of the model-based presentation he later published. Again, part of his hesitation and indecision may have related to the difficulty of linking the spin/speed articulation to his real theory of light. In January 1632 he sent to Golius what he termed 'the first portion of the *Dioptrique*', dealing with 'refractions without touching upon the rest of philosophy'. (AT i. 235) This, too, tends to indicate that Descartes still contemplated publishing in the *Dioptrique* more of his dynamics and real theory of light than we find in the publication of 1637. If so, he was probably then still facing the problem of the relevance of the spin/speed articulation to the real theory.

In the end Descartes' problems were solved on a pragmatic basis, motivated by external events. When he learned of the condemnation of Galileo and decided to withhold *Le Monde* from publication, he reorganized his publication program, producing within three years the *Discours* and three *Essais* in the form with which we are now familiar. The reorganization allowed him to design the *Dioptrique* and the optical portions allotted to the *Météores* around the tennis ball model, without having to face up to the problem of whether the model in its articulated form could represent aspects of the real theory of light. In this respect, perhaps, he came to see the demise of *Le Monde* as something less than a complete disaster, since it allowed him to resolve the problem of presenting and justifying his optical achievements. Again, from this perspective, he may well have viewed the tennis ball model as a qualified success.

of refraction might all have been discovered by using his method. This part of the rule dates from 1626 to 1628: it obviously post-dates the discovery of the law of refraction, the first elaboration of lens theory and the initial attempts to provide a physical rationalization of the law.[87]

Descartes' story in Rule 8 of the methodological investigation of the anaclastic and other problems unsurprisingly contains an initial analysis and a concluding, demonstrative synthesis, and follows the general lines of the method doctrine extractable from the early *Regulae*.[88] The analysis consists in the discovery of that ordered series of questions upon the solution of which the resolution of the anaclastic problem ultimately depends. If, Descartes begins, one were going to search for the anaclastic curve using the method, the initial step would be to see that the solution depends upon first discovering the law of refraction, 'the relation which the angles of refraction bear to the angles of incidence'. At this point, Descartes observes, a mathematician would have to give up the search, for all he can do is assume some relation and work out the consequences. Further analysis shows that the problem of the law of refraction in turn depends upon knowledge of 'physics' as well; for the relation between the angles of refraction and incidence depends in some way upon the manner in which light passes through media. But the answer to that question would be seen to depend on the more general issue of 'what is the action of light', and the answer to that question would be seen to depend in turn upon the answer to the ultimate question in this series, 'what is a natural power?' One would have to determine, by a 'mental intuition' what this 'absolute nature' is.[89] This would be the last step in the analysis and the first in the deductive synthesis.

Unfortunately, Descartes does not inform us as to the content of this 'intuition'; but, we can presume that light and all other natural 'powers' are to be explained mechanically, by corpuscular motion, impact or tendency to motion. In any case, having discovered this by 'intuition' one would have to pursue the rest of the synthesis by proceeding back along the chain of questions, deducing the more relative natures from the less relative ones. However, our deduction might stall at some point, for example, at the step of trying to deduce the nature of light from the nature of natural powers in general. In such cases one would have to proceed by 'analogy'. The investigator must 'enumerate all the other natural powers, in order that the knowledge of some other of them may help him, at least by analogy…to understand this one.' Again, we are not told anything more here about the analogies, but we are acquainted with one of Descartes' favorites from this period, the bent arm balance he was soon to expound to Beeckman.[90] Allowing for such occasional and unpredictable

[87] It can also be shown that it is the first of the passages added to the *Regulae* in Paris and leads directly to the core of the third stratum of the text. See below, Sects. 7.2 and 7.3. Cf Schuster (1980) 58–9.

[88] AT x 393–5.

[89] We shall learn more about Descartes' methodological terms, 'absolute' and 'relative natures' in Chap. 5, where we examine his dream of a universal method and the opening portions of the text of the *Regulae*.

[90] Perhaps he also had in mind other analogies for the action and refraction of light, for example, a rudimentary and unarticulated kinematic model, a tennis ball model; we simply do not know.

recourse to analogy, the synthesis would ultimately lead from a theory of natural powers, via a theory of light to a deduction (and physico–mathematical explanation) of the law of refraction, and thence to a theory of lenses.

As one would expect, Descartes' methodological tale about how he 'could have done optics' bears no relation to the complex trails of research that we have reconstructed in this chapter. In Chap. 6 we shall argue that what when such tales of particular researches are woven out of the discursive cloth of a grand doctrine of method (Descartes' or anyone else's) some characteristic effects follow. On the one hand, the 'thick', sui generis conceptual and procedural density of the field of inquiry in question is necessarily suppressed and lost from view. This entails that the method story really cannot accurately describe any actual or even possible course of genuine practice in that field; it necessarily structurally mystifies the dynamics of knowledge production and evaluation in that field. On the other hand, the little methodological story bears structural similarities to other such stories which can be generated within the same method discourse. To the methodologist, therefore, the story seems to be true, or at least possibly true, and his belief in the unity and efficacy of his method are enhanced by this further 'evidence' of its value.[91]

Given all this, Descartes' story is to be construed as a rationalization of the complex and sometimes abortive course of his researches up through 1626–1628; as an attempt to show that, since the results could in principle have been produced by using method, they should enjoy certain epistemological and methodological accolades. After all, our reconstruction indicates that Descartes' lived experience of 'being an optician and physico-mathematician' had not been entirely happy or tidy. On the one hand, there was the tortuous and none too orderly course of his researches, which had, at long last, produced some results of note. On the other hand, despite or indeed because of these results, he confronted a confusing array of resources, theories, programs and commitments—the disorderly residues of 8 or 9 years of endeavor. Among these we can number (a) a law of refraction discovered using the possibly discredited image locating principle; (b) an unarticulated theory of light as mechanical impulse; (c) two dynamical premises read out of (a) in the light of (b); (d) a body of lens theory in the process of refinement and alteration; and, (e) at least one analogy for the deduction of (a) from (c). Upon this chaos of personal history and conceptual baggage the method tale imposes a double order. There is the diachronic order of an ideal course and flow of research, and conflated with, or contained within, that diachronic order is a logical/explanatory order, revealing the deductive relations holding amongst his theories and principles.[92] From the perspective of a

[91] Schuster (1984, 1986, 1993), Richards and Schuster (1989), and Chap. 6 below.

[92] Like a myth viewed in a Lévi-Straussian perspective, the method discourse provides a structure which imposes order on this jumble of biographical and in part contradictory conceptual meaning-tokens, by means of a narrative of particular events and actions which is, at bottom, yet another instance of his core myth of method. Lévi-Strauss (1972), 216, 224. Alternatively, if one prefers Roland Barthes' view of myth, we might say Descartes' account amounts to a none too convincing rational reconstruction, motivated by a host of personal, philosophical and ideological concerns, and posing as a true story of the discovery. Barthes, 'Myth Today' in (1973), 109–59. We return to these theoretical reflections on 'method-talk' as akin to mythopoeic talk in Chap. 6.

believer (or at least a promoter of method), the virtues of this story are considerable (if only it could have been done this way) and, although virtues here are displayed in a particular case; but they are generic to method-talk, as we shall learn in Chap. 6.

This interpretation further allows us to make sense of two otherwise peculiar aspects of Descartes' tale in Rule 8: (1) his appeal to the use of analogies, and (2) his reticence about the nature of light and natural powers in general.

1. It would appear likely that Descartes introduced an analogy when moving to the step of deducing the nature of light, because in 1626–1628 he simply did not quite know what else to say about the issue. At the time he possessed an unarticulated theory of light as mechanical impulse, two roughly hewn premises read from the Mydorge diagram, and the bent-arm balance beam analogy. The theory of light was not closely articulated to a system of mechanistic natural philosophy; he simply did not have such a system. Similarly, the dynamical premises were not yet part of a systematically theorized dynamics, explicitly forming the causal register of such a larger system of natural philosophy. Leaving aside the Mydorge diagram, read 'physico-mathematically', the only thing holding together the theory of light and the premises was the bent-arm balance analogy: it modeled light as an impulse and it modeled the two premises; and, it could be used to explain/deduce the law of refraction, as we know it was used in October 1628 to explain the law to Beeckman. In rule 8 Descartes is probably simply echoing this as yet unsystematized and unresolved state of affairs. His only alternative would have been to begin discoursing about the Mydorge diagram; his program in physico-mathematics; how to read Kepler; as well as admitting to having used the now arguably superseded traditional image location rule—all amounting to a most unmethodical undertaking, if our reconstructions are to be believed.
2. A similar sort of explanation applies to the question of why Descartes was coy and reticent about the 'nature of natural powers' in general and about the 'nature of light' in particular. We may surmise that Descartes preferred to be non-committal, because as of 1626–1628 he had not yet committed himself to articulated theories on either topic. The beauty of the method tale is that it can accommodate this vagueness and hide it by enfolding it in 'orderliness'. Certainly he had a sketch theory of light, a mechanistic outlook on nature and premises from which to deduce the law of refraction; but none of this was settled or elaborated. Since he had to hand a workable analogy for deducing the law and modeling light, it was better in such circumstances to inject into the tale a sub-discourse on the use of analogy, than it was to imply that any of his currently unsettled ideas might have the status of products of 'intellectual intuition' or 'deduction' therefrom.

A final point about this may now be troubling scrupulous readers, who may feel something has been missed in this account: Does not Descartes' physico-mathematics, as presented thus far, entail a kind of methodological skeleton that Descartes could have exploited here in *Regulae* 8? His physico-mathematical work in hydrostatics, and eventually in optics can be given the following sort of 'method-talk' gloss: *'First in the manner of mixed mathematics, find a simple, workable, and on this superficial level "true" geometrical rule for the phenomenon in question. Then move "analytically" to discover the natural philosophical bases for the law, by inspecting the geometrical representation of law and intuiting or "seeing the causes" in it.* [93] *Finally, in an explanatory synthesis, start with those natural philosophical [matter and cause] premises and deduce the law in question.*' Why in 1626–1628 didn't Descartes give us this sort of methodological tale about (physico–mathematical) optics?

This query can be answered by the following observations. First, one must note again that the above method-gloss, like all such little tales cut from the cloth of a larger doctrine of method, misses the conceptual and practical density of the work that actually produces the results so cavalierly paraded at each methodological step. It cannot capture what Descartes did in mixed mathematical optics, and it cannot tell anyone else how to do that work. Similarly, it cannot capture how and why Descartes selected and designed his natural philosophical commitments, or explain to anyone else how to do so. *A fortiori*, it cannot explain how Descartes or anyone else might 'analyze' natural philosophical results out of mixed mathematical findings.[94] So, even if one prefers the above 'physico-mathematical' method tale to the one Descartes actually tells in Regulae 8, no real grip is going to be obtained on his actual decisions, actions and practices in his course of research. With that point understood, it can still be asked, 'Why did not Descartes report his optical work the following way (which at least conforms to the traditional methodological movement of analysis followed by synthesis)?'[95] (Fig. 4.11).

In Descartes' *Regulae* 8 story, we need natural philosophical insight before we get the law of refraction, and the mixed mathematical discipline of geometrical optics is not mentioned at all. But why not say what he knew (and practiced) very well, that a mixed mathematical optician can find an instrumentally useful and descriptively fairly accurate geometrical expression of the law of refraction, and that this law can be used to design lenses embodying the anaclastic curve? Again, I think we have to say Descartes avoids the actual story of his discovery of the law of refraction because his geometrical optics involved the increasingly dubious traditional image location principle; that is, a bit of outmoded mixed mathematics. Accordingly, there is no mention of geometrical optics as such, and the law of refraction is found

[93] This conceit of 'seeing (natural philosophical) causes inside well grounded mixed mathematical results' emerged in discussion of 'Baroque Optics' with my colleagues, Dr. Ofer Gal (Unit for HPS, University of Sydney) and Dr. Sven Dupré (then of the Department of History of Science, University of Ghent). We have put this notion to work in research on the physico-mathematization of optics in the work of Kepler and Descartes, brought together in a dedicated issue of *Synthèse*. See Schuster (2012).

[94] See below Chap. 6. On the specific issue of *the necessary vacuity of the rules of grand methods* see Schuster (1984, 1986) and, as noted therein the very important, and little noted paper of Paul Feyerabend (1970) on exactly this issue, which is to be preferred to his wider ranging and better known works on method in relation to this critically important point.

[95] And where the arrows in the figure, of course, do not, and cannot, represent strictly valid logical movements.

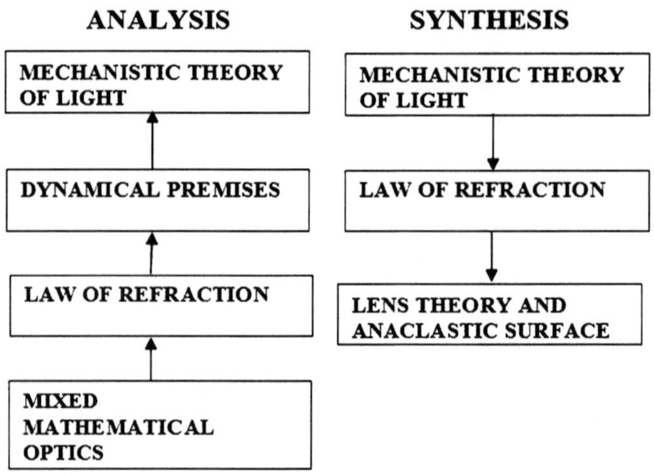

Fig. 4.11 A traditional analysis–synthesis methodological schema for Descartes' optical researches

in the second, synthetic phase of the method story, not at the beginning of its first or analytical phase. Moreover, he did not want to say he could 'analyze' all the way from the law of refraction to natural philosophical premises, because as yet—in 1626–1628—he did not have a settled and systematized natural philosophy to put forward as the outcome of that process. Accordingly, the opening analysis does not consist in findings at all, but in discovery of a sequence of appropriately ordered questions, ending with the ultimate natural philosophical question of 'what is a natural power'. Then, that question is left vague, and the synthetic movement back down toward the discovery of the law of refraction is made to depend on intervening use of analogy, as described above. The law of refraction does not appear early in the (analytical) game based on geometrical optics; and even in the synthesis phase, it is not reached by a conclusive deduction from natural philosophical premises, but emerges from some auxiliary play with models and analogies, rather than firm natural philosophical principles and commitments.

4.10 Conclusion: Looking Forward—Mathematics and Method: 1618–1629

This completes our two chapter reconstruction of Descartes' work in physico-mathematics and embryonic corpuscular-mechanism from 1618 to 1628. If the reconstruction offered carries some degree of plausibility, it brings into relief complex diachronic and conceptual relations amongst Descartes' early enterprises, and demonstrates the centrality of geometrical optics, and optical concerns in general, in the evolution and cross fertilization of his agendas in physico-mathematics and corpuscular-mechanism. His work in physico-mathematical optics had reached a notable climax, but in no way did this signal the onset of some smooth, linear course of work and endeavor, leading to the more mature Descartes we recognize in his published work from the *Discourse on Method* onward. Our next three chapters will address additional complexities and layers of endeavor spanning the period as far back as 1618 and reaching forward to the composition of his first system of corpuscular-mechanical natural philosophy in *Le Monde (1629–1633)*. This part of our story will be complex in itself, since it deals with both solid mathematical work, and rising, concatenating, and increasingly unrealistic and unrealizable aspirations of a general methodological type. We shall also see that Descartes' tortured trajectory in mathematics and method intersected and articulated with the story of physico-mathematics and natural philosophizing which we have told so far, and which in fact cannot be fully understood on its own, but only when this second dimension of the young Descartes' struggles is brought to light.

References

Works of Descartes and Their Abbreviations

AT = *Oeuvres de Descartes* (revised edition, 12 vols.), edited by C. Adam and P. Tannery (Paris, 1964–76). References are by volume number (in roman) and page number (in Arabic).

SG = *The World and Other Writings*, edited and translated by Stephen Gaukroger (Cambridge,1998).

MM = *René Descartes, The Principles of Philosophy*, translated by V. R. Miller and R. P. Miller (Dordrecht, 1991)

MSM = *Rene Descartes, Le Monde, ou Traité de la lumière*, translated by Michael S. Mahoney (New York, 1979).

CSM(K) = *The Philosophical Writings Of Descartes*, 3 vols., translated by John Cottingham, Robert Stoothoff, and Dugald Murdoch, and (for vol. 3) Anthony Kenny, (Cambridge, 1988) References are by volume number (in roman) and page number (in arabic).

HR = The Philosophical Works of Descartes, vol I translated by E.S. Haldane and G.R.T. Ross (Cambridge, 1968 [1st ed. 1911])

Other

Alquié, F. (ed.). 1963. *Oeuvres philosophiques de Descartes*, t.1. Paris, Garnier Frères.

Barthes, Roland. 1957, 1973. *Mythologies*. Paris, Editions du Seuil. English Trans. St. Albans: A.Lavers.

Beeckman, I. 1939–53 *Journal tenu par Isaac Beeckman de 1604 à 1634*, 4 vols. ed. C. de Waard. The Hague: Nijhoff.

Bossha, J. 1908. 'Annexe note', *Archives Neerlandaises des Sciences Exactes et Naturelles*, ser 2 t. 13, pp.xii–xiv.

Buchdahl, G. 1972. Methodological aspects of Kepler's theory of refraction. *Studies in History and Philosophy of Science* 3: 265–298.

Buchwald, Jed Z. 2008. Descartes's experimental journey past the prism and through the invisible world to the rainbow. *Annals of Science* 65: 1–46.

Clarke, Desmond. 2006. *Descartes, a biography*. Cambridge: CUP.

deWaard, C. 1935–6. Le manuscrit perdu de Snellius sur la refraction. *Janus* 39–40: 51–73.

Dijksterhuis, F.J. 2004. Once Snell breaks down: From geometrical to physical optics in the seventeenth century. *Annals of Science* 61: 165–185.

Fermat, Pierre de. 1891–1922. *Oeuvres de Fermat.*, 5 vols. eds. Charles Henry and Paul Tannery. Paris: Gauthier-Villars et fils.

Feyerabend, P.K. 1970. Classical empiricism. In *The Newtonian heritage*, ed. R.E. Butts and J.W. Davis, 150–170. London: Blackwell.

Gabbey, A. 1980. Force and Inertia in the seventeenth century: Descartes and Newton. In *Descartes: Philosophy, mathematics and physics*, ed. S. Gaukroger, 230–320. Sussex: Harvester.

Gaukroger, S. 1976. Bachelard and the problem of epistemological analysis. *Studies in History and Philosophy of Science* 7: 189–244.

Gaukroger, S. 1995. *Descartes: An intellectual biography*. Oxford: OUP.

Gaukroger, S. (ed.) and Trans. 1998. *Descartes, the world and other writings*. London: Cambridge University Press.

Gaukroger, S. 2000. The foundational role of hydrostatics and statics in Descartes' natural philosophy. In *Descartes' natural philosophy*, ed. S. Gaukroger, J.A. Schuster, and J. Sutton, 60–80. London: Routledge.

Gaukroger, S., and J.A. Schuster. 2002. The hydrostatic paradox and the origins of Cartesian dynamics. *Studies in History and Philosophy of Science* 33: 535–572.

Korteweg, D–.J. 1896. Descartes et les manuscrits de Snellius d'après quelques documents nouveau. *Révue de Métaphysique et de Morale* 4: 489–501.

Kramer, P. 1882. Descartes und das Brechungsgesetz des Lichtes. *Abhandlungen zur Geschichte der Mathematischer (Natur) Wissenschaften* 4: 235–278.

Knudsen, O., and K.M. Pedersen. 1968. The link between "Determination" and conservation of motion in Descartes' dynamics. *Centaurus* 13: 183–186.

Lévi-Strauss, Claude. 1972. *Structural anthropology*. Trans. C. Jacobson and B.G. Schoepf. Harmondsworth: Penguin.

Lohne, J. 1959. Thomas Harriot (1560–1621) The Tycho Brahe of optics. *Centaurus* 6: 113–121.

Lohne, J. 1963. Zur Geschichte des Brechungsgesetzes. *Sudhoffs Archiv* 47: 152–172.

McLaughlin, P. 2000. Force determination and impact. In *Descartes' natural philosophy*, ed. S. Gaukroger, J.A. Schuster, and J. Sutton, 81–112. London: Routledge.

Mahoney, M. 1973. *The mathematical career of Pierre de Fermat 1601–1665*. Princeton University Press: Princeton.

Mersenne, M. 1932–88. *Correspondence du P. Marin Mersenne*, 17 vols. eds. C. de Waard, R. Pintard, B. Rochot and A. Baelieu. Paris: Centre National de la Recherche Scientifique.

Milhaud, Gaston. 1921. *Descartes savant*. Paris: Alcan.

Mouy, Paul. 1934. *Le développement de la physique Cartésienne*. Paris: Vrin.

Prendergast, T.L. 1975. Motion, action and tendency in Descartes' physics. *Journal of the History of Philosophy* 13: 453–462.

Richards, E., and J.A. Schuster. 1989. The myth of feminine method: A challenge for gender studies and the social studies of science. *Social Studies of Science* 19: 697–720.

Sabra, A.I. 1967. *Theories of light from Descartes to Newton*. London: Oldbourne.

Schuster, J.A. 1977. *Descartes and the scientific revolution 1618–1634: An Interpretation,* 2 vols. unpublished Ph.D. dissertation, Princeton University.

Schuster, J.A. 1980. Descartes' Mathesis Universalis: 1619–28. In *Descartes: Philosophy, mathematics and physics*, ed. S. Gaukroger, 41–96. Sussex: Harvester.

Schuster, John. 1984. Methodologies as mythic structures: A preface to the future historiography of method. *Metascience: Review of the Australasian Assoc. for the History, Philosophy and Social Studies of Science* 1–2: 15–36.

Schuster, J.A. 1986. Cartesian method as mythic speech: A diachronic and structural analysis. In *The politics and rhetoric of scientific method*, ed. J.A. Schuster and R.R. Yeo, 33–95. Dordrecht: Reidel.

Schuster, J.A. 1993. Whatever should we do with Cartesian method: Reclaiming Descartes for the history of science. In *Essays on the philosophy and science of René Descartes*, ed. S. Voss, 195–223. Oxford: OUP.

Schuster, J.A. 2000. Descartes *Opticien*: The construction of the law of refraction and the manufacture of its physical rationales, 1618–29. In *Descartes' natural philosophy*, ed. S. Gaukroger, J.A. Schuster, and J. Sutton, 258–312. London: Routledge.

Schuster, J.A. 2005. "Waterworld": Descartes' cortical celestial mechanics: A gambit in the natural philosophical contest of the early seventeenth century. In *The science of nature in the seventeenth century: Changing patterns of early modern natural philosophy*, ed. Anstey Peter and Schuster John, 35–79. Dordrecht: Springer.

Schuster, J.A. 2012. 'Physico-mathematics and the Search for Causes in Descartes' Optics—1619–37', *Synthèse* 185: 467–499. [published online Dec. 2011 DOI 10.1007/s11229-011-9979-4].

Shea, W. 1991. *The magic of motion and numbers: The scientific career of René Descartes*. Canton, MA: Science History Publications.

Smith, Russell. 2008a. 'Optical reflection and mechanical rebound: The shift from analogy to axiomatisation in the seventeenth century', Part 2. *British Journal for the History of Science* 41(2): 187–207.

Stevin, Simon. 1955–66. *The principal works of Simon Stevin*, 5 vols. eds. Ernst Cronie et al. Amsterdam: Swets & Zeitlinger.

Turbayne, C. 1959. Grosseteste and an ancient optical principle. *Isis* 50: 467–472.

Vollgraff, J.A. 1913. Pierre de la Ramée (1515–1572) et Willebrord Snel van Royen (1580–1626). *Janus* 18: 595–625.

Vollgraff, J.A. 1936. Snellius notes on the reflection and refraction of rays. *Osiris* 1: 718–725.

Westfall, Richard. 1971. *Force in Newton's physics: The science of dynamics in the seventeenth century*. New York: Elsevier.

Chapter 5
Analytical Mathematics, Universal Mathematics and Method: Descartes' Identity and Agenda Entering the 1620s

5.1 Introduction: The Struggle Over Mathematics, Universal Mathematics and Method

The young Descartes, physico-mathematicus, whom we have studied in the previous two chapters was certainly not a builder of natural philosophical systems, nor had he even consistently applied himself to underwriting his physico-mathematical results in corpuscular-mechanical terms, although he had clear leanings in that direction. Even at the time of the optical triumph of the 1620s, the future author of *Le Monde* and the *Principia* was nowhere in sight, and it would be drawing a long bow indeed to characterize Descartes at that stage as having the vocation or identity of a bold would-be conqueror of the field of systematic natural philosophizing. Our problem is that thus far we have, necessarily, taken too narrow a view of the young Descartes. We have to trace in this and the next two chapters the wider trajectories of intellectual endeavor, and shifting self-definition, that Descartes had been pursuing during the very period we have canvassed in the previous two chapters. These projects interacted with, and constantly promised—at least in his view—productively to subsume the 'mere' physico-mathematics we have thus far observed him practicing. And, unsurprisingly, these projects did not concern systematic natural philosophizing, its neo-Scholastic hegemons or increasingly numerous challengers. In these years, Descartes was not simply a *physico-mathematicus*, but he certainly was no serious natural philosophical player; that is, a competitive builder of systems. We need to discover what, in fact, he was doing, and what he thought his identity and agenda might be, so that, in the end, we can understand how, why and with what aims and resources to hand, he turned to natural philosophical systematics and the composition of *Le Monde*, from 1629.

We are going to see that since his early days with Beeckman, Descartes had pursued a set of projects related to physico-mathematics, but far outstripping even it in potential scope and invested hopes. From 1618, Descartes had pursued an analytical, problem-solving oriented agenda in mathematics, which in these respects

J. Schuster, *Descartes-Agonistes: Physico-mathematics, Method & Corpuscular-Mechanism 1618-33*, Studies in History and Philosophy of Science 27,
DOI 10.1007/978-94-007-4746-3_5, © Springer Science+Business Media Dordrecht 2013

resembled his physico-mathematics, or so he thought. Indeed, the parallels he perceived between his mathematical and physico-mathematical work triggered in 1619–1620 his dream of a unified analytical approach to all mathematically based disciplines—practical, pure and physico-mathematical—to which he appropriated the already circulating name 'universal mathematics'. We shall see that this conception was somewhat overblown and overheated, and that despite that fact, or perhaps because of it, this project quickly gave way to the even more encompassing mirage of a universal method. He remained committed to this idea from 1619 right through to the mid and late 1620s, when, after his optical breakthrough, he picked up universal mathematics and method again in detail. It is these compounding enlargements of his mathematical and physico-mathematical agenda—and of his self-understanding of his intellectual identity—that we now have to trace in this and the following two chapters.

We shall see in the present chapter that Descartes' analytical mathematics, and his dreams of universal mathematics and a universal method, involved their own complicated genealogy, which interacted in intended and unintended ways with his work in physico-mathematics and (piecemeal) natural philosophy. Descartes *agonistes*, it turns out, was not just struggling to work out a physico-mathematics with possible corpuscular-mechanical bearings. He was also a master analytical mathematician and dreamer of gigantic and seductive methodological fancies, all of which arguably affected his shifting and evolving self-understandings and agendas. Then, in the next chapter, we shall pause to consider just what Descartes' youthful dream of a universal method entailed, and to what degree, if any, the method in fact could have guided and facilitated his intellectual work. On this important issue, which has constituted a pitfall to generations of Cartesian scholarship, we shall conclude that Descartes' method—like all grand discourses of method, then or now—could not accomplish what it promised, but that it easily created in the minds of willing believers the impression that it could do so. Clearing away epistemological and historiographical obstacles on this issue is essential to any serious work on the history of science and natural philosophy involving Descartes' career, practices and achievements. Accordingly, in Chap. 7 we shall be able to return to the story of Descartes' trajectory in these projects in the 1620s. We shall learn that his concerns with physico-mathematics, universal mathematics and method came to a climax and inflection point in the late 1620s. Working partly in the shadow of Marin Mersenne and his cultural battle against both radical scepticism and radical (religiously heterodox) natural philosophies, Descartes launched out, trying to realize his earlier dream of a methodologically sound 'universal mathematics'. Riding on his physico-mathematical and more purely analytical mathematical results and the confidence they fed, he worked himself into an intellectual and vocational dead end. We shall see that this project, inscribed in the latter portions of his unfinished *Rules for the Direction of the Mind*, did not blossom into a magisterial 'post–Mersennian' work of method and universal mathematics, but collapsed under its own weight of self-generating problems and contradictions. Descartes now had to struggle to redefine his projects and his vocation, and it was only at this point, from 1629, that he set out to become something we have not seen him intending to become at any previous moment. He became the author of a systematic, radical, pro-Copernican and corpuscular-mechanical, new philosophy of nature, embodied first in *Le Monde*, which we shall study in Chaps. 8, 9,10, and 11.

The pivot and basis of our reconstruction will be a careful textual and contextual reading of the surviving text of Descartes' *Rules for the Direction of the Mind*. The text is generally taken to have been composed between 1626 and 1628, having been conceived on a coherent plan as a unified exposition of Descartes' method. Here I develop my earlier work, partially based upon and modifying that of J.-P. Weber, who argued that the *Regulae*, in fact, were composed in stages between 1619 and 1628, and that different strata in the text correspond to different stages in the development and reformulation of Descartes' methodological ideas.[1] The extent of my borrowing from Weber and my sometimes drastic revision of some of his claims are apparent in my previous work and on display here.[2] The fundamental point for my argument is that certain strata in the text can be identified with stages in the development of universal mathematics, which emerged in embryonic form in 1619, just before the idea of method, but which was only articulated later, between 1626 and 1628, under and within the framework of Descartes' ideas about method.[3]

On my modification of Weber's findings, the *Regulae* really consist in three main textual strata, written at different times between 1619 and 1628, with rather different aims in view.[4] The *first stratum*, consisting in a portion of rule 4, is the remnant of a

[1] Schuster (1980) and Weber (1964) '…ce n'est pas une Méthode que les Regulae exposent, mais plusieurs, qui se succèdent, se perfectionnent ou s'annulent mutuellement.' (p.2).

[2] As Schuster (1980) argued in detail, I fully concur with Weber's three main findings:

(1) That Rule 4 of the *Regulae* consists of two autonomous and chronologically skewed segments: one dealing with universal mathematics and dating from mid-1619; the other dealing with method and dating from November 1619.

(2) That a substantial portion of the first eight rules dates from Descartes' earliest period of work on the method in 1619–1621.

(3) That an important break in the aim and content of the text occurs in the middle of Rule 8 (a point we have already touched upon in Sect. 4.9).

Weber deploys these results, and his identifications of other 'strata' in an attempt to show that the text was composed fairly continuously between 1619 and 1628, and that its various layers contain different and often contradictory versions of the method. Because of issues of textual interpretation and dating, discussed below, I do not accept this picture of various methods sedimented into the text. I believe the text teaches one method (in Rules 3–7 essentially), which dates from 1619 to 1620. The bulk of the later portion of the text (Rule 12 forward) was written much later, after 1626, that is after the discovery of the law of refraction, but before 1629, and deals essentially with an elaborated version of the universal mathematics (now taken as framed within and constrained by the method). All these points will be reaffirmed in the course of this and the next two chapters.

[3] In addition to the modified version of Weber's thesis, the most important studies guiding my interpretation of the text have been: P. Boutroux (1900), Brunschvicg (1927), pp. 277–324; Klein (1968), and Buchdahl (1969). What these works have in common is a serious concern with relating Descartes' mathematical and scientific (let us say physico-mathematical!) practices to his methodological pronunciamentos, and realistic views of the conceptual tensions holding within and between various of his intellectual pursuits.

[4] Schuster (1980).

treatise which Descartes planned to compose in mid 1619 on the subject of 'universal mathematics'. Descartes conceived of this 'discipline' in mid 1619, viewing it as some sort of synthesis of his physico-mathematical project and his more purely mathematical researches, in particular his recent work in the generalization of analytical procedures applied to classes of geometrical and algebraic problems. Universal mathematics was supposed to embody general analytical methods applicable to all genuinely mathematical fields, whether pure or physico-mathematical. It was more an enthusiastic post-adolescent dream rather than a practical reality. Descartes overestimated the generality and power of his analytical findings, and, as has been seen, his physico-mathematics was itself a loose assemblage of embryonic concepts and protocols for 'figuring up' and resolving problems of a physico-mathematical type. These difficulties most likely did not become clear to Descartes at the time, for by November 1619 his horizons widened even farther. The half-baked project of universal mathematics was superseded by and encysted within the main lines of his method, the dream of a general analytical machinery suitable for all rational disciplines, mathematical or not. We shall see that Descartes' constructed his doctrine of method by analogically extending concepts embedded in his none too efficacious discourse about universal mathematics. This was done in the winter of 1619–1620, the results being recorded in the *second stratum* in the *Regulae*, rules 1–3, part of 4, and 5–11, excluding some material in rule 8. After we deconstruct Descartes' method and its illusory efficacy in the next chapter, we shall in Chap. 7 see that Descartes arrived back in Paris in 1625, with his apparently effective method in hand, and there was led into the composition of the *third and final stratum* of the *Regulae*. Under the influence of Marin Mersenne, Descartes' *Regulae* project now took the form of returning to the universal mathematics of 1619, which he would attempt to construct in detail, by expanding and extending his 1619/1620 text on method, that is, roughly rules 1–11 of the *Regulae*, in such a way that the now articulated universal mathematics would both express the terms of the method and be shaped by them. It was the collapse of this major initiative in late 1628 that induced the crisis through which Descartes emerged as the system-building natural philosopher of his later years.

In this chapter, we therefore proceed as follows: First we unpack the problematical Rule 4 of the *Regulae*, which teaches both universal mathematics and method. In Sect. 5.2 we look at exactly what Rule 4 tells us about universal mathematics and then, in Sect. 5.3, we compare the passages on universal mathematics with the slightly later passages in Rule 4 which present, in a parallel manner, the idea of a universal method. Next, in Sect. 5.4, we survey Descartes' 1619 projects and aspirations in analytical mathematics, which, we argue in Sect. 5.5, were crossed with the idea of physico-mathematics to produce, before November 1619, the idea of the universal mathematics. Section 5.6 examines what we may term the core of Descartes' method doctrine, as presented both in the *Regulae* and even later in the *Discours*. This sets the stage for Sect. 5.7, where it is shown how the elements of the method doctrine, first mooted in November 1619, themselves emerged from rather overheated and impractical extension and inflation of the vague concept of the universal mathematics.

Woven through these developments, as we follow them in this and later chapters, will be issues about the young Descartes' own understanding of his intellectual agenda and role, his intellectual identity if you will. Each step in the accelerating, indeed dizzying genealogy of projects in 1619 presumably signals, if only momentarily, a state of Descartes' aspiration and self-understanding, as a bold seeker after varieties of intellectual fame. Clearly the physico-mathematics begun with Beeckman, and continued, as we have seen, very successfully into the 1620s was not in itself sufficient or definitive. From very early on, Descartes was an aspiring analytical mathematician, seeking general analytical insights and the powerful problem solving command they would yield. The *mathesis universalis* of 1619 bespeaks a hope to generalize and transcend both physico-mathematics and the enterprise in pure analytical mathematics, rendering Descartes some sort of super mathematician, the man who would finally show what the idea of *mathesis universalis* might really mean. But, in 1619, this was only a momentary flirtation, since he was almost immediately to construct the idea of a universal method through another round of enthusiastic analogical extension of elements of his half–developed universal mathematics. Thus, by late 1619, Descartes' rapidly concatening intellectual and identity affairs had reached a remarkable condition—superheated, yet simultaneously raw and underdone. We are presented with a twenty-two year old Descartes, self-styled prince of analytical mathematicians, practitioner of a new physico-mathematics, and owner of the synthetic meta-discipline of universal mathematics, who, not to stand on any modesty regarding his former endeavors, further thought that he had found a general method that could control work not only in these his pet disciplines, but in all disciplines governed by reason. The issue of René's identity and agenda had certainly reached a post-adolescent stage of megalomaniacal incandescence, and correspondingly, as it stood in late 1619, it lacks any real interest for us in terms of concrete intellectual endeavors and products.

Indeed, the only way by which we can make any sense of the position into which Descartes had enthused himself by late 1619 is to follow the ensuing paths along which he played upon and pursued both universal mathematics and method over the next decade. Here, at least, there are interesting initiatives, practices, successes and failures to explore, above and beyond his apparently continuing identification with his grandiose dreams. (Recall that we uncovered his trajectory in physico-mathematical optics in the 1620s without having previously to explore his mounting enthusiasms of 1619.) As we shall see, in the 1620s the universal mathematics came to represent to Descartes some sort of practicable agenda and hence a serious anchor to his identity and vocation. For, despite its underdeveloped state and transient place at the forefront of his thinking in 1619, his universal mathematics would later resurface with a vengeance, when in the mid to later 1620s he attempted to construct it in great detail. This would for a time become of the focus of his work, and presumably his sense of his identity as some new sort of natural philosophically relevant mathematician (in the radical sense that what would remain of the traditional field of natural philosophizing would be dictated by what could be taken up into universal mathematics via physico-mathematics). But, the commitment to the method would be entangled in this, because as we shall see, the detailed universal mathematics of the later 1620s was clearly meant to take shape within the bounds of his beliefs about method, and to express its principles. So, the dream of the method from 1619 also remained with Descartes and framed his self-understandings. Indeed, it was to do this in a complex way: As we shall later see, at first, in the period 1619–1629, he took the method quite genuinely and seriously as a guide to self-understanding about who he was and what he was doing. But later, after the failure of the intellectual project of the *Regulae* in the late 1620s, the method would remain not as a genuinely held source of identity and agenda, but rather would be more cynically wielded as a tool of presentation and public rhetoric. By that time, and for that reason, Descartes' agenda and identity will have shifted to the ground of 'systematic and boldly innovative philosopher of nature', the guise in which he clearly intended to step forth publicly in *Le Monde*, had it not had to be withheld from publication for other emergent reasons. Throughout all these twists and turns, we shall have occasions to note, in this and later chapters, that the published *Discours* of 1637 offers only dim or occluded views of all this—Descartes' supposed life plan; his actually shifting vocations and agendas; the nature of the method; the engagement with mathematics; and the claim that his post-1629 systematic natural philosophy is 'mathematical' in some serious sense.

5.2 The Universal Mathematics of 1619: Rule 4 of the *Regulae*

We begin with rule 4 of the *Regulae.* This is the only place in Descartes' work or correspondence where he both names and describes (however sketchily) the discipline of universal mathematics. I shall argue that this is an early text, dating from between March and November 1619, and that it contains something close to Descartes' earliest vision of universal mathematics.

Introducing the discussion of universal mathematics in rule 4, Descartes laments the disorderly pursuit of study in the mathematical fields. Often people are satisfied with 'superficial demonstrations', discovered more frequently by chance than by skill. It is therefore no wonder that many abandon mathematics as empty and childish.[5] Pondering this state of affairs, he continues, he was struck by the fact that the ancient Pythagoreans and Platonists had held mastery of mathematics to be a prerequisite for the study of wisdom. They surely, therefore, must have commanded a

[5] *Regulae*, rule 4, AT, X, pp.374–5; CSM, p.18, HR, pp.11–12. References to the *Regulae* in English translation will generally rely on the now standard Cottingham, Stoothoff and Murdoch version, in CSM vol. 1, but occasionally we shall prefer the much older Haldane and Ross in HR vol I. Neither translation was made in the light of the reconstruction of Descartes' trajectories offered here, and so they occasionally need modification or comment, always within the scope of the tenor of the Latin text of AT. Additionally, appeal will be made from time to time to Jacques Bruchschwig's French translation in t. I of the Alquié (1963) edition of the *Oeuvres philosophiques*.

sort of mathematics very different from that now extant. Not that they had a complete knowledge of it, rather merely some traces of it springing from 'primary germs of truth implanted by nature in the human mind', which 'had a very great vitality in that rude and unsophisticated age of the ancient world'.[6] In particular, traces of this true mathematics are to be discerned in Pappus and Diophantus, and may even be glimpsed today in 'the art which goes by the outlandish name of "algebra"', which, if its symbolism and canons of procedure could be rationalized, would display 'that abundance of clarity and simplicity which I believe the true mathematics ought to have'.[7]

Such reflections, claims Descartes, recalled him from particular mathematical studies to the question of what precisely is meant by the term 'mathematics', and why not only arithmetic and geometry, but also astronomy, music, optics, mechanics and several others are termed 'branches of mathematics'.[8] The problem was thus to disengage the exact character of the underlying unity which held together the various mathematical sciences and arts and so fitted them for the name.[9] This, he asserts, will not be too difficult, because anyone with the least schooling easily recognizes matters relating to mathematics and can distinguish them from non-mathematical matters.[10] The answer is that:

> When I considered the matter more closely, I came to see that the exclusive concern of mathematics is with questions of order or measure and that it is irrelevant whether the measure in question involves numbers, shapes, stars, sounds or any other object whatever. This made me realize that there must be a general science which explains all the points that can be raised concerning order and measure irrespective of the subject–matter, and that this science should be termed *mathesis universalis*—a venerable term with a well–established meaning—for it covers everything that entitles these other sciences to be called branches of mathematics. How superior it is to these subordinate sciences both in utility and simplicity is clear from the fact that it covers all they deal with and more besides; and any difficulties it involves apply to these as well, whereas their particular subject–matter involves difficulties which it lacks.[11]

Though the text is tantalizingly cryptic, certain aspects of this universal mathematics are tolerably clear. Perhaps most striking is the limited character of the discipline. Universal mathematics somehow subsumes and is superior to *properly mathematical fields only.* There is no claim to mathematicize all knowledge

[6] *Ibid.* AT, X, pp.375–6; HR, p.12; CSM, p.18.

[7] *Ibid.* AT, X, pp.376–77; CSM, pp.18–19; HR, pp. 12–13.

[8] *Ibid.* AT, X, p.377; CSM, p.19; HR, p.13.

[9] 'Sciences and arts' because Descartes names here, besides geometry and arithmetic, the 'mixed mathematical sciences', such as music, astronomy and optics, as well as mechanics and algebra, which some might not have admitted unconditionally as such 'sciences'. In addition, of course, all the mixed mathematical sciences had practical dimensions, so one arguably can assume that universal mathematics was meant to encompass all domains of mathematics—pure, mixed and practical.

[10] AT, X, p.377; CSM, p.19; HR, p.13.

[11] *Ibid.* AT, X, pp.377–78; CSM, p.19; HR, p.13

(whatever that might mean) and subordinate it to universal mathematics; nor is it even hinted that in some metaphorical sense all knowledge is to be rendered 'mathematics like' and commanded through a suitably extended notion of universal mathematics. Descartes insists that '...almost anyone with the slightest education can easily tell the difference in any context between what relates to mathematics and what to the other disciplines'.[12] So, when he refers to numbers, figures, stars and sounds as among the objects about which questions of measurement arise, it is unlikely he intended a metaphorical extension of 'measure' (or its correlative term 'order') to encompass any and all objects of rational knowledge.[13] Moreover, Descartes goes on to state quite clearly the relation of universal mathematics to 'higher disciplines'. Universal mathematics does not subsume or displace such higher studies; rather, it is to be pursued as a moderately useful introduction to them.[14] There is no indication here that universal mathematics offers methods, tools or concepts directly, or even indirectly, applicable to the actual practice and cultivation of higher studies. But, given that universal mathematics is limited to properly mathematical fields, it was certainly intended to include and subsume Descartes' early physico-mathematics, such as we have unveiled it, since Descartes clearly believed that physico-mathematics involved properly mathematical procedures and protocols for the stating and resolving of problems within its domain.

It is also obvious that Descartes was acquainted with earlier discussions about the possible existence, scope and content of a 'universal', 'general' or 'common' mathematics. Ultimately deriving from passages in Aristotle's *Metaphysics* and more especially in Proclus' *Commentary on the First Book of Euclid's Elements,* these issues were widely canvassed in the sixteenth century against the background of the revival of the study of classical mathematics, debates over the place of mathematics in the scholastic curriculum, and—in some quarters—promotion of algebra as an important

[12] Ibid.

[13] The interpretation advanced here differs from that advocated by Marion in his *Ontologie grise* (1981, 55ff), and in the notes to his admirable French translation of the *Regulae* prepared with the help of the computer-assisted Latin–French Descartes lexicon (Marion 1977, 155–7). Given that Descartes was a working mathematician, the straightforward meaning of the passage cited at Note 11 is the following: Anyone who knows the least mathematics, in the narrow sense of the term, can tell the fields of pure and mixed mathematics from any other non-mathematical discipline. Why the fields of mixed mathematics are called 'parts' or 'branches of mathematics cannot be explained in terms of the meaning of the Greek word mathesis, or 'discipline', for then all those 'more physical branches of mathematics' (Aristotle, *Physics* II, 194a 7ff) would be called mathematics, not parts of mathematics.

[14] *Regulae*, Rule 4, AT, X, p.379; CSM p.20; HR, p.14: Descartes states that his order of study has been to start with the 'simplest and easiest' of disciplines and to master them before moving on. Therefore, he has hitherto cultivated only universal mathematics rather than more advanced or profound sciences. Prior to undertaking higher studies, as he 'hopes to do soon', he will 'try to bring together and arrange in an manner whatever I thought noteworthy in my previous studies'. These findings, collected in 'this little treatise' will serve as an aid to memory so that he may be free to concentrate his mind on his future studies.

(or unique) source of analytical insight in mathematics.[15] The very term 'universal mathematics', concedes Descartes, is a traditional one, and he goes on to castigate earlier writers for pursuing the subordinate mathematical fields, despite the fact that they understand the name 'universal mathematics' and what its object ought to be.[16] However, the degree of Descartes' acquaintance with the traditional literature, and his precise sources within it, cannot at present be specified.[17] It is at least very likely that he had read Proclus, perhaps in Barozzi's Latin edition of 1560, and, as a result of his Jesuit training, was familiar with the relevant Aristotelian texts. Nevertheless, the lack of more precise information will not be absolutely crucial for our present purpose.

Beyond this point, the text does not speak with much clarity about Descartes' views on issues central to the earlier discussions of universal mathematics. For example, in the sixteenth century, much of the debate about the existence and scope

---

[15] Aristotle had alluded to the existence of such a general field, intimating that it consisted in the Euclidean axioms, taken as applicable to any sort of quantity whatever, as well as the Eudoxian theory of proportion which appears in Euclid, Book V. *(Metaphysics E* 1, 1026a 25–7; *M* 2, 1077a 9–10; 174, 1005a *19–22; Posterior Analytics Γ* 10, 76a 37–41; cf. Klein (1968) 158–9. Proclus, in his commentary on Euclid, Book I, had discussed a 'general mathematics' prior to arithmetic and geometry, as well as to the more subordinate fields of astronomy, mechanics and optics. See Proclus (1970), Prologue, Part I, Chapters III, VII and XIV. General mathematics would have provided the principles and procedures constitutive of all the mathematical subjects. Not only the theory of proportion, but also the 'methods' of analysis and synthesis were included. Similarly, some sixteenth-century algebraists, starting with Gosselin and Bombelli, had seized upon Proclus' conception and identified the general science with algebra in the sense of a general analytical discipline dictating the art of discovery in the mathematical fields. (Klein 1968, 148–9, 181). More recently, in the generation before Descartes, Adrianus Romanus (van Roomen) had advanced a conception of universal mathematics reminiscent of that of Proclus; he did not stress the role of contemporary algebra in the field in his first version of a mathesis universalis, but as Paul Bockstaele (2009) has shown, did so in his second attempt following his acquaintance with the work of Viète. It was a rare work (and remains so now) printed but not properly published. An extremely useful survey of the sixteenth-century debates about the existence, content and extent of universal mathematics and its relation to 'metaphysics' and 'dialectic' is contained in Crapulli (1969). Bockstaele (2009, 435) points out that (quite understandably) Crapulli's analysis of van Roomen was based on knowledge of the first version only.

[16] *Regulae,* Rule 4, AT, X, p. 378 ; HR, p. 13; CSM, pp.19–20.

[17] Some scholars have claimed Romanus (Note 15) as the proximate source for Descartes' knowledge of the ideal of universal mathematics. See Weber (1964) 'Appendix A'; and J. Brunschwig, in a note to his useful French translation of the *Regulae* in Alquie (1963, t. I, 98, note 3). The text does seem to derive from Proclus, perhaps by way of Romanus or Ramus. (Klein 1968, 182)) that Descartes seems to have Proclus in mind when he dismisses the importance of inquiring into the origin of the term 'mathematics'. See Proclus (1970) Prologue, Part I, Chapter XV, and *Regulae,* Rule 4, AT, X, p. 377; CSM, p.19; HR, p. 13. Recent work on Romanus includes the very valuable study by Bockstaele (2009) which discusses van Roomen's two attempts to devise a *mathesis universalis,* and briefly (pp. 464–68) canvasses his quite possible influence upon Descartes. It would appear it was van Roomen's much more widely distributed first attempt, lacking an emphasis on algebra, that Descartes might have seen, although there are tantalizing hints that Descartes may have also seen the second version (Bockstaele 2009 467–68). However, it should be noted that these hints occur in portions of the *Regulae* which we will argue were written in Paris 1626–1628, so that, again, van Roomen's second attempt may not have been known to the young Descartes when, in 1619, he formulated his conception of universal mathematics embracing algebra).

of universal mathematics turned on the question of whether the Euclidean common notions, or axioms, and the Eudoxean theory of proportion should be taken to apply to the study of continuous quantities only (and hence solely to geometry and its subordinate disciplines), or whether they also applied to the study of 'multitudes' or discontinuous quantities (and hence to arithmetic and its subordinate fields).[18] Both Proclus and Aristotle had held the common notions and Eudoxean theory to be essential elements in the discipline in question, though they did not exhaust its content. Descartes would clearly have been committed to the broadest view of the applicability of the common notions and theory of proportion, for he held that universal mathematics embraced the classical geometrical analysis preserved in Pappus, Book Seven, as well as the 'arithmetic' of Diophantus and the doctrines of contemporary algebra. But beyond that, it is not at all obvious what role he would have assigned to the theory of proportion in relation to algebra. Did he intend to identify the two by stressing, in the manner of Stevin and Vièta, the interconvertibility of proportions and equations? Or, as some modern commentators insist, did he intend that an improved symbolic algebra be identified with universal mathematics *tout court.*[19] One's doubts in this connection are compounded by the fact that Descartes makes no comment about the overall content of universal mathematics. Does it extend, as Proclus insisted, to methods of analysis and synthesis, or does Descartes' implied stress on analysis exhaust the field?

Similar ambiguity surrounds the connotation of the apparently central terms 'order' and 'measure'. These might signify discontinuous and continuous quantity respectively, and so signal the subsumption of arithmetical and geometrical fields.[20] Alternatively, 'measure' might connote 'quantity in general', regardless of whether it has been abstracted from continuous or discontinuous quantities; and 'order' might connote a concern for discovering systematic ways of unfolding the orders of relations which can hold among such abstracted quantities. There is warrant for this reading in Descartes' later assertion that his teaching is concerned solely with the unfolding of relations among 'measures', so that the problems they present can be viewed as ones of order.[21] Even in rule 4 he remarks that there is 'no difference' whether the 'question of measurement' arises in 'numbers, figures, stars, sounds or any other object'.[22] 'Order' could then refer to the business of studying the relations which can hold among the 'measures' of any and all quantitative objects. Ultimately, it is this interpretation of 'order' and 'measure' which will be vindicated by our reconstruction of the universal mathematics of 1619, as well as our interpretation of the developed version of universal mathematics, constructed in 1626–1628, which

[18] See Crapulli (1969) *passim.*

[19] For example Klein (1968), Boutroux (1900) and Liard (1880) 591, 593.

[20] There is warrant for this later in the text at Rule 14, AT, X, p. 450; CSM, p.64; HR, p. 63.

[21] *Regulae,* Rule 14, AT, X, pp. 451–2; HR, p. 64 is much to be preferred to CSM pp.64–5 which compresses the long paragraph involved and seems to ignore the key phrase 'de quibus evolvenda', in making no mention of 'unfolding' relations.

[22] *Regulae,* Rule 4, AT, X, p. 378; CSM, p.19; HR, p. 13.

we shall examine in Chap. 7. But, to go further in this direction, we must try to date the text we have been examining and attempt to place it in the context of Descartes' work and aims at the time of composition.

5.3 Reading Rule 4: Method and Universal Mathematics

The discussion of universal mathematics occupies the latter two paragraphs of rule 4 (AT, X, p. 374 1.16 to the end of the rule). The first four paragraphs of the rule (AT, X, p. 371 1.1 to p. 374 1.15) describe Descartes' conception of his general method of discovery and they are continuous with the immediately surrounding text (rules 1–3 and 5–7). Jean-Paul Weber maintained that the two portions of the rule are divided by a clear boundary of conception, intention and chronology, and he argued that a proper understanding of the stratigraphy of the rule is crucial to historical reconstruction of the development of Descartes' thoughts on method. In this section we shall follow and amplify Weber's views, using his denotation of the opening portion of the rule as 'rule 4A' and the latter portion as 'rule 4B'.[23]

The differences between the two sections are indeed very striking. At no point does 4A mention universal mathematics, nor does 4B mention the method.[24]

[23] Weber (1964) 7ff. The initial occasion for Weber's division was probably the fact that in the Hanover ms. of the text rule 4B is displaced to the end, after rule 21. Weber argues that Descartes intended the separation, but this becomes much less plausible in the light of Crapulli's republication of the Dutch edition of the *Regulae* of 1684, see Crapulli (1977). This text, like the Latin edition published at Amsterdam in 1701, has no such displacement. Nevertheless, textual and contextual evidence will support Weber's basic contention that 4A and 4B need to be treated as conceptually and chronologically separate. It should be noted that Dr Richard Serjeantson of Cambridge University has recently announced the discovery of a hitherto unknown manuscript of the *Regulae*, to be called the Cambridge Manuscript. He informs me (private correspondence, February 2012) that he hopes to publish an edition of this document in the near future. The new manuscript is characterized by, amongst other things, the fact that it is about forty per cent shorter than the other versions; does not contain rule 4B; ends at rule 16, rather than with the mere title of rule 21, and omits the discussion in rule 12 of 'simple natures'. An initial conjecture about the dating and intent of this document is offered below, Chap. 8, note 73, after we learn more about the composition of the *Regulae* and the reasons for its abandonment in 1628.

[24] The CSM translation uses the term 'method' at least six times (on pages 18–19) in the text we, following Weber, have denominated rule 4B: AT X 375 l.17 to 377 l.15. The Latin text employs in these loci the term 'art' or unambiguous reference to a preceding use of that term. Correctly, HR, and Brunschwig in the Alquié (1963) French edition use the terms 'art' and 'invention' in the corresponding loci. Clearly, it is reasonable for a translator of rule 4 to use the term 'method', if he believes the rule is unitary and devoted to method only. But, if, following Weber, one accepts the conceptual, and temporal disjunction between 4B and 4A, then it is inappropriate to interject 'method' multiple times into rule 4B, where the term does not appear in the Latin. Without the benefit of Weber's arguments, HR and Brunschwig had arrived at very good translations, simply by following the Latin text more literally in this context. These matters are, to put it mildly, of the utmost importance for Anglophone students attempting to decode the *Regulae* in technical detail, whilst relying in whole or in part on the CSM translation.

Universal mathematics is not explicitly said to issue from the method, nor is method explicitly said to derive from universal mathematics. Nowhere are the two enterprises identified.[25] Whereas 4B describes universal mathematics as a discipline of limited scope, applicable to properly mathematical fields only, and bearing only a modest propaedeutic relation to 'higher' studies, rule 4A presents the method in grandiose tones. The scope of method, Descartes writes, should 'extend to the discovery of truths in any field whatsoever'.[26] Waxing enthusiastic he concludes that, 'Frankly speaking, I am convinced that it is a more powerful instrument of knowledge than any other with which human beings are endowed, as it is the source of all the rest'.[27]

Furthermore, as we shall see in more detail below in Sect. 5.6, rule 4A fits precisely into the flow of argument of the first 7 or 11 rules of the *Regulae*.[28] Indeed, it is the very fulcrum of the early portion of the text. By contrast, rule 4B with its universal mathematics produces no echo elsewhere in the early portion of the text, although, as we shall see in Chap. 7, the later portion of the text, composed in Paris in the mid and late 1620s, deals almost entirely with an articulated version of the discipline of universal mathematics. Method, Descartes explains in 4A, consists in

> ...reliable rules which are easy to apply, and such that if one follows them exactly, one will never take what is false to be true or fruitlessly expend one's mental efforts, but will gradually and constantly increase one's knowledge till one arrives at a true understanding of everything within one's capacity.[29]

---

[25] Weber (1964) 5–7. This does not mean that the two projects are not related; they are, as will be argued below. The present point is textual; they are introduced and discussed independently in the text, and that will provide an important key to their chronology and natures.

Though it is nowhere stated in the *Regulae* that universal mathematics issues from the method, Descartes does claim in *Discourse II* (AT, VI, p. 20; HR, p. 93) that a discipline which would seem to be identical with universal mathematics was developed after the discovery of the four rules of method. Weber takes account of this in a footnote (1964, 9 note 34) by pointing out that at least in the Latin version of the *Discourse* (AT, VI, p. 551) the discipline in question demands that all quantities be represented *by straight* line lengths. Since rule 4B makes no such stipulation, Weber concludes that the discipline evolved later and so does not prove the priority of 4A over 4B. My interpretation will tend to support Weber's line, because, on the basis of our findings in Chap. 7, it will be possible to identify the discipline in *Discourse II* more precisely with the mature, explicated form of universal mathematics worked out after 1626 in rules 12–21; for in this explicated form the representation of all quantities in terms of straight lines and rectangles plays a crucial legitimatory role. This interpretation therefore preserves and deepens Weber's claim and explains the residual similarity between 4B and the discipline in *Discourse II.* Traditionally, this passage in the *Discourse* has been the basis of rather speculative assertions about: (a) the priority of the method over universal mathematics; and/or (b) the identification of the discipline (a product of the method) with analytical geometry, simply because of the mention of representation by (straight) lines. On these various options, see the literature cited in Schuster (1980, 41 notes 1 and 2).

[26] Rule 4, AT, X, p. 374; CSM, p.17; HR, p. 11; see Weber (1964) 7–8, 40, 43.

[27] *Ibid.* AT, X, p. 374; CSM, p.17; HR, p. 11.

[28] Weber (1964) 5–6.

[29] *Regulae,* Rule 4, AT, X, pp. 371–2; CSM, p. 16; HR, p. 9.

This, then, as we shall further explicate later, is a two-fold conception of the method: first, there is a rule or rules describing inborn human cognitive faculties productive of true judgments and inferences; second, there are additional largely heuristic rules which offer aid to the inquirer in preparing for or checking-up after inquiries. In the *Regulae* this distinction corresponds to the discussion of 'intuition' (and 'deduction') in rule 3, and the heuristic rules 5–7, of which Descartes wrote that they exhaust the essential content of the (heuristic part of the) method.[30] The former rule gives us a basis for 'distinguishing the true from the false', while the latter ones give a non-exhaustive set of what we may term 'tips' about 'how not to waste our mental efforts to no purpose'. The placement of rule 4, or, to be precise, rule 4A, then becomes clear. Entitled 'There is need for a method for finding out the truth', it comes directly after the discussion of intuition and deduction, and right before a series of heuristic guidelines.

It is a curious fact that rules 4A and 4B, which differ so much in their content, tone and linkage to the surrounding text, are nearly identical in structure and form of argument.[31] In 4A, as in 4B, Descartes points out the futility of disorderly studies (now studies in general, not simply mathematical ones in particular); he alludes to the inborn seeds of truth, from which can grow the discipline in question (now the method, formerly universal mathematics); and he gleans intimations of the discipline from the history of mathematics, in particular, Greek geometrical analysis and contemporary algebra. Neither discipline is derived from the other, instead both are independently derived by parallel arguments.[32]

The structural similarities between 4A and 4B, combined with their contrasting contents and tones, raise the issue of their relative dates of composition, since it is rather implausible that they were composed simultaneously with an integrated rule 4 in mind. It is highly plausible that 4B pre-dates 4A and provided a model for its composition. Loosely following Weber, one need first assume that Descartes initially developed some conception of universal mathematics and intended to write a 'small treatise' about it.[33] Assume secondly, that Descartes subsequently hit upon the grander idea of the universal method of discovery. Descartes might then have tried to model the pivotal passages of his methodological treatise upon important draft sections of his discussion of universal mathematics. This would

---

[30] *Regulae,* Rule 7, AT, X, p. 392; CSM pp.27–28; HR, p. 22. In the *Discourse on Method* this distinction corresponds to that between rule 1, the rule of evidence, and the three following heuristic rules.

[31] These points derive from Weber (1964) 1–4; Marion (1981) 55ff. has also noted the parallel construction between 4A and 4B, but he argues for the essential unity of the two as moments in the elaboration of a unified method.

[32] Again, these arguments are textual, see Note. 25. They do not purport to show that in fact there was no genetic relation between universal mathematics and the method. Weber argues from textual autonomy to genetic autonomy; Marion argues from a supposed textual unity to an underlying identification of the two projects.

[33] Weber (1964) 8, 9, 15. (On the 'small treatise' on universal mathematics, see above Note. 14).

explain the content and tone of rule 4A, its integral relation to the surrounding text, as well as the curious isolation and 'provincial' character of rule 4B, which would then constitute a fossilized relic of a treatise (the 'little treatise' he mentions in rule 4B)[34], the project of which was now subsumed within the scope of the elaboration of the method.

While such textual arguments suggest that 4B preceded 4A, they cannot tell us much about the precise dates of composition. I take it as not contentious in Cartesian circles that Descartes first hit upon the main themes of the method and began to work some of them out in detail in the winter of 1619–1620, following on from his initial insights and self-justifying dreams of November 1619: After all, Part 2 of the *Discourse on Method* describes the origin of the method and its guiding insights in the winter of 1619–1620; and the *Olympica*, some early notes of Descartes partially preserved by Baillet, describe the Descartes' famous three dreams (and his own interpretations) of the night of 10 November 1619, during a period of work on the 'foundations of a marvelous science'.[35] The 'marvelous science' is not explicitly identified as the method, but the dreams can be interpreted (by us) as bespeaking a recent concern with the basic premises of the method.[36] Rule 4A cannot then pre-date November 1619. Weber has gone so far as to suggest that it does date from around that time and that it records Descartes' initial enthusiastic aspirations for the method. Rule 4B, he further concludes, therefore dates from the days or weeks just preceding his experiences of early November.[37] Nevertheless, it must be conceded that these remain only plausible conjectures so long as one attends only to the text of the *Regulae* and to the collateral evidence in the *Discourse* and *Olympica*. Even if one agrees that Descartes began to work out his vision of the method in the winter of 1619–1620, and that universal mathematics very probably preceded it, there is still a real possibility of rule 4A having been written at any time between 1619 and 1628, with 4B preceding it at some distance in time and intention. We therefore first concentrate on the possibility of dating rule 4B to 1619. This can be further confirmed by looking for contextual evidence plausibly bearing on its content and composition. In the next two sections, it will be argued that Descartes' earliest mathematical work in 1619 provides a likely context in which universal mathematics was developed and rule 4B in fact composed. Then, once we know more about the universal mathematics of 1619 embedded in rule 4B, we will be able further to discern that the method discussed in rule 4A and the surrounding rules arose from an attempted analogical extension of that only partly elaborated universal mathematics, and that this enthusiastic and hopeful extension arguably did take place from November 1619, hard on the heels of the inscription of the notion of universal mathematics in rule 4B.

---

[34] See Note 14 above.

[35] AT X pp.179–88.

[36] See Weber (1964) 16. On the status of the dreams and their interpretations see Gouhier (1958) 37ff. and Rodis-Lewis (1971) vol I, 46ff.

[37] Weber (1964) 16–17.

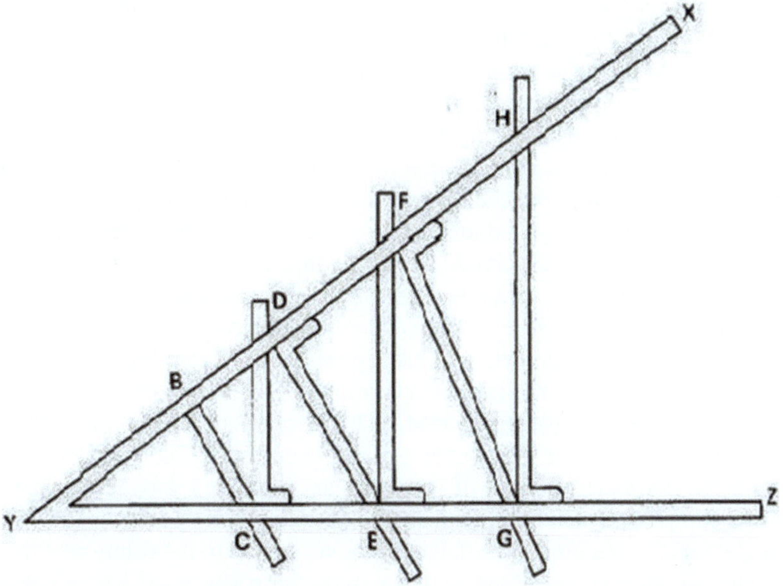

Fig. 5.1 The chief proportional compass 1619

5.4 Straining at the Classical Bit: Descartes' Early Work in Analytical Mathematics

In addition to the physico-mathematical investigations we examined in Chap. 3, Descartes also pursued mathematical researches during the year prior to November 1619.[38] Surprisingly, his focus was not upon algebra nor upon geometrical analysis, nor indeed upon the attempt to relate the one to the other in ways anticipating his mature mathematical thought. He took a rather instrumentalist tack, devoting much effort to devising compasses which would generalize and solve geometrical as well as algebraic problems. One compass in particular deserves notice (Fig. 5.1). It is basically a device for generating magnitudes in continued geometrical proportions, and it is the same instrument introduced twice into the *Geometry* nineteen years later.[39]

---

[38] *Cogitationes Privatae*, AT, X, pp. 234–41.

[39] The compass, which will be termed Descartes' proportional compass, is described in Books II and III of the *Geometry* in terms corresponding to the more crude figures and implied mode of use in the *Cogitationes privatae.* The lettering in the figure is based on that in the *Geometry.* The compass consists of two main branches, YX and YZ, pivoted at Y. Set inside the branches are a series of rulers, of which BC, DE and FG are set at right angles to YX, while CD, EF and GH are set at right angles to YZ. BC is fixed to YX at B, but the bases of the rest of the rulers can slide along the inner side of the branch to which they are set. *As* the compass is opened BC pushes CD along YZ, and CD in turn pushes the base of DE along YX and so on.

In the *Geometry* Descartes uses the compass to show that the curves generated by the motion of points D, F and H are of increasing orders of complexity and can be represented and analyzed by means of algebraic equations reflective of those orders.[40] In 1619 Descartes does not attend to the curves, though he names some of them, nor does he represent them algebraically. Instead, he looks at the actual articulation of the limbs and branches of the physical compass. He sees that many geometrical and algebraic problems can be reduced to problems of determining magnitudes in continued geometrical proportions, and that those so reduced can be modeled to the architecture of the compass and solved. I contend that this manner of exploiting the compass was to serve as the veritable 'exemplar' for his emerging doctrine of universal mathematics. Let us explore in some detail how this came to pass.

Descartes' use of the compass and his aspirations for extending analytical techniques in mathematics emerge in a letter written to Beeckman on *26* March 1619.[41] During the previous six days, he reported, he had found four remarkable 'demonstrations' with the aid of several compasses of his own devising. The first demonstration dealt with the classical problem of trisecting an angle and was accomplished by means of a compass, which in principle could be elaborated to produce the *n*-section of an angle.[42] The remaining three demonstrations related to the solution of the three general classes of cubic equations i.e. those in the form $x^3 = \pm ax^2 \pm c$; $x^3 = \pm bx \pm c$; and $x^3 = \pm ax^2 \pm bx \pm c$[43] Descartes' working notes of the time, preserved in the *Cogitationes privatae*, show how he hoped to use the proportional compass of Fig. 5.1 to produce these demonstrations.[44]

The proportional compass first appears in connection with the solution of the equation $x^3 = 7x + 14$. Descartes commits a curious algebraic error, arguing that after reducing the equation to the form $x^3/7 = x + 2$, he will solve $x^3 = x + 2$, and then 'multiply x^3 by 7'.[45] A similar error runs through the notes in this section of the *Cogitationes*.

The compass is 'a machine for generating series of magnitudes (line lengths) in continued geometrical proportion' (Vuillemin 1960, 112)], for, by similar right triangles CYB, DYC, EYD, FYE, GYF, and HYG it is the case that:

$$\frac{YB}{YC} = \frac{YC}{YD} = \frac{YD}{YE} = \frac{YE}{YF} = \frac{YF}{YG} = \frac{YG}{YH}$$

The compass was obviously designed to construct this series, most probably in the first instance to solve the problem of inserting two mean proportionals between two given line lengths (Milhaud 1921 41).

[40] *Geometry,* III, AT, VI, pp. 442–4.

[41] To Beeckman, 26 March 1619, AT, X, pp. 154–60.

[42] Schuster (1977) 116–118, 124–127; Shea (1991) 40–1; Gaukroger (1995) 93–95.

[43] AT, X, pp. 154–6. Descartes claims to be able to solve '13 species' of cubic falling under these classes; that is, of the 16 possible types of cubic, he rules out those of the form $x^3 = -ax^2 - C$; $x^3 = -bx - c$; $x^3 = -ax^2 - bx - c$. Throughout Descartes' cossic symbolism has been modernized.

[44] Cogitationes Privatae, AT, X, pp. 234–9.

[45] *Ibid.* AT, X, p. 234.

It seems to arise from Descartes' desire to generalize his solutions, and from his still considerable naiveté concerning the principles of the art of algebra. The remainder of the note, however, shows that Descartes' drive for analytical generality may have been the decisive cause here, for he proceeds to show how $x^3=x+2$ can be solved on the compass. YB is taken as a unit and the compass is opened until CE is equal to two units; then YC will be the root.[46] Descartes probably envisioned this procedure as a model for the general 'demonstration' of equations of the form $x^3=bx+c$, as mentioned in the letter to Beeckman of 26 March 1619. But he apparently did not notice that the compass cannot be applied to any 'species' of the equation in which a negative term appears on the right-hand side.[47] He seems rather to have been intrigued by the possibility of generalizing his method of solution. This is apparent both from his haste to eliminate the coefficient of x^3 in the example given, and from his injudicious claim in the letter to be able to extend his 'demonstrations' to all thirteen permissible cases of the cubic, despite the fact that in this example the cases $x^3=-bx+c$, $x^3=bx-c$ and $x^3=-bx-c$ cannot be solved by his instrumental method.

Another entry in the *Cogitationes* confirms this line of interpretation, for it shows Descartes reducing cubics of the form $x^3=ax^2+bx+c$ to the form $x^3=b_1x+c_1$, suitable, he thought, for solution on the compass.[48] Having obtained the latter form, Descartes asserts that the root is extracted 'according to our invention' *(ex invento nostro),* which no doubt means that the compass is to be employed as explained above.[49] Here again Descartes concentrates on the search for general solutions to the exclusion of certain otherwise obvious difficulties.[50]

In the first instance, therefore, the proportional compass had been devised to solve a classical problem in geometry, the insertion of two mean proportionals between two given lines. In constructing the compass so as to generalize the problem

---

46 *Ibid.* AT, X, pp. 234–5. This is obvious from the geometry of the compass. Let $YC=x$, $YD=x^2$ and $YE=x^3$, because: $\frac{1}{YC}=\frac{YC}{YD}=\frac{YD}{YE}$ then, $\frac{1}{x}=\frac{x}{x^2}=\frac{x^2}{x^3}$

And, since $YE=YC+CE$, or $x^3=x+CE$, when CE is opened to two units, the root x can be read off the compass.

47 This is clear from the figure. YE or x^3 can only be constructed as the sum of $YC=x$ and $CE=b$.

48 *Cogitationes Privatae,* AT, X, pp. 244–5 and Enestrom's Note *a* to p. 245.

49 Cogitationes Privatae, AT, X, p. 245, 1.3.

50 For example, his method of reducing the original form was of limited value and was subject to the same errors of algebraic manipulation that he committed above. In addition Descartes continued to fail to see the limitation of his use of the compass to 'species' of this equation with positive terms only. Descartes also applied the compass to the first listed class of cubics in the form $x^3=ax^2+c$ (AT, X, pp. 238–9). Equations of this sort with positive terms can indeed be solved on the compass if $a=1$. (One sets $YB=1$, $YC=\sqrt{x}$, $YD=x$, $YE=x^{3/2}$, $YF=x^2$, $YG=x^{5/2}$, $YH=x^3$. Then one opens the compass until $FH=c$, for then $YF+FH=YH$, or $x^2+b=x^3$ and YD is the root x sought). Descartes, however, erred in assigning the powers of x to parts of the compass. (In effect he set $YC=x$, $YD=x^2$ and $YF=x^3$ so that $DF=c$). The text is so garbled that it is not possible to determine just how much he understood about the possibility (and limitations) of a correct manipulation of the compass.

to the determination of any number of mean proportionals, Descartes aimed to recapitulate and surpass the achievements of the ancients by means of rough-and-ready instrumental solutions. By March 1619 he had also moved hopefully beyond the purely 'geometrical' use of the compass for finding mean proportionals. He had seen, in part mistakenly, that the compass also permitted the solution of certain types of algebraic equations, provided they could be interpreted as proportions and unfolded on the compass. Descartes' enthusiastic use of the compass in order to transcend the immediate geometrical or algebraic statement of problems, and to reduce them to common forms of relation among proportional magnitudes, was soon to constitute the technical core of universal mathematics (in so far as it had one) and to provide exemplars for its discursive elaboration. This, it will now be shown, must have occurred between late March and November of 1619.

5.5 Genesis and Dating of Universal Mathematics

The letter to Beeckman of 26 March 1619 shows that Descartes did not then possess the notion of a universal mathematics as expressed in rule 4B. In the first place Descartes was envisioning, not a unified analytical discipline, but a loose compendium of analytical techniques. His proportional compass and his compass for the sectioning of angles were just two of the tools which he would admit to his compendium.[51] Second, he failed to mention the subsumption of physico-mathematics or

[51] In the letter of 26 March, Descartes says he is undertaking 'not an Ars brevis of Lull but a fundamentally new science, by means of which may be solved all questions which can be proposed about any sort of quantity you wish, both continuous and discrete' (AT, X, pp. 156–7). But, it transpires that this 'science' is only a compendium of techniques, for the passage continues: 'But each one according to its nature: for just as in arithmetic certain questions are solved by means of rational numbers, others only by irrational numbers, others finally can be imagined but not solved; thus I hope to demonstrate that in regard to continuous quantity, certain problems can be solved with only straight lines and circles; others cannot be solved except with other curved lines (than circles), but which are produced by one motion, and therefore they can be drawn by means of new compasses which I do not judge to be less certain and geometrical than the ordinary compass which is used to draw circles; other problems, finally, can only be solved by means of curves generated by motions not subordinated to one another, which (curves) certainly are only imaginary, such as the quadratrix, which is well known. And I think nothing can be imagined which cannot in any event be solved by means of such lines.'

G. Milhaud (1921) 43 commented on this text, 'C'est une sorte de classification complète de toutes les questions relatives à la quantité, selon leur nature, leur solution devant chaque fois y être adoptée.' Note that among the curves used to make constructions Descartes includes the quadratrix, a curve banned from geometry in the *Geometry* of 1637, because it is not produced by a motion 'subordinated to others'; that is, it is not describable by means of a polynomial equation. For a different interpretation of this passage, based, I believe, on a misconstrual of Descartes' attitude toward the quadratrix, see Gäbe (1972) Anhang 1, 113–20.

the mixed mathematical fields within his program. Finally, there was as yet no hint that he was acquainted with the traditional discussions of universal mathematics.[52] For universal mathematics to emerge, therefore, Descartes would have to envision a more unified analytical discipline, one which embraced 'physico-mathematics', and was designed in the light of Proclus' speculations.

Little can be said about how or when Descartes acquired his knowledge of the traditional notion of universal mathematics. It is simply clear from rule 4B that he probably would not have been motivated to design a universal mathematics without the stimulus directly or indirectly owing to Proclus. The case is different as regards the other two factors; the move to an integrated discipline, and the subsumption of physico-mathematics. Even on the basis of the evidence of March 1619, one can begin to conjecture that Descartes was then very near to realizing them and could not have been very long delayed in so doing.

The move to subsume physico-mathematics within some sort of analytical program is not difficult to understand. Although in March 1619 the physico-mathematical researches still appear to have been independent of the compendium of analysis, Descartes already firmly believed that such work should, in principle, depend upon properly mathematical modes of representation, analysis and demonstration. He need only have realized that his compendium had to include tools suitable for physico-mathematics. Although we have no evidence of this score between March 1619 and rule 4B, the small but logical move to include physico-mathematical problems would have constituted the key step between the letter to Beeckman and rule 4B, as far as physico-mathematics were concerned. The central problem in reconstructing the genesis of universal mathematics, therefore, is to understand how the loose and relatively untheorized notion of a compendium came to be transformed into the project for creating an integrated analytical discipline. It is in this respect that the researches with the proportional compass may have become very significant in the period after March, especially if and when they were viewed, as it were, through newly acquired Proclean 'spectacles'.

The early work with the proportional compass showed Descartes that certain sorts of algebraic and geometrical problems were subject to identical types of analytical treatment, provided that they could be rendered in terms of relations between proportional magnitudes. If so, the proportional magnitudes could be represented or instantiated

[52] At the time it seems Descartes' thoughts about a universal science were dominated by a vague interest in the Lullian art (see Note 51, and To Beeckman, 29 April 1619, AT, X, p. 165). Beeckman advised him in effect to stick to 'mechanics' (To Descartes, 6 May 1619, AT, X, p. 168). In the letter of 26 March 1619, Descartes had also confessed that the project of his compendium was an 'incredibly ambitious' one, and that it was 'infinite, not to be accomplished by one person'. This is a far cry from the tone of rule 4B, where universal mathematics is not said to be 'infinite' in this sense, despite the fact that it clearly is intended to subsume all properly mathematical fields. All this suggests that early in 1619 Descartes rightly judged his proposed program to be infinite, or at least *very* demanding, precisely because he well knew that it consisted in a compendium of disparate techniques, and he either had not envisioned or seriously entertained the conception of a unified discipline, such as universal mathematics was intended to be.

on the limbs and branches of the compass, and general solutions could be discovered in the form of procedures—literally manipulations—for the unraveling of the relations between known and unknown line lengths. To extrapolate further from such observations would be to speculate that perhaps problems about *all* types of quantities could and should be rendered in the abstract in terms of relations between some 'magnitudes in general', so that, as a consequence, very general schemas for unpacking classes of relations might be derived and applied. Such a speculation is, of course, very close to the meaning of rule 4B, at least on our preferred reading of 'order' and 'measure'. If Descartes read such meanings into his compass researches, he would have been well on the way to displacing the earlier idea of a 'grab bag' of analytical techniques with the idea of a universal mathematics.

That the work with the compass, though still part of the compendium, was moving toward the centre of his concerns, is evident in the careless enthusiasm of Descartes' notes in the *Cogitationes privatae.* But the notes themselves are not sufficient evidence to carry the weight of my conjecture. It is obvious that if his route to universal mathematics were anything like this, he would sooner or later have had to acquire more specific motivation and direction for his speculations by reading or re-reading Proclus' discussion. Descartes' personal vision of universal mathematics most likely crystallized when, reflecting on Proclus, he thought he saw expressed in his compass work some specific elements of a general analytical discipline which could subsume physico–mathematics, as well as arithmetic, geometry and algebra.

On the one hand, the work with the compass, and the aspirations surrounding it in early 1619, could be interpreted in a new and more ambitious way in the light of Proclus' notion of a universal mathematics. In the mathematical researches could then be seen concrete exemplars for certain otherwise ambiguous, or at least unarticulated, aspects of Proclus' teaching. For example, the compass could be seen to materialize the idea that there are operations and axioms applicable to all species of quantity, a notion traditionally vested by proponents of the existence of a universal mathematics in the Euclidean axioms and Eudoxean theory of proportion. In addition, the representative straight lines realized on the compass were a veritable materialization of the vague idea of a 'magnitude in general', which would be the object of the axioms and operations.

On the other hand, those implications of the use of the compass focused through Proclean spectacles could be promoted to become central elements in Descartes' personal vision of universal mathematics. This, I think, is just what rule 4B illustrates in regard to the concepts of 'order' and 'measure': If rule 4B were written in the light of notions highlighted by the now 'exemplary' work with the compass, it would seem that 'order' and 'measure' should be interpreted in the second of the two manners suggested at the end of Sect. 5.2 above. 'Measure' would then denote quantity or magnitude in general, that which one measures regardless of whether the specific object originally in question was a 'number, figure, star or sound'. 'Order' would then denote a concern for finding and employing very general procedures of analysis. The discovery of these procedures would depend upon insight into the characteristic 'orders' or structures of relation that can obtain amongst the 'magnitudes in general', into whose terms problems have been cast. Read this way, 'order' and 'measure' would render into general terms the methodological advantages and imperatives detectable in concrete instances in the compass researches.

It is just such a reading which Descartes could have performed whilst bearing in mind the pre-existing debates about the nature of a 'common', 'general', or 'universal' mathematics. Descartes' personal vision of universal mathematics most likely therefore crystallized when, reflecting on the culturally available but vague theme of a 'universal mathematics', he saw expressed in the compass researches concrete exemplars for the central concepts of that discipline: [1] an exemplar for the concept of 'magnitude in general', the abstract, generalized quantity capable of subsuming and representing the objects of any mathematical field, whether 'pure' or 'physico–mathematical'; and [2] an exemplar for the conception of generalized analytical procedures consisting in the identification of the classes of structure (or 'order') of 'relations' holding amongst such generalized magnitudes, when they represent various classes of problem. Rule 4B would therefore seem to be the culmination of a course of research, speculation, aspiration and discursive elaboration which led Descartes from physico–mathematics and his early compass researches, through the letter of 26 March 1619, down to the fabrication of universal mathematics prior to November 1619.

Genetic arguments such as these can illumine other aspects of rule 4B. Why, for example, does 4B generally suggest that universal mathematics is mainly or solely concerned with analysis, while at the same time defining the discipline as one about 'order' and 'measure'? The association of the two sets of commitments can be explained by the fact that 'order' and 'measure', articulating and promoting the core of the compass researches, were *ipso facto* delimiting the general conception of universal mathematics as an analytical discipline. Consider also Descartes' rejection of the traditional Aristotelian and Proclean 'philosophical' settings of universal mathematics as respectively subsumed by metaphysics or propaedeutic to dialectic.[53] A route to universal mathematics out of his working experience and aspiration as a mathematician would make sense of the limited and pragmatic tone of universal

[53] Proclus (1970) claimed operative value for his proposed general mathematics, but he stressed even more the role of general mathematics as an object of contemplation and philosophical edification serving as an introduction to higher reaches of philosophy and theology. *(Prologue,* Part II, Chapter II; Part I, Chapters VII, XIV). He had drawn additional metaphysical insight from the analogy between the cosmic architectonic of emanations and the genesis of discursive mathematical knowledge by the unrolling of ideal mathematical concepts from Nous through Intellect down to Imagination *(Ibid.* Part I, Chapter VI; Part II, Chap. 1). Aristotle's allusions to a general mathematical science in the *Metaphysics* point not so much to a generalized technical procedure for use in lower mathematical sciences as toward a philosophical denouement in which the metaphysician co-opts the subject as part of his inquiry into 'being as such'. *(Metaphysics* Γ 4 1005a 19–25).

mathematics in rule 4B.[54] Finally, there is the question of the place of algebra in universal mathematics. As a practicing algebraist, Descartes could have placed the art at the very centre of the discipline. But he did not. Algebra, just like geometrical analysis and Diophantine arithmetic, merely bespeaks the underlying discipline, it is not identified with it. Once again, this position makes sense on the hypothesis that Descartes' universal mathematics arose from valued examples, in which he thought he saw both algebra and geometry being transcended, with new, overriding exemplars for what analysis should be coming into view.

All these arguments serve to link rule 4B to a likely context for the development of universal mathematics between March and November 1619. In effect, we have now argued for the dating of rules 4A and 4B from two different directions. In Sect. 5.3 on 'Reading Rule 4', rule 4A was plausibly—but not definitively—linked to the winter of 1619/1620. Rule 4B was shown on rather stronger textual evidence, derived from Weber, to be a precursor of rule 4A and a model for it. So, if 4A dated from around November 1619, 4B certainly came before. In this section, we have seen independent contextual arguments for the genesis of 4B around the middle of 1619. To all of this a final set of considerations can be added, although we simply note them here, since they will be argued in detail below in Sect. 5.7, in relation to our analysis of the articulation of Descartes' doctrine of method in the period immediately after November 1619. We shall see that certain crucial elements of the method as discussed in rules 5–11 probably arose from the analogical extension of ideas originally embedded in the sorts of problems typical of universal mathematics. A process of enthusiastic analogical extension may well have led Descartes from problems characteristic of universal mathematics, problems about series of proportional magnitudes, to his 'enchainment' vision of knowledge and to his set of heuristic 'tips'.[55] Such a conclusion further suggests, of course, that when the method was developed in the winter of 1619–1620, it was being elaborated hard on the heels of the formulation of the notion of universal mathematics. It also reinforces the idea

[54] More generally, one might note that Descartes had not been 'recalled to study' by Beeckman for the purpose of engaging in school disputes about the explication and articulation of first principles of philosophy, including natural philosophy. Their relationship was colored by a sense of on-going discovery and progress, resting on the basis of the resolution of piecemeal problems. Descartes' impetus to generalization did not arise from the imperatives of a system, metaphysical or natural philosophical: he and Beeckman had, or thought they had, technical grounds for belief in the value of a unified 'physico-mathematics', and Descartes' researches seemed to indicate similar sorts of opportunities. Universal mathematics took shape as a proposed working discipline to be directed toward the practice of its subordinate fields, not as a cog in a philosophical system.

[55] The core notion of 'enchainment' also occurs in the *Discourse, II,* in the famous passage at AT, VI, p. 19 'The long chains of reasonings, every one simple and easy, which geometers habitually employ to reach their most difficult proofs had given me cause to suppose that all those things which fall within the domain of human understanding follow on from each other in the same way…' (Maclean, 2006, 17–18) This implies that the notion is abstracted from the deductive character of geometry; but the very much more elaborate discussion in the *Regulae* is both earlier, and, as we have seen, derived from very particular models in universal mathematics.

that rule 4A and its surrounding rules actually date from 1619 to 1620, and hence that rule 4B, in fact, dates from earlier in 1619; for just as 4A was shown to have been modeled on 4B, so much material in rules 5–11 was possibly modeled upon ideas essential to universal mathematics.

In accordance with the historiographical principles and caveats given in Chaps. 1 and 2 concerning the handling of grandiose agendas that arise from belief in a universal method (or similarly enthusiastically embraced but under theorized programs of huge trans–disciplinary conquest and command), let us conclude this section with some observations on both the objective nature and possibilities of the universal mathematics of rule 4B, as well as Descartes' likely self-understanding of his agenda and identity as a *mathematicus universalis*.[56] To create universal mathematics in 1619, Descartes combined what must have been a hopeful gloss on his existing physico-mathematical case studies with partial, optimistic extrapolations of his compass researches. This amalgam was finished and polished through creative deployment of the culturally available discourse on a 'common' or 'universal' mathematics. Thus, did he construct his initial version of universal mathematics; or to be precise, thus did he cobble together a text on the theme 'universal mathematics'. In a word, the universal mathematics of rule 4B is a textual artifact, and unlike other things sometimes discussed in texts, only a textual artifact.[57] Accordingly, the universal mathematics of rule 4B was not a viable discipline with real possibilities of development and articulation; it was a stillborn dream: As we have seen, it had only the most tenuous links to the curious exercises which Descartes apparently believed to be genuine instances of it. On the plane of actual mathematical practice, from whence universal mathematics would presumably draw its power, its examples and its applications, the state of play was quite parlous: Descartes' researches with the proportional compass were really quite limited; they were laced with errors recognizable by other contemporary algebraicists, and with proclaimed techniques of analytical manipulation that did not even work for all the algebraic problems he had attempted. The wider hope that such generalized techniques would work in the treatment of the sort of physico-mathematical problems he had attempted concerning hydrostatics and fall, was, of course, also a non-starter, once it is looked at with an unblinking, technically attuned eye.

---

[56] See Sects. 1.3.1, 1.3.3 and 2.6. As noted, there is a need to watch both sides of this issue lest pitfalls ensue.

[57] Of course, any gloss of a discipline is a discursive construct; that is not quite the point. Descartes was glossing a discipline, universal mathematics, which had no social and technical/practical density in itself, and which owed its entire two paragraph 'existence' in rule 4B to the literary devices of extrapolating and generalizing from glosses of his 'physico–mathematical' exercises and of the purported import of the proportional compass. Later, in the 1620s, Descartes would indeed return to the project of a universal mathematics, and set out in much more detail, in the latter portions of the *Rules for the Direction of the Mind,* what this discipline would look like. Going beyond the mere two paragraphs of youthful exuberance of rule 4B, this would be a serious and sustained intellectual construction, aimed at articulating the method; explicating the grounds and procedures of universal mathematics; and addressing the cultural politics of scepticism and radical natural philosophizing as seen by the Mersenne circle. These developments will be examined in Chap. 7.

From the perspective of the young Descartes, however, the situation was quite different. He did have before him a few successful special cases and a grand idea, much bandied about in vague terms by others, but now capable of being pinned down, and developed by none other than himself! He no doubt saw universal mathematics as a promising, realizable project, emerging naturally from his previous course of endeavor and reflection, and holding out the opportunity to subsume and rationalize those efforts while setting them in their proper disciplinary matrix, a matrix which would demonstrate, as against previous inconclusive speculation, just what the prized universal mathematics was to be. For a moment in 1619, before the whole undertaking was taken up within the even more grandiose vision of the method, Descartes must have thought he, and he alone, was on the way to showing what this great field of hyper-mathematics would be, and hence may have thought of this as commanding his agenda and identity, in much the same ways the dream of the universal method was soon going to do.[58] He certainly had a right to think that since his March 1619 letter to Beeckman about a 'compendium' of mathematical analysis, his ideas about the subject had both expanded in scope and crystallized into an elegant and powerful machinery.

In the event, however, the initial enthusiasm for universal mathematics, and for the agenda it indicated and the identity it might confer, were soon overridden and subsumed by a second, similar access of grand programmatics, when from November 1619, he hit upon the vision of the universal method, which, we are about to see, was in fact, a vast analogical extrapolation of notions embodied in universal mathematics, notions themselves underdone and overextended. So, we turn now to explore the core of Descartes' teaching on method, particularly as expressed in the *Regulae*. This will help us date the origin of the method as embodied in rules 4A, 1–3 and (most of) 5–11, and hence, as noted, further solidify our dating of rule 4B and the construction of the universal mathematics. It will also make it possible for us to understand the full scope of the delusions of identity and agenda that Descartes' engendered for himself through his emersion in the project of method.

5.6 The Core of Descartes' Method Discourse in the Early *Regulae*

The structural analysis, and demystification, of Descartes' method depends upon the identification of his central methodological claims, the core of his method discourse. This core is to be found in the *Discours de la méthode* and the *Regulae*

[58] When, much later in 1637, Descartes told the reading public about his early encounters with 'mathematics', his puzzlement about its lack of philosophical standing and existence in diffuse, merely useful pockets and domains, despite its unique truth finding and binding capabilities, he may well have been sanitizing, and smoothing out in regard to nearly twenty years of additional thought and struggle, several key episodes in his mathematical and physico-mathematical career, including this excited episode with universal mathematics in 1619. Chapter 6 will offer a full explication of these matters.

ad directionem ingenii, where Descartes offered formal, systematised versions of the method. The texts proclaim themselves to be the pre-eminent vehicles of his method message; and a vast consensus of historical research further supports the claim that these texts were central to Descartes' intentions for systematizing and presenting his method. In what follows, we are of course much more interested in the *Regulae*, since we are concerned with the argument that Descartes' methodological ideas emerged early in his career, from November 1619, and strongly reflect his aspirational trajectory up from analytical mathematics and physico–mathematics, through the ambitious but unelaborated universal mathematics, and ending with the grandiose and self-inflating dream of the universal method. We shall, however, start our examination with the *Discours* 1637. This is not because the *Discours* teaches some more mature, more elaborate and considered method; this is far from the case. Rather, the *Discours* offers an accurate but very much simplified version of the method first taught in the early *Regulae*, and so it is a useful initial object of study, pointing toward that more interesting and important text. In addition, as we shall learn later, the *Discours* offers this thin, almost half hearted exposition of the method. By the time Descartes began to contemplate writing the *Discours*, he had lost confidence in the idea that his method was truly efficacious, and was selling method–talk largely as public packaging for his startling achievements, accomplished, of course, by quite other means. From 1637 the method may have provided some of his public persona, but by that time his self-understanding of his own work in optics, mathematics and natural philosophy had nothing to do with these achievements being 'products of his method'.

I am, of course, well aware that some Cartesian scholars may be sceptical of my identification of the core of Descartes' method discourse—whether in the *Discours*, *Regulae* or both—because they know that Descartes produced disparate and varied informal remarks about methodological matters which are scattered in his correspondence and published works. These remarks include, among other topics, meta-reflections on the explanatory structure of his mechanistic natural philosophy, comments on the argumentative tactics in his *Meditations,* and methodologically relevant debate with critics of his optics and mathematics, particularly after the publication of the *Discours* and accompanying *Essais* in 1637. Commentators have often sifted these texts in search of 'the method'. I would suggest that such projects cannot succeed: one finishes with an account of Descartes' method skewed to the particular selection and weighting made of these texts. For some, the method is an account of how Descartes discovered and/or deployed his arguments in the *Meditations;* for others, it is a supposedly definitive account of how he went about explaining things in his mechanistic natural philosophy or his optics.

There are two profound problems implicated here. The first is how all these different special pleadings, for this or that text to exemplify the use of the method, in this or that special case of application, can be brought together to yield some unified sense of what the method was. The answer is that either there is no core method doctrine, or that some attempt must be made to find those core principles, as we do here. But, beyond this lurks the second question, the more fundamental one which derails any and all attempts to seek to capture a glimpse, here or there, of 'the method truly at work, producing results'. The issue here is the idea, to be argued in Chap. 6, that

Descartes' grand method—and indeed anybody's grand method—is a species of mythic discourse, in the sense that it cannot, for structural reasons, accomplish the feats it claims for itself, whilst that very structure tends to produce textual effects, or illusions to the effect, that such methodological work can be accomplished.[59] So, all the labor expended to find snippets and hints of Descartes using his method amounts to this: First of all, any such text either does, or does not, show Descartes articulating the discursive fabric of the core of his method. This presupposes we locate that core, and it implies that in many instances where it has been claimed Descartes is talking about his method, he is not talking about the core of his method doctrine at all. Secondly, and again more importantly, it also means that even if we find Descartes arguably talking about a case or example, by deploying material from the core of his method, it still remains quite simply impossible that the work being discussed was actually produced by using that method. At best, we are left with certain examples of the rhetorical and legitimatory use of method–talk to package and publicize claims reached by quite other ways and means. So, in the cases alluded to in the previous paragraph, we can conclude that all that happens is that the commentator produces a more or less compelling account of how Descartes went about glossing and defending a particular text, either the *Meditations,* the *Principia philosophiae* or the *Dioptrique*. If this sounds a bit over stated, let us recall that in Chap. 4 we have already seen a very good case, in optics with the discovery of the law of refraction, where Descartes tells a methodological story about the course of his research which in fact, and in principle, had nothing to do with the actual optical practices by which he accomplished that feat. Just that small case should begin to signal to us the abysmal depths to which our historical accounts can dive if we wittingly or unwittingly take at face value the sort of things Descartes, or some of his scholars, say about the method–driven basis of actual courses of his technical work. Let us therefore return to our more promising sequence of tasks: first, in the remainder of this chapter, to analyze the core of Descartes method–talk and to seek its genesis in the period from November 1619; then, in the next chapter to understand how the method functions to produce illusions of efficacy so that, as a consequence, we can improve the way we approach the problem of understanding and narrating the trajectory of Descartes *agonistes*.

So, turning first to the method as set out in the *Discours*, we find in *partie* II four rules of method surrounded by a text which purports to narrate the history of the method. The narrative includes rhetorical elaborations of certain themes, as well as what might be judged more thematically central explications of technical portions of the method.[60] Leaving aside, for treatment in the next chapter, the question of the historical accuracy of the narrative (and hence the further question of the historical

[59] The material in Chap. 6 will build upon my previous work along these lines: Schuster (1986, 1993, 1984), Schuster and Yeo (1986), and Richards and Schuster (1989).

[60] AT, VI, pp. 11–2. Gilson's commentary on these passages and their surroundings runs to seventy pages, bringing out their metaphorical and rhetorical elaboration Gilson (1947) 155–228.

relevance of the elaborations and explications), one can, I think, isolate three related propositions, well founded in the text, which form the context of Descartes' presentation of the four rules of method.

1. All rationally obtainable truths subsist in a network of deductive linkages, and this is the meaning of the 'unity of the sciences'.[61] (This will henceforth be termed Descartes' 'latticework' vision of the unity of the sciences.)
2. As rational beings, humans possess two divinely given faculties for the attainment of truth; the power of intuiting individual truths, and the power of deducing valid links between them.[62]
3. A single mind, exercising intuition and deduction, could in principle traverse the entire latticework; but, some help is required in the form of practical hints or suggestions—heuristic rules—to aid in the preparation of inquiries, the ordering of inquiries, and the checking up after inquiries.[63]

Not surprisingly, therefore, there are two complementary moments or aspects within the statement of the rules of the method. Firstly, there is a doctrine of truth. On the one hand, it informs us of what we presumably already know—that we can intuit and deduce truths. On the other hand, it adduces some negative heuristic advice from this fact: trust not in a authority, nor in unclear, indistinct belief, will or emotion; avoid precipitation and hasty judgment; go only as far as intuition and deduction reveal the truth. All this is essentially contained in rule 1 of the *Discours.*[64] Secondly, there is an open ended set of heuristic rules, initially gathered from easy excursions around the latticework of knowledge. These are contained in part in rules 2, 3, and 4 of the *Discours,* and they advise the inquirer to divide each problem into as many simpler parts as possible; to resolve each sub-problem in due order, starting with the simplest and rising by degrees to the most complex; to assume a fictitious order in a problem when there is no natural one; to assess ('enumerate') relevant aspects and materials of a problem beforehand, and afterwards carefully review ('enumerate') one's steps so that nothing is omitted, no falsehood admitted or truth overlooked.[65] These heuristic rules are not to be applied mechanically. They require sagacity and practical experience. As inquiry proceeds they can be refined and additional rules added.

---

[61] AT VI 11 l.13 –12 l.16 and Gilson's commentary (1947) 157–62; AT VI 19 l.6-16.

[62] AT VI 17 l.11–18 l.23 (the appeal to logic and mathematics as models and the statement of the first rule of the method, cf. below Note 64).

[63] AT VI 18 l.24–19 l.5 (rules 2, 3 and 4); *ibid.* 20 l.25–21 l.6 (elaboration of precepts in the development of the method through mathematical applications).

[64] AT VI 18 l.16–1.23: 'Le premier était de ne recevoir jamais aucune chose pour vraie, que je ne la connusse évidememment être telle: c'est-a-dire, d'éviter soigneusement la précipitation et la prévention: et de ne comprendre rien de plus en mes jugements, que ce qui se présenterait si clairement et si distinctement à mon esprit, que je n'eusse aucune occasion de le mettre en doute.' Cf. Gilson's commentary (1947) 197–204.

[65] On 'enumeration' see the last five paragraphs of this section, and on 'fictitious order' Note 68.

The two moments or aspects of the rules of method are closely interrelated, especially when they are set against the backdrop of the vision of the latticework of rational truths and unity of the sciences. On the one hand, there is little point in developing heuristic rules in the absence of at least some preliminary picture of the cognitive terrain (latticework) the conquest of which they facilitate; on the other hand, the capacity for intuiting and deducing truth is likely to be insufficient *in practice;* that is, heuristic aids are necessary if we are to move about the latticework in an efficient manner, and if we are to learn and profit methodologically from the experience.

Turning now to the *Regulae,* one can say that the opening sections of the text, roughly rules 1 and 11, simply parallel the *Discours*. Rules 1 and 2 present and develop the notion of the unity of the sciences. Rule 3 discusses intuition and deduction, and hence it corresponds to rule 1 of the *Discours* and to the first moment of the method. Rules 5–11 contain detailed heuristic advice, and so they correspond to rules 2, 3, and 4 of the *Discours,* the second moment of the method.[66] What is of interest here, however, are the ways in which the *Regulae* provide a clearer and yet more detailed picture than the *Discours* of the core of Descartes' method doctrine. For our purposes two aspects of this greater elaboration must be considered.

First of all, in rule 6, the reader is treated to Descartes' most elaborate (and we shall find the *earliest)* explication of the vision of the latticework of rational truths. The logical chains of truths consist in 'absolute' terms linked to a 'series' of 'relative' terms through a greater or smaller number of rationally specifiable 'relations' *(respectūs)*.[67] Absolute terms are the initial terms in particular deductive series, and they are themselves relative to a small set of what might be termed 'absolutely

---

[66] Without wishing to foment pedantic controversy and splitting of hairs, one might set out the following parallels: *Discours,* rule 1, corresponds to *Regulae,* rule *3; Discours,* rule 2, corresponds to *Regulae,* rules 5 and *7; Discours,* rule 3, corresponds to *Regulae,* rules 5 and 6; *Discours,* rule 4 corresponds to *Regulae,* rule 7. Cf. F. Alquié (1963) t.l, 587 Note 1; and Weber (1964) 64. With reference to rules 8–11, one should perhaps say 'in material preserved in rules 8–11' and not the literal rules themselves, for it will be shown below in Chap. 7 that Descartes only worked out the shape of the material after the first two paragraphs in rule 8, after 1626. But this does not preclude material intimately linked to the idea of a heuristic method having been initially developed in 1619–1620. Weber (1964) 205, dates rules 9–11 from 1628. Clearly, some sections do post-date the discovery of the law of refraction, and hence, as we know, post-date 1626, for example, the second last paragraph of Rule 9, AT X 402 1.9-28; CSM 34; HR 29–30). But, the overall aim and structure of these rules seem continuous with rules 5, 6, and 7, because they seem to present straightforward commentaries and addenda to the basic heuristic rules offered in the latter rules. Even if incontrovertible evidence appeared for the dating of rules 9–11 after 1626, it would still remain true that these rules are well within the confines of the early heuristic method and have little direct bearing upon the new direction taken in part of rule 8 and in rules 12–21 in the later 1620s, which we shall uncover in Chap. 7.

[67] AT X 381–382; HR 15; CSM 21: 'I call absolute whatever has in it the pure and simple nature in question; that is whatever is viewed as being independent, a cause, simple, universal, single, equal, similar, straight, and so forth; and the absolute I call the simplest and easiest of all, so that we can make use of it in the solution of questions...The 'relative', on the other hand, is what shares the same nature, or at least something of the same nature, in virtue of which we can relate it to the absolute and deduce it from the absolute in a definite series of steps.'

absolute' terms.[68] Relative terms, properly so called, are those occurring further down deductive series. In some degree, they 'participate in the same nature' as their antecedents, the absolutes; but, they are ordered according to the increasing number, and hence complexity, of the 'relations' *(respectūs)* linking them to the absolute term in question. That is, relatives are distanced from their absolute 'to the degree that they contain more relations subordinated one to another'.[69] By keeping in mind this explication of the latticework, we shall find it easier to reconstruct the development of Descartes' method discourse and to subject the discourse to a structural critique.

The second important aspect of the early *Regulae* follows from the first. It is the deceptively simple point that in the *Regulae* the articulated vision of the latticework is the template against which Descartes forms the heuristic rules of the method. Even in the *Discours* the three heuristic rules find their rationale in large measure in the latticework vision. But, in the *Regulae,* thanks to the elaboration of rule 6, the process of manufacture of the heuristic rules against the backdrop of the latticework is in plain view. Indeed, it could be suggested that Descartes' main intention in articulating the latticework was to facilitate the formulation and presentation of the heuristic rules. This may surprise knowledgeable Cartesian scholars who are aware of a large literature concerned with the correct interpretation of the intended ontological and epistemological statuses of Descartes' absolute and relative terms.[70] What must be realized is that Descartes did not stop to wrangle over these issues when composing these early rules. In 1619–1620 he was an aspiring methodologist and not yet the metaphysician and builder of a system of natural philosophy of the post–1628 period. One must read his text keeping in view the nature and trajectory of his methodological

[68] The grounds for introducing the non-Cartesian term 'absolutely absolute' are as follows: Later in rule 6 Descartes indicates that what is absolute and what is relative may vary depending upon the task in question: 'Herein lies the secret of this whole art, that in all things we should diligently mark that which is most absolute. For some things are from one point of view more absolute than others, but from a different standpoint are more relative.' (AT. X 382; CSM 22; HR 16) This implies that Descartes intends a notion of 'methodological' order, created by the subject in furtherance of his cognitive interests, and apparently independent of any 'ontological' order there may be of' absolute and relative terms. J.-L. Marion and others take this as Descartes' definitive view of the matter. (Cf. Marion 1981, 'La fiction de l'ordre', 71–8.) But, one should note that Descartes clearly credits the existence of an objective, let us say 'cosmic' order of truths, given for the subject to explore. In the very next passage, he discusses the existence of some terms which are absolute in respect of any series in which they play a part (AT X 383; CSM 22; HR 16). They are, as one might say, 'absolutely absolute'. Consider, furthermore, that Descartes has already said that 'absolute' applies to such things as 'causes' and 'essences' (pure and simple nature) in the text cited above in Note 67.

[69] Relatives are 'whatever is said to be dependent, or an effect, composite, particular, many, unequal, unlike, oblique etc. (AT X 382 l.8-11 : CSM 21–22 HR 15–16). The translation in the text is neither CMS nor HR but my own conditioned by a preference here for the tenor of the HR translation.

[70] There is an immense literature concerned with the interpretation of Descartes' absolute and relative terms. See, for example, Hartland-Swan (1947); Keeling (1937); Le Blonde (1937); Beck (1952), chapters IV–VI; Gibson (1932), chapter V. For debate between 'realist' and 'neo-Kantian' interpretations of 'absolute' and 'relative' terms and 'simple' and 'compound' natures, see Marion (1981) and O'Neil (1967). Cf. also Ree (1974), 24–5, 36–8.

project (such as we can reconstruct it). Descartes, at this early stage, simply proceeded to use his articulated version of the latticework, along with the doctrine of intuition and deduction, as a basis for constructing some specimen heuristic tips in rules 6–11. This point, too, will be of use in our reconstructing of the making of Descartes' method discourse, and it will be partially confirmed in the process.

Before turning to the heuristic rules in the *Regulae*, we must briefly examine rule 5. Strictly speaking, this rule does not contain any detailed heuristic advice. Occurring before the articulation in rule 6 of the latticework vision, rule 5 simply states that any given question ('proposition') is to be reduced to simpler sub-problems; these are to be resolved in due order, and the overall solution should then be assembled by reintegrating, in order, the sub-solutions.[71] In a word, rule 5 tells us that problem solving involves analysis as well as synthesis. Descartes does not elaborate this point, because it is, after all, highly traditional, and because it represents the least innovative portion of his discourse.[72] The real discursive work in Descartes' presentation, and presumably the main locus of its novelty in his view, resides in the way he tries to elicit concrete heuristic rules from the 'analysis/synthesis' couple, *after* it has been articulated onto the details of the latticework vision.[73] We shall find that Descartes presupposes the existence and necessity of analysis and synthesis as loosely sketched in rule 5. His method consists in a further set of suggestions as to how analysis and synthesis are to be carried out. In his view, these suggestions and aids are discovered in practice and can be ploughed back into practice to facilitate the solving of ever more 'complex' or higher 'order' problems.

In rule 6 one finds what is arguably the principal heuristic rule: reflection upon already mastered deductive series will lead to the acquisition of skill in pursuing new inquiries. It is the 'chief secret of the art' always to note the absolute term in question and the order of relations binding the relatives to it.[74] Here the entire heuristic

---

In the main, this literature sees the problem of interpretation as one of specifying the ontological and epistemological statuses of these entities and their relations to Descartes' later conceptions of 'innate' and 'simple' 'ideas'. Such attempts almost inevitably construe 'absolute' and 'relative' terms in relation to some picture of Descartes' overall 'system', as supposedly expounded in the *Meditations* (1641) and *Principia Philosophiae* (1644). Although this approach can have some merit in elucidating Descartes' later position in metaphysics, it is ill-suited to the project of viewing the text of the *Regulae* in a strict biographical and diachronic framework, for the purpose of analysing the genesis and structure of the method. And, with respect to rule 6 in particular, it would be useful to notice the essentially unexplicated character of these terms in this early context. The vision of cosmic 'enchainment' of knowledge points the way to the formulation of heuristic rules; it is not itself an object of interpretive elaboration, as it was to become in the twentieth century.

[71] AT X 379–80; CSM 20; HR 14–5.

[72] On Medieval and Renaissance concepts of analysis/synthesis, resolution/composition see Crombie (1953); Randall (1940); Gilbert (1960); Jardine (1974). On the relation of these concepts to geometrical and algebraic problem solving and theorem proving procedures see Klein (1968), and Mahoney (1980).

[73] This illustrates, again, the greater elaboration of the heuristic rules in the *Regulae* over those in the *Discourse,* and it also hints at the vacuity of taking off from rule *5* to trace the 'influences' upon it of the traditional methodological discussions of analysis/ synthesis, resolution/composition.

[74] AT X 381 1.7; CSM 21; HR 15.

method first gets off the ground with the suggestion that one study the contexture of some already known and available segment of the latticework. Many of the specific suggestions in the heuristic method follow from this admonition. For example, one will acquire insight into the classes of problems which can arise about the series under scrutiny, and, from that, one will develop sagacity in ferreting out the simplest routes of solution to given classes of these problems.[75] This sort of heuristic insight will be of use when one is again confronted with these sorts of problems, whether by themselves or as sub-portions of a more complex problem.[76]

Rule 7 is perhaps the most complex portion of the text on heuristic method, and for that reason contains important hints about how Descartes composed and inscribed his rules. Ostensibly, it sets out a number of heuristic suggestions, presented as coherent, well motivated variations and extensions of the procedure called 'enumeration'. This surface order is quite specious, however. A close analysis of rule 7 would show that the text was composed as it were 'on the run', and that it contains traces of several shifts and alterations in Descartes' thinking as he came to grips with the necessity of writing down some heuristic rules. A full analysis of the inner dynamics and tensions in the text cannot be undertaken here, where we are mainly concerned to show how the heuristic suggestions, as stated, relate to the background of the latticework vision and doctrine of intuition and deduction. But, to discuss the latter issue requires some attention to the internal complexities of the text. What follows is therefore an uneasy, and, I concede, unsatisfactory mixture: A catalogue of the heuristic advice (and its background) mixed with allusions to some of the discursive dynamics of the text.

[75] AT X 382 1.17-1.18; 383 1.16-1.26, and the extended example which Descartes gives at AT X 384 1.20–387 1.7, and which is discussed below in Sect. 5.7. In this connection, it is worth noting that even when Descartes first veers toward a notion of 'methodological order' and then subordinates it to a notion of 'given', 'cosmic' absolutes (see above Note 68*)*, he still is mainly interested in wresting heuristic capital from this version of the latticework, rather than in exploring the precise ontological statuses of the absolute and relative terms: 'These (few pure and simple natures) we say should be carefully noted, for they are just those facts which we have called the simplest in any series. All the others can only be perceived as deductions from these, either immediate and proximate, or not to be attained save by two or three more acts of inference. *The number of these acts should be noted in order that we may perceive whether the facts are separated from the primary and simplest proposition by a greater of smaller number of steps.*' (AT X 383: HR 16–7; CSM 22) (emphasis on Descartes' heuristic advice added).

[76] A more exhaustive and textually meticulous analysis of rules 6 and 7 would produce the following picture of how Descartes' perceived the 'flow' of methodical procedure, leaving specific heuristic suggestions aside:

1. reflection upon finished series leads to new heuristic insight;
2. analysis and solution of slightly higher order problems mobilizes those insights;
3. synthesis of the problem in *(2) is* followed by
4. reflection upon the newly finished series, leading to more heuristic insight.

Much of the complexity of the text can be explained as arising from Descartes simultaneously elaborating this crude schema, whilst progressively becoming aware of new or differentiated types of heuristic advice required by the schema. See the related comments in the next four paragraphs.

'Enumeration' first appears when Descartes advises us to conduct numerous continuous reviews of completed deductive series. He hopes this process will lead, in many cases, to some sort of unitary validating intuition of the link between the widely separated first and last terms. This is needed to overcome the fact that since long chains of deduction take time, the certainty of the procedure to some degree relies upon the *memory* of previous deductive steps. Enumeration in this sense will minimize the role of memory, and, as indicated, perhaps eliminate it entirely in some lengthy deductions.[77] What seems to be happening here is that Descartes has discovered some problems about the 'certainty of synthesis' which are conditioned by the very terms of his discourse upon 'intuition', 'deduction' and the 'lattice-work'. He proposes to 'solve' these discourse-generated puzzles by adducing some additional heuristic advice, in this case the advice to try to telescope a 'series' of 'deductions' into a 'memory-evading' unitary 'intuition'.

Enumeration, now differentiated as 'sufficient enumeration or induction', also appears as the performance of a non-linear inference based upon the grasp of several independent chains of deductions.[78] This might be necessary, for example, in the final assembly of a synthesis (or in the sub-assembly of the solution of a sub-problem). Here, again, a problem of procedure emerges against the horizon of the discourse on latticework, intuition and deduction; and once again the solution is generated from within the discourse, in the ostensible form of 'further heuristic advice', this time by adducing a new species of the genus 'enumeration'. Thirdly, from the first sense of enumeration, Descartes literally slides into a new sense of enumeration (called 'enumeration or induction') as preliminary categorization of the materials and means relevant to the solution of a problem.[79] And this usage, taken in the context of later portions of the rule, appears to subdivide further into senses of enumeration as: (1) the specification (during analysis) of the relevant sub-problems (and assessment of their solvability!), and as (2) the solution of such sub-problems.[80]

By this point, Descartes is rather neatly entrapped by the momentum of his own discourse: little puzzles and issues about problem–solving procedure keep popping up from within his grand discourse on intuition/deduction, analysis/synthesis, and latticework. The puzzles are laid to rest with the triumphant 'discovery' of further heuristic insights, a trick accomplished by multiplying and ramifying senses of 'enumeration' in terms of the resources of the puzzle generating discourse. One should note, for future reference, *just how far these developments take Descartes*

[77] AT, X, p. 387 1.16- p. 388 1.9 (This would also advance the function of 'reviewing for heuristic insight' (cf. above), but here Descartes is centrally concerned with this 'problem' about memory and certainty.)

[78] AT, X, p. 389 1.8-15.

[79] AT, X, p. 388 1.18- p. 389 1.7 should be carefully examined in this regard.

[80] Subdivision occurs, for example, at AT, X, p. 390 1.6-20; cf. rule 11, ibid. p. 407 1.8- p. 409 1.10. Alquié, op. cit., p.110 Note 1 and p.111 Note 1 comments pertinently on the relations between these senses of 'enumeration'.

from any arguably significant and useful concern with the real practice of any concrete, living field of inquiry. He is happily churning heuristic rules out of his discourse; but, rules which enjoin us to identify and solve sub-problems hardly begin to tell us *how* to identify and solve them in any actual domain of inquiry. Here, again, we begin to sense the illusory, indeed mythic, character of this discursive enterprise, an issue explored in systematic terms in Chap. 6.

Returning to the text of the *Regulae,* one finds more general advice in rules 8–11. For example, to train oneself to intuit well, one should start with simple matters (rule 9). Similarly, to learn to discern the orderly deductive texture of series, one should start with simple and to-hand examples (rule 10); one should not wander where deduction cannot lead, and one must learn to recognize when enumerative reviews need to be 'complete' and when merely 'sufficient' (rule 8). Descartes, insists at the end of rule 7, that virtually the whole of the (heuristic) method consists in these profundities.[81] Descartes has had little real difficulty in finding and formulating his dazzling collection of rules, because, as has been suggested, the rules are epiphenomena of his discourse on analysis/synthesis, intuition/deduction, latticework, series, absolutes, relatives and relations; that is, they are answers to puzzles the discourse might seem to entail; or, they are straightforward, textually conditioned remarks on what 'analysis/synthesis' involves when articulated onto the wonderful, new, method-relevant discourse.

5.7 The Making of Cartesian Method–Talk, Winter 1619–1620

In 1619 Descartes apparently did no further work on universal mathematics beyond what we have identified above in rule 4B. This was because, by November 1619, universal mathematics was overtaken and subsumed by the grander project of the method. Now, we already know (Sect. 5.5, above) that the central elements of his discourse on universal mathematics had been derived as optimistic metaphorical extensions of aspects of his earlier work in physico–mathematics and analytical mathematics. And, we have just seen that at the core of Descartes' teaching on that method in the *Regulae*, one finds a discourse about intuition and deduction, about 'absolute' and 'relative' terms, about 'series' and 'relations', and about the latticework. Then rules 6–11 announce heuristic rules, forged against the template of that discourse. We shall now discover that Descartes' core discourse on method, as well as a number of the heuristic rules, had themselves resulted from the analogical extension of terms and rules constitutive of his discourse on universal mathematics.

In rule 6, there is a remarkable remnant of the sort of process of analogical extension which seems to have produced the method discourse. Near the end of rule 6, the rule which articulates the latticework vision, there is a little mathematical example concerning a series of numbers in a continued geometrical proportion:

---

[81] See AT, X, p. 392, l.1-7

$$\frac{3}{6}=\frac{6}{12}=\frac{12}{24}=\frac{24}{48}$$

Such a series is, of course, typical of the sort of object to be treated in universal mathematics and it was the sort of relation embodied by, and manipulated upon, the proportional compass of 1619. Indeed, it would seem that in universal mathematics all problems were supposed to be reducible to this type of structure. In rule 6, Descartes uses the series to illustrate some of the general heuristic rules he is in the process of unveiling. So, one is advised to intuit the basic defining ratio of the series, and to inspect the order of numbers sequentially generated by reiterated application of the ratio. From this procedure will follow, predictably, insight into how problems about the series may be classified, and hence how to choose the simplest routes of solution: Given that we are dealing with a basic ratio of 2:1, applied first to 3, Descartes points out that there are three orders of difficulty of problem—[1] To find 12 given 3 and 6; or, to find 24 given 3,6, and 12; [2] To find 6, given 3 and 12; [3] To find 6 and 12, given 3 and 24. But to find 6 and 12, given 3 and 48 is really easier than it might look, for it is of order [2]. The route to solution is: First find 12, the mean proportional between 3 and 48, and then 6 and 24, the mean proportionals between 3 and 12 and 48 respectively.[82] By implication, one could go on to apply other heuristic rules to the tasks of posing and solving problems about this series, and increasingly complex series of which it forms a part.

Descartes offers this example as an illustration of the method, and indeed it serves this purpose well. The mathematical structure of the series illustrates the more profound latticework of truths. The intuition of the defining ratio and the inspection of the structure of the resulting proportions mirror the suggestion to inspect the structure of an already mastered deductive series, that is, to mark the absolute and the order of relatives subordinate to it. From inspection of the mathematical series, there flow methodological insights which parallel some rules of the heuristic method. But, the very perfection of the example raises the issue of whether it (or another similar case) was the very model upon which Descartes erected the core of his method discourse. Consider these translations between 'methodological' terms and 'mathematical' terms involved in the example (or in any putative instance of universal mathematics): For the methodological concept of the 'absolute term', *read* 'defining ratio applied to an initial number'; for 'relative terms' *read* 'subsequently generated numbers in continued geometrical proportion'; for the grand 'latticework of rational truths', *read* 'orderly interlinked series of numbers in continued geometrical proportions'; finally, for the heuristic rules of the method, read 'concrete but fairly trivial pieces of advice concerning the solution of problems arising about such series in continued proportion'.[83] Everything we have previously seen about the genesis and structure of universal mathematics reinforces the conclusion that the

[82] AT, X, pp. 384–7; CSM, pp. 23–24; HR, pp. 17–9.

[83] I am not suggesting that all the heuristic rules arise this way; see Note 84 and our earlier discussion in Sect. 5.6 on derivation of the heuristic rules.

'methodological' terms were inscribed by analogical articulation of the 'universal mathematical' terms. Descartes, moving beyond the exciting idea of universal mathematics in rule 4B (itself the product of an ambitious crossing of his mathematical and physico-mathematical interests), analogically extended universal mathematics and happily concluded that *all* knowledge consists in structures of logically related elements; that all such structures have basic and (orderly generated) derivative terms; and that all the heuristic insights adhering to typical examples in universal mathematics can thereby be transformed into generally applicable heuristic rules of method.[84] The method discourse was not directly abstracted from successful practice in some area of mathematics.[85] It was produced by analogical extension of the terms of a discourse, universal mathematics, which itself could not do what it purported to do.

This reconstruction can perhaps be reinforced by the following considerations: Descartes' insights of November 1619—the doctrine of intuition/deduction, the vision of the latticework and the perceived need for heuristic rules—mutually imply each other as interrelated elements in the overriding programme of developing the method. But, as with any system of concepts, the obvious structural relations amongst the elements do not fully account for their respective contents. They did not assume shape and content solely in relation to one another. Pre-given materials, concepts, resources and goals were moulded to give the desired interrelations of elements. For example, although the latticework is correlative with 'intuition', the latter concept does not dictate much about the precise explication of the former, beyond the assertion of valid deductive links between terms. It cannot account for the specific and problematical discourse which emerges concerning 'absolutes', 'relatives', 'series' and 'degree of relation'. Similarly, the correlated notions of intuition and deductive latticework might or might not suggest the notion of heuristic aides. They certainly do not narrow the field of specific candidates for the title 'an official heuristic rule of the method of Descartes'. The point is that the latticework and the idea of heuristic rules were explicated by Descartes with precise exemplars in mind, exemplars, I suggest, which were drawn from reflection upon universal mathematics, or to be precise, from reflection upon the structure of purported examples of that discipline.[86]

All this permits us to look again at the question of dating. In Sect. 5.3, it was shown that rule 4A is closely related to rule 4B, being in fact modeled upon it. In Sects. 5.4 and 5.5, we saw that rule 4B fits precisely into the context and flow of Descartes'

---

[84] Strictly speaking, it is only necessary that the general notion of 'heuristic rules' arise in this connection along with sketches of some few of the detailed rules. Close study of rules 6–11 would show a second rule-generating phenomenon in which emerging problems and tensions in Descartes' method discourse invited and conditioned the formulation or reformulation of rules. We have seen something of this process in the analysis of rule 7 above.

[85] This, of course, is the conventional view, deeply entrenched in Cartesian studies since the advent of their modern phase in the later nineteenth century. Cf. for example, Gibson (1896) and Liard (1880).

[86] See Chap. 6 on the further implications of this view for the requirements of a new, 'non-believer's' history of grand doctrines of method in the history of modern science.

work and aspiration in mid 1619. It has just been argued that rules 1–3 and 5–11 are closely linked to the universal mathematics which emerged during the course of 1619. Just as 4A was modeled upon 4B, so the other rules appear to be based upon the analogical extension of core concepts of the very universal mathematics which 4B discusses. It therefore appears likely that rule 4A, and the text of which it is the pivot, rules 1–3, 5–11, date from the winter of 1619/1620 and constitute Descartes' first detailed version of his method as it grew out of, and beyond, the project of universal mathematics.[87]

5.8 Conclusion: Descartes' Unfolding Agendas and Identities 1618–1620

We have seen that during 1619 Descartes was swept forward on a mounting wave of rapidly evolving ideas about what he would take to be his paramount intellectual project, and accordingly about his agenda and identity as a high cultural player. He started with attempts to advance Beeckmanian physico-mathematics, itself a project displaying both promise of radical and important success (in the hydrostatics manuscript), and worrying signs of limitations (regarding the attack on the problem of local fall). Along with physico–mathematics, he had pursued some interesting moves in mathematical analysis, characterized by a drive for unification and materialization of techniques. He was then driven on, first around mid 1619, to the dream of a universal mathematics, which promised consolidation and elevation of his role as a master (perhaps 'the' master) mathematician of the age—including the conquest of the vast empire of natural philosophy, by virtue of the reduction of its annexable territories to physico-mathematics and the envelopment of physico-mathematics within universal mathematics.[88] And then, in November 1619, he was carried onward, to the founding insights of the method, which implied an agenda as master of all the rationally based disciplines.

[87] Cf Weber (1964) 15–7, 40–7.

[88] This really would have meant the destruction of the hitherto largely discursive realm of natural philosophizing. Only those parts amenable to physico-mathematical treatment would have survived, having been translated and 'shanghaied' to universal mathematics. It was a move not so much within the game of natural philosophizing as over against it. Those historians of science who in recent years have claimed to discern in the Scientific Revolution a mathematisation of natural philosophy, or even the destruction of that field by mathematics and mathematicians, would find here, in Descartes' fantasy program, their best example of the larger supposed phenomenon, except for one trivial problem: Descartes' gambit failed, as he himself acknowledged by all his relevant decisions and actions after 1628. And, of course, no such thing as the larger putative process occurred at all, as this entire volume illustrates, and our sketch of stages and phases in the Scientific Revolution in Chap. 2 foreshadowed. The long term relation of mathematics and mathematicians to natural philosophizing was not murder and displacement, no matter what may presently pass for conventional wisdom amongst some enthused inmates of North American graduate programs.

If our reconstruction carries conviction, we can perhaps recapture some of the excitement which must have gripped Descartes in November 1619, as he worked his way toward his idea of the method. Recalled to study in late 1618 by the vision of Beeckmanian physico-mathematics, he had, by mid 1619, begun to work out his own version of that project, instantiated, by then, in at least two case studies, regarding hydrostatics and accelerated free fall. Shortly thereafter, he started to imagine he could merge this physico-mathematics with his work in compass mediated analytical mathematics to formulate, in the text known as *regulae* 4B, the promising, if only briefly explicated, idea of what the long sought after 'universal mathematics' might be. Then, musing by his stove in the late Bavarian autumn of 1619, he had thought he had seen how to conquer all rational knowledge by further generalizing his earlier revelations. It is little wonder, therefore, that his famous three dreams of 11 November 1619 display a nearly mystical state of enthusiasm over his recent insights, and that he interpreted his third dream of that Saint Martin's Eve as a divine consecration of this the latest and most grandiose of his proposed peak projects (and hence of his imagined future identity and mission). [89]

From Descartes' perspective his path to the method, therefore, would have seemed a marvelous and triumphal progress. However, for us, his trajectory, the excitement it generated and senses of agenda and identity it encouraged, all pose obstacles to historical reconstruction and narrative. As we shall learn in the next chapter (and saw exemplified in Sect. 4.9), we need to exercise extreme care, distinguishing between what Descartes believed his method might accomplish and what in fact such grand method doctrines can achieve. These two issues are linked by the further fact, also to be established in the next chapter, that grand method doctrines, such as that of Descartes, are very good at creating illusions as to their own efficacy—a matter all historians of science need to bear in mind. We have already glimpsed, for example, how Descartes' heuristic rules probably unfolded and branched out during the course of their inscription, and we suggested that, as a consequence, Descartes was probably beginning to be enmeshed in the momentum of his own writing and dreaming about method. Now, this entrapment of a believer, in the enticing discursive toils of a grand method doctrine, is in fact a typical effect of such discourses. It arises, we shall see, from the way that general method discourses generate great powers of textual persuasion as a direct consequence of the very way they are generically structured. In the language we shall develop, the young René was probably beginning to fall for the 'literary effects' of his own method discourse. This entails that we need to exercise great care in dealing with Descartes' method,

[89] See above Note 36 and main text related thereto. In addition, Gouhier (1958) 53–55. A refreshingly commonsensical reminder of Descartes' youthfulness and self-deception as regards these projects was contained in Alice Browne's comment on the ambitions displayed in the letter of 26 March 1619: Descartes' works 'merely express the sort of vague and megalomaniac intellectual ambition many people have in youth' (Browne 1977, 256–7). Browne went on to assert that no one really knows what the 'marvelous science' of November 1619 was (p.258), a perhaps too modest conclusion given the evidence that it was the method.

as well as his own responses to it (not to mention those of later scholars who might also believe, at some level, in the real efficacy of such a grand method). Unless we do so, our understanding of Descartes' technical work in natural philosophy, mathematics and the subordinate sciences will be rendered both ahistorical and epistemologically suspect, and the project of a critical reconstruction of his projects in the various disciplines producing knowledge of nature will be fatally compromised. Therefore, before proceeding further, in Chap. 7 with the narrative of Descartes' projects in the late 1620s, we must pause briefly to consider, in more theoretical and textual detail than we mooted in Chap. 1, the problem of 'method and the search for an historical Descartes'.

References

Works of Descartes and Their Abbreviations

AT=*Oeuvres de Descartes* (revised edition, 12 vols.), edited by C. Adam and P. Tannery (Paris, 1964–76). References are by volume number (in roman) and page number (in Arabic).

SG=*The World and Other Writings*, edited and translated by Stephen Gaukroger (Cambridge,1998).

MM=*René Descartes, The Principles of Philosophy*, translated by V. R. Miller and R. P. Miller (Dordrecht, 1991)

MSM=*Rene Descartes, Le Monde, ou Traité de la lumière*, translated by Michael S. Mahoney (New York, 1979).

CSM(K)=*The Philosophical Writings Of Descartes*, 3 vols., translated by John Cottingham, Robert Stoothoff, and Dugald Murdoch, and (for vol. 3) Anthony Kenny, (Cambridge, 1988) References are by volume number (in roman) and page number (in arabic).

HR=The Philosophical Works of Descartes, vol I translated by E.S. Haldane and G.R.T. Ross (Cambridge, 1968 [1st ed. 1911])

Other

Alquié, F. (ed.). 1963. *Oeuvres philosophiques de Descartes*, t.1. Paris: Garnier Frères.

Barthes, Roland. 1957, 1973. *Mythologies*. Paris, Editions du Seuil. English Transl. St. Albans: A.Lavers.

Beck, L.J. 1952. *The method of Descartes: A study of the Regulae*. Oxford: Oxford University Press.

Bockstaele, Paul. 2009. Between Viète and Descartes: Adriaan van Roomen and the *Mathesis Universalis*. *Archive for the History of the Exact Sciences* 63: 433–470.

Boutroux, P. 1900. *L'imagination et les mathématiques selon Descartes*. Paris: F. Alcan.

Browne, A. 1977. Descartes' dreams. *Journal of the Warburg and Courtauld Institutes* 40: 256–73.

Brunschvicg, L. 1927. Mathématique et métaphysique chez Descartes. *Révue de Métaphysique et de Morale* 34: 277–324.

Buchdahl, Gerd. 1969. *Metaphysics and the philosophy of science*. Cambridge, MA: MIT Press.

Crapulli, G. 1969. *Mathesis Universalis*. Rome: Genesi di un'Idea nel XVI Secolo.

Crapulli, G. (ed.). 1977. *Rene Descartes Regulae ad directionem ingenii: Texte critique etabli par Giovanni Crapulli avec la version Hollandaise du XVII Siecle*. La Haye: Martinus Nijhoff.

Crombie, A.C. 1953. *Robert Grosseteste and the origins of experimental science, 1100–1700*. Oxford: Clarendon.

Gäbe, L. 1972. *Descartes Selbstkritik, Untersuchungen zur Philosophie des Jungen Descartes*. Hamburg: F. Meiner.

Gaukroger, S. 1995. *Descartes: An intellectual biography*. Oxford: OUP.

Gibson, A.Boyce. 1896. La géometrié de Descartes au point de vue de sa méthode. *Révue de métaphysique et de morale* 4: 386–98.

Gibson, A.Boyce. 1932. *The philosophy of Descartes*. London: Methuen & Co.

Gilbert, N.W. 1960. *Renaissance concepts of method*. New York: Columbia University Press.

Gilson, E. (ed.). 1947. *René Descartes, Discours de la Méthode: Texte et Commentaire*. Paris: Vrin.

Gouhier, H. 1958. *Les Premières Pensées de Descartes*. Paris: Vrin.

Hartland-Swan, J. 1947. Descartes 'simple natures'. *Philosophy* 22: 139–52.

Jardine, Lisa. 1974. *Francis Bacon. Discovery and the art of discourse*. Cambridge: CUP.

Keeling, S.V. 1937. Le réalisme de Descartes et le role des natures simples. *Révue de métaphysique et de morale* 44: 63–99.

Klein, Jacob. 1968. *Greek mathematical thought and the origin of algebra*. Cambridge, MA: MIT Press.

Le Blonde, J. 1937. Les natures simples chez Descartes. *Archive de philosophie* 13: 163–80.

Lévi-Strauss, Claude. 1972. *Structural anthropology*. Trans. C. Jacobson and B. G. Schoepf. Penguin: Harmondsworth.

Liard, Louis. 1880. La méthode de Descartes et la mathématique universelle. *Révue Philosophique* 10: 569–600.

Maclean, I. Ed. and Trans. 2006. *René Descartes: A discourse on method*. Oxford: OUP.

Mahoney, M.S. 1980. The beginnings of algebraic thought in the seventeenth century. In *Descartes: Philosophy, mathematics and physics*, ed. S. Gaukroger, 141–55. Sussex: Harvester.

Marion, Jean-Luc. 1977. Ed. and Trans. *René Descartes, Regles utiles et claires pour la direction de l'esprit en la recherche de la verité*. La Haye: Martinus Nijhoff.

Marion, Jean-Luc. 1981. *Sur l'ontologie grise de Descartes*, 2nd ed. Paris: Vrin.

Milhaud, Gaston. 1921. *Descartes savant*. Paris: Alcan.

O'Neil, B.E. 1967. 'Epistemological direct realism in Descartes' Philosophy'. Unpublished Ph.D. dissertation. University of California, Berkeley.

Proclus. 1970. *A commentary on the first book of Euclid's elements*. Trans. G. Morrow. Princeton: Princeton University Press.

Randall, J.H. 1940. The development of scientific method in the school of Padua. *Journal of the History of Ideas* 1: 177–206.

Ree, J. 1974. *Descartes*. New York: Allan Lane.

Richards, E., and J.A. Schuster. 1989. The myth of Feminine method: A challenge for gender studies and the social studies of science. *Social Studies of Science* 19: 697–720.

Rodis-Lewis, Geneviève. 1971. *L'Oeuvre de Descartes*, 2 vols. Paris: Presses Universitaires de France.

Schuster, J.A. 1977. *Descartes and the scientific revolution 1618–34: An interpretation*, 2 vols. unpublished Ph.D. dissertation, Princeton University.

Schuster, J.A. 1980. Descartes' Mathesis Universalis: 1619–1628. In *Descartes: Philosophy, mathematics and physics*, ed. S. Gaukroger, 41–96. Sussex: Harvester.

Schuster, John. 1984. Methodologies as mythic structures: A preface to the future historiography of method. *Metascience: Review of the Australasian Assoc. for the History, Philosophy and Social Studies of Science* 1–2: 15–36

Schuster, J.A. 1986. Cartesian method as mythic speech: A diachronic and structural analysis. In *The politics and rhetoric of scientific method*, ed. J.A. Schuster and R.R. Yeo, 33–95. Dordrecht: Reidel.

Schuster, J.A. 1993. Whatever should we do with Cartesian method: Reclaiming Descartes for the history of science. In *Essays on the philosophy and science of René Descartes*, ed. S. Voss, 195–223. Oxford: OUP.

Schuster, J.A., and Richard R. Yeo. 1986. Introduction. In *The politics and rhetoric of scientific method*, ed. J.A. Schuster and R.R. Yeo, ix–xxxvii. Dordrecht: Reidel.

Shea, W. 1991. *The magic of motion and numbers: The scientific career of René Descartes*. Canton: Science History Publications.
Vuillemin, J. 1960. *Mathématiques et Métaphysique chez Descartes*. Paris: Presses Universitaires de France.
Weber, J.-P. 1964. *La constitution du texte des Regulae*. Paris: Société d'édition d'enseignement supérieur.

Chapter 6
Method and the Problem of the Historical Descartes

6.1 The Way Forward: Between Naïve Belief and Pure Debunking

As we learned in Chap. 2, there are important traditions in the interpretation of the Scientific Revolution that have been committed to narratives of the discovery, perfection and application of the scientific method. Many pioneering professional historians of science of the past century were persuaded, along with the bulk of the educated reading public since the Enlightenment, that Descartes, Bacon, Galileo, Harvey, Huygens and Newton variously contributed to the invention of a single, transferable and efficacious scientific method, the advent of which was the central achievement and event in the rise of modern Western science. We also know, from Chap. 2, that serious questions have been raised about the existence of such a unique, efficacious and transferable method, and that other traditions in the fields of history and philosophy of science, deriving sustenance from the writings of Bachelard, Koyré and Kuhn, have cast serious doubt upon the idea that any general method commands and explains the actual practice of living fields of scientific inquiry.

Now, the entire present study has been conceived and executed under quite post-Kuhnian commitments about the sui-generis character of natural knowledge–making traditions, whether natural philosophy itself or its subordinate and cognate specialist disciplines. At no point has method, whether some general scientific method, or the method of Descartes, been invoked in our reconstruction of the practices, agendas and outcomes of Descartes' pursuits in mathematics, natural philosophy, physico-mathematics or mixed mathematics. We have now come to a critical turning point in our inquiry, however, because we have reconstructed how and when Descartes conceived of the core of his method, and seen the likely enthusiasm and belief it ignited in him. The issue is, how shall we deal with a the young Descartes' 'discovery' of his method? Shall we now cave in to traditional, and still popular, belief and proceed to explain his subsequent work as the product of that method? Or, shall we simply ignore Descartes' method claims, in the manner implied by the debunking historiographies of Koyré and Kuhn?

J. Schuster, *Descartes-Agonistes: Physico-mathematics, Method & Corpuscular-Mechanism 1618-33*, Studies in History and Philosophy of Science 27,
DOI 10.1007/978-94-007-4746-3_6, © Springer Science+Business Media Dordrecht 2013

In fact, I propose to do something different, neither falling back into a Whiggish historiography of efficacious method, nor simply marginalizing and ignoring method claims in the manner of Koyré and Kuhn. What we must do, on the one hand, is understand how and why any general method discourse, Descartes' included, persuades believers as to its unity, efficacy and transferability, whilst also, on the other hand, show why in fact generalized methods must necessarily fail to work in the sciences in the ways they literally claim to work. It will turn out that these issues are two sides of the same coin: It is the generic structure of grand method doctrines, Descartes' included, that both promotes the illusion of their efficacy and guarantees that no such efficacy is possible by literal application of the method in question. Only if we clear the ground in this fashion, can we continue to pursue the historical Descartes in his mathematical, physico-mathematical and natural philosophical endeavors, for only in this way can we do justice to the undoubted influence the discovery of the method had over time on Descartes' self-understandings, agendas and modes of public presentation. As such we still remain true to the post-Kuhnian axiom that general methodologies cannot and do not explain how work is accomplished in living traditions of making natural knowledge. We begin by surveying how Descartes' method has been dealt with in the literature and where our alternative approach fits in this picture

6.2 The Cult of Method in Descartes Studies

The treatment of Descartes' method by historians of science, intellectual historians and historians of philosophy has largely conformed to the general pattern of understanding the Scientific Revolution to be found in past and present popular accounts, as well as in the writings of early professional historians of science. That is, the *Discours de la méthode*, has been seen as one of the most important methodological treatises in the Western intellectual tradition, and Cartesian method has been viewed as doubly successful and significant within that tradition. Firstly, Descartes' method has been taken to mark an early stage in that long maturation of the scientific method resulting from interaction between application of method in scientific work and critical reflection about method carried out by great methodologists, from Bacon and Descartes down to Popper and Lakatos.[1] Secondly, Descartes' own considerable

[1] This sort of work, therefore, tends to ignore the long Scholastic tradition of methodological debate, founded upon Aristotle's works, and reaching back to the foundation of the great European universities, which reached new heights of sophistication and density in the universities of the late sixteenth and early seventeenth century. More recent work in history of science, history of philosophy and Cartesian studies as such, have begun to rectify this older origin tale of modern method arising only from the early seventeenth century. Even in such historically sophisticated work, however, there remain the historiographically crucial and logically independent questions of whether method is efficacious, and why people believe in method, even if it is not efficacious. It should be noted, moreover, that to question the wider, more audacious claims of general methods, as we do in this chapter, is not to deny the tremendous importance to early modern thinkers of their university

achievements in the sciences and in mathematics during the crucial stage of the Scientific Revolution of the seventeenth century have been taken to have depended upon his method.

It can be said, in general, that the aim of much research on Cartesian method is serious, scholarly, 'apologetic' exegesis: the analysis and explanation of how and why Descartes' well-omened methodological enterprise came to pass. Just as all Christian apologists believe in God, so apologists for Cartesian method agree on the basic aim of elucidating, historically and philosophically, what was, in principle and in practice, a triumph of an efficacious method. To be sure, differences over minor points of interpretation and emphasis have arisen. Just as Christian apologists differ over points of biblical exegesis, so, as I have argued elsewhere, apologists for Cartesian method fall into broad camps: there are naive literalists, sophisticated hermeneutical exegetes, and those whose belief takes a dry rationalist and sceptical turn.[2]

1. *Literalists* accept, at more or less face value, Descartes' epistemological and autobiographical claims for his method. Accordingly, their scholarly task is simply to explicate and clarify the essential truth of Descartes' tale about his own life and method.[3]
2. *Sophisticated hermeneutical exegetes* are found almost exclusively amongst leading historians of philosophy. Combining careful textual scholarship with at least tacit belief in the method, they include most of the recognizedly great Descartes' scholars of the last hundred years, for example, Gilson, Gouhier, Sirven, Hamelin, Liard and latterly Jean-Luc Marion. These men are great scholars because they pose a serious scholarly question and try to answer it using the highest standards of philosophical and textual criticism. That question is, 'What do the *Discours* and the *Regulae ad directionem ingenii* really say about method, and how much of these texts corresponds to historically recoverable fact?' It is widely accepted that you cannot study Descartes without taking their work as a starting point. Exegetical apologists are all united on one axiom, despite a wide range of other differences. They virtually all believe at some level in the efficacy

training in concepts of method, Aristotelian logic, and related tools of thought. This was the bedrock from which prophets of new grand methods launched their programs with any degree of plausibility amongst the rank and file of educated men.

[2] Schuster (1986) 38–40.

[3] Lest this seem improbable, especially to younger historians of science, the reader is referred, for example, to Vrooman (1970), an informative, readily accessible English language treatment from the generation before the latest round of popularizations, and sensationalizations of Descartes' life. We read, on pages 66 and 67, what is actually an amplification of Descartes' own rhetoric in the *Discours:* The method is real and efficacious; it was tested and used in particular technical fields; it dictated the order of study; it is an epoch making cultural achievement ('. .. the method that would be adapted *[sic]* by virtually the entire civilized world, the method that would he accepted as a monument in the history of Western thought'). None of this is questioned as to evidence or meaning. Indeed, the text is a continuation of what we shall below term the 'mythic speech' of Descartes himself.

of the method.[4] True, they do not spend their time reconstructing Descartes' scientific work in terms of the method; they are not 'Whig' historians of science. Rather, tacitly taking the method as efficacious, they pursue other goals: explicating the content of the method or reconstructing its development. Like devout Christian biblical hermeneuts, these 'apologists' are engaged in a difficult, scholarly search for *grains of truth* hidden in a selection of canonical texts. But, as we shall see, you do not have to be a believer in method in order to pursue its history; and if you are a sceptic about method, then all sorts of new interesting questions appear, such as, 'what is method discourse', 'how does it affect believers', 'how can we save believers from their illusions'? etc.

3. Now, if literalist readings of Descartes on method correspond to biblical fundamentalism, then the ultra sophistication of the exegetical hermeneuts perhaps corresponds to the intricacies of Catholic theology. This seems to leave a place for a dry, *rationalizing and slightly sceptical apologetic* approach to Cartesian method similar in tone to the more modish variants of advanced Protestant theology. One striking example of this type was the redoubtable and very learned E. J. Dijksterhuis, one of the last, and certainly one of the most proficient positivist historians of science. Clearly having little time for 'metaphysics' or for empty rhetoric, the exasperated Dijksterhuis took the view that the four rules of method in the *Discours* had no relevance for the bits of hard science to be found in the *Dioptrique* and *Météores*. However, for Dijksterhuis a serviceable, positivistically conceived scientific method assuredly exists, and since Descartes did produce some good physical science, it must have been the product of his possession of such a method. Dijksterhuis therefore concluded that Descartes possessed the method shared by all real scientists, a method bearing no relation to the pap Descartes rhetorically spoons out in the *Discours*.[5]

My approach here is motivated by some news which we canvassed in detail earlier in Chap. 2, and which will probably be unwelcome amongst method cultists: For almost three generations now we have had excellent grounds for being 'atheists' about method. Although the message has not yet spread extensively through the world

[4] The sceptical reader should consult the following: L. Liard (1880) 573, Sirven (1928) 349–53; Gilson (1947) 196, 222, 180–1, H. Gouhier (1958) 62, Gadoffre (1961) 'Introduction historique', p, xxxviii; Lefèvre (1956) 145, 149, 152; Allard (1963) 28, 30, 143, Chevalier (1937) 6–7, Beck (1952) 198, Röd (1971) 18, Note 8.

[5] Dijksterhuis (1950) 22–44. In a similar vein, Elie Denissoff (1970) pp. 28, 30, 96–98, dismissed the claimed universality of the four rules of the method as a literary sop intended to hold together the disparate parts of the *Discours*. Yet, the chief burden of his study was to read the *Discours* as a coded message from Descartes concerning his real method, which is limited to 'mathematical physics', and which Denissoff clearly believes was indeed efficacious. Such dry debunking, in the ultimate service of belief in some sort of scientific method, offers us perhaps the worst of two worlds: On the one hand, there is a dismissive scepticism about Descartes' explicit methodological discourse which defuses critical historical inquiry into whatever it could be that Descartes thought he was talking about. And, on the other hand there is a cloying, now much outdated faith in the existence of a unique, efficacious, rather positivistically conceived scientific method, of which Descartes was one of the first 'discoverers'.

of history of philosophy and intellectual history (let alone the satellite zones of popularization), we have seen that some historians, philosophers and sociologists of science have established that no doctrine of method, whether Descartes' or anybody else's, ever has guided and constituted the actualities of scientific practice—conceptual and material—in the literal ways that such methods proclaim for themselves From this perspective it follows that apologetic scholarship directed to Descartes' method is misguided, not so much in its separable scholarly detail, but certainly in its view of science, of method, and of their intertwined histories. And it further follows that in so far as biographical writing about Descartes is a function of the larger historiographies of method and of science, it too requires reformation. As an historian of science and natural philosophy of this peculiarly atheistical bent, my intention is to reclaim Descartes as a de-mystified object of study in my field. Since the cult of method and the apologetic Cartesian scholarship block that possibility, I seek the tools of demystification within those developments in the historiography of science and the related field of sociology of scientific knowledge canvassed in Chap. 2.

The challenge for us is that the strategy of Koyré and Kuhn, as described in Chap. 2, of pure debunking of method will not fully suffice for our project. It certainly immunizes us against the seductions of any form of apologetic, whether literal, dryly sceptical or sophisticatedly exegetical. But, it also runs a serious risk of encouraging us to ignore the problem of method entirely, because the Koyré–Bachelard–Kuhn position tends to reject method–talk as simply not worth taking seriously. This is not the way forward, for the simple reason that it has become perfectly obvious, through the work first of all of Paul Feyerabend, but more thoroughly in the work of some social and contextual historians of science, as well as post–Kuhnian sociologists of scientific knowledge, that political and rhetorical deployments of method claims are important in the life of the sciences, that is, in the weave of tradition dynamics that a post-Kuhnian view would encourage.[6] But even more to the point is a deeper problem raised by the debunking tendency of Koyré, Bachelard and Kuhn. It is this. If method talk is complete nonsense and of no account whatsoever in the life of traditions of scientific research, we may reasonably ask, 'How, then, can it possibly be that throughout the history of science methodologists and their audiences have often genuinely believed in the efficacy of method doctrines which we 'post–Koyréans' 'know' cannot have worked?' No historian wishes to accuse his subjects of being fools or mad persons, just because they appear to disagree with him. Therefore, we are obliged to discover just what it is about systematic method doctrines that creates and sustains their plausibility to believers, past and present. We must, in short, become more like anthropologists of method, seeking to understand how belief in various types of putatively unified, efficacious methods is sustained amongst certain groups and what are the consequences of those beliefs (and differences of opinion about them) for players in living traditions of research—despite the fact that we cannot possibly subscribe to the substance of their beliefs.

---

[6] Schuster and Yeo (1986).

My answer to this question, previously developed not only in relation to Descartes' method, but to any grand method doctrine past or present is this:[7] All systematic method doctrines belong to a definite species of discourse. The species is characterized by the presence of a certain discursive structure common to all instances of the type. This structure is such that it necessarily defeats the ability of any methodology to accomplish what it literally announces itself to be able to accomplish. At the same time, this same discursive structure easily sustains or creates a set of illusions (in the form of literary effects) to the effect that the method in question can indeed accomplish what it claims to be able to do. In other words, all grand, set piece method doctrines have the same underlying discursive structure which explains their lack of efficacy *as well as* their ability to create the literary effect that they are efficacious. This, I suggest, is the way forward in dealing with Descartes' colossal claims about his method—claims he apparently genuinely believed in (at least up to the late 1620s), but claims we should never literally accept as explanations of his technical achievements (let alone their order and trajectory, as he asserted in the *Discours*). Therefore, my study of Cartesian method in this chapter is both a contribution to the historical study of Descartes, and at the same time a working example of how we might address the deeper historiographical problem of method. It seems to me that we have little chance of comprehensively understanding Descartes historically, according to now state of the art standards in the history of science, until we know what to do about his method; that is, until his method discourse is demystified and historicized.[8]

6.3 Descartes' Method as Mythic Speech: Where 'Myth' Is Not a Colloquial Term of Abuse

The title of this section speaks of Cartesian method as 'mythic speech'. These words are not chosen lightly. In part they are intended to have a certain shock value for those readers who do not share in the not entirely uncommon, but I think largely tacit, post-Koyréan opinion that 'method is myth' in the usual dictionary sense of the term. More importantly my choice of words intends something more theoretically precise. My use of the terms 'myth' and 'mythopoeic' (myth making) derive, at one or two removes, from Roland Barthes' early essay 'Myth Today'.[9] Barthes

[7] I began to generalize from Descartes' case to systematic method discourses of any type in Schuster (1984).

[8] By the same token, whilst acknowledging the value of studies of the tactical uses of method–talk in the life of the sciences, I would suggest any future social and political historiography of method must confront the phenomenon of 'methodology' head on, and must not concede too much by claiming to treat only the 'external' or merely 'social' deployment (or 'abuse') of method–talk. If we are largely agreed that grand theories of method are bunk, we should be prepared to theorize about why that is so, and what that means for writing the history of method.

[9] Barthes (1973) 109–59. Also relevant here was the work of Claude Levi–Strauss (1972), mentioned earlier in Chap. 4 (Note 92), when we first mooted the issue of the mythopoeic character of Descartes' story in the *Regulae* concerning how the law of refraction of light and anaclastic surface might have been discovered using his method.

claimed to identify a peculiarly modern ('bourgeois') form of myth, in which the trick is to naturalize values, interests and socially negotiated outcomes, making them appear to be factual, natural and inevitable. It seemed to me that one characteristic myth, operating in exactly this manner, is the myth of scientific method, even though, in this case, the mythopoeic discourse in question can be traced back to Aristotle and up to our methodological prophets of modernity, such as Bacon and Descartes, through the high medieval and renaissance methodological debates of the Scholastics. After all, if the naturalizing of human commitments of theory, value and aim is the mark of modern forms of myth, then we have to accept that perhaps the first example of a characteristically modern Western myth is indeed the myth of scientific method. Although this myth did not start with the heroes of the Scientific Revolution, it certainly was given new force and cultural cachet as a result of being attached to the novelties of natural philosophy and the sciences emergent in the seventeenth century. Now, according to Barthes these effects are brought about by the structure of the discourse in question. Even though I do not deploy the kind of semiotic techniques which Barthes advocated, I do claim to have identified certain structural levels in any systematic method discourse, and I locate the persuasive power and naturalizing force of such discourses in the relations holding generally amongst these levels. In my account, grand methodologies are discourses so structured that they necessarily lie about their own powers and capabilities in the interest of turning *culture* (how the natural sciences are actually practiced) into *nature* (a simple outgrowth of human rationality and nature's amenability to it) In these precise senses, then, methodologies earn the (Barthian) title of myths.

My approach also owes much to the anti-methodism of Paul Feyerabend, although in a precise way which requires clarification. Post-Kuhnian debunkers of method can, I think, perceive two rather distinct initiatives in Feyerabend's work. On the one hand, there is Feyerabend's historical critique of methodology, consisting mainly in case-study illustrations of the non-binding character of any and all systematic methodologies. 'Progress' in science, he persuasively argued, has always broken the pat rules laid down by methodologists, and it has always *had* to do so. New standards are constructed and refined in the act, through the very processes of major scientific change; and, in the cases studied, Feyerabend tended to show that rigid adherence to the rules of contemporary (or later) methodologies would have aborted or obstructed the course of development.[10] Feyerabend's efforts in this direction must seem brilliant and historically revealing to any Koyré– or Kuhn–influenced debunker of method. Nevertheless, this work did not really constitute a great advance in our ability to theorize, seriously, about the nature of methodological *discourse.* On the other hand, in a small and rather neglected corner of his work, Feyerabend, in my opinion, offered the first sustained demonstration of the structural sources of the mythic character of method discourse. In his important paper, 'Classical Empiricism', Feyerabend laid bare the discursive mechanisms by which Newton's methodological claims in physical optics present a systematically distorted picture of his actual practice, producing a convincing fairy tale about the

[10] Feyerabend (1975, 1978).

genesis and status of his claims in that field.[11] Although Feyerabend did not pursue his analysis in explicitly structural terms, his approach, viewed through my semi-Barthian spectacles, catalyzed my own structural schema for methodologies.[12]

We can now return to the question posed toward the end of Sect. 6.2, transposing it down to the case of Descartes, so that it takes the form: 'How can it be that Descartes and others apparently believed in the reality and efficacy of a method which most post-Koyréan historians of science are convinced cannot have worked'? My fundamental claim is that the vacuity and sterility of the method *and* its appearance of efficacy are both effects of a common cause. That cause is the way Descartes' method discourse is structured onto several interacting levels. Descartes' method cannot possibly do what it claims to be able to do, because, as discourse, it has a particular structure; and yet it is that very structure which can create and sustain illusions or literary effects about the efficacy, applicability and unity of the method.

The analysis begins from a naive but fundamental premise: In order for the rules of the method to be considered *efficacious* in the practice of a given field of research, the rules have to be applied and deployed within inquiry in the target field in ways adequate to the proclaimed goals and foci of the method.[13] Granting this point, it would seem that Descartes must give some arguably adequate account or redescription of the contents and workings of the target field. Such an account must be couched in terms supplied by the core of his methodological discourse, his talk of 'absolutes' and 'relatives', of 'relations' and 'series', which are aspects of the latticework, as discussed above in Chap. 5. This is necessary because the heuristic rules are claimed to apply to entities of this type. The heuristic rules, after all, were formulated directly in terms of, and are clearly 'relevant' and 'applicable' to the discourse about 'absolutes', 'relatives', 'series' and 'relations'. Approaching a target field with his method, Descartes must be able to construe that field in terms relevant to the use and application of his rules. Descartes, as we shall soon see, accepted these conditions and worked within them. He thought his results exemplified the efficacy of the method.

However, following the dictates of Chap. 2, we subscribe to a post-Kuhnian understanding of the dynamics of living traditions of research. Any grand method discourse aims to produce 'arguably adequate accounts of the contents and working

[11] Feyerabend (1970).

[12] How Feyerabend's argument maps onto my structural schema for method discourse is shown below, see Note 47. My earliest suspicion that there was a specific mechanism of mystification involved in methodological accounts was aroused in the mid 1970s by reading Bachelard (1949), where he deals with the systematic role of traditional philosophical perspectives (not particularly methodologies) such as empiricism, rationalism, conventionalism, in producing a structured series of illusory pictures of how theory and practice (or 'applied rationalism' and 'technical materialism') interrelate in the constitution of mature mathematico-experimental sciences. (Cf. LeCourt 1975, 41 ff.) I read Barthes and Levi-Strauss at that time as well. Up to that point I was a (Kuhn–trained) Kuhnian debunker of method.

[13] This eliminates claims for the efficacy of a method that rest on an arguable misconstrual of its own proclaimed resources and goals, for example, claims that Descartes' method facilitated the discovery of articles of *faith,* or that Popper's method was of use in the *discovery* of a fact or law.

of its target fields'. We hold that such methodological accounts, adequate in the sense required, cannot be achieved. The post-Kuhnian understanding of research tradition dynamics sees the conceptual structures and modes of practice of living fields as both sui generis, and in constant re-negotiated flux. We saw in Sect. 2.6 that to redescribe, gloss or translate these conceptual structures and modes of practice into some other idiom or discourse, is simply to translate them, on paper, for some 'outside' purpose. Research proceeds from within each field's proper, and evolving framework, and not in terms of glosses provided by putative single, transferable methods.[14] So, in what follows here, we are not concerned with further supporting the initial premise that method discourses, such as that of Descartes, must fail adequately to gloss or redescribe target fields; rather we explore the mechanisms which *simultaneously explain both the necessity of that failure and the creation of the illusion or literary effect that no such failure has in fact occurred*—terming such curious and important mechanisms 'mythopoeic' in a precise and considered sense of the term, inspired by Barthes: In short, Descartes' method discourse, like any grand method discourse, produced only literary effects of its own efficacy, applicability and unity, effects that tended to convince Descartes (and other believers) that the method actually possessed these virtues.

6.4 The Failure of Adequate Redescription: An Example of Descartes Attempting to 'Methodologize' a Field of Inquiry

We have, in fact, already studied in passing such a case of inadequate redescription. In Chap. 4, we looked closely at the distance separating, on the one hand, Descartes' actual path of discovery of the law of refraction and his subsequent struggle to find an adequate mechanistic rationale for it, and, on the other hand, his methodological account or redescription in rule 8 of the *Regulae* of how one might accomplish these ends. We shall return to our findings in that case later in this chapter, using them to

[14] One reservation must be registered to this claim. It is perfectly true that bits and pieces of 'methodological discourse', including putative glosses of the field in question, can be deployed in practice as resources in debate, negotiation and adjudication of the content and acceptability of knowledge claims. In Sect. 6.8, we shall identify these as the 'rhetorical' uses of method discourse in debates and negotiations about knowledge claims inside the living fabrics of disciplines and fields. Scientists can appeal to methodological principles to attempt to substantiate or undermine such claims. However, such deployments of method discourses within scientific debates are merely small portions of the total structure of action and belief through which knowledge is made and unmade. Such deployments do not represent clear and accurate meta-level versions of the specific practice of that field. Indeed, the deployment of method claims in scientific debate in no way whatsoever constitutes even prima facie evidence of the efficacy of that method. The issue must be turned on its head—how are such claims, as discursive phenomena, shaped by the resources of method discourses and what literary effects of genuine efficacy are thereby created? Classic sociological studies of methodological discourse in scientific debate and negotiation include: Gilbert and M. Mulkay (1980, 1981); Mulkay and Gilbert (1981, 1982).

articulate further the structural model of method discourse we are developing. For the moment, however, it will be useful to start our analysis with a new and different example. Our case deals with what we may term for ease of expression 'the science of magnets', which Descartes discusses twice in the *Regulae*.[15] In fact, what we are dealing with here is better described as one typical domain of explanation within the larger realm of corpuscular-mechanical natural philosophizing as a whole. What we are going to find out about this one domain arguably holds across any and all regions of phenomena one would wish to cover with corpuscular-mechanical explanations. Although Descartes first discussed 'magnet science' in the *Regulae* at a time before he was committed to constructing a system of corpuscular-mechanical natural philosophy, we can still treat this case as involving, in the end, an attempt at such corpuscular-mechanical explanation. There are two reasons for this: Firstly, as in the case of the explanation of light in the *Regulae*, we know that although Descartes was no system builder in the 1620s, he nevertheless clearly preferred to ground exercises in his brand of physico-mathematics in piecemeal corpuscular-mechanical explanations. Secondly, later in the *Principles of Philosophy*, his second and definitive system of corpuscular-mechanism, magnetism took pride of place as an object of study and example of explanatory success.[16]

So, let us now consider the fate of the 'science of magnets' when Descartes tries in the *Regulae* to explain how to 'do' this science according to his method. We need to compare Descartes' methodological tale, or redescription of magnet science, with what he, in fact, had to do to produce corpuscular-mechanical explanations of magnets and their phenomena within the living field of natural philosophizing, where actual corpuscular-mechanical explanations had to be thought up and inscribed. Methodologically speaking, Descartes instructs us first to isolate a fixed set of experimental data about magnets, in practice the experiments reported in Gilbert's *De magnete (1600)*. We are then to inquire into the 'intermixture' of 'simple natures' which will explain the magnet.[17] Here the absolute natures or terms surely are primitive geometrico-mechanical elements, corpuscles, with their properties of size, shape, hardness and state of motion or rest. What, then, are the 'relatives' in this case? Descartes terms them 'intermixtures' of absolutes. That conveys the image of some set of complex corpuscular-mechanical models, models for the structure *of* lodestones, magnetic 'effluvia', magnetizable bodies etc.[18] Hence, in this case Descartes' methodological gloss or redescription of 'magnet science' does

[15] *Regulae*, Rule 12, AT, X, p. 427; Rule 13, AT, X, pp. 430–1.

[16] *Principles of Philosophy*, Part IV arts 133–183. See below, Sect. 12.5 on the key role of 'cosmic' magnetism in the systematizing strategy of the *Principles*.

[17] I leave aside the problem, obvious to anyone familiar with the post–Kuhnian sociology of scientific knowledge literature cited in Note 14, of the criteria by which Descartes selects as 'adequate and 'reliable' Gilbert's own selection of a set of experiments, their performance and their glossing in his book.

[18] On the complexity of interpreting Descartes' remarks see Buchdahl (1969) 85–8, 126–47 and Schuster (1980) 74–5, and notes 150, 151 thereto.

depend in a loose sense upon the required specification of 'absolutes' and their studied 'complexification' into sets of 'relatives'. And these absolutes and relatives are arguably the sorts of entities which did eventually enter into his detailed explanation of magnetism, set forth later in his *Principia philosophiae* of 1644. The mythologist of method, however, must ask an embarrassing question: 'Has Descartes provided here an adequate redescription or gloss of what it took to construct corpuscular-mechanical explanations of magnetism (or of anything else for that matter)?'

Modern Cartesian scholarship has given us answers to the question of what really was involved in Descartes' formulating and inscribing *of* corpuscular-mechanical explanations, and something approaching a consensus has existed in the literature for a considerable time:[19] In non-methodological contexts and later in his career, after 1628, Descartes increasingly came to see that although there are some absolutely certain metaphysical principles, for example, that the essence of matter is extension, neither the details of particular corpuscular-mechanical explanatory models, nor the facts to be explained, can be *deduced* in the strict sense from such absolutely certain metaphysical principles. *A fortiori* there is no question of the full details of the corpuscular-mechanical world system being fully deduced from such 'first principles'. Nevertheless, the absolutely certain metaphysical principles do place constraints upon what can and cannot be asserted of any detailed corpuscular model designed to explain a particular class of phenomena. For example, nothing should be asserted in a particular explanatory model that contradicts any of the metaphysical principles. Additionally, available empirical evidence, and in particular, the 'facts' to be explained, also need to be considered in the formulation of the detailed explanatory models. By the time he published the *Principles of Philosophy* in 1644 his position became very clear: We may know with certainty from metaphysical deduction that the essence of matter is extension, as well as certain laws of motion and collision, but we cannot deduce from these truths more detailed explanatory models for such diverse phenomena as gravity, light, magnetism, planetary motion, sensory perception and animal locomotion. The best one can say is that such models should not contradict metaphysically derived certainties and that relevant facts must also be considered in shaping explanatory models. Hence, such lower level models are necessarily hypothetical and can achieve at best only 'moral certainty'. When, in his later works, Descartes spoke of 'deducing' phenomena from his principles, he did not mean the strictly mathematical deduction envisioned in his central methodological texts, but rather 'deduction' in the looser contemporary acceptation of 'plausibly explain'.

Such, then, were Descartes' own later and more considered views about the production of corpuscular-mechanical models and explanations. Although they show that his strict methodological views bore little relation to the procedure, they do not quite do full justice to what we might now term the interpretational complexity and fluidity of his project and the indexical character of virtually every

---

[19] Buchdahl (1969) 97, 118–26, Sabra (1967) 21–45, Clarke (1977) and (2006) 154, 161–68, Schuster (2000c).

move within it. Imagine a sociologist or anthropologist of science transported back in time to observe Descartes as he attempted to produce and inscribe a piece of corpuscular-mechanical discourse, about magnets for example. Our temporal interloper would probably have identified three interacting moments in Descartes' performance. His field notes might read as follows:

1. *Logically and temporally prior to the construction of any particular explanation, Descartes tries to devise and legitimate his basic metaphysical principles which will constrain and condition the formulation of specific corpuscular-mechanical models. Such principles include his fundamental definitions of matter and mind, and his basic laws of motion, collision and the behavior of directional tendencies to motion.*[20] *Needless to say, the production of Cartesian conclusions in metaphysics and dynamics, and their legitimation, are not amenable to clear, consistent, rule-bound procedural glossing. One might parody M. Descartes' own account of his procedures and say that 'God only knows how he does it'. Nor is it clear how and in precisely what sense the models should be 'constrained' in any given case. That, too, can only be a matter of on the spot interpretation and 'negotiation', if only with himself!*
2. *'Relevant' empirical evidence has to be selected, weighed and 'appropriately' deployed and described. Evidence can include 'facts' needing explanation, or 'facts' lending credibility to the explanatory model offered (including 'facts' purporting to weaken the credibility of competing explanations). It is not clear that M. Descartes has procedures for accomplishing these tasks which are any more rule-bound than those ongoing negotiations and 'constructions' of facts and arguments revealingly studied from the later twentieth century by post–Kuhnians such as Latour, Collins, Pickering, Pinch or Shapin.*[21]
3. *In the light of the 'evidence' and the metaphysical 'constraints', a specific corpuscular-mechanical model for the phenomena in question has to be constructed. Given the un-methodological character of the proceedings under (1) and (2), it is not to be expected that M. Descartes' inscribing of characterizations of particular models is a method-bound activity. Consider, additionally, that in Descartes' usages the meaning of 'deduce' in the phrase 'deduce the phenomena from the model' is fluid and reinterpretable. As if this were not enough, there is also the point that each specific model has ultimately to 'fit' into a 'system' of natural philosophy. This raises a host of additional interpretive challenges which reflect back upon the way in which a model is to be constructed. For example, Descartes, the 'systematizer' always asks, 'To what degree does a particular model 'comport' (itself a fraught word) with other specific models*

[20] The principles and laws of Descartes' dynamics are included here, not because there is scholarly agreement that Descartes intended all of them to be deducible from his metaphysics, but rather because they are foundational for all his detailed model building and particular explanations, and because he often gives strong indications that they were meant to be deducible from first principles.

[21] Latour and Woolgar (1979), Pinch (1985), Shapin (1982).

within the system in respect of (a) consistency of mode of metaphysical constraint;[22] *(b) similarity (or difference) of explananda in view,*[23] *and (c) the degree of structural, cosmological 'interplay' intended to hold between these models in the overall system of the world machine.*[24]

Not to put too fine a point on it, in the field notes of our sociologist or anthropologist, (1), (2) and (3) together constitute a complex undertaking—it's not easy (or method–bound) to think up and write down corpuscular-mechanical explanations of things, let alone under seventeenth century conditions of pursuit of 'systematic' completeness. Each of the three steps involves discursive practices, interpretations, weightings and selections for which no rules were ever given, and from which it is not plausible to imagine any workable and consistent full set of rules could ever be elicited by *post facto* glossing.[25] *A fortiori* these moments elude Descartes' own recommendation, in effect, to 'find the absolutes and the structure of relatives and use my heuristic rules in so doing'. Descartes' methodological discourse on how to 'do' magnet science bears no significant relation to whatever it is he must have been 'doing' in order to accomplish magnet science; that is, to think up and inscribe his typical corpuscular-mechanical texts on the subject. The recommendations embedded in his method–talk could never have led to the construction of this corpuscular-mechanical discourse about magnets. Here, the methodological recommendations which he gives in the *Regulae* simply batten upon the prior accomplishment of (1), (2) and (3), and upon the uncodified and arguably uncodifiable body of discursive practices which underpin them at every step. The methodological version of what it is to do magnet science drains the practice of magnet science of its actual *sui generis* procedural density (whatever that might have been); and, having drained that density, the method discourse poses as the real basis of Descartes' practice.

---

[22] How, for example, can the behavior of Descartes' 'first element' comport with the metaphysical principle that extension is the essence of matter, since it seems able to change density instantaneously, as it instantaneously changes shape to fully 'fill' interstitial 'spaces'; or, how does Descartes' account of the internal sensations and passions of the soul comport with his ontological dualism? (Descartes seems not to have attended to the former but devoted much effort to the latter.)

[23] Descartes' judgments about these similarity/difference relations would, at least in part, depend upon already established patterns of interpretation. For example, Descartes thoroughly accepts the broad difference Gilbert had sought to draw between magnetic and electrical phenomena, although he does not accept this as an ontological distinction (between a spiritual and a corporeal cause respectively), and yet *à la* Gilbert, he still promotes magnetism as a phenomenon of 'cosmic' scope and importance, contrasting it to the rather trivial role of the known electrical phenomena. (Obviously, our anthropologist of method has taken a position reminiscent of Barry Barnes in these remarks, Barnes 1982, especially Chap. 2.)

[24] As we shall see in Chap. 11, especially on the issue of the systematicity of *Le Monde* and the possible criteria of the goodness of that systematicity. For example, in Cartesian mechanical philosophy in the *Principia*, aspects of the theory of magnetism and of the theory of light have crucial bearing on the structure of the theory of celestial motions (cf. Note 23): but, in a different mechanical world picture this particular linkage might not have been valued or sought.

[25] Cf. Garfinkel (1967), Barnes and Law (1976).

Nobody should really be surprised by this. Very few serious post-Kuhnian students of the history of scientific practice would expect that Cartesian method discourse (or any systematic method discourse) could adequately gloss the content and practice of a field of research.[26] And yet, an important fact presents itself at this point. Descartes' methodological story about doing magnet science has a curious property: *Certainly it fails adequately to redescribe its target; but, inadequate as the story is, the rules of the method do seem to apply to it and mesh within it.* Consider that in the *Regulae* Descartes discusses the problem of explaining magnetism, and his discourse proceeds by using the terms 'absolute', 'relatives', 'relations' and 'series' ('intermixture'). So, as far as 'just discoursing' goes, one can gear the rules of the method right in, so as to facilitate the investigation; that is, to facilitate the continuation of our little methodological story about how we really should 'do' magnet science. Descartes' redescription of magnet science, his little account of what we should do, is just the sort of text into which additional talk about the rules and their use can be inserted 'convincingly' and 'coherently'.

What, then, should we make of the fact that the rules of the method can be applied to and deployed within Descartes' account of doing magnet science? The answer depends upon whether we are believers in the method, or we are post-Kuhnian mythologists trying to demystify it. Believer and mythologist can agree that the rules 'go together' with the methodological account. But, the mythologist sees that the redescription is a phantom, a parody of what must have gone on when Descartes 'did' corpuscular-mechanical natural philosophy. The situation of the believer, however, is very different, especially if the believer is an early modern methodologist, like Descartes. René did not have the benefit of *our* contemporary critical tools, through which to articulate what we mean by saying that the dynamics of the inscribing corpuscular-mechanical explanations were very complex, and, quite probably ultimately elusive. Our early modern methodologist (or our surviving contemporary methodologists of various ilks) might have easily overlooked the slide between, on the one hand, the practices involved in thinking up and writing down corpuscular-mechanical explanations, and, on the other hand, the methodological glossing of those practices. Or, better expressed, he is unlikely to have overlooked it, because, he probably had no discursive tools through which to

[26] Analogous remarks apply to that supposed case of application of the method, the discovery of the explanation of the formation and geometrical properties of the rainbow. This was indeed an exceedingly good piece of normal science, the solution to a classic puzzle in geometrical optics. But it was also highly traditional, conditioned by the aims, concepts, tools and standards of the discipline. Descartes' recourse to a water filled flask as a model rain drop was not novel, and even had it been, it could be interpreted as having been mediated by a very commonsensical, rather than methodological rationale. Descartes' sole advantage over others was possession of an exact law of refraction, which now served, as laws often do, as a tool in facilitating further research. An exact tool, a standard model, some sufficiently accurate data, and laborious calculation resolved the problem. To invoke the rules of the method here is to glide over the rich, tradition bound dynamics of the research. Moreover, on hitherto little noticed problems, and successes, of this research, see Buchwald (2008), whose brilliant reconstructions further undermine the idea that Cartesian method actually controlled Descartes' course of work in this area.

thematize it in the first place—except for the resources of a *methodological* discourse of one sort of another, a type of discourse, which enjoyed cultural prestige and was accorded epistemological precedence in attempts to 'account for' intellectual practices. Now, if you overlook the structural slide; or if it is just unthematizable for you, then you are bound to be impressed by Descartes' method story, and in particular, you are bound to be impressed by the wonderful way the rules of his method apply beautifully within the story.

Any believer who reaches such a level of conviction would then be in a position to engage in a pair of characteristic behaviors, quite reasonable from his viewpoint, which reflect his belief in method and would serve to reinforce it. Having constructed the method story about magnet science, or having consumed it from an 'authoritative' source, the believer could then do the following:

1. He could practice corpuscular-mechanical natural philosophy—thinking up and writing down explanations—more or less proceeding in the way discovered by our time traveling anthropologist, while monitoring his actions to himself and/or to others in terms of the little method story. This would lend credence to the method, by seemingly attaching it to segments of the practice as they occur in the flow described by the anthropologist. Such 'voicing over' or real time monitoring of practice could be institutionalized in pedagogy to entrench both the method in question and the theories embedded in the routines so glossed.[27]
2. He could practice corpuscular-mechanical natural philosophy, and then after the fact claim in good faith that, although what he had done might not fully reflect the dictates of the method, he could, in principle, have accomplished the same things by strictly adhering to it.[28]

Both (1) and (2) are very easy to accomplish. The method story which the methodologist recites while working, or after the fact (and 'in principle' or 'in fact'), is simple to construct, because it mainly utilizes the core of the method discourse,

[27] Feyerabend (1970) effectively first identified this phenomenon and isolated its mythopoeic character through what amounted to a structural delineation of Newton's methodological discourse. I introduced the term 'voicing over' in this context in Schuster (1984) 21ff in articulating Feyerabend's argument.

[28] If this move were accompanied by the claim that the results of research are justified only if such *a post facto* gloss is possible, then we would have the typical modern methodologist's tactic of appealing to the possibility of 'rational reconstruction'. On the view to be advanced here, 'rational reconstructions' guided by systematic methodologies are simply a species of methodological story or account, and share in their mythopoeic character. Existing studies which suggest this criticism of Popperian-Lakatosian conceptions of method include: Feyerabend (1978) 201–2; Schuster (1984) especially. Section VIII; and Mulkay et al. (1983) 172–82 (where the authors produce a brilliant ethnomethodological critique of a quasi-Lakatosian attempt to offer an historical 'rational reconstruction' of some developments in particle physics). Reasons for the mythic character of Lakatosian reconstructions are also discussed in Schuster (1979) 301–17. As for Descartes' own *post facto* glossing of practice, we have already studied the wonderful example in the *Regulae* (Rule 8, AT, X, pp. 393–5) where he discusses (in the carefully chosen subjunctive mood) how one *might have discovered* the law of refraction of light by following the method.

its generalized statement of rules; and, it is easy to accept, because the tame methodologist (and his tame audiences), almost by definition, either have no alternative critical frameworks available for theorizing about practice, or, if they have them to hand, they are for one reason or another unwilling to deploy them.

6.5 The Structural Levels and Underlying Metaphors in Descartes' [or Anybody's] Method Discourse

By reflecting on the example of Descartes' 'magnet science', we are about to uncover the basic structural units of his method discourse, indeed, the basic structural units of any systematic method discourse. Knowledge of the structural levels in a method discourse, such as Descartes', will, in turn, allow us to locate and explain the four characteristic illusions or textual effects which method doctrines exercise upon those who believe in them.

Before we look at this structure and its characteristic effects, we must, however, remind ourselves of the fact that all method doctrines encountered in the Western tradition, from Aristotle to Popper and beyond, are structured around two intertwined metaphors: (1) to acquire knowledge is a matter of establishing a correct subjective grasp, or more typically, vision, of independently existing, objective objects of knowledge; (2) method, drawing on the literal Greek meaning of the term, is the subject's 'way through' to the objects of knowledge, a set of prescriptions as to the path to be followed, by the subject, in the pursuit of knowledge. All particular method doctrines are attempts to explicate the key metaphors. Indeed, the history of method doctrines is in large measure the history of various and competing attempts to dress these notions in conceptual vestments deemed appropriate to each methodologist's perception of the context of debate and structure of socio-cognitive relevances holding in his time and place. Typically, a new doctrine is fabricated out of bits of older method doctrines, as well as pieces of neighboring varieties of discourse—theological, natural philosophical, ethical, mathematical, psychological, and so on.

Let us now turn to the generic structure of method discourses (Fig. 6.1). Reflection upon the discussion of 'magnet science' in the previous section, shows that we are dealing with three levels of discourse: The systematic 'core' of the method discourse, the *sui generis* target fields, and the methodological accounts, glosses, redescriptions and stories which can be manufactured about the latter, using the resources of the former. Figure 6.1 is a map of these levels, which will also become a guide to the places where the characteristic illusions or literary effects of Descartes' (or anyone else's) method discourse are generated, between and across levels.

Level I is that of explicit, 'systematised' discourse about the core of any given method doctrine. In any particular method doctrine, Level I will consist in (1) generalized (non-discipline specific) statements of the rules of that method, and (2) explicit, more or less systematised, abstract and generalized discourse concerning the canonical themes, 'knowing subjects' and 'objects of knowledge', and how the

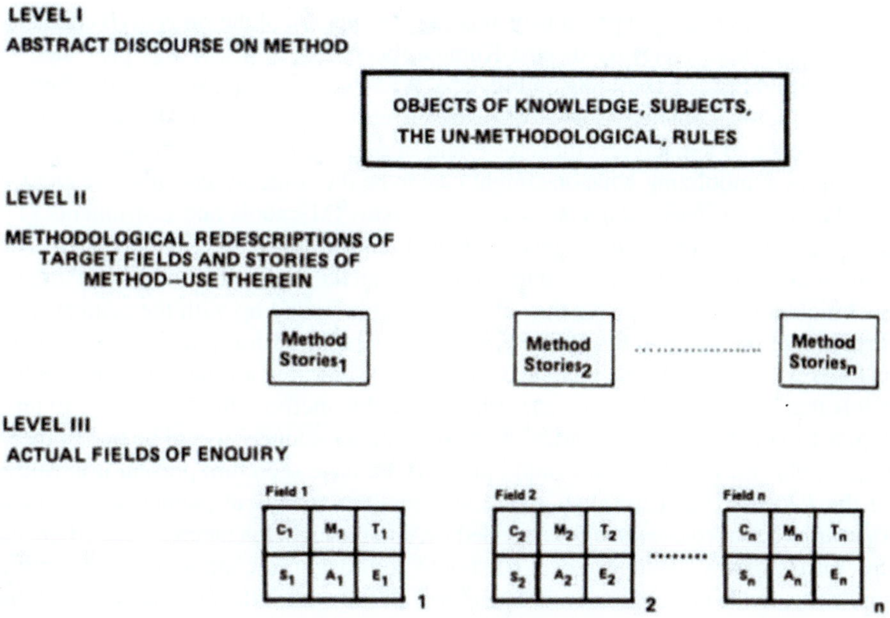

Fig. 6.1 The structural levels in any grand method doctrine of method

rules help them to get together. There is typically present also (3) some discourse on the 'pitfalls', 'obstacles' and 'sources of error' which can deflect a subject, mask or distort the objects or lead to misapplication of the rules. In some grand method doctrines Level I is itself packaged within a metaphysical or even theological framework. In Descartes' method we have already discussed this Level I core, as presented in the *Discours* and *Regulae*. It includes his teaching concerning intuition and deduction, the conception of the latticework with its intertwined concepts of absolutes, relatives, series and relations, the statement of the rules of the method, and, in addition, a discourse on pitfalls and sources of error, which above, in Sect. 5.6 we termed his 'negative heuristic advice'.[29]

---

[29] For example, trust not in a authority, nor in unclear, indistinct belief, will or emotion; avoid precipitation and hasty judgment; go only as far as intuition and deduction reveal the truth. Descartes also has, as many methodologists do, a catchall 'saving clause' to the effect (using our terminology) that somebody might indeed point to gaps between methodological stories of practice and what it really is like to pursue the discipline in question. Systematic methodologists, being reasonable people, do notice gaps between stories and counter-accounts, when these are brought to their attention. But, this does not induce them to become mythologists of method. Consider Descartes, writing to Mersenne (27 February 1637) '...je n'ai su bien entendre ce que vous objectez touchant le titre; car je ne mets pas *Traité de la Méthode*, mais *Discours de la Méthode*, ce qui est le même que *Préface* ou *Avis touchant le Méthode*, pour montrer que je n'ai pas dessein de l'enseigner, mais seulement d'en parler. Car comme on peut vois de ce que j'en dis, elle consiste plus en pratique qu'en théorie... (Alquie 1963, t.1, 521–2)

Level III has been met previously. It is a representation of the *sui generis* character of scientific fields, viewed in post-Kuhnian perspective, which was presented in Sect. 2.6 and Fig. 2.6 (building on Fig. 2.5). What we have here, again, is the domain of living fields, depicted as a set of matrices, each one schematically representing the existence of a discrete, efficacious field of inquiry. Inside each matrix there are six spaces symbolizing what one might take to be the contents and structure of such a field—its own basic concepts [C]; metaphysics [M]; tools and instruments [T]; standards of relevance and adequacy [S]; goals and aims [A] and exemplars [E]. The aim, as in Chap. 2, is to represent the neo-Kuhnian thesis that the 'method(s)' of any such field are *sui generis,* for they are inextricably bound up with the contents and structure of the relevant disciplinary matrix. The point of thus placing Fig. 2.6 here in Fig. 6.1, as Level III of a method doctrine, is to show that it is the set of such living fields that are the 'targets for redescription' in this method. So, if we were to take Fig. 6.1 to be depicting Descartes' method, 'magnet science' would be one of these target fields on Level III, so would optics, as we have seen him present it in rule 8 of the *Regulae,*[30] and we shall soon meet another, analytical mathematics, which Descartes believed could be commanded by, and through, his method. Indeed, in the case of Descartes' method, Level III should be thought to contain all the scientific traditions, fields or disciplines, because they all fall within the claimed scope of the method, along with all mathematical disciplines, and, indeed, all domains of rational inquiry, as opposed to those controlled by faith.

Finally, *Level II* consists of a set of 'methodological versions' of the corresponding fields of inquiry represented on Level III. Here one finds methodological accounts, redescriptions or stories, which purport to describe or capture the essence of the practice of the corresponding Level III fields. These stories or accounts are structured in terms of the elements provided by Level I, by the core discourse on 'subjects', 'objects' and rules characteristic of the particular method discourse in question. Such stories or accounts analytically proceed as follows: the 'target' field, the corresponding Level III field, is redescribed or glossed in terms of the elements provided by Level I of this particular method discourse, and an account or story of practice is woven, by reference to a subject (conceived in Level I terms) applying the rules within the glossed field. Hence, Level II stories and accounts can only exist in so far as they are shaped by deployment of the conceptual resources of Level I of that method. In any given method, the stories on Level II are specific, episodic unfoldings of the conceptual resources provided on Level I as elements in the core discourse of the method. In the case of Descartes' method, we have just met one example of a Level II methodological story corresponding to the target field of 'magnet science', having seen his Level II story for optics in rule 8 of the *Regulae,* and shall soon meet others. Each of these stories or accounts is couched in terms of the core methodological terms and rules available on Level I of Descartes' discourse on method.

[30] Section 4.9.

6.6 The First Two Structural 'Effects': Adequate Redescription' and 'Application'

If we return to the case of magnet science, bearing in mind the structure in Fig. 6.1, we can begin to locate the sites at which a method discourse generates illusions or creates literary effects concerning its own efficacy, and we can specify its *modus operandi* at those sites. Thus, we can examine how the structure of a method discourse contributes to its dynamics as a mode of mythic speech; that is, we can explain how method discourses, Descartes' included, succeed in creating literary effects of their own efficacy, whilst, in fact, being structurally incapable of doing what they literally claim to be able to do.

The key to the mythological operation of Descartes' (or anybody's) method discourse resides in getting the audience, potential believers, to operate on Level II, where they bask in the methodological version(s) of the target field(s) with which they are concerned. On Level II, the rules of the method do 'apply to the redescriptions offered'; ongoing work on Level III can be glossed (or 'voiced over') in terms of Level II, or *post facto* 'accounting' for practice on Level III can be offered in terms of Level II stories. It is crucial that the Level II story both embodies the rules of the method and disastrously misses all the cognitive and organizational density of actual Level III practice. Indeed, these two characteristics of Level II stories may be seen as interacting in the very constitution of such stories: Level II stories eviscerate actual Level III practice because, failing to engage the density of that practice at all, they are episodic, fabular versions of the core methodological terms available on Level I. They are fairy tales of methodological comportment, spun out of materials and scenarios available on Level I. But, of course, the rules of the method only mesh into such stories because the stories derive from Level I.

Recall our case study of Descartes' practices for formulating corpuscular-mechanical explanations of magnets, and his methodological tale in the *Regulae* purportedly corresponding to that activity. According to our new terminology, there is a Level III field of natural philosophical practice and a corresponding Level II methodological account of this target domain. Descartes' methodological tale about 'magnet science' eviscerates and suppresses the specific content and dynamics of his practice in corpuscular-mechanical explanation of magnets, the target field, while the tale itself is spun out of the Level I cloth of core discourse about rules, series, absolutes, relatives, etc. In fact, Descartes' inscription of his method tale is dependent upon those two processes: (1) the suppression of the real content of 'magnet science'; and (2) the fabular rendition of the core discourse of Level I as a Level II story, to replace that content as the methodologically sound 'essence' of the target field. However, whilst post-Kuhnian mythologists of method know all this, historical actors living in the early modern culture of method most probably did not, for they, *ex hypothesi*, had virtually no discursive resources for explicating and accounting for successful practice in a discipline other than those offered by some method discourse or other—either a version of the neo-Scholastic discussions of method acquired at university, or some alternative of their own or other's manufacture.

Such a 'believer' is likely to miss the slide between Level III and the method accounts on Level II; indeed, he might not even be aware of it, since 'method talk' is his preferred (or only) way of thematizing practice.[31] Once on Level II, however, he is likely to be impressed by the way the Level II account (1) 'applies' the rules of the method (and generally articulates the core concepts of the method), whilst (2) (apparently) constituting an adequate account of what the disciplinary practice is about.

One may, therefore, state that when a reader or listener is confronted with a Level II redescription or story of rule-following, he is in danger of succumbing to two structurally produced illusions or literary effects characteristic of systematic method discourses (Fig. 6.2). Firstly, he may be taken in by the 'adequate redescription effect', producing the illusion that Level II redescriptions are in any sense adequate to Level III (target field) contents and practices. Secondly, he may be taken in by the 'application effect', producing the illusion that the application of the rules in the Level II story is (or could be) the application of the rules to the practice of the target field (Level III). These effects are *structural* in the sense that they are made possible and sustained by the relationships amongst the three levels of discourse. That structural arrangement also simultaneously explains why a method discourse, such as Descartes', must be inadequate and ineffective in real practice. In an appropriate cultural environment its upper two levels marginalize or displace the discursive thematizing of the Level III field as such, and pose in its place a desiccated phantom of its actual structure and practice. That phantom, the Level II redescription or account, is then solidified and underwritten by its 'obvious' congruence with the grandiose, self-proclaimedly authoritative core discourse on Level I. Since the very process of manufacturing, solidifying and underwriting phantoms on Level II is the source of the effects, we may well reassert the central thesis of this chapter: *Descartes' method discourse is a species of mythic speech whose discursive structure renders it vacuous, whilst simultaneously sustaining powerful illusions that it is not.*

It may well be asked, how a reader/listener ever gets to Level II in the first place; what types of reader/listeners are at risk of falling victim to these effects; and what exactly are the dynamics of their acquiescent reading/listener? These are crucial problems. The first point to grasp is that the seduction of an historical actor is greatly facilitated if he or she is a member of a culture in which 'scientific method' is generally believed to exist, in practice or in principle. Early Modern figures, such as Bacon and Descartes, moved in an intellectual culture permeated by this belief. The in principle existence of efficacious methods of discovery and proof in mathematics and the subordinate sciences was largely unquestioned (except by some sceptics). The task was to devise and enforce *the* 'correct' method. Beyond this, it must be stressed that two broad lines of inquiry are involved: In my view, the structural study of the dynamics of method discourse always must be joined to social historical and biographical enquiry into the expectations, aims and discursive resources concerning method available to and/or enforced upon actors, in their particular historical

[31] Even in his activities on Level III, Cf above Note 14 and see below Sect. 6.8 on the rhetorical functions of Cartesian method.

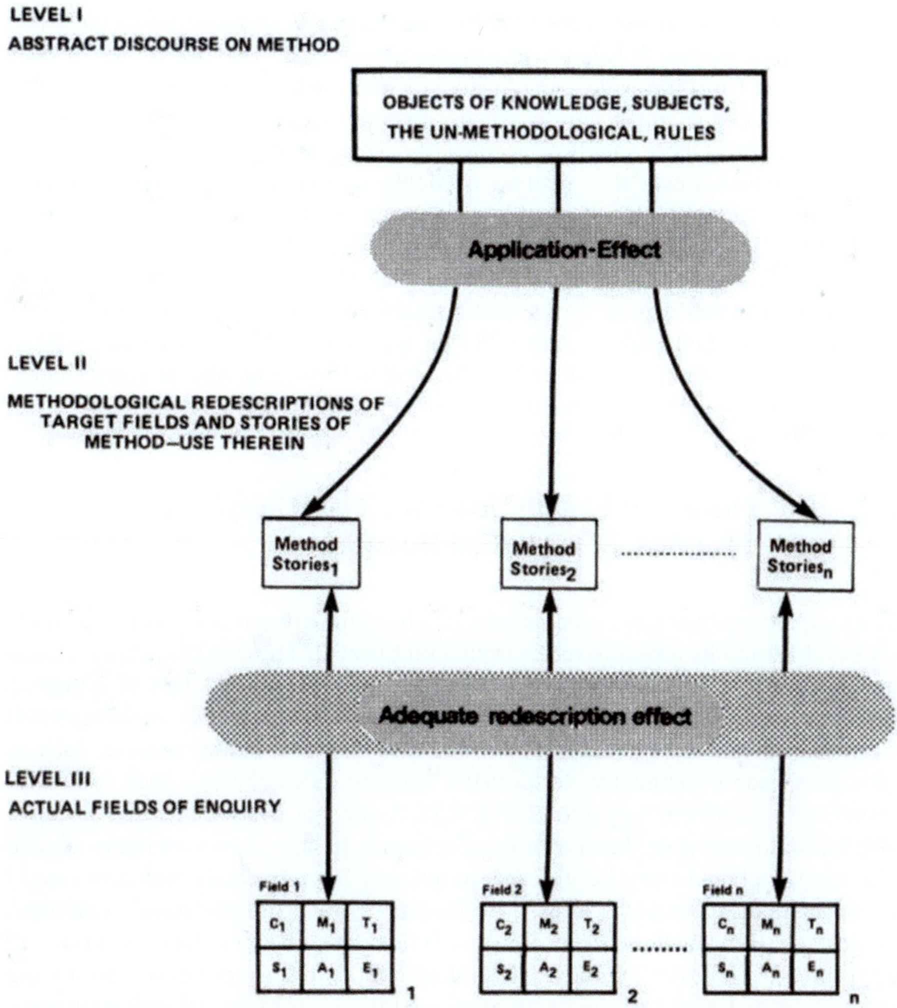

Fig. 6.2 Location of the first two structural effects of a grand method discourse

circumstances. So, firstly, one can ask, in general, how stories or accounts of putative matters of fact come to be heard as true even prior to, and possibly completely independently of any attempt at further inquiry? I suggest that ethnomethodological investigations of the strategies used in making sense of everyday accounts and narratives are relevant here.[32] Secondly, the general theory of construing of accounts needs to be supplemented, in any given historical case, by an inquiry into the conditions

[32] Schuster (1984) especially Section V. In this connection, I particularly rely on the researches of Harvey Sacks (1972); read against the background of Schutz (1970) and Schutz and Luckmann (1974).

and determinations, which turned the historical actor in question into a particular sort of reader/listener. A believer in a method is someone who has chosen or been led to choose (in some wide social-historical senses of the term) to have no other resources for explicating scientific practice than those offered by the method of his 'choice'. We must try to supply an account of the historical context and biographical trajectory which made that method available to him and 'preferred' by him. Analytically speaking, there is the historical problem of explaining the election of a method by an actor, and then there is the general problem of explaining how a method discourse functions upon an actor, once he is 'inside' it.[33] In the case of Descartes, how and why he formulated his particular method is an historical problem; how his method could be sterile and yet appear not to be is a structural problem it shares with other method doctrines. To explain how and why Descartes could believe in such a method is a function of both inquiries taken together.

6.7 The Third and Fourth 'Effects': The 'Unity' and 'Progress' of a Method Discourse

Thus far, the analysis has concerned the relationship between any given target field on Level III and its phantom redescription on Level II. The first two textual effects are produced by vertical relations holding amongst the three levels of a method discourse. A third and fourth textual effect are created horizontally, across Level II of a method discourse. This is due to the fact that a method discourse, such as Descartes', can generate across Level II a range of redescriptions, each one corresponding to a different target field (Fig. 6.3). A general method, after all, has to be able to command more than one area of research. In any given systematic method discourse, each and every Level II redescription will be couched in terms of Level I elements. In the case of Descartes, these involve reference to 'absolutes', 'relatives', 'series' etc. Hence, all the Level II redescriptions in a given method discourse will appear to be 'similar', although we shall see that serious equivocations are introduced as Level I elements are deployed in several Level II redescriptions. In addition, any Level II story then produced about the use of the method will involve an account of the application of the rules of the method to the redescribed field. Unsurprisingly, the believer in Descartes' or some other method will find that the rules of his favorite method gear into each and every redescription offered across its Level II. With that realization, the application effect gives birth to what I term the 'unity effect'—the illusion that the rules of the method are efficaciously applicable to some set of discrete fields of inquiry.

[33] In some cases, the second task is involved in the first. For example, in cases where actors were arguably indoctrinated into a method, part of the account of the indoctrination process has to do with the mythological mechanics of method discourses. Early modern figures, such as Descartes, present a more complex picture. As I just proposed, their intellectual culture widely accepted the, in principle, existence of efficacious methods of discovery and proof. The challenge was to devise, select and enforce the 'correct' one.

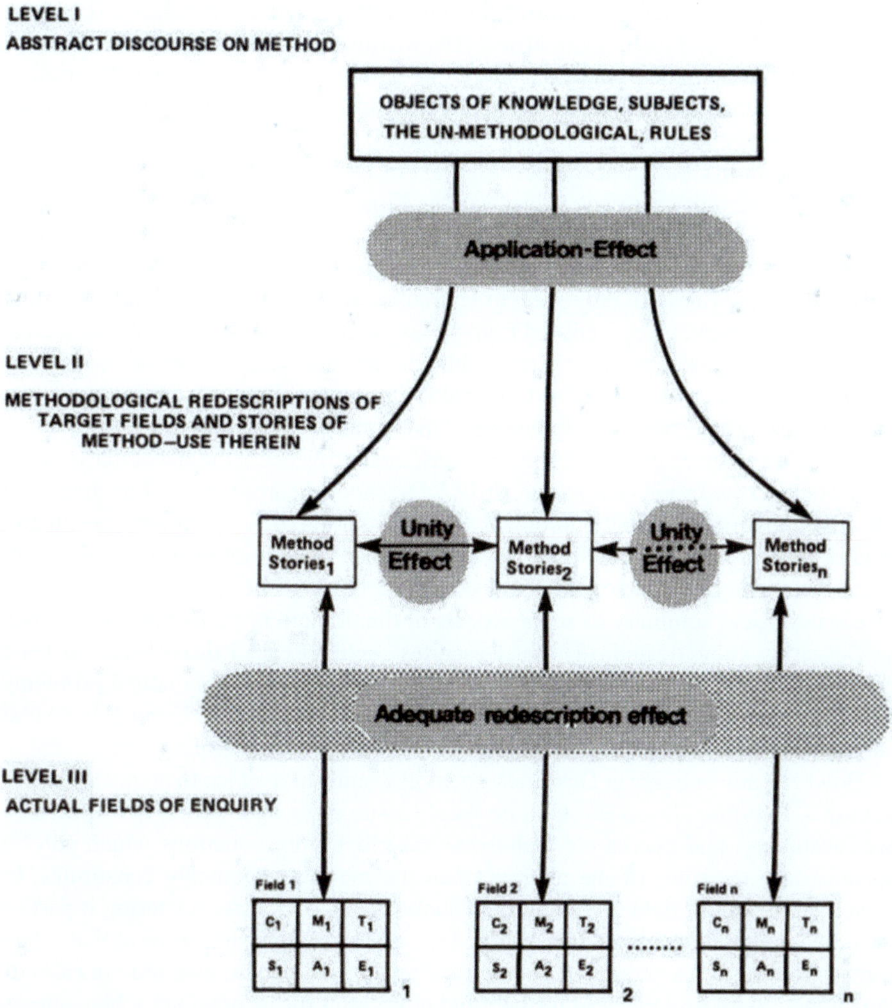

Fig. 6.3 Adding the unity effect to the application and adequate redescription effects

In the *Regulae*, Descartes provides splendid illustrations of these points. He not only offers his methodological version of magnet science, but also his methodological version of mathematical analysis. He writes that, in this case, the 'absolute' is the straight line, the coordinate, the 'relative' is a curve.[34] Presumably, we are to refer a curve to straight lines by means of equations expressing the nature and degree of its 'relativity' to the coordinates. These clear and distinct algebraic representations of curves then become the vehicles for further inquiry into the properties and relations of curves. The entire discipline is pursued under the aegis of an emerging algebraic

[34] AT, X, p. 381 1.25, p. 382 1.9.

theory of equations. Before exploring the unity effect, by comparing Level II magnet science with Level II mathematical analysis, we must first consider how this *Regulae* version of mathematical analysis fares, in relation to the adequate redescription effect (and hence in relation to the application effect). This will further illustrate points made above in Sect. 6.6, and, in addition, it will show the relevance of our reconstruction in Chap. 5 of the development of Descartes' method discourse.

Recall Descartes' heuristic rules, such as 'note always the absolute in question and the order of relations binding the relatives to it'; or, 'break down questions into simpler parts and resolve them in due order proceeding from the simpler to the more complex'. These rules not only *seem* to be transcriptions of maxims used in resolving equations, they were, indeed, elicited from parts of Descartes' mathematical practice. As we have seen, the concepts of the 'latticework', of 'absolute' and 'relative' terms, of 'relations' and 'series', were all derived by analogy from the elements of the sort of little mathematical problems informing 'universal mathematics'; that is, from problems about quantities in continued geometrical proportions, such as the one at the end of rule 6, or the ones which so fascinated Descartes in 1619.[35] Descartes' heuristic rules of method were likewise constituted as methodological analogues of maxims which were useful and readily at hand in dealing with such little problems. So, mythologists of method can agree with believers in Descartes' method: (1) that the heuristic rules of the method do give *a post facto* summary of *some* aspects of the domain of mathematical analysis as Descartes knew it; and, (2) that Descartes' entire Level I discourse is at least minimally adequate and applicable to the small domain of mathematical problems which formed the basis of 'universal mathematics', because the Level I discourse and the rules are simply further analogical developments of that base.

However, any believer in Descartes' method would surely wish to go much further than this. Echoing Descartes' text, he would issue a stronger claim: *If* coordinates are 'absolutes' and curves are 'relatives' describable by equations which can be solved using the rules of the method, *then* the method adequately constitutes, in principle, the entire field of geometrical analysis, by, of course, rendering it part of the domain of application of the method. Descartes would have us believe that if one were to set up such a version of geometrical analysis, he could then use the rules of the method to expand the field further (and to account for its previous achievements as well). This, in historical and mathematical point of fact, is not the case; and, as our structural analysis shows, cannot be the case.

The heuristic rules of Descartes' method were simply too vague and too limited to be of use in the further expansion of the domain of geometrical analysis. The heuristic rules did not, and could not help Descartes, or anyone else, move beyond the resolution of just the sorts of trivial problems which formed the basis of 'universal mathematics' and of Level I of the method discourse. Nor can they explain the prior development of the field in any way other than through a tendentious and implausible Level II account, a methodologist's 'rational reconstruction', of the sort Imre Lakatos and his followers used to advocate.[36] Consider, for example, that in Book

[35] See above Sect. 5.7.

[36] Schuster (1979)

Three of the *Géométrie* (1637) we have Descartes' own mature and considered statement of the procedures and techniques of contemporary algebra and theory of equations. He summarized, rationalized and extended the rapidly growing early seventeenth century 'tool kit' of algebraic techniques. He showed how to evaluate the number and signs of the roots of an equation; discussed real and imaginary roots of an equation; showed how to alter roots by common factors; how to reduce varieties of equations to canonical forms, etc.[37] It is obvious that the heuristic rules of the method are mildly reflective of a small amount of this material. But, in themselves, they cannot account for the full richness of the theory of equations or its development to that contemporary point of perfection. Algebra, which was increasingly seen as a general theory of equations, was developing within a largely *sui generis* and rapidly changing realm of properly algebraic conception, expression and technique (it was a Level III discourse and practice).[38] The heuristic rules redundantly echo some few of the findings. However, they furnish no significant insight into how the algebraic art had developed, nor how it could be further extended.[39]

In short, Descartes did not formulate an adequate Level II version of mathematical analysis. This was despite the fact that his Level I rules were indeed applicable to a small portion of that domain, and the fact that his entire Level I discourse was an analogical transcription of that small portion. So, although Descartes' method discourse was based upon a sub-domain of mathematical practice, this did not prevent his Level II version of mathematical analysis from suffering from the adequate redescription and application effects. This has led to much misunderstanding of the real scope and efficacy of his method.[40]

---

[37] *Géométrie,* Liv. III, AT, VI, pp. 444–61

[38] Mahoney (1973, 1980).

[39] Consider a parallel case, Descartes' discovery of a general constructive solution for all cubic and quartic equations, presented in the third book of the *Géométrie* as the solution of all the 'solid' problems of the ancients. I have reconstructed Descartes' path to this construction (Schuster, 1977, 131–49), showing how this achievement grew out of Descartes' early compass researches, recounted in part above in Sect. 5.4, and how the analytical skills involved elude pat generalizations, in terms of the simple heuristic rules of the method.

[40] Let us consider how this line of analysis affects our understandings of (1) Descartes' posture toward his own method, and (2) the relevant traditions of Descartes scholarship on these issues:

(1) In 1619–1620, Descartes might well have believed his method did, in fact, or could, in principle, subsume all of the analytical mathematics he knew. The point is that his own work (for example, in the mid 1620s on 'all the solid problems of the ancients'—cubic and quartic equations—or, in the early 1630s, the problem of Pappus which focused the research issuing in the *Géométrie* itself) and that of others would eventually be seen to outstrip the method-talk of 1619–1620, elaborated as it had been from tiny bits of analytical practice. These points concern his likely belief in the relevance of his method to his high level mathematical practices. Below, in Chap. 7, we discuss the likely processes (and timing) by which Descartes became, in a more general sense, a non-believer in his own method (at least in private, if not in public!).

(2) Considering these issues more generally, I suggest that this line of analysis offers closure to several generations work of heated and contentious speculation amongst Cartesian scholars, concerning 'the relation' between his method and his mathematics, particularly his mature analytical geometry and theory of equations.

Now we can compare Descartes' Level II version of mathematical analysis to his Level II version of magnet science. Everyone immediately sees that the rules of the method 'apply' to both Level II stories, and that the stories have a certain resemblance: They are both about the 'relations' of 'absolutes' to 'relatives' etc. Mythologists realize, however, that this resemblance is superficial and misleading. It is based on deeply worrying equivocations about the terms 'absolute', 'relatives' and 'series'. The words are the same, but they are badly abused. Depending upon how one uses the term 'absolute', as coordinate or corpuscle, we then suitably frame our 'relatives' as algebraic equations or corpuscular–mechanical models. But, coordinates bear to curves or equations vastly different 'relations' than basic corpuscles bear to physical conjunctions thereof.[41] *Within* each story, there is some unity of expression about absolutes, relatives and relations, whilst across versions there is nothing but an almost comical equivocation.[42]

How can it possibly be that a believer does not see these equivocations? How is it that he falls for the unity effect? In Sect. 6.6, it was suggested that a believer in a method tends not to see, or at least not to thematize, the slippage between Level III and Level II, what was termed the evisceration of Level III in favor of its Level II redescription. Once located on Level II, and lacking any motive or machinery for removing himself from that comfortable perch, the believer may be confronted with two (or more) Level II redescriptions, corresponding, respectively, to two (or more) target fields. He is now likely to be impressed by the observation that these Level II versions are 'rather similar', if not identical. To the extent that he can think of the target fields in independent, Level III, terms, he is likely to say that the methodological versions of the fields render them 'more alike' than the non-methodological accounts (which accounts may be further stigmatized for their lack of 'methodological articulation').

Now, it is worth noting that, even for mythologists, there is some truth in this. The Level II versions are, indeed, more similar than any serious accounts which might be given of their Level III matrices.[43] Level II accounts do not lie about

---

[41] The only way out of this problem would be to maintain, with Cassirer (and some of his latter day acolytes), that Descartes' (discursive) corpuscularianism was some sort of mistake or illusion, and that Descartes did (or should) have entertained an 'ontology' for physics of coordinates and functions of physical quantities thereof. This, however, would hardly establish the efficacy of Descartes' method. If true, it would only show that mathematics provided genuine tools for the pursuit of Descartes' physics. In any case, after 1628 the historical Descartes resolutely stuck to grinding out qualitative corpuscular-mechanical discourse. See Cassirer (1902), Einleitung, 'Descartes' Kritik der mathematischen und naturwissenschaftlichen Erkenntnis'.

[42] This may well help explain Gerd Buchdahl's acute observation (1970, 126–7) that Descartes seems to use terms such as 'analysis' and 'absolutes' in several contrasting ways; for example, the latter may be (a) elements to which explananda are to be reduced; (b) elements into which explananda are decomposed; (c) elements to which explananda are referred; (d) elements from which they can be physically generated, or (e) logically derived. The cause of this proliferation is not so much a philosophical profundity struggling with inadequate terminology, or underdeveloped conceptual distinctions. Rather, it is the structural product of the way the method discourse devours and denatures domain-specific discourses in the interests of (a pseudo) generality.

[43] 'Serious' here denotes the accounts given by 'properly trained' historians and sociologists of science, of course!

each other in quite the way they each lie about their respective Level III targets. The Level II accounts are honest (that is both leading and misleading) analogues of each other. They are of the same discursive fabric, Level I, and there are legitimate things to be said about their analogies. All this, of course, greatly aids the unity effect, which in itself is no better, but certainly no worse, than any other categorical grouping of instances by 'similarities'.[44] It is the structural location of this effect which makes it an accomplice of the really delusional two prior effects.

Returning to the believer, we find that he is much impressed by the discoveries that the Level II versions have a wondrous similarity and that the rules of his method gear into each and every one of these accounts. The method truly is 'unitary': For him, it reveals the underlying similarity of the fields in its domain, and it empowers him, *via* its widely applicable rules, to command and pursue, efficaciously, any and all of these fields. To the extent, if any, that such a believer becomes aware of any contrasts or tensions between any Level II version and its independently characterizable target field, he can mobilize a discourse on the 'practical problems of applying the method'; for, as Descartes and other methodologists are wont to say, 'the method consists more in practice than in abstract statement after all'.[45]

As in the case of the first two structural effects, the reason for this third effect is the obverse of the reason why the method cannot actually work in the ways it claims to work. There is a similarity between Level II versions and a gearing of the rules of the method to them *because* those versions are woven out of the cloth of the Level I discourse; but, for that very reason these Level II versions cannot hope to be adequate glosses of the structure and dynamics of living, Level III fields; they eviscerate those fields in the interests of Level I and still necessarily equivocate amongst themselves.

Finally, for much the same reasons, there is even a fourth literary effect of method, which I term the 'progress effect'. Reading across Level II of Fig. 6.3 over time, methodologists can proudly point to 'progress' as the method is 'extended'; that is, as *new* Level II accounts of *new* target domains are added. Methodologists can also label as 'progress' the revising of existing Level II accounts of old domains in order to grasp and 'explain' new developments in those already methodologized fields. 'Progress' can also be discerned in the discovery and resolution of certain internal problems set in train by the very structure of the method discourse. Often, this takes the form of adding to or revising the rules, such as we have seen the young Descartes doing in 1619–1620.[46]

In general, then, the literary effects of Cartesian (or any other) methodology relate to each other in this manner: In any systematic method discourse the adequate redescription effect is fundamental, and it ultimately depends upon the plausibility of Level II stories within a context of cultural precedence accorded to the Level I

[44] The allusion here is again (cf Note 23) to Barry Barnes, explicating Kuhn, on 'finitism' in regard to the acquisition and ongoing negotiation of meaning and reference of concepts. Barnes (1982), Chap. 2.

[45] Cf Note 29 above.

[46] See Sect. 5.6

discourse as the way of thematizing practices in the disciplines in question. The application effect depends upon the adequate redescription effect, for it fosters the illusion that the application of the rules, within stories on Level II, is the application of the rules in actual practice. The unity effect results from the iteration of the application effect across the spectrum of fields thought to be commanded by the method in question, and it is facilitated by the fact that Level II entities bear some analogical relations to each other, despite inevitable and dangerous equivocations. The progress effect is the unity effect experienced or accounted over time as more Level II stories are added to the collection, or, as Level I rules are explicated, new ones added and 'problems' of the discourse identified and 'resolved'.[47]

One can conclude that any believer, seriously engaged in the business of grand prescriptive general methodology, will probably stumble into this hall of discursive effects. The believer will then happily expatiate on the unity, applicability, efficacy and progress of his method, or that of his master; he will refine and explicate Level I,

[47] In Sect. 6.3, it was mentioned that my approach to methodology as mythic speech has been heavily influenced by Feyerabend's dissection of the mechanics of Newton's methodological claims (Feyerabend 1970). We can now examine Feyerabend's argument in an attempt to bring out the implicitly structural character of his analysis, and thereby also to show how his analysis can be mapped onto my general schema: Feyerabend is concerned with the ways in which Newton's proclaimed empiricist method enabled him to package and defend his theory of physical optics. In Feyerabend's version of Newton's empiricism, the essential methodological rule is: 'The sole basis for the derivation (induction) of theory is "experience"' (p.150) 'Experience' is the stable, objective, repeatable, public foundation from which theories may be rigorously induced (hence this is part of Newton's Level I). Experience, moreover, requires no interpretation or justification. A methodological corollary, expressed by Newton in versions of his 'fourth rule of philosophizing', states that theories derived from experience are not to be challenged on the basis of 'speculative alternative hypotheses' (p.159 Note 7). (This is a Level I counter-discourse on the 'unmethodological').

What interests Feyerabend is the essential vacuity of Newton's central methodological rule (the parallel of our first two 'effects'), for the rule itself states nothing about what experience is; what is to count as experience; how it is to be located; how, if located, it is to be interpreted. Newton's (Level I) pronunciamentos name a privileged basis for theory without being able to specify how that basis is to be produced, recognized, evaluated or inferred from. This produces a situation in which any institutionalized teaching about what is to count as 'experience' can be used by believers to fill the criteriological void (pp.155, 168–9). Feyerabend then shows how Newton summons 'experience' into existence by a two stage semantic slight of hand. First, we have the paradigmatic experimental illustrations of the theory which are, of course, thoroughly theory- and standard-laden. (These being his presentation of the famous elongated prismatic spectrum, and his *'experimentum crucis'*.) The 'results' of these experiments, what Newton or we might perceive in them, are first identified with what Newton calls 'phenomena'. But 'phenomena', in his usages, are to us highly generalized, idealized and differentially weighted observation reports. Then, Newton claims that such 'phenomena' are the 'experience' which we seek as the sole basis of theory (pp.161–5). That a theory does seem to follow from such 'experience' is hardly surprising, Feyerabend observes, since such experiences already have the logical form of laws (p.163 Note 11), being general propositions well loaded with the terms of the very theory to be 'induced'. (This, I would suggest, can he seen as a very neat instance of the application effect, the convincing 'realization' that the rules of one's method do apply in the realm of the 'authoritative' method story, which one has spun about one's practice using the resources of Level 1.) (Cf. Feyerabend's remarks at p.165).

his Level II stories and the rules; he will castigate other methodologists, and those who do not believe in methodology at all; and, he will comment upon all these matters at ever higher levels of meta-discourse. Like other believers, René Descartes got lost in this hall of discursive effects, only to be followed there by many of his loyal scholars. In order to understand the trajectory of Descartes in natural philosophy and its subordinate disciplines, and to understand his role in the Scientific Revolution, one must leave the hall of effects and subject it to the sort of critique begun here. Accordingly, we have two further items on our agenda before concluding the present chapter. We must, in the next section, examine more closely the ways the young Descartes deployed method stories rhetorically, that is, as tools of self or public persuasion within the Level III practice of given disciplines. Then, we need to reconsider the issue of whether Descartes ever stopped believing in his method as a guide to his intellectual agenda, and source of his intellectual identity, choosing thereafter to maintain a rather cynical public posture as to its efficacy, and its role in his agenda and identity.

6.8 The Rhetorical Functions of Cartesian (and Other) Method Discourses

Whilst the sceptical historiographies of Koyré and Kuhn effectively debunked method as having no role in the dynamics of the sciences, our proposed discursive model of grand method doctrines entails that methodologies can play some roles in the formation and negotiation of knowledge claims in scientific disciplines. However, they cannot play the definitive roles they claim for themselves. Methods do not capture the (non-existent) essences of their target fields; but, they are certainly rather useful resources in the rhetorical combats and political struggles, through which knowledge claims come into being, prosper and/or die. This section explores these political and rhetorical functions of method and suggests some ways in which they apply to Descartes' work. This advances our overriding interest in reclaiming him, and his method, for an historiography of natural philosophizing and its subordinate sciences, which neither merely debunks method, nor falls victim to its literary effects.

As already noted, the work of Paul Feyerabend, on the rhetorical and propaganda functions of Galileo's and Newton's methodological pronouncements, began to point toward the political functions of method discourse in the life of the sciences.[48] His initiative was extended in the literature within the history and sociology of science that capitalized on the 'post-Kuhnian' challenge to explain what method discourse does in the sciences, if it does not and cannot do what had traditionally been claimed for it. Broadly speaking, this work suggested that method discourses are often deployed as rhetorical weapons in those negotiations and struggles over

[48] Feyerabend (1975, 1978). Cf. Schuster (1986) 36–37, 79–80.

the framing and evaluation of knowledge claims, which go on at all levels of scientific activity, from the laboratory bench, through published texts, to disciplinary debate and its necessarily associated micro-politics of groups and institutions.[49]

Let us first consider what the 'rhetorical' function of method discourse means at the level of the formulation of technical arguments and knowledge claims. Some historians of science and sociologists of scientific knowledge plausibly claim that technical scientific arguments, even in published form, are pieces of practical, rather than formal reasoning, more akin to legal briefs than to chains of strictly valid inferences. The burden of a scientific argument is, typically, to promote some novel, or revised, claim about the 'objects of inquiry' within a given field. To that end, various resources may be deployed: Appeals are made to theory- and standard-laden data; claims are made about the objects, tools and techniques currently accepted in the field; and, implicitly, at least, field-specific standards of adequacy and relevance guide the assemblage of these resources into a 'compelling' but not rigorous argument. Hence scientific argument, as essentially persuasive argument, may rightly be termed 'rhetorical', in the sense defined by students of 'the new rhetoric', denoting the entire field of discursive structures and strategies used to render arguments persuasive in given situations.[50]

Now, all the various grand doctrines of scientific method, as well as the particular stories derivable from them, form a reservoir of discursive resources available to scientists for use in the formulation of such essentially rhetorical arguments. Hence, to this extent it is correct to say that methodological doctrines can be *partially* constitutive of knowledge claims in the sciences. That is, in terms of our model, Level I and II method discourse, especially Level II stories, can be deployed on Level III in the cut and thrust of scientific practice and debate, and hence, in that sense, can be said to be partially constitutive of socially negotiated outcomes within the Level III matrices. Methods do not command, explain or grasp the essence of Level III practices; but, they can be deployed by players, on that level, as resources in the struggle to establish claims. Historians and sociologists of science have observed that all such rhetorical deployments of method discourses are highly flexible and context dependent, scientists sometimes giving different methodological accounts in different argumentative contexts, and sometimes even contradicting themselves, by offering contradictory interpretations of their own methods or those of famous methodologists.[51]

René Descartes certainly practiced such rhetorical deployments of method, mobilizing Level II accounts in order partly to constitute knowledge bids he was advancing on Level III—where the Level III fields might be construed as natural

[49] Schuster and Yeo (1986) and Yeo (1986).

[50] Perelman (1979), Perelman and L. Olbrechts-Tyteca (1971), J.R. Ravetz (1971), Yearley (1981), Weimar (1977).

[51] Mulkay and Gilbert (1981), Feyerabend (1975), Miller (1986), LeGrand (1986), Wood (1980), Richards and Schuster (1989).

philosophy and other cognate or subordinate fields such as mathematics, medicine, optics, mechanics and the like. His methodological account in the *Regulae* of the discovery of the law of refraction and of its mechanistic explanation is just such a gambit. As we saw in detail in Chap. 4, the story bears no relation to Descartes' 'bench practice'; yet, it structures a presentation of his work and therefore is partly constitutive of it as a knowledge claim proffered to the intended audience of the *Regulae*.[52] Moreover, as we already know, Descartes' method story about his optical work served other subordinate functions, in the overall interest of facilitating the acceptance of his claims. First, it occluded the dependence of his actual work upon the traditional image principle rendered dubious by Kepler's findings in his new theory of vision. Second, it provided a (method-)logical connection between the geometrical optical and physico-mathematical explanatory stages in his work. Thirdly, the vagueness of Descartes' methodological language about 'natural powers', and his methodological reflections about 'analogy' covered what was, in 1628, hesitation and ambivalence about the best direction to take in articulating a mechanistic model of light.[53] In other words, our analysis in Chap. 4, in effect, discovered the rhetorical uses of the method in this case: The method story was a very valuable way of framing, constituting and presenting his knowledge claims while finessing these secondary problems. When one additionally considers that Descartes probably believed that the work could have been done the way the story tells, the power and utility of the method become very clear. Descartes, one suspects, was probably getting the benefit of his own 'just so' story (by virtue of the literary effects), just as his readers were (honestly, rather than cynically) intended to do.

All the foregoing points are based upon our model of method discourse. Taken together they also reinforce and articulate that model, because they allow us to see additional reasons why actors quite reasonably fall for the apparent efficacy and applicability of any method doctrine: For believers in a particular method, any deployment on Level III of its Level II stories will be highly privileged and impressive. These stories will probably be the only resources in play on Level III which label themselves as 'methodological'. Participants debating and negotiating claims on Level III will generate and hear these method stories as the only elements in the cluttered landscape of debate which are of a 'methodological' character. Hence believers will see method-talk 'in action' as a crucial, or *the crucial* element in the debate. This will lend more support to the truth of the Level II stories. The stories say 'practice proceeds just thus and so', and here is 'practice', that is the social

[52] Of course, given the fact that Descartes left off the *Regulae* unfinished, the actual audience has consisted not of his natural philosophical and mathematical contemporaries, but mainly of modern historians of philosophy.

[53] We recall that Descartes was then probably playing with models of light involving bent arm balances, balls, as well as crude versions of his ontological model—mechanical disturbance in a medium. (Sects. 4.7.3 and 4.7.4).

world of the laboratory, conference, published debate etc., in which method discourse is a crucial resource in the fray.[54]

Method claims on Level III need not be consensually accepted by all parties and can be contested. This can be understood in terms of the post–Kuhnian work in the sociology of science which further established that the evaluation and negotiation of knowledge claims is a social and political process, and that any and all of the tools or weapons used in constructing or evaluating a claim can be questioned.[55] The recourse to methodological discourse on Level III is simply one possible tactic in this knowledge-making/knowledge-breaking game, and so deployments of method discourse can become objects of contention within it.[56] Hence, for a contestant like Descartes, not only did particular claims need to be woven out of the sturdy cloth of method discourse; but, the method itself, the ultimate legitimizing weapon, required support and justification. So, when Descartes presented his optics in terms of his method, he not only tried to legitimate the optics, in the ways we have indicated; he was also legitimizing the method by the 'evidence' of concrete application and success. (In the *Regulae*, the optics case illustrated a text on method, not vice versa.)

All this was particularly important, because the method in turn was going to have to bear the weight of legitimating any and all of his projects. Descartes, like others contending for natural philosophical pre-eminence, was not concerned simply with particular claims and arguments. He wanted to group together and package a certain family of results ranging over a spectrum of specialties, from mathematics to medicine. So, when Descartes grouped together otherwise widely disparate pieces of research as products of *his* method, he was staking out a series of political claims in the economy of natural philosophy, its cognate and subordinate disciplines.

---

[54] From all this, we can derive two laws in the 'anthropology of method', which help to explain why method-talk is deployed in certain ways in scientific debate. Consider a specialist scientific community engaged in debate over two divergent knowledge claims:

(1) To the extent that all debaters share elements of the same method discourse, their debate will tend to take the form 'to which claim does the method story attach' *not* 'how can one credit stories generated in our method discourse?'

(2) If there are differences about preferred method discourse, debates about method will take center stage away from debate about the divergent claims per se. That is, debate about the claims will be carried on to a large extent by means of debate about which method is to be followed.

In either case, all sides will still share the method believer's view that the crucial element in debate is method.

[55] Bourdieu (1975), Latour and Woolgar (1979), Callon (1980), Shapin (1982), Mulkay and Gilbert (1981).

[56] As noted earlier in Chap. 2, we can draw some controlled analogies between recent findings about the competitive dynamics of modern sciences, and the situation in the highly contested realm of natural philosophizing in the generation of Descartes. Not only were natural philosophical systems contested, along with access to important institutional bases, but the fields cognate and subordinate to natural philosophy were also in flux, in terms of their own content and practices and the relation of the latter to natural philosophies in conflict. These circumstances invite the *controlled heuristic use* of insights gained about the competitive dynamics of modern scientific traditions.

Not only was he endorsing his results individually, he was also linking them under the claim that they were all to be accepted as a piece, because they all fell within, and followed from, his method, *the method*. He was claiming methodological hegemony over these and other fields, positioning himself in relation to practitioners within and across those fields. The literary effects of method, especially those of unity and progress, probably provided him with a great deal of honestly held confidence about taking this posture.

In the final analysis, after 1628 when he began to work on *Le Monde*, the key issue for Descartes was the status of this, his emerging first system of mechanistic natural philosophy. In this regard, he now became a symptomatic, leading player in the central dynamic of this stage of the so-called Scientific Revolution, which focused precisely on the clash of opposing systematic visions of natural philosophy.[57] His method functioned on this peak level of struggle, by supposedly underpinning his entire project in natural philosophy, underwriting, that is, his claim to pre-eminence in resolving the clash of natural philosophies of his day. This is intimated in the way the *Essais* of 1637, themselves appetizers for the natural philosophical system, are subordinated to the overarching tale of the method in the *Discours*; and in the way the metaphysical grounding for his natural philosophy is also offered as a triumph of method. Descartes even carried this method-rhetorical shaping of his claim to cognitive dominance to a higher, more personal, heroic, indeed Baroque level in the *Discours*, when he claimed that his life as a natural philosopher, mathematician and metaphysician had itself been shaped and lived, in order, according to the method.

But, whether Descartes himself believed these wider claims, especially after he abandoned the *Regulae* in 1628, is another matter. Method discourses may systematically delude believers, but there may also be particular circumstances, social and biographical, in which actors cynically exploit the rhetorical power of a method discourse, in which they have cause not to believe. In the next section, we will examine the possibility that Descartes' career in methodology conforms to a narrative in which the honest delusion of youth later gave way to cynical opportunism. Our next two chapters will then be devoted to showing why and how that shift occurred and what it had to do with the emergence of Descartes, the systematic corpuscular-mechanical philosopher of nature and metaphysician.

6.9 Rethinking Method and the Career of Descartes

6.9.1 The Original Inscription of Descartes' Method: Bricolage, Self-Deception and Self-Definition

We have spoken about the need for a new, non-Whiggish historiography of method. The Western tradition of method discourse certainly extends back to Aristotle, and debate about method and other instruments of thought, such as rhetoric and dialectic,

[57] See Chap. 2 and Schuster (1990, 2002).

were endemic in early modern universities and the experience of educated men, not only in the bachelor's degree courses, but also in the higher faculties of law and medicine. The critical period of the Scientific Revolution in the early and mid seventeenth century, with its heightened natural philosophical contention, also saw the proliferation of new, competing anti– or post–Aristotelian doctrines of method, which in turn mark the real starting point for the dynamic of methodological debate in Western science, down to the present day. The bold methodologists of the period, such as Bacon, Descartes, Gilbert, Kepler and Galileo, first of all, always operated with (and against) the available formal discourses on method, meaning, in turn, that they were always elaborating upon the core metaphors and structure of discursive levels that, as discussed at the beginning of Sect. 6.5, shape the very possibility, and limits, of the field of method discoursing. Additionally, each new methodology was constructed by its author in the light of problems and goals, which might relate to the tradition itself, to the perceived state of one or more of the contemporary sciences, or to other discourses believed to be relevant, such as natural theology, political theory, and moral philosophy. The perception and weighting of such concerns by a methodologist was a complex function of his biography, social location, institutional affiliations and perceived interests. Moreover, it seems that a certain biographically and contextually conditioned *bricolage* of available cultural resources governed the manufacture of any particular 'great' methodologist's brand of method.[58] In this chapter and the preceding one, we have reconstructed the young Descartes' trajectory of methodology-building *bricolage*, and his related self-reflections on his agenda and identity qua possessor of a great method.

We have seen that the core of Descartes' method doctrine was constructed in late 1619 and early 1620; that his enthusiastically constructed method doctrine marked the third and final step in a series of youthfully over-ambitious and under-articulated enterprises—physico-mathematics, universal mathematics and method—each one more grandiose and general than the previous one, each one inscribed partly by analogical extension of its predecessor. In particular, we have seen that what the over-excited young Descartes thought, wrongly, was true of universal mathematics, he daringly extended into the realm of all rational enquiry. The method discourse was not abstracted from successful practice in some genuine area of mathematics; it was produced by performance of operations of analogical extension upon the terms of a discourse, universal mathematics, which itself could not do very much of what it was purported to be able to do.

What we seem to have, therefore, in Descartes' path to his initial inscription of his method is a trail of somewhat confused and over-enthusiastic *bricolage*. Bits of his own work were assembled with elements of culturally available discourse on 'general mathematics', and then 'method', in a series of analogical extensions and subsumptions of previous discourse, issuing in the manufacture of his method. And, as we now know from our study of the enticing discursive dynamics of method

[58] See e.g. Yeo (1979) and Richards and Schuster (1989). The term 'bricolage' derives from Levi–Strauss' usage in his account of mythopoeic behavior. (Cf above Note 9 and Chap. 4 Note 92; Chap. 3, Note 86).

discourses, Descartes was probably beginning to succumb to the literary effects of his discourse. Yet, from Descartes' perspective his path to the method would have seemed a marvelous and triumphal progress. Recalled to study in late 1618 by Beeckman and his program of physico-mathematics, Descartes had, by mid 1619, merged that project with his own work in mathematics to formulate the intoxicating dream of universal mathematics. Then, musing in the late autumn of 1619, he had seen how to conquer all rationally obtainable knowledge by generalizing his earlier revelations. No wonder, then, that on St. Martin's eve 1619, Descartes, enthused by his skill in thus transforming one discourse into another, dreamt that the project he had glimpsed had been consecrated by God himself.

Let's therefore take stock of what has been discovered in the last two chapters: In this chapter we have stressed the mechanisms by which method discourses, Descartes' included, mislead believers as to their efficacy. In the previous chapter we focused on Descartes' shifting early agendas and self-understandings, firmly grounded for him in what he lived and perceived as a sequence of marvelous intellectual accomplishments and revelations. As was established in Chap. 5, during 1619, Descartes worked sequentially through physico-mathematics, universal mathematics and arrived at the core of his method, so that by late 1619 things had reached an overheated and underdeveloped state of affairs. The twenty-three year old Descartes, analytical mathematician, practitioner of a new physico-mathematics, and owner of the meta-discipline of universal mathematics, now thought that he had found a general method that could control work in all disciplines governed by reason. Descartes' senses of his own identity and agenda were bound up in these experiences, and their wonderful outcome. And now, with the present chapter, we know more about how such belief and commitment were generated, due in part to the mythical power of any well formed methodological discourse. *Self-awareness*, commitment to an heroic identity and agenda, and *self-delusion*, product of precisely the intellectual achievements so valued, are simply two sides of the same coin. Each side needs to be fully understood and related to the other, as we have tried to do. Accordingly, and finally, we come to the issues of how long Descartes' belief in his own method manifested itself in its original, simple and naïve form, and whether Descartes ever became aware of the limits of his method, and hence more calculating, indeed cynical, about its public deployments—whether, in short, the naïve enthusiasm of 1618–1620 ever turned to opportunistic public posturing about a method known or suspected to be of dubious efficacy.

6.9.2 *The Failure of the* Regulae, *the Birth of the System and the Problem of the Cynical Discours de la méthode*

I have been suggesting, all along, that Descartes' project of method is crucial to understanding his career as a physico–mathematician, mathematician and natural philosopher, but not in the senses that he (or approving scholars) claim. Nowhere is this more apparent than in the decisive fifteen years following the methodological frenzies of 1619/1620. Unless we maintain a cool, sceptical approach to method, we are likely to get hopelessly lost in Descartes' own mystifications, and so lose the key to reclaiming him as a realistically conceived actor in the history of natural philosophy and the sciences. To this end, a proper, demystified understanding is required of Descartes' activities in the 1620s, in particular his attempt in the later portions of the *Regulae* to flesh out and partially redirect his method project of 1619.

If we reflect upon the young Descartes' position in 1620, we can confidently assert that he was neither a builder of systems of natural philosophy, nor a systematic metaphysician. He was, a practicing mathematician, and, following Beeckman, a physico-mathematician with corpuscular-mechanical leanings, as well as, from November 1619, a self-appointed methodological prophet. We can also now understand how he was convinced that his method subsumed universal mathematics; that it was efficacious; and that it could guide his researches in every field of rational inquiry. What we know of his intellectual activities over the next few years, gives us little reason to think he, in any way, doubted his method was real and efficacious.

By the time he settled in Paris in the mid 1620s, Descartes had produced a further genuine analytical mathematical triumph with his construction of all the 'solid' problems of the ancients, using only a circle and parabola (equivalent to a general construction for all cubic and quartic equations).[59] And, as we know, shortly thereafter, sometime in 1626 or 1627, he produced his master stroke in physico-mathematics: the construction of the law of refraction, followed by the development of a theory of lenses and the attempt to subsume the law under a mechanistic theory of light. Although, we modern method atheists have no grounds whatsoever for believing these achievements had been produced by application of the method, Descartes no doubt conceived the method to be relevant to his triumphs, and they in turn reinforced his belief in his method, according to the mechanisms we have described. Even in the case of the discovery of the law of refraction and its mechanistic rationales, whilst our reconstruction makes it clear that the method had virtually nothing to do with the trajectory of discovery, and that Descartes would have known that, he still wrote (in the subjunctive mood), in rule 8 of the *Regulae*, as though, in principle, the method could have led to the discovery of the law and its explanation. So far, so good, therefore, concerning Descartes' original belief in his method. Nevertheless, things were about to get more complex and interesting. We signal these events here and will pursue them in detail in the next two chapters.

Whilst in Paris in the mid and late 1620s, Descartes decided, for reasons we canvass in the next chapter, to articulate in detail his universal mathematics of 1619, under the guise of extending his 1619/1620 text on method, roughly rules 1–11 of the *Regulae*. Rules 12–21, and parts of rule 8, were written in Paris for this purpose. Unfortunately for the young Descartes, now intent upon establishing a public reputation, there was only one thing wrong with this newly articulated universal mathematics: It did not work. The new text collapsed under the weight of internally

[59] Schuster (1977) 127–149. Needless to say, this reconstruction of Descartes' discovery owes nothing to invoking the rules of his method.

generated problems which revealed, amongst other things, that his analytical mathematics could not be subsumed or produced by his method, and that whilst he now needed to be systematic and explicit about his corpuscular-mechanical commitments in natural philosophy, such a system was not, and could not be, a product of the method. Descartes, we shall see, realized all this by late 1628, when he abruptly abandoned composition of the *Regulae* and moved to the United Provinces, there to work on the metaphysics and systematic mechanistic natural philosophy which could answer and transcend the difficulties upon which the *Regulae* had foundered.

After this 'inflection point' (as we shall term it), pivoted on the failure of the later *Regulae,* Descartes never again was to advocate his elaborated, method-based, universal mathematics. He quickly evolved into the metaphysician and systematic natural philosopher he was to remain during the later and more public part of his career. We shall trace all these developments in the next two chapters. The problem for the moment, however, is that he was still to write the *Discours* and publish it in 1637. There he proclaimed to the public that his method had guided his life and work—thereby in effect announcing that none of the messy post-1628 history just outlined ever happened. Can Descartes have seriously still believed in his method? In this period of his life, after the collapse of the later *Regulae*, we have to suspect that he no longer had a genuine personal conviction as to the reality and efficacy of the method, although he exerted himself to appear, at long last, in public as an advocate of that method. Surely he was now being cynical, playing for the public a method card he now knew to be flawed or counterfeit.

Virtually everything Descartes states in the *Discours* about the provenance, use and development of the method, and its role in his career, is a fiction—indeed given our mythopoeic model of method, must have been fiction, because nothing he claimed for the method could have been accomplished by using it (or any other general method). Looking at the *Discours*, it should by now be patently obvious: [1] that Descartes did not elicit his method by abstracting out and synthesizing the best aspects of scholastic logic, Greek geometrical analysis and algebra (his construction being more fraught and opportunistic);[60] [2] that applying his method did not generate an ever enlarging collection of rules for mathematical analysis;[61] [3] that, after 1618, the method in no way offered a full account of 'everything that gives the rules of arithmetic their certainty';[62] and, finally, [4] that the method, in 1619, did not dictate the subsequent course of his career, the preparatory years spent in lower studies, before he was ready to assay metaphysics after 1628.[63]

---

[60] AT. VI.p.18; CSM I, p.19

[61] AT VI. pp.20–21; CSM I p.121.

[62] AT VI, p.21; CSM I, p.121

[63] ibid., pp.21–22; CSM I pp.121–22 Since, for example, his move to systematic natural philosophy and metaphysics was a response to the failure of the later *Regulae*, not some unfolding of methodological imperatives built into it. By the same token, we shall also see that he did not develop his elaborated version of universal mathematics in 1620 (AT VI. p.20), nor did he actually do it in 1628 by applying his method.

We can, I think, conclude that down until the collapse of the renewed project of the methodologically based universal mathematics in the later *Regulae* in 1628, Descartes was probably under the sway of his method discourse, generally believing, for the reasons already discussed, in its de facto or in principle relevance to his physico-mathematical and mathematical projects. After 1628, one cannot be so confident that Descartes was so firmly in the grip of the discursive dynamics of method, nor, accordingly can one be so charitable about his likely beliefs and intentions. It would seem likely that when he used the method to articulate his autobiography in the *Discours*, he was largely covering the tracks of his abortive enterprise of the late 1620s and was cynically exploiting the kind of rhetorical uses of method-talk we discussed above. The method, he insisted to the public, governed his life, his order of study, the content of those studies, including the metaphysics, medicine, optics, meteorology and geometry published in the *Discours* and its accompanying *Essais*—all appetizers for his yet to be unveiled system of corpuscular-mechanism. That he actually believed what he was saying beggars our own belief; that he wished he could still believe what he was saying seems more plausible; that he wished the public to believe him seems indisputable. Additionally, in the years following 1637, as we mentioned earlier in Sect. 5.6, Descartes produced disparate and varied informal remarks about methodological matters scattered in his correspondence and published works, dealing with meta-reflections on the explanatory structure of his mechanistic natural philosophy, comments on the argumentative tactics in his *Meditations*, and methodologically relevant debate with critics of his optics and mathematics, as presented in the *Essais*. In these cases, too, it becomes increasingly difficult to believe that Descartes genuinely believed what he was saying about the efficacy of his method.

So René, the believer in his own method, became, arguably, René the public spin doctor of his own method. And yet, this matter requires one more modulation, softening our black and white conclusions: It must be admitted that the discursive mechanisms of (any) method are such that no amount of experience must dissuade a believer. The fact that Descartes was probably both a cynical manipulator of the method, and the first of its many victims, may explain the air of ambiguous ambivalence that seems to surround many of his later methodological pronouncements. After 1628, Descartes may have feared that the method did not work, and feared and resisted coming to grips with that suspicion. The psychology of a crisis of belief in a method may bear similarities to the better known contours of crises of religious belief, especially if methods are indeed powerful species of mythic speech.

Whatever one makes of these problems, it should, at least, be clear that the sorting out of Descartes' method discourse, the reconstruction of its genesis and the identification of its discursive structure and dynamics, are all necessary conditions for our recovery of an historical Descartes, the mathematician, physico-mathematician, methodologist, and later systematic corpuscular mechanical natural philosopher. Although the mature Descartes posed in public behind his method, as a lone prophet of a new natural philosophy, in reality—as an exponent of mechanism, practitioner of the mathematical sciences and advocate of new values in natural philosophy—he was, as this entire book is intended to show, a figure highly symptomatic of the contextual forces in play and opportunities at hand, at this crucial moment in the process of the Scientific Revolution. His method explains neither his manner of work, his achievements, nor the course of his symptomatic career. Rather, his absorption in method, his succumbing to its effects, and even his later suspected manipulation of it, are simply a part, an essential part, of that very contextual weave, a weave the method deceivingly claims to command and explain. Now, with the tangled problem of 'method and the historical Descartes' hopefully laid to rest, we can return to our more chronological focus, beginning with a full dissection of Descartes' attempt to articulate his universal mathematics in the later portions of the *Regulae*, written in the mid and late 1620s, and the reasons for, and large consequences of, the failure of that project.

References

Works of Descartes and Their Abbreviations

AT=*Oeuvres de Descartes* (revised edition, 12 vols.), edited by C. Adam and P. Tannery (Paris, 1964–76). References are by volume number (in roman) and page number (in Arabic).

SG=*The World and Other Writings*, edited and translated by Stephen Gaukroger (Cambridge,1998).

MM=*René Descartes, The Principles of Philosophy*, translated by V. R. Miller and R. P. Miller (Dordrecht, 1991)

MSM=*Rene Descartes, Le Monde, ou Traité de la lumière*, translated by Michael S. Mahoney (New York, 1979).

CSM(K)=*The Philosophical Writings Of Descartes*, 3 vols., translated by John Cottingham, Robert Stoothoff, and Dugald Murdoch, and (for vol. 3) Anthony Kenny, (Cambridge, 1988) References are by volume number (in roman) and page number (in arabic).

HR=The Philosophical Works of Descartes, vol I translated by E.S. Haldane and G.R.T. Ross (Cambridge, 1968 [1st ed. 1911])

Other

Allard, J -L. 1963. *Le mathématicisme de Descartes*. Ottawa: Éditions de l'Université d'Ottawa.

Alquié, F. (ed.). 1963. *Oeuvres philosophiques de Descartes*, t.1. Paris: Garnier Frères.

Bachelard, Gaston. 1949. *Le rationalisme appliqué*. Paris: Presses Universitaires de France.

Barnes, Barry. 1982. *T.S. Kuhn and social science*. London: Macmillan.

Barnes, Barry, and John Law. 1976. Whatever should be done with indexical expressions? *Theory and Society* 3: 192–222.

Barthes, Roland. 1957, 1973. *Mythologies*. Paris, Editions du Seuil. English Trans. A. Lavers: St. Albans.

Beck, L.J. 1952. *The method of Descartes: A study of the regulae*. Oxford: Oxford University Press.

Bourdieu, Pierre. 1975. The specificity of the scientific field and the social conditions of the progress of reason. *Social Science Information* 14: 19–47.

Buchdahl, Gerd. 1969. *Metaphysics and the philosophy of science*. Cambridge, MA: MIT Press.

Buchwald, Jed Z. 2008. Descartes's experimental journey past the prism and through the invisible world to the rainbow. *Annals of Science* 65: 1–46.
Callon, M. 1980. Struggles and negotiations to define what is problematical and what is not: The sociologic translation. In *The social process of scientific investigation. Sociology of the sciences*, vol. IV, ed. K. Knorr, R. Krohn, and R. Whitley, 197–219. Dordrecht: Reidel.
Cassirer, Ernst. 1902. *Leibniz system in seinem wissenschaftlichen Grundlagen*. Marburg: Elwert.
Chevalier, J. 1937. Le Discours de la méthode. *Archives de Philosophie* 1: 1–13.
Clarke, Desmond. 1977. 'Descartes' use of "Demonstration" and 'deduction. *Modern Schoolman* 54: 333–344.
Clarke, Desmond. 2006. *Descartes, a biography*. Cambridge: CUP.
Denissoff, E. 1970. *Descartes, premier théoricien de la physique mathématique*. Louvain: Universitaires de Louvain.
Dijksterhuis, E.J. 1950. La méthode et les essais de Descartes. *In Descartes et le cartesianisme hollandaise*. Ed. E. J. Dijksterhuis, 22–44. Paris: Presses Universitaires de France.
Feyerabend, P.K. 1970. Classical empiricism. In *The Newtonian heritage*, ed. R.E. Butts and J.W. Davis, 150–170. London: Blackwell.
Feyerabend, P.K. 1975. *Against method*. London: New Left Books.
Feyerabend, P.K. 1978. *Science in a free society*. London: New Left Books.
Gadoffre, G. 1961. *Descartes' Discours de la méthode*, 2nd ed. Manchester: Manchester University Press.
Garfinkel, Harold. 1967. *Studies in ethnomethodology*. Englewood Cliffs: Polity Press.
Gilbert, N., and M. Mulkay. 1980. Contexts of scientific discourse: Social accounting in experimental papers. In *The social process of scientific investigation. Sociology of the sciences*, vol. IV, ed. K. Knorr, R. Krohn, and R. Whitley, 269–294. Dordrecht: Reidel.
Gilbert, G.N., and M. Mulkay. 1981. Warranting scientific belief. *Social Studies of Science* 12: 383–408.
Gilson, E. (ed.). 1947. *René Descartes, Discours de la Méthode: Texte et Commentaire*. Paris: Vrin.
Gouhier, H. 1958. *Les Premières Pensées de Descartes*. Paris: Vrin.
Latour, Bruno, and Steve Woolgar. 1979. *Laboratory life, the social construction of scientific facts*. London: Sage.
Lefèvre, R. 1956. *La vocation de Descartes*. Paris: Presses Universitaires de France.
Lecourt D. 1975. *Marxism and epistemology: Bachelard, canguilhem, foucault*. Trans. B. Brewster. London: New Left Books.
LeGrand, H.E. 1986. Steady as a rock: Methodology and moving continents. In *The politics and rhetoric of scientific method*, ed. J.A. Schuster and R.R. Yeo, 97–138. Dordrecht: Reidel.
Lévi-Strauss, Claude. 1972 *Structural anthropology*. Trans. C. Jacobson and B.G. Schoepf. Harmondsworth: Penguin.
Liard, Louis. 1880. La méthode de Descartes et la mathématique universelle. *Révue Philosophique* 10: 569–600.
Mahoney, M. 1973. *The mathematical career of Pierre de Fermat 1601–1665*. Princeton: Princeton University Press.
Mahoney, M.S. 1980. The beginnings of algebraic thought in the seventeenth century. In *Descartes: Philosophy, mathematics and physics*, ed. S. Gaukroger, 141–155. Sussex: Harvester.
Miller, David P. 1986. Method and the "micropolitics" of science: The early years of the geological and astronomical societies of London. In *The politics and rhetoric of scientific method*, ed. J.A. Schuster and R.R. Yeo, 227–257. Dordrecht: Reidel.
Mulkay, Michael, and Nigel Gilbert. 1981. Putting philosophy to work: Sir Karl Poppoer's influence on scientific practice. *Philosophy of the Social Sciences* 11: 389–407.
Mulkay, M., and N. Gilbert. 1982. Accounting for error: How scientists construct their social world when they account for correct and incorrect belief. *Sociology* 16: 165–183.
Mulkay, M., J. Potter, and S. Yearley. 1983. Why an analysis of scientific discourse is needed. In *Science observed*, ed. K. Knorr-Cetina and M. Mulkay, 171–203. London: Sage.

Perelman, C. 1979. *The new rhetoric and the humanities*. Dordrecht: Reidel.
Perelman, C., and L. Olbrechts-Tyteca. 1971. *The new rhetoric: A treatise on argumentation*. London: University of Notre Dame Press.
Pinch, Trevor. 1985. Towards an analysis of scientific observation: the externality and evidential significance of observational reports in physics. *Social Studies of Science* 15: 3–36.
Richards, E., and J.A. Schuster. 1989. The myth of feminine method: A challenge for gender studies and the social studies of science. *Social Studies of Science* 19: 697–720.
Röd, W. 1971. *Descartes' Erste Philosophie*. Bonn: Bouvier.
Sabra, A.I. 1967. *Theories of Light from Descartes to Newton*. London: Oldbourne.
Sacks, Harvey. 1972. On the analysability of stories of children. In *Directions in Sociolinguistics*, ed. J.J. Gumperz and D. Hymes, 325–345. New York: Holt, Rinehart and Winston.
Schuster, J.A. 1977. *Descartes and the Scientific Revolution 1618-34: An Interpretation*, 2 vols. unpublished Ph.D. dissertation, Princeton University.
Schuster, John. 1979. Kuhn and Lakatos revisited. *British Journal for the History of Science* 12: 301–317.
Schuster, J.A. 1980. Descartes' Mathesis Universalis: 1619–1628. In *Descartes: Philosophy, mathematics and physics*, ed. S. Gaukroger, 41–96. Sussex: Harvester.
Schuster, John. 1984. 'Methodologies as mythic structures: A preface to the future historiography of method. *Metascience: Review of the Australasian Association for the History, Philosophy and Social Studies of Science* 1–2: 15–36
Schuster, J.A. 1986. Cartesian method as mythic speech: A diachronic and structural analysis. In *The politics and rhetoric of scientific method*, ed. J.A. Schuster and R.R. Yeo, 33–95. Dordrecht: Reidel.
Schuster, J.A. 1990. The scientific revolution. In *The companion to the history of modern science*, ed. R.C. Olby, G.N. Cantor, J.R.R. Christie, and M.J.S. Hodge, 217–242. London: Routledge.
Schuster, John. 2000c. René Descartes. In *Encyclopedia of the scientific revolution*, ed. W. Applebaum. New York: Garland Publishing.
Schuster, John. 2002. L'Aristotelismo e le sue Alternative. In *La Rivoluzione Scientifica*, ed. D. Garber, 337–357. Rome: Instituto della Enciclopedia Italiana.
Schuster, J.A., and Richard.R. Yeo. 1986. Introduction. In *The politics and rhetoric of scientific method*, ed. J.A. Schuster and R.R. Yeo, ix–xxxvii. Dordrecht: Reidel.
Shapin, S. 1982. History of science and its sociological reconstructions. *History of Science* 20: 157–211.
Schutz, Alfred. 1970. *Reflections on the problem of relevance*. New Haven: Yale University Press.
Schutz, Alfred, and Thomas Luckmann. 1974. *The structures of the life-world*. Trans. R.M. Zaner, and H.T. Engelhardt. London: Heinemann.
Sirven, J. 1928. *Les années d'apprentissage de Descartes*. Albi: Imprimerie Coopérative du Sud-Ouest.
Vrooman, J.R. 1970. *Rene Descartes, a biography*. New York: Putnam.
Weimar, W. 1977. Science as rhetorical transaction: Toward a nonjustificational conception of rhetoric. *Philosophy and Rhetoric* 10: 1–29.
Wood, Paul. 1980. Methodology and apologetics: Thomas sprat's *History of the royal society*. *British Journal for the History of Science* 13: 1–26.
Yearley, S. 1981. Textual persuasion: The role of social accounting in the construction of scientific arguments. *Philosophy of the Social Sciences* 11: 409–435.
Yeo, Richard. 1979. William Whewell, natural theology and the philosophy of science in mid-nineteenth century Britain. *Annals of Science* 36: 493–516.
Yeo, Richard. 1986. Scientific method and the rhetoric of science in Britain, 1830–1917. In *The politics and rhetoric of scientific method*, ed. J.A. Schuster and R.R. Yeo, 259–297. Dordrecht: Reidel.

Chapter 7
Universal Mathematics *Interruptus*: The Program of the Later *Regulae* and Its Collapse 1626–1628

7.1 Introduction—Toward the Renewed Project of *Mathesis Universalis* in the Later *Regulae*

When, in 1625, Descartes settled in Paris, he had reason to be confident in his intellectual accomplishments and prospects.[1] We already know that he had to hand a draft treatise on the method, roughly rules 1–11 of the *Regulae.*[2] Encysted within the text, in rule 4B, was the remnant of his earlier initiative in universal mathematics. The discipline had not been further developed in its own right. It had been subordinated to the method, but could be read into any, and all, technical successes in mathematics, physico-mathematics and natural philosophy. And such success there certainly had been since 1619. Sometime between that date and 1625 Descartes had produced his construction for all the 'solid' problems of the ancients, using only a circle and parabola. This was equivalent to a general construction for all cubic and quartic equations and was later to form a centerpiece in Book III of the *Geometry*.[3] As we know from Chap. 4, within a year or two he discovered the cosecant form of the law of refraction, a triumph of mixed mathematics, and he quickly moved on to a theory of lenses and the anaclastic surface, as well as the attempt, as a physico-mathematician,

[1] Between 1619 and 1625 Descartes travelled extensively, stopping in France briefly in the winter of 1622–1623. He settled in Paris in 1625 and remained there between trips to the countryside until late 1628, when he moved to the United Provinces and launched his projects in metaphysics and systematic corpuscular-mechanical natural philosophy.

[2] As we also know, some material in rule 8, dealing with the refraction of light and its law, can only date from 1626 or later; and as we shall see, the overall design of the text after the second paragraph of rule 8 also dates from 1626 or later. By late 1628 the entire enterprise had been abandoned for reasons we shall uncover in this chapter.

[3] *Geometry, III,* AT, VI, pp. 464–85. For a reconstruction of the path to the solution and its dating see Schuster (1977) 124–49.

J. Schuster, *Descartes-Agonistes: Physico-mathematics, Method & Corpuscular-Mechanism 1618-33*, Studies in History and Philosophy of Science 27,
DOI 10.1007/978-94-007-4746-3_7,

to subsume the law under a mechanistic theory of light. Working in the mathematical circles around Marin Mersenne, he was very much a rising star in the emerging community of French mathematical savants. As we also now understand, the method itself, expressed in vague rules, seemingly significantly applicable to any and every field of practice, could have appeared to have produced these triumphs. One suspects that Descartes could have continued on with his special studies in mathematics, mixed mathematics and physico-mathematics, resting content in the apparent knowledge that his triumphs confirmed the method and instantiated his universal mathematics.

Descartes, however, became enmeshed in the wider intellectual life of the capital; and this involvement led him back to serious concern with his universal mathematics proper.[4] At this time Parisian intellectual circles, including theologians, savants, courtiers, literary figures and cultured lawyers and bureaucrats, were gripped by a mounting wave of religious, political and philosophical debate. Tensions centered on apologetical issues, both within the fragmented Catholic camp and between Catholic apologists and their real, or apparent, unorthodox challengers. But debate spilled over into areas of literature and moral philosophy, and for some, natural philosophy and its foundations.[5]

It has long been accepted that Descartes was party to this turmoil. For the most part, it has been his turn to dualist metaphysics after 1628 which has been linked to

[4] It is of course well known that in the years 1625–1628 Descartes associated with establishment literary figures like Guez de Balzac, an apologist for the Jesuits and fierce anti-sceptic and anti-stoic; religious apologists such as Silhon and Mersenne; and, with the neo-Augustinian fathers of the Oratory, including Gibieuf, Condren and, briefly, Cardinal Bérulle himself-the founder of the Order and chief figure in the French Counter–Reformation in that generation. See e.g. Adam, *Vie de Descartes,* AT, XII, 66–98; Sirven (1928) 313–37, Espinas (1906), Gadoffre (1961) 'Introduction', especially pp. xxff.

[5] Historians have frequently employed the term 'crisis' to deal with the period. Spink (1960) discerned the 'crisis of 1619–1625', which he saw primarily in institutional terms as a 'repressive reaction on the part of the authorities, namely the Parlements' against 'libertinage' and 'free thought'. René Pintard, in his massive and now dated study of *Le libertinage érudit* (Pintard 1943), described 'la crise de 1623–1625', which was characterized as a breaking point when nascent scepticism and free thought began to meet increased resistance, mainly in the form of apologetical writings by thinkers as diverse as Mersenne, Garasse and Silhon. Thereafter, free thought became more a private affair of well-placed scholars such as Gassendi, La Mothe le Vayer and Naudé, rather than an aggressive public movement. Henri Gouhier (1954), starting with an assessment of the apologetical aims of Descartes' metaphysics, described a 'theological crisis' in the 'era' of Descartes, fought out between Catholic proponents of 'mystical' and 'positive' theology. Popkin (1964) pointed to a generalized sceptical 'crisis' of the early seventeenth century, which came to a head in France in the 1620s and 1630s, and there elicited constructive new attempts at resolution, first in the form of Mersenne and Gassendi's 'mitigated scepticism', and then in Descartes' dogmatic metaphysics. A well-rounded account of the institutional and ideological conflicts of the period remains to be written to synthesize this set of largely single-factor accounts. For present purposes it need only be granted that there was a 'common context' of theological, political, ethical and epistemological turmoil in the period, and that different actors had differing perspectives on it. But, see more recently, Staquet (2009) and Torero-Ibad (2009).

the Parisian context.[6] While accepting that view, I wish to suggest that there was an earlier and more clear-cut stage in Descartes' involvement, which, in fact, is causally related to his eventual turn to systematic corpuscular-mechanical natural philosophy as well as metaphysics. For in 1625 Descartes was not yet a metaphysician or a systematic natural philosopher. As we concluded earlier, he was following Beeckman, practicing physico-mathematics in relation to a piecemeal corpuscular-mechanical philosophy, whilst also pursuing analytical mathematics and luxuriating in the self-appointed if as yet private role of methodological prophet. He was, moreover, closely associated with Mersenne, who not only shared his interests in mathematics and the mathematical disciplines subordinate to natural philosophy, but who was also pursuing a highly personal strategy of apologetic which made essential use of those interests. Mersenne's example significantly channeled and focused Descartes' view of the larger contemporary ideological turmoil, and importantly influenced his initial response.

In his first two major works Mersenne started as a mathematically and natural philosophically literate Catholic apologist.[7] His special calling was to deploy in a piecemeal manner findings from optics, acoustics, mechanics, mathematics, astronomy and medicine to undermine the specifically natural philosophical claims of unorthodox systems, particularly those of alchemical, neo-Platonic or 'Hermetic' inspiration. In the mid-1620s, as his own thought matured and the Parisian turmoil deepened, he became aware of the sceptical threat to the grounds of his own position, and he turned—in his *La vérité des sciences* (1625)—to meet the challenge. His answer, which was to be further developed in five important treatises of the early 1630s, depended upon acquiescing in the sceptical critique of the possibility of knowledge of principles or essences, while holding that a more modest form of descriptive knowledge was possible. This would consist in the mathematical correlation of appearances,

---

[6] See, for example, Blanchet (1920), Gilson (1913), especially. Chapters IV, V; Gouhier (1924), especially. pp. 54–62 and Popkin (1964). One can, with Gouhier, stress the properly apologetical aims of the metaphysics, or argue that the main role of the metaphysics was to ground the mechanistic physics, which itself is to be seen as aimed at resolving the natural philosophical conflicts of the time. Alternatively, with Popkin one can stress a supposed sceptical core of the contemporary malaise and so identify the anti–sceptical thrust of the metaphysics as Descartes' response to the situation. In any of these sub-theses, Descartes' association with the Oratorians, with Bérulle, and with Silhon and Mersenne can take on special significance, in which one can stress the general apologetical aims of these figures: the special role of the Oratorians as proximate—if not sole or original—sources of Descartes' neo-Augustinian leanings; the views of Mersenne on Voluntarist theology; and the anti-sceptical tenor of Silhon's rationalistic apologetic. My thesis here is that no matter what precise position one takes on the apologetical role of the metaphysics and its intellectual sources, one crucial determinant of its aims, problems and content is the very failure of the ideological charged project of the later *Regulae*.

[7] On Mersenne see the immensely significant Lenoble (1943), which locates Mersenne's interests as not simply scientific and anti-sceptical, but as apologetical in the sense of seeking the scientific refutation of apparently unorthodox natural philosophies, especially those of neo-Platonic, 'Naturalist' or Rosicrucian inspiration. Also, most importantly in the Anglophone literature, see Dear (1988), Popkin (1964) and Hine (1967).

as already illustrated in the sciences of optics and acoustics. Such knowledge would systematically overcome the deficiencies and illusions of sense, by showing how appearances are regularly related to one another. As a consequence, this knowledge would serve the pragmatic purpose of facilitating and progressively improving humankind's transactions with its immediate physical environment. Hence, by this stage Mersenne was engaged in a two-sided struggle. On the one hand he hoped to outflank, rather than refute, the sceptical arguments against the possibility of certain knowledge of essences. On the other hand, he hoped to secure bits of mixed mathematical knowledge to be deployed against the claims of natural philosophical systems he considered to be unorthodox.

Though he was younger than Mersenne, a relative newcomer to Paris and no cleric, Descartes was in several ways similarly placed in the contemporary turmoil. Like Mersenne, Descartes had been trained by the Jesuits of La Flèche into their special mix of Counter-Reformation neo-scholasticism, humane letters and mathematical arts. Both men nurtured the scholastic ideal of the relevance (within limits) of true natural philosophy to correct theology; but neither man found Aristotelian–Thomist natural philosophy and metaphysics a credible basis for sound theology or morals. Both had, in different ways, felt their ways toward an ideal of the relevance of mathematics in the pursuit of knowledge of nature. And hence both were sensitive to the ways in which scepticism and alternative unorthodox visions of nature each threatened the relationship of sound natural philosophy to orthodox Catholic theology. Accordingly, Descartes, like Mersenne, came to focus not so much on formal apologetics *per se,* but rather on natural philosophical and increasingly 'epistemological' issues which bore, at one remove, upon apologetics.

Entering the intellectual fray, Descartes was probably eager to exploit his technical achievements, and to deploy and to win public recognition for his personal methodological illuminations of 1619. So placed, he found a concrete (and surpassable) model in the strategies and concerns of his friend Mersenne. He decided to return to his text on the method, the *Regulae,* to build his universal mathematics in detail as an apparent outgrowth of it; to show in detail how universal mathematics would function, and how it could be constructed so that it could elude sceptical attack, while itself precluding neo-Platonic and magical views of mathematics and of nature. In short universal mathematics, expanding upon Mersenne's tactics, would appear to grow out of the doctrine of method, and carry the fight, under the banner of method, versus scepticism and radical natural philosophizing alike. The elaborated universal mathematics would sidestep and finesse scepticism while showing what could be accomplished in mathematics and physico-mathematics.

In the fashion of Mersenne, the approach was non-dogmatic, relying not upon systematic metaphysics, but upon the exploitation of attractive piecemeal natural philosophical claims. But, unlike Mersenne, Descartes was not content merely to parade disparate physico-mathematical and mathematical results before the sceptics as examples of our ability to overcome sense illusions and fashion a science of appearances. Descartes intended to weld his actual and his hoped-for technical results into a new account of perception and of mental function, which would ground and legitimate universal mathematics and show in detail how its machinery was to work. This gave Descartes' project a more sharply defined epistemological edge than that of Mersenne, and that in turn unexpectedly raised a series of dangerous new difficulties. These problems forced the abandonment of the project of constructing universal mathematics under the aegis of the method and conditioned Descartes' turn after 1628 to systematic metaphysics and systematic corpuscular-mechanical natural philosophy—the crucial inflection of his entire intellectual trajectory which we shall treat in Chaps. 8, 9, 10, and 11 below.

The argument of this chapter therefore will proceed as follows: the dating of the initiative of the later *Regulae* will be established in the next section. Then Sect. 7.3 will show exactly where and how the project of the later *Regulae* emerges in the surviving text. In Sect. 7.4 the core conceptual and logistical machinery of the new universal mathematics will be described. The findings of Sects. 7.3 and 7.4 will be reprised and diagrammatically represented in Sect. 7.5, with particular regard to how universal mathematics was intended to carry out its 'Mersenne–like' legitimatory functions. Further textual evidence for our dating of the later *Regulae* and for our reading of its aims will also be presented. In Sect. 7.6 the unintended difficulties raised by the project will be examined and an indication will be given of they way they shaped Descartes' post-1628 enterprises. Finally, Sect. 7.7 will underline the place of the aborted universal mathematics of the later *Regulae* as an inflection point in the career of Descartes, where his identity and agenda begin to shift decidedly toward the systematizing corpuscular-mechanical natural philosophy and metaphysical legitimation we recognize as the beginning of the emergence of the mature Descartes.

7.2 Rule 8: The Emergence of the Project of the Later *Regulae*

The new initiative of the mid and late 1620s can first be detected in portions of rule 8. The first two paragraphs of the rule advance general heuristic advice in the manner of the surrounding rules 5–7 and 9–11.[8] This portion of rule 8 probably dates from the earlier period of construction of heuristic rules in 1619–1620, although it most probably did not exist in this form.[9] The dating of the new initiative of the later *Regulae* depends on the next three paragraphs; for they clearly pre-date the remainder of the rule, and in that latter portion of the rule is found a definitive plan for the remainder of the work, just as it is executed in part in the extant rules 12–21.[10]

In the three paragraphs in question Descartes tries to illustrate the heuristic pointers given in the opening paragraphs. He offers his famous 'methodological' story of the

[8] Rule 8, AT, X, p. 392 1.14 to p. 393 1.21.

[9] See above Sects. 5.6 and 5.7.

[10] Cf. Weber (1964) 88–103.

discovery of the law of refraction and derivation of the anaclastic curve.[11] As we know from our study in Chap. 4, this 'cover story' bears no relation to the actual path of discovery. But Descartes had every reason to believe he could have made his discoveries in this fashion, and, in addition, he probably desired to cover the tracks of his actual path, which depended at its crucial point upon the dubious traditional optical principle for image location.[12] At the moment, it is only important to note the extreme unlikelihood that Descartes would have written this section before his optical triumphs of 1626 and 1627. This places these passages well within the Parisian years, and so also dates rules 12–21.

In the remainder of the rule Descartes introduces a second methodological example, which he terms 'the most splendid example of all'.[13] At first it is presented simply as the application of the method to the discovery of the capabilities of the mind itself.[14] But as the closing passages of rule 8 unfold, the 'most splendid example' is articulated and extended. It becomes less an example of the method of rules 3–7 and more a program for the construction and legitimation of universal mathematics, a program which dominates and controls rules 12–21. The closing sections of rule 8 seem to have been written in great haste and with mounting excitement: Descartes formulated and reformulated the 'most splendid example' three times.[15] He eventually decided that two problems were involved—an inquiry into the nature and limits of the mind, and a concomitant inquest into the character and status of the objects of knowledge.[16] He devised two programs for accomplishment of the project: the first was left half-stated, but the second was actually carried through in the sequel.[17]

What is most striking about his third and last setting of the example is Descartes' sudden realization of the unique epistemological role of such an inquiry:

---

[11] Rule 8, AT, X, p. 393 l.22 to p. 396 l.25. Apart from illustrating the overall use of the method, the 'cover story' exemplifies in particular two points made at the beginning of rule 8: [1] Do not proceed where deduction cannot take one, as the pure mathematician can proceed only so far in the search for the law of refraction without physical premises about matter and cause. [2] Learn to know when your 'enumerations' need to be 'complete' and when merely 'sufficient', as in enumerating the types of 'natural power'. On enumeration see above Sect. 5.6.

[12] See Sect. 4.9 for the several other issues about the path of discovery that Descartes' method tale seeks to hide.

[13] The CSM translation (p.29) renders this as 'the finest example of all'. I prefer here for dramatic effect the Haldane and Ross expression 'most splendid example of all', to underscore the import of this phrase, and shift in the text. Brunschwig in the Alquié edition (p.118) renders this as 'l'exemple de tous le plus éclatant...' See below Note 18 for a further preference for HR over CSM over another significant line in this part of the rule.

[14] Rule 8, AT, X, pp. 395–6; CSM p.30; HR, I, pp. 24–5.

[15] The second and third formulations occur at p. 396 l.26 to p. 397 l.3 and p. 397 l.26 to p. 398 l.5.

[16] Rule 8, AT, X, p. 398 l.10–25. See also Descartes' enumeration of only three faculties of mind in the first setting (AT, X, pp. 395–6) and his enumeration of four faculties in the third setting of the example (AT, X, pp. 398–9).

[17] Rule 8, AT, X, p. 398 l.26 to p. 399 l.21. First Descartes promises to deal with the question of the faculties of mind in the succeeding proposition, but no such discussion occurs in rule 9. He proceeds to discuss the objects of knowledge on p. 399 but then pulls up short and introduces the second plan for a work in 36 rules divided into three books or 12 rules each.

> Now no more useful inquiry can be proposed than that which seeks to determine the nature and scope of human knowledge. This is why we state this very problem succinctly in the single question, which we deem should be answered at the very outset *with the aid of the rules which we have already laid down.* This investigation should be undertaken once at least in his life by anyone who has the slightest regard for truth, *since in pursuing it the true instruments of knowledge and the whole method of inquiry come to light.* (emphasis added) [18]

Here Descartes reaches a qualitatively new stage in his reflection on the status and use of his original method. Back in rule 7 he had asserted that the (heuristic) method was virtually complete with rules 5, 6 and 7 and would only be worked out in more detail in the rules to follow.[19] But now there is no question of simply adding a few more rules; now, guided by the rough methodological insights of 1619 (which had been lauded in nearly mystical terms in rule 4), a new understanding of mind, its instruments and its objects is going to be produced. As Descartes worked and re-worked rule 8 a considerable shift took place. At first the 'most splendid example' was an illustration of one of the sub-rules of the heuristic method, now it had become a program, not only for a deeper account of 'method', but also for a new teaching on epistemology.[20]

It is true that in rule 8 Descartes still expresses himself in the idiom of the method, but we are about to see that Descartes now understood the 'most splendid example' in terms of an inquiry into cognitive function and perception, which—in the first

---

It is incorrect to think, as is widely held, that rules 13–24 were meant to pertain to mathematics and rules 25–36 to physics, or that the former were to pertain to 'synthesis' and the latter to 'analysis'. Problems of physics occur in 'book two', provided they are 'fully determinate' and 'book three' could contain mathematical, as well as physical, problems in which the relevant terms and data have to be elicited from a larger body of raw material. In addition, as we shall see, the entire thrust of the procedure of universal mathematics is to reduce problems to solution in algebraic form, in which an 'analysis' takes on a deductive character. On 'determinate' sorts of problems see the Sect. 7.4 below, on rules 14–18 and also AT, X, pp. 429–30.

[18] Rule 8, AT, X, p. 397 1.27 to p. 398 1.5; CSM p.31; HR, I, p. 26. Here again the HR translation is preferred, but not for any significant reason in the opening two sentences which are essentially similar in CSM. The issue is with the last sentence, which CSM render as 'This is a task which everyone with the slightest love of the truth ought to undertake at least once in his life, since *the true instruments of knowledge and the entire method are involved in the investigation of the problem.*' [p.31] (emphasis added) This loses the sense that the instruments of knowledge and method *are going to be progressively and iteratively uncovered by means of pursuing this, and later inquiries*, a significant, and I submit misleading watering down of Descartes' statement. Brunschwig in the Alquié edition renders the last phrase as '…parce que c'est dans cette enquête que se trouvent les véritables outils du savoir, et la méthode tout entière.' (p.120) The Marion (1977) translation of the *Regulae*, in contrast, has 'parce que cette recherche contient les vrais instruments du savoir et la méthode toute entière' (p.30). Now, the Latin verb in question is '*continentur*' amongst whose main meanings can be not only 'involved in' but 'depend upon' (according, for example, to several Ciceronian usages). Hence, again, taking the full contest into account we can construe Descartes intended message to be that, *[ultimately uncovering] the true instruments of knowledge and whole method of inquiry depend upon [undertaking this investigation at least once in one's life].*

[19] Rule 7, AT, X, p. 392; CSM pp.27–8; HR, 1, p. 22.

[20] Between the second and third statements of the 'example' (AT, X, p. 397) Descartes introduces a telling metaphor in which the development of the method beyond the early rules, but on their basis, is likened to the origin of practical arts, in which first the tools themselves must be fashioned in a rough form before the art is practiced and perfected and its fruits produced.

instance—would ground the machinery of an elaborated version of universal mathematics. The growing urgency and significance of the 'most splendid example' in rule 8, and its unfolding after rule 11 as an articulated version of universal mathematics, literally show Descartes thinking himself into the 'crisis' of the 1620s—or at least into a Mersenne-like take on the challenge to justify mathematics and mathematically based knowledge of nature—and moving to meet it with the tools and talents at his disposal.[21]

7.3 Rule 12: From 'Most Splendid Example' to the Articulation of the Machinery of Universal Mathematics

In rule 12 Descartes develops a theory of psychology and perception that will provide the basis for the machinery of universal mathematics. With this theory he will be able to lend ontological certification to the objects of universal mathematics and display precisely in what the certainty of its operations consists.

Descartes denies the existence of separate scholastic faculties of the soul, such as the common sense, imagination, memory and understanding. Rather, he posits, first, the natural light of reason, or *vis cognoscens,* which is the unique and purely spiritual agency of the cognitive apparatus and which carries out intuition and deduction, the two fundamental intellective functions. Second, he posits certain physical loci in the brain where mechanically delivered corporeal impressions or patterns are registered, thus providing the content of sensation, imagination and memory.[22] Two of the loci are named (but not localized), the common sense and the imagination (or phantasy). According to Descartes, the *vis cognoscens,* acting alone, apart from awareness of corporeal patterns, constitutes the 'understanding'. 'Applying itself' *directly* to the

---

This might, in a different context, be read simply as one of Descartes' broad and empty claims that the method consists 'mainly in practice'. But coming here it indicates a consciousness of the fact that his project of method (read 'methodologically grounded universal mathematics') is going to be vastly deepened and widely articulated.

[21] It might be useful here to note the cash value of our proposed dating of the earlier rules (in Chap. 5) in the light of these findings about rule 8. One can now see that even if that dating proves untenable, the overall thesis of a change in aim and content of the text in rule 8 can be maintained, and the change can be dated from around 1626. One could even assume that rules 4A, 4B and 1–11 (excluding parts of 8 and with the caveat about the material following those passages in rule 8, given above in Note 2) were composed in Paris before 1626 or 1627. It would still be the case that universal mathematics was very likely first developed in 1619 and that universal method was even more probably initially worked out in the winter of 1619/1620. Rules 4B and 4A would still reflect at a distance the character of these projects. Moreover one could still demonstrate in rule 8, and then in rules 12–21, the very shift in aim and content which we have uncovered.

[22] The *vis cognoscens will* be identified with the 'understanding' *(intellectus)* as used both earlier and later in the text. Context always indicates whether the term is used to denote the one spiritual faculty attending to purely intellectual matters (Descartes' technical definition of the understanding), or whether it is attending to corporeal patterns in the brain.

corporeal phantasy and common sense, the *vis cognoscens is* said to sense; applying itself to the phantasy, in so far as the latter is stocked with formerly impressed patterns, the *vis cognoscens is* remembering; finally, in imagining, the *vis cognoscens* applies itself to the phantasy to create new corporeal impressions.[23]

Norman Kemp Smith aptly described the action of the *vis cognoscens* as a 'cognitive awareness' directed toward either purely spiritual or purely corporeal entities, which thus become the immediate objects of consciousness. There is no doctrine of representative perception in rule 12, no postulation that physical entities can be known only by means of mental duplicates.[24] Regardless of the nature of the object of consciousness, whether sense impression, memory pattern or purely intellectual conception, the mode of action of the *vis cognoscens is* always a direct cognitive awareness. Of course, Descartes' position rests upon an unexplicated ontological dualism—a single spiritual 'power' 'applies itself' to corporeal loci.[25] Moreover, a visual metaphor lies at the very heart of this account, for it is as though the *vis cognoscens* constituted a second, spiritual pair of eyes within the brain, there to attend to patterns delivered up on the corporeal screens of the brain loci.[26]

It is absolutely crucial to understand precisely how Descartes conceives of the brain loci and their contents. The loci are, as already noted, physical locations in the brain, 'genuine parts of the body'.[27] They are, moreover, macroscopic, 'The phantasy is a genuine part of the body, and is large enough to allow different parts of it to take on many different figures, and generally, to retain them for some time....'[28] Typically, Descartes terms the impressed patterns or figures 'ideas',[29] because they are the immediate objects of consciousness. They are not, however, to be identified

[23] Rule 12, AT, X, pp. 415–16: 'Atque una et eadem est vis, quae, si applicet se cum imaginatione ad sensum communem, dicitur videre, tangere etc; si ad imaginatione solam ut diversis figuris indutam, dicitur reminisci; si ad eamdem ut novas fingat, dicitur imaginari vel concipere ....' (HR, I, p. 39; CSM, p. 42).

[24] Norman Kemp Smith (1952), 51–2, writes that what Descartes offers in a 'quite unqualified way' is: 'an empirical realist view of the data available to the mind. The only 'objects' which he allows to the mind—all of them directly apprehended—are obtained, he [Descartes] holds, from one or other of two sources. (1) The self is aware of itself as thinking, i.e. as doubting, affirming, desiring etc .... (2) The self ... is no less aware of the physical patterns which external objects, by way of their action on the bodily sense organs, imprint on the brain ....' Cf. O'Neil (1967).

[25] See text cited in Note 23.

[26] In his mature metaphysics Descartes explicitly rejects the metaphor of the spiritual 'helmsman' in the 'ship' of the body, and that is a measure of the changes which overtake his epistemology in the wake of the difficulties created by the doctrine in the *Regulae* (see the discussion on the problems of perception in the *Regulae*, below Sect. 7.6.2) and also *Discourse on Method, V,* AT, VI, p. 59 (HR, I, p. 118); *Sixth Meditation,* AT, VII, p. 81 (HR, I, p. 192).

[27] *Regulae,* Rule 12, AT, X, p. 414; CSM pp.41–42; HR, I, p. 38.

[28] *Ibid.*

[29] Rule *12,* AT, X, *p. 414 1.17;* cf. rule 14, p. 441 1.10 to *13:* 'sequitur ex dictis ad regulam duodecimam, ubi phantasiam ipsam cum ideis in illa existentibus, nihil aliud esse concepimus, quam verum corpus reale extensum et figuratum'; Rule *14, p. 450* 1.10 to 11: 'Quod attinet ad figuras, iam supra ostensum est, quomodo per illas solas return omnium ideae fingi possint ...' For further citations see Jean-Luc Marion's translation of the *Regulae*, pp. 231–2. (Marion 1977).

with the immaterial impressed and expressed *species* of scholastic psychology; they do not convey or consist in the 'form' of the object perceived, including what we would now term its secondary qualities. These figures or ideas are purely mechanical impressions, congeries of geometrical shapes, impressed upon the sense organs and conveyed mechanically via the nerves to the common sense.

Descartes writes that the external senses 'perceive in virtue of passivity alone, just in the way that wax receives an impression (*figuram)* from a seal'. This is no mere analogy: just as the wax is physically impressed with the image of the seal, 'the exterior figure of the sentient body is really modified by the object'. All sensations, those of light, color, odor, savor and sound, and not merely the tactile sensations, are ultimately caused by the mechanical disturbance of the external sense organs.[30] From the sense organs the impressed 'figures' are transmitted to the common sense via the nerves. This occurs 'instantaneously' by the passing of a pattern of mechanical disturbance. 'No real entity travels from one organ to the other', just as the motions of the tip of a pen are instantaneously communicated to its other end. For, as Descartes rhetorically concludes,

> who could suppose that the parts of the human body have less interconnection than those of the pen? And what simpler way of explaining the matter could be devised?[31]

Patterns so registered in the common sense can then be imprinted in the imagination, there to be stored in memory for the future 'attention', we might say, of the *vis cognoscens,* or to be immediately attended to in sense perception.[32]

The model of the pen is virtually identical to that of the blind man's staff which Descartes was to use in the *Dioptrique (1637)* to illustrate the claim that light consists in an instantaneously transmitted mechanical impulse.[33] Here, in rule 12, despite the hypothetical tone of some of his remarks, including the last part of the above quotation,[34] he seems to be saying that whatever the details of anatomy may turn out to be, it is very likely that the nervous system in its sensory aspects is nothing

[30] *Ibid.* AT, X, pp. 412–13; CSM pp.40–41;HR, 1, pp. 36–7.

[31] *Ibid.* AT, X, p. 414; CSM, p.41; HR, 1, pp. 37–8.

[32] *Ibid.* AT, X, p. 415 1.16 to 24; CSM p. 42; HR I, pp.38–39

[33] *Dioptrique 1,* AT, VI, *pp. 83–6.*

[34] Descartes does indeed introduce this material in a seemingly hypothetical tone. There is not space, he contends, to present all the material upon which the truth of the account depends; one need not believe 'the facts are so' unless one prefers to. Yet, despite the hypothetical tone, he also insists that his suppositions 'do no harm to the truth', that they 'promote his purpose' and that they 'render the truth more clear'. He has already stated that the wax and seal offers an exact model for the impression of patterns *on* sense organs and their transmission through the nerves to the common sense and thence to the imagination. He also deploys the pen-analogy which derives from his seriously held mechanistic theory of light. Furthermore, he clearly implies that valid reasons could be advanced for the more detailed mechanical theories upon which the wax and seal model and pen-analogy trade (rule 12, AT, X, pp. 411–12; CSM p. 40; HR, I, p. 36). It seems likely, therefore, that Descartes wished the explicit physiological and psychological account to be taken as true in its main lines.

but a mechanism for the instantaneous transmission of impulses from organ to brain through continuous matter in or of the nerves.[35]

Throughout the later *Regulae* it is clear that the impressions or ideas are two-dimensional figures. This is initially suggested by the strict analogy of the seal and wax and by the overall picture of sense perception. Later, Descartes gives an example of the reduction of the differences between colors to differences between two-dimensional patterns.[36] More importantly, the entire mechanism for carrying out the operations of universal mathematics will consist in the manipulation of lines and rectangles in the imagination. The third dimension is never represented and is always avoided.[37] Presumably, perception of three dimensions is caused by the impression of a perspective rendering in two dimensions. This raises difficulties to which we shall return in Sect. 7.6 below.

The template for the entire mechanistic theory of perception, and in particular for the idea that the objects of perception are two-dimensional mechanically impressed patterns was, very probably, Descartes' mechanistic theory of optics. As we saw in Sects. 4.7.3 and 4.7.4, there is little doubt that by the time he composed rule 12 in the mid to late 1620s, Descartes was already committed to a mechanistic theory of light as a mechanical impulse instantaneously conveyed through a continuous optical medium, even though the details remained unexplicated. Commitment to such a mechanical theory no doubt powerfully influenced his choice and design of the mechanical theory of perception in rule 12; but, to see the precise relevance of mechanistic optics to the physiology and psychology of rule 12 one has to look in particular at Descartes' theory of vision.

In the *Dioptrique* and in the *Treatise of Man* Descartes was to take over Kepler's revolutionary theory of vision in a suitably mechanized form. Kepler had shown in *Ad Vitellionem Paralipomena* (1604) that the eye is a dioptrical instrument which focuses incoming rays to form on the retina an inverted image of the visual field. As against the long-dominant theory of Alhazen, Kepler established that *all* the rays entering the eye from a point on the visible object are focused to a single point on the retina. The image so constructed consists in the summation of these points, each

---

Descartes also takes a decidedly hypothetical tone in introducing the idea that colors should be represented as figures and the difference between them taken as differences between figures (AT, X, p. 413; CSM p .41; HR, I, p. 37*)* This does not necessarily mean that the theory of mechanical sense impression is hypothetical *per se,* but only that any particular claim about the correlation of certain figures with certain colors must at present be conjectural.

[35] Working out the details in the *Treatise of Man* a few years later with the aid of some practical anatomical experience, Descartes devised a complicated mechanical account of the sensory and motor aspects of nervous function. But sense impression still depended upon the instantaneous passage of a mechanical impulse, now conceived to be conveyed along continuous filaments running in the centers of the nerves from sense organs (and sites of internal sensory excitation) to the central brain locus surrounding the pineal gland (AT, XI, pp. 141–6, 151–8).

[36] Rule 12, AT, X, p. 413; CSM p. 41; HR, I, p. 37.

[37] See the discussion of rules 14–18 in the following Section.

corresponding to a unique point source on the visible object.[38] Kepler's brilliant reconstruction of the geometry of visual perception was set within an immaterialist theory of light of neo-Platonic inspiration: light was a spiritual emanation, propagated spherically and instantaneously from each luminous point. Color was, of course, a real entity, of the same ontological genus as light and able to be borne by it from object to observer.[39] He spoke of a full-color pictorial perspective representation of the visual field being 'painted' onto the surface of the retina.[40] By contrast, Descartes mechanized the theory of light. As the *Dioptrique* and *Treatise on Man* show, this entailed that the images formed on the retina can consist only in patterns of mechanical disturbance, which are in turn conveyed to brain loci by the instantaneous passage of the disturbance along the continuous solid filaments that supposedly run through the cores of the nerves.[41]

It is likely Descartes possessed the outline of this mechanical version of the theory by 1628 and that it was the implicit basis for the manifest new theory of perception and mental function.[42] The retina, a sense organ receiving two-dimensional patterns of disturbance from the external world, would then have provided the model for the parallel construal of all senses. The removal of retinal patterns via the optic nerve to the brain would very plausibly be the model for the generalized account of the transmission of 'figures' 'without the passage of any real entity'. The central metaphor of spiritual eyes applying themselves to patterns lodged in the brain substance would merely have generalized to all the senses the sort of account that would probably have had to be given of the psychology of vision, once the physical process had been rigorously mechanized. Since there is a good *prima facie* reason to think that an achieved mechanization of the theory of vision lies behind the physiology and psychology of rule 12, and is implicitly maintained as their complement, we shall from now on term Descartes' entire teaching in rule 12, both explicit and implicit, the '*optics-psychology-physiology nexus*', or '*o-p-p nexus*' for short. It is on the basis of the o-p-p nexus that the truth of the operations of universal mathematics will be grounded, as well as the ontological reference of its objects of inquiry.

---

[38] Johannes Kepler, *Ad Vitelionem paralipomena*, in (Kepler, 1938ff) Vol. II 151–4. See Straker (1970), Lindberg (1976), Chap. 9; and Crombie (1967), reprinted as Chap. 9 of Crombie (1990).

[39] Kepler, *Ad Vitelionem*, (Kepler, 1938ff) Vol II, Chap. 1, Propositions I-V, XV, XVI.

[40] *Ibid.* Chapter V, Section 2.

[41] *Dioptrique*, V, AT, VI, pp. 114–29; *Treatise of Man*, AT, XI, pp. 133–4, 142–6, 151–60, 170–88.

[42] Acting in his 'Beeckmanian' style of mechanizing Kepler's speculations, Descartes may have meditated about a mechanical theory of vision in 1620, upon reading Kepler's optics. His notes from the time (AT, X, p. 243) contain some remarks on image formation which very plausibly derive from Kepler's theory of vision and the new theory of image formation it entailed: a matter he clearly he did not keep in mind, or wish to be reminded about in the course of his discovery of the cosecant law of refraction! Alternatively, the mechanistic theory of vision may have awaited the discovery of the law of refraction and the formulation of a more precise covering mechanical theory of the action of light. In a sense the best evidence for Descartes' possession of the theory in 1626–1628 is its implied role in the later *Regulae.*

As already indicated, the o-p-p nexus was seriously intended as a correct account of perception and mental function, notwithstanding Descartes' occasionally hypothetical tone.[43] Just for this reason one must be extremely careful in interpreting Descartes' position. There is a danger of reading into the text too much of his mature, post-1628 epistemology and ontology and hence missing the internal dialectic which is to lead to those later positions. On the one hand, Descartes is certainly not claiming that the essence of corporeal substance is extension. He speaks of 'extension', 'corporeal nature' and other attributes of matter.[44] And it will be shown that the particular epistemological slant of the o-p-p nexus leads him to claim that extended, shaped, mobile 'figures' can be known with certainty in the common sense or imagination, *but not* that extension therefore exhausts the essence of corporeal reality. On the other hand, he is not yet fully aware of the deep epistemological puzzles latent in his mechanistic account of perception, and in the unarticulated dualism of his account of mental function. This, again, is due to his peculiar epistemological commitments here, his focusing upon what *can* be claimed to be known with certainty on the basis of the o-p-p nexus, *not* upon the problems created by the o-p-p nexus in regard to what will later be termed secondary qualities.

For example, near the end of rule 12 Descartes insists, in full accord with his o-p-p nexus doctrine, that imaginations *qua* imaginations are veridical. In imagining, one is intuiting a corporeal state of affairs, just as it is in the imagination.[45] This is crucial, for it will be shown that a good deal of the machinery of universal mathematics is grounded in the imagination and hence is rooted in corporeal reality, even if in the first instance it is merely the corporeal substance of a brain locus. Descartes then goes on to assert that one is 'liable to go wrong' in judging 'that the imagination faithfully represents the objects of the senses, or that the senses take on the true shapes of things, or in short that external things *always* are just as they appear to be'.[46] Far from necessarily meaning that sense perception is always illusory, this could mean, in the context of the o-p-p nexus, that some aspects of sense deliverances may be veridical. The remainder of the passage confirms this. The 'wise man', Descartes writes, will not think that the patterns in his common sense and imagination 'have passed complete and without alteration from the external world to his senses and from his senses to his imagination, *unless he already has some other grounds for claiming to know this'*.[47] This astounding statement would appear to

---

[43] See above Note 34.

[44] *Regulae,* Rule 12, AT, X, p. 418 1.7 to 10; CSM p. 44; HR, I, p. 40. On the absence in the *Regulae* of any of the specifically Cartesian metaphysical theses see, for example, Alquié (1950), 71 ff. and Gäbe (1972), 54 and *passim.* The present study differs from Alquié on the issue of just when Descartes' characteristic mature metaphysical theses began to be developed, and it differs from Gäbe on the reasons for the abandonment of the project of the *Regulae.*

[45] *Regulae,* Rule 12, AT, X, p. 423 1.1 to 5, 13 to 16; CSM p.47; HR, I, p. 44.

[46] *Ibid.* AT, X, p. 423 1.1 to 7; CSM p.47; HR, I, p. 44. emphasis added

[47] *Ibid. AT, X, p.* 423 1.13 to 20; CSM, p.47; HR, I, p. 44. Emphasis added. The translation combines elements of both CSM and HR.

mean that some patterns can at least come unaltered directly from the surface of the sense organs to the forum of the *vis cognoscens,* there to be intuited. But the 'other grounds' cannot be faith or authority, ruled out by the method, nor can it be the metaphysical system only designed after 1628 and of which there is no hint in the text. The 'ground', in fact, must be the o-p-p nexus and the articulation of the old 1619 doctrine of intuition (rule 3) accomplished in the reform of psychology.

Descartes has already claimed, after all, that impressions pass 'instantaneously and without transfer of any real body from the sense organs to the common sense and imagination', there to be scrutinized by the *vis cognoscens.* The automatic and instantaneous imprinting of sense patterns already gives some 'prior ground' for believing that the patterns are accurately delivered from the surface of the sense organ to the brain. But, of course, all that is delivered, and delivered accurately (in a healthy individual), are patterns of disturbance impinging on sense organs and ultimately derived from objects in the outside world. What those objects are really like, what their complete natures or 'forms' are, cannot be known with certainty, for we know them only in respect of their geometrical-mechanical patterns of effect upon us, patterns further mediated in the cases of sight and hearing by the mechanical transmission of disturbances through intervening media. On the other hand, in sensation we *are* in touch with some directly registered aspects of the external corporeal world and not with some 'spiritual' object, nor with some corporeal object of our own manufacture (an imagination) or stored in our brain from prior experience (a memory). It may seem that on this basis we know desperately little of the corporeal world for certain, but the little we can know—along with the machinery of imagination—will prove sufficient to ground universal mathematics, as we shall now see.[48]

7.4 Rules 14–18: The Machinery of Universal Mathematics

After some preliminaries in rule 14 Descartes points out that there is a single mental operation, a direct inspection or comparison, by which shapes, magnitudes and figures can be judged to be equal or similar. Deduction, he intimates, is merely a series of such comparisons, whereby at each step in the logical chain the equality of a pair of quantities is intuited.[49] Quantitative reasoning, in short, is to be seen as consisting in iterated steps, each involving the immediate inspection of the equality of quantitative objects implanted in the imagination.

---

[48] It is worth noting the 'Mersenne-like' and 'Mersenne-transcending' aspects of this position. That we know desperately little of the outside world for certain echoes Mersenne; what goes far beyond Mersenne's proposals is that according to Descartes we have more than a piecemeal collection of reliable bits of knowledge, because we have a procedurally coherent, general discipline, universal mathematics, providing physico-mathematical knowledge of nature (as well as grounding the objects and procedures of all of mathematics).

[49] *Regulae,* Rule 14, AT, X, pp. 439–40; CSM p.57; HR, I, p. 55.

Descartes next suggests that only figures and shapes should be used to represent in the imagination all the quantities to be compared in any question. The critical passage, central to the machinery of universal mathematics in the later *Regulae*, and to the way Descartes hoped to ground the legitimacy of that machinery, must be cited at length:

> ...when the terms of a problem have been abstracted from every subject in accordance with the preceding Rule, then we understand that all we have to deal with here are magnitudes in general.
> The final point to note is this: if we are to imagine something, and are to make use, not of the pure intellect, but of the intellect aided by images (*speciebus*) depicted in the imagination, *then nothing can be ascribed to magnitudes in general which cannot also be ascribed to any species of magnitude.*
> It is easy to conclude from this that it will be very useful if we transfer what we understand to hold for magnitudes in general to that species of magnitude which is most readily and distinctly depicted in our imagination. But it follows from what we said in Rule Twelve that this species is the real extension of body considered in abstraction from everything else about it save its having a shape. In that Rule we conceived of the imagination, along with the ideas existing in it, as being nothing but a real body with a real extension and shape. That indeed is self-evident, since no other subject displays more distinctly all the various differences in proportions.[50] (emphasis added)

This passage epitomizes Descartes' constructive strategy for universal mathematics: Earlier in rule 14 he established that all well-defined problems, regardless of their subject matter, consist in unraveling structures of relations between magnitudes by means of imaginative comparison between impressed patterns. He now invokes the o-p-p nexus of rule 12 to remind the reader that even in the imaginative rendering of quantity, one is dealing with a real body. The imagination and the patterns in it are corporeal entities, bearing the same ontological certification as the mechanical deliverances of sensation registered in the common sense. Then, the argument moves forward in two well-planned stages. First he says that any particular sort of magnitude may be chosen to depict in the imagination the conditions of the problem previously construed in terms of magnitudes in general. His point is that any sort of magnitude depicted in the imagination will be ontologically certified as corporeally present, and, in addition, it will be susceptible to the same sort of manipulations as would have applied to the original quantity from which the magnitude in general was abstracted. Next, a criterion of intuitive simplicity specifies the sort of magnitude to be used—'figures' or shapes. But simplicity is not a merely heuristic criterion. The point about clear and simple intuition is, as always, that it is self-verifying. The aim of representing magnitude in general by figures is to lend ontological grounding to the objects, and logical certainty to the manipulations, which in this case are immediate acts of comparison.[51]

---

[50] *Ibid.* AT, X, pp. 440–1; CSM p.58; HR, I, p. 56.

[51] Boutroux (1900), 32, seems to have been the first to notice the ontological import of this passage. However, he did not see the justificatory aim, but rather stressed Descartes' falling back on the use of imagination after an attempt to found a purely intellectual universal mathematics (p. 25). There is no evidence for this. On the interpretive conflations involved see below Sect. 7.7 and the accompanying notes.

In specifying the sorts of figures to be employed in order 'most readily' to 'express differences of relation or proportion', Descartes mentions only 'numerical assemblages and magnitudes'.[52] The former, exemplified by a genealogical tree and a Pythagorean representation of 'triangular' numbers, do not appear again in the text as integral parts of the machinery of universal mathematics. Even arithmetic, it will transpire, is to be pursued in terms of 'magnitude', which Descartes now quickly limits to geometrical figures.[53] In those figures, he continues, one should attend only to their length and width.[54] This, he claims, will best facilitate the pair-wise comparison of magnitudes, upon which each step of a universal mathematical procedure depends.[55] By way of illustration, Descartes asserts that the solution of problems in geometry will no longer involve complex figures and constructions. Instead, lines and rectangles will somehow stand for the given and sought quantities, and they will be manipulated in the imagination to determine the latter in terms of the former.[56] To understand fully what Descartes intended by this representing function of lines and rectangles, one must return to an earlier part of his discussion—his definitions of 'dimension' and 'unit'.

A 'dimension' for Descartes is 'a mode or aspect according to which a subject is considered to be measurable'. Division of an entity into several identical parts constitutes a dimension, according to which numbers are applied to things. Dimensions include, for example, the three dimensions of extension; weight, a dimension according to which heaviness is measured; and speed, a dimension of motion.[57] Furthermore, dimensions can also be founded on mere distinctions of reason having no real basis in the object measured, for example, the division of the day into hours and minutes.[58] The central point is the conflation of length, width and depth with such physical dimensions as weight and speed. This indicates Descartes' intention to integrate geometry (and arithmetic) with a particular brand of 'physico-mathematics' in the conception of universal mathematics (an entirely unsurprising move given our excavation of his physico-mathematics and initial concept of universal mathematics in 1619). Geometry deals with the three dimensions of extension (and particular dimensions based on distinctions of reason and applied to figures in respect of chosen reference frames). Physico-mathematics, whatever it might prove

[52] *Regulae,* Rule 14, AT, X, p. 450; CSM, p.64 (HR, I, p. 63). The HR translation of 'numerical assemblages' is preferred to CSM's 'sets'.

[53] *Ibid.* AT, X, p. 452; CSM p.65; HR, I, p. 65. Descartes represents some discontinuous quantities in rule 15, but by rule 18 *all* mathematical operations are being carried out upon lines and rectangles, just as p. 452 1.22 to 26 suggests.

[54] *Ibid.* AT, X, p. 452; CSM, p.65; HR, I, pp. 64–5.

[55] *Ibid.*

[56] *Ibid.*

[57] *Ibid.* AT, X, pp. 447–8; CSM, p.62; HR, I, p. 61.

[58] AT X p. 448; CSM 62-63; HR I 61. Descartes may have had in mind dimensions measured in respect of conventionally selected co-ordinate frames in the solution of geometrical construction problems. See *Geometry,* I, AT, VI, pp. 382–3, 372.

to be in detail, is generally meant to deal with other measurable physical dimensions.[59] It is a 'particular brand' of physico-mathematics, because unlike his earlier physico-mathematics, as we have studied it in Chaps. 3 and 4, in this 'late *Regulae*' physico-mathematics there can be, strictly speaking, no attempt to move from mixed mathematical findings to underlying corpuscular-mechanical causes. The reasons for this are intimately tied up with the design and legitimatory intent of the later *Regulae* machinery, as will soon become more obvious.

We have now arrived at the point at which the basic machinery of the later *Regulae* version of universal mathematics finally comes to light. The magnitudes or figures with which universal mathematics has to deal have been revealed to be, in fact, 'dimensions' measured out according to appropriate given or selected units. One can therefore see more clearly what Descartes meant in asserting that geometry was to be about the manipulation of lines and rectangles, and not about the analysis and construction of complex figures. The relevant given geometrical magnitudes are to be represented in the imagination by extensional measures. The sought magnitude is to be expressed in terms of a relation among these extensional magnitudes (on analogy with the procedures of algebraic analysis), and the unknown is to be determined by an unfolding and simplification of the relations among the given extensional measures. All this is carried out in the imagination on real extensions in the simplest possible way before the validating intuitional gaze of the *vis cognoscens*.[60] Below we shall see that the schemas for unfolding the relations will be given within a theory of equations, expressed in terms of an improved symbolic algebra.

As regards physico-mathematics, it will now be that part of universal mathematics which deals with problems about the relations holding between given and sought dimensions of physical properties measured in or between bodies. An example given in rule 13 provides the perfect illustration of what Descartes intends. The problem deals with an inquiry into the nature of sound, and, very tellingly, it draws upon Mersenne's recent work on the basic acoustical laws. Descartes wrote:

> … the question may be, what is my conclusion as to the nature of sound, founding my judgment merely on the precise fact that the three strings A, B and C give out an identical sound, when by hypothesis B, though twice as thick as A, but not longer, is kept in tension by a

[59] 'Unit' is simply the element through which a given 'dimension' is measured. If a unit is not given for a sort of dimension involved in a problem, Descartes is perfectly willing to have the unit represented by any arbitrarily chosen magnitude of that type. Hence, he allows for units applicable to each type of figure which might be employed in a problem, whether, for example, collections of points, rectangular figures or straight lines, whose units would be a point, square or unit length respectively. *Regulae,* Rule 14, AT, X, pp. 449–50; CSM, pp.63–64; HR, 1, p. 63.

[60] Descartes gives the example of a triangle to be analyzed in terms of its 'dimensions', … ut in triangulo, si illud perfecte velimus dimetiri tria [dimensiones] a parte rei noscenda sunt, nempe vel tria latera, vel duo latera et unus angulus, vel duo anguli et area, etc; item in trapezio quinque, sex in tetraëdro, etc; quae omnia dici possunt dimensiones' (AT, X, p. 449; CSM, p.63; HR, I, p. 62). See the treatment in terms of 'simple natures' in rule 12 prior to the transformation of his old 'methodological' terminology into the technical language of the new universal mathematics, rule 12 (AT, X, p. 422; CSM, p.46; HR, 1, p. 43).

> weight that is twice as heavy; while C though no thicker than A, but merely twice as long, is nevertheless kept in tension by a weight four times as heavy.[61]

Here a correlation of vision and hearing permits a direct visual intuition and measurement of appropriate 'dimensions'. Measures of length, cross-section and weight can be read off macroscopic objects and correlated with the tonal properties perceived.[62] The science of tone then consists in the correlation of extensional measures of observable properties, a mathematical science correlating appearances, as Lenoble so aptly characterized Mersenne's scientific ideal.[63]

If one should ask how weight is determined, it is clear that that depends upon the prior achievement of a science of statics, built up through direct visual inspection of equilibrium conditions of standard weights and lever arms, all representable by surfaces and lines.[64] In a certain sense, therefore, physico-mathematics differs from elementary geometry in that the relevant dimensions are not straightforwardly available in experience. They have to be selected and devised in order progressively to get purchase on more complicated sorts of scientific objects—for example, geometry is prior to statics, and statics, as we have seen, is prior to the science of tone. However, the higher reaches of geometry also depend upon the progressive devising of appropriate schemas of 'dimension-formation'; for, in complex locus and construction problems the reference frames for the production of extensional measures (and their algebraic symbols) must be chosen, they are not given; and as Descartes insists, the solution of higher-order problems depends on the prior mastery of simpler cases.

Beyond all this there is a striking methodological unity in the vision of universal mathematics.[65] The solution of any problem in any properly mathematical field is held to consist essentially in an unfolding in the imagination of relations holding among extensional measures of dimensions. The art of unfolding those relations, to be given in an algebraic theory of equations, presumably holds for every properly mathematical field. And, to repeat and underscore our earlier point, this late *Regulae*

---

[61] *Regulae*, Rule 13, AT, X, p. 431; HR, I, pp. 49–50. The HR translation is preferred for the following reason: CSM may be misleading as to Descartes' intention and state of knowledge here, with their translation of the conditions on string C being 'C is twice as long as A, *though not so thick*, and is tensioned by a weight four times as heavy.' [emphasis added, p. 52] Surely Descartes idiomatic Latin was not meant to be conveying a mistake about Mersenne's quite exact, and mathematically simple, results.

[62] For the time being we overlook the problem that according to Descartes' account of perception in rule 12, sound also is delivered as a mechanical disturbance in the brain loci and is directly attended to by the *vis cognoscens*. See Sect. 7.6.2 below.

[63] Lenoble (1943) 272–6, 313–17, 319–21.

[64] The general implication, not spelled out by Descartes, is that all relevant physical properties can somehow come to be expressed as geometrical extensions by means of sub-procedures constitutive of each of the 'physico-mathematical' fields subordinate to universal mathematics. So, expressed in terms of extensional measures, these properties can then become the objects of general analytical procedures, worked out in the corporeal imagination, according to rules given by the theory of equations.

[65] As we expect, given our interpretation of the universal mathematics of the later *Regulae* as both articulating, and being constrained by, Descartes' methodological ideas in the earlier strata of the text.

version of physico-mathematics, taken with strict attention to its logistical machinery and legitimatory intentions, in no way aims at or allows micro-corpuscular causal reductions of firm mixed mathematical findings. This issue will shortly reappear in our discussion of the tensions and problems within the text of the later *Regulae*.

To summarize, then, we may say that in rule 14 Descartes shows how straight lines and rectangular figures may function as *symbols* to be used in signifying any and all magnitudes which enter into a problem to be solved within universal mathematics. The machinery of universal mathematics is thus essentially a 'logistic' of 'extension-symbols', as I shall term them, specifically designed and related to the o-p-p nexus in order to assure the truth of the operations and their ontological reference.[66] If one links the o-p-p nexus of rule 12 with the introduction of extension-symbols in rule 14, one can see that Descartes is claiming a natural philosophical basis; that is, an o-p-p basis, for the following points:

1. The corporeal world is indeed the ultimate object of universal mathematics; but, it is known only under the category of the two-dimensional shapes and patterns registered in sensation and delivered up to the validating gaze of the intellect.
2. For this reason the hierarchy of physico-mathematical fields and pure mathematics (geometry and arithmetic) can, and must, consist in the construction and manipulation of relevant dimensions given in, or manufactured from, these data—both the data and the dimensions being extensional objects really present in the imagination.
3. As a corporeal locus, the imagination is an ontologically suitable 'screen' upon which extension-symbols can be manipulated; and the operations performed on the symbols have the certification of being clearly intuited in the 'real extension of bodies'—they are true and true of the world.

All that remains for the construction of the machinery of the newly elaborated universal mathematics is that Descartes show how an improved symbolic algebra can be put to the service of the discipline. Symbolic algebra, and the theory of equations whose construction it facilitates, are absolutely necessary for the functioning of universal mathematics as a general analytical discipline. This is because the techniques

[66] It was Jacob Klein in his brilliant study (Klein 1968), who first attained the fundamental insight that Descartes was offering a mathematics expressed in and manipulated through line lengths functioning as operative symbols (pp. 198, 202, 208). Klein saw that in rule 14 Descartes was trying to ground his universal mathematics, a general science of proportions, in a symbolism consisting of real, concrete line lengths depicted in the corporeal imagination (pp. 197–8). Descartes wanted to realize, indeed materialize, abstract algebra in concrete, intuitively clear, objects and operations, and he wanted to show how a mathematical physics falls [actually a species of physico–mathematics, as we can now see] under the analytical procedures that algebra provides (p. 198). My only reservation with Klein's reading arises from his tendency to say that Descartes intended the theory of mind and perception to give insight into the real structure of the world (p. 210). On my reading, Descartes is saying in 'Mersenne-like' fashion that we have access to certain aspects of the world, not that we have insight into the essential structure of it. To understand why Descartes later came to claim the latter through his metaphysically backed theory of matter-extension, one must comprehend the nature of the epistemological position in the *Regulae* and the reasons for its demise (see Sect. 7.6 below).

for unraveling complex structures of relations are best pursued and recorded in algebraic terms. Nevertheless, and this is crucial, Descartes must also exercise extreme care to show how all operations dictated by a theory of equations can be grounded in the logistic of extension-symbols and so be certified as true.

In rule 16 Descartes gives a cautious introduction to his improved symbolic algebra, which was far superior to the cossic abbreviations he had been using in 1619, but which was not yet fully developed into the classical notation introduced in the *Geometry* of 1637. Here only very modest functions are accorded to the symbolism,[67] and it is only in rule 17 that its profound value in facilitating mathematical analysis comes to light and is mobilized for universal mathematics. In this sense rule 17 serves as an introduction to the theory of equations which Descartes planned to annex to universal mathematics and which he began to sketch in rules 19–21 before abandoning the text.

Employing his terminology of proportions, which had loomed so large in the original 1619 notion of universal mathematics (and in the development of the method), Descartes distinguishes between 'direct' and 'indirect' problems.[68] Direct problems take the form of straightforward proportions, e.g. $1/a = b/x$. The solution is a simple case of 'deduction', a direct manipulation of the known terms a and b. Indirect problems, in contrast, involve the search for mean proportionals, given the first and last terms, and this, Descartes well knows, is equivalent to solving an equation of a corresponding degree of complexity. For example if $1/x = x/a$, the form in which a and unity are connected is known, and one must find the mean proportional x which connects them. Here x is not immediately revealed through a direct manipulation of the knowns. To unravel the proportion one must solve a second-degree equation in x.

It is precisely through this translation into the terms of a theory of equations, Descartes now insists, that 'indirect' questions may be rendered into 'direct' form:

> ... if from the fact that we know the first (term) and the last to be connected with each other in a certain way, we should want to deduce the nature of the middle terms which connect them, we should then be following an order that was wholly indirect and upside down. But because here we are considering only involved inquiries, in which the problem is, given certain extremes, to find certain intermediaries by the inverse process of reasoning, the whole of the device here disclosed will consist in treating the unknowns as though they were known, and thus being able to adopt the easy and direct method of investigation even in problems involving any amount of intricacy'.[69]

[67] *Regulae,* Rule 16, AT, X, pp. 454–9: the improved algebra aides memory by facilitating the recording of the results of the comparison and manipulation of magnitudes. All attention can then be directed to the comparison at hand. Second, the recording of the steps preserves the distinctions amongst the relevant quantities and reveals at a glance the operations performed upon them.

[68] *Regulae,* Rule 17, AT, X, pp. 459–60; CSM, p.70; HR, 1, pp. 70–1.

[69] *Ibid.* AT, X, p. 460; CSM, pp.70–1; HR, 1, pp. 71. Again as on occasion above, the HR translation is preferred to CSM, not on overall diction or accuracy, but for a matter of technical precision. CSM, whilst utilizing the word, 'term', earlier in the passage, before the portion we have quoted, nevertheless consistently render the above passage in terms of talk of extreme and intermediate 'propositions'. In contrast HR keeps to the strongly implied technical mathematical context, and speaks of extreme and intermediate 'terms'.

When we assume the unknown as known, that is, provide a symbol for it, and relate it by means of equalities to the data of the problem, we obtain an algebraic equation which can be solved for x in an 'easy and direct' 'deductive' manner. Here, then, symbolic algebra attains its full role within universal mathematics, a role masked by the exposition of rule 16. Algebra allows complex problems to be reduced to the ease of 'direct' deduction; and the theory of equations, symbolically expressed and manipulated, permits the derivation, articulation and codification of general schemas of solution for increasingly broad classes of problems. All the contemporarily perceived fruits of algebra are imported into the heart of universal mathematics.

It is crucial to note, however, that universal mathematics is not to be identified with symbolic algebra *tout court.* Algebra and the theory of equations are an indispensable *element* in the actual working of the machinery of the new universal mathematics, but they are carefully subordinated to the overriding aims of grounding the logical validity and ontological grip of the discipline. In rule 18 Descartes tries to show how all the operations expressed and commanded in terms of abstract symbolism and theory of equations must be grounded and certified at each step by the intuitively certain and ontologically validated 'logistic of extension-symbols'. Only in this way can the analytical power of algebra be harnessed to and certified by the o-p-p nexus, which guarantees the truth and reference of universal mathematics. Quite simply, Descartes attempts to provide geometrical-intuitive interpretations for the four operations needed in solving problems—addition, subtraction, multiplication and division.[70] The first two operations present obvious cases of laying-off of line lengths to generate sums or differences.[71] The intuitional rationale for these operations is supposedly apparent in the immediate inspection of the very diagrams which illustrate them. Descartes intends the same to be the case with multiplication and division. Multiplication of two magnitudes *a* and *b* presented as straight-line segments (extension-symbols) is to be accomplished by fitting them together at right angles to form a rectangle. If rectangle *ab* has to be multiplied by a third quantity *c*, represented by a straight-line segment, then one constructs a line of length *ab* units and then constructs again a rectangle *ab(c).*[72] Similarly, in divisions, where the divisor is given, one takes the magnitude to be divided as a rectangle, the divisor as one side and the quotient as the other.[73] In general, then, Descartes assumes that any

[70] Raising to a power and extracting a root are considered to be species of multiplication and division respectively. Difficulties arise from this in the case of root extractions. See Sect. 7.6.3 below.

[71] *Regulae*, Rule 18, AT, X, pp. 464–5; CSM, p.73; HR, I, p. 73.

[72] *Ibid.* AT, X, pp. 465–6; CSM, p.74; HR, 1, pp. 74–5. One determines the line *ab* by constructing a rectangle of area *ab,* one side of which is of unit length.

[73] Ibid. AT, X, pp. 466–7; CSM, p.75; HR, 1, pp. 75–6. The procedure as stated would assume the result is known beforehand. One can 'reconstruct' Descartes' view of the operation as follows: represent the divisor by a line of length *a*; then normal to one end of a lay-off line *b*, the quotient, initially of unknown length. Box-off unit squares in the resulting rectangle until *ab* units, the dividend, has been obtained, thus specifying the actual length of *b* (and indicating any remainder). The complexity of this procedure weighs heavily against the notion that Descartes intended it as a practical aid to working calculations (see also below, Sect. 7.5): his aim was legitimatory.

power of any quantity can be represented by a straight line or rectangular surface. He insists, however, that multiplication and division always respectively produce a rectangle from two lines, or decompose a rectangle into straight lines.[74] The reason for the restriction of the logistic of extension-symbols to two dimensions would seem to be the technical *and hence legitimatory* constraints of the o-p-p nexus. Such complicated flipping back and forth between lines and rectangles is necessary in order to retain the operations within the narrow bounds of the clearest and simplest intuitions, and therefore the most certain imaginative-intuitive performances.

7.5 The Structure of Universal Mathematics in the Later *Rules* and Its Legitimatory Functions

Figure 7.1 summarizes the foregoing interpretation of the universal mathematics of the later *Regulae.* The o-p-p nexus, that is, Descartes' mechanistic theory of light and vision, his sketch of a mechanistic theory of perception, and his reformulation of faculty psychology, provided natural philosophical grounds for holding the deliverances of sense to be geometrical-mechanical alterations of certain *loci* in the brain. Whatever else the external world may consist in, it is knowable under two broad categories of 'dimension' given in, or manufactured from, such deliverances as follows:

1. Measures of length, width and depth, making up the objects of geometry (and arithmetic, as we have seen).
2. Extensional measures of certain observable physical properties, such as motion, weight, tone, tension, color etc. which are the objects of the physico-mathematical disciplines.

Once constituted, both orders of dimension are directly and infallibly known, and hence there is no difference between the objects of geometry and those of physico-mathematics in respect of their ontological status or mode of presentation to the *vis cognoscens.*

The immediate clarity and simplicity of intuitions of dimensions impressed in the common sense and imagination is the ground of the truth of the 'comparisons' made between them, whether this consists in a single 'intuitive' comparison, or in an iterated series of them, that is, a deduction. To aid in achieving such legitimatory clearness and simplicity, only straight lines and rectangles are used to symbolize the

---

[74] Ibid. AT, X, pp. 4678; CSM, pp.75–6; HR, I, p. 76. Descartes continues by asserting that these transformations between lines and rectangles can always be performed by geometers: 'provided they recognize that whenever we compare lines with some rectangle, as here, we always conceive those lines as rectangles, one side of which is the length that we took to represent the unit. For if we do so the whole matter resolves itself into the following proposition: Given a rectangle to construct another rectangle equal to it upon a given side'. (AT, X, p. 468; CSM, p.76; HR, I, pp. 76–7.)

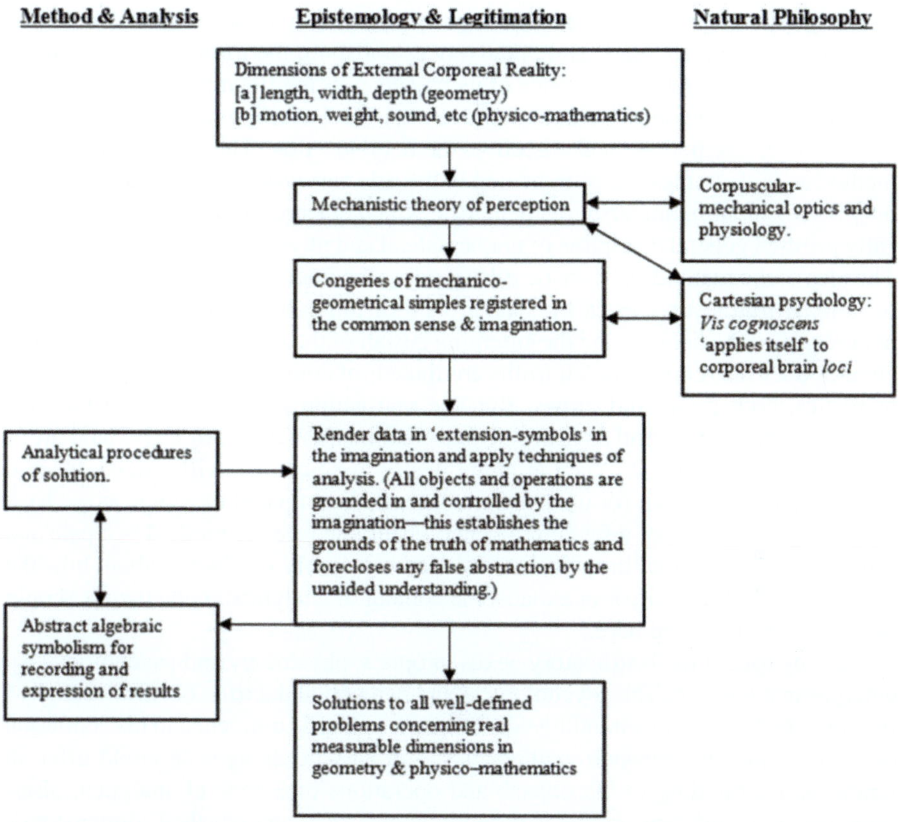

Fig. 7.1 The Structure of Universal Mathematics 1626–1628: Natural Philosophical, Epistemological and Methodological Elements

given and sought dimensions. All operations to be performed on these dimensions are to be controlled in the imagination by the construction and decomposition of line lengths and rectangles. The procedures or schemas for unraveling relations among dimensions are elaborated in a theory of equations, expressed by means of an improved operative algebraic symbolism; but, every move dictated by the algebraic theory must be representable to the intellect upon the corporeal screen of the imagination in precisely the ways set down in rules 14 and 18. So interpreted, universal mathematics is not to be identified with symbolic algebra, with Descartes' later views on analytic geometry, with corpuscular-mechanical natural philosophy, nor even with a properly mathematical physics *tout court* (on the model, say, of Galileo's mechanics).[75] It is a general mathematical discipline, providing machinery

[75] The best analyses of exactly what Galileo's mechanics amounted to as the first species of a mathematico-experimental science remain those of Clavelin (1974) and Gaukroger (1978).

for the analysis of all problems occurring in properly mathematical fields, including physico-mathematical ones, and putatively establishing the truth of its own procedures and the ontological reference of its objects—its legitimatory functions.[76]

The figure underscores the contention that the later *Regulae* are not a mere collection of methodological directives or heuristic tips. The original notion of method as 'aids to reasoning' indicated in the title survives in the 'mature' work as only one element in universal mathematics, which emerges as a natural philosophically justified general discipline of mathematical and physico-mathematical inquiry. The idea of the method as heuristic rules guiding the intellect around the latticework of rational truths survives in the background of the universal mathematics of the later *Regulae*: the field behind the emerging gestalt of the elaborated universal mathematics doctrine. Of course, all truths are linked; of course, method as a tool kit of heuristics, both exists and grows. But, the real action and focus are on the now articulated mechanics and logistics of universal mathematics. An elaborated art of algebra, mainly expressed via a growing theory of equations, will provide the real problem-solving tool kit for universal mathematics, compared to which, as we have argued, the statement of a few more general heuristic rules of method is a pale and limited shadow. All the limitations mentioned previously in Chap. 6 about how the method cannot fully mirror or exhaust the content of analytical mathematics should be recalled and applied here.

The key role of the justificatory nexus of optics, physiology and psychology also emerges in the figure. The psychological and perceptual doctrine of rules 12 and 14 provides the key to understanding why Descartes placed confidence in his conflation of mathematics and physico-mathematics, and why he thought he could offer an ontological grounding to the objects and operations of a general analytical discipline which would treat problems posed in geometrical and physical 'dimensions'. The justification for the entire edifice of the elaborated universal mathematics of 1626–1628 rests on the degree of credence one can give to Descartes' theory of perception (conjoined to his mechanical theory of light) and the faculties of the mind.

There may yet linger a suspicion, however, that the o-p-p nexus was intended merely to provide heuristic devices useful for working mathematicians, not in order to help establish the truth of mathematical operations and the ontological status of their objects. For example, some commentators have claimed that the machinery of lines and rectangles was introduced because Descartes, in 1628, did not yet know how to represent geometrically the product of two line lengths as a third line length. So, in the manner of classical Greek geometry, he had to construct rectangles representing

[76] Noting again that in the physico-mathematical part of this universal mathematics, all inquiry remains on the macroscopic level of correlations of physical dimensions and problem solving about them. Strictly speaking, in view of these legitimatory ends and their supporting machinery, reduction to corpuscular–mechanical explanations is neither sought nor allowed, unlike the case in Descartes' 1619 physico-mathematics, or in his recent optical work. As a result, problems lurk for the *Regulae* project, which we canvass in the next Section.

Fig. 7.2 AT VI p. 370

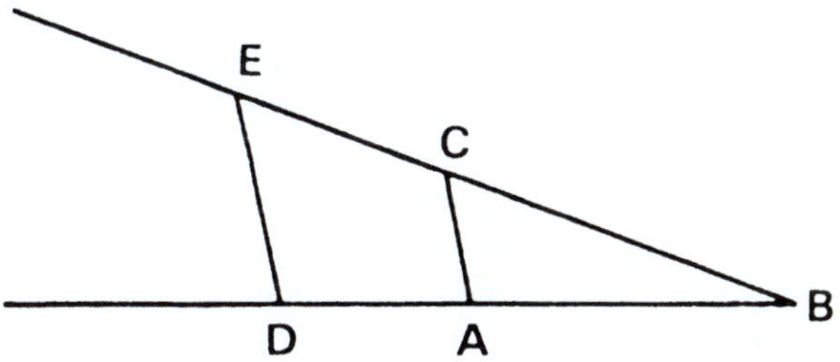

the products as areas.[77] Only in the *Geometry* of 1637, they observe, did Descartes publish a geometrical illustration of how the result of the multiplication or division of two line lengths can be construed as a third line length.[78] But, in fact, Descartes did know very well how to represent the product or quotient of two homogeneous quantities, say line lengths, as a third homogeneous quantity. In the *Regulae* he explains how multiplication and division (and raising to a power and root extraction) can be interpreted as operations upon homogeneous proportional magnitudes.[79] (Indeed this insight is the basis of his construction in the *Geometry.)* Nor is it difficult to see that the insight is closely related to his use of the proportional compass, and so goes back to 1619.[80] Therefore, I conclude that the reason Descartes devised the elaborate machinery of lines and rectangles in rules 12–18 was *legitimatory* and not *procedural* or *heuristic*.[81] After all, a sceptic might scoff at the interpretation of mathematical operations as manipulations of proportional magnitudes, even when modeled on the compass (the rationale of the manipulation of the compass presupposes a good deal of geometrical knowledge, the validity of which is in question). But, in contrast, the laying-off of line lengths and the composition and decomposition of rectangles has commonsensical intuitive appeal, and, in addition, could now be backed with an attractive account of perception and imagination, which showed just where, how and upon what these most simple of operations occurred.

[77] P. Boutroux (1900) 43, Milhaud (1921) 70–2, L. Brunschvicg (1927) 283–9, Mahoney (1971) Vol. IV, pp. 56–7.

[78] *Geometry, I,* AT, VI, p. 370: In Fig. 7.2 BA is the unit; to multiply BD by BC, join points A and C and then draw DE parallel to AC, then BE is the product. For, by similar triangles:

(BC/BA)=(BE/BD); or BE=BC x BD. To divide BE by BD one reverses the process.

[79] He does this in rule 18, prior to introducing the logistic of extension symbols, AT, X, pp. 463–4; CSM, pp.72–3; HR, 1, pp. 72–3.

[80] See above Sect. 5.4.

[81] Descartes used his logistical machinery in solving a problem only once in his extant corpus of writings. This occurs in a report he gave to Beeckman in 1628 concerning his researches over the previous ten years (*Algebrae Des Cartes Specimen Quoddam*, AT, X, pp. 334–5). But in this case Descartes was illustrating the teaching of the *Regulae* rather than showing Beeckman how he ordinarily solved quadratic equations. Put bluntly, for the practicing mathematician familiar with the methods of arithmetic, algebra and geometry, the reconstruction of each step in terms of imaginative manipulation of straight lines and rectangles is heuristically otiose.

That Descartes designed the o-p-p nexus with legitimatory aims in view may be further confirmed by looking at a portion of rule 14, which we did not examine earlier. Descartes opens by insisting that when the *vis cognoscens* reasons about corporeal objects, and that, of course, includes mathematical objects, it must not act alone without making continual reference to corporeal images.[82] Again, the suggestion is not merely heuristic. The corporeal images are the very objects of universal mathematics—geometrical dimensions and extensional measures rendered in terms of extension-symbols. Then, in an absolutely critical passage, Descartes insists that unless reasoning is so constrained the unaided understanding will tend to manufacture fantastic and illusory properties, attribute them to its abstract notions, and thence to their corporeal referents. The delusions of neo-Platonic number harmonies and numerology flow precisely from this pathology of cognition.[83] Next, turning to the sceptical threat, he grants, in a manner similar to Mersenne, that the stock sceptical arguments against the foundations of geometry are valid. Geometers have undermined their own position by founding their science on illicit manipulations of wrongly abstracted notions of geometrical objects.[84] But, Descartes continues, he is going to expound a doctrine which will outflank the objections by giving a 'proof of whatever is true in arithmetic and geometry'.[85] *This, to reiterate, can only mean the legitimatory doctrine of the o-p-p nexus and the logistic of extension symbols,* and it shows Descartes' intention of using the new machinery to rescue matters considered in some circles to be under serious sceptical threat.

To summarize, then, the dating of the text of the later *Regulae* and the reconstruction of its content and aims seem to confirm the larger thesis that the renewed attempt to construct universal mathematics in detail was triggered and conditioned by the Parisian turmoil and the immediate example of Mersenne. Rather than standing firm on his methodological visions (and illusions) of 1619, Descartes returned to the idea of universal mathematics and tried to work it out in detail toward very particular ends using very particular means. Both the ends, and means, bear the stamp of the Paris of the 1620s.

Like Mersenne's limited conception of natural knowledge, the physico-mathematical part of universal mathematics was to consist in the mathematical correlation of quantitative measures of observable properties. Like Mersenne, Descartes eschewed, outright, systematic metaphysical construction. Rather, like Mersenne, he ventured into the theory of knowledge by exploiting presumed natural philosophical achievements and by appealing to the dictates of 'intuitive' good sense.[86] The aim, as with

[82] *Regulae*, Rule 14, AT, X, pp. 442–5; CSM, pp.58–61; HR, I, pp. 57–60. This paragraph gives only the briefest sketch of this rich and significant material, in which Descartes displays a striking sarcasm toward the claims of the 'naked intellect' in these matters. See Schuster 1977, pp. 501–10.

[83] Ibid. AT, X, pp. 445–6; CSM, pp.60–61; HR, 1, pp. 59–60.

[84] Ibid. AT, X, pp. 446–7; CSM pp.61–2; HR, I, p. 60.

[85] Ibid. AT, X, p. 447; CSM, p.62; HR, I, p. 61. See Alquié (1950) 64. Once again the HR translation is preferred. CSM have '...our aim being to provide the easiest possible demonstration of such truth as may be found in arithmetic and geometry.' (p.62)

[86] Gäbe (1972) 39, Note 45, points to the similarity of this tactic to that of Mersenne (1625).

Mersenne, was to walk an ideological knife-edge by, on the one hand, outflanking (and possibly seducing) sceptics, while, on the other hand, securing a brand of physico-mathematical science of nature immune from the dangers of seemingly ascendant neo-Platonic and mystical approaches to nature. The turn toward a Mersennian conception of physico-mathematics is obvious in the later *Regulae*, where, as already noted, the physico-mathematical part of universal mathematics eschews any kind of micro-corpuscular reduction, being a discipline strictly limited to treatment of measurable, macroscopic 'dimensions'. As previously noted, this 'late *Regulae*' version of Descartes' physico-mathematics differed from the physico-mathematics he had produced in hydrostatics and optics since 1619, because the latter tended to envision the discovery of corpuscular-mechanical causes through analysis of well formed mixed mathematical results.

Nevertheless, the parallels to Mersenne ultimately break down. For perfectly comprehensible biographical reasons Descartes' project, unlike Mersenne's, grew out of a doctrine of method and was expressed in the form of a methodological treatise intended to show how a unified mathematical science could be constructed, which would be subordinate to and illustrative of that method. Descartes' mode of appropriation of pre-existent and hoped-for results also differed from that of Mersenne. Where the Minim made a diffuse appeal to bits and pieces of achieved mixed mathematical science, Descartes attempted to fuse real and imagined physico-mathematical and corpuscular-mechanical results into the o-p-p nexus, which would provide a systematic new account of perception, mental function and knowledge. Mersenne had tended to assume both the self-evident truth of mathematical propositions and their applicability to the world.[87] Descartes, too, had always placed crucial methodological weight upon 'intuition'; but in the later *Regulae* he displays what for him at that date was an unprecedented sensitivity to the depth of the challenge to the grounds of mathematics. Accordingly, much of the new universal mathematics is concerned with explicating, as it were, where, upon what, and in what sense this intuition occurs. So, despite broad similarities to Mersenne's approach, Descartes' project was philosophically more profound, for it united his serious concern with method to a daring bid to construct a type of what we must term a 'scientistic epistemology'. In conclusion we are going to see that Descartes' concern with method and his deployment of the o-p-p nexus and logistic of extension symbols led him into new and unexpected difficulties, well beyond the horizon of Mersenne's problematic. These problems undermined the project of the later *Regulae,* and the attempt to resolve or finesse them, while preserving the old legitimatory aims, led directly to the beginning of the constitution of the main lines of mature Cartesianism—a turn to systematic corpuscular-mechanical natural philosophy, with attempted metaphysical grounding.

[87] Mersenne (1625) 226–7, and the tenor of the entire argument, in which Mersenne basically assumes and asserts that we are well advanced in acquiring a mathematical knowledge of appearances, regardless of sceptical doubts about the knowledge of essences.

7.6 The Instability of the Later *Regulae* and the Beginnings of the Origins of Cartesianism

The program of the later *Regulae* broke down in three areas. First, tensions emerged between the methodological vision of a properly mathematical science dealing with macroscopic physical 'dimensions' and Descartes' long standing, albeit underlying and piecemeal commitment to corpuscular-mechanical explanation. Second, what we may legitimately term epistemological problems arose concerning the status of perceptions which are not purely 'intellectual' and yet do not reasonably consist in immediately intuited congeries of shapes and figures. Third, and finally, difficulties emerged in showing how all the objects and operations of mathematics could possess the desired sort of intuitive grounding in the imagination. In general, one can say that Descartes, by his over-concern with constructing the machinery of universal mathematics and with securing its legitimatory benefits, backed himself into a set of unintended difficulties which aborted the project and set the stage for his post-1628 enterprises in systematic metaphysics and corpuscular-mechanical natural philosophy. The development of the instabilities in these three areas are treated in the following three sub-sections.

7.6.1 Discursive Corpuscular-Mechanism in Tension with Genuine Mathematization and the Aims of the O-P-P Nexus

The later *Regulae* ripened, and hence brought to notice for the first time, a pre-existent but latent tension in Descartes' thought between his methodological ideal of a properly mathematical and deductive science and his underlying commitment to corpuscular-mechanical explanation within his physico-mathematical enterprise. The problem, of course, was that Descartes' corpuscular-mechanical discourse was just that, *discourse*, the explanations being always verbal, qualitative and discursive. In this Descartes' natural philosophizing differed not at all from that of any other player. As we observed in Chap. 2, this was the nature of the game of natural philosophizing. Every species and instance of a natural philosophy—from the time of Aristotle, Plato and the atomists, down through the time of Descartes and to the era of the dissolution of natural philosophy as a cultural form in the late eighteenth and early nineteenth century—was a discursive entity, and never a mathematical one per se.[88]

[88] Of course concepts of natural philosophical provenance could be mathematicized (tending to the production of fields increasingly independent of the culture of natural philosophizing); and work in the traditional mixed mathematical fields could be read 'physico–mathematically' as bespeaking issues of matter and cause, hence of natural philosophical relevance. But, no sustained system of natural philosophy ever was, or could be, mathematical; that would be to mistake an instance of natural philosophizing for a bit of mathematical science.

In Descartes' life and agenda, the method based deductive-mathematical ideal and the commitment to corpuscular mechanism both date from 1619 when they subsisted in a loose unarticulated amalgam. Descartes' early physico-mathematics had envisioned the corpuscular-mechanical explanation of well defined mixed mathematical results, whilst simultaneously advocating a genuinely mathematical sense of problem setting and solving. For example, the physico-mathematical treatment of Stevin's work on the hydrostatic paradox was supposed to involve both analytical style problem solving and yet also lead to corpuscular-mechanical explanations of matter and cause. The unexplicated universal mathematics of 1619 preserved this unarticulated amalgam, under the idea that properly mathematical schemas of problem solution would be evolved for mathematical and physico-mathematical questions, whilst the latter inquiries would somehow also lead to natural philosophical explanations. Descartes' early dreams of method further served both to entrench and occlude the problem—it was simply assumed that physico-mathematics (with its encysted bent toward causal corpuscular-mechanical discourse) would be enveloped by the method, along with every other rational discipline. In short, so long as Descartes refrained from constructing in detail his universal mathematics, which was meant to subsume 'physico-mathematics', his optimistic rhetoric about method and a properly mathematical universal mathematics masked tensions about how or whether physico-mathematical inquiries should or could end with 'seeing' the corpuscular-mechanical causes in well formed geometrical representations of macroscopic regularities. But, the building of a version of universal mathematics in the later *Regulae* virtually dictated that the tensions would emerge under two related modalities: first when Descartes tried to appropriate typical objects of corpuscular-mechanical explanation to the machinery of universal mathematics, a difficult enough problem as it stands, but one compounded, secondly, by the fact that the o-p-p nexus grounding and legitimating that universal mathematics was itself a tissue of claims ultimately based in corpuscular-mechanical natural philosophy. This raised the question, 'What exactly grounds such corpuscular-mechanical explanations'?

Let us consider the first form of the difficulty, by reprising an example we studied from a slightly different angle in the previous chapter: Descartes' attempt to fit the study of magnetism to the procrustean bed of the procedures of universal mathematics.[89] From a finite set of observations we are to 'deduce' the character of the 'intermixture' of 'simple natures' necessary to produce the effects of the magnet. The vocabulary is that of the new universal mathematics (with vestiges of the abstract methodological language of rules 4–11).[90] The science of magnets is about the inter-relation of

[89] See Sect. 6.4 where we looked at this matter from the standpoint of 'what, really, was involved when Descartes rendered corpuscular-mechanical explanations of phenomena such as magnetism—did his construction of corpuscular-mechanical discourse follow his method?' Here we ask whether he could really treat magnetism under the precise protocols for the 'science of dimensions' outlined in the universal mathematics of the later *Regulae*.

[90] *Regulae*, Rule 12, AT, X, p. 427; Rule 13, AT, X, pp. 430–1.

appropriately designed and measured 'dimensions'. The term 'deduce' carries with it all the rigorously mathematical apparatus supposedly contained in the analytical procedures of universal mathematics. One should deal with macroscopic measures of observable properties, empirically derived, but known with certainty, to which one would apply properly mathematical methods to produce necessarily true results. Explanation, of course, takes the form of deductions simplifying and revealing the relations among 'dimensions'. In methodological terms, therefore, this case is assimilated to the study of sound, discussed above in Sect. 7.4, or to the analysis of the properties of triangles, to which Descartes alludes twice in the course of the later rules.[91]

However, the three cases cannot really be congruent on the methodological plane as precisely dictated by the universal mathematics of the later *Regulae*. Physical dimensions arise from the measurement and representation of observable properties. In the case of the triangle, the 'dimensions' are immediately given in sense-experience or imaginative intuition. In the case of the science of sound, all 'dimensions' are observable in the sense that one can correlate tones and consonances, perceivable qualities, with equally observable extensive measures of tension (weight), length and cross-section of the vibrating strings. In the study of the magnet it is not at all clear what sort of observable properties and measures are appropriate.[92] And, compounding this obvious failing, it is also clear that the overall aims and style of Descartes' physico-mathematics, as we have come to know it *before* the later *Regulae*, called for reductive corpuscular-mechanical explanations of matter and cause to be arrived at on the basis of clear mixed mathematical results. Just as in 1619 he wanted to know how the corpuscular make up of fluids explains the general phenomenon, representable geometrically, of the hydrostatic paradox, so arguably even here, in the later *Regulae*, Descartes ultimately would have wanted to know how magnetic effluvia push bits of iron about, on the basis of 'reading' such causal knowledge out of some neat geometrical representation of the phenomenological law(s) of magnetism.[93] But, and here is the absolutely critical point, even had

[91] See above note 60, Rule 12, AT, X, p. 422; CSM p.46; HR, I, p. 43 and Rule 13 AT, X, p. 449; CSM, p.63; HR, I, p. 62.

[92] A 'Whiggish' 'rational reconstruction' of Descartes' aims might suggest that he intended, in the light of his methodological ideal, to be a sort of Baroque Coulomb, applying measures of mathematically well-defined 'force' (determined through engineering applications of a science of mechanics) to fully determinate experimental conditions. This, of course, gets us nowhere, because Descartes could neither have conceived nor executed such a project, though it is precisely what the methodological ideal of universal mathematics demands in the strict sense. On Coulomb's determination of the law governing the attraction and repulsion of electrostatic charges see Gillmor (1971) and King (1964).

[93] That is why we used this case in Chap. 6 to illustrate the distance separating Descartes' methodological story about magnet science from his actual discursive practices in constructing corpuscular-mechanical explanations of magnetism. In other words, there is a charitable reading of the relevant passages in the *Regulae*, according to which the 'intermixture' of simple natures might refer to a package of geometrico-mechanical properties to be ascribed to magnetic corpuscles, and that is how we proceeded back in Chap. 6: As he wrote out these passages Descartes may have intended such a gloss, for he may have been wearing his customary rosy-tinted methodological spectacles.

Descartes obtained a macroscopic law of magnetism, the articulated machinery of universal mathematics of the later *Regulae* is not only not interested in corpuscular-mechanical explanation, but also, more importantly, it is not adequate to that task, provided we take it seriously and precisely.

Descartes, in this serious and precise sense, could not offer a micro-mechanical account of the magnet in the later *Regulae*, for that would have destroyed any chance of grounding the certainty of the results in the imaginative intuition of macro-geometrical dimensions manipulated mathematically. Conversely, though universal mathematics could provide a language, and—in the proper cases—actual procedures, for unraveling relations between dimensions, it could not account for the efficacy of the magnet in the required sense of specifying matter and cause in the corpuscular-mechanical terms ultimately at stake in physico-mathematical proceedings *as Descartes had always previously understood them.* Working out universal mathematics in the later *Regulae* had unveiled the long standing though previously latent tension, between, on the one hand, purely discursive traditional natural philosophizing, and, on the other hand, the genuinely mathematical aspects of the method and universal mathematics (including now as well the Mersenne-like 'brand' of physico-mathematics on offer in the later *Regulae*). Moreover, the articulation of universal mathematics in the later *Regulae* had crystallized this tension as a clear, pressing and arguably insoluble problem. In the event, in the strongly 'methodological' context of the *Regulae,* the attractive epistemological and logical aspects of newly elaborated universal mathematics prevailed and Descartes slipped into its vocabulary to describe what his 'science of magnetism' would be like, thus managing to express the worst of both worlds: that is, to be ambiguous both about how a mixed mathematical law might be obtained, and whether he really desired a causal, corpuscular-mechanical account to round off such a significant inquiry.

Even though this aspect of the problem clearly pointed to the conclusion that discursive corpuscular-mechanical claims do not fit within the machinery of universal mathematics, Descartes perhaps did not immediately see or respond the difficulty when it was thus presented in the limited form of 'try to formulate micro-mechanical explanations of particular phenomena using universal mathematics'. In this he was perhaps helped along both by his now long standing posture about physico-mathematics, where corpuscular-mechanism lurked, un-grounded in itself, as the explanatory repertoire, and by the lure, and textual effects, of his own method discourse, as examined in Chap. 6.

However, the second mode of the problem did arguably register with him and helped persuade him to drop work on the *Regulae*, and turn to systematic corpuscular-mechanism with dualist metaphysical grounding. This is because the second aspect

This also clearly was part of Gerd Buchdahl's incisive reading (Cf Sect. 6.4 Note 18 and text thereto) — a very persuasive reading, as well, provided one is not also factoring in a picture of the specific logistical and legitimatory machinery of the later *Regulae*, as we are here. Hence, on our own present strict reading of what the universal mathematics of the later *Regulae* is about, we must conclude that its procedures will not really stretch so far, because the 'dimensions' then charitably in question would be neither observable, nor measurable, nor could the explanation take a properly mathematical form.

of the problem cut to the very rationale of the later *Regulae*, that is, to ground and legitimate the new, articulated universal mathematics itself. Here the issue was, in extension of the first, the simple fact that the o-p-p nexus doing that grounding and legitimating, consisted in a set of corpuscular-mechanical natural philosophical claims which, obviously, were not in the nature of findings in universal mathematics, and which could not be produced by means of its machinery. This raised the issue in the more pointed form of 'what grounds the o-p-p nexus', or more generally, 'what grounds corpuscular-mechanical discourse, corpuscular-mechanical natural philosophizing'. Descartes quite simply could not fail for long to see that the central ontological postulates and modes of procedure of his favored form of micro-mechanical natural philosophy were not guaranteed by the putative justificatory devices of the later *Regulae*, which work only at the 'Mersennian' level of phenomenological appearances grasped in mixed mathematical ways.

His theory of light and vision, and (implicitly) his mechanistic interpretation of neuro-physiology, depended upon the precise but hypothetical ascription of mechanical properties to unobservable corpuscles. Although a good deal of practical support was lent to these theories by their internal consistency, accordance with phenomena, and analogical basis in macro-mechanical contexts, one could not claim deductive certainty for them on the basis of the justificatory nexus of the *Regulae*. As we have seen, that justificatory machinery implicitly presupposed the practice of corpuscular-mechanical natural philosophy in order to ground universal mathematics, thus the latter could come to be seen as a methodological phantom posing in the place of the actual mechanistic natural philosophy which Descartes practiced and needed to be able to justify. In short, Descartes' previous physico-mathematics simply assumed that well grounded mixed mathematical results can be achieved—as in hydrostatics or optics—and that corpuscular-mechanical causes can be revealed via the geometrical representation of those mixed mathematical results. Now, in the later *Regulae*, he went to extremes to ground and legitimate (beyond Mersenne) a strictly phenomenological (Mersennian) physico-mathematics. However, his legitimating and grounding machinery was discursive, corpuscular-mechanical natural philosophy, which had no place in the revised physico-mathematics, there being no stateable procedures in the revived and expanded universal mathematics/method for generating such natural philosophical findings.

Thus, as we shall see, Descartes was soon to shift gears to escape these binds. He would jettison the universal mathematics he had just worked out in partial detail, along with any serious claim that what he was now doing derived from the method. Instead, he would begin to work on the composition of *Le Monde,* a system of inevitably qualitative and discursive corpuscular-mechanical natural philosophy and cosmology. He soon stepped forth as a systematic natural philosopher, a dealer in corpuscular-mechanical stories, which he now hoped he could draw together in a system, whilst bolstering the new whole, or at least its more fundamental parts, with some deeper, metaphysical grounding. At the same time, as we shall also see, key parts of his discursive natural philosophy would have physico-mathematical roots, for example, the principles of his dynamics, and even the conceptual core of his vortex theory.

In *Le Monde,* in the metaphysical speculations which he began to pursue around the same time, as well as in the *Principles* (1644) and the later correspondence, one detects a multi-level response to the difficulties emergent in the later *Regulae.* On the rhetorical and propagandistic level—that is, mainly in contexts dominated by talk about 'method'—Descartes sometimes continued to assert that his natural philosophy was 'geometrical' and that it had a deductive structure of explanation.[94] But in practice (that is, in constructing natural philosophical discourse), and sometimes by overt admission, he conceded that his system of corpuscular-mechanism had to be verbal and qualitative and that there was a necessarily hypothetical dimension in all explanations of particular phenomena, as well as in such important concepts as those of the elements.[95] The metaphysics and Voluntarist theology which he began to pursue after 1628 were intended, in part, to ameliorate the problem of the necessarily hypothetical status of detailed corpuscular models: Though the details of the system could not be rigorously 'deduced' from the principles demonstrated in metaphysics, metaphysical findings, for example, that the essence of matter is extension, placed important constraints upon what could, and could not be, asserted of corpuscular models, and they leant important extra legitimatory weight to necessarily hypothetical particular explanations.

7.6.2 Ambushed by the Unexpected Manifestation of the Problematic of Modern (Cartesian) Epistemology

The second set of problems in the newer portion of the *Regulae* grew out of the daring claim that the intellect directly intuits two-dimensional patterns carved into the corporeal substance of the common sense and imagination. The problem with this is probably obvious to just about anybody who is not busy perseverating on the wonderful

[94] For example, in the *Discours, II,* when he writes of the deductive inter-linking of all truths (AT, VI, p. 19; HR, I, p. 92) or, when he claims that parts of his physics were deduced from first principles (*Discourse, V,* AT, VI, p. 41, 63–4, HR, I, p. 106). Typical of such contexts are also: to Mersenne, 16 March 1640, AT, 111, p. 39; and *Principles III,* art 43 (but see the hypothetical tone of III, art 44), and IV, art 206 which makes strong deductivist claims but also wavers and waffles on their extent.

[95] See Chap. 6 note 19 for the consensus view on this matter in the literature. When dealing with this literature it is important not to slide into simply assuming that the young Descartes held the same sophisticated 'probabilist' position. That is, some may perhaps wish to say that Descartes already recognized in 1626–1628 the necessarily hypothetical status of his corpuscular models and was using the words 'deduce' and 'intermixture' in the loose sense (which he definitely adopted later) of 'plausibly explain'. The problem with this is that there is no evidence for such an interpretation in the work of the younger Descartes in physico–mathematics, universal mathematics or method. Surely it is therefore preferable to say that Descartes only began to see the problem of the status and grounding of corpuscular-mechanical explanations in anything like his later fashion as a result of the inability of the later *Regulae* to give him a physico-mathematical practice that was really mathematical and corpuscular-mechanical at the same time.

legitimatory results of the o-p-p nexus; that is, to almost anybody other than the enthused Descartes, during his likely excited inscription of the later *Regulae*: Human beings have immediate sense perception of certain 'objects' that seem to be in, or derived from, the physical world and that are not simply congeries of two-dimensional patterns. There is, for example, the perception of depth in three dimensions, which, despite the feats of Renaissance painting, can generally be differentiated in common experience from two-dimensional perspective representations. More generally there were all the qualities later called 'secondary', such as colors or the 'tones' perceived in the 'science of sound', which are immediately perceived as such, not as patterns.[96] If sense perception is, strictly speaking, the 'application' of the *vis cognoscens* to brain *loci,* then one cannot argue that some (or all) perceptions are not of patterns unless the epistemological implications of the theory are developed further than they are in the *Regulae.* That development begins to take place immediately after the abandonment of the *Regulae,* for example, in the first chapter of *Le Monde* and in the main lines of the new dualist metaphysics. There classical epistemology starts to take shape as a response to the overstated and vulnerable theory of perception and mental function in the *Regulae.*

Although classical atomism and Beeckman and Descartes' earlier corpuscular-mechanism harbored similar problems about the causes, ontological status and reference of 'secondary qualities', they remained latent for the most part. The focus of theoretical concern lay elsewhere, for example, in the struggle to strip the world of 'unintelligible' forms, qualities and powers. The *Regulae* are crucial in the development of the classical epistemological puzzles just because of their *unintended* precipitation of these issues. Descartes elaborated the o-p-p nexus in order to lend legitimatory weight to the machinery of universal mathematics; but the unexpected consequence of this strategy was to present the latent epistemological problems of atomism or corpuscularianism in a new light. If, in sensing, imagining and remembering, we (that is, *vires cognoscentes)* apply ourselves directly to mechanically impressed patterns and shapes, then it is not at all clear what the perception of 'secondary qualities' is or how it comes about. The theory assumes an unarticulated dualism of spiritual *vis cognoscens* and material brain loci. In its unarticulated state it left no room to account for the 'given-ness' of colors *qua* colors, tones *qua* tones, etc. *By mechanizing Kepler's theory of vision and building the o-p-p nexus, Descartes went so far in pursuit of his methodological and legitimatory goals that he unintentionally actualized the latent epistemological difficulties of his assumed and as yet unexplicated dualism of* vis cognoscens *and material brain loci in a way, and in a context, in which they could hardly be ignored.*

After dropping the *Regulae* in 1628, Descartes moved to meet these problems and in so doing unwittingly began to work on lines that led to his prominent but elusive place in modern philosophy. During his first six months in the United

[96] See N. Kemp Smith, op. *cit. pp.* 229–31.

Provinces in 1629 he began to work on systematically elaborating a strict ontological dualism of mind and body which allowed him to drive an ontological, and hence epistemological wedge between mental acts and their purely mechanical occasioning causes.[97] The doctrine was only fully presented, and hotly debated, with the publication of the *Meditations* and *Objections and Replies in 1641;* but its origins do go back to 1629, and very full hints of its epistemic import can be detected in the *Dioptrique,* published in 1637 but apparently written in large part by around 1630, and in the first chapter of *Le Monde,* presumably written at the same time.[98] In claiming that perceptions of secondary qualities are purely spiritual entities, modes of thinking substance to be precise, and that they are caused or 'occasioned' by certain corpuscular-mechanical states of affairs and that they do *not* arise from the direct inspection of corporeal patterns, Descartes was undermining the explicit *Regulae* doctrine of direct cognitive awareness, by developing its nascent and defocalized ontological dualism. This began to open up the universe of modern epistemological discourse.

Consider, for example, this striking passage from the *Dioptrique, partie iv:*

> ...we must note that it is only a question of knowing how they (the mechanical patterns which are formed in the brain) can enable the mind to perceive all the diverse qualities of the objects to which they refer, not of (knowing) how the images themselves resemble their objects.[99]

In part six of the *Dioptrique* this epistemological schema of purely mechanical disturbances 'instituted by nature' to 'occasion' purely mental 'ideas' is extended to the explanation of the perception of all the 'qualities' of vision, including location, distance and color.[100] A similar clear break with the later *Regulae* appears in the first chapter of *Le Monde.* After tentatively suggesting that there can be a difference between the sensation of light and the external cause of that sensation, Descartes leads the reader to the epistemological abyss:

> You well know that words bear no resemblance to the things they signify, and yet they do not cease for that reason to cause us to conceive of those things, indeed often without our paying attention to the sound of the words or to their syllables....Now, if words, which signify nothing except by human convention, suffice to cause us to conceive of things to

---

[97] To Mersenne, 25 November 1630, AT, I, p. 182.

[98] Descartes had begun to compose *Le Monde* in the fall of 1629 (To Mersenne, 8 October 1629, AT, 1, p. 23; see To Mersenne 13 November 1629, AT, 1, p. 70). In November 1630 he termed the *Dioptrique a* 'summary' of *Le Monde* (*To* Mersenne, 25 November 1630, AT, I, p. 179), though neither text was yet ready.

[99] ... nous remarquions qu'il est seulement question de savoir comment elles [patterns which are formed in the brain] peuvent donner moyen à l'ame de sentir toutes les diverses qualités des objets auxquels elles se rapportent, et non point comment elles ont en soi leur ressemblance.' (AT, VI, p. 113) LLA translation p.113

[100] AT, VI, pp. 130–1, 137–40.

which they bear no resemblance, why could not nature also have established a certain sign that would cause us to have the sensation of light, even though that sign in itself bore no similarity to that sensation?[101]

Here the naive, natural philosophically buttressed theory of direct intuitive 'application' is swept aside with the analogy:

$$\frac{\text{natural sign}}{\text{idea or perception of light}} = \frac{\text{word sign}}{\text{conception of object signified}}$$

The metaphysical arguments for ontological dualism are not given, but an ontology of mental events occasioned by mechanical states of affairs which they do not resemble must be in play, and with it the modern epistemological problems in principle begin to enter the field of possible discourse. Now, although here in Chap. 1 of *Le Monde* Descartes struggled to gloss this problem without getting embroiled in a full scale doctrine of mechanically occasioned, purely mental acts or ideas, it will be seen below in Chap. 8 that a good deal of Descartes' metaphysical construction in the period starting early in 1629 was indeed aimed at carrying through this epistemological reorganization, by erecting a new justificatory doctrine upon a now self-consciously metaphysical basis. In sum, he began to work out his dualism in detail right after dropping the *Regulae* project, although in the text of *Le Monde* he eschewed a full metaphysical buttressing, for reasons we shall canvass later, and only nodded and winked in the direction of the dualism residing below and propping up the text.

Finally, we note that there is considerable irony in these developments. The later *Regulae* had depended upon an implicit and untheorized dualism, presupposed in the application of the *vis cognoscens* to each of the brain *loci.* The o-p-p nexus, built on this implicit dualism, sinned against obvious experiential facts temporarily defocalized in the drive to create the machinery for universal mathematics. Descartes' mature position grew up around the attempt to accommodate those facts. He was driven to a formally elaborated, metaphysically enforced dualism, a dualism of mental and material substances, pushed far enough to detach 'ideas' from their corporeal-mechanical grounds. To make a long and continuing story short, this alleviated the difficulties emergent in the later *Regulae* but at the cost of virtually inventing the problems of classical epistemology. Hence the characteristic epistemological concerns of the mature Descartes arose neither from his corpuscularianism *per se,*

[101] Vous scavez bien que les paroles, n'ayant aucune ressemblance avec les choses qu'elles signifient, ne laissent pas de nous les faire concevoir, et souvent meme sans que nous prenions garde au son des mots, ni a leurs syllables ... Or, si des mots, qui ne signifient rien que par l'institution des hommes, suffisent pour nous faire concevoir des choses, avec lesquelles ils n'ont aucune ressemblance: pourquoy la Nature ne pourra-t'elle pas aussi avoir estably certain signe, qui nous fasse avoir le sentiment de la Lumière, bien que ce signe n'ait rien en soi, qui soit semblable à ce sentiment? (AT, XI, p. 4.) MSM 3 (cf SG 3-4)

nor from his early method; they arose as an unexpected consequence of the attempt to prop up the new, late *Regulae* version of universal mathematics by means of the o-p-p nexus and a related, implicit, and unarticulated ontological dualism.

7.6.3 Analytical Mathematics Works and Is Useful, But Resists Mapping onto the Legitimatory O-P-P Nexus

Descartes' endeavor in the later *Regulae* to immunize mathematics from sceptical doubts also faltered on the daring but clumsy legitimatory machinery. Descartes sought with the o-p-p nexus and logistic of extension-symbols to establish the ontological reference of the objects of mathematics and to justify the truth of its operations. However, some of the objects and operations of mathematics eluded the legitimatory machinery and thus helped to undermine the program of the later *Regulae.* Of the three areas of difficulty discussed in this section, this one is best exemplified in the text, and, in fact, the location of the precise point at which the text breaks off can be explained by reference to the mathematical difficulties of the justificatory program.[102]

The *Regulae* end with the mere enunciation of the titles of rules 19–21, where Descartes was about to embark on a discussion of the theory of algebraic equations. He abandoned the text at this point because he probably realized that the solution of quadratic and higher-order equations, or the extraction of square- and higher-order roots, would elude the excessively simple manipulation of lines and rectangles demanded by his legitimatory aims and doctrine. Such operations, he now probably recalled, require constructions utilizing circles or higher-order curves respectively, or—what is equivalent—a device such as the proportional compass, which is essentially a tool for drawing the curves and making the constructions. Descartes' difficulties in this regard, and his ineffectual evasions of them, are clearly apparent in rule 18.

We have already examined how, in rule 18, Descartes treats the operations of mathematics in terms of the manipulation of extension symbols, and how, in particular, multiplication and division were to consist respectively in the composition and decomposition of rectangles out of, or into, straight lines. While raising to a power bears a straightforward analogy to the process of multiplication, the extraction of roots is not so easily analogized to division, and Descartes falters at precisely this point. He writes:

> But in those divisions in which the divisor is not given, but only indicated by some relation, as when we are bidden to extract the square or cube root, then we must note that the term to

[102] For interesting articulation and reinforcement of the general tenor of the claims made in this section, compare Henk Bos' informative recent study, 'Descartes' Attempt, in the *Regulae*, to base the certainty of algebra on mental vision—A Conjectural Reconstruction' in *Proceedings of the 13th International Congress of Logic, Methodology and Philosophy of Science (Beijing August 9–15, 2007)* forthcoming.

be divided and all the others *must be always conceived as lines in continued proportion,* of which the first is unity, and the last the magnitude to be divided. The way in which any number of mean proportionals between this and unity may be discovered *will be disclosed in its proper place.* At present it is sufficient to have pointed out that according to our hypothesis those operations have not yet been fully dealt with here, since to be carried out they *require an indirect and reverse movement on the part of the imagination,* and at present we are treating only of questions in which the movement of thought is to be direct.[103]

Significantly, Descartes then goes on in the next paragraph to reassert that all the 'direct' operations (addition, subtraction, multiplication, powering and simple division) are to be carried out by composition and decomposition of line lengths and rectangles.[104]

Descartes is struggling here against the collapse of his justificatory enterprise (but not, assuredly, the collapse of his ability to 'do' this sort of mathematics). Note, first, how he slides back into the interpretation of root extraction as the finding of proportional magnitudes. This begs the question, since the point of rule 18 has been to show how multiplication, powering, and simple division can be construed as operations upon proportional magnitudes *and* as manipulations upon extension-symbols, the latter being the important justificatory maneuver. One can judge the degree of Descartes' discomfort by recalling that in rule 17 he claimed that algebraic symbolism permits 'indirect' problems to be treated in 'direct' form. However, when in rule 18 he attempts to legitimate the mathematical operations, he explicitly defers treatment of the extraction of roots which is now taken as an 'indirect' operation of the imagination. This contradicts rule 17, and it seems to concede that root extraction eludes the simple machinery of rule 18, because it requires complex constructions using circles or higher-order curves. Seeing the futility of seriously trying to fit the theory of equations to the legitimatory machinery, Descartes simply abandoned the text after jotting down the titles of his first few algebraic' rules.

Here again, as in the previous two cases, the narrow legitimatory machinery revealed its bankruptcy, and once again Descartes was soon driven back upon metaphysical construction to meet the difficulty. Henceforth the truth of mathematics was vested in metaphysical arguments concerning God's guarantee of the truth of clear and distinct intuitions. Algebra and arithmetic, as well as geometry, could then have direct grounding in metaphysics, and mathematical objects such as negative or imaginary roots would be justified by their clear and distinct structural relations to other sorts of objects, notwithstanding the fact that they had eluded the justificatory machinery of the imagination in the later *Regulae.*

[103] *Regulae,* Rule 18, AT, X, p. 467; CSM p.75; HR, I, p. 76; HR translation preferred, emphasis added.

[104] *Ibid.* AT, X, pp. 467 1.17 to 468 1.6 (Division is mentioned explicitly in this passage).

7.7 Conclusion—The Project of the Later *Regulae* and the Inflection of Descartes' Agenda and Identity Toward Systematic Natural Philosophy and Metaphysical Grounding

Commentators, led by the perceptive studies of Brunschvicg, have often noticed the contrast between the heavy reliance upon the imagination in the mathematics of the *Regulae* and the more abstract, algebraic and structural tone of the theory of equations in Book III of the *Geometry.*[105] While the contrast is valid, it is not quite what needs to be explained. To understand what happens after the demise of the *Regulae,* one must separate Descartes' justificatory aims from his actual mathematical practice. In technical terms, Descartes' mathematics always had a strongly algebraic flavor. This was less pronounced in his first studies, surveyed in Chap. 5, but was increasingly the case in his mathematical work from 1619 to 1625, and then from 1628 to 1635, when crucial developments leading to the *Geometry* occurred. In the *Regulae* Descartes was not trying to supplant or alter his modes of mathematical practice. Rather, he was trying to show how they could be legitimated, and where they would fit into universal mathematics. When that enterprise failed, he turned to a new legitimatory strategy in his systematic metaphysics; but his mathematical work remained, before and after the demise of the *Regulae,* strongly and increasingly algebraic in its analytical orientation.

To conclude, then, I would suggest that the later *Regulae* were related to Descartes' subsequent work in metaphysics and systematic corpuscular-mechanism in the following ways: In the later *Regulae* Descartes had attempted, in accord with his own view of 1619, to conflate pure, practical, mixed- and physico-mathematics in 'universal mathematics', now to be constructed in detail. But the newly articulated universal mathematics was in an important sense a phantom discipline. It neglected or distorted Descartes' actual practice in analytical mathematics. Additionally, it brought into view the fact that his underlying commitment to corpuscular-mechanical explanation in physico-mathematics—going back to 1619—would not fit the pattern and machinery of the elaborated universal mathematics. The distortions occurred largely because of the way universal mathematics had to be formulated in the light of the legitimatory aims. The o-p-p nexus and the logistic of extension symbols were designed to forge the unity of universal mathematics, and to forestall scepticism while immunizing the new discipline from the dangers of neo-Platonism and mysticism. The daring attempt failed with the appearance of unintended new problems about the relations between method and corpuscular-mechanical explanation; about 'epistemology', that is about how the o-p-p nexus really was supposed to work; and about the legitimation of mathematics and its operations. The phantom

[105] For example, L. Brunschvicg (1927) and (1922) 106–123; Mahoney (1971).

discipline of universal mathematics disappeared and Descartes' practice was more clearly re-directed back toward the corpuscular-mechanism and the analytical mathematics which had always interested him. But the legitimatory aims, originally spurred by the Parisian turmoil, remained and were intensified by the discovery of the new difficulties. Descartes turned to the grand style of constructive metaphysics, beginning in the next few years to elaborate the dualism implicit in the later *Regulae*. Hence, it should be clear that we are not saying that Descartes pursued dualist metaphysics after 1628 in order to save the *Regulae* project and its articulated universal mathematics. That program was now rejected. Over the next few years it became increasingly obvious to Descartes that he practiced a discursive corpuscular-mechanism in natural philosophy, and an increasingly abstract and analytical style of pure mathematics, and that both of these very real, and very difficult pursuits required a daring new type of metaphysical grounding. Additionally, he came to realize that he might be able to provide that metaphysical grounding from a common source in a dualism, to be worked out in detail, beginning in part from the tacit rudiments of it embedded in the *Regulae* doctrine itself

To a considerable extent, and perhaps in Descartes' own view, the new metaphysical buttressing of mathematics and corpuscular-mechanism constituted an advance on the program of the later *Regulae*.[106] Though dogmatic and not commonsensical in tone, and so abrasive of some Parisian tastes, it not only carried out the legitimatory intentions and met (or seemed to meet) the new problems; but it also subsumed more accurate representations of Descartes' researches—his actual and now systematically pursued corpuscular mechanism, and his analytical mathematics in all its challenging algebraic abstraction. Descartes' original physico-mathematics envisioned the 'seeing' or eliciting of causes from sound mixed mathematical results. Now he would be a fully committed, systematizing corpuscular-mechanical natural philosopher. His natural philosophy would have metaphysical warrant and would benefit from the presence within it of fruits of physico-mathematical research in optics and hydrostatics, ripened in his concepts of the dynamics of corpuscles and his theory of vortex celestial mechanics. Accordingly, in our final five substantive chapters we focus on how and why Descartes came to compose an initial system of corpuscular-mechanism in *Le Monde*; what that system was intended to accomplish; just how far it succeeded, compared to the later *Principia philosophiae*; and, what both *Le Monde* and the far more ambitious *Principia* meant within the overheated natural philosophical contest of the second generation of the seventeenth century.

---

[106] Brunschvicg (1927) 292, held that Descartes' metaphysics may be seen as an attempt to mediate between the increasingly divergent views of space which corresponded respectively to his newly extended abstract mathematics and to his mechanical corpuscular physics. As such, the metaphysics would have served to integrate the mathematics and physics on the justificatory plane, just as had been attempted in a more 'scientific', that is, natural philosophical, guise in the later *Regulae*.

References

Works of Descartes and Their Abbreviations

AT=*Oeuvres de Descartes* (revised edition, 12 vols.), edited by C. Adam and P. Tannery (Paris, 1964–76). References are by volume number (in roman) and page number (in Arabic).

SG=*The World and Other Writings*, edited and translated by Stephen Gaukroger (Cambridge,1998).

MM=*René Descartes, The Principles of Philosophy*, translated by V. R. Miller and R. P. Miller (Dordrecht, 1991)

MSM=*Rene Descartes, Le Monde, ou Traité de la lumière*, translated by Michael S. Mahoney (New York, 1979).

CSM(K)=*The Philosophical Writings Of Descartes*, 3 vols., translated by John Cottingham, Robert Stoothoff, and Dugald Murdoch, and (for vol. 3) Anthony Kenny, (Cambridge, 1988) References are by volume number (in roman) and page number (in arabic).

HR=The Philosophical Works of Descartes, vol I translated by E.S. Haldane and G.R.T. Ross (Cambridge, 1968 [1st ed. 1911])

Other

Alquié, F. 1950. *La découverte métaphysique de l'homme chez Descartes*. Paris: PUF.

Blanchet, L. 1920. *Les antécédents historiques du 'Je pense, donc je suis'*. Paris: Felix Alcan.

Boutroux, P. 1900. *L'imagination et les mathématiques selon Descartes*. Paris: Felix Alcan.

Brunschvicg, L. 1922. *Les Etapes de la Philosophie Mathématique,* second edition. Paris: Felix Alcan.

Brunschvicg, L. 1927. Mathématique et métaphysique chez Descartes. *Révue de Métaphysique et de Morale* 34: 277–324.

Clavelin, M. 1974. *The natural philosophy of Galileo*. Cambridge, Mass.: MIT.

Crombie, A.C. 1967. The mechanistic hypothesis and the scientific study of vision. In *Historical aspects of microscopy*, ed. S. Bradbury and G.E. Turner, 1–112. Cambridge: CUP.

Crombie, A.C. 1990. *Science, optics and music in medieval and early modern thought*. London: Hambledon.

Dear, Peter. 1988. *Mersenne and the learning of the schools*. Ithaca: Cornell University Press.

Espinas, A. 1906. Pour l'histoire du cartesianisme. *Révue de Métaphysique et de Morale* 14: 265–293.

Gäbe, L. 1972. *Descartes Selbstkritik. Untersuchungen zur Philosophie des Jungen Descartes*. Hamburg: F. Meiner.

Gadoffre, G. 1961. *Descartes' Discours de la méthode*, 2nd ed. Manchester: Manchester University Press.

Gaukroger, Stephen. 1978. *Explanatory structures*. Harvester: Hassocks.

Gillmor, C.S. 1971. *Coulomb and the evolution of physics and engineering in eighteenth century France*. Princeton: Princeton University Press.

Gilson, E. 1913. *La liberté chez Descartes et la théologie*. Paris: Alcan.

Gouhier, H. 1924. *La pensée religieuse de Descartes*. Paris: J. Vrin.

Gouhier. H. 1954. 'La Crise de la théologie au temps de Descartes', *Révue de Théologie et de Philosophie,* ser. 3, 4: 19–54.

Hine, W.L. 1967. *The interrelationship of science and religion in the circle of Marin Mersenne.* Unpublished Ph.D. dissertation. University of Oklahoma.

Kepler, Johannes. 1938ff. *Gesammelte Werke*, ed. M. Caspar. Munich: C.H. Beck.
King, W.J. 1964. The quantification of the concepts of electrical charge and electric current. *The Natural Philosopher* 11: 107–125.
Klein, Jacob. 1968. *Greek mathematical thought and the origin of algebra*. Cambridge, MA: MIT Press.
Lenoble, Robert. 1943. *Mersenne ou la naissance du mécanisme*. Paris: J.Vrin.
Lindberg, D. 1976. *Theories of vision from al-Kindi to Kepler.* Chicago: University of Chicago Press.
Mahoney, M.S. 1971. Descartes. In *Dictionary of scientific biography*, vol. 4, ed. C.C. Gillespie, 56–57. New York: Scribners.
Marion, Jean-Luc. 1977. (ed.)and Trans. *René Descartes, Regles utiles et claires pour la direction de l'esprit en la recherche de la verité.* La Haye: Martinus Nijhoff.
Mersenne, Marin. 1625. *La vérité des sciences*. Paris: Toussainct du Bray.
Milhaud, Gaston. 1921. *Descartes savant*. Paris: Alcan.
O'Neil, B.E. 1967. Epistemological direct realism in Descartes philosophy. Unpublished Ph.D. dissertation. University of California, Berkeley.
Pintard, *R*. 1943. *Le libertinage érudit,* 2 vols. Paris: Boivin.
Popkin, Richard. 1964. *The History of Scepticism from Erasmus to Spinoza*. Berkeley: University of California.
Schuster, J.A. 1977. *Descartes and the scientific revolution 1618–34: An interpretation,* 2 vols. unpublished Ph.D. dissertation, Princeton University.
Sirven, J. 1928. *Les années d'apprentissage de Descartes*. Albi: Imprimerie Coopérative du Sud-Ouest.
Smith, Norman Kemp. 1952. *New studies in the philosophy of Descartes*. London: Macmillan.
Spink, J.S. 1960. *French free thought from Gassendi to Voltaire.* London: Athlone Press.
Staquet, Anne. 2009. *Descartes et le libertinage*. Paris: Hermann.
Straker, S. 1970. *Kepler's Optics*. Unpublished Ph. D dissertation, Indiana University.
Torero-Ibad, Alexandra. 2009. *Libertinage, science et philosophie dans le matérialisme de Cyrano de Bergerac*. Paris: Honoré Champion.
Weber, J.-P. 1964. *La constitution du texte des Regulae*. Paris: Société d'édition d'enseignement supérieur.

Chapter 8
Reinventing the Identity and Agenda: Descartes, Physico-Mathematical Philosopher of Nature 1629–1633

8.1 The Problem of Descartes' Career 'Inflection Point' and How to Approach It

> Perhaps you find it strange that I have not persevered with some other treatises I began while I was at Paris. I will tell you the reason; while I was working on them I acquired a little more knowledge than I had when I began them, and when I tried to take account of this I was forced to start a new project rather larger than the first.[1]

Thus Descartes wrote to his friend Marin Mersenne from the United Provinces early in 1630. There are as many ways of decoding Descartes' customarily understated and partially masked revelation as there are views of his career and trajectory. For example, if one still cared to indulge in the 'method-centric' myth of Descartes' career, which he himself initiated first in private correspondence and conversation and later publicized in the *Discours*, then we might well say that this letter shows him moving from the articulation of the method in the *Regulae*, toward application of the method in his embryonic projects in corpuscular-mechanism and dualist metaphysics, both products of exactly this period. Or, perhaps, following the perceptive Léon Brunchschvicg, as cited at the end of the previous chapter, we might speculate that Descartes was commenting upon how his mathematical practice was moving beyond the strictures of the *Regulae*, which tied operations down to manipulations of geometrical representations of quantities, toward the more purely analytical outlook of Book Three of the *Géométrie*. Clearly, on the basis of our findings thus far, neither of these options is correct, nor do they satisfactorily describe Descartes' trajectory, identity and

[1] To Mersenne 15 April 1630 AT I, 137–8. 'Que si vous trouvés estrange de ce que j'avais commence quelques autres traités estant à Paris, lesquels je n'ai pas continués, je vous en dirai la raison: c'est que pendent que j'y travaillais, j'acquerais un peu plus de connaissance que je n'en avais eu en commencent, selon laquelle me voulant accommoder, j'estais contraint de faire un nouveau projet, un peu plus grand que le premier'.

J. Schuster, *Descartes-Agonistes: Physico-mathematics, Method & Corpuscular-Mechanism 1618-33*, Studies in History and Philosophy of Science 27,
DOI 10.1007/978-94-007-4746-3_8, © Springer Science+Business Media Dordrecht 2013

agenda leading up to the attempt to write the later *Regulae*, nor, *a fortiori*, what became of that trajectory, identity and agenda, with the obvious failure of the project of the later *Regulae*, due to the internally generated contradictions, tensions and problems we have just canvassed.

Between late 1629 and 1633 Descartes was engaged in constructing his first system of natural philosophy, *Le Monde*, as well as devising the first skeletal lines of his dualist metaphysics and some elements of a voluntarist theology meant to buttress parts of that natural philosophical system.[2] He had never previously engaged in projects of this type. By the same token, work on the *Regulae* stopped, never to be resumed again, meaning that both the method and the physico-mathematics supposedly contained within the *Regulae*, were never again further articulated. Additionally, Descartes never again represented himself simply as a 'physico-mathematician' or practiced the style of piecemeal, problem oriented physico-mathematics that had marked his earlier years. Things had changed for René, although to be sure, it is not as though his mature intellectual persona emerged all at once in these years. His career after 1633 is marked by numerous further twists and reorientations, increasingly keyed to public debate and controversy—he remained Descartes *agonistes* to the end.[3]

Moreover, continuities persisted, as they do in real life, as opposed to the over dramatized narratives of historians of crisis and rupture. The Descartes of *Le Monde* in 1633 is not the Descartes of the *Meditations* and *Principles of Philosophy* of the mid 1640s; yet, that later Descartes was indeed a systematic philosopher of nature and metaphysician, just what Descartes became between 1629 and 1633. Similarly, and more pertinently for the present study, the inflection of Descartes' agenda and identity between 1629 and 1633 had not completely erased or cancelled his earlier incarnations, practices or products. They lived on, sifted, revised and retranslated, as we shall see. His new systematic natural philosophy bore definite marks of his aims and results in physico-mathematics; his dream of method, now taken more cynically as a useful mode of public packaging and expression, survived to become, misleadingly, his veritable intellectual trademark on the public stage; and his physico-mathematical optics, and the principles of the dynamics of corpuscles it suggested and supported, became central pillars of his system of mechanical philosophy. In constructing *Le Monde*, he borrowed and transformed so much, in terms of style and content, from his older physico-mathematics, that the title of the present chapter signals this point, while Chaps. 10 and 11 dealing with the conceptual structure of *Le Monde*, will confirm this claim in detail.

---

[2] Given the limitations and foci of the present study, we reiterate that the medical, anatomical and physiological endeavors of Descartes in these years, very much part of his larger natural philosophical project, will not be canvassed, apart from brief mentions.

[3] Works that make clear the continuing patterns of change in Descartes career after 1633 include most notably Gaukroger (1995), Clarke (2006) and most recently Machamer and McGuire (2009). Of these only Gaukroger gives sustained and useful attention as well to the earlier period, down to 1633. Cf. Schuster (2009).

The issue is, How shall we go about describing and explaining this inflection in Descartes' career? Since this study is not a full intellectual biography of Descartes, our focus remains upon Descartes' evolution from physico-mathematician into a systematic philosopher of nature (of physico-mathematical coloring); our main goals being to analyze that first system of corpuscular-mechanism, and its systematic aims, strengths and weaknesses as part of the natural philosophical agon of the age, and to understand its emergence from his earlier projects, agendas and senses of intellectual identity. Hence, we are not going to examine every documentable source and event in the years in question, nor certainly enter into debate about other large interpretations of that evidence, such as, for example, those which focus chiefly on the emergence of Descartes the metaphysician, the theologian, the conqueror of scepticism, or the triumphant methodologist. But, since good intellectual biographies, and topical studies of the sort mentioned, do exist, we may rely upon them about basic evidential and chronological issues, even if we do not always share their aims and conclusions. The present analysis will remain consistent with the known evidence, of which much more certainly survives for this period than for the earlier years we have studied thus far. In turn, it is hoped that our more analytical approach, and focus on the emergence and structure of Descartes' first system of natural philosophy, will perhaps help others read, interpret and select the relevant primary and secondary sources for the benefit of studies with other aims and foci.

Our strategy involves what might be termed a structural and layered approach to historical understanding, rather than either a sustained single narrative, or univocal thematic argument (about, for example, a sceptical, metaphysical or theological crisis or break, or methodological continuity, in these years). Simple chronological narrative, made possible by the relative wealth of evidence we have for Descartes in these years, compared to his earlier career, simply does not explain how this inflection relates to our picture of his earlier endeavors, and it threatens constantly to elevate some particular event into the decisive moment in the inflection (or crisis), supplying either the causal motivation or intellectual content, or both. Univocal thematic arguments about this or that crisis or break, to which I myself had an inclination in earlier work on Descartes, simply do not get to grips with the inner dynamic of Descartes, physico-mathematician, mathematician, methodologist and natural philosopher, which we have been trying to unpack, describe and explain. Such thematic accounts also tend to batten upon select bits of the chronological narrative, to find the 'crunch' events where motive and or content were produced or elicited. These approaches lose touch with the reconstructable lived reality of practicing simultaneously in several contested, fluid and dynamically interrelated intellectual fields—the situation of Descartes and other contending natural philosophers of his generation. Similarly, if we are going to place Descartes' (and others') natural philosophical struggles into a larger pattern of, for example, sceptical crisis and resolution, or some general crisis of the seventeenth century, then that should be done after we have sufficiently conceptually and historiographically nuanced studies of natural philosophical dynamics and contention in these years, rather than by imposing very grand templates from the outside onto selected slices of evidence about particular natural philosophers.

There are three layers in our structural approach to the problem of explaining and describing Descartes' career inflection 1629–1633. They are:

[1] A set of fundamental intellectual agendas and projects built into, or providing framework for, his emerging systematic natural philosophy. These concerned dualist metaphysics, voluntarist theology, and what we shall term Descartes' 'holistic, or 'plenist–realist' style of explanation in natural philosophy. This style will find its central exemplar and achievement in his vortex theory of celestial mechanics in *Le Monde*, and contrasts with the abstracting and idealizing style of explanation typical of the mixed mathematical sciences and of the mechanics of Galileo, which was to appear in 1638.

[2] A set of events or interpersonal encounters which are quite disparate and not capable individually or collectively of explaining his career inflection and its products, but which arguably did shape moments in the inflection in particular directions rather than others.

[3] A careful chronology of the evolution of the text of *Le Monde* in relation to the events picked out in [2] as well as the agendas and projects picked out in [1].

An important aim of this strategy of explanation is to come to grips with the somewhat paradoxical facts that, on the one hand, the surviving correspondence suggests that Descartes slipped and stumbled his way toward the composition of *Le Monde*, as we shall see when dealing with [3] below, whilst, on the other hand, the resulting work displays a considerable, and often underestimated degree of systematic design and conceptual coherence, to be demonstrated in Chaps. 10 and 11.

8.2 Fundamental Intellectual Agendas and Projects 1629–1633

We turn first to three interrelated intellectual agendas/projects initiated in the same period during which *Le Monde* was germinating: explicit work on dualist metaphysics, excursions in voluntarist theology, and examples of—and reflection upon—'plenist realism'. Each of these gives explicit or implicit framing or grounding to one or another central element of the emerging system of corpuscular-mechanism. They are each new to Descartes' concerns in this period, and unsurprisingly, they all reflect attempts to deal with the problems which had scuttled the project of universal mathematics in the later *Regulae*. As such they link the failure of the *Regulae* to the creation of the system of natural philosophy, and they, along with the composition of *Le Monde* itself, provide the very matter and form of Descartes' career inflection—they are the inflection in process and substance at the deepest intellectual level.

Two of the three agendas ran in chronological tandem with the emergence and execution of the project of *Le Monde*. The voluntarist theology was initiated in 1630, in parallel with the natural philosophical moves leading directly toward *Le Monde*. It had clear and explicit legitimatory roles in the newly emergent natural philosophy, as well as for mathematics, both in a very well defined 'post-*Regulae*' tenor, as we shall see in Sect. 8.2.2 below. The plenist realism emerged with the composition of *Le Monde*, being part of its very explanatory style and substance, as discussed in Sect. 8.2.3. The third agenda and project, explicit work in dualist metaphysics, differs in that Descartes began work on it before the events of mid and late 1629 which crystallized into the project of *Le Monde*. It is the only one of the three agendas whose initiation pre-dates the processes leading to *Le Monde*. The dualism was therefore initially aimed at addressing problems emergent in the *Regulae*; but it soon came to play an important grounding role behind *Le Monde*, once it began to form.[4] That is, Descartes' embryonic dualist metaphysics eventually resided not so much inside the text of *Le Monde*, as immediately behind it (whilst also being an indispensible response to the problems that derailed the *Regulae*), and it is to this crucial agenda and project that we first turn.

8.2.1 The Emergence of Cartesian Metaphysical Dualism

We know that an embryonic ontological dualism lay behind the justificatory machinery of the later *Regulae*, involving, as we described it in the previous chapter, 'an implicit and untheorized dualism presupposed in the application of the *vis cognoscens* to each of the brain *loci*'. We also know that in this unarticulated state, as embedded in the doctrine of the o-p-p nexus, the dualism actually created difficulties for Descartes' universal mathematics, difficulties we foreshadowed he would address during his period of career inflection by explicit construction of a systematically articulated dualism. In November 1630, he wrote to Mersenne that he had indeed begun work on what he termed 'a small treatise of metaphysics' during his first few months in the United Provinces. The 'principle points' of the work he claimed to be to prove the existence of God and the immortality of the soul, when separated from the body.[5]

[4] Let us put this into perspective by first noting how Gaukroger (1995) persuasively and innovatively argued with regard to the mature Descartes that the *Meditations,* with their formal presentation of the dualist ontology, were integral to, and motivated by, Descartes' natural philosophical project, fully exposed to the public shortly afterward in the *Principles*. The strategy was to 'establish (metaphysically) the unique legitimacy of a particular way of pursuing natural philosophy without raising a single natural philosophical question' [p.345, Cf. *ibid.* p.352 and Schuster (1995) p.135]. Now, *Le Monde* does not come forward with its metaphysical legitimation on display in the text, but Descartes was assuredly working on it, and had it in reserve behind the text. And, even if we note that the early metaphysical work was stimulated by the failures of the *Regulae*, recall that one of those failures had to do with the problem of grounding the corpuscular-mechanistic models that had been deployed within what we have termed the o-p-p nexus. For a firm, scholarly statement about the significance of Descartes' attempted metaphysical grounding of his mechanical philosophy, and the early provenance of his first attempts in that direction, see Henry (2004).

[5] Descartes to Mersenne 25 November 1630, AT I 182, 'J'eprouverai en la Dioptrique si je suis capable d'expliquer mes conceptions, et de persuader aux autres une verité, après que je me la suis persuadé; ce que je ne pense nullement. Mais si je trouvais par experience que cela fût, je ne dis pas que quelque jour je n'achevasse un petit Traitté de Metaphysique, lequer j'ai commencé étant en Frize,

In discussing Descartes' metaphysical work of the late 1620s, Charles Adam suggested that Descartes then possessed something approaching a complete exposition of the *Meditations*.[6] However, Descartes' own allusion to this early work would indicate that he then possessed at most the rudiments of a metaphysical system sufficient in his opinion to support the propositions he mentions and their logical precursors. So, on the basis of the later argument of the *Meditations* we might conjecture that in 1629 Descartes' metaphysics embodied in draft form the *cogito,* the criterion of clearness and distinctness, the proof of the existence of God, a doctrine of innate ideas and the mind-body dualism linked to the identification of matter and extension. In addition, for reasons discussed below, Descartes had probably begun to articulate a distinction between 'judgments' and the immediate intuitions of appearances which had played a central justificatory role in the *Regulae.* No evidence whatsoever supports the assumption that the full text of the *Meditations* was then in hand. This is especially true in regard to the eventual contents of the Sixth Meditation. Although Descartes could well have entertained the demonstration of the existence of material bodies at this time, the detailed analysis of internal sensations and passions of the soul, as well as the 'metaphysicalization' of the mind-body union, appear to be concerns which only matured in the 1640s. The best way to reinforce these conjectures, and their main purpose in any case, is to try to identify the potential justificatory roles of those pieces of the metaphysics which, on the basis of Descartes' letter of November 1630, seem to date from the period around late 1629–1630.

In the first place, let us recall the damaging ontological dilemma Descartes uncovered in articulating his universal mathematics in the later *Regulae.* The legitimatory optics–physiology–psychology (or o-p-p) doctrine and logistical machinery of universal mathematics envisioned a science of correlation of macro-geometrical measures of physical 'dimensions' or properties, whilst the o-p-p doctrine itself, and Descartes' preferred mode of matter/cause explanation in physico-mathematics, involved claims about corpuscular-mechanical states of affairs, indeed a tacit commitment to an unsystematized corpuscular-mechanical natural philosophy. Now, to work out in more detail the mind/body distinction with the concomitant assertion of a metaphysically grounded doctrine of matter-extension, would seem to signal an attempt by Descartes to resolve the ontological dilemma of the *Regulae* in

---

et dont les principaux points sont de prouver *l'existence de Dieu* et celle de nos ames, lors qu'elles sont separées du cors, d'ou suit leur immortalité. Car je suis en colère quand je vois qu'il y a des gens au monde si audacieux et si impudens que de combattre contre Dieu.' It is clear from Descartes' last sentence that he may well have been thinking of his 'treatise' as a direct response to the apologetical aspect of the sceptical challenge. That does not detract from the overriding argument of this chapter that the dualist metaphysics, and voluntarist theology, were being explored for their value in grounding and legitimating the emergent corpuscular-mechanical natural philosophy, once by mid 1629 Descartes began his work leading in that direction.

[6] Adam's biography of Descartes in the first Adam and Tannery (1897ff) AT XII 130–42. Adam's summary of the putative content of the treatise would virtually exhaust the eventual content of the *Meditations*.

favor of corpuscular-mechanism, rather than continuing with the project of universal mathematics per se. The more explicitly worked out metaphysical dualism would firmly commit Descartes to a corpuscular-mechanical ontology, by establishing that reality beyond the mind consists merely of matter-extension. Henceforth the seeking of knowledge of nature would proceed as 'natural philosophy', not as 'universal mathematics'; that is, one proceeds by reducing phenomena to the consideration of the shapes, sizes and motions of corpuscles of this matter-extension: And that is to say [1] that the explanatory mode of his original physico-mathematics—where the causal register is corpuscular–mechanical—visible as early as 1619, had won out over the elaborated machinery of universal mathematics in the later *Regulae* or what we termed his new 'Mersennian' version of physico–mathematics; and [2] that the needed grounding for corpuscular-mechanical discoursing about matter and cause would come from the dualist metaphysics. The issue thereby arising would have been this—how far and in what way does the dualist metaphysics ground corpuscular-mechanical natural philosophizing at all? This question would follow Descartes the rest of his career, receiving different answers in different rhetorical contexts, whilst slowly maturing to a considered statement of the matter in the latter portions of the *Principles of Philosophy* of 1644.[7] Here we need to consider how far matter-extension could ground the micro-mechanism of *Le Monde*, and why, if it had some legitimatory role, intended and designed by Descartes, was it not systematically and explicitly inscribed into the text of *Le Monde*, rather than being systematically left out?

The doctrine of matter-extension grounds Descartes' corpuscular-mechanism in two general ways. First it bids to exclude from consideration any alternative theory of matter, be it Aristotelian, Stoic or neo-Platonic. Secondly, it strongly suggests that corpuscles, micro-fragments of extension, can only possess the geometrico-mechanical properties ascribed to matter-extension in general. The problems, tensions and ambiguities of the metaphysical grounding strategy arise at the next level, when one asks about the logical status of particular assertions about the corpuscles and their behaviour: are the laws of motion derivable from the doctrine of matter extension; are Descartes' claims about the existence of three types of particle, or elements so derivable, or are they more speculative, hypothetical and merely plausible assertions. The latter point holds even more strongly for very particular claims about corpuscular-mechanical models meant to explain particular types of phenomena. Can such detailed explanatory models for particular genres of phenomena be deduced from the nature of matter-extension? As mentioned earlier, Descartes displayed an ambivalence and ambiguity about these issues, only throwing more definite light on them in the *Principles of Philosophy* of 1644. He had an unchecked proclivity, in certain rhetorical contexts, to assert the strict deducibility of the above matters from his metaphysics.[8] Yet, the fluidity and hypothetical nature of particular

7 See Chap. 6 Note 19.

8 *Discours de la méthode*, partie 5, AT VI 43: 'Further, I revealed what were the laws of nature; and basing my reasoning on no other principle than the infinite perfections of God, I set out to prove all those laws about which one might have had some doubt, and to show that they are such that even if God had created many worlds, there could not be any in which they could have failed

corpuscular-mechanical claims would have been obvious to him, since he created and modified them and experienced shifts, doubts and changes about them. (And he certainly knew that his machinery of method and universal mathematics in the later *Regulae* did not permit deduction of them.) In the *Principles of Philosophy* he finally conceded the merely 'moral certainty' of particular corpuscular-mechanical models, asserting of course that they were constrained by, and consistent with, the underlying metaphysical principles.[9] Though modern scholars tend to agree that this was his mature position on detailed models, there is still debate about whether Descartes continued to think that higher level corpuscular-mechanical claims, say about the elements, the formation and structure of vortices and the like, could in principle be deduced from his metaphysics.

Le Monde marks an early stage in these considerations. As we shall see later, Descartes strongly hints that the nature and existence of the three elements follows directly from the nature of matter extension, given that God has chosen to inject a certain quantity of force of motion into it; but, of course, no deduction is offered. Similarly, he suggests that the formation of an indefinitely large number of stellar-planetary vortices also follows directly, given the existence of a conserved quantity of force of motion, and the nature of the elements.[10] So, the implied grounding role of the doctrine of matter-extension is quite striking. However, as is well known, *Le Monde* does not contain any strictly metaphysical argument about dualism; it does not begin with a précis of that 'little treatise of metaphysics'; and indeed, as we learn in the next chapter, the entire work eschews anything looking like Scholastic pedagogy and structure, opting at first for a breezy, commonsensical '*honnête homme*' style, which then gives way to an exercise in explication of the core of the natural philosophy and cosmology under the guise of a fable. We will examine both these moments in the text in the next two chapters. What should be resisted is the temptation to conclude that *Le Monde* has no metaphysical backing or grounding, just because the most fundamental steps in Descartes' dualism, which he arguably already possessed and had worked out for this purpose, do not appear in the text.

to be observed. After that, I demonstrated how the greater part of the matter of this chaos must, in consequence of these laws, be disposed and arranged in a way which made it similar to the heavens above us; how. at the same time, some of its parts had to compose an earth, some others planets and comets, yet others a sun and fixed stars' (Maclean 2006, 36–7). Cf. Descartes to Mersenne 15 April 1630, AT I 143–44, 'Or j'estime que tous ceux à qui Dieu a donné l'usage de cette raison, sont obligés de l'employer principalement pour tâcher à le connaître, et à se connaître, eux-mêmes. C'est par là que j'ai tâché de commencer mes études; et je vous dirai que je n'eusse jamais su trouver les fondements de la physique, si je les eusse cherchés par cette voie.' Also, Descartes to Mersenne April 1634, AT I p.285, 'Or, je vous dirai que toutes les choses que j'expliquais en mon traité, entre lesquelles était aussi cette opinion du mouvement de la terre, dépendaient tellement les unes des autres, que c'est assez de savoir qu'il y en ait une qui soit fausse, pour connaître que toutes les raisons dont je me servais n'ont point de force; et quoique je pensasse qu'elles fussent appuyées sur des demonstrations très certaines, et très évidentes, je ne voudrais toutefois pour rien du monde les soutenir contre L'autorité de l'Eglise'.

[9] See Chap. 6 Note 19 and accompanying text; Chap. 7 Note 95 and accompanying text.

[10] These matters are discussed below Sect. 9.5.3 with textual reference to *Le Monde*.

The doctrine of matter-extension certainly does appear, doing quite heavy work, and in turn its own grounding in metaphysical argument is tacit, held in reserve in the work Descartes had already done in parallel with the construction of *Le Monde*. Moreover, as we learn in Sect. 8.2.2 below, Descartes' second new set of legitimatory tools, deriving from his version of voluntarist theology, are quite obviously present in *Le Monde*, underpinning crucial parts of the natural philosophy, and they too were worked out in the United Provinces, during the period of career inflection when *Le Monde* also took shape.

The second way in which he mind/body dualism would have provided Descartes with needed aid was in regard to the epistemological difficulties which surfaced in the course of the later *Regulae*. Earlier, in Sect. 7.6.2 containing part of our discussion of the instability of his elaborated universal mathematics, we saw that the first chapter of *Le Monde* deals with the problem of the status and role of secondary qualities, as that problem had become framed by the account of perception and mental function in the later *Regulae*. We witnessed in chapter one of *Le Monde* the initial emergence of a theory of ideas as purely mental entities triggered by, but not necessarily representative of, mechanical states of affairs. In particular, Descartes dealt with the distinction between the act of perceiving light and the pattern of mechanical impressions which must be the cause of the percept, but not its directly intuited object. An emergent theory of ideas allowed Descartes to deal with the perception of secondary qualities, while not necessarily elaborating the doctrine so far that one would be forced to question the direct empirical realism of primary qualities. Through an explicit doctrine of mind/body distinction Descartes could metaphysically enforce the mental/mechanical distinction involved in the theory of knowledge of the *Regulae*. Metaphysical stress would thus be placed on the distinction of mind and body, rather than upon the mind's direct intuitive inspection of body as in the *Regulae*.[11] Support would thereby be lent to the construal of the objects of thought, ideas, as strictly mental entities, and metaphysical 'room' would be provided for the perception of secondary qualities. At the same time direct realism in regard to geometrical qualities would not be explicitly ruled out, but rather, as in the first chapter of *Le Monde*, it would be sidestepped for the moment.

---

[11] It is worth underscoring once again the dialectic involved in this movement of conceptualization: The later *Regulae* already depended upon a clearly stated but unarticulated ontological dualism of *vis cognoscens* and brain loci (as well all the rest of the material world). It was this very approach to grounding universal mathematics which brought into focus new, proto-modern epistemological problems, the answer to which, according to Descartes in his period of career inflection, was further to articulate that dualism on an explicit level of metaphysical construction, now stressing not the 'directness of application' of mind to body, but the metaphysical gap between mind and body, so that ideas as states of mind could be separated from their material, not necessarily representative, causes. (As we know, much later in his career, in the Sixth part of the *Meditations*, the debates that followed, and his own drift toward issues of psychosomatic medicine and ethics, Descartes focused once again on the nature of the mind/body union, its functions and nature, rather than upon the conceptual gap between mind and body.) The issues at stake, as we would now realize, are not capable of being fully and finally resolved, and the claims and counter-claims of those debating them are shaped by the needs and opportunities of the context and process of debate.

Descartes' first steps in the metaphysics of this problem were probably only intended to lend support to a conceptual juggling act of the kind he would perform in the opening chapter of *Le Monde*. The deeper problems entailed by this mode of hedging the problems of the *Regulae* would only have emerged later in the published *Meditations*, and the disputes to which they would give rise.[12] Nevertheless, this is not to belittle the 'little treatise of metaphysics' or the conceptual shadow it was casting on Chapter 1 of *Le Monde*, for it is crucial to realize that Descartes had indeed seen the need to move beyond the doctrine of the later *Regulae*, and to do so in the interest of providing epistemological living space for his now focused program in corpuscular-mechanical natural philosophy.

The third and final way in which the emerging metaphysical dualism could be deployed by Descartes in the wake of the collapse of the *Regulae* and during his period of career inflection concerned, of course, the status and grounds of mathematics. Just as physico-mathematics (as a part of the collapsed universal mathematics) was giving way to a definite program in systematic corpuscular-mechanical natural philosophy, so pure mathematics, the other component of universal mathematics, also had to find a new, post-*Regulae* grounding. As discussed in Sect. 7.6.3, it undoubtedly became increasingly apparent to Descartes that the thrust of his mathematical work was to reduce consideration of complicated geometrical figures and curves to the analysis and classification of abstract sets of algebraical relations. With the collapse of the *Regulae* on the very issue that he could not justify these relations by grounding them in an ontology of directly intuited macroscopic geometrical elements, there arose the following difficulty: It was no longer possible to assume that with regard to their manner of grounding, pure mathematics and physico–mathematics could be treated the same way—that both could be subjected in a single elegant movement to the same justificatory doctrine, as had occurred in the later *Regulae*. (This was true quite apart from the further complication that physico–mathematics was in any case giving way to corpuscular-mechanical systematics.) Pure mathematics needed a new mode of justification, and so of course did the now focalized corpuscular-mechanical natural philosophizing. Was it possible to devise a new justificatory doctrine that could still grasp both endeavors (even if separate and detailed elaboration was required to achieve each result, compared to the compact justificatory machinery of universal mathematics?). The answer Descartes found early in his sojourn in the United Provinces was 'Yes'—the dualism not only propped up the natural philosophizing,[13] but also the more analytical and abstract parts of his mathematical practice. Descartes could employ his metaphysical

[12] These later difficulties helped to constitute the problem nexus of classical epistemology as it emerged from Descartes' mature writings. The questions include: How are ideas caused by mechanical states of affairs; how, given the mind/body dualism, can the realism of primary qualities be asserted; how can the mind entertain ideas of extended objects at all; what are the ontological status and cognitive function of such 'ideas' as internal sensations and passions?

[13] We say here 'propped up' to remain neutral amongst the various difficulties mentioned earlier about what exactly Descartes meant and included at any moment in the concept of supporting his natural philosophy with his metaphysics.

teachings on God's existence and veracity, and the criterion of clearness and distinctness to supply the justification for mathematical propositions no longer offered by the doctrine of the *Regulae*. In that work direct intuition into the necessary truth of mathematical propositions was based in part on the operation of the optics-physiology-psychology nexus. Appeals to such natural philosophical grounds were now replaced by metaphysical (ultimately theological) guarantees of mathematical propositions. We will have more to say about this in relation to Descartes' deployment of voluntarist theology, to which we turn in Sect. 8.2.2.[14]

We might summarize by reflecting that Descartes' post-*Regulae* problems and their attempted resolution by means of metaphysical construction contained the seeds of what Kemp Smith termed Descartes' 'later theory of knowledge', an apt label, since it points toward the importance of the career inflection following the collapse of the *Regulae.*[15] By 1630 Descartes seems to have realized that if one grants the justificatory nexus of the *Regulae*, secondary qualities (as immediate perceptions) threaten to be ontologically certified, while mathematical truths seem to elude the entire justificatory net. The recourse to metaphysics and the contents of the first chapter of *Le Monde* suggest that the basic philosophical strategy of Descartes in the post-*Regulae* period was to try to drive a logical and psychological wedge between the intuitive inspection of 'appearances' and assertions of 'truth'. For obvious reasons the *Regulae* had conflated these elements; to split them apart Descartes articulated his metaphysical dualism and a doctrine of 'judgments' as mental acts over and above appearances.[16] As Kemp Smith showed, in the *Meditations*

[14] Several subsidiary considerations probably helped conduce to this reformulation. In the first place Descartes' own heightened insight into the power of the tools of sceptical thought, consequent on certain events we will canvass in Sect. 8.3 below, may also have helped motivate a metaphysical, rather than natural philosophical, justification of mathematical procedures. In addition, dawning awareness of the very need for a theory of ideas, and the concomitant assertion of mind/body dualism, may itself have further rendered problematic the relation between mathematical propositions and the deliverances of external reality—a relation seemingly so clear, 'mechanical' and direct in the *Regulae*. All these considerations may have further suggested that metaphysical support had to be introduced to supply the justificatory arguments surrendered by the weakened theory of direct empirical realism of the *Regulae*. Similarly, the now definite commitment to a corpuscular-mechanical ontology precluded the type of direct macro mathematical intuition asserted in the *Regulae*. Mathematical relations might still be knowable in external reality, but the exact nature of the process of abstraction of such truths from the play of matter in motion became more problematical, and hence hardly suitable for grounding the very truth of these propositions.

[15] Kemp Smith (1952), chapter 9.

[16] That Descartes had begun to consider a theory of judgment as early as his first metaphysical speculations in 1629 is suggested by his letter of to Mersenne of 27 February 1637 (Alquié 1963, I, p.522). Descartes concedes that his proof of the metaphysical distinction between mind and body in the *Discours* was not as clear as it could have been, but that he could only have improved his presentation, 'en expliquant amplement la fausseté ou l'incertitude qui se trouve en tous les jugements qui dépendent du sens ou de l'imagination, afin de montrer ensuite quels sont ceux qui ne dépendent que de l'entendement pur, et combien ils sont évidents et certains.' He goes on to claim that this was shown in a Latin treatise of metaphysics written 'eight years ago'.

and *Replies* to *Objections* Descartes would still hold that ideas *qua* objects of thought are directly intuited, but that it is the role of judgment to determine whether and to what degree these immediately intuited appearances are true and hence have ontological relevance.[17] Therefore the direct apprehension of secondary qualities need not entail the Aristotelian conclusion that these qualities are real constituents of the physical world. Judgment, especially natural philosophically trained and experienced judgment, can decipher the true nature of these appearances as mere ideas, triggered by certain mechanical states of affairs. Furthermore, the distinction of judgment and appearance rests on the explicit formulation of a doctrine of innate ideas against which appearances are judged.[18] This same metaphysical construction putatively solves the problem of justifying mathematical procedures by arguing that God's benevolence guarantees the truth of clearly and distinctly intuited innate mathematical relations.

8.2.2 Some Voluntarist Theology and Its Strategic Uses

Descartes' second enterprise of justificatory construction in this period involved the highly personalized employment of aspects of Voluntarist theology. This enterprise, best expressed in Descartes' letters to Mersenne in the late spring of 1630—when he was already committed to the course of work which would lead to *Le Monde*—was an attempt to put forward justificatory doctrines bearing at several points on details of corpuscular-mechanistic natural philosophy.[19] Descartes' work on Voluntarist theology does not form an integrated justificatory doctrine with his emergent dualist metaphysics. There are, of course, points of congruity between the teachings, but virtually no logically necessary links.[20] For our purposes, in regard to the question of Descartes' justification of corpuscular-mechanical natural philosophy, Voluntarist theology can be considered as a specialized justificatory tool, summoned to support certain key points in a way consistent with, but not necessarily derived

[17] Kemp Smith (1952) pp.155–60; 224–37; Cf *Meditations* AT VII 71–3; 37; *Replies to Objections*, AT VII 387.

[18] Kemp Smith (1952) pp.237–447; Descartes to Mersenne 15 April 1630 where Descartes speaks of principles of geometry 'mentibus nostris ingentiae'. AT I 145.

[19] It is a fundamental fact in Scientific Revolution studies that Mersenne and other later mechanists such as Boyle shared a strategy similar to that of Descartes in seeking to exploit the reciprocal support that mechanical philosophy and Voluntarist theology seemed to lend to each other. For example, classically stated by Oakley (1961) and McGuire (1972); modern views are synthesised and criticised in Osler (1994); Harrison (2002a, b), correctly cautions against assuming a necessary logical link between the Voluntarism of such figures, and the empiricism many of them display. Cf. Note 28 below.

[20] One key point of communication is the defense of the veracity of clear and distinct ideas, resting upon God's decision to consistently adhere to a world order in which clearly and distinctly perceived ideas are indeed true.

from, the broader covering doctrines of the metaphysics.[21] Apparently only during times of intellectual creation or combat did Descartes have recourse to the further reaches of his Voluntarism. For example, he sought to exploit voluntarist tenets in 1630 during the composition of *Le Monde*, and in the early 1640s in constructing the *Replies* to *Objections* to the *Meditations*.

Descartes' dual recourse to metaphysics and Voluntarist theology becomes explicable on the basis of our arguments about the fate of the *Regulae*. We have taken Descartes to have come to the brink of creating a systematic corpuscular-mechanism by way of a very elaborate, yet ultimately abortive attempt at a methodological and epistemological justification of universal mathematics. Descartes was able to build upon this failure by attacking the problems and tensions raised by the *Regulae* from the standpoint of a new metaphysics. In order further to secure strategic points in his metaphysics and natural philosophy, Descartes drew upon certain elements in the tradition of Voluntarist theology. Thus Voluntarism did not constitute as central a justificatory machinery for Descartes' mechanism as it perhaps later would for the chief English mechanists. The reason for this lies in Descartes' idiosyncratic route to his corpuscular-mechanistic system, and its justification, through the sophisticated but failed enterprise of the later *Regulae*, a text which created problems inviting more concerted elaboration of a metaphysical dualism which, we have argued, had existed in a tacit, embryonic yet strategically important form in that very text.

Descartes and other seventeenth century mechanists could draw upon rich resources of Voluntarist thought stemming originally from the writings of Ockham and Duns Scotus.[22] Before the marriage of voluntarist theology and corpuscular-mechanical philosophy in the seventeenth century, Voluntarism had primarily served the ends of particular strains of theological, ethical and legal thought. Like any intellectual tradition, it had suffered alteration and modulation, depending upon the various contexts, aims and motives of its upholders. With the application of Voluntarism to mechanical philosophy in the seventeenth century, the process of articulation continued, due to the necessity of each mechanist subtly adjusting elements of the tradition to render it congruent with the particular stresses and aims of his mechanism. Nevertheless, it is possible to abstract out the main lines of the Voluntarist doctrine.

As the term implies, Voluntarist theology stresses the primacy of God's creative will over His intellect. Despite variations in their exact formulations, Voluntarists from Scotus to Descartes seem to have been trying to avoid the conclusion that God's intellect may be necessitated by truths existing in some sense independently of himself.[23] Voluntarist stress on the primacy of God's will was naturally entailed

[21] As Gilson and Kemp Smith have pointed out, in his published works Descartes was perfectly willing and able to adhere closely to what have been termed the properly metaphysical teachings, and to stress within those teachings the more traditional neo-Thomistic elements Gilson (1913) 177–8; Kemp Smith (1952), 182–4.

[22] Oakley (1961)

[23] Gilson (1913) 38–41. This was a conclusion one could draw from propositions in Thomist theology, although, in fact, as Gilson argued, no Thomist ever seems to have allowed himself to be pushed into such a position.

by the fundamental Christian doctrine of creation as the absolutely free act of God: The natural world, ethical and natural law, are all regarded as free creations of God, not necessitated by pre-existing logical or normative considerations, which could have imposed themselves on His understanding. From this it follows that creation does not necessarily embody ends rationally ordered by God, and indeed it becomes likely that divine ends cannot be read out of creation at all. Since the cosmos and the laws ordering it were freely decreed by God, human reasoning based on teleological or analogical procedures is powerless to plumb the nature of creation. Such types of reasoning cannot decipher God's ends, for He had no ends in anything approaching a human sense of the word; nor can these procedures reveal the laws which God has in fact decreed.

Voluntarist theology, running along these rails set down in the doctrine of creation, therefore emphasizes the absolute power, and freedom of God and the complete dependence of creation upon His will. The conclusion follows that although creation is absolutely dependent upon God, His essence and ends cannot be known through his works, yet, as was generally stressed in the Voluntarist discourse, there is an important sense in which God's free creative action is not arbitrary. This argument rests on the Voluntarists' distinction between God's 'ordinary' and 'absolute' power. By his ordinary power God has freely willed an order for the universe. Though he maintains the option (in the case of miracles and the like) of suspending the ordinary concourse by an extraordinary act of his absolute power, in the normal course of events, he continues to maintain creation according to the freely willed dictates of his original ordinary dispensation. The ordinary concourse is open to inspection by human reason. We can learn, usually by experiential means, the manner in which God has chosen to rule in the ordinary course of events; but, to reiterate, knowledge of the ordinary concourse of creation does not entail knowledge of God's essence or ends.

The usual concomitant of the above doctrine was a nominalist interpretation of ontology. The universe of the Voluntarist was most often one of unrelated particulars. Since the sole creative and causal agency is God's free will, the unrelated particulars have no inherent or immanent laws, entelechies, potencies or actions. Laws of their action and development are freely imposed upon them by God. This ontology, conjoined with the doctrine of freely imposed law, and the prohibition of analogical or teleological reasoning, forced many Voluntarists, starting with Ockham, to the position that natural knowledge can only be acquired by the study of efficient causes, based on the observation of what God in fact has decreed in nature.

8.2.2.1 Descartes' Articulations of Voluntarism

Descartes' own version of Voluntarism, as revealed in the letters of 1630, articulates the account given above in three main respects. The most striking innovation of Descartes' within the tradition is his identification of the will and understanding of God, which forecloses the possibility of introducing any consideration or contemplation on the part of God distinct from acts of His will. For example, in regard to mathematical truths Descartes wrote to Mersenne in early May 1630.

> As for the eternal truths, I say once more that they are true or possible only because God knows them as true or possible. They are not known as true by God in any way which would imply that they are true independently of Him. If men really understood the sense of their words, they could never say without blasphemy that the truth of anything is prior to the knowledge which God has of it. In God willing and knowing are a single thing in such a way that by the very fact of willing something, he knows it, and it is only for this reason that such a thing is true.[24]

Hence, following Gilson, one observes that Descartes rejects the Thomistic distinction between God's contemplation of an eternal truth, such as 'man is an animal', and God's possible willing to create a man fulfilling the definition.[25] To Descartes the distinction between essence and possible existence threatened to lead to the conclusion that necessarily true essences can somehow subsist independently of God, forcing themselves upon His intellect, and thus constraining or conditioning His will. We shall shortly see that through such considerations Descartes attempted to vindicate the truth of mathematical propositions and the laws of nature. In addition, he hoped to render material nature and its principles dependent upon God, and thus oppose the tendency of so-called Naturalist philosophies of nature to put forward a notion of a self-sufficient nature activated by immanent principles.[26]

Descartes· second main articulation of the Voluntarist position follows immediately from his view of divine attributes. Having purposely conflated divine will and intellect, thus demolishing the distinction between necessarily true essence and contingent existence, Descartes proceeded to assert that not only material existents, but also natural laws and mathematical truths are created by God through acts of

---

[24] Descartes to Mersenne, 6 May 1630, AT I 149; Kenny (1970) 13–4. 'Pour les veritez eternelles je dis derechef que *sunt tantum verae aut possibiles, quia Deus illas veras aut possibiles cognoscit, non autem contra a Deo cognosci quasi independenter ab illo sint verae.* Et si les hommes entendaient bien le sens de leurs paroles, ils ne pourraient jamais dire sans blaspheme, que la vérité de quelque chose precede la connaisance que Dieu en a, car en Dieu ce n'est qu' un de vouloir et de connaitre; de sorte que *ex hac ipsa quod aliquid velit, ideo cognoscit, et, ideo, tantum talis res est vera.*'

[25] Gilson (1913) 46, Citing Suarez, *Metaphysicae Disputationes* (1597) for a contemporary exposition of this view.

[26] There is a further argument tending to show that it was just the perceived need to vindicate the freedom of God's will from logical necessitation which motivated Descartes' identification of the divine will and intellect. Gilson expended much effort to show that Duns Scotus was not the source of Descartes' Voluntarism. Much of the argument rested upon the acceptance by Scotus of a distinction between God's will and understanding. (Gilson 1913, 132–47) But, on Gilson's own showing, Duns Scotus' adherence to this doctrine tended to reduce his Voluntarism toward more properly Thomist positions. (*Ibid.* p.144) Although there are major doctrinal differences between Descartes and Scotus, this does not foreclose the possibility of a drawing upon these sources. It is still possible to view Descartes as borrowing from a Scotist Voluntarist tradition while at the same time radically altering the teaching on divine faculties to secure a major weak point he perceived in the Voluntarist approach—the quasi Thomist loophole seemingly allowing for the logical necessitation of God's will.

His free will. As such, these laws are only true by virtue of God's fiat, rather than decreed by God because He recognizes them as true. In addition, Descartes held that God is related to his freely willed laws as an efficient and total cause, not as a final cause. About three weeks later, he wrote to Mersenne,

> You ask me *by what kind of causality God established the eternal truths.* I reply: *by the same kind of causality* as He created all things, that is to say as their *efficient and total cause....* You ask also what necessitated God to create these truths and I reply that He was just as free to make it untrue that all the lines drawn from the centre of a circle to its circumference are equal as he was free not to create the world. [27]

The third articulation of Descartes followed more closely from traditional Voluntarist—or more properly Creationist—tenets.[28] Starting from the proposition that creation utterly depends upon God's will, he elaborated a doctrine of continuous creation, or at least a doctrine of the necessity of God's continuous sustenance of creation from instant to instant. In *Le Monde* he stresses the necessity of God's acting at each moment both to conserve the existence of natural bodies and their modes (including force of motion and rest), [29] and, as we shall see, to enforce the rules according to which natural change occurs. These rules, or laws of motion, become nothing but the freely willed consistent modes of operation of the divine will in governing the exchanges of force of motion in nature. Creation is so utterly dependent upon God, and so utterly devoid of any immanent principles, that God must reiterate his active maintenance of it at each instant of time.[30]

---

[27] Descartes to Mersenne, 27 May 1630, AT I 151–2; Kenny (1970) 14–5 (slightly modified). 'Vous me demandez *in quo genere causae Deus disposuit aeternas veritates.* Je vous reponds que c'est *in eodem genere causae* qu'il a crée toutes choses, c'est a dire ut *efficiens et totalis causae*...Vous demandez aussi qui a necessité Dieu à créer ces veritez; et je dis qu'il a été aussi libre de faire qu'il ne fut pas vrai que toutes les lignes tirées du centre a la circonference fussent égales, comme de ne pas créer le monde.'

[28] Harrison (2002a) has importantly pointed out that one should not simply conflate the notion that God's will is primary with the idea that nature is totally dependent upon God: 'the doctrine that places God as the direct cause of what takes place in nature is thus independent of a voluntarism according to which the divine will is above reason.' (p.69). Harrison also uses this point to show that there is no necessary link between Voluntarism and the typical late seventeenth century versions of natural philosophical empiricism, as the case of Descartes' Voluntarism of course also illustrates.

[29] Why the term 'force of motion' is used in preference to 'motion' should be apparent on the basis of the discussions of Descartes dynamics in Chaps. 3 and 4.

[30] There is a fourth articulation or difference of Descartes from many other later seventeenth century mechanist voluntarists, which we have mooted already—Descartes is not a radical empiricist. Harrison (2002a) threw light on this, as we have seen, by rejecting the necessary connection that many assume between Voluntarism and empiricism, pointing out that it is based on a confusion between two different aspects: the notion that nature is dependent upon God, and the idea of God's will being primary. According to Harrison (p.66), 'the reason that voluntarism does not issue in empiricism in Descartes's scheme of things is that God, having created the eternal truths by his sheer will, then proceeded to stamp them onto the human mind. We can know the laws of nature without recourse to empirical investigation because these truths are in our minds.' Harrison indeed argues that the religious sources of empiricism reside elsewhere than in Voluntarism. (2002a)

8.2.2.2 Descartes' Uses of Voluntarism—Mathematics and the Concepts of Dynamics

Turning to the uses to which Descartes put his Voluntarism in the period immediately following abandonment of the *Regulae*, let us note first the function of his teaching that mathematical relations are true because willed by God and not vice versa. Descartes' key point is that mathematical truths are not ultimately self-validating.[31] Regardless of how true they may appear to us, or how valid their application and manipulation seem, one begs the question of their justification unless it is recognized that their truth derives from God's creative fiat.[32] The question for Descartes thus turns on his fundamental problem of how we are to know that a proposition is true. To assert that God decrees such propositions because they are true begs the question. Descartes tries to plumb the ultimate grounds of truth, coming to the position that the truth of these propositions is only certified through recognition of their absolute dependence on the free creation of God.

Descartes' problem here is the same one which motivated him in the *Regulae*. There he hoped to justify the truth of mathematics by demonstrating through natural philosophical arguments that we have immediate infallible intuition of mathematical relations true of the external world, and that the very operations of mathematics enjoy the same ontological grounding. Of course, Descartes came to realize that his

---

To further support Harrison, we have seen that Descartes has his early metaphysics in view, as well as his hopefully deductivist rhetoric of method, and, more to the point, his interest in 'seeing the causes' through his style of physico-mathematics. Nevertheless, Harrison is talking mainly about Descartes and laws of nature, not the workaday demands of natural philosophical explanation. There we have seen, and will see again in our study of *Le Monde,* that, all method-rhetoric aside, Descartes had a healthy respect for empirical reports in the fashioning of corpuscular-mechanical explanations, and indeed at one stage, as we shall see, waxed lyrical about the need for a full natural history to complete his natural philosophy. Cf. Sect. 8.6 below and our analysis of the actual demands of corpuscular-mechanical explanation in Sect. 6.4.

[31] Kemp Smith (1952) 177–9.

[32] Granted, of course that God exists, but since Descartes claims to have worked on such proof before 1630, there is no need to conclude as did, for example, James D. Collins (1971, 9) that, '… Descartes assigns a role to God in the ordering of the universe, even considered as having a mechanical genesis and structure. But *Le Monde* employs a set of presuppositions about God, without being able to supply the philosophical basis for accepting them. Since these theistic presuppositions figure quite prominently in the Cartesian conception of nature, the latter is proposed to students only as a well-conceived but incompletely established theory.' (cf. *Ibid.* pp. viii, 10) Collins pinned his contention on the fact that the text of *Le Monde* does not contain explicit metaphysical justifications for the physical and theological conceptions set forward. But surely Descartes' reports on his metaphysical work and his insistence on the metaphysical grounding of his natural philosophy allows us to conclude that he put forward his Voluntarist tenets in full confidence of his ability to supply the needed metaphysical support, in this case a proof of God's existence and analysis of his attributes. The pedagogy of *Le Monde* is intended to convince the sceptical, uninitiated but commonsensical; Descartes himself thought he had certain grounds for his rhetorically styled presentation. Thus the fundamental error of Collins was to mistake *Le Monde's* lack of metaphysical support for Descartes' actual state of preparedness. See below Chap. 9 Note 3 and accompanying text.

doctrine of direct realist intuition grounded in the o-p-p nexus was marred by the problem of perception of secondary qualities. A new level of doubt was thus generated, rooted in the realization that immediate, direct inspection of appearances may be ontologically deceiving. This ruled out a blanket appeal to intuitions of external reality as a justificatory device. In addition, as we also know, the increasing use of algebra and theory of equations in Descartes' mathematics rendered justificatory appeals to geometrical intuition irrelevant, if not impossible.

With the justificatory procedures of the *Regulae* in doubt, Descartes in effect was left in the position of Mersenne in *La Vérité des Sciences*, a position the *Regulae* had probably been intended to avoid. Mersenne had been inclined simply to appeal to mathematics as true and then proceed to show how one could wield mathematics against scepticism. From the standpoint of Descartes, such a position would have been inadequate, for when examined it amounted to asserting that mathematics appears true to us. In the *Regulae* Descartes had in part tried to anchor that appearance in external reality via articulation of the o-p-p nexus. With that tactic bankrupt, his justificatory procedures are marked on the one hand by his dualism, theory of ideas and criterion of clearness and distinctness, and, on the other, by the Voluntarist interpretation of God's creation of truth.

In sum it is well to remember Richard Popkin's observation to the effect that the contemporary sceptics did not dispute that certain propositions appear true, but rather questioned whether there is adequate evidence that they are in fact true.[33] Starting in the later *Regulae* as a methodologist and proto-epistemologist, and after 1628 as a budding metaphysician and Voluntarist, Descartes was always sensitive to just this distinction in ways which Mersenne, his less profound colleague in anti-scepticism, was not. Nothing could be more indicative of the progressive inflection of Descartes' post-*Regulae* justificatory enterprise than his daring and paradoxical attempt to ground mathematical truth in the arbitrary will of God.

The second function of Voluntarism for Descartes was to elevate to the status of so-called 'laws of nature' the principles of the dynamics of corpuscles he had elaborated in several stages since 1619. In setting himself the task of composing the full system of corpuscular-mechanical natural philosophy, Descartes ran up against the problem of the ultimate rationale for those principles of his dynamics of corpuscles embedded in his physico-mathematical work, and emergent in quite developed form from his work on optics in the mid 1620s, as we have seen in Chap. 4. What was new for Descartes in 1630 was the dual attempt to construct a systematic corpuscular-mechanism, partly out of elements of his previous physico-mathematical exploits, and to guarantee, as far as possible, the truth of that construction. We must always remember that Descartes' progressive development of the concepts of his dynamics between 1619 and 1633, and the construction of its legitimation as part of a systematic articulation of a natural philosophy 1630–1633, were two different but possibly related things. The latter enterprise is where the specification of the ultimate metaphysical, rather than pragmatic and operational grounds of his dynamics would

[33] Popkin (1964) 167.

have come into question. Accordingly, it was just in the course of writing *Le Monde* that Descartes' principles of dynamics reappear as divinely decreed laws of nature, whose expression is intimately tied to the idiom of Voluntarist theology.

We have already canvassed a fair amount of this material, under slightly different angles. First, in Chap. 3 we saw how Descartes' earliest formulations of his physico-mathematical program involved the provision of natural philosophical explanations in terms of matter and cause, wherein he preferred corpuscular-mechanical discourse, and demonstrated an explicit concern with elaborating the principles according to which corpuscles move and interact—what we have consistently termed Descartes' 'dynamics' of corpuscles throughout this work. This dynamics, even in 1619, was focused on issues of instantaneously exerted force of motion, whether the body was moving or only tending to motion, and its analysis according to graphically represented components.

Then, in Chap. 4, we explored in considerable detail the genealogy of Descartes' mature principles of dynamics in his physico-mathematical optical work of the mid 1620s, and their presentation, as laws of nature, in *Le Monde*. We proceeded in that manner because of the detective work needed to uncover how Descartes discovered the law of refraction. We needed to understand and how and why he occluded his mechanical theory of light and his path to the law of refraction when presenting his 'tennis ball' demonstration of the law in the *Dioptrique*. And, to do that we needed first to understand his principles of dynamics as present in *Le Monde*, including their theological tonality. In Sect. 4.2 our explication of the principles of the dynamics, his concepts of instantaneously exerted force of motion, and its analysis into determinations or directional magnitudes of such force, could not have proceeded without including the basic Voluntarist theological points Descartes insisted upon regarding God's instant to instant maintenance of the cosmos and everything in it. Then, having uncovered how the law of refraction had been discovered by Descartes, we completed the detective work by returning to our first stop, the laws of nature in *Le Monde*. It will be recalled that we showed in Sect. 4.8.1 how Descartes forged his statement of the principles of his dynamics in *Le Monde*, by superimposing Voluntarist ideas and the conceptual constraints of his plenist matter theory upon a technical exemplar, provided by a particular representation of the law of refraction which had been instrumental in his very discovery of the law.[34] So, while we looked in Chap. 4 at the how Descartes' Voluntarist conception of God's relation to nature was shaping the formulation of his dynamics in terms of laws of nature, we turn here to the obverse side of that formulation, looking at the Voluntarist theology as providing grounding and legitimation for the principles of Descartes' dynamics, allowing them to be expressed as divinely decreed laws of nature. Since a good deal of the content has already been discussed, we limit ourselves here to generalizations not previously made, and to a few remarks on the fine structure of the Voluntarist discourse woven into and behind the laws of nature.

[34] We also argued that it was this genealogy in physico-mathematical optics, and the combination of these two further shaping factors, which explain why Descartes' laws of corpuscular mechanics came to differ so much from those of his original mentor, Beeckman.

Three principles of Descartes' dynamics of corpuscles, originally emergent in his optical work, appear in *Le Monde* with the status of 'laws of nature'. These laws assert: (1) the moment to moment conservation of bodies and their modes, including 'force of motion' and rest, in the absence of external disturbing factors; (2) the conservation of the total quantity of 'force of motion' in collisions between bodies;[35] and, (3) the 'determination' of the force of motion conserved in the first law to act at each moment in a straight line tangent to the path of the body at the point under consideration.[36] These laws, along with the assertion of God's conservation of the total quantity of force of motion in the universe, are interpreted as rules God has decreed according to which change is to occur in nature. The Voluntarist character of Descartes' justification of the laws can be grasped through consideration of his interpretation of their ontological status. The first two laws can be treated together.

The laws of nature are the modes of conserving activity through which God has freely elected to act from instant to instant since the moment of creation. Descartes claims,

> ... it is the case that these first two rules manifestly follow from this alone: that God is immutable and that, acting always in the same way, He always produces the same effect. For supposing that He placed a certain quantity of motions (*'certaine quantité de mouvemens* [sic]) in all matter in general at the first instant He created it, one must either avow that he always conserves the same amount of it there or not believe that He always acts in the same way. Supposing in addition that, from that first instant, the diverse parts of matter, in which these motions are found unequally dispersed, began to retain them or to transfer them from one to another according as they had the force to do, one must of necessity think that He causes them always to continue the same thing. And that is what those two rules contain.[37]

The two conservation laws follow, granted God's immutability and the conditions He decreed and supported at the instant of creation[38]; that is, the existence of a definite summed total quantity of force of motion and the conservation of the force

[35] On our insistence upon the term 'force of motion' rather than 'motion' see above Sects. 4.2, 3.3.3, 3.4. and 3.5.4

[36] *Le Monde* AT XI 37–48. Let us recall here the explication of the meaning of the third law offered in Sect. 4.2: 'The third law of motion in *Le Monde* specifies the direction in which the Divinely conserved quantity of force of motion is to act. The force of motion is directed along the tangent to the path of motion at the point under consideration. We have to be careful here. The third law does not say that merely a direction is conserved. Rather, it asserts that a quantity of force of motion is annexed to a privileged direction. That is, the law specifies a directional quantity of force of motion. It says that in the absence of external constraint, this directional quantity of force of motion would be conserved by God from instant to instant. This directional quantity of force of motion is, of course, that determination mentioned above. Let us call the directional quantity of force of motion directed along the tangent to the path of motion at a given instant the principal determination of a moving body; following Descartes one can decompose that directional quantity into components, also called determinations. In any given case, mechanical conditions and the spatial relations of bodies dictate which components of the principal determination come into play.'

[37] AT XI 43; MSM.69–71.

[38] To speak of 'God's immutability' is to use a shorthand for the full Voluntarist understanding to the effect that God has freely willed to exert his ordinary concourse immutably, but can decide to suspend it for miracles under his extraordinary concourse.

of motion redistributed in those instantaneous collisions occurring at the moment of creation. Since God is immutable, he conserves in each succeeding instant the conditions which he decreed at the first instant. Hence all instants of time are assimilated to the instant of creation. God's willing to be immutable in his ordinary concourse of nature, becomes the source of the metaphysical guarantee that His instantaneously reiterated conservation will give rise to an orderly and law like ordinary concourse of nature.

Descartes' Voluntarist vision of the universe emerges in these passages of *Le Monde*. Matter and its modes are utterly dependent from moment to moment on God's reiterated conserving action. In particular, God acts in each instant to conserve motion in a self-prescribed manner, according to rules He has freely set down. These rules or laws of nature are thus nothing immanent in creation; they merely express the particular manner of operation through which God has chosen to exercise His ordinary conserving concourse. In the final analysis neither force of motion, rest, nor any other mode of body, nor indeed matter itself, has an independent capacity of subsisting from instant to instant. Without explicitly saying as much, Descartes seems to slip here toward a doctrine of continuous re-creation. After all, it is hard to discern a distinction between saying that the motion of a body is nothing if not conserved at each moment by God, as opposed to saying that God recreates the motion at each instant.[39] Descartes did not proceed in *Le Monde* to deal with these issues at a further level of theological analysis. He was primarily interested in the reduction of the analysis of phenomenal translation to the consideration of divinely governed instants of time. This procedure was critically important to his enterprise of justifying the principles of his dynamics (which we know had always *in practice* been focused on the instantaneous status of corpuscles in motion or tending to motion and in terms of their instantaneously possessed quantities of force of motion and their 'determinations'). Descartes' discussion of the first two laws of nature thus aimed to impart theological backing to his species of dynamics, and to this end he selectively employed resources of Voluntarist theology, articulated with special attention to the punctiform character of God's conserving concourse.

This pattern of selective and targeted theological construction, reaching to the conceptual heart of Descartes' dynamics, is further exemplified in his discussion of the third law of nature. As we know, this law specifies the unique direction in which

[39] Cf. Kemp Smith (1952) 195–6. These points were foreshadowed in Sect. 4.2, where we commented as follows on the reduction of phenomenal spatio-temporal translation to discrete instants of divine (re)-creation or support of a body with a given quantity of force of motion: 'God must continually support (or re-create) bodies and their attributes from moment to moment. This implies that in the final analysis a body in phenomenal translation, in motion, is really being recreated or continually supported at successive spatial points during successive temporal instants. In addition, and this is the key point, in each of those instants of re- creation, it is characterized by the Divine injection of a certain quantity of 'force of motion'. We should view the instantaneously conserved 'force of motion' as a kind of quantity of efficacy (the phenomenal mirror of the instantaneously injected Divine action).'

God conserves the quantity of force of motion at each instant during the motion of a body,

> I shall add as a third rule that, when a body is moving, even if its motion most often takes place along a curved line and, as we said above, it can never make any movement that is not in some way circular, nevertheless each of its parts individually tends always to continue moving along a straight line. And so the action of these parts, that is the inclination they have to move, is different from their motion. (...leur action, c'est à dire l'inclination qu'elles ont à se mouvoir, est different de leur mouvement).[40]

For a body moving independently of all external constraints, this law, conjoined with the first, would entail the principle of inertia of classical mechanics, and, as we have already seen, one version of Beeckman's statement of the principle. Of course, such externally unhindered rectilinear motion is impossible in the plenum universe of Descartes' *Le Monde*. That, however, does not seem to have been the only reason Descartes framed his law in terms of instantaneous tendencies to rectilinear motion along tangents to the trajectory. Rather, as we have seen in the case of his work in physico-mathematical optics and the two previous laws, Descartes' dynamics focused on instantaneous collisions of non-elastic bodies and thus was concerned with the instantaneous force of motion predicated of moving bodies at discrete instants of their motion. Operationally speaking, in terms of the application and practice of Descartes' dynamics, the enunciation of a 'Beeckman-like', fully kinematic principle of inertia would have been irrelevant, whereas it was important to be able to deploy a principle specifying the instantaneous 'determination' (or quantity and direction of the tendency to motion) of a moving body. For example, in *Le Monde*, in what we shall later term the 'cosmological' theory of light (theory of light as produced in and by stellar vortices) Descartes wanted to be able to derive rectilinear propagation as a centrifugal action, by resolving into components the instantaneous tangential 'determination' of the 'second matter' constrained to rotate in the celestial vortices.[41]

Therefore, one may conclude, consistent with our findings in Sect. 4.8.1 that the punctiform character of the third law was dictated by the exact technical and conceptual requirements of Descartes' dynamics as applied to physico-mathematical and natural philosophical uses, while the theological justification of the law exploited the doctrine of continuous re-creation or conservation in order to rationalize that punctiform character.

Let's examine a sample of that strategy. Descartes argued concerning the third rule of nature that,

> This rule rests on the same foundation as the two others and depends only on God's conserving everything by a continuous action and, consequently, His conserving it not as it may have been some time earlier, but precisely as it is at the same instant that He conserves it.

---

[40] AT XI. 43–44; SG. 29.

[41] See below, Chap. 10, also cf. Descartes' treatment of the instantaneous tendencies to motion, or determinations of a stone in a sling, described above, Sect. 4.2.

> Now it is the case that, of all motions, only the straight is entirely simple; its whole nature is understood in an instant. For, to conceive of it, it suffices to think that a body is in the act of moving in a certain direction (*en action pour se mouvoir ver un certain coté*), and that is the case in each instant that might be determined during the time that it is moving. By contrast, to conceive of circular motion, or any other possible motion, one must consider at least two of its instants, or rather two of its parts, and the ratio between them.[42]

Now, this argumentation concerns God and His grounding of a dynamical principle, not the technical or operational meaning or use of that principle per se: The instantaneous determination of motion is rectilinear, *because* only a straight line can be grasped entirely in an instant without God having to calculate or observe the path of the body at one or more other instants past or future. Descartes argument would seem to be that a straight line can be defined in any instant of the body's motion through God's consideration of its present position and the implicit endpoint of the (straight) line along which one would point in saying that, 'At this instant the body is in the act of moving in ***that*** direction.' By contrast, a circular path would have to be defined by at least one other point in addition to the present position of the body. To conserve such a curved determination God would have to recalculate the determination at each instant based on memory or prediction of one other point, rather than as above by the instantaneous ostension of the implicit endpoint of a straight line. Thus, to press the theological point, God would not conserve the body 'precisely as it is at the same instant that He conserves it', but rather in a manner also dependent upon consideration of its past or future path.[43]

In what we might term a further 'justification' of his justification of the third rule, Descartes continues in the same paragraph by warning against disciplinary transgression by meddling non-theologians (philosophers and 'sophists').

> But in order that philosophers, or rather the sophists, may not take occasion here to exercise their superfluous subtleties, you should note that I do not say that rectilinear motion can take place in an instant, but only that everything which is requisite to produce rectilinear motion is found in bodies in each instant which can be determined during their motion, and not everything which is requisite to produce circular motion.[44]

The implication is that the Voluntarist grounding of the third law can withstand mere philosophical meddling, provided we continue to focus, as honorary and temporary theologians, on those notions of the just articulated manner of God's moment to moment (re–) creative concourse.

Finally, before ending this section, some of the more general uses of Voluntarism in the system of *Le Monde* should also be noted. The corpuscular-mechanical universe of *Le Monde* was to be stripped of all Aristotelian forms, qualities, potencies and entelechies. Impact and pressure became the sole causes of natural change.

---

[42] AT XI 44–5; MSM pp.71–73

[43] This argument was first worked out in informal conversation with the late Professor Michael S. Mahoney.

[44] AT X p.45. MSM p.73

Accordingly, as Norman Kemp Smith observed, the last vestiges of the qualitative approach had to be excised from the account of impact. A universe of perfectly hard particles foreclosed consideration of the problematical quality of 'elasticity'; but, in such a universe impact and pressure were rendered inexplicable by natural causes.[45] At just this point a recourse to metaphysics and theology became necessary in order to provide a rationale for the laws of nature, explicating how bodies interact at the mechanically opaque instant of collision. The answer on this legitimatory level is that material bodies do not in fact interact physically and causally with each other. They only appear to do so on a phenomenal level, this being the expression or effect of the rule bound ways in which God from moment to moment maintains or alters their forces and determinations of motion.[46]

On an even more general level Descartes also recognized that his mechanical universe was a strictly inanimate created entity, devoid of any and all immanent sources of activity and change. The principle of natural change, motion (or rather 'force of motion' as we have argued throughout), and the laws of its transfer had therefore to be continually impressed upon (or conserved in) creation by God at each instant. It was out of the necessity of meeting this nexus of theological and conceptual constraints that Descartes elaborated his Voluntarism to include an explicit doctrine of continuous conservation or re-creation.

Descartes' theological construction in *Le Monde* was certainly one of the earliest attempts at a Voluntarist rationalization of the laws of nature in the seventeenth century. The mechanical, hence non-teleological, character of his principles accorded well with the potential of Voluntarism to provide justificatory devices. Mechanical laws of efficient causes do not permit entry into God's ends, or analogical argument from nature to God's attributes. With its denial of the efficacy of teleological reason and its concomitant stress on the radical separation of creator from creation, Voluntarism formed a most suitable theological complement to Descartes' system of corpuscular-mechanical natural philosophy and its 'causal register' of dynamical principles.

---

[45] Kemp Smith (1952) p.194. One is reminded of Beeckman's wranglings with the problem of the elastic impact of ultimate particles

[46] The conceptual entanglements of this position are obvious and begin to show up in Descartes' own thinking later with the *Meditations* and *Principles of Philosophy*. A possible way through is suggested by the ground breaking work of Martial Geuroult (1954, 1980). It is to work with the following distinctions: force of motion (causal) is identified with God's own causal action and is discussed in the language of theology and metaphysics; it is equated with a force of motion (modal) which is the manifestation of that divine action in the material world, potentially observable in some of its effects, and taken by us humans, on the level of technical and useful natural philosophical discourse of dynamics, as a possession of a body moving or tending to motion. Finally, our commonsense notions of motion in space and time, unenlightened by either theology or natural philosophy, are seen to be merely appearances caused by God's law like, moment to moment causal actions upon matter. (Bodies themselves, of course, having to be supported in existence moment to moment by God.)

8.2.3 *Plenist or Holistic Realism*

In Sect. 8.2.1 we focused on the possible roles of Descartes' doctrine of matter-extension, a principle of his dualism, in grounding his newly emerging system of corpuscular-mechanism. We did not explore the further implications of a commitment to matter-extension for the actual design and construction of a corpuscular-mechanical natural philosophy, beyond the obvious points that the corpuscles are, as it were, fragments of extension, partaking in its properties (as well as being moveable) whilst it is insisted that the universal plenum is maintained. But, it is one thing to produce metaphysical argument for the nature of matter as extension, further claiming that such genres of corpuscles (elements) that exist, are shattered fragments of that matter-extension. It is another thing altogether to be guided, or forced, by these claims into formulating a style of natural philosophical explanation applicable to this system, and further to arrive at certain highly technical exemplars for the key ranges of phenomena to be explained. This indeed is what happened in the composition of *Le Monde* and its is what we examine here under the label 'holistic' or 'plenist' realism, which will denote the explanatory style of *Le Monde*.

8.2.3.1 Defining the Style and Its Genealogy

Plenist realism as an explanatory style in natural philosophizing may be characterized by the kinds of explanatory moves, exemplars and tools that are encouraged or discouraged. Descartes' plenist-realist style prohibits mathematical abstraction or idealization of the sort characteristic of the traditional mixed mathematical sciences, favoring instead explanations which, arguably, are inclusive or holistic about the factors taken into account, and in doing so arguably reflect the 'real complexity' of phenomena in the plenist universe, and the 'real set' of causes in play, not some 'abstract' or 'fictitious' picture.[47] The realism and the holism are intertwined defining characteristics of this style. By the time *Le Monde* was well under way Descartes

[47] The scare quotes here are intentional, as is the introduction of the word 'arguably'. They signal that we are dealing here with matters that are up for interpretation and judgment by contemporary proponents and opponents of such claims and labelings, whilst we, the observer-analysts of these debates, need not necessarily enter into any agreement or disagreement with the historical actors about them. The intent here is—in the style of the leading 'post-Kuhnian' sociologists of scientific knowledge and experiment, such as Collins (1985), Pinch (1985), Barnes (1982) and Shapin (1992)—to point to how actors negotiate the application of such terms. Descartes' implied claim to grasp the full range of factors in play in any explanatory problem may usefully be viewed through the cautionary spectacles provided by Collins and Pinch in their respective classic studies of the negotiation of 'relevant factors' in instrumental and experimental controversies in modern science. More generally, as was the case for these post-Kuhnian sociologists, the best conceptual preparation for dealing with this sort of issue in 'actors' construction of natural philosophical claims' resides in the resources of phenomenological sociology, as in the original dispensations of Schutz and his followers, e.g. Schutz and Luckmann (1974).

was convinced that matter-extension, the doctrine of elements and his principles of dynamics all have metaphysical-theological grounding, and hence this doctrine of matter and cause in its higher reaches is true to reality. Natural philosophical explanations thus need immediately and completely to grasp the tangle of causes and conditions in play behind any phenomenon in this plenum universe. Explanations must not abstract away from some or most of these causes, issuing in over simplified (strictly not real) models of phenomena under study. *Le Monde* and the correspondence leading to it reflect Descartes' growing awareness of and commitment to this style.

To further specify Descartes' plenist-realism, we should also note that physico-mathematical procedures are not banned, and hence that the technique of 'figuring up' phenomena to be explained and then appending presumed corpuscular-mechanical explanations to diagrams so produced, continues to be central in *Le Monde*, as we shall see in Chap. 10. Additionally, within the realm of plenist-realism, the exemplary physico-mathematical discipline remains optics, meaning, of course, Descartes' physico-mathematical optics, not traditional mixed mathematical optics. Where the latter abstracts and idealizes away from the complexity of plenist reality, Cartesian physico-mathematical optics, at least in René's view, actually cuts to the core of that plenist reality, revealing the underlying dynamics of corpuscles that runs the cosmos. So, the resulting Cartesian plenist-realist natural philosophy remains—like all natural philosophies—merely discursive, not truly mathematical, but in Descartes' case makes use of figures and diagrams to express, as needed, the physico-mathematical genes also present, remembering that for Descartes ever since 1619, physico-mathematical explanations might invoke diagrams, but end in stories about corpuscles and forces and determinations of motion.

Hence we arrive at three key points to bear in mind about the genealogy and content of Descartes' plenist-realism, before we explore it in the correspondence during the years of the composition of *Le Monde*:

[1] Metaphysically grounded matter-extension was part of the justification of natural philosophy, but also part of the articulation of the natural philosophical system 'out of' metaphysical construction. We see this by noting that matter-extension emerged in the *Regulae* problematic, where Descartes' dualism was implicitly, rather than systematically, inscribed in the *vis cognoscens*/brain loci pairing, the embryo of the mind/body distinction. The more formally derived matter-extension doctrine then arose with a now specifically designed dualism, and played a role in grounding the corpuscular-mechanical system in process of development.

[2] The matter-extension doctrine, as articulated with the growing design of the system, then highlighted further requirements and opportunities: displacement circuits of matter were required for any movement to take place; vortices became imaginable, and in turn invited detailed dynamical description. So, the vortex mechanics, central to the entire content and structure of *Le Monde*, was elicited from the matter-extension plenum which itself followed from the initial metaphysical work. Descartes might well have thought of this as an admirably fruitful course of intellectual discovery, from the metaphysics down to the intricacies of the vortex mechanics.

[3] The emerging sophisticated vortex celestial mechanics, in turn became the core exemplar for working out wide swathes of the natural philosophy—not just the celestial mechanics of planets, but of comets as well, plus the theory of light in cosmic setting, the theories of local (planetary) gravity, and tides, and the behaviour of planetary satellites, as we shall explore in great detail in Chap. 10. Indeed, the complexity of phenomena and of explanations in plenist-realism turns out to be the just the obverse side of the ubiquity and import of vortices (and their dynamics) in the entire system of natural philosophy.

Hence, in sum, we find that our third fundamental intellectual agenda or project in the making of *Le Monde*, the style of holistic realism, was itself a distant product of attempts to resolve issues which arose with the project of the later *Regulae*, but which flowered and articulated well beyond those origins as part and parcel of the process of writing *Le Monde.*

8.2.3.2 Articulating Plenist Realism as Composition of *Le Monde* Proceeds

We find evidence in Descartes' correspondence of the crystallization of the realist-plenist register of natural philosophical explanation consequent upon the continuing composition of *Le Monde*. Descartes increasingly expressed the view that the only valid and meaningful approach to natural philosophical explanation of particular phenomena is one which immediately grasps and deploys as a whole the full range of corpuscular-mechanical factors involved, without any idealization or abstraction, which are now seen as sources of error and distortion.[48] Thus in two letters to Mersenne in the fall of 1631, as *Le Monde* was taking mature shape, Descartes refused to discuss as serious natural philosophical matters such idealized problems as the determination of the law of free fall in a void, whereas in 1629, whilst beginning his treatise, he had discoursed at length about both free fall and the motions of pendula under such ideal conditions.[49] Descartes now insisted that the computation concerning free fall he had previously sent to Mersenne, and which echoed his work with Beeckman in 1619, was of no value, because it was based on two falsehoods: first that there could be void space, and second that the motion in the first instant of fall was 'the slowest that can be imagined' and that the motion increased thereafter in a equal manner (in each succeeding interval of time). If these propositions were true, the increase in speed over time would follow the numbers he had calculated. But now, he admits he does not know the 'true proportion' according to which a body falls 'in air', and he will be seeking over the next few days to explain the

[48] Cf. Tannery (1896) who was the first to note this trait in Descartes' work. He did not specifically link it to the composition of *Le Monde*, although he recognised that the *Dioptrique* presents a different style of what he saw as a more properly mathematical physics (p.486).

[49] Of course Descartes could do this sort of mixed mathematical work, as we see from this work in 1629 as well as from his 1619 physico-mathematics of fall and, of course, his physico-mathematical optics. Garber has excellently captured and documented this point in Garber (2000).

cause of weight.[50] In his next letter he says that under the idealized conditions the result '*est demonstratif*', but one cannot assume a void without committing an error, meaning an error in natural philosophy, rather than an error in an abstract mathematical analysis (where there is no error other than of unrealistically framing the problem in the first place).[51] He continues, declaring that that to assume the existence of void space and that 'the force which moves the falling body always acts equally' openly violates 'the laws of nature' ('*ce qui repugne apertement aux lois de la Nature*') because

> ...all natural powers act more or less, according to how the subject is more or less disposed to receive their action; and it is certain that a stone (in the act of falling) is not equally disposed to receive new movement, or an increment of speed, when it is already moving very fast, and when it is moving slowly. [52]

In short, the earlier arguments, based on abstractions such as the assumption of the existence of the void and of uniform instantaneous increments of speed, are dismissed as false *in natural philosophy and hence unable to be accepted or applied within it.* Descartes is articulating the sorts of doubts about the abstract analysis of fall that we conjectured might have been worrying him back in 1619.[53] Now, however, he has in hand an almost complete system of plenist, corpuscular-mechanical natural philosophy, and an elaborated set of dynamical principles for dealing with the instantaneous states of motion, or tendency to motion, of corpuscles and their necessarily instantaneous changes. In contrast to 1619, he now stands not on dimly awakening doubts, grounded in common sense intuitions such as 'bodies always fall through air', and linked to a vague commitment to corpuscular-mechanical causes of weight and fall; rather, he now stands on articulated natural philosophical truth, and a strong sense of proper explanatory protocols for such natural philosophizing. To hypothetically paraphrase his message to Mersenne:

> Our world, the real world, is a material plenum, consisting of corpuscular-mechanical elements, whose motions and changes of motion are dictated by known, divinely sanctioned and enforced, laws of nature. Local fall in our world is caused by impacts of corpuscles of the second element; the laws of nature dictate that the instantaneous increments of velocity are not and cannot be uniform under these real circumstances. Moreover, since the real world is a plenum, falling bodies are resisted not simply by (particles of) the air as even the uninstructed sensibly conclude, but by the fact that if any body whatsoever is to move at all, in any direction, not just in local fall, a volume of matter-extension of the same volume

---

[50] Descartes to Mersenne, October 1631, AT I 221–2: 'Pour ce qui est de la vraye proportion selon laquelle s'augment ou diminue la vitesse d'un poids qui descent dans l'aer, je ne la scay pas encore. Il me faudra dans peu de jours expliquer la cause de la pesanteur dans mon traité: si en l'escrivant je trouve quelque chose de cela, je vous le manderai.

[51] Descartes to Mersenne, October or November 1631, AT I 228.

[52] Descartes to Mersenne, October or November 1631, AT I 230

[53] See Sect. 3.5.5. Recall also our suggestion that in 1619 Descartes' qualms also extended to the issue of the lack of relevant empirical evidence (as Garber 2000, also argued regarding Descartes' attitude to work of this mixed mathematical type, as discussed in correspondence with Mersenne later in the 1630s and 1640s).

(including as circumstances determine quantities of all three elements) must be displaced in some sort of circuit, since there is no void space for anything to enter freely. The principles of the dynamics of corpuscles dictate that complex exchanges of force of motion and changes in its 'determinations' will occur in this process.

The treatment of ideal cases, as in this case of the analysis of fall, or in traditional mechanics generally, as he further explains to Mersenne in May 1632 under subsequent questioning, is irrelevant to the new dispensation in natural philosophy. This is because in mechanics one abdicates the task of immediately and fully grasping the real in terms of its natural philosophical truth, that is, the full complexity of the plenum universe arising from the conjunction of the principles of matter (full yet divided into elements) and cause (the laws of nature/principles of dynamics), applied to the particular problem or circumstances in view. If one relinquishes the 'real' at the start of a problem, there is no compensating possibility of a true or fruitful later return from the idealized realm of mathematics or traditional mechanics to the concrete case, because the ideal assumptions are flatly false, although mechanics does allow in many cases for useful practical application.[54]

8.2.3.3 Aero-Statical Theory and Experiment in the (Partial) Shadow of Plenist-Realism

In June 1631 Descartes wrote his budding follower and student Reneri a detailed letter on fluid mechanics and air pressure which opens a window on the ways in which his plenist holism was developing, with important continuities and contrasts both back to the hydrostatics manuscript of 1619, and forward to portions of *Le Monde* we shall survey in the next two chapters.[55] We may conjecture that Reneri had addressed two questions to Descartes: [1] why the great height of the atmosphere does not occasion a crushing weight at the surface of the Earth; and [2] why

[54] Descartes to Mersenne, 3 May 1632, AT I 246–7. Mersenne seems to have asked Descartes what is the status of results in mechanics given his friend's now staunch defense of his holistic-plenist natural philosophical stance about truth and explanation. Mersenne's chosen example was the law of the inclined plane. Descartes' response, in which we should note his distinction between 'Mechanics' and 'Nature' was as follows: 'Si on suppose qu'un poids poli, estant trainé sur un plan poli horizontal, ne le touche qu'en un seul point indivisible, et que l'air n'empesche point du tout son mouvement, la moindre force sera sufficant pour le mouvoir, tant grand qu'puisse estre. Et quoy que ces deux suppositions soient tousjours fausses en la Nature, et que les plus gros poids el les plus pesans soient plus empeschez par l'air, et appuyent en plus de parties sur le plan our ils se mouvent, que les plus legers et plus petits; toutefois cela empesche de si peu leur mouvement que, lors qu'on examine en Mechanique combien il faut de force pour lever un poids, ou pour le trainer sur un plan incline…on suppose que l'air, ny l'attouchement du poids sure le plan incline, n'empesche rien du tout.' In short, in relation to the real world, there is no doubt that the assumptions of mechanics, made for the purposes of mathematical analysis and demonstration, are false. But in simple mechanical situations, where some useful result is in view, the error or falsity introduced by idealization is not sufficient to prevent our practical reliance upon the results.

[55] Sections 9.3, 10.5.2, 10.6.1.

mercury poured into a pipe or tube does not flow out then the pipe is inverted and its upper end closed off.

Descartes begins with an analogy, asking Reneri to consider the third element (earthly matter) making up air to be like wool, and the interstitial aether of his first and second elements moving around the corpuscles of air to be like little puffs of air moving about around the fibers of wool.[56] His discussion then switches back and forth between the natural philosophical basis—air corpuscles and interstitial aether corpuscles—and their commonsense analogues—fibers of wool and puffs of air amongst and around them.

Next, Descartes proceeds to answer the first question, portentously mixing two different approaches. Introducing, in a somewhat backhanded way, the important concept of the weight of a column of air reaching from the Earth's surface to the top of the atmosphere, Descartes at first claims that the weight of such a column of air particles is somewhat alleviated by the particles being agitated by the aether, and consequently disjoined from one another. Speaking to Reneri in terms of his model, he writes that Reneri should (Fig. 8.1),

> ...consider that this wind, which plays in all directions amongst the small fibers of the wool, prevents them from pressing as strongly against one another as they would if there were no wind. This is because the wool fibers are all heavy and press upon each other as much as the agitation of the wind permits them, so that the wool which is near the earth is pressed by all the wool above it, right up to and beyond the clouds, which makes a large weight; so, if it were necessary to raise that part of the wool which is, for example, at spot marked O, with all the wool above it in the line OPQ, it would take a very considerable force...[57]

This explanation, it should be noted, can hold regardless of whether or not, in switching from the model to the natural philosophical discourse, one accepts an interstitial void amongst the particles of aether.

Next Descartes literally slides into his second explanation of this problem, as if he were only explicating the first, as we see by picking up from the last part of the previous quotation (in italics):

> *...so that the wool which is near the earth is pressed by all the wool above it, right up to and beyond the clouds, which makes a large weight; so, if it were necessary to raise that part of the wool which is, for example, at spot marked O, with all the wool above it in the line OPQ, it would take a very considerable force.* Now, this weight is not usually felt in the air when

[56] Descartes to [Reneri] 2 June 1631, AT I 205: 'Pour resoudre vos difficultez, imaginez l'air comme de la laine, et l'aether qui est dans ses pores commes des tourbillons de vent, qui se meuvent çà et là dans cette laine...'

[57] *Ibid.* '...et pensez que ce vent qui se joüe de tous costez entre les petits fils de cette laine, empesche qu'ils ne se pressent si fort l'un contre l'autre, comme ils pourraient faire sans cela. Car ils sont tous pesans, et se pressent les uns les autres autant que l'agitation de ce vent leur peut permettre, si bien que la laine qui est contre la terre est pressée de toute celle qui est au dessus iusques au dela des nues, ce qui fait un grande pesanteur; en sorte que s'il fallait élever la partie de cette laine, qui est, par exemple, à endroit marque O, avec toute celle qui est au dessus en la ligne OPQ, il faudrait une forece tres-considerable.'

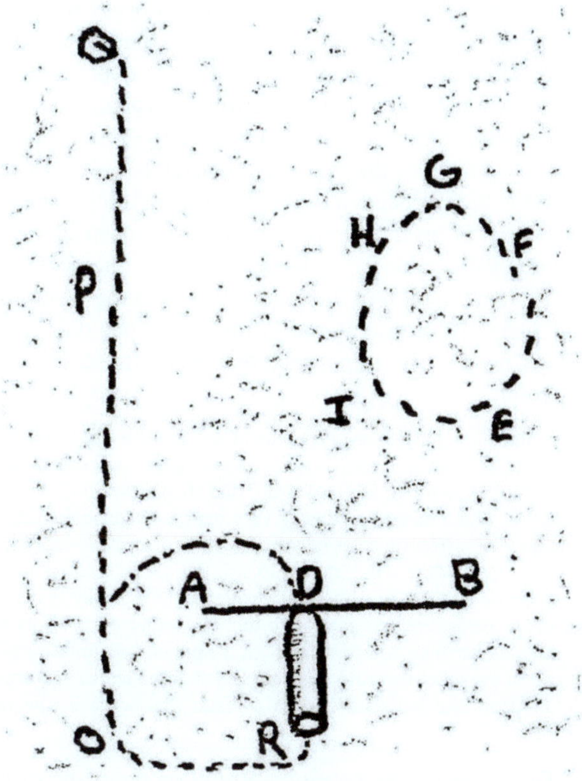

Fig. 8.1 Descartes' aerostatics explained to Reneri, AT I, p.206

one pushes it upward, because if we raise up a part of the air, say that at point E toward F, then the air at F goes circularly toward GHI and returns to E; and thus its weight is not felt any more than that of a wheel which one turns when it is perfectly balanced upon its axle.[58]

We may view this second explanation as embodying his mature, plenist matter theory: Gross matter (air particles) and the interstitial aether (particles of first and second element) form a corporeal plenum; therefore, all motion must occur by means of simultaneous mutual replacement of particles along a displacement circuit. Although Descartes does not explicitly insist upon a plenum in the passage cited, it seems called for both by the fact that he would subsequently employ this argument to meet the objection to the possibility of motion in a fluid plenum, and because the

[58] *Ibid.* 'Or cette pesenteur ne se sent pas communement dans l'air, lors qu'on le pousse vers le haut; pour ce que si nous en élevons une partie, par exemple celle qui est au point E, vers F, celle qui est en F va circulairement vers GHI et retourne en E; et ainsi sa pesanteur ne se sent point, non plus que serait celle d'une roüe, si on faisait tourner, et qu'elle fût parfaitement en balance sur son aissieu.'

displacement circuit obviously need not occur if there are void spaces between the particles of air.[59] Hence, despite the fact that Descartes was clearly capable of imagining a line of explanation articulated from the notion of the weight of a column of air in the atmosphere—the idea Reneri arguably put before him—Descartes needed to avoid the conclusion Reneri wished to draw. Agitation by an all pervasive aether can help (that was the first explanation), but what Descartes really wants is to introduce the idea of necessary displacement circuits (the second explanation). They are called for by his actual commitment to a material plenum, and they are going to be of great use in answering Reneri's second question, to which we now turn.

The explanation of the mercury held in the inverted tube follows from the material plenum considerations, with a few peculiar argumentative twists and turns which need to be unpacked, but only after a first reading of the passage in its entirety:

> ...in your example, involving the tube DR, closed at D where it is attached to the board AB, the mercury inside cannot all at once begin to descend, unless the wool at R were to move toward O, and that at O toward P and (that) toward Q, thus raising all the wool in the line OPQ, which taken together is very heavy. Because the tube is closed at the top, no wool, that is to say (corpuscles of) air, can enter the tube to replace the mercury when it descends. You say that the wind, that is to say, the aether, can enter through the pores of the tube. I concede that; but you should consider that the aether that would enter can only come from the heavens; because, although there is aether everywhere in the pores of the air, there is only as much as is necessary to fill the pores; and consequently if there were a new space to fill in the tube, it would be necessary that the aether required should come from above the air, from the heavens, and as a result that some air be raised into the space it would vacate.[60]

Descartes is now arguing entirely on the plane of natural philosophy. One must distinguish between the role of the (third element) air (corpuscles) and that of the interstitial aether in this explanation. As for the air, Descartes first concedes a role to the weight of the air in holding the mercury in place, alluding to the weighty

[59] Cf. *Le Monde*, AT XI, p.20; SG 15; and Descartes to Reneri, 24 July 1634, AT I 301. '...it faut considerer qu'il n'y a point de vuide en la nature, et que par consequent lors qu'un cors se meut, il doit necessairement entrer en la place de quelque autre, de laquelle celui qu en est chassé, doit au mesme instant occuper celle d'un autre...jusque a ce que le dernier occupe la place qui est laissee par le premier, de façon que tous les mouvemens qui se font au monde sont en quelque façon circularires.'

[60] Descartes to [Reneri], 2 June 1631, AT I, pp.206–7. '...dans l'exemple que vous apportez du tuyau DR, fermé par le bout D par òu il est attaché au plancher AB, le vif–argent que vous supposez etre dedans, ne peut commencer à descendre tout à la fois, que la laine qui est ver R n'aille vers O, et celle qui est vers O n'aille vers P et vers Q, et ainsi qu'il n'enleve toute cette laine qui est en ligne OPQ, laquelle prise toute ensemble est fort pesante. Car le tuyau estant fermé par le haut, il n'y peut entrer de laine, je veux dire l'air, en la place du vif–argent, lorsqu'il descend. Vous direz qu'il y peut bien entrer du vent, je veux dire de l'aether, par les pores du tuyau. Je l'avouë; mais considerez que l'aether qui entrera ne peut venir d'ailleurs que du ciel; car encore qu'il y en ait part tout dans les pores de l'air, il n'y en a pas toutefois plus qui'il en faut pour les replir; et par consequent s'il ya une nouvelle place à remplir dans le tuyau, il faudra qu'il y vienne de l'aether qui est au dessus de l'air dans le ciel, et partant que l'air se hausse en sa place.'

column of air corpuscles QPO. But he only views the weight as having a role because there does not exist a realizable displacement circuit RODR in the medium. (One might parenthetically note that if the displacement circuit of air particles is not possible because of the material plenum, and the closure of the top of the tube to air particles, the issue of the column of air surely is irrelevant; but let us continue because the next stage of the argument is extremely telling.)

Next, as regards the particles of aether, Descartes claims that, similarly, there cannot be a realizable displacement circuit of aether particles. True, in principle, he agrees with Reneri that aether particles can enter the closed end of the tube at any time, but Descartes has an elaborate and tendentious argument about why this cannot occur. First he claims that any displacement circuit in the aether must reach above the region of air to the region of pure aether above. This is because any aether entering the tube cannot be replaced from aether dispersed amongst the air particles of the atmosphere. This, in turn, is because although aether is everywhere in the 'pores of the air', there is only enough there to fill those pores, and if any aether were to move into the tube, the replacement aether would ultimately have to be recruited, in a circuit, from the repository of pure aether in the heavens, above the atmosphere. But why can't that just happen? Well, Descartes adds with an ad hoc flourish, if some aether were removed from the above the atmosphere, it in turn would have to be replaced by some air particles (rather than circulating aether particles), and that would involve lifting a very heavy column of air (as rhetorically mooted for the reader in the previous lines!). All this supposedly rules out the possibility of an actual aether displacement circuit, and does so by combining what we shall term a 'cosmic' injunction that new aether must be recruited above the atmosphere, with the 'aerostatic' injunction, that only air particles lifted up a heavy column can replace such a loss. Hence we see that it was no wonder that back at the beginning of his discussion Descartes was eager to acknowledge to Reneri, if only briefly, the idea of weighty columns of air, which it would seem the latter had posed to Descartes in the original question.[61]

Finally, therefore, granting all this, what happens if one actually opens the hole at D? The mercury will descend, of course, but not because of the weight of the column of air above D, nor because of an actual circulation of aether (which presumably still cannot occur because the column of air QPO would have to be lifted, and it presumably is much heavier than the mercury) but simply (on plenist principles) because the mercury has a greater tendency to descend than the air and thus would now be able to initiate the (short) displacement circuit ROD.[62]

[61] Indeed in the very next paragraph of the letter Descartes returns to the idea of the weighty column of air particles, thus again acknowledging in some fashion the force of Reneri's original suggestion. He continues (Ibid. p. 207 l 10–14) 'Et afin que vous ne vous trompez pas, il ne faut pas croire que ce vif-argent ne puisse estre separé du plancher par aucune force, mais seulement qu'il ya faut autant de force qu'il en est besoin pour enlever tout l'air qui est depuis là iusqu'au dessus des nues.'

[62] The implication is that some finite force is required to initiate a circular displacement. Cf Descartes to Reneri 2 July 1634, AT I 302 l 7–10.

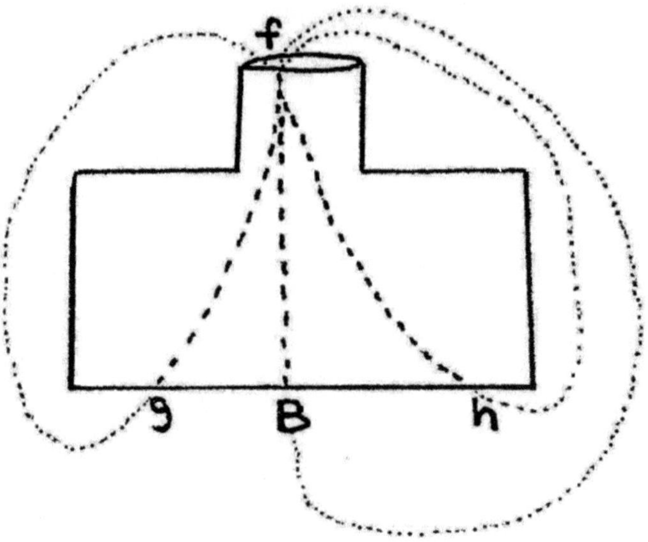

Fig. 8.2 Didactic modification of part of Fig. 3.3 (Hydrostatics Ms. 1619)

However, more is going on in this text than even the above discussion implies. We must first note that the explanation is structured in part on the lines of the explanation of the hydrostatics paradox in the physico-mathematical exercises of 1619 (Cf. Sect. 3.3). Descartes poses his argument as if it arises synthetically from the known properties and behaviour of air and aether particles. In actuality, it is the primitive datum that the mercury does not descend which is primary. Descartes merely adduces a corpuscular-mechanical explanation of the fact, primarily, it turns out by drawing the line of tendency to motion ROPQ to illustrate the long and weighty series of particles which must be displaced if motion is to occur. So far the style of argument parallels the hydrostatics manuscript of 1619. But, on closer inspection, it is clear that Descartes has to some extent altered his previously favored mode of analysis by potential descents, because of the precise nature of the problem to be explained, and because of the added complication of his new natural philosophical commitment to a plenum of gross and aetherial fluids.

That the mercury possesses a tendency to motion of descent greater than that of the nearby air is an obvious fact necessary to Descartes' explanation, for it alone can explain why mercury falls when D is opened. Were we to try to represent the tendency to motion of the mercury in the style of the hydrostatics text of 1619, we would consider the path of displacement RODR which would appear upon opening the hole at D. We can see this by recalling the hydrostatics manuscript in Fig. 8.2, which is a modified version of part of Fig. 3.3.

Had Descartes wanted to bring the hydrostatics manuscript up to date, he could have merely completed the explanation of the hydrostatic paradox by drawing in the circuits of displacement shown in the new figure, thus representing how surface f can supply sufficient total tendency to descend to points on the bottom of the vase. Of course, what makes the present case different is that the problem now is not merely to trace out or 'figure up' the tendencies to motion, but to explain the absence of motion, despite the undoubted existence of a tendency to motion in the mercury and the 'open possibility' of its falling, since the tube has no bottom, unlike the basins in the hydrostatics manuscript. That is, what really needs explaining is what holds the mercury in when R is open, a problem that did not arise in the closed basins of 1619. This is surely the reason Descartes embraces, before the fact, the 'Torricellian sounding' idea of the weighty column of air particles; but ironically, his tactic only makes sense when one accepts that he has in effect ignored some of his own prior assumptions and modes of explanation in physico-mathematics.

Descartes may have reasoned as follows: There is a sense in which the air in the path ROD cannot plausibly hold the mercury in the tube, even when D is closed. Mercury is 'heavier' than the air and therefore one might maintain that even when D is closed, the mercury should flow out at R, pushing the air below in all directions, whilst aether flows in the top to complete the circuit. Certainly no occult resistance to the formation of a vacuum can be asserted to make up the needed extra measure of 'retentive' force lacked by the air. Hence, in order to hem in the mercury, it seemingly became necessary to posit [1] the 'cosmic' injunction on the circulation of aether particles, and [2] the weight of the column OPQ of air particles.[63] *But he did this despite the fact that as against the hydrostatics text of 1619, one was now asserting that the tendency to motion is exercised along a different path than that which would be realized if motion were to occur.* In other words, the cosmic injunction was probably not implausible at all, in Descartes' view, given his new plenist matter theory. After all it is an obvious fact, indeed the very core of the problem at hand, that the mercury does not fall out of the inverted, closed tube. Hence, if aether is real, aether must not be flowing into the closed end of the tube! However, Descartes was also still reaching back to his physico-mathematical hydrostatics of 1619 for conceptual resources to address the problem: As in the earlier case, he still did not possess, and hence could not introduce, a generalized notion of fluid pressure—he does not even say to Reneri that QPOR presses back on the mercury. Rather, as in 1619, he refers to ROPQ as the line of particles which must be raised if the mercury is to fall at R. That is, a tendency to motion is illustrated by a potential line of displacement constructed through an ad hoc distribution of properties to particles of a fluid, even though, as we just noted, contra the hydrostatics manuscript, the imputed, 'figured up', line of tendency to motion, is *not* the path that the particles in question would follow if their tendency to motion were actualized. [64]

[63] And that is just what Descartes also reinforced in the last passage we cited above at Note 61, when he suggested to Reneri that the measure of the force it would take to initiate the descent of the mercury out of the closed tube would be the weight of the column of air OPQ.

[64] On this and the entire document, see Gaukroger (1995, 235–6) whose interesting discussion itself in part extended my own earlier account (Schuster 1977, 594–601), which in turn I am further extending here, given the aims and findings of the present volume.

Viewed this way, the letter to Reneri and its incoherences reveal the flow of Descartes' concerns from the physico-mathematics of 1619 and toward the systematic, plenist-holist, corpuscular-mechanism of *Le Monde*. He still creatively exploits his physico-mathematical approach of 1619, with its rather ad hoc style of figuring up, while striving to blend it with the needs of a newer plenist fluid mechanics, involving displacement circuits and considerations of the properties and movements of earthy and aetherial particles making up that plenum. And, these newer considerations in turn point forward to the style and content of *Le Monde*, in particular, as we shall see, to the central vortex mechanics and the accounts of celestial motions and production of light which it facilitates.

We have now completed the first stage in our structured analysis of Descartes' post-*Regulae* career inflection, by looking at three new agendas which both address problems of the later *Regulae* and eventually play roles in the assemblage of *Le Monde*. We turn now to the second moment in our analysis—consideration of some of the specific events and interactions which have been taken to have shaped the 'mature' post-*Regulae* career of Descartes.

8.3 Events and Interactions Partially Shaping the Motives for and Content of *Le Monde*

8.3.1 Abandonment of the Later Regulae Actually the Most Important 'Event' of All

The events or episodes we are about to discuss have sometimes been taken, individually, as the keys to understanding the character of mature Cartesianism, especially when focused on Descartes in his role as a dualist metaphysician and first of the modern epistemologists. They are worth pondering, but in a cooler frame of mind, keeping in view, on the one hand, the coherent model of his trajectory up to the failure and abandonment of the *Regulae* we have developed thus far, and, on the other hand, the sheer magnitude and import of the three intellectual agendas we identified in Sect. 8.2. In addition we shall eventually need to factor in the flux and drift of Descartes' activities 1629–1633, as he wrote and organized *Le Monde*, which we shall survey below in Sect. 8.4. In these circumstances, no one subsequent event or interaction in the years 1628–1633 is likely to have set, confirmed, given definitive content to, or even in itself triggered mature Cartesianism, let alone the smaller target of his next big product, *Le Monde* as a system of natural philosophy.

In order to sort out these events, and place them in a more adequate historiographical light, we must remind ourselves of the facts of Descartes' situation around 1628–1629 as established so far in this study: In the manner of Mersenne, Descartes had addressed the *Regulae* to a perceived need to counter, or side-step, a scepticism corrosive of both natural philosophical and religious belief, while also blunting the threat—cognitive and religious—of unorthodox philosophies of nature. But the

Regulae project of articulating his method through an ontologically validated universal mathematics had failed. Now, this certainly did not prevent a somewhat cynical—and sometimes self-deludedly hopeful—Descartes (as we have diagnosed him) from subsequently displaying and advocating the method in public. Nevertheless, on the levels of actual day to day work, larger agendas, and self-identity as a man of knowledge, the collapse of the project of the later *Regulae* left Descartes with no larger programs than his bits and pieces of analytical mathematics, and his piecemeal style of physico–mathematics linked to matter-cause discourse about corpuscles and their dynamics. Yet, within two years of the downfall of the *Regulae* he would be well at work on a system of natural philosophy, having also already done non-trivial work on dualist metaphysics and special items in voluntarist theology designed to back it up. We have seen the articulations between the problems that crippled the project of the later *Regulae* and the emergent dualism and voluntarist theology, which aimed to solve some of them, whilst also grounding the now crystallizing natural philosophical system in *Le Monde*. This internal dialectic of intellectual struggle, and resultant trajectory of creative work are arguably much more important in understanding Descartes' inflection toward the making of *Le Monde* than any particular event or interaction within the period in question. Suitably prepared, and cautiously sceptical of quick fix pictures of 'entire destinies in one event', we now turn to three events of the years 1628–1633, seeking more precisely to gauge their impacts, if any, on the inflection of Descartes' intellectual identity and agenda in these years of germination of *Le Monde*.

8.3.2 The Chandoux Episode and Relations with Cardinal Bérulle: Method and/or Metaphysics and the Defeat of Scepticism

In his biography of Descartes, the first comprehensive one ever produced, Adrien Baillet related that Bérulle was quite taken with Descartes' dialectical demolition of a new natural philosophical system presented by Chandoux at the residence of the Papal Nuncio.[65] This event most likely occurred in December 1628 after Descartes had arrived back from the United Provinces.[66] Descartes and other philosophical and theological dignitaries had gathered that evening at the residence of the Papal Nuncio in order to hear the alchemist Chandoux present a new system of natural philosophy meant to replace that of Aristotle. Urged on by his fellow listeners, who suspected that he had not been entirely pleased with Chandoux's performance, Descartes put on a rare public display of his dialectical skill. He assailed Chandoux's claims to certainty and then proceeded to argue persuasively for the falsity of

[65] Baillet (1691) liv II.

[66] Mersenne (1932–1988) II, 163. De Waard's note puts the Chandoux episode in December 1628.

several propositions commonly taken as true, and conversely, for the truth of some propositions usually accepted as false. This dialectical tour de force captured the interest of Bérulle, who wanted to hear more about Descartes' projects. Bérulle subsequently had a least one interview with Descartes, during which Descartes outlined his hopes for a revised mechanics and medicine, and was encouraged in his work by the Cardinal, who exhorted him as bound by a duty to God to use his gifts to erect a new system of natural philosophy conveying medical and mechanical benefits upon the public.[67]

Of course we next see Descartes bemoving himself back to the United Provinces and becoming involved in elaborating his dualism and bits of voluntarist theology, just before sliding into the complex process that issued in *Le Monde.* Beginning as early as Baillet in the late seventeenth century, it has struck commentators that we have here the beginning of the emergence of the real or mature Descartes, the dualist metaphysician, modern looking epistemologist (and not least of all, the systematic mechanistic natural philosopher), no longer the mere piecemeal 'scientist' or practicing mathematician. The deeper 'philosophical' activities consequent on the move to Netherlands seem somehow possibly motivated, triggered, even shaped in content by the Chandoux episode and interaction with Bérulle. There have been two main variants of this investing the Chandoux/Bérulle episode with career shaping meaning, which we shall call the religio-apologetic and the anti-sceptical. Both focus on the fact that immediately following these events Descartes began his explorations of dualist metaphysics, in the former case because the dualism hinges on the proof of the existence of God and in turn His guarantee of the truth of clear and distinct intuitions, in the latter case because it is this metaphysics which claims to defeat scepticism about the possibility of the truth of mathematics, and of the correct form of natural philosophy.

It should be clear, whatever else we make of these claims, that there is no difficulty in seeing apologetical and anti-sceptical overtones in Descartes' dealings with Bérulle, even though they seem to have focused on natural philosophy, not metaphysics or theology. For example, in analysing the interaction, Gouhier observed that properly apologetical subjects were not reported to have been discussed.[68] He had been motivated by Blanchet's concern about this point: the omission of apologetical topics having so bothered Blanchet that he struggled to show that Bérulle had actually encouraged Descartes to pursue metaphysics and theology in order to crush the 'libertines'.[69] But, as Gouhier pointed out, we do not have to believe that Bérulle lacked interest in non-theological matters and that he could not have discussed natural knowledge matters with Descartes. It is true that throughout his career Bérulle chided the vanity of the human sciences. That still does not entail that he was uninterested in the possibility of a new, well grounded, non-Scholastic natural philosophy.[70] What, after all, was he doing at the Nuncio's

[67] Baillet (1691) liv II.

[68] Gouhier (1924) 58–9.

[69] L. Blanchet (1920) 86–7.

[70] Gouhier (1924) 60.

residence in the first place except auditing the presentation of a new alternative to Aristotle? So, we may draw this conclusion: Providing a firm basis for natural philosophy could easily have been seen as a devout act in a period when, in the private view and/or public rhetoric of some, sceptics and libertines were attacking orthodox knowledge and theologically suspect natural philosophies were proliferating. Accordingly, Gouhier concluded that,

> ...il n'etait pas indispensible que le Cardinal de Bérulle ordonnat expressment à Descartes à la fin de 1628 de trouver des preuves de l'existence de Dieu, pour que Descartes eut le sentiment de faire une oeuvre chretienne.[71]

Descartes and Bérulle certainly both could have perceived the challenge of constructing a new system of natural philosophy as an eminently apologetically relevant enterprise. Furthermore, what Gouhier perceived about the Bérulle interview holds *a fortiori* regarding Mersenne's work and the general intellectual atmosphere in which he and Descartes had moved in the mid and later 1620s. Bérulle was not pointing out anything about the apologetical valencies of natural philosophizing that Descartes could not have known for himself, especially as a result of talking with Mersenne, and, in the later *Regulae*, trying to emulate and surpass him in these regards. The defense of the truth of mathematics and some bits of natural knowledge would protect the orthodox from destructive scepticism and the encroachments of unsavory natural philosophies, and it would renew and reform the central Scholastic teaching that natural philosophy is a propaedeutic to theology, a result much desired by the two former students of the Jesuits of La Flèche.

Similarly, Richard Popkin was able to launch an argument about the centrality of the Chandoux-Bérulle episode in motivating and shaping Descartes' career as a metaphysical warrior against scepticism. Popkin did not discuss the significance of the *Regulae* as Descartes' first attempt to resolve the sceptical crisis (albeit as we have seen, in a way avoiding explicit metaphysical argument and natural philosophical systematization). Popkin did, however, focus on the events of late 1628 as the triggers to Descartes' mature metaphysical campaign against scepticism, due to his now awakened alarm at what Popkin depicted as an all encompassing 'crise pyrrhonienne' infecting all fields of knowledge. Descartes' own sense of the sceptical threat to the possibility of certain knowledge might have been heightened as a result of the Chandoux episode. After all, what he really had done at the Nuncio's residence was to give a masterful exercise in the strategies of destructive scepticism. Skillful dialectics had systematically reduced firm conviction to paralyzing doubt. So, on Popkin's well known telling, it is just possible that Descartes' forthcoming strenuous attempt to defeat scepticism with the tools of dualist metaphysics owed its proximate motivation to his sudden re-acquaintance with the imminent danger posed by a general 'sceptical 'crisis' engulfing all culture. [72]

---

[71] *Ibid.* p.61

[72] Popkin (1964) 178–9, 176; We have placed 'crise pyrrhonienne' and 'sceptical crisis' in scare quotes because Popkin's hypostatization of a totalizing sceptical crisis seems implausible by contemporary standards of intellectual history and historiography. This does not mean we must reject his claim that the Chandoux/Bérulle episode deepened Descartes' arguably already existing concern

Nevertheless, considerable care, both historical and historiographical, must be exercised in assessing these sorts of claims. The Chandoux-Bérulle episodes may well have been important, but not in such automatic, totalizing ways, and not by means of such straightforward psychology of motivation. To understand why, we need to start with some already established facts about these matters: We know that Descartes' *Regulae* project was in trouble and that he knew it, which means that the efficacy of his method and his ways of grounding mathematical and physico-mathematical knowledge against sceptical attack were in question. Nor *a fortiori* could his proposed *mathesis univeralis* offer an alternative to, and immunization against, putatively unorthodox alternative natural philosophies. This in turn certainly throws a poignant light on his appeal to his method against Chandoux! However, we also know that despite all this, the wistful hope lingered that the method existed and could work. Behind that hope stood the very real facts of his concrete mathematical and physico–mathematical accomplishments, not produced by his method, as we know, but which might appear to have been able to have been so. We also know that Descartes had already gone to the United Provinces in October 1628 and clearly may have been considering moving there, Chandoux and Bérulle not withstanding. He of course had not envisioned *Le Monde* by this stage, and would not do so for almost another year, hence arguably quite apart from any dealings with Chandoux and Bérulle.

Finally, to this accounting of reasonably certain facts, let us consider, in a more speculative, but still defendable vein, Descartes' likely situation as regards motivation and self-understanding at this juncture. He was not some 'cultural dope', possessed of a real method that really worked, and aware of a real 'sceptical crisis' and hence ready to be prompted by an authority figure to go off into exile to defeat scepticism or save orthodox natural knowledge by means of metaphysical construction. No, Descartes had good reasons to doubt his ability to address any such large agenda. He was offered the encouragement of a respected and authoritative cultural figure. But Descartes knew that, to put it mildly, he had given Bérulle a somewhat optimistic gloss about the state of his intellectual tools and agendas, prompted most likely by the tone and excitement of the interview itself. We can grasp the likelihood of this if we contrast Descartes own 'post-*Regulae*' doubt and malaise against Bérulle's encouragement and the overstatements and optimistic spins elicited from Descartes by the interview. Just picture Descartes in the interview with Bérulle: he is a little out on limb; his public display and posturing about method[73] against

about making out knowledge claims in ways that contemporaries might see as avoiding or overcoming the usual topoi of sceptical attack, now so popular in some (not all!) cultural circles. If sceptical posing was popular and a way to attract attention, patronage or fame, so might be attempts to overcome or defeat it. The later *Regulae* already display Descartes' sensitivity to being able to play some kind of 'post-Mersennian' hand in this culture war, whose epicenter seemed to be the very Paris in which he was living.

[73] Although Descartes had hinted to the Chandoux gathering that appeal to his 'natural method' provided a way out of cognitive dilemmas, given that he was dubious about the *Regulae* at this point, and had just abandoned them, this confident assertion would have rung hollow in his ears on that very occasion. This observation in turn creates the appropriate point to return to the issue of

Chandoux had attracted the attention of Bérulle and raised *his expectations*; but, of course, Descartes does not reveal, cannot reveal, the true 'history of his spirit' (climaxing in the unraveling of the *Regulae*) to Bérulle (or ever to anyone); he makes some big claims to Bérulle which he knows are not quite right; yet the mythopoeic power of method is such that he still hopes they still might be; and he is urged on by Bérulle, as if his own claims to Bérulle were much less risky than he knew they were. Some cultural dope!

So, the intellectual and psychological situation of Descartes was more complex than the usual accounts of Chandoux-Bérulle acknowledge. Descartes was not, as some serious biographers assert, simply re-awakened to the task of building a philosophical system—for apologetic and/or anti-sceptical reasons—a task for which he was already intellectually equipped, but which he had long delayed.[74] Bérulle's exhortations to system building, or to metaphysical construction, if such they were, certainly could have had a positive, reinforcing effect at this juncture, but without giving much guidance as to where and how Descartes' projects would next evolve. In short the intervention of Bérulle may have given a boost to Descartes' resolve and confidence, perhaps convincing him that he should shake off a mounting sense of

the newly discovered 'Cambridge manuscript' of the *Regulae*, mentioned above in Chap. 5, note 23. We now have seen the content, intent and failure of the later *Regulae*, and have studied the most important issues involved in Descartes' career inflection consequent upon that failure. The 'Cambridge ms.' is reported to be 40 % shorter than the other versions; to end at rule 16; to contain the material on the anaclastic curve in rule 8; and to lack rule 4B, as well as the material on 'simple natures' in rule 12. All these reported facts, combined with our overall interpretation to this point, suggest the following conjecture about the Cambridge ms., which I advance pending its full publication sometime in the next few years: The abridged Cambridge ms. looks to be a version concocted as a holding action, after the discovery of the law of refraction in 1626 and before the final abandonment of the project in 1628. The ms. seems to have been sculpted to avoid overt revelation of the obvious difficulties arising from rules 17–22, (See Sect. 7.6.3) so that it might seem to a reader that the project was still on course. I suspect that it was intended for a friend, Mersenne or Beeckman, or potential patron, like Bérulle, to whom, we have seen, Descartes had probably sounded off about his 'method'. In the context of Descartes' awareness of the failure of the project, and his growing public profile, the idea behind this shortened document would perhaps have been something like this: *'People expect something about method/universal mathematics from me after my recent private and public posturing; but the* Regulae *are not going to work out, as I have recently discovered; however, until I find my way forward, a streamlined, less confused and confusing version of the* Regulae *can circulate. (When I see how to get around what stymied it, my new projects and results will drive this out of people's minds anyway.)'* A *Regulae* text, lacking the confusingly redundant rule 4B, and ending with rule 16, could still look like a promising and yet to be fully completed project. Had the remaining rules been present (amounting to about ten pages of mathematical material in the Adam and Tannery edition), smart mathematicians at the very least, would have seen Descartes' difficulties and textual squirming, and so suspected something was seriously amiss. Judicious cutting made the text look as though it were still representing a living project, and that clear sailing would still be ahead, should Descartes move past the intriguing rule 16.

[74] Vrooman (1970) 74–5. 'This meeting with Chandoux and the subsequent conversations with Bérulle served to rekindle the enthusiasm and sense of mission he had received from his friendship with Beeckman and the revelation of his dreams'. So much for the tortured trajectories of Descartes' physico-mathematics, method and corpuscular-mechanism we have traced thus far in our own account.

doubt, and awareness of the limitations of his previous large projects, in order to press on with the constructive work he still believed might be possible on the basis of his past record of accomplishment. But none of that alone could or would have created the motive, let alone the content, of his next moves. Descartes' agenda and self-identity were already in flux; he was already beyond the dream of method and universal mathematics in the later *Regulae*, but on what roads exactly, were not obvious to him—nor to us beyond the facts that he was soon to work in metaphysics and theology in ways evoked by the failure of the *Regulae*, and that by mid 1629 he was also going to embrace eagerly challenges and problems that eventually led to *Le Monde* as a system of nature.

The traditional accounts of why Descartes soon left Paris for the United Provinces fit into this view of the pattern of events and our non-linear reading of the Chandoux/Bérulle affair.[75] Having become something of a cultural celebrity himself, Descartes found much of his time wasted with what to his mind were fruitless and inane interruptions. His move to the Netherlands was primarily motivated by the need to escape the annoyances of Paris in order to maintain privacy for his work. But, given the considerations set out above, we need not simply say that after nine years Descartes finally saw that the time had come to construct his physics or his metaphysics in the United Provinces instead of in Paris. Rather, it now seems plausible to suggest that at least part of the reason Descartes decided to isolate himself was because he had just passed through a period of deepened doubt and disappointment concerning his own projects. More celebrity and public posturing around town would not have seemed inviting prospects. With the encouragement of Bérulle he might have come to see that some kind of new start had to made beyond the project of the *Regulae*, yet still embodying its concerns with foundations of natural and mathematical knowledge and defense from unorthodox natural philosophical systems. But, what that start would be; where it would lead; and whether and how his intellectual took kit actually could be refurbished after the debacle of the *Regulae*, could hardly have been known by him in Paris at the end of 1628, with or without his discussions with Bérulle, themselves bound up in a confused and confusing dialectic where defeated hope and agendas intertwined with publicly expressed over–confidence and in private, we theorize, an almost delusional skating on thin ice of programmatic claim and hope.

8.3.3 Challenge of Renewed Interaction with Beeckman

During the period of his career inflection Descartes had two phases of significant interaction with his old mentor Isaac Beeckman. The first occurred in the autumn of 1628, when Descartes paid a short visit to the Low Countries prior to his settling there permanently early the next year.[76] It was mutually productive, and in some

[75] Adam's biography of Descartes in the original AT XII p.106; Baillet (1691) livre II.

[76] Beeckman (1939–1953) iii. 114 note 3; Mersenne (1932–1988) ii. 222, 217–8, 233–44; AT x 341–3. Beeckman (1939–53) iii. 103.

ways very important in eventually shaping the aims and content of *Le Monde*. The second occurred at a distance through direct and indirect correspondence, and took place over a least a year, leading to a break in their relations and a difficult period for Descartes in the midst of his trying to compose *Le Monde*.

As to the first phase: on 4 October 1628 Descartes met with Beeckman for the first time since early 1619. As we already know, he sketched for Beeckman some of his discoveries of the previous nine years, including the work on lens theory. This was prefaced by a statement of the (sine) law of refraction which Beeckman recorded in a short memorandum, along with the interesting bent arm balance analogy for the physical causes of the law of refraction, which we analyzed in detail in Chap. 4. Beeckman contributed to this intellectual exchange, for example, by providing Descartes with an improved demonstration of how the law of refraction permits the geometrical specification of the anaclastic problem in the case of a plano-hyperbolic lens.[77] Beeckman was also involved at this time in a sustained intellectual program which we did not mention in our earlier study of his exchanges with Descartes about optics. On his visit Descartes found Beeckman systematically working through the astronomical works of Kepler, focusing on passages where Kepler invokes immaterial celestial forces, which Beeckman sought in each case to re-write in corpuscular-mechanical terms. Beeckman, in other words, saw in realist Copernican astronomy, especially as transformed by Kepler, a target for natural philosophical explication; that is, a problematic to be addressed in terms of matter and cause, since it presented the problem of explaining the motion of the planets around the sun, and the nature of the sun and planets (including the Earth) such that these celestial motions can occur. Beeckman was indulging in what he called a *restitutio astronomiae*,[78] and we may term it an exercise in celestial mechanics, of a discursive, natural philosophical type. This project of Beeckman was of the utmost import for Descartes. For that reason, we defer discussion of the details here, preferring to introduce them later, in Chap. 10, after we have explored in detail the content of Descartes' own vortex celestial mechanics in *Le Monde*. We note this here for future reference as one of the most important stimuli, and indeed models for what Descartes would soon attempt in *Le Monde*. But more was at stake than this specific stimulus to a focal point of Descartes' own emergent system. With these celestial mechanical and Copernican realist speculations, Beeckman was also beginning a project to publish his own version of the mechanical philosophy, a fact which was going to shape then next, turbulent period of his relations with Descartes.[79]

The second phase of the renewed Descartes-Beeckman relations began in the summer of 1630, when Mersenne paid a short visit to the United Provinces. Mersenne also saw Beeckman, who invited Mersenne to examine his *Journal*, Mersenne thus becoming only the second savant, after Descartes, to have done so.

---

[77] Section 4.5.2 and Appendix 1.

[78] Beeckman (1939–53) iii. 103. In the period July 1628 to June 1629 roughly 21 out of 59 pages of Beeckman's journal deal with celestial mechanical and related matters.

[79] van Berkel (2000)

One of Beeckman's aims was to have Mersenne see for himself who had been the author of many natural philosophical ideas Descartes had recently been casting about in his correspondence. Curiously, or perhaps not so curiously given his agenda, Beeckman included a report on this meeting in a subsequent letter to Descartes, and Descartes in turn cited it back to Mersenne, to the effect that Beeckman had told him that Mersenne had realized in perusing the *Journal* that Beeckman deserved credit for a number of achievements he, Mersenne, had previously been led to attribute to Descartes.[80]

The inelegant, but by no means groundless implication that much of Descartes' work derived from the speculations contained in Beeckman's diary was the proximate cause of the agitation and lassitude about pushing on with *Le Monde* that Descartes displayed in the late summer and fall of 1630, which we discuss below in Sect. 8.4.7. Given our study of the matter in Chap. 3, we are perhaps better placed than even Mersenne to appreciate the deep lines of filiation that connect Beeckman's corpuscular-mechanism, and his dream of a physico-mathematics, with the early agenda and identity of Descartes. In addition, as just flagged for later detailed analysis, Descartes' new enterprise in 'celestial mechanics', in the vortex mechanics he was just then devising, had a strong affinity to the celestial mechanical speculations Mersenne would have found in the more recent parts of Beeckman's *Journal*. Descartes surely thought his approach was better, as we shall discuss in Chap. 10, but he had not finished his work, and the perusal of the *Journal* by informed third parties, especially his other chief friend and mentor Mersenne, was bound to raise alarm bells in Descartes' mind, and indeed his conscience.

In fact, for some time difficulties had been simmering with Beeckman. Their origins date the late summer of 1629 when Beeckman had intimated to Mersenne that the *Compendium of Music* only contained material Descartes had drawn from his own work.[81] Hearing of this from Mersenne, Descartes had expressed mild displeasure with Beeckman,

> I am very obliged to you for calling the ingratitude of my friend to my attention: I think the honor I have done him of writing to him has dazzled him, and he thought that you would have an even better opinion of him if he wrote to you that he had been my master ten years ago. But he is completely mistaken, for what glory is there in having taught a man who knows very little and freely admits this, as I do?[82]

[80] Descartes to Mersenne 4 November 1630, AT I 171. 'Cumque Mersennus tuus totas dies in Libro meo manuscripto versaretur, atque in eo pleraque, quae tua esse existimabat, videret, et ex tempore illis addito, de illorum Authore merito dubitaret, id quod res erat illi liberius fortassis, quam tibi aut illi placuit, aperui.'

[81] Descartes to Mersenne 8 October 1629, AT I 24 and the Note to this passage on p.30. Beeckman had written to Mersenne concerning Descartes, 'Ipsius, inquam, is est cui ante decem annos ea quae de causis dulcedinis consonantiarmum scripseram communicavi'.

[82] Descartes to Mersenne 8 October 1629 AT I 24. 'Vous m'avez extremement oblige de m'advertir de l'ingratitude de mon ami; c'est je croi, l'honneur que vous lui avez fait de lui escrire, qui l'a eblouy, et il a cru que vous auriez encore meilleure opinion de lui s'il vous ecrivait qu'il a ete mon maitre il y a dix ans. Mais il se trompe fort; car quelle gloire y a-t-il d'avoir instruit un homme qui ne scait que tres peu de chose, et qui le confesse librement comme je fais?' I cite here Gaukroger's well atuned translation (Gaukroger 1995, 223)

Although Descartes was willing to admit to considerable ignorance in the past, he clearly was sensitive to Beeckman's potential claims upon their mutually held opinions. Soon afterward he took an opportunity to belittle Beeckman's intellectual progress in the years since 1618. Hearing from Mersenne that Beeckman had presented copies of his doctoral theses to Gassendi when the latter visited Dordrecht, Descartes sarcastically remarked how surprised he was that 'mon Docteur' had given Gassendi these decade old works, let alone that he had kept them that long, apparently having nothing better subsequently written to display.[83]

Clearly, even before Mersenne's personal inspection of the *Journal*, Descartes was on the verge of a serious confrontation with Beeckman. What lay behind Descartes' denigration of Beeckman was a very well founded awareness of an imminent clash over intellectual property, now that he had decided to offer a system of natural philosophy to the public, just as Beeckman had also decided to do.[84]

In any case Descartes' concern over Beeckman's possible claims reached a climax in October 1630 when Mersenne apparently sent him an account of some of Beeckman's more recent pronouncements.[85] Writing to Beeckman, Descartes unleashed a storm of sarcastic invective noteworthy even in the annals of Baroque histrionics. Beeckman, contends Descartes, surely suffers from some sort of illness, rather than from consciously malevolent intents. Therefore he is more to be pitied than blamed. Descartes, in a parody of his later style of humanitarian psycho-therapy, is prepared to offer the 'cure' in the form of an admonition, based upon a curious distinction, and of dubious applicability. Even though one might have gathered some information from another individual, he has not learned it from the other unless persuaded by his valid arguments or authority (presumably in matters of faith only). In addition and accordingly, several people might be said to know something without any of them having learned it from another.[86] These contentions, and Descartes' snide rhetorical play in the letter on merely 'sensing' versus 'knowing', imply that Descartes has the honor of having logically deduced and articulated a body of natural philosophical knowledge out of disparate pieces of information, some of which may have been culled from Beeckman, but which Beeckman had accepted on false or non-existent grounds. We might say that Descartes' polemical line here invokes his doctrine of demonstrative *Scientia* in order to secure his interest

[83] Descartes to Mersenne 25 February 1630 AT I 122. 'Vous m'etonnés de dire que mon Docteur ait donné ses Theses a Mr Gassendi: je n'eusse pas cru qu'il les eut gardées si longtemps, et c' est bien a dire qu'il n'a rien fait depuis qui sont meilleur.'

[84] Beeckman, for his part, may have even started to push his claims against Descartes even more vigorously now that he knew of Descartes' new project and publication plans. On Beeckman's own plans and work toward his own system of natural philosophy, see van Berkel (2000).

[85] Evidence in Descartes to Mersenne 25 November 1630 AT I 177–78. The opening lines of this letter imply that Descartes had been informed by Mersenne and that he had sought to hide Mersenne's role when writing to Beeckman.

[86] Descartes to Beeckman 17 October 1630, AT I 158 'Si quis vero nullius auctoritate nec rationibus adductus aliquid credit, quamvis hoc ipsum a plerisque audiverit, non tamen ab illis didicisse putandus est. Imo potest fieri ut sciat, quia propter veras rationes ad credendum adducitur; alii autem, quamvis prius idem senserint, non tamen scierunt quoniam ex falsis principiis deduxerunt'.

in a budding priority dispute triggered by the impending completion of *Le Monde*.[87] Beeckman and Descartes did meet again, but relations were never the same as they had been in 1618–1619 or, apparently in October 1628. Beeckman never published his system (a version, still in the form of an assembly of notes, being published by his brother Abraham in 1644, well after Beeckman's death in 1637) and indeed he did not work toward it after 1630 as this dispute grew.[88] Descartes therefore may have succeeded in browbeating the modest and controversy-shy Beeckman into retreat to his previous style of private, piecemeal diarization of his work. That probably explains his openness to a (still cool) reconciliation of sorts, marked amongst other things, by lack of emotional engagement at the time of Beeckman's death.[89] Finally, returning to the situation in late 1630, we can see that Descartes was definitely on the hunt for priority in systematic corpuscular–mechanical natural philosophy. Later we shall see just how much Beeckman's celestial mechanical speculations additionally aided and spurred him on in this period.[90]

8.3.4 The Galileo Affair and Its Perceived Meanings

Only in November 1633, as *Le Monde* was very close to completion (Descartes claims he had it ready for Mersenne as a New Year's gift)[91], did he learn of the trial and condemnation of Galileo five months earlier. He had been found 'vehemently suspect of heresy' for having held that the 'sun is the centre of the world and does not move from east to west and that the Earth moves and is not the centre of the universe.'[92] Descartes, inquiring in Leyden and Amsterdam for a copy of Galileo's *Dialogues Concerning the Two Chief World Systems*, the work for which Galileo had been prosecuted, had been informed that 'all the copies had been burned at Rome and Galileo condemned'.[93]

If we are going to assess the impact of the Galileo affair on Descartes' agenda and the project of *Le Monde*, we need to be very clear about certain facts, the chief of which is this: Properly read, *Le Monde* is a system of natural philosophy embodying a full and frank endorsement of realist Copernicanism. Despite Descartes' deployment

[87] Just as he would always, when pressed or threatened intellectually, invoke his method and the fairy tale of deductive certitude for key parts of his work. The further psychological reaches of this episode are well canvassed by Gaukroger (1995, 224), invoking also Floris Cohen's analysis, to remind us that even this baroque storm blew over, with Descartes and Beeckman being to a degree reconciled, although they did not ever after see much of each other and the genuine warmth of 1618–1619 had completely evaporated.

[88] See again the admirable account of van Berkel (2000), especially p.57.

[89] Gaukroger (1995) 224, citing Descartes to Colvius 14 June 1637, AT I 379–80

[90] Section 10.3

[91] Descartes to Mersenne, end of November 1633, AT I 270–1.

[92] Langford (1971) 152.

[93] Descartes to Mersenne, end of November 1633, AT I 270.

of the presentational conceit of a fable (about which we shall hear more in the next chapter), there can be no serious doubt that *Le Monde* teaches realist Copernicanism, and a radical form of realist Copernicanism at that. Descartes joins with Bruno, in opposition to Copernicus, Galileo and Kepler, in asserting an infinite universe of infinitely many star and planets systems, of which our solar system is an undistinguished ordinary member (or if you like to follow Descartes' spin doctoring, an indefinitely large universe containing an indefinitely large number of stars with orbiting systems of planets). In addition, as we shall see in detail later, the universe of stellar vortices is not some add on to a generic mechanistic system of natural philosophy whose centre of interest resides elsewhere. Rather, the vortex mechanics is the system's centerpiece, in two senses—it is the pivot of the content and it contains the conceptual core, because the complex, but exemplary vortex celestial mechanics, an amalgam of a number of strands of Descartes' physico-mathematical and natural philosophical thinking, is the intended template and basis for a wide range of explanations across the scope of the system.

Indeed, the fable form of explication was present in the text before the condemnation of Galileo, and it was not meant seriously to deflect an informed reader from taking the realist meaning of the cosmology. If one were interested in Descartes' principles of matter, and of dynamics, and the letter of his laws of nature, there could have been little doubt amongst intelligent and natural philosophically literate readers that this was not a sceptical fancy, but rather an intended best realist explanation. Its limits are meant to be precisely those commonly asserted of natural philosophizing; that is, it is putatively based on experience and reason, not divine revelation or orthodox theological dictums, nor the poet or dramatist's imagination. At best the fable form fits with the *honnête homme* style of the opening, more pedagogical, chapters. This might influence the educated amateur, and perhaps seduce the non-expert cleric or censor, to think something different from natural philosophizing is going on. But, regardless of whether they would have accepted the claims of *Le Monde*, the idea that a Galileo, Beeckman, Gassendi, Hobbes or Mersenne would not have taken it as intended as a bid for 'best explanation in natural philosophy' beggars the imagination, provided *Le Monde* is read with the same expertise, and natural philosophical and cosmological concern they would have exercised. This means that the fabular narrative in *Le Monde* is irrelevant to the issue of Descartes' response to the Galileo affair. *Descartes' problem with the condemnation of Galileo was precisely that he too was teaching realist Copernicanism and, unlike Galileo, an 'infinite universe and planetary systems' realist Copernicanism as part of a completely new system of natural philosophy.* Even putting aside all possible issues and niceties of Church law, political geography, publishing licensing, censorship and the distribution of ones' friends and enemies, he had a right to think he had a serious problem on his hands.

Writing to Mersenne at the end of November 1633, Descartes recounted first hearing of the condemnation,

> I was so surprised by this that I nearly decided to burn all my papers, or at least let no one see them. For I could not imagine that he, an Italian, and I believe, in favor with the Pope, could have been made a criminal, just because he tried, as he certainly did, to establish that

> the earth moves. I know that some Cardinals had already censured this view, but I thought I had heard it said that all the same it was being taught publicly even in Rome.[94]

Descartes and other savants certainly knew of the decree of the Congregation of the Index of 5 March 1616 to the effect that Copernicus' *De Revolutionibus* be suspended until corrected; that is, rendered more hypothetical. The 'Pythagorean doctrine' had been declared false and contrary to Holy Scripture, but only books attempting to prove the truth of the doctrine and its concordance with the Bible had been prohibited and condemned. Descartes correctly surmised that the original decree had not amounted to the creation of an article of faith, and, accordingly, he had not doubted that Copernican theory had subsequently been discussed publicly, even in Rome itself. But, if Galileo, a presumed intimate of the Barberini Pope, Urban VIII, could subsequently be condemned, what point would there be for Descartes to offer to the Catholic public his system of natural philosophy, with its deeply Copernican cosmology? Descartes was thirty-seven years old; he had still not published any of his work; if his desire to burn his papers was a genuine but momentary emotional response, his confession of a wish to keep his work secret seems perfectly in character and an understandable response to the shock the Galileo affair offered to his own agenda.[95]

So, it would be wrong to see in Descartes' initial response, written to Mersenne (and hence in knowledge that his views might well be broadcast further in the Minim's correspondence net and immediate Parisian circle) some kind of disingenuous, even dishonest, posturing, particularly on the theological issues and matters of personal belief at stake. Descartes was justifiably and genuinely discouraged by this turn of events; he continued to worry and write about it to Mersenne over the next ten months, and even beyond.[96] Descartes had good reason to wonder about the fate of his highly pro Copernican system of natural philosophy, and about what he, a genuine lay adherent of the Catholic Church should or could do about it.

The remainder of this letter makes clear that three considerations lay behind Descartes' initial response: desire to be, after his own fashion, a devout Catholic; concern for the possibility of establishing truth over mere opinion in natural philosophy; and concern over the possibility of henceforth procuring an audience for his work in France and other Catholic lands. These points were probably closely linked in Descartes' view, and they all ultimately refer back to his now fundamental problem: that the Copernican theory was the basis of his cosmology and unified theory of light and vortex celestial mechanics. He continues to Mersenne,

> I must admit that if this view is false, then so too are all the foundations of my philosophy, because it can be demonstrated from them very clearly. And it is so closely tied to all the parts of my treatise, that I would not be able to detach it without making the whole work defective.[97]

---

[94] Descartes to Mersenne, end of November 1633, AT I 270–1.

[95] But see also the interesting remarks of Gaukroger concerning Descartes' very likely misreading of the situation in Paris at this stage (Gaukroger 1995, 291).

[96] Cf. Grayling's interesting review and commentary about Descartes' prolonged mulling over of the implications, and his less than bold stance, compared to the immediately subsequent publication activities of Mersenne and Gassendi Grayling (2005, 170–173).

[97] Descartes to Mersenne, end of November 1633, AT I 271.

This assertion is quite accurate, because we shall see that the vortex celestial mechanics, and realist Copernicanism, subtending innumerable stellar-planetary systems, was the very core of *Le Monde.* If Copernicanism, particularly in the enlarged sense Descartes was advocating it, could not be asserted in natural philosophy as true, let alone, as he contended, shown on his principles to be true, then his teaching was no better than any other, and a theologically dubious one as well—his system would wind up on the kind of cultural scrap heap where he and Mersenne had intend to park a host of other contenders. 'There are already so many opinions in philosophy', he added,

> that have a plausible appearance, and which can be maintained in debate, that if my claims are no more certain, and cannot be approved of without controversy, then I have no wish ever to publish them.[98]

A system with these handicaps was not going to secure an audience and pedagogical foothold in France, especially amongst his former preceptors, the Jesuits. And to this one can add the issue of what kind of Catholic would be indulging in such work. Descartes was in no personal danger from the Catholic authorities in the Dutch Republic, he was therefore expressing more a personal belief than a public relations stance when he also said to Mersenne directly after his remark about rendering the whole work defective,

> But for all the world I did not want to publish a discourse in which a single word could be found that the Church would have disapproved of; so I preferred to suppress it rather than to publish it in mutilated form.[99]

In the United Provinces Descartes had been engaged in building a new system of natural philosophy, underpinned by his dualist metaphysics and selectively designed bits of voluntarist theology. He was continuing, in a much inflected and enlarged sense, his original Mersenne-inspired program for an anti-sceptical, anti-libertine, arguably orthodox form of natural knowledge, begun in the abandoned *Regulae*, at least in the parts written after 1626, as we have seen. *Le Monde* was going to be sent to Mersenne, and published with his help. It would be perverse to maintain that Descartes had to feign obedience to the Church under these circumstances and given the facts of his upbringing and behaviour. Lacking any knowledge of the 'conflict' of science and theology trumped up in some quarters in the nineteenth century, he was, like his Jesuit preceptors, simultaneously on the side of natural philosophical truth and of the Catholic Church. His problem was that these were not simply passive beliefs on his part. He was bidding to present a form of natural philosophy to replace that which the bulk of educated Catholics believed, or at least had been taught. The Galileo affair drove a wedge between natural philosophical truth, which in his view now had to embrace and articulate the most radical form of realist Copernicanism, and Catholic theological doctrine. He was forced to avoid what

---

[98] *Ibid.*

[99] *Ibid.* This, of course, does not mean Descartes was immune from scrutiny and harassment from the Dutch Calvinist authorities. Descartes' repeated concerns in this regard during his long years in the United Provinces are well narrated and explained by Clarke (2006).

would now appear bound to be not only a public breach with the Church, but a renunciation of his identity as a Catholic.

Nevertheless, Descartes still seemed to hold out hope that the Galileo case had not fully and finally closed the door on realist Copernicanism and on his natural philosophical ambitions. He ends, perhaps surprisingly, by promising Mersenne he will send *Le Monde* within a year, and then somewhat querulously asks what Mersenne himself knows of the Galileo affair.[100] While some might see this as a sign of a disingenuous attitude on the part of Descartes toward his professions of religious orthodoxy and obedience,[101] it can be suggested that Descartes is showing quite human ambivalence: hope, fear and emotional investment in his own work are all in play. He is indeed a genuine, obedient Catholic, but he still wishes that somehow his masterwork might appear: (*'one day soon Père Mersenne will surely have my full text in hand...'* he seems to speculate, wistfully adding, *'and perhaps the decree against Galileo will not totally and forever ban discussion of the truth of Copernicanism...indeed what does it really mean and how will it really be applied?'*)[102]

There is some evidence for these conjectures, provided by a letter Descartes wrote to Mersenne five months later in April 1634, when he learned that his original letters on the Galileo affair had been lost en route.[103] In general Descartes reiterated the positions he had taken in November 1633, by way of bringing Mersenne up to date on these lost observations. We, who possess all these letters, can see that Descartes' sense of the tension, between the natural philosophical truths he had discovered and assembled over the past four years, and their denial by the Church, had not decreased. In a striking passage he juxtaposed the faith of an orthodox natural philosopher (who had recently produced the best system to date) and the good conscience of any obedient lay son of the Church.

> I must tell you that all the things I explained in my treatise, which included the doctrine of the movement of the earth, were so interdependent that it is enough to discover that one of them is false to know that all the arguments I was using are unsound. Though I thought they were based on very certain and evident proofs, I would not wish, for anything in the world, to maintain them against the authority of the Church.[104]

Moreover, as in the letter of November, and again with no less barely disguised plaintiveness, he shows he was not entirely certain that the condemnation of Galileo was the final word on the matter in terms of Church doctrine. And yet, while seeming

---

[100] AT I 271.

[101] Cf. Grayling (2005) 163–4.

[102] In his next letter, February 1634, AT I 281, Descartes asks Mersenne whether the condemnation of Copernican realism has been made an article of faith, and what is the state of debate about the affair in France.

[103] Descartes to Mersenne, April 1634, AT I 284. Descartes states 'les dernieres' had been lost. The letters would be those of November 1633 February 1634, if Descartes meant all his previous correspondence to Mersenne on the subject. Or perhaps he meant only that the intervening letter of February 1634 (AT I 281–2) had been lost, or not sent, or had not yet arrived in Paris at the time Mersenne last wrote him.

[104] Descartes to Mersenne, April 1634, AT I 285. Kenny (1970) 25–26.

to hope there might still be scope to move ahead, he also expresses his orthodox loyalty in a particularly depressed and resigned manner.

> I know that it might be said that not everything which the Roman Inquisitors decide is automatically an article of faith, before it is decided upon by a General Council. But I am not so fond of my own opinions as to want to use such quibbles to be able to maintain them. I desire to live in peace and to continue the life I have begun under the motto (from Ovid) *to live well you must live unseen*. And so I am more happy to be delivered from the fear of my work making unwanted acquaintances than I am unhappy at having lost the time and trouble which I spent on its composition.[105]

Nevertheless, further reflecting the ambivalent play of hope and dejection, he ends the letter by expressing the possibility that the decision of the Church might be reversed, and *Le Monde* eventually appear.[106]

Of course, *Le Monde* was never sent to Mersenne, nor did it appear during Descartes' lifetime; but that does not mean that his post 1629 work was not put before the public under a revised strategy. By 1635 Descartes had decided to detach the *Météores* and *Dioptrique* entirely from *Le Monde* and to publish them as samples of his philosophy.[107] He published the two essays in 1637, adding to them the *Géométrie* and a hastily pieced together preface, the *Discours de la Méthode*. The *Discours* hinted at the content of *Le Monde*, and, as we have seen, placed all this work, and his supposed life-long strategy and trajectory leading to it, under the aegis of a ludicrously truncated version of that method which he hoped to teach in the *Regulae*, before its implosion and collapse had set in train the actual, rather than mythical, saga of *Le Monde* and its intended metaphysical and theological underpinnings. As for the system of corpuscular-mechanism, it appeared, significantly improved and augmented, as the *Principles of Philosophy* in 1644 in a vastly extended Scholastic textbook form, integrated with the now fully articulated dualist metaphysics, which in any case had already been elaborated as a preparatory move in the *Meditations* of 1641.[108]

8.4 The Chronology of *Le Monde*

8.4.1 Introduction: Sliding and Tinkering Toward a System of Natural Philosophy

The key events of late 1628, discussed above, such as the collapse of the program of the *Regulae* and the Chandoux/Bérulle episode, as well as Descartes' re-acquaintance with Beeckman, surely helped to mould the emotional and intellectual environment

105 AT I 285; Kenny (1970) 26.

106 AT I 285.

107 Cf. Descartes to * * *, Fall 1635, AT I 322.

108 The strategies of Descartes in the *Meditations* to deploy the dualist metaphysics to legitimate the natural philosophy to come without revealing any of that natural philosophy in detail are best

in which *Le Monde* was to be composed over the next four years. Suppose one wanted vastly to simplify the picture, ignore the mass of correspondence, and seek instead a 'thirty-second grab' version of Descartes' situation in the winter of 1628–1629. One might describe Descartes, settled back in the United Provinces in early 1629, realizing that a new framework of metaphysics would have to be set down, and convincing himself that such a new framework might permit the construction of a coherent, systematic corpuscular-mechanical natural philosophy. This new system would in turn be the preferred path forward from the wreck of the *Regulae* and would also win the implied challenge to surpass the speculations of Beeckman concerning the corpuscular-mechanical theory of light and celestial mechanics. Unfortunately, this gloss cannot adequately explain the content of *Le Monde* or its diffuse path of composition. The evidence of Descartes' correspondence shows that the conjuncture of events in 1628–1629 was not sufficient to motivate *Le Monde*, account for the timing of its composition or explain its structure. To understand these issues we need to attend to Descartes post 1628 intellectual processes, some key subsequent events, as well as the detailed chronology of composition of *Le Monde*.

Indeed, the evidence, now more abundant, from Descartes' correspondence starting in 1629, shows that he did slide and tinker toward *Le Monde* as a systematic project and as a set of contents. Yet, in the end, the project crystallizes and its content settles into a quite systematically and strategically composed text on natural philosophy (which also had a carefully calculated public face and rhetoric of presentation). Hence, we must not lose sight of that product forming below the surface turbulence. In this task we are helped by remembering the key agendas discussed in Sect. 8.2, as well as taking on board in a cautious way the possible shaping toward *Le Monde* by the key episodes discussed in Sect. 8.3. Still, the very rich evidence of Descartes slipping, sliding and tinkering toward *Le Monde*, the project and the content, also shows the unlikelihood that a particular event or single project can account for the path of inflection and its outcome in *Le Monde*.

8.4.2 *Optics—Physico-Mathematical, Mixed and Practical*

Descartes' first several months in the Dutch Republic in early 1629 were devoted to metaphysical construction and to some new initiatives in his optical work, rather than to any properly natural philosophical concerns.[109] We have already dealt with the metaphysical exertions above in Sect. 8.2.1. It is worth briefly considering the

described in Gaukroger (1995), see above Note 4. On consequential changes and improvements in the natural philosophy of the *Principles* over that in *Le Monde*, see Machamer and McGuire (2009), Biro (2009) chapter 3 part 3, and Chap. 12 below, where some new findings in this regard are presented.

[109] Descartes to Gibieuf, 18 June 1629, AT I 17; to Mersenne 15 April 1630, AT I 144; to Mersenne 25 November 1630, AT I 182.

optical work, since it casts light on Descartes' practical and intellectual concerns at the moment just prior to his beginning work on what was to become *Le Monde*.

Writing to the artisan Ferrier from Franeker in June 1629, Descartes announced that he had recently learned much concerning their lenses.[110] He invited Ferrier to join him so that they could work together on optical matters. Ferrier, however, apparently ignored this invitation. Descartes decided to commit his new ideas to writing in order that Ferrier might pursue the proposed technical work in Paris by himself. From letters exchanged in October 1629, it is clear that Descartes had envisaged a new design for machine which they had previously planned to build in Paris for cutting more accurate hyperbolic lenses.[111] In three extremely detailed letters, they explored the technical difficulties of the new design.[112] Apparently, because of a combination of practical obstacles and Ferrier's preoccupation with other matters, nothing came of this initiative. Descartes was later to complain to Mersenne that he had not heard from Ferrier for some time.[113]

What do these transactions say about how Descartes was perceiving the relations between practical mathematical and instrumental pursuits and his more high cultural program in physico-mathematical optics? First, Descartes was, in his fashion, making a play inside the field of practical mathematics—albeit a stillborn play, but a real play nonetheless. He did indeed want to make and 'show' lenses that would embody his law of refraction, and control a more standard and improved telescope. Such behaviour is indistinguishable from that of a mathematical practitioner, at least

---

[110] To Ferrier 18 June 1629, AT I p.13. These transactions are not to be confused with the work Ferrier actually undertook with Descartes and Mydorge regarding refraction earlier in the 1620s. (Schuster 2000, 274–5; Shea 1991, 150–2.)

[111] Cf. Descartes' report of December 1635 to Constantijn Huygens (AT I 336–7) on the original production of a convex hyperbolic lens by Mydorge and Ferrier in 1626–1627. In the letter Descartes mentioned the difficulty Ferrier had encountered in trying to produce a concave lens. One may conclude that a prime motive for devising an improved lens grinding machine was to facilitate the production of a concave hyperbolic lens. This conjecture is supported by the fact that throughout this exchange of letters in 1629, Descartes and Ferrier drew diagrams of the machine in positions which would lead to the production of convex cutting plates, to be used to cut a concave grinding wheel, which in turn would produce convex lenses. However, they consistently drew the cutting plates supposedly produced by the machine as *concave*, and thus suitable for producing a concave lens. (See Descartes to Ferrier, 8 October 1629, AT I 34–5; Ferrier to Descartes 26 October 1629 AT I 39 and compare the *Dioptrique* Discourse 10). Only later, in his letter of 26 October did Ferrier adjust the figure of the machine, reversing it. The machine depicted in the published *Dioptrique* is a later hybrid between Descartes original design mentioned as having been discussed in Paris, and the machine of the letters of 1629. With the Parisian design Descartes would have cut the lens directly from the motion of the machine; in the 1629 plan, he would have used the machine to cut out individual hyperbolic plates, which in turn would be used to cut the grinding wheel. In the *Dioptrique* the hyperbolic motion of the machine is transferred to a cutting instrument placed at the end of a beam. Several cutting plates are produced at once. The plates and the cutting end of the machine are then used to shape the grinding surface of the wheel.

[112] To Ferrier, 8 Oct. 1629, AT I, p.32; To Ferrier 13 Nov 1629, Ibid.; Ferrier to Descartes 26 Oct 1629, AT I pp.38ff.

[113] To Mersenne 18 March 1630, AT I p.129.

a rather elite and well educated one. Had René Descartes actually needed to make a living this way, presumably he would have been more seriously and steadfastly engaged in this work.

But, secondly, he was maneuvering for primacy at a more elite level, in what he called physico-mathematics, but which, since he was to become fully involved with natural philosophical systematics within weeks, we might as well say was maneuvering for primacy within the culture of natural philosophy. This would be the case when as he continued to pursue the issue of lens grinding machines down through the publication of the *Discours* and *Dioptrique* in 1637. The reason his proposed moves with Ferrier had significance in the higher cultural *agon* of natural philosophy was this: The lens grinding machine was also a physical/mechanical instantiation of the law of refraction—it was not just a machine for pride and profit of the results, as it would have been for a more humdrum type of mathematical practitioner. Indeed, it was a natural philosophical signifier, and what it signified was a concrete achievement in natural philosophy of a specially valued kind, at least in avant garde circles. For this lens machine, guided by natural philosophical principles, was much better than anything that could have been produced by crafty trial and error. It was both illustrative of the truth and maximally productive. Truth and utility are attained simultaneously. As Descartes undoubtedly knew, Ramus and Bacon would have approved of that.[114] Descartes might have been undergoing the early phases of his period of career inflection; he might have had the failure of the *Regulae*, and the challenge of his new metaphysics on his mind. But he was also ready and willing to push on in his most successful arena, physico-mathematical optics, trying to improve his position in the realm of practice and practitioners, and the domain of natural philosophy at the same time. Had the *Regulae* not collapsed, he might have been writing up this sort of work as further illustration of the power of the method and of universal mathematics. And that leads directly to our final historiographical observation about all this.

It should also go without saying—but needs to be said in the light of recent claims in the literature—that Descartes was not forsaking natural philosophy, or physico-mathematics, in the interest of some new, modern experimental 'method' or 'science'. Ferrier, who had worked with Descartes and Mydorge in the 1620s, came again into potential play regarding the new machine after the law of refraction was discovered; after the construction of the central concepts of Descartes' dynamics; and just as *Le Monde* was starting to be written. Hobnobbing, or wanting to

[114] Rossi (1970) masterfully established this general perspective. My points here relate to the putative signification of the lens grinding machine as such. Neil Ribe interestingly widens this perspective, by demonstrating that for Descartes the ultimate aim of optical knowledge, practically embodied in telescopes and microscopes, is the improvement of (inherently limited) unaided human vision, in aid of the improvement of genuine knowledge to the purpose of generalized human mastery of nature. To that end, Ribe reminds us, Descartes called at the conclusion of the *Dioptrique* for a new kind of artisan, from amongst the ranks of the 'more curious and skilful persons of our age…' (Ribe 1997, 61).

hobnob, with Ferrier was not driving Descartes' natural philosophical agenda, inscriptions or strategies at all. Inside natural philosophy the instrument was FOR natural philosophical agendas and actions. Descartes was not a different natural philosopher or a new kind of 'scientist' because he played with instruments and instrument makers. He played with instruments and instrument makers because this fitted his evolving agenda as a natural philosophical contender.[115] The general historiographical lesson here follows from our cultural process model of natural philosophy developed in Chap. 2: Suppose we ask, 'What were instruments FOR inside natural philosophy; what was the élan of instruments and their makers FOR inside natural philosophy?' The answer is, they were FOR natural philosophizing, FOR natural philosophers' agendas and actions. If we forget that, essence and origin stories will loom up and overwhelm our historiographical imaginations.[116]

---

[115] Put this way our conclusion becomes entirely consistent with, and further articulates Ribe's illuminating finding that for Descartes the discipline of optics aims at improving upon nature (Ribe 1997, 60–61): 'In effect, Cartesian optics masters nature by making it possible for human vision to serve a purpose higher than nature originally intended. Moreover, this mastery has a double aspect, corresponding to Descartes' specific and general senses of the word Nature. In the more specific sense, the 'Nature' mastered by Cartesian optics is the human visual apparatus, whose naturally given capabilities are extended by means of artificial optical instruments. But mastery in this limited sense makes possible in turn a more encompassing mastery of nature as a whole. The instrument of this wider mastery is no longer an optical device such as a telescope or microscope, but, rather, the knowledge of nature that such devices reveal. It is precisely the generality and infinite perfectibility of this new instrument that allows Cartesian mastery to extend itself beyond the optical domain into all realms of human endeavor.' Translating this conclusion into the framework and language of the present argument, we see that Descartes had a studied (indeed brilliant and daring) strategy to position himself as a dominant philosopher of nature, by means of tactically crucial articulation with, and appropriation of, the rhetoric, as well as the findings, practices and artifacts of practical mathematics.

[116] Similarly there is a lesson here, consistent with our conceptual tool kit in Chap. 2, for handling claims about the 'influence' of the rhetoric and values of mathematical practitioners. This is because we are dealing with concrete 'cultural process' transactions in natural philosophy in a concrete case. We can temper any large claim that Descartes was 'influenced' by mathematical practitioners by seeing how the values and rhetoric he appropriated geared into the process of work in a specific natural philosophical project of his. Hence we can calibrate what can and cannot be attributed to such a vague 'influence' as the rhetoric, values or ideology of the mathematical practitioners. So, first of all, it is entirely possible Descartes appropriated practitioners' rhetoric—as Paolo Rossi long ago suggested—and that this was used to express to others, and even to himself, what he was doing and why. (Note, however, to sample this rhetoric, he didn't actually need real practitioners, or their writings, he could catch the rhetoric buzz from elite broadcasters like Bacon, whom he had indeed read by this stage.) That is neither here nor there. The fundamental point is that Descartes was doing more than buying and selling rhetoric—to himself or others—he was also 'doing' physico-mathematical optics, and 'doing' natural philosophy in specific technical ways. Those 'doings' are not deducible from the practitioners' rhetoric, caused or influenced by it. All Descartes does is appropriate the rhetoric to wrap his results in culturally attractive understandings, attractive and persuasive to his audience, and importantly to himself as well, for we should never ignore the question of Descartes' own resources for thematizing his own roles and strategies. After all, we have seen throughout this book the importance to him of his personal twist on the contemporary avant garde category of *physico-mathematicus*.

8.4.3 *'Tous les Phainomenes Sublunaires'—The Parhelia: Grasping Opportunities and Pregnant Problems*

Whilst Descartes was working on the metaphysics, and embroiled with Ferrier in the strategies just mentioned, a completely unexpected event drew his attention to a set of problems, triggering inquiries which would crystallize into the project of writing *Le Monde*. Some time in the summer of 1629, most likely in late July, Descartes' recent Dutch acquaintance, Reneri, requested an explanation of the parhelia, or false suns, observed at Rome the previous March.[117] Reneri also forwarded a copy of Christopher Scheiner's description of the phenomenon which he had received from Gassendi, who in turn had obtained it from Peiresc.[118] Descartes quickly became engrossed in the problem to the point of setting aside his metaphysical work and expanding his inquiry more generally to other problems of meteorology (as understood of course in terms of Aristotle's treatise on the subject of sub-lunar, terrestrial atmospheric phenomena, in his geo-centric conception). Descartes wrote Mersenne on 8 October 1629 that he had received a fairly adequate account of the parhelia two months previously, and having been asked his opinion, he had found he needed to suspend all other work and 'examine in order all of meteorology' Now, he asserted, he had found the explanation and was writing a 'small treatise' which would not only also contain the explanation of the colors of the rainbow, a problem that had caused him more difficulty than all the rest, but generally speaking the explanation of 'all sub-lunar phenomena'.[119]

Not surprisingly Mersenne took an immediate interest in Descartes' project. At Descartes' request he had recently sent him a slightly different version of the description which had come into his possession, and he soon offered to have the treatise published when completed (recall that Descartes had still not published any of his work).[120] Therefore, Mersenne's surprise must have been considerable when, about a month later, Descartes side-stepped the offer, announcing that the work would not be finished for a year, because he now contemplated a complete text of physics, that is natural philosophy, going beyond the original project on '*phainomenes sublunaries*' to explain '*tous les Phaenomenes de la nature'*.[121] The challenge of the false suns had crystallized within a month into the program

---

[117] AT I 29, Note to p. 23 l.2; Cf Descartes to Mersenne 8 October 1629 AT I 23.

[118] AT I 29, Note to p. 23 l.2.

[119] Descartes to Mersenne 8 October 1629, AT I 23. '...il ya plus de deux mois qu'un de mes amis m'en a fait voir ici une description assey ample, et m'en ayant demandé mon avis, il m'a fallu interrompre ce que j'avais en main, pour examiner par ordre tous les Meteores, auparavant que je m'y sois pû satisfaire. Mais je pense maintenant en pouvoir rendre quelque raison, et suis resolu d'en faire un petit Traitté qui contiendra la raison des couleurs de l'Arc-en-Ciel, lesquelle m'ont donné plus de peine que tout le reste, et generalement de tous les Phainomenes sub-lunaires.

[120] Descartes to Mersenne 8 October 1629 AT I 23; to Mersenne 13 November 1629 AT I 70.

[121] Descartes to Mersenne 13 November 1629, AT I 70. '...depuis le temps que je vous avais ecrit il y a un mois...je me suis resolu d'expliquer tous les Phaenomenes de la nature.'

for a complete Cartesian natural philosophy, sub-lunar and celestial, as the term implies. Given everything we now know about both Descartes' intellectual pursuits and achievements since 1619, and the state of his *post-Regulae* intellectual affairs, we can understand his abrupt change of plans, and some of the challenges and difficulties the newly crystallized program would now entail for him.

8.4.4 'Tous les Phaenomenes de la Nature'— Le Monde Begins to Crystallize

The invitation to explain the parhelia appealed directly to Descartes' greatest strength as a physico-mathematician. His best work had been in optics, and, as we have seen, after discovering the law of refraction he had sought to embed it in a mechanical theory of light. Now, in the letter of 8 October 1629, Descartes linked the problem of the parhelia to that of the rainbow, just as he would later in the published version of the *Météores*. He obviously thought the explanation of both the rainbow—a classical puzzle in natural philosophy and in mixed mathematical optics—and the suddenly controversial parhelia, would constitute a dazzling exhibition of his talents and results (and we may conjecture, all the more enticing because of the fluid state of his intellectual affairs in this immediate post-*Regulae* stage). This dual achievement was conceivable because these problems involved identical sorts of theoretical tasks: the geometrical analysis of the refractive and reflective aspects of the phenomena; and, the subsidiary question of explaining the production of spectral colors, which, as it were, lay further out along the (methodically conceived) trajectory of problem solution. Given his previous results, Descartes probably saw the parhelia and rainbow as phenomena whose dioptrical and catoptrical aspects would be amenable to complete analysis, while the explanation of the spectral colors would be the crowning touch, an achievement hitherto elusive but surely within his grasp now that he had the law of refraction.

These sorts of considerations help to explain Descartes' initial enthusiasm for a meteorological treatise, but they do not account for the sudden and wide extension of the project to all of natural philosophy. The proposal for a 'complete physics' requires some additional examination. Between the eighth of October and the thirteenth of November 1629, it probably became apparent to Descartes that to attempt to explain the optical puzzles would necessarily demand a mechanical theory of light more detailed than any he had yet envisioned. It was certainly true that the dioptrical aspects of the phenomena required a mechanical theory of light in so far as Descartes wished to justify and explain the use of the law of refraction. But, as we have already seen him doing in the wake of the discovery of the law, this would only have required the introduction of an optical medium, broadly construed as mechanical, whose nature was sufficiently well explicated to be able to bear the conceptual strain of the demonstration of the law of refraction. In this case the detailed corpuscular structure of the medium would not become an issue, for Descartes could 'prove' the law on the basis of those macroscopic properties of the medium which putatively arise from its underlying corpuscular-mechanical structure. Using, for example, the bent arm balance analogy he had shown to Beeckman a few months earlier, and without offering any specific corpuscular model for optical media, he could have moved to an explanation in terms of the differential light-bearing properties of media of different densities and employed his principles of the constant ratio of the absolute quantity of the force of motion of the light and the conservation of the horizontal component of the determination of its movement.[122] After all, this was the strategy he later followed in the *Dioptrique*, where, as we have discovered, his tennis ball model, properly decoded, deals with precisely these principles and properties of a light ray, at the instant of impact with a refracting interface. That is to say, in the *Dioptrique* he did not deal with corpuscular-mechanical models for interface phenomena, and via the tennis ball model expressed a set of construction rules for deriving ray paths from the principles of his dynamics applied to the gross, macroscopic optical properties of the media (their differential ability to convey the force of light).[123]

So far so good, it must have seemed to Descartes. The conceptual situation would have become much more complex, however, as soon as Descartes contemplated the next puzzle, production of spectral colors. This would have required the articulation of claims concerning the micro-structure of optical media, their corpuscular-mechanical make up—not that this in itself would have worried Descartes. After all, he was a corpuscular-mechanist by natural philosophical preference, and this was now a high stakes, yet arguably solvable challenge in that domain. From the contents of the published *Météores*, as well as the text of *Le Monde*, we know almost certainly that Descartes now started to contemplate the idea of corpuscular *boules*, constituting optical media, and whose differential rotary motion would account for the production of different spectral colors.[124] Such direct contemplation of the corpuscular make-up of optical media would have further alerted him to the need to postulate an additional interstitial (inter *boule*) mechanical aether of some kind

[122] Recalling our discussion earlier, Sect. 4.7.4.

[123] The only model for the micro-structure of optical media in the *Dioptrique* occurs near the end of the second discourse (AT VI 102–3) where Descartes tries to explain deflection toward the normal in water at an air/water interface, by invoking the relatively greater hardness of the water particles and their more rigid interconnections. Hard particles, rigidly joined to one another, do not absorb the force of motion of light, just as a hard smooth table does not absorb much of the motion of a ball rolled upon it. We have also seen in Sect. 4.4 why Descartes' avoided dealing with the micro-structure of optical media—such considerations threatened to raise conceptual dilemmas for his explanation of the law of refraction based on his two dynamical principles, principles he had been prompted to articulate on the very basis of how he had discovered the law of refraction in the first place.

[124] Descartes never explained this problem any other way, and the *Dioptrique*, *Météores* and *Le Monde* are direct genealogical descendants of conjectures and plans he was now in the process of formulating, in November 1629 or shortly thereafter. It is possible that he already had hit on the tennis ball conceit for 'proving' the law of refraction, and in that case need only have begun to think in terms of the spin/speed ratios of the ball, thence translated, in an initial set of bold, crude strokes, to the micro level.

to preserve the fullness of the matter-extension he had been working out in his metaphysical work. This need to 'anatomize' optical media at the corpuscular-mechanical level, therefore, would have brought Descartes face to face with the need to work out an explicit corpuscular-mechanical ontology, a theory of elements if you like, if not yet a cosmology. In addition, as the 'actual' corpuscular texture of optical media emerged as a question, Descartes might have found it hard to resist the temptation to promote his construction rules for refraction of light rays to the parallel ontological dignity of 'laws of nature', as we have argued he certainly did in the composition of *Le Monde*. This reconstruction, if plausible at all (and it does not assert anything outside the soon to be realized contents of *Le Monde*, the *Dioptrique* and *Météores*), offers a plausible working account of the stage Descartes had reached by November 1629, showing how the project of a small treatise on meteorology could have become envisioned as a year long project in natural philosophy. Descartes need not have worked any of this out in detail, merely sketched, if only in his mind, the heads of issues as we have canvassed them. Such a sketch surely amounted to the plan for 'Cartesian' natural philosophy, a project that would take at least a year of concerted work, just as he told Mersenne.

In fact we can go one step further in reconstructing the genesis of *Le Monde* between October and November 1629. Recall that Descartes alludes to a treatise on 'all the phenomena of nature', contrasting it to the earlier topic of 'sub-lunar' effects. This plausibly indicates his awareness of the cosmological implications of his plans. We can speculate that around this time Descartes began to realize that a corpuscular-mechanical theory of elements and of light, when cast on a cosmic level, might offer an entrée to the questions of celestial motion and causation with which his now erstwhile friend Beeckman had been wrestling.[125]

Given all this we should also note for reference below the following point: Descartes would now have had in hand a proposed project for a system of natural philosophy. His planned very specialized treatments of meteors, optics and lenses either might in the end be contained in a treatise on that system, or they might be hived off as separate but related works. We shall see that organizational questions of this type crop up all along the course of composition of *Le Monde*, and that the latter solution was arrived at for both conceptual and quite contingent reasons.

8.4.5 Reprise: In a Spin Over Light and Color

Thus far we have taken a 'Descartes' eye view' of the unfolding challenge and opportunity, and hence a rather optimistic prospect has appeared. Before we move

[125] As noted above we have deferred detailed analysis of Beeckman's celestial mechanical speculations of the late 1620s, since they are of the utmost significance in understanding the genesis and details of Descartes' vortex celestial mechanics, the very conceptual engine room of natural philosophy, as we shall see in detail in Chap. 10.

on through the next stages in the composition of *Le Monde*, we need to register a serious problem, indeed a considerable pitfall which Descartes was going to encounter along the way, given our picture of his likely aims and intentions in November 1629. It, too, was going to play into the more general problem of how to apportion material amongst the projected treatises. We have already encountered this problem in Sect. 4.8.2, where we saw that in his *Dioptrique* and *Météores*, Descartes articulated the tennis ball model for light, used in the proof of the laws of reflection and refraction, to develop his model for the production of color dependent upon the speed/spin ratio of the balls. Unfortunately, this model could not seriously be transferred to the level of the corpuscular-mechanical theory of light, because there, light is taken really to consist in the tendency to motion of the *boules* of second element, whilst there is no sense in talking about a tendency to spin, let alone differential ratios of tendency to motion/tendency to spin.

We found that Descartes realized this was a problem, and never introduced the spin theory of color into his corpuscular-mechanical natural philosophical treatises, either *Le Monde* or later the *Principles of Philosophy*. The reader is always referred to the *Dioptrique* and *Météores*. Indeed in *Le Monde* he refused to list color as an essential property of light, and passed up the opportunity to present his theory of refraction, referring the reader to the *Dioptrique*. We used these findings to suggest, in addition, that this was the reason Descartes persevered with the tennis ball model for refraction and reflection in the *Dioptrique*, even though, as we also discovered, the tennis ball model somewhat occludes and masks the underlying logic and conceptualization of his corpuscular-mechanical theory of light and his actual dynamical principles for explaining refraction. Descartes was making a calculated choice: at least the tennis ball model in the *Dioptrique* would allow for and comport with the spin/speed ratio model for colors in the accompanying *Météores*. We concluded that he contented himself with the reasonably adequate job done by the speed/spin ratio theory of color in the latter two works, whilst avoiding raising in public the issue of whether his 'real' theory of light could sustain this elaboration.

These findings give us another perspective on Descartes' path of composition of *Le Monde*. We know he will encounter these problems and we shall see him in the act of dealing with them, or at least we will observe statements and actions on his part which can be explained by his encountering these matters. Part of his problem of apportioning material amongst the emerging texts of *Le Monde*, the *Dioptrique* and the *Météores* (or even having any distinction amongst them) therefore revolves around this issue.

8.4.6 Descartes at Work, November 1629 to April 1630

The correspondence of Descartes over the next several months reveals a strenuous tempo of work on the natural philosophy treatise. On 18 December 1629 Descartes thanked Mersenne for forwarding queries about the explanation of phenomena relevant to the treatise and urged him to continue to do so.[126] He went on to discuss a wide variety of natural philosophical issues, including Beeckman's explanations of accelerated free fall in a void and terminal velocity in resisted free fall; the relations of length and period in pendula; the motion of sunspots; and the production of consonances by vibrating strings. Apparently Descartes was beginning to sift through the possible subject matters of his treatise, because none of these topics appear in *Le Monde.*[127] However, some strong hints toward the eventual content, and structure, of *Le Monde* do appear. For example, Descartes noted that in his opinion fire struck from flint is of the same nature as all other fire.[128] This was a matter that would appear prominently in the opening chapters of *Le Monde*, where, as we shall see in the next chapter, Descartes placed considerable pedagogical stress upon it and related points, hoping to use such commonsensical examples to motivate the general idea of a corpuscular-mechanical reduction of qualities. In addition, as if preparing for the task of explicating his principles of dynamics into laws of nature, Descartes jotted down a reminder of Beeckman's old inertial law for motion in a void, a move we have already seen in Sect. 4.8.1, was preparatory to his transforming Beeckman's principle into conceptual terms suited to his punctiform dynamics of a plenum universe.[129] Writing to Mersenne in January 1630, he continued to hint at the articulation of his dynamical principles into laws of nature [130], and by late February 1630 he was linking the inertial principle to the concept of conservation of quantity of motion under alteration of direction.[131] A related question about his dynamical principles had also appeared in the letter of January. Descartes inquired whether in fact Mersenne had any empirical evidence supporting the proposition that a stone thrown from a sling or shot from a musket moves faster or has more force at some point in its trajectory beyond the point of origin.

---

[126] Descartes to Mersenne 18 December 1629, AT I 84

[127] On 13 November 1629 (AT I 71) Descartes asserted to Mersenne that free fall should be discussed in his treatise. Sunspots play no role whatsoever in *Le Monde*. However, as we shall see in Chap. 12, Descartes' attempt to co-opt them into the system of the *Principia* marked a pivot of the new systematizing strategy which resides at the heart of his second natural philosophical text.

[128] Descartes to Mersenne 18 December 1629, AT I 88. '…mais il faudrait un long discours pour l'expliquer, ce que je tascherai de faire en mon petit traité.

[129] Descartes to Mersenne, 18 December 1629, AT I 90. Also see above Sect. 4.8.1.

[130] Descartes admitted that air-filled balloons rebound from impact because of the elastic reaction of the enclosed air (to Mersenne January 1630, AT I 107). He insisted, however, that to some degree the phenomenon depends upon another cause, 'the continuation of movement' (*Ibid.*). One is reminded of Descartes' eventual tennis ball model for the reflection of light in which the rebound of the non-elastic, perfectly hard 'tennis ball' from a perfectly hard reflecting surface is attributed to the conservation of the ball's own quantity of force of motion, conjoined with an altered normal 'determination' of that force of motion (Sect. 4.3 above).

[131] Descartes to Mersenne February 25 1630, AT I 117. '…pour le rejaillissement des balons, je n'ay pas dit que toute la cause en deust estre attribuée a l'air enfermé, dedans, mais principalement à la continuation du mouvement, de qui a lieu en tous les corps qui rebondissent, c'est a dire ex hoc ipso quod una res coepit moveri, idea pergit moveri, quamdiu potest; atque si non possit recta pergere, potius in contrarias partes reflectitur quam quiescat.'

> Because that is the common belief, although my reasoning does not agree with it. I find that bodies which are pushed and which do not move of their own accord, must have more force at the commencement of the motion than they have (even) immediately afterward.[132]

Descartes was indeed confident of his dynamical principles, but even having started composing his system of natural philosophy, he clearly was a little concerned lest some well attested evidence undermine a central tenet of his dynamics.[133]

In addition to his inquires into this range of possible issues for his natural philosophy, Descartes was occupied during this period with his first sustained investigations into medicine. In January 1630 he remarked to Mersenne in a rather offhand manner that he was seeking a 'demonstrative medicine'.[134] Descartes would renew this utopian aspiration in subsequent years, increasingly modulating it with discouraged remarks about he technical difficulty and his lack of sufficient evidence. For the moment his medical enterprise apparently consisted of untold hours of anatomical observation. Given the surviving version of *L'Homme*, the treatise of corpuscular-mechanical physiology accompanying *Le Monde*, one can only conclude that Descartes· anatomical research was done for the purpose of delimiting the field of macro-phenomena for which micro-mechanical explanations had to be postulated. In April 1630 Descartes was still pursuing chemistry and anatomy 'tout ensemble', claiming to learn each day something not to be found in books.[135] By May 27 he informed Mersenne was ready to explain the 'souls of brutes', an explanation which proceeded by reducing them to complex patterns of mechanically mediated reflexes.[136]

8.4.7 Spring to Autumn 1630—Writer's Melancholy, Visioning, Spinning and Dealing with Beeckman

Not surprisingly, this initial burst of research and writing eventually brought on a spell of lassitude and discouragement. In an important letter to Mersenne dated 15 April 1630 Descartes appeared to have lost enthusiasm for the project.

---

[132] Descartes to Mersenne, January 1630, AT I 113–4.

[133] This query may have been related to Descartes' speculations about what was to become the second law of nature in *Le Monde*, for, in considering the translation of bodies in a plenum universe of resisting media, he would conclude that, '...when one of these bodies pushes another it cannot give the other any motion except by losing as much of its motion at the same time....[this rule] tells us that the motion of one body is not retarded by its collision with another in proportion to how much the latter resists it, but only in proportion to how much of the latter's resistance is surmounted, and to the extent that, in obeying the law, it receives into itself the force of motion that the former gives up.'(AT XI, 41; SG 27–28; MSM 65–67)

[134] Descartes to Mersenne, January 1630, AT I 105. 'Je suis marry de vostre eresypele....je vous prie de vous conserver, au moins iusqu'a ce que je sçache s'il y a moyer de trouver une Medicine qui soit fondée en demonstrations infailibles, qui est ce que je cherche maintenant.'

[135] Descartes to Mersenne 15 April 1630, AT I 137.

[136] Descartes to Mersenne 27 May 1630, AT I 153.

> If anybody has the idea that I plan to write, please try to remove this impression, not to confirm it; I swear that if I had not already told people I planned to do so, so that they would say I have not been able to carry out my plan, I would never undertake the task at all.[137]

He continued to work largely because of the promise made to Mersenne to send him the finished treatise, and he understandably admits to finding more enjoyment in his research than in actually organizing and writing his treatise.

> My work (on the treatise) is going very slowly, because I take much more pleasure in acquiring knowledge than in putting into writing the little that I know....Altogether, I pass the time so contentedly in the acquisition of knowledge that I never settle down to write any of my treatise except under duress, in order to carry out my resolution, which is, if I am still living, to have it ready for posting to you by the beginning of the year 1633.[138]

Some of the reasons for Descartes' hesitation and dejection concerning the writing, if not his researches, are apparent from the text of the letter. No doubt the sheer scope and difficulty of the composition were taxing his patience. After all, it had only been four months since he committed himself to the grand project. In the interim the conceptual, organizational and evidential demands of the work must have begun to erode his initial confidence in his ability to bring the work to term.[139]

In addition, beyond the purely natural philosophical aspects of the treatise, there now increasingly loomed the enterprise of metaphysical and theological justification. We know, of course, that even before launching *Le Monde* Descartes had worked on the foundations of his metaphysics. Now, in the course of composing *Le Monde*, he was encountering quite specific difficulties requiring metaphysical and theological justification. An early instance of this had occurred in December 1629 when Descartes queried Mersenne about theological doctrines concerning the infinite, as opposed to indefinite extension of created beings.[140] The correspondence of the spring of 1630 further supports this view. We learned earlier that in letters of April and May Descartes worked out his Voluntarist theology in response to certain critical aspects of his principles of dynamics *cum* laws of nature.

Some further insight into Descartes' state of mind and its relation to his greatly expanded enterprise can be gathered from his remarks in the same letter of 15 April

---

137 '... je vous jure que si je n'avais pas ci-devant tesmoigné avoir ce dessein, et qu'on pourrait dire que je n'en ai sceu venir a bout, je ne m'y resoudrai jamais.' Descartes to Mersenne, 15 April 1630 AT I 136; Kenny (1970) 7.

138 'Au reste je passe si doucement le temps en m'instruisant moi-meme, que je ne me mets jamais a escrire en mon traitté que par constrainte, et pour m'acquiter de la resolution que j'ay prise qui est, si je ne meurs, de le mettre en estat de vous l'envoyer au commencement de l'annee 1633.'Descartes to Mersenne, 15 April 1630 AT I 137; Kenny (1970) 8.

139 On March 18 1630 Descartes still thought the treatise would include issues in music theory and the nature of sound. AT I 134.

140 Descartes to Mersenne 18 December 1629, AT I 86. 'Je vous prie me mander s'il n'y a rien de determine en la religion, touchant l'étendue des choses créés, sçavoir si elle est finie ou plutost infinie, et qu'en tous les pais qu'on appele les especes imaginaires il y ait des cors créées et veritables.'

1630 concerning some treatises started in Paris but now left off unfinished. We cited some of these lines as the epigraph to this chapter; we now cite in full:

> Perhaps you find it strange that I have not persevered with some other treatises I began while I was at Paris. I will tell you the reason; while I was working on them I acquired a little more knowledge than I had when I began them, and when I tried to take account of this I was forced to start a new project rather larger than the first. It is as if a man began building a house and then acquired unexpected riches and so changed his status that the building he had begun was now too small for him. No-one could blame such a man if he saw him starting to build another house more suitable to his condition. I am sure that I shall not change my mind again; because what I now possess will stand me in good stead no matter what else I may learn; and even if I learn nothing more I shall still carry out my plan.[141]

We can now appreciate, even better than when we first encountered it, the coded message in this passage. Descartes is glossing the collapse of the *Regulae*, and legitimating his new program, which Mersenne well knows has to do with systematic natural philosophy and its metaphysical and theological grounding. So confident is the tone—or so tightly wound the spin—of what we would today term Descartes' 'vision statement', that he asserts that never again will he switch tack, implying that everything he is learning and recording is simply quantitatively rather than qualitatively progressing the enterprise. We have called this a process of inflection of his agenda and identity; but he is making it out to his friend (and potential spokesman) Mersenne as a more radical break—albeit, thankfully, a final, definitive, and never to be altered one. However, one can say in Descartes' favor that if our interpretation is correct, then he was entitled to put this more optimistic gloss on his situation, having in a sense overcome the no doubt originally dispiriting collapse of the project of the later *Regulae*. Even if the path ahead correctly appears to him likely to be arduous, he is claiming that at least he has found the way forward.[142]

The summer and autumn of 1630 appear to have been a period of slackened work and increasing outside interference. No letter of natural philosophical import survives from late May to early November 1630. The hiatus is only partly explained by Mersenne's short visit to the United Provinces in the summer of 1630, which obviated Descartes' need to write. There were other matters disturbing Descartes at this time, traceable to the unintended consequences of Mersenne's visit to Isaac Beeckman, which, as we have seen above, fomented turbulence and a virtual breakdown of intellectual relations between Descartes' and his old friend and mentor. Now that we can view that controversy in the light of what we have learned about Descartes' course of work on *Le Monde*, we can appreciate even more than before his concern about priority claims by Beeckman, in turn possibly to be backed up by

[141] Descartes to Mersenne 15 April 1630, AT I 137–8; Kenny (1970) 9.

[142] It should also be noted that this interpretation would make even more sense if, as conjectured in note 73 concerning the 'Cambridge ms.' of the *Regulae*, Descartes had left Mersenne with that hastily concocted place holder, which momentarily seemed to promise the eventual completion of a project Descartes had already known was defunct before he left Paris. Mersenne, as a recipient of this abbreviated and optimistically edited version of the *Regulae* would indeed have been surprised by the recent shifts in Descartes' projects and agenda.

the publication of his own system of corpuscular-mechanical natural philosophy. It is easy to be critical of Descartes' reactions and rhetoric, as we tended to be in our earlier discussion. Now, perhaps, this perspective should soften just a little, as many readers can understand, even if not fully sympathize with, the highly sensitized proprietary worries of an author in full flight. At this precise juncture, René, in modern parlance, 'really did not need' Beeckman to be carrying on about intellectual property to his friends and possible patrons, and he certainly would have been far from happy about the idea of Beeckman at long last publishing a system (although the same could easily have been said by others about Descartes himself, and not least of all, by Beeckman).

8.4.8 Organizing the System and Himself— Investigations, Decisions, Hesitations

Descartes' frustration and anger with Beeckman peaked around October 1630, and it was only later that he could return more full attention to the treatise of natural philosophy, as problems of organization and exposition soon become the focus of concern. Writing to Mersenne in November, Descartes noted that he was working on the section of the *Dioptrique* dealing with the nature of color and light. This section, he admitted, would be longer than originally planned and would contain 'quasi une physique toute entière'.[143] We may deduce that Descartes was wrestling with the two problems of how much of his corpuscular-mechanical theory of light, as opposed to mere models or analogies thereof, should be incorporated into the *Dioptrique*, and whether or how the explanation of color via the spin/speed ratio model could be annexed to it. We know this because, on the one hand, he says that with this material the *Dioptrique* is virtually a work of natural philosophy (and hence surely had the real, or ontological, corpuscular-mechanical theory of light); and, on the other hand, he asserts that the treatise will also deal with color. Additionally, we know that this was an unpromising mix for Descartes, because the real theory of light could not bear the weight of the spin/speed ratio model. Of course we have the benefit of also knowing the answer at which he arrived: His natural philosophical texts would never articulate the theory of light into the theory of color, whilst his texts giving the spin/speed ratio model for color, would not be natural philosophical texts. They would deal with the tennis ball and other analogies for the mechanical action of light, or at most pretend that only in isolated systems do light-causing *boules* actually *move and spin*. The *Dioptrique* was not destined to be 'quasi une physique toute entière', and *Le Monde* was not destined to deal with phenomena of spectral colors.

What can therefore be said is that it looks as though Descartes momentarily believed that, regardless of what the treatise on natural philosophy might contain, the November 1630 version of the *Dioptrique* could emerge as more of a text of natural philosophy, perhaps with the exposition of the corpuscular-mechanical

[143] Descartes to Mersenne, 25 November 1630, AT I 179.

theory of light articulating to a theory of elements and perhaps even some of the vortex cosmology. So as late as November 1630 Descartes was still experimenting with various ways of allocating his material between what would in the end become *Le Monde* and the *Dioptrique*, perhaps even thinking of the *Dioptrique* as a separable portion of *Le Monde*, to appear in or with it, or not. No principle of differentiation or decision had yet emerged. Two overlapping drivers shaped his eventual decision: first the fact that the spin/speed model does not sit well with the real theory of light as corpuscular tendency to motion, and secondly, the fact that in the end, the Galileo affair forced him to avoid publishing *Le Monde*, with its infinite universe version of realist Copernicanism, thus reinforcing the choice of the non-cosmological presentation of the tennis ball model in the *Dioptrique*, elaborated into the spin/speed ratio model in the *Météores*.

Further difficulties of exposition were reported a month later, in December 1630, reflecting a quite advanced state of development of the vortex celestial mechanics and accompanying theory of production of light in its cosmic setting—the very core of the system of *Le Monde*. Descartes was now ready to sketch the emergence of the universe out of a primordial chaos of created matter in motion, but he was at a loss to find a mode of exposition which would not 'shock the imagination of his readers' or scandalize received opinion.[144] The pedagogical and cultural political constraints on his composition were coming to fore, as he apparently became more settled about his vortex celestial mechanics. We shall explore his pedagogical answers below in Chap. 9, and the system-binding vortex celestial mechanics will be the key topic in Chap. 10.

Relatively little correspondence survives from 1631 dealing with the natural philosophy project. We have seen above in Sect. 8.2.3 that the bulk of what does exist from that period strongly bespeaks Descartes' maturing sense of the technical and epistemological contours of his plenist cosmology: his increasing sense of the irrelevance of mathematical abstraction in the treatment of natural philosophical problems, and the crystallization of what we have termed his plenist holism in regard to natural philosophical explanation. *Le Monde* was nicely taking shape, and Descartes was well aware of where he was being taken by his inflection from 'physico-mathematician with corpuscular-mechanical leanings' to 'systematic corpuscular-mechanical, and plenist, natural philosopher of physico-mathematical lineage'.

In any case, by the spring of 1632 Descartes was very nearly finished with the treatise. He had finally decided on the format for the *Dioptrique* and had sent Golius a draft of the first section, where he 'explains the matter of refractions without touching on the rest of philosophy'; that is the element theory and cosmological theory of light—just the solution we find in the published version and which we would expect, given the conceptual dilemmas posed by trying to connect the ontological theory of light and the theory of spectral colors.[145] However, in April he

[144] Descartes to Mersenne, 23 December 1630, AT I 194.

[145] Descartes to Golius, January 1632, AT I 235. Note that this occurred before the Galileo affair, which presumably further reinforced Descartes' earlier decisions about limiting discussion of the real, corpuscular-mechanical theory of light in the *Dioptrique*, and *Météores*.

asked Mersenne to wait a few months longer while he edited the text, completed the diagrams and, it now transpired, explored a number of additional chemical and meteorological phenomena, in order to 'learn more and add to his knowledge'. He explains that, having in hand

> …the general description of the stars, the heavens and the earth, I did not originally intend to give an account of particular bodies on the earth but only to treat of their various qualities. In fact, I am now discussing in addition some of their substantial forms, and trying to show the way to discover them all in time by a combination of experience and reasoning. This is what has occupied me these last days; for I have been making various experiments to discover the essential differences between oils, ardent spirits, common and strong waters, salts, etc.[146]

Here Descartes was engaged in serious experimental work, at least to the extent of generating evidence reported in everyday language as to the properties and qualities to be explained; suggesting comparisons and discriminations; and eliciting matters of fact that might strongly confirm or disconfirm his intended corpuscular-mechanical explanations. We need to remember, however, that as in the case of any and all phenomena to be explained, all Descartes could do was make up corpuscular-mechanical stories which, he hoped, did not contradict his deeper story about elements and laws of nature; which arguably 'comported well' with the stories offered about arguably 'similar' phenomena; and for which 'striking' factual evidence existed, and 'striking', putatively 'negative' factual evidence did not exist.[147] In the event, Descartes decided to suppress *Le Monde* before this material could be added to the text, but some of these issues are treated in the *Météores* and listed in the *Discours de la Méthode* as integral portions of *Le Monde*.[148]

This hands-on empirical work, in the interest of prompting explanatory stories, was delaying the dispatch of *Le Monde* to Mersenne, and as if to compensate for this Descartes at this time tantalized Mersenne, and perhaps himself, with his daring speculations about an 'a priori' science of nature, beginning with deduction of the cause of the distribution of the fixed stars.[149] One can imagine René, mixing substances in his cabinet, jotting down the odd corpuscular-mechanical hypothesis about what he observed, and wandering off in internal conversation into more grandiose glosses of his already achieved vortex celestial mechanical and cosmology: *'does it not amount, after all, to a deductively tight movement of thought from God and matter-extension, through God's injection of motion, the necessarily formation of the elements and vortices, thence stars, and à fortiori, their nature and spatial distribution? This needs to be written down, and sent off to Père Mersenne; this will keep him engaged and on edge until all this chymical work can be completed and patched into my "World".'*

---

146 Descartes to Mersenne, 5 April 1632, AT I 242–3; Kenny (1970) 22.

147 See Sect. 6.4, on the 'phenomenological description' of what one is doing when one constructs discursive corpuscular-mechanical explanations of things.

148 Cf. Discours, AT VI 64; CSM I 143–4; Maclean (2006) 52–3.

149 To Mersenne, 10 May 1632, AT I 250ff. See below, Sect. 8.6 on these speculations and their meaning in relation to Descartes' agenda and self-identity.

Thus, the work dotting the i's and crossing the t's of *Le Monde* dragged on to July 1633. Descartes had in fact spent much of this period putting together the *Treatise on Man*.[150] And finally, by 22 July 1633, he announced *Le Monde* was virtually complete, but still only promised Mersenne that he would have it at the end of the year.[151] In November, as we know, Descartes finally heard about the trial and condemnation of Galileo months earlier, and his voluntary suppression of *Le Monde* followed, as we have seen, followed shortly thereafter by the revised publication strategy, leading to the *Discours* and *Essais* of 1637.

8.5 An Exercise in Counter-Factual History: If the *Regulae* Had Not Failed...

Our chronological investigation of the composition of *Le Monde* completes our description *cum* causal analysis of Descartes' career inflection, carried out in the previous three sections. We are therefore ready to explore the text of *Le Monde* itself in the next three chapters. But, before we proceed to that task, we have the opportunity to shed more light on both the inflection process and the system of *Le Monde*. This we shall do by indulging, in this section, in a controlled piece of counter–factual history, and in the following section by delving into a remarkable piece of evidence concerning Descartes' state of intention, self–understanding and strategy at exactly the time he was finishing *Le Monde*.

The premise for our counter–factual history exercise is this: Imagine that there had been no failure of the later *Regulae* and no abandonment of the text by Descartes. Thinking this through in, as Max Weber would say, 'objectively possible' terms will help solidify our analysis of the inflection process and further set the stage for considering Descartes' actual situation by the time *Le Monde* was largely complete.[152] Now, if the *Regulae* had not failed, one would surely first ask why would Descartes have needed to engage in the new intellectual agendas we discussed in Sect. 8.2? Indeed it is entirely possible he would not have done so, at least in the period we have been surveying. And, beyond that, assuming no failure of the *Regulae*—and hence no launching into explicit metaphysical construction, exercises in Voluntarist theology or explorations of plenist–realism—what then would have been the impacts, if any, of the key episodes and interactions treated in Sect. 8.3? Finally, since the chronology toward *Le Monde* as such therefore might not have happened, what might have been the consequences for Descartes, physico-mathematician, piecemeal corpuscular-mechanist and possessor of the *Regulae*, of hearing about the parhelia in mid 1629? Let's look at these possibilities sequentially and in more detail.

---

[150] Descartes to Mersenne, November or December 1632, AT I 263.

[151] Descartes to Mersenne, 22 July 1633, AT I 268.

[152] Max Weber, 'Objective Possibility and Adequate Causation in Historical Explanation' in (Weber 1949) 164–88.

If the attempt at construction and legitimation of universal mathematics in the *Regulae* had not failed, why indeed would Descartes have ever switched to exploring dualist metaphysics and writing *Le Monde* (including its overt invocations of voluntarist theology)? Never previously had Descartes shown the slightest sign of being a systematic natural philosopher, let alone the dualist metaphysician. Indeed the first fruit of his moving in the larger cultural context, and trying to play a role in public, had been precisely the later *Regulae*, not the natural philosophical system and its metaphysics. So, most likely he would have continued as a physico-mathematician, in his piecemeal, brilliant way. The *Regulae* would have been *the* vehicle for presenting the method, and the universal mathematics it taught would have been put forward as the legitimated synthesis of his work in mathematics and in physico-mathematics. Or, if, the *Regulae* had come to seem too unwieldy, or to suffer from problems—but not ones sufficient to set off the major inflection process that did occur—we can further speculate that a simplified 'sampler' might have been produced, and that it would have looked like a cousin of the *Discours* of 1637, the counter-factual *Discours* now being closer to the *Regulae* in design.

In turn, under such circumstances, with the *Regulae* either published, or at least represented by a proxy, counter-factual *Discours*, we can imagine some texts resembling the *Dioptrique*, *Météores* and *Géométrie* would arguably also have eventuated, appearing, as in reality, in public under the legitimatory covering rhetoric of the method. What really would have been missing in this version of the public premiere of 1637 would have been *Le Monde* (and *L'homme*) written and lurking in the background as the work ultimately to be introduced by the method and essays. So, part of the motive behind the actual *Discours* would not have been there, and the passages therein flagging and hinting toward the system in the background also would not have been present. In short, the big picture lurking in the background would have been something like the *Regulae*, whether published by that stage or not.

Consider, finally, some possible details in these counter–factual developments. Would not Descartes' color theory and related material in the *Météores* have required natural philosophical framing somewhere in the background? The answer is, 'Well, yes and no'. They would not have required the production of a system, but certainly would have demanded some additional lines of articulation of the mechanistic theory of light and matter theory. And, if one reads the relevant existing portions of the *Dioptrique* and *Météores*, without knowledge of *Le Monde* preceding them, or of the *Principia* to come, that is just what one gets—piecemeal, somewhat speculative and disparate corpuscular-mechanical claims about the structure of matter and real mechanics of light, just far enough to facilitate the relevant bits of optics and meteorology. As for the parhelia episode, it could very well have triggered this counter–factual trajectory, with Descartes having been pushed further along it by the need to out perform Beeckman, not just on paper, but in public. Beeckman's just commenced search for a system could have been haughtily dismissed, not so much because Descartes too would have then been building a mechanistic system, but because our counter-factual Descartes, physico-mathematician and universal mathematician, would have no interest in the textbook style systems of his contemporaries and would have believed he had something new, more worthy of a mathematician and more relevant for inquiring into nature. His interaction with Beeckman might have been a little less fraught—Beeckman was not going to produce a competitor for the triumphant *Regulae*, after all!

Further to all this, one can imagine Descartes, under the pressure of some future circumstances, eventually moving to a system of natural philosophy, from the directions of light theory, matter theory and speculation about element theory and celestial mechanics, stimulated by Beeckman and under the shadow of Kepler. But, even then, why would he necessarily have had to carry out legitimatory activities in metaphysics and voluntarist theology?[153] The answer is that our counter–factual Descartes might well have not needed to do that—which brings us back to the actual history (that is, our arguably satisfactory description/explanation of his trajectory). We see with more force than ever that it was precisely with the failure of the *Regulae* that he was driven toward corpuscular-mechanism on a systemic basis, if he were to have anything solid in the realm of knowledge of nature; and, that because of this, he knew exactly which problems needed to be addressed, if he were going to consolidate his now inflected agenda. He was back in the situation in which we depicted him at the beginning of this chapter, with the failure of the project of the later *Regulae*. Let's therefore return finally to the question of Descartes' situation near the end of his inflection process, and in particular to a remarkable and hitherto little commented upon piece of self-revealing evidence.

8.6 Aspiration, Identity and Strategy at the Birth of *Le Monde*: Between Natural History and A Priori 'Science'

As mentioned near the end of Sect. 8.4, in May 1632 Descartes wrote Mersenne a curious letter, a window on his state of mind concerning the nearly completed *Le Monde*. Properly deciphered the letter displays clues and traces concerning Descartes' opinion about, and aspirations for, his vortex mechanics-centric system of natural philosophy. Speculating upon the possibility of deciphering the cause of the distribution of the fixed stars (something about which one could say that *Le Monde* does offer an explanation, although it is not the distribution we seem to observe from the Earth) Descartes suggests that such knowledge could lead on to an *a priori* science of nature (something he certainly does not have in *Le Monde*, but

[153] It will be noted that the odd man out in these scenarios so far is any reference to the Chandoux/Bérulle episode. Had the *Regulae* not failed, Bérulle might have been taken by Descartes as reinforcing the project of the later *Regulae*, with the entire episode becoming less fraught and ambiguous for Descartes than we depicted it earlier. Or, if we need more out of the episode, it could be factored in alongside the 'future circumstances' just mentioned, as eventually helping to contribute to Descartes' desire to build a more traditional style system of natural philosophy, and to ground it metaphysically and theologically. But, even on this view, he would not have had to have agonized about the Galileo affair, as no realist Copernican natural philosophy would have been to hand at that stage, and the eventual natural philosophical product might have been more in the traditional textbook style adopted in the *Principles*. One might conclude that had the *Regulae* not failed, Descartes might well have had a smoother public and private trajectory, still being deluded about method, but more 'naively' and with less overlay of overtly cynical manipulation of the public.

which his current articulations from the vortex celestial mechanics might have suggested as some sort of possibly attainable ideal):

> For the last two or three months I have been rapt in the heavens. I have discovered their nature and the nature of the stars we see there and many other things which a few years ago I would not even have dared to hope; and now become so rash as to seek the cause of the position of each fixed star. For although they seem very irregularly distributed in various places in the heavens, I do not doubt that there is a natural order among them which is regular and determinate. The discovery of this order is the key and foundation of the highest and most perfect science of material things which men can ever attain. For if we possessed it we could discover *a priori* all the different forms and essences of terrestrial bodies, whereas without it we have to content ourselves with guessing them *a posteriori* from their effects.[154]

He immediately goes on to ask about factual reports about comets, which would be a very great help toward discovery of the sought order, and then he writes at length about the need for a Baconian type natural history of the heavens, 'without arguments or hypotheses' dealing with the position, sizes, colors and brightnesses of the fixed stars (which in any case he is convinced constantly change position), as well as repeating a request for observations, and, in the manner of Tycho, recorded tracks of comets.[155] Having swung first toward *a priori* science and then to demands for Baconian natural history, he reaches a kind of closure near the end of the letter, when he confesses that he does not really think anyone will do the factual work on comets he has envisioned,

> ...just as I do not hope to discover the answers to my present questions about the stars. I think that the science I describe is beyond the reach of the human mind; and yet I am so foolish that I cannot help dreaming of it though I know that this will only make me waste my time as it has already done for the last two months. In that time I have made no progress with my treatise.[156]

Much might be said about this curious missive to Mersenne. Descartes was certainly letting his feelings, aspirations—and hesitations—flow out on to the page. His alternating expression of desire for, and disillusionment with, the possibility of an *a priori* science conveys the image of man seriously involved in constructing a new and remarkable system of nature, who is overtaken first by a lust for even deeper grounds of certainty about his work, which in turn is deflated by recalling the slightly mad character of that ambition. After all, in what to us should be an unsurprising break with the comic book Descartes of some philosophy textbooks, he expatiates at length about the lack of hard records of matters of fact, the natural history of the heavens, comets included, which would provide the material to be explained, supply the controls on wild explanations, and help shape the formulation of those explanations. He knows he has worked without such ideally complete natural historical archives; and he also knows from bitter experience the implausibility of the grand vision of deductive knowledge produced by a method.

[154] Descartes to Mersenne, 10 May 1632, AT I 250–1; Kenny (1970) 23–4. Note also the curious astrological/alchemical tone of this aspiration.

[155] Descartes to Mersenne, AT I. p. 252; Kenny (1970) 24.

[156] Descartes to Mersenne 10 May 1632, AT I 252; Kenny (1970) 24.

Interesting as this picture of Descartes' possible psychological roller coaster ride undoubtedly is, with its echoes of the dreams of 1619 and the deflations of 1628, the main issue for us here has to do with what all this shows about Descartes' state of mind about *Le Monde*, and in particular, the crystallization of Descartes' sense of a powerful, systematically explanatory, plenist holism, pivoted on the vortex celestial mechanics and its articulations. By this point it probably appeared to Descartes that within the field of natural philosophizing, *Le Monde* offered a uniquely coherent and fluid passage of argument from matter-extension and the laws of nature; down through the elements; vortex formation and dynamics; the cause and nature of light; to the nature and distribution of stars; the nature and reasons for planetary systems; the behaviour of comets and planetary satellites; the cause of weight on and near planets and the rest as glossed in the letter—the forms and essences of terrestrial bodies (i.e. the substances found on any planet).[157] But how did *Le Monde* sit between the two ideals and ideally related projects, the *a priori* science and its complete natural historical base? This presumably was what Descartes was really reflecting upon: Well, *Le Monde*, to be sure, was not *a priori*, nor was it strictly deductive, but its very coherence and systematicity, and the scope of explanation allowed by the powerful and elegant vortex mechanics, might make one dream, again, of an even tighter logical weave, but only provided that—he reminds himself—we can work from a massive natural historical base. The sting in the tale of this round of ruminations was perhaps not the fact that *Le Monde* was not deductive or *a priori*, but rather that even the excellent and very real *Le Monde* probably needed more natural history as well.[158]

[157] Cf. Gaukroger's comment on this letter, which he quotes at length at the beginning of a section dealing with the technical contents of *Le Monde*. 'When Descartes says that we can discover all the different forms and essences *a priori*, he means that we can discover them from their causes, not that there is some non-empirical way of discovering them' (Gaukroger 1995, 249 and Note 65 thereto). This is a sensible way of dealing with this strange letter, by extracting a philosophically defendable position from it, and no doubt had Descartes been challenged about his meaning, he might well have retreated a step to Gaukroger's reading. Nevertheless, one might note that Gaukroger's comment still allows for Descartes to be speculating more wildly here, to be dreaming in fact of a 'methodological fix', a rigorous movement of analysis from empirical information contained in a near perfect natural history up to knowledge of causes and then down again deductively in explanation, a protocol epistemologically superior to his actual procedure of messy corpuscular-mechanical model building, in the light of metaphysical constraints, and available empirical evidence. In the letter, Descartes seems to be hankering after something beyond the workaday practices that have yielded even the powerful and innovative *Le Monde*—to wit, a natural philosophy based on applying the method to a complete natural history, yielding in the end *a priori* knowledge of causes and true explanations. But, as usual, he also knows that the method does not work to produce such ideal natural philosophical knowledge of causes and explanations. Hence, also as usual, his strange, ambiguous and aspirational discourse. The obviously Baconian resonances of the letter need not be belabored.

[158] Speaking loosely and with larger historical lessons in mind, one might imagine here that Descartes, the systematizer, has an inkling of the world of the more collective research in experimental natural philosophy and emerging experimental domains which was to develop in the realm of natural philosophizing in the next two generations. At least our prescient René would not have committed the modern historiographical blunder of assuming that in the opening generations of this process the field and agon of natural philosophizing simply and suddenly died, giving birth to 'Modern Science[s]'. Unfortunately, however, our René probably would share with some modern historiographers a debilitating belief in the complete separability of fact and theory. Cf. on these problems Schuster and Watchirs (1990), Schuster (1990), Schuster (2002), and below, Sect. 12.12.

In short, the letter shows that Descartes still in his heart of hearts dreamt of the deductive certainty of the method, or even perhaps the grand cosmic intuition of the neo-Platonists, whilst he also knew this to be something he ridiculed and disdained in others, and had found not possible in his own program. But beyond both of these points was an even deeper and more immediately real personal truth: *the only reason he could push his dreams this far once again was that on another level he knew* Le Monde *was a remarkable, and remarkably coherent, system of natural philosophy.* The problem was that in this dialectic of thought, emotion and recollection, such a proud and sober thought could at any moment kick off the both wilder variant… '*what a shame, after all, that nobody's natural philosophy, even my own admirable* Le Monde, *can live up to the methodologist's dream of an a priori and deductively certain knowledge of nature,* as well as the more sober worry…*if ideal* Scientia *requires perfect natural history, as my own (and Bacon's) method doctrine dictates, what does my admirable yet not ideal* Le Monde *require in that line*?' Just because Descartes knew in a sober, commonsensible way that the *Regulae* had failed; that method was dead except as window dressing; and that, qua natural philosophy, *Le Monde* was very good; does not mean that he, any more than any other mortal, would not be hostage to inflated hopes and fears shaped by past experience as recollected for internal narrative, spilling out on paper to a trusted friend and mentor.

It is time, therefore, to look closely at *Le Monde* as both a systematic natural philosophy and as the genealogical outcome of Descartes' struggles and achievements since 1618 qua *physico-mathematicus*, natural philosopher and proponent of universal mathematics and method. This task will take three chapters, dealing respectively with the didactic and pedagogical style of the opening portions of *Le Monde* (Chap. 9); its systemic heart in the vortex celestial mechanics and cosmological theory of light (Chap. 10), and finally with its status as a gambit in the natural philosophical *agon* and the ways its systematicity was secured (Chap. 11). At that point, having exhausted the study of Descartes' first system, we will be a position to look in Chap. 12 at some new findings about how his second, fully mature, system in the *Principia philosophiae* was structured and compares to *Le Monde*.

References

Works of Descartes and Their Abbreviations

AT = *Oeuvres de Descartes* (revised edition, 12 vols.), edited by C. Adam and P. Tannery (Paris, 1964–76). References are by volume number (in roman) and page number (in Arabic).

SG = *The World and Other Writings*, edited and translated by Stephen Gaukroger (Cambridge, 1998).

MM = *René Descartes, The Principles of Philosophy*, translated by V. R. Miller and R. P. Miller (Dordrecht, 1991).

MSM = *Rene Descartes, Le Monde, ou Traité de la lumière*, translated by Michael S. Mahoney (New York, 1979).

CSM(K) = *The Philosophical Writings of Descartes*, 3 vols., translated by John Cottingham, Robert Stoothoff, and Dugald Murdoch, and (for vol. 3) Anthony Kenny (Cambridge, 1988), References are by volume number (in roman) and page number (in arabic).

HR = The Philosophical Works of Descartes, vol I translated by E.S. Haldane and G.R.T. Ross (Cambridge, 1968 [1st ed. 1911]).

Other

Alquié, F. (ed.). 1963. *Oeuvres philosophiques de Descartes*, vol. 1. Paris: Garnier Frères.

Baillet, Adrian. 1691. *Vie de Monsieur Descartes*. Paris: Paris Daniel Horthemels.

Barnes, Barry. 1982. *T.S. Kuhn and social science*. London: Macmillan.

Beeckman, I. 1939–1953 *Journal tenu par Isaac Beeckman de 1604 à 1634*. 4 vols, ed. C. de Waard. The Hague: Nijhoff.

Biro, Jacqueline. 2009. *On earth as in heaven: Cosmography and the shape of the earth from Copernicus to Descartes*. Saarbrucken: VDM Verlag.

Blanchet, L. 1920. *Les antécédents historiques du 'Je pense, donc je suis'*. Paris: Alcan.

Clarke, Desmond. 2006. *Descartes, a biography*. Cambridge: Cambridge Univeristy Press.

Collins, James. 1971. *Descartes' philosophy of nature*. Oxford: Blackwell.

Collins, Harry. 1985. *Changing order*. London: Sage.

Garber, Daniel. 2000. A different Descartes: Descartes and the programme for a mathematical physics in his correspondence. In *Descartes' natural philosophy*, ed. S. Gaukroger, J.A. Schuster, and J. Sutton, 113–130. London: Routledge.

Gaukroger, S. 1995. *Descartes: An intellectual biography*. Oxford: Oxford University Press.

Geuroult, Martial. 1954. Métaphysique et Physique de la force chez Descartes et chez Malebranche. *Revue de Métaphysique et de Morale* 59: 1–37.

Geuroult, Martial. 1980. The metaphysics and physics of force in Descartes. In *Descartes: Philosophy, mathematics and physics*, ed. S. Gaukroger, 196–229. Sussex: Harvester.

Gilson, E. 1913. *La liberté chez Descartes et la théologie*. Paris: Alcan.

Gouhier, H. 1924. *La pensée religieuse de Descartes*. Paris: Vrin.

Grayling, A.C. 2005. *Descartes: The life of René Descartes and its place in his times*. London: Free Press.

Harrison, Peter. 2002a. Voluntarism and early modern science. *History of Science* 40: 63–89.

Harrison, Peter. 2002b. Original sin and the problem of knowledge in early modern Europe. *Journal of the History of Ideas* 63: 239–259.

Henry, John. 2004. Metaphysics and the origins of modern science: Descartes and the importance of laws of nature. *Early Science and Medicine* 9: 73–114.

Kenny, Anthony. Trans. and ed. 1970. *Descartes, philosophical letters*. Oxford: Clarendon Press.

Langford, Jerome J. 1971. *Galileo, science and the church*. Ann Arbor: University of Michigan.

Machamer, Peter, and J.E. McGuire. 2009. *Descartes' changing mind*. Princeton: Princeton University Press.

Maclean, I. Trans. and ed. 2006. *René Descartes: A discourse on method*. Oxford: Oxford University Press.

McGuire, J.E. 1972. Boyle's conception of nature. *Journal of the History of Ideas* 33: 534–542.

Mersenne, M. 1932–1988. *Correspondence du P. Marin Mersenne*. 17 vols, ed. C. de Waard, R. Pintard, B. Rochot and A. Baelieu. Paris: Centre National de la Recherche Scientifique.

Oakley, Francis. 1961. Christian theology and the Newtonian science: The rise of the concept of the laws of nature. *Church History* 30: 433–470.

Osler, M. 1994. *Divine will and the mechanical philosophy*. Cambridge: Cambridge University Press.

Pinch, Trevor. 1985. Towards an analysis of scientific observation: The externality and evidential significance of observational reports in physics. *Social Studies of Science* 15: 3–36.

Popkin, Richard. 1964. *The history of scepticism from Erasmus to Spinoza*. Berkeley: University of California.

Ribe, Neal. 1997. Cartesian optics and the mastery of nature. *Isis* 88: 42–61.

Rossi, Paolo. 1970. *Philosophy, technology and the arts in the early modern era*. New York: Harper and Row.

Schuster, J.A. 1977. *Descartes and the scientific revolution 1618–34: An interpretation.* 2 vols. unpublished PhD dissertation, Princeton University.

Schuster, J.A. 1990. The scientific revolution. In *The companion to the history of modern science*, ed. R.C. Olby, G.N. Cantor, J.R.R. Christie, and M.J.S. Hodge, 217–242. London: Routledge.

Schuster, John. 1995. Descartes *Agonistes:* New tales of Cartesian mechanism. *Perspectives on Science* 3: 99–145.

Schuster, J.A. 2000. Descartes *Opticien*: The construction of the law of refraction and the manufacture of its physical rationales, 1618–29. In *Descartes' natural philosophy*, ed. S. Gaukroger, J.A. Schuster, and J. Sutton, 258–312. London: Routledge.

Schuster, John. 2002. L'Aristotelismo e le sue Alternative. In *La Rivoluzione Scientifica*, ed. D. Garber, 337–357. Rome: Instituto della Enciclopedia Italiana.

Schuster, J.A. 2009. Descartes—Philosopher of the scientific revolution; Or natural philosopher in the scientific revolution. *Journal of Historical Biography* 5: 48–83.

Schuster, J.A., and Graeme Watchirs. 1990. Natural philosophy, experiment and discourse: Beyond the Kuhn/Bachelard problematic. In *Experimental inquiries: Historical, philosophical and social studies of experimentation in science*, ed. H.E. Le Grande, 1–47. Dordrecht: Kluwer.

Schutz, Alfred, and Thomas Luckmann. 1974. *The structures of the life-world.* Trans. R. M. Zaner and H. T. Engelhardt. London: Heinemann.

Shapin, Steven. 1992. Discipline and Bounding: The History and Sociology of Science As Seen Through the Externalism-Internalism Debate. *History of Science* 30: 333–369.

Shea, W. 1991. *The magic of motion and numbers: The scientific career of René Descartes*. Canton: Science History Publications.

Smith, Norman Kemp. 1952. *New studies in the philosophy of Descartes*. London: Macmillan.

Tannery, Paul. 1896. Descartes physician. *Révue de Métaphysique et de la Morale* 4: 478–488.

van Berkel, Klaas. 2000. Descartes' debt to Beeckman. In *Descartes' natural philosophy*, ed. S. Gaukroger, J.A. Schuster, and J. Sutton, 46–59. London: Routledge.

Vrooman, J.R. 1970. *Rene Descartes, a biography*. New York: Putnam.

Weber, Max. 1949. *The methodology of the social sciences.* Trans. Edward Shils and Henry Finch. New York: Free Press.

Chapter 9
Reading *Le Monde* as Pedagogy and Fable

9.1 Introduction

In *Le Monde* Descartes offers a bold exposition of realist Copernican cosmology and corpuscular mechanism. Indeed, with *Le Monde* the corpuscular-mechanical approach to nature first aspires to the level of natural philosophical systematics which had characterized the Aristotelian teaching and its neo-Platonic competitors.[1] *Le Monde* can be approached in several ways. For example, one might emphasize its mode of presentation, for it is far from a formal natural philosophy textbook and more an appeal to the commonsense and everyday rationality culturally attributed to an *honnête homme*—that is, an intelligent, practical, trustworthy gentleman, well educated, but not embroiled within—or attracted to—the controversies and trivialities of the Schools.[2] Or, one might ponder in detail the curious mode of presentation of the middle sections—the well known fable of the world. In this chapter, we shall be doing both, but only as means to a deeper end, explored in the next chapter. In this book our focus ultimately remains on the process of emergence of Descartes,

[1] *Le Monde* is termed 'more original than any of [Descartes'] other works' by no less an expert than Theo Verbeek (Verbeek, 2000, 149).

[2] Ranea (2000) As the Editors comment (Gaukroger et al., 2000, 12) in their Introduction: Ranea 'shows that Descartes treated experience and experiment as something problematic that had to be regulated, thus demonstrating the existence of an earlier and continental variant of the English controversy over how one defines the 'experimental life', studied by Shapin and Schaffer. Ranea focuses on Descartes' dialogue, *La recherche de la vérité par la lumière naturelle*, in which one of the interlocutors, Poliandre, is cast as the *honnête homme*, relying on his natural faculties and not on scholastic training. In addressing himself to the *honnête homme*, Descartes is identifying an audience of practical gentlemen whom believes he can educate to adjudicate (in his favor) in natural philosophical controversies. Ranea therefore argues that Descartes intended his natural philosophy as something that might close controversies stirred by the endemic variability and unreliability of factual reports.'

J. Schuster, *Descartes-Agonistes: Physico-mathematics, Method & Corpuscular-Mechanism 1618-33*, Studies in History and Philosophy of Science 27,
DOI 10.1007/978-94-007-4746-3_9, © Springer Science+Business Media Dordrecht 2013

corpuscular-mechanical natural philosopher, out of the original carapace of Descartes, physico-mathematician. We shall therefore in the end concentrate on the central natural philosophical topics, claims and constructs in *Le Monde*, linking them to the genealogy of Descartes' development, and to our understanding of what natural philosophical contestation was chiefly about in his generation. In other words we are setting up the next chapter to offer a synthetic reading, explicating *Le Monde* as the point where Descartes' long evolving project of physico-mathematics gave way to an interest in systematic corpuscular mechanical natural philosophy, albeit still conceived to some extent as physico-mathematical in tenor and underlying conceptual weave. These matters constitute the very climax of the argument of this work.

Before that, however, we need, to deal in this chapter with the *honnête homme* style and flow of *Le Monde* and Descartes' famous narrative conceit of the fabular world of corpuscular mechanism. We need to get some sense of the rhetoric and style of *Le Monde*, because whatever canonical natural philosophical topics we take it to be explicating, as examined in the next chapter, they certainly are expressed in this loose and discursive manner. This affected Descartes' framing of what actually are quite sophisticated and technical natural philosophical concepts and explanations, and accordingly, the style of the text must be brought into play in our reading, as we attempt to unearth and explicate those natural philosophical elements. Moreover, as to Descartes' explication of his system within the confines of a fable, this certainly can mislead us into thinking that in some sense *Le Monde* is not a committed staking of very serious claims in the natural philosophical agon of the day. So, we need to see exactly what the fabular exposition does and does not mean in this regard, especially given that we already know Descartes had in place justificatory and grounding machinery in the form of an embryonic metaphysical dualism, and voluntaristic theology, some of which was brought explicitly into play in *Le Monde*, whilst all that machinery, certainly in his view, remained in reserve for possible articulation and use.

9.2 The Ground Plan of *Le Monde*

Close consideration of the structure of *Le Monde*, and Descartes' probable aims regarding the completed work, lead to the conclusion that he intended *Le Monde* as a work of pedagogical artistry, even if, as we shall see in both this chapter and the next, the extant text falls short of that aim in many respects. It was Charles Adam, in his notes to *Le Monde*, who first sketched the outline of Descartes' strategy.[3]

[3] We should recall that this strategy was not forced upon Descartes by the lack of a metaphysical framework—a common if often unstated premise, explicitly articulated with considerable force years ago by James Collins (1971). (Cf. above Chap. 8 Note 32). The claim that Descartes did not possess a metaphysical doctrine equal to the task until 1637 or later cannot be sustained by the evidence and argument we have advanced earlier in Chap. 8.

According to Adam, Chapters 6 and 7, dealing with the laws of nature and the properties of matter of his fabular world, lay down the foundation upon which Descartes constructs his cosmological edifice in Chapters 8–14.[4] In the latter chapters Descartes explains the formation of the stars, comets and planets; the vortex celestial physics determining the celestial motions, local gravity and weight, and the tides; and the nature and properties of light. Adam went on to suggest that the following chapters, of which only part of Chapter 15 survives, would have contained an exercise in matching the observable consequences of this postulated mechanical corpuscular world against the phenomena of our real world.[5] Presumably the final chapters would have drawn heavily upon the material Descartes mentioned in his letter to Mersenne of April 1632 concerning the explanation of the 'substantial forms' of some selected terrestrial bodies.[6]

The structural plan set out by Adam seems convincing in its main lines. It accounts for the order and content of the later chapters and makes room for the missing final sections of the text. Adam, however, did not have much to say concerning the first five chapters of *Le Monde* other than to point out that they serve as a kind of introduction.[7] This is a remarkable omission, given the obviously pedagogical and at times tendentious nature of the opening sections. It should be possible to integrate these chapters into a view of *Le Monde* as a self-consciously didactic work. This will be attempted in the remainder of this chapter, starting in Sect. 9.3 with the opening chapters of *Le Monde*, before turning to the problems of the fable of the world in Sect. 9.4. Then in Chap. 10, we will be able to explore the vortex celestial mechanics, and other systematic and canonical natural philosophical claims which follow from first portion of *Le Monde*, including the issue of the matching of appearances in our world to those in the putatively fabular world of *Le Monde*, a matter to which we shall indeed attend closely throughout our attempted synthetic reading of the key elements of Descartes' 'physico-mathematically' tinged natural philosophy.

By extension, there is no need to grant that Descartes constructed *Le Monde* as a didactic work because he could do no better. He freely chose his format. In fact Descartes' letter of 25 November. 1630 to Mersenne makes explicit his strategy: 'J'éprouveray en la *Dioptrique* si je suis capable d'expliquer mes conceptions, et de persuader aux autres une verité, après que je me la suis persuadée: ce que je ne pense nullement. Mais si je trouvais par expérience que cela fût, je ne dis pas que quelque jour je n'achevasse un petit Traitté de Métaphysique, lequel j'ay commencé éstant en Frize et dont les principaux points sont de prouver *l'existence de Dieu*, et *celle de nos ames*, lors qu'elles sont separées du cors, d'ou suit leur immortalité. Car je suis en colère quand je voy qu'il y a des gens au monde si audacieux et si impudens que de combattre contre Dieu.' (AT. I. 182)

[4] AT XI 698–700.

[5] Ibid., p. 701.

[6] AT I 243, see also above Sect. 8.4.8. The *Discours*, 5ie partie, also contains a summary of *Le Monde* which mentions topics absent from the extant text, but which appear in the *Météores* and *Principia Philosophiae*.

[7] AT I 698.

9.3 The Common Sense of Corpuscular-Mechanism pour les honnêtes hommes—The Opening Chapters of *Le Monde*

Consider the titles which Clerselier gave to the first five chapters of *Le Monde*: 'On the difference between our perceptions and things that produce them'; 'In what the heat and light of fire consists'; 'On hardness and liquidity'; 'On the void and how it happens that our senses do not perceive certain bodies'; and 'On the number of the elements and on their qualities'. Upon reading this list one might conclude that the five chapters merely contain some simple exercises in corpuscular-mechanical explanation. That is, Descartes might merely be showing the reader that in the mechanical philosophy one must abstract from secondary qualities, that phenomena such as heat and fire can be easily explained on a corpuscular basis, and that such traditional problems as the existence of the void and the constitution of liquids and solids can be clearly resolved in corpuscular terms. Having motivated the reader through such simple examples, Descartes could then set out in detail his corpuscular-mechanical natural philosophy, especially its cosmology. Nevertheless, I believe that a close reading of these chapters from the standpoint of one who must be systematically won over to mechanism leads to a somewhat different interpretation. On the latter view these chapters are constructed in an eminently pedagogical manner. Descartes intends to interweave appeals to common experience, putatively intuitive common sense truths and mechanical analogies in order gradually to unfold to the reader the central precepts of his mechanism, finally leading him to the very brink of the cosmological construction by gently opening the question of the nature of matter and the elements.

We have gotten a taste of this, starting in Chap. 7, Sect 6.2, when we looked at Chapter 1 of *Le Monde* as part of our exploration of the collapse of the project of the later *Regulae* and Descartes' response thereto. Recall that Descartes was addressing the emergence, out of the optics-physiology-psychology nexus of the *Regulae*, of what we can with Whiggish convenience term the problem of secondary qualities. Here we must examine the didactic function of the chapter in the overall structure of *Le Monde*. Descartes does not appeal to mechanical principles in order to justify his distinction between perceptions and their physical causes. Rather, he hopes to establish *prima facie* grounds for the distinction by appealing to what he takes to be common sense notions, which are implicitly tied to mechanistic principles. One might therefore say that the aim of Chapter 1 is to stimulate awareness of a distinction between 'the mental' and the 'non-mental', or physical, while Chapters 2 and 3 will involve arguments favoring the purely mechanical-corpuscular constitution of the non-mental.

This interpretation of Chapter 1 is confirmed from the very first line,

> In setting out to deal here with light, the first thing I want to make clear to you is that there *can be* a difference between our perception of light (i.e.: the idea that is formed in our imagination through the intermediary of our eyes) and what is in the objects that produce that perception in us ...[8] (emphasis added)

[8] AT XI 3; MSM 1; SG 3.

Descartes appeals first to the presumably obvious fact that words bear no necessary resemblance to the conceptions they cause us to entertain. In like manner he then asks,

> Now, if words, which signify nothing except by human convention, suffice to cause us to conceive of things to which they bear no resemblance, why could not nature also have established a certain sign that would cause us to have the perception of light, even though that sign in itself bore no similarity to that perception?[9]

But, compared to the example drawn from words and ideas, it is difficult to grasp the necessity of believing that the cause of the perception of light does not resemble that perception. After all, Descartes has not revealed the theory of perception of the *Regulae* which had recently rendered problematical the entire issue of the perception of non-geometrical sense qualities. Accordingly, he poses his common-sense argument more firmly upon consideration of the sense of touch. Surely 'ideas' of pain or tickling (that is, the sensation of being in pain or being tickled) bear no resemblance whatsoever to the objects which must impinge upon our bodies in order to cause those ideas.[10] Here the case rests at its most intuitively obvious level. Of course, what the argument gains in appealing to common sense it loses in generality.[11] But, in keeping with his pedagogical aims Descartes does not need to insist upon the immediate generalization of his argument. It will be the task of his next chapter to unfold further arguments for the construal of all physical actions as forms of touch or contact. For the moment it is sufficient merely to have stimulated the reader's interest in the possibility of a mental/non-mental distinction.

> Now, I see no reason that forces us to believe that what is in the objects from which the perception of light comes to us is any more like that perception than the actions of a feather and of a strap are to tickling and pain. Nevertheless, I have not adduced these examples to make you believe absolutely that this light is something different in the objects from what it is in our eyes, but only so that you will doubt it and so that, forbearing from being preoccupied by the contrary, you can now better examine with me what light is.[12]

In Chapter 1, Descartes raised the possibility that states of mind are not identical to their causes. In Chapter 2, he aims to confirm and articulate this distinction by suggesting that those causes are material, arising from the principles of matter and motion, and in particular from the motion and collision of unobservable corpuscles. He focuses upon the phenomena of heat and fire, in which the conceptual transition from empirical conditions to putative corpuscular causes is rather easy and relatively convincing. This strategy affords the rhetorical and common-sense base from which he can move to the conclusion that material corpuscles and their motions are sufficient explanatory principles for heat, fire, flame and perhaps other phenomena.

---

[9] AT XI 4; MSM 3; SG 4.

[10] AT XI 5–6; MSM 5; SG 5.

[11] Although we might grant that touch is the least misleading sense.

[12] AT XI 6; MSM 7; SG 5–6.

Consider his account of the action of a flame,

> When a flame burns in wood or some other similar material, *we can see with our eyes* that it moves the small parts of the wood and separates them from one another, thus transforming the subtler parts into fire, air, and smoke, and leaving the grosser parts as ashes. Hence, someone else may, if he wishes, imagine the form of 'fire', the quality of 'heat', and the action that 'burns' it to be completely different things in this wood. For my part, afraid of misleading myself if I suppose anything more than what I see must of necessity be there, I am content to conceive there the motion of its parts. For, posit 'fire' in the wood, posit 'heat' in the wood, and make the wood 'burn' as much as you please: if you do not suppose in addition that some of its parts are moved or detached from their neighbors, I cannot imagine that it would undergo any alteration of change. By contrast, remove the 'fire', remove the 'heat', prevent the wood from 'burning': provided only that you grant me that there is some power that violently removes the subtler of its parts and separates them from the grosser, I find that that alone will be able to cause in the wood all the same changes that one experiences when it burns.[13] (emphasis added)

Descartes first appeals to empirical grounds: from a common-sense point of view one 'sees' the flame gouging out small pieces of the wood. The changes resulting from burning—the disintegration of the wood into smoke, flame and ash—seem to depend on the dissociation and translation of visible particles of the wood. Therefore, what needs to be explained is the removal and motion of pieces of the wood. This accounts for Descartes' sarcastic rejection of the scholastic descriptive and causal jargon about 'fire', 'heat' and 'burning'. In a manner still reminiscent of Mersenne's mitigated scepticism, he is not directly denying the reality of substantial forms and qualities, but rather is insisting upon their conjoint conceptual and ontological redundancy. They are simply irrelevant to the experiential kernel of the problem, the motion and removal of parts, from which all the effects seem to follow.

Only after thus structuring the conditions of the problem of explaining fire, and presumably winning the reader's tentative assent thereto, does Descartes postulate mechanical corpuscular causes. Understandably, then, this is the rhetorical moment to invoke the touchstones of the mechanist's position—belief in the primitive and irreducible intelligibility of local motion, and of collision as the cause of change of motion.

> Now, insofar as it does not seem to me possible to conceive that one body could move another unless it itself were also moving, I conclude from this that the body of the flame that acts against the wood is composed of small parts, which move independently of one another with a very fast and very violent motion. Moving in this way, they push and move with them the parts of the body that they touch and that do not offer them too much resistance.[14]

Another mechanical intuition is added to flesh out this account. Surely, Descartes suggests, the extremely small corpuscles of flame must move very quickly in order to compensate for their lack of size. Their 'force to act against other bodies' thus somehow arises conjointly from their speed and magnitude. The reader versed in the science of machines (surely a small sub-set of the *honnête homme* readership)

---

[13] AT XI 7–8; MSM 7–9; SG 6–7.

[14] AT XI 8; MSM 9; SG 7.

might well have imagined a gloss based upon the 'dynamical' interpretation of the balance or lever, in which speed (length) of the lever arm compensates for the smaller magnitude of the moving force.

Nevertheless, all this is merely intended as a foretaste of the formal system of dynamical principles to be presented later, and whose central tenets we have already discussed in earlier chapters.[15] The dictates of common-sense discourse suitable for an *honnête homme* interlocutor require that Descartes quickly descend from the realm of natural law to the ground of logical adequacy and empirical confirmation. It is sufficient, he reiterates, to 'conceive' of the motion of particles in order to understand how flame consumes wood.[16] In addition, the explanation of other effects of flame, such as heat and light, require nothing more than the same agitation of unobservable parts,

> ... as regards heat, the perception that we have of it can, it seems to me, be taken for a type of pain when it is violent, and sometimes for a type of tickling when it is moderate. Since we have already said that there is nothing outside of our thought that is similar to the ideas we conceive of tickling and pain, we can well believe also that there is nothing that is similar to that which we conceive of as heat; rather, anything that can move the small parts of our hands, or of any other part of our body, can arouse this sensation in us. Indeed many experiences favor this opinion; for merely by rubbing our hands together we heat them, and any other body can also be heated without being placed close to a fire, provided only that it is shaken and rubbed in such a way that many of its small parts are moved and can move with them those of our hands.[17]

Descartes' explanation telescopes and conflates logical, empirical and hypothetical elements, but the substance of his argument is clear. The sensation we have of heat can be taken for a species of pain when the heat is intense and for a kind of tickling when the heat is moderate. In the opening chapter Descartes suggested that the perceptions of pain and tickling are nothing similar to their material causes and that those causes depend on a physical disturbance or impingement upon the surface of one's body. Now, if some of the mental effects of flame are of the same nature as pain or tickling, then we can suggest that the physical impingement of the flame consists in violently agitated unobservable corpuscles; therefore, the perception of heat may depend in particular upon the physical impact of micro-particles against the surface of one's body. Descartes does indeed admit that this conclusion is merely an 'opinion'.[18] As yet the mechanistic program rests on this rhetorically suggestive mixture of logical and empirical conditions and analogical postulation of particles. Still, one must admit that Descartes has carried off a very neat little didactic exercise, which leaves the reader with the limited but tantalizing notion that corpuscular matter and motion are sufficient explanatory principles for such an important agent of natural change as fire.

[15] See above, Chap. 4, Sect. 8.1; Chap. 8, Sect. 2.2.2.

[16] AT XI 9; MSM 11; SG 8.

[17] AT XI 9–10; MSM 13; SG 8.

[18] AT XI 10; MSM 13; SG 8 has '... this view.' rather than 'opinion'.

Chapter 3, 'On Liquidity and Hardness', is in the nature of a transition, binding the opening chapters to the remainder of the introduction. On the one hand Descartes extends the conclusions of Chapter 2 concerning mechanical-corpuscular explanation to a theory of the constitution of liquid and solid bodies. On the other hand, the theory of liquidity and solidity leads to the claim that all bodies are full, a contention which is developed further in Chapter 4 with particular reference to the air. By the end of the latter chapter the reader will be brought to the point of conceding the existence of unobservable particles filling the interstices between air corpuscles, thus setting the stage for the postulation and description of the elements in Chapter 5.

Descartes' notorious doctrine concerning the causes of hardness and liquidity is based on the recognition that it is necessary to apply force to a resting body to set it in motion.[19] If two bodies at rest are in contact, a small but finite effort is required to separate them by moving one or the other aside. To shove both apart from each other at once seems to require an even greater effort. Surely then, he contends, if a gross body is made up of millions upon millions of contiguous corpuscles mutually at rest, then a sensible force is required to break or tear the body apart, or to separate off some part of it from the rest.[20] Thus, the criterion of the hardness of a body is its resistance to penetration and consequent division, interpreted as arising from the summed resistances to being set in motion of the resting particles which make it up. By contrast, liquids are characterized by their relative lack of resistance to penetration and separation. Interpreted at the corpuscular level, this means that the constituent particles are in a constant state of agitation relative to one another, so that they offer no resistance to the attempt to penetrate the volume of the liquid by pushing them apart from one another.[21] These criteria, and their corpuscular interpretations, suggest the possibility of classifying all material substances along a continuum from the hardest to the most liquid:

> Thus, to constitute the hardest body imaginable, I think it is enough if all the parts touch each other with no space remaining between any two and with none of them being in the act of moving. For what glue or cement can one imagine beyond that to better hold them one to the other. I think also that to constitute the most liquid body one could find it is enough if all its smallest parts are moving in the most diverse ways from one another and as quickly as possible, even though in that state they do not cease to be able to touch one another on all sides and to arrange themselves in as small a space as if they were without motion. Finally, I believe that every body more or less approaches these extremes, according as its parts are more or less in the act of moving away from one another.[22]

This certainly is one of the most problematical doctrines in Cartesian natural philosophy. Descartes leaves himself open to several objections which even contemporary critics could have made. In the first place Descartes conflates, or rather

[19] This intuition appears later, of course, as an aspect of the first law of nature.

[20] AT XI 12–13; MSM. 17; SG 10.

[21] AT XI 13–14; MSM 17–19; SG 10–11.

[22] AT XI 13–14; MSM 19; SG 10–11.

does not explicitly distinguish, density and hardness. A plausible implication of his teaching is that the harder a body is, the more dense it will be, for relatively more of its constituent particles will be at rest and mutually contiguous to each other. Indeed, as Paul Mouy long ago observed, Descartes made precisely this conflation in the *Dioptrique* when he identified denser, more strongly refracting media, with those whose parts are most rigidly joined.[23] Furthermore, the theory leaves little room for explaining any relative difference between the response to tensile and shearing stress in the same material (especially very dense materials). Nor can it account easily for the viscous properties of 'soft' materials such as wax, honey and oil.

Granted these obvious difficulties, Descartes' motivation for advancing a mechanical-corpuscular theory of hardness and liquidity must have gone beyond a mere desire to resolve some interesting problems in natural philosophy. The deeper motivation lies in the didactic structure of *Le Monde* and thus explains how Descartes could side-step the incoherencies of the doctrine, when it is taken in isolation. The key to the pedagogical function of the theory appears in Descartes' claim, cited above, that all types of material substance can be located along axes determined by the corpuscular properties of relative rest or motion, and relative contiguity or separation. After all, on Descartes' view all material bodies are either solid or liquid. Therefore, the simple intension or remission of particulate motion and the consequent rearrangement of parts is sufficient to explain the make-up of all material bodies, as well as the changes of 'state' they may undergo. This is tantamount to asserting that all forms of matter, their qualities and their transformations, arise from the size, shape, arrangement and motion of corpuscles. Thus a general theory of matter and material change unfolds from the original attempt to specify the natures of the most fluid and most solid bodies.

Mere generalization of earlier results, however, does not exhaust the pedagogical and rhetorical function of the theory of hardness and liquidity. Descartes also desires to set up his discussion of the existence of the void and eventually motivate his theory of elements. He does this through a series of examples chosen solely from the phenomena of liquids, because of their pedagogical value, and, no doubt, because it would be difficult to imagine a plausible empirical example directly supporting the theory of solids.[24] Indeed, most experiences which come to mind, such as ordering the relative densities and hardnesses of materials, or their tensile and shearing strengths, would falsify, or at least greatly strain Descartes' explanation.

[23] Mouy (1934, p. 59).

[24] The freezing of water would be one good example, provided one overlooks the expansion of volume involved. In 1612 in his *Bodies That Stay Atop Water or Move in It* Galileo had challenged the previously widely accepted natural philosophical topos according to which ice results from the condensing of fluid water, a 'fact' whose natural philosophical causes were in turn the subject of increasing dispute in the late sixteenth and seventeenth century amongst Scholastics and their challengers. (Cf. Boschiero, 2007, Chapter 6, which goes on to document the experiments on expansion of freezing water, the force of this expansion and artificial freezing conducted later in the century by the *Accademia del Cimento*.)

First, Descartes relativizes the melting of metals by flame to the burning of wood discussed in Chapter 2,

> Since, as I have already said, all the parts of flame are perpetually agitated, not only is it liquid, but it also renders most other bodies liquid. Note also that when it melts metals, it acts with no different power than when it burns wood. Rather, because the parts of metals are just about all equal, the flame cannot move one part without moving the others, and hence it forms completely liquid bodies from them. By contrast, the parts of wood are unequal in such a way that the flame can separate the smaller of them and render them liquid (i.e. cause them to fly away in smoke) without agitating the larger parts.[25]

As Alquié correctly observed, from Descartes' standpoint combustion and liquefaction are perceived as the same type of process.[26] If corpuscular motion and consequent negation of rest-cohesion are sufficient to explain the violent dissociation and diffusion of particles of wood, surely it suffices to explain the liquefaction of metals. When wood is burned, extremely fluid substances—flame and vapors—fly off into the surrounding air. When a metal is melted, a somewhat less fluid material is produced, whose constituent particles do not quite possess sufficient agitation to enable them to fly off. The slight difference between the effects produced arises from the difference between the corpuscular make-up of wood and metal. At a more commonsensical level Descartes next appeals to the random fluttering of macroscopic dust particles in the air as an indication of the incessant agitation of unobservable particles making up the air.[27] As in the case of flame devouring wood, Descartes' operates at the interface of the observable and unobservable realms, arguing from the perceptible motion of small parts to the underlying, causal agitation of imperceptible parts.

The corpuscular theory of matter forms a basis upon which in the following Chapter 4, 'On the Void and How It Happens that Our Senses Do Not Perceive Certain Bodies', Descartes seeks to establish that air constitutes a material plenum. The argument proceeds in the following way: We already find it plausible that air consists of a mixture of particles, probably deriving for the most part from earthly exhalations, vapors, the fluid products of burning, etc. In addition it is obvious and a commonplace of Scholastic Aristotelianism that air is not as dense or solid as water or earth. Therefore, if it can be established that the air is a material plenum, then we will be forced to postulate the existence of other genres of unobservable particles completely filling the interstices which must exist between the grosser particles of air.

Descartes puts forward the problem in a highly discursive manner: we will discover that matter is a plenum by divesting ourselves of a vulgar error and prejudice,

> … we must examine in greater detail why air, although it is as much a body as the others, cannot be felt as well as they. By doing so, we will free ourselves from an error with which we have been preoccupied since childhood, when we believed that there were no other bodies

[25] AT XI 14; MSM 19; SG 11.

[26] Alquié (1963) t.I. p. 328, Note 1.

[27] AT XI 14–15; MSM 21; SG 11.

> around us except those that could be perceived, and thus that, if air were one of them, then because we perceived it so faintly, it at least could not be as material nor as solid as those we perceive more clearly.[28]

Next Descartes notes that all particles of any body, liquid or solid, consist of the same matter, having the same particulate solidity. No particle can become more or less solid or take up any more or less space; therefore, the most solid gross body would be made up of particles each of which is completely surrounded on all sides by other particles.[29] It would then follow that,

> if there can be a void anywhere, it ought to be in hard bodies rather than liquid ones; for it is evident that the parts of the latter can much more easily press and arrange themselves against one another (because they are moving) than can those of the former (which are without motion).[30]

If a solid body is something short of the hardest possible body, it is reasonable to assume that some void spaces are present between its rigidly fixed particles.

For the moment Descartes can operate with this rather mild claim, because he only hopes to establish that fluids, in particular air, contain no vacua. As a matter of pedagogy and exposition the question of the existence of vacua in solids can be bracketed for the time being. Descartes' tactic is to retail some examples which play rhetorically upon the implausibility of positing interstitial vacua in the air:

> ... pray tell me what explanation [*apparence*] would there be for nature's causing the heaviest bodies to rise and the most solid to break (as one experiences her doing in certain machines), rather than to suffer that any of their parts should cease to touch one another or to touch some other bodies, and for her nonetheless permitting the parts of air (which are so easy to bend and to arrange in all manners) to remain next to one another without being touched on all sides, or even without there being another body among them that they touch? Could one really believe that the water in a well should mount upward against its natural inclination merely in order that the pipe of a pump be filled, and think that the water in clouds should not fall in order that the spaces here below be filled, if there were even some little void among the parts of the bodies that they contain?[31]

It should be noted that this difficult passage is not directed toward disproving the existence of vacua in solids. Rather, it functions rhetorically, suggesting the absurdity of implicitly attributing vacua to fluids by playing upon the reader's commonsense experience of the behavior of solids and dense fluids in fluid environments. In the opening sentence Descartes ironically asks how plausible it is to attribute void spaces to the air when we commonly see that in mechanical devices solid bodies are raised up against their 'natural' inclination to fall or are ruptured, rather than suffer dissolution of the contiguity of their parts. The sentence operates in two modes: (1) it affects a posture of ingenuous surprise and wonder that nature should be so disordered that the most unnatural things occur in solids in order to avoid that which

---

[28] AT XI 16–17; MSM 25; SG 13.

[29] AT XI 17; MSM 25; SG 13.

[30] Ibid.

[31] AT XI 18; MSM 27; SG 14.

is presumably natural to fluids; (2) it implicitly functions as a more properly natural philosophical argument by asking why solids should so unnaturally resist dissolution of continuity if it is the case that the surrounding air, permeated as it is by interstitial vacua, stands ready to receive the disjoined or unnaturally elevated portions of solids. Put more positively, statement (2) means, 'only if air is full and offers no void spaces to intruded bodies does the behavior of solids make sense'. The alternative explanation by *horror vacui* is not disproved, but ignored as Descartes presses on with his exposition. Given this interpretation, one can go a step further, because the concluding sentence of the passage then falls into place as a detailed commentary on statement (2). Water does indeed mount unnaturally in pipes as the incumbent air is removed. Certainly, this would not occur only when the air is removed if it were the case that the air contained interstitial vacua into which the water could seep at any time. Therefore, there is all the more reason to think that water drops in clouds would continually descend *naturally* to fill any interstitial vacua in the underlying air. This, of course, does not occur, thus it is generally implausible to grant interstitial vacua in the air.

Lurking behind Descartes' involuted passage, of course, is his theory of fluid mechanics, presumably the immediate descendant of the theory he discussed in the letter to Reneri of June 1631 in the course of composing *Le Monde*.[32] In that letter Descartes implied that the world is full and thus concluded that motion must take place by means of instantaneous mutual circular displacements of parts. The mechanical difficulty attending such a displacement explains the phenomena commonly ascribed to the *horror vacui*. The passage from *Le Monde* which we have been discussing reveals just how obscure and convoluted Descartes' text could become when he insisted on couching his exposition in an *a posteriori* and (supposedly) rhetorically pleasing manner. In the passage the phenomena of rising and breaking appear as evidence for the conclusion, rather than as effects demonstrated from causes. The non-existence of the vacua in air is to be established, not used as a starting point for demonstration of effects; and the rhetorical exigencies of arguing from experience and common sense require the use of the scholastic notions of natural and unnatural motions.[33] Similarly, Descartes only introduces his doctrine of circular displacement in *Le Monde after* the labored passage cited above.[34] Evidence is supplied by the phenomenon of fish swimming in water not too near the surface. Their motion does not disturb the surface, supposedly because of the smooth circular displacements set up under the water.[35] Another experience involves a sealed wine vat punctured at the bottom. The case parallels that of the inverted tube of mercury in the letter to Reneri:

> When wine in a cask does not flow through an opening at the bottom because the top is completely closed, it is improper to say (as one ordinarily does) that this takes place due to

[32] See above, Sect. 8.2.3.3.

[33] Cf. Alquié (1963) t.I, p. 332

[34] AT XI 19; MSM 27–29; SG 14.

[35] AT XI 19–20; MSM 29; SG 14–15.

> *horror vacui*. One knows that the wine has no mind to fear anything; and, even if it did have one, I do not know under what circumstances it would apprehend that void, which is in fact nothing be a chimera. Rather, one should say that the wine cannot leave the cask because outside everything is as full as it can be and that the part of the air, whose place the wine would occupy should it descend, cannot find another place to put itself anywhere in the rest of the universe, unless one makes an opening in the top of the cask, through which this air can rise circularly to its place.[36]

Despite these arguments and the implicit theory upon which they rest, Descartes is well aware of the fact that the non-existence of the void is far from having been demonstrated. He readily admits that the chapter is sufficient if it has persuaded the reader that some spaces which we perceive as being empty are really filled with as much matter as equal spaces filled by perceptible bodies.[37] Nor is there anything strange in his suggestion that some matter cannot be perceived. He recalls the thesis of the first chapter concerning the mediation of perception by physical impingement in order to suggest that not all matter can be perceived,

> ... for it is certain that we cannot perceive any body unless it is the cause of some change in our sensory organs .... But if those that continually touch us ever had the power to produce any change in our senses, and to move some part of their matter, in order to move them they had perforce to separate them entirely from the others at the beginning of our life, and thus they can have left there only those that completely resist their action and by means of which they cannot be perceived in any way. Whence you see that it is no wonder that there are many spaces about us in which we perceive no body, even though they contain one no less than those in which we perceive it the most.[38]

Returning finally to the prime question, the constitution of the air, Descartes can concur in the 'common opinion of philosophers' that air is rarer than water or earth, and that therefore air particles cannot be as closely packed together in a given space as those of earth or water. Consequently, there must be 'a great quantity of small intervals among the parts of which the air is composed'.[39] After all, 'there is no other way to conceive of a rare body'.[40] But, as has been shown, the air is most likely full. This enables Descartes to conclude,

---

[36] AT XI 20; MSM 29–31; SG 15 Note that at this stage in the exposition Descartes is ignoring the possible role of an interstitial aether filling all the spaces between the grosser air particles, and hence the possibility that particles of this aether could enter the closed cask and drive out an equal volume of the wine. Descartes only introduces the types of interstitial aether—first and second element filling all spaces between air particles of third element—in subsequent passages. Recall that in the letter to Reneri about aerostatics discussed in Sect 8.2.3.3, Descartes already allowed for the possibility of aether particles entering the closed top of the inverted, mercury filled tube. To prevent this he invoked what we called his 'cosmic injunction': if any aether were to enter the tube, an equivalent volume of the aether would have to be recruited from above the atmosphere, requiring a long and weighty column of atmospheric air to be lifted to that height.

[37] AT XI 20–21; MSM 31; SG 15.

[38] AT XI 22; MSM 33; SG 15–16.

[39] AT XI 23; MSM 35; SG 16.

[40] AT XI 23; MSM 35; SG 16.

> that of necessity there are mixed with the air some other bodies, either one or several, which fill as exactly as possible the small intervals left among its parts.[41]

These 'other bodies', of course, are the other as yet not specified types of 'aetherial' matter, taken in addition to the grosser particles of air, water and earth which we have been implicitly discussing all along. Thus Descartes' introduction debouches on to his theory of the elements, which in turn will be the basis of the cosmology and theory of light to follow.

The final arguments of Chapter 4 bring the introductory didactic section of *Le Monde* to a close. A ramifying network of mechanical intuitions, appeals to common sense, and shrewdly stated examples have lent credence to the principles of mechanical-corpuscular explanation and a plenum theory of matter, pointing on to the necessity of a theory of elements.

In Chapter 5, 'On the Number of the Elements and on Their Qualities', Descartes sets aside the indirect didactic style of the opening chapters, launching into a rather ad hoc postulation of the three genres of particle, or elements, which will make up his mechanistic universe:

> I conceive of the first, which one can call the element of fire, as the most subtle and penetrating fluid there is in the world. And in consequence of what has been said above concerning the nature of liquid bodies, I imagine its parts to be much smaller and to move much faster than any of those other bodies. Or rather, in order not to be forced to imagine any void in nature, I do not attribute to this first element parts having any determinate size or shape; but I am persuaded that the impetuosity of their motion is sufficient to cause it to be divided, in every way and in every sense, by collision with other bodies, and that its parts change shape at every moment to accommodate themselves to the shape of the places they enter ....
>
> As for the second, which one can take to be the element of air, I conceive of it also as a very subtle fluid in comparison with the third; but in comparison with the first there is need to attribute some size and shape to each of its parts and to imagine them as just about all round and joined together like gains of sand or dust. Thus, they cannot arrange themselves so well, nor press against one another, that there do not always remain around them many small intervals, into which it is much easier for the first element to slide in order to fill them. And so I am persuaded that this second element cannot be so pure anywhere in the world that there is not always some little matter of the first with it.
>
> Beyond these two elements, I accept only a third, to wit, that of earth. Its parts I judge to be as much larger and to move as much less swiftly in comparison with those of the second as those of the second in comparison with those of the third. Indeed, I believe it is enough to conceive of it as one or more large masses, of which the parts have very little or no motion that might cause them to change position with respect to one another.[42]

One should not think that these elements are totally conventional constructions. Certain constraints of a methodological and theoretical order do condition Descartes' discussion. In the first place we should note Descartes' continual interjection of

[41] AT XI 23; MSM 35; SG 16.

[42] AT XI 24–26; MSM 37–39; SG 17–18. What shall later term the cosmographical overtones of this passage will be taken up in due course in Sect. 12.2 at Note 10.

phrases such as 'I conceive', 'I accept' or 'I judge'. An epistemological constraint is involved, implicitly harking back to the doctrine of the *Regulae*. Supposedly, nothing is conceived or imagined of these elements which is not clearly intuitable. The description involves only considerations of motion, size, shape and arrangement.[43] Although it cannot be proved that elements exactly like these exist, the discussion moves within the discursive limits set out in the *Regulae* on the basis of a theory of perception, and further employed in Chapters 1–4 of *Le Monde*. In addition, there is a second set of constraints arising jointly from the requirements of the theory of vortex mechanics and cosmological theory of light annexed to it. The three elements are designed to account for the three kinds of matter minimally needed for a theory of light as mechanical pressure: that which produces light by mechanical agitation, that which conveys light-pressure, and that which reflects light and is opaque to it. If Descartes started in the late 1620s with an unexplicated real theory of light as tendency to motion in a bearer medium, not very much imagination would have been needed to see that at the very least two other types of matter would be necessary, one providing the cause of the tendency to motion in luminous bodies—the sun, stars and flame, and the other constituting reflecting materials. These distinctions have obvious cosmological parallels which Descartes exploits with ease. The sun and stars produce light and thus are identified with the first matter; the vortex heavens propagate light and so are identified with the bearer medium of second element; and, the Earth, moon, planets and comets reflect received light and thus consist of the gross opaque third matter.[44] To be sure, the elements are hypothetical and lack independent evidence or argument for their existence. But they are not arbitrary, for they speak to logical constraints and conceptual possibilities which grow directly out of Descartes' earlier work.

Despite all this, however, it can still be said that the overall argument of *Le Monde* is becoming very tenuous at this juncture. Indeed it seems that it cannot help but become even more tenuous and ad hoc, for it is clear that the remainder of the work must consist in a natural philosophy built with these three elements. Descartes recognizes this, although he does not put the issue squarely as having to do with the necessarily hypothetical character of the elements. Rather, he abruptly ends the chapter by half-heartedly asserting that this logical and epistemic problem is really one of exposition. Turning a logical embarrassment into a rhetorical triumph, he proclaims,

> Many other things remain for me to explain here, and I would myself be happy to add here several arguments to make my opinions more plausible. In order, however, to make the length of this discourse less boring for you, I want to wrap part of it in the cloak of a fable, in the course of which I hope that the truth will not fail to appear sufficiently and that it will be no less agreeable to see than if I were to set it forth wholly naked.[45]

---

[43] Although, to be sure, the behavior of the first element is quite inexplicable. How can it continually change shape and adapt itself to the ever shifting interstices of the second element without experiencing a change in density?

[44] AT XI 29–30; MSM 45–47; SG 19–20.

[45] AT XI 31; MSM 49; SG 21.

Rather than consistently having to face the nagging question of the existential status of the elements, Descartes liberates his discussion by freely calling it fable. The cosmology, cosmogony, celestial mechanics and theory of light will be presented in this guise and it is to the nature and status of this fable that we must turn in the following section.

9.4 Why the Fable of the Mechanistic World?

The remainder of the mechanical construal of the world will be presented in the form of a fable, an imaginative construction occurring in the indefinite reaches of matter-extension well beyond the region of the real world. Our initial question must be, why does Descartes take this new approach which differs so much from the straightforward didactic style of the first four chapters? The considerations mentioned at the end of Sect. 9.3 offer part of the explanation; but, there are several other aspects to this problem.

We know from our exploration in Chapter 8 of Descartes' metaphysical and voluntarist theological work in 1629–1630 that it is not true that much of *Le Monde* is cast in the form of fable because Descartes then lacked the materials for what he would later take to be a metaphysically grounded natural philosophy.[46] More agreeably to the original views of Charles Adam, I argued that Descartes' metaphysical enterprise cannot be dated from the composition of the *Discourse* or the more extensive works of the 1640s. Rather, Descartes had worked on metaphysics and the theological justification of corpuscular-mechanical natural philosophy since 1629, and, had he so desired, he could have linked metaphysics and natural philosophy at least to the degree achieved in the *Discourse*. We can agree, however, with the tactical reasons some scholars have adduced for the fabular format. These include considerations of exposition, persuasiveness and prudence. Indeed, now that we have uncovered the highly didactic structure of the opening chapters, it is even more likely that the fable was framed for tactical reasons, rather than for lack of a more synthetic, metaphysical mode of presentation. From this perspective we can examine some of the advantages of the fabular form.

---

[46] It is worth noting that Alquié (1950) made an even stronger claim, to wit, that the doctrines of creation of the eternal truths and continuous creation had so 'de-realized' the world that human science must take something like this fabular form (p. 125). This is one element in his grandiose attempt to read Kantian problems and themes into Descartes' work after 1629–1630. Alquié's approach in this instance led to a rather ahistorical account, in which, for example, Descartes' grounding of physics in metaphysics in the *Principia* is seen as a 'retrograde' step (p. 115). Indeed it was for Kant, but not for Descartes, who was and remained a seeker of ontological truth in natural philosophy, or at least for the upper conceptual reaches of his natural philosophy. The entire issue of Descartes' ongoing struggle to ground an increasingly hypothetical science in metaphysics (and Descartes' long term relation to Kant) was then much more convincingly treated in Buchdahl (1969).

In the first place, description of an explicitly imaginary cosmos avoids for the moment any hint of direct conflict with the Church over the issue of the reality of the Copernican system.[47] However, one should stress the term 'for the moment', because it is quite difficult to believe that Descartes actually thought he could side-step the issue of realist cosmological Copernicanism permanently, or that he even desired to do so. In the long run of the entire argument of *Le Monde*, Descartes certainly wants to suggest that his world matches the real world effect for effect, appearance for appearance. Presumably, in the strongest statement of his position, the metaphysics and voluntarist theology could have later been brought to bear to guarantee the truth of the major tenets of the system. Thus, the use of a fable is intended to gain time, both rhetorically and politically, in order to permit the full elaboration of the system in fabular guise before the process of matching is to begin.

These considerations help to explain another aspect of the fable. As we shall see, Descartes intends to produce not only a mechanistic cosmology but also a cosmogony describing the evolution of the mechanistic universe out of a divinely created chaos in accordance with the laws of nature. By this exercise Descartes probably did not mean to deny the scriptural account of creation. Rather his point was pedagogical: he could drive home the mechanistic approach by using it to explain the very genesis of the cosmic system. To accomplish this maneuver while avoiding insulting the religious sensibilities of his readers, he places the cosmogony in a fable. In this way he can offer the mechanistic genesis without claiming ontological validity or connection to our world. Whilst it can certainly be argued that Descartes' cosmogony is if anything even more ad hoc and unsatisfying than his cosmology, this observation does not weigh against the above argument, and indeed it tends to support it. From Descartes' perspective the relative inadequacy of the cosmogony may not have been apparent. The very possibility of matching the temporal development of the universe to the logical movement from principles to effects may have presented rhetorical and methodological advantages which hindered a clear appreciation of the vacuity of most of the cosmogonical descriptions.[48]

Finally, it should be noted that although the fable embodies and hence concedes the necessarily hypothetical nature of much of Descartes' world, this would not have appeared to Descartes to license unrestricted speculation. As he understands it,

---

[47] James Collins (1971) agreed with this (p. 8) but curiously went on to say that Descartes used a fable 'lest his own theory of the world suffer the same fate as befell that of Galileo.' But the trial of Galileo only took place in 1633, by the time *Le Monde* was virtually complete. Descartes' putative earlier caution would have been in relation to the condemnation of realist Copernicanism in 1616.

[48] Jacques Roger (1973) contended with respect to the cosmogonical passages in the *Principia* about element and vortex formation, taken together with the detailed Earth history in Book IV of the *Principia,* that we are dealing primarily with a logical rather than genealogical-historical exposition—a movement from principles to effects, more than a seriously intended history to rival *Scriptures*. On Peter Harrison's brilliant sequel to this sort of line of argument, see Chap. 12, Note 83 below. The role of Descartes' cosmogonical claims and Earth history in the *Principia* are discussed in detail in Chap. 12, with regard to his systematizing goals for that text and the ways they surpass those he had set for *Le Monde.*

the fable, a form of imagining in the sense of the *Regulae*, allows him to apply certain epistemological conditions to his universe, even before it has been ontologically certified. At the end of Chapter 6, he insists that nothing obscure or vague has been posited in the new world.[49] The fable consists only in that which is clearly imaginable, that is, on the doctrine of the *Regulae*, only that which is geometrical-corporeal and mechanical. In addition, since anything clearly imaginable is possible to God,[50] nothing strictly impossible or fantastic has been assumed. This is therefore a natural philosophy into which nothing occult, incorporeal or vacuously numerological has been injected, and that is a great deal to accomplish without an ideologically disruptive excursion into legitimatory metaphysics or theology.

9.5 Working out the Fable: Chapters 6–8 of *Le Monde*

We are now going to follow Descartes fabular exposition into Chapters 6, 7 and (the opening parts of) 8. Here we find his cosmogony articulated through the introduction of motion and its laws, the formation of cosmic vortices and the emergence of elements—which he has already described in Chapter 5—as the chief components respectively of stars (first element); planets, comets and planetary satellites (third element) and the fluid heavens themselves (spherical *boules* of second element.). Our account will continue to be rather more descriptive than interpretive or critical, because we want to reserve the final state of the cosmic vortices for very detailed conceptual dissection and genealogical study in the next chapter, as the key to our opening up the middle and concluding sections of *Le Monde* to intense scrutiny as an elaborate and daring system of natural philosophy. Accordingly, our discussion of these three chapters concludes our introduction to *Le Monde*, its style and pedagogy, and forms a bridge to the synthetic analysis in Chapter 10 of the system it contains.

9.5.1 Cosmogony, Matter–Extension and the Introduction of Motion and Its Laws

Descartes opens the fable in Chapter 6 by asking us to imagine that in the indefinitely large spaces beyond our real world God has created a uniform, space-filling continuous matter. This stuff is devoid of all secondary qualities and is conceived solely in terms of its solidity and continuous extension in three dimensions,

> … since we are taking the liberty of imagining this matter to our fancy, let us attribute to it, if you will, a nature in which there is absolutely nothing that anyone cannot know as perfectly

[49] AT XI 36; MSM.57; SG 24.

[50] Ibid.

as possible. To that end, let us expressly assume that it does not have the form of earth, nor of fire, nor of air, nor any more particular form (such as of wood, of a stone, or of a metal); nor does it have the qualities of being hot or cold, dry or moist, light or heavy, or of having some taste, or smell, or sound, or color, or light, or suchlike, in the nature of which one could say that there is something that is not clearly known by everyone.

Let us also not think, on the other hand, that our matter is that prime matter of the philosophers, which one has so well stripped of all its forms and qualities that nothing remaining of the rest can be clearly understood. Let us rather conceive of it as a true, perfectly solid body, which uniformly fills the entire length, breadth, and depth of the great space at the centre of which we have halted our thought. Thus, each of its parts always occupies a part of that space and is so proportioned to its size that it could not fill a larger one nor squeeze itself into a smaller one, nor (while it remains there) suffer to find a place there.[51]

Note the epistemic condition on this postulation, mentioned above, which derives from the act of 'distinctly imagining'. In the manner of the *Regulae*, that which is fully and distinctly intuited about matter is its geometrical extension. No person in his right faculties can deny that he clearly understands what this matter-extension is.

The matter created by God is utterly lifeless and without internal efficacy. The creation of matter, therefore, does not bring into being a Nature of self-sufficient causal processes, forces or active principles. Indeed, as yet, there is no Nature in the full sense of the term, only as it were, a block of impenetrable matter-extension. Local motion, which will be the principle of all natural change, must be injected into this dead block-universe by a second—but none the less simultaneous—creative act of God. By imparting diverse motions to portions of the block God thereby constitutes particles of different sizes and shapes.

Let us add further that this matter can be divided into any parts and according to any shapes that we can imagine, and that each of its parts is capable of receiving in itself any motions that we can also conceive. Let us suppose in addition that God truly divides it into many such parts, some larger and some smaller, some of one shape and some of another, as it pleases us to imagine them. It is not that He thereby separates them from one another, so that there is some void in between them; rather, let us think that the entire distinction that He makes there consists in the diversity of the motions He gives to them. *From the first instant that they are created, He makes some begin to move* in one direction and others in another, some faster and others slower (or indeed, if you wish, not at all): thereafter, He makes them continue their motion according to the ordinary laws of nature.[52] (emphasis added)

Motion, like geometrical extension, is clearly imaginable, being in some sense a simple term, incapable of explication. Thus, Descartes insists, the world, even in its initial chaotic state, contains nothing that cannot be perfectly known by the reader,[53] which is a very useful state of affairs, given that the above passage continues with one of Descartes' most daring gambits in the rhetoric of *Le Monde*:

For God has so wondrously established these laws that, even if we suppose that He creates nothing more than what I have said, and even if He does not impose any order or proportion on it but makes of it the most confused and most disordered chaos that the poets could

[51] AT XI 33; MSM 53–55; SG 22–23.

[52] AT XI 34; MSM 53–55; SG 23.

[53] AT XI 35; MSM 55–57; SG 23–24.

> describe, the laws are sufficient to make the parts of that chaos untangle themselves and arrange themselves in such right order that they will have the form of a most perfect world, in which one will be able to see not only light, but also all the other things, both general and particular, that appear in the true world.[54]

In short, we are being asked to accept suppositionally for the sake of argument the conceit that even if the omnipotent creator God somehow failed or reneged upon his instant to instant ordinary concourse of the universe and its laws, the system, left to its own devices after the instant of creation, would eventually settle into an analogue of our very own observable cosmos. Now, let us be clear, Descartes' voluntarist theology, and his not to be doubted Catholic orthodoxy, do not really envision this sort of deistic outcome, wherein a creator God first fashions matter and its laws but then leaves the machine to settle by its own mechanics into its final, stable form. His remark here is not meant to falsify or deny his underlying voluntarist commitment; but rather, as usual in *Le Monde*, to underscore the full conceptual contours of his systematic claims. No doctrinal denial or insult to orthodox Christianity (viewed through voluntarist theological spectacles) is literally intended; but rather a dramatic overstatement of his natural philosophical claims: If we, the *honnêtes hommes* readers, think through (that is imagine) this supposed deist cosmogony along with Descartes step by step, we shall establish a very firm grip on what Descartes' natural philosophical principles of matter and (laws of) motion are. As mentioned above in Note 48, Jacques Roger long ago saw this kind of expository logic at work eleven years later in the analogous cosmogonical passages, as well as the extensive, new Earth history of the *Principia*. But expository logic in natural philosophizing is one thing, and a legitimatory, even worshipful, commitment to voluntarist theology is another, and Descartes' voluntarist commitments to undergirding actual parts of his system are immediately on show—more clearly than anywhere else in *Le Monde*—in the very next chapter, on 'the laws of nature of this new world'.

9.5.2 The Laws of Nature

Descartes' statement and explanation of the laws of nature in Chapter 7 of *Le Monde* have been discussed above in detail with respect to the mechanical and metaphysical-theological problems involved; that is, we have explored both their origin and the factors shaping their content.[55] Therefore, we need only touch briefly on a few issues here.

It is important to keep in mind that motion is a real entity in Descartes' world. The total (scalar) amount of motion, or rather force of motion, in the universe is conserved by God from instant to instant. Most often he refers to motion as a real

[54] AT XI 34–35; SG 23; MSM 55.

[55] Sects. 4.2; 4.8.1; 8.2.2.2.

'state' or 'quality' of bodies which is to be conserved, and which causes their being 'in the act of moving'. Furthermore, his analysis of circular motion, which leads to recognition of the centrifugal tendency arising from constrained motion along a curve, clearly rests on the view that the 'principal' tendency, as we have termed it, and the component tendencies into which it may be resolved are really in the body, in the sense of internal moving efficacies.[56]

Although motion is real, we have no physical knowledge of how, in Descartes' world, it is maintained or transformed. As Descartes himself explains, the laws of nature are divinely decreed rules according to which natural change, i.e. transfer of motion, takes place.[57] They dictate how bodies will behave under certain conditions. In Descartes' view the motion of a body can be treated in respect to a series of temporal instants at which God recreates the body at a series of sequential spatial points and endows it with a certain quantity of force of motion linked to a certain principal determination (as we called it), or directional tendency to motion, characterizing it at those instants.[58] Since collision takes place in an instant, and the ultimate particles are perfectly hard and inelastic, there can be no physical interpretation of the process of impact and transfer of motion. Rather, on the occasion of an impact, God readjusts and redirects the motions of the bodies, according to the rule or law of motion governing that particular case. The rule describes the disposition of the force and direction of motion in the instant of time immediately following the collision. Thus the rule records what the outcome of God's law-like adjustment will be, while on the physical plane the phenomenon is conceptually opaque.

As we have already seen in Chapters 4 and 8 above, in large measure the actual form the laws of nature assume in Chapter 7 of *Le Monde* can be explained by Descartes' earlier work in optics, as it was conditioned by the new 'plenist realism' he came to espouse while composing *Le Monde*. The laws involve precisely the dynamical principles involved in the demonstration of the optical laws. Indeed, they are theoretical transcriptions of the geometrical rules of construction for refracted and reflected rays which Mydorge and Descartes had employed soon after the discovery of the law of refraction. The first law guarantees the conservation of quantity of motion or rather the quantity of force of motion, and thus it reflects the construction of the circular locus about the point of incidence upon which a refracted or reflected ray will be found. This law is intimately connected to the second, which

---

[56] As has been observed by several commentators, Descartes' notion of inertial motion or tendency has strong overtones of an impetus theory and centrally involves the idea of a real force or power of motion present in the body. For example, Cohen (1964), Gabbey (1980) and Westfall (1972) Note that all these claims attach at the level of a physical, or natural philosophical, understanding of Descartes' dynamics and laws of motion. On the theological plane, the moment to moment causal efficacy of a body in motion or tending to motion is to be attributed directly to the immediate action of God, without which neither the body nor its force of motion (and 'determinations' thereof) could exist or subsist. (Cf. Chap. 8, Note 46).

[57] AT XI 37; SG 25; MSM 59.

[58] Cf. Chap. 4, Sect. 2; Chap. 8, Sect. 2.2.2.

dictates the conservation of total quantity of motion in collisions, or, that the loss of motion by one body entails a corresponding equal increase in quantity of motion in the other. This conception is implied in the proof of the law of reflection, where the absence of transfer of motion to the perfectly hard reflecting surface necessitates that the total quantity of force of motion of the incident ray is preserved unchanged and thus that the ray is reflected at the appropriate angle. The third law asserts that the instantaneous tendency to motion is rectilinear, directed along the tangent to the path of motion at the point of question. This law guarantees the rectilinear transmission of light. In addition, this law provides a rationale for the optical proofs, which, after all, proceed on the basis of the resolution and composition of tendencies to motion characterizing the light ray or 'tennis ball' at the critical instant of impact with the optical surface. Put more generally, the optical proofs are based on a distinction between instantaneously possessed 'quantity of force of motion', a magnitude, and instantaneously possessed 'tendency to motion', or determination, a directed magnitude, both magnitudes being exercised and analyzed instant by instant. These two quantities respectively represent, and grow out of, the circular locus and horizontal component (or line) which were used in the original geometrical construction of refracted rays in Mydorge's letter to Mersenne of 1626. But, the forms of the first and third laws especially can also be linked to Descartes' desire to have principles adequate to the complex corpuscular-mechanical reality of a plenum universe. We have seen that they seem to derive from a bifurcation and transformation of Beeckman's original inertial law for actual rectilinear motion in a void. No actual inertial translation can ever occur in Descartes' world, but the concepts which underlay Beeckman's original law are still true of bodies in and of themselves. Thus, although bodies will always be constrained to move along curves and will always transfer motion to the omnipresent resisting media, in and of themselves, they will tend at each instant to move off along the tangent to the curve at the point under consideration, and conserve their total force of motion.

9.5.3 Vortex Formation, Stability Principle and [Re-]Introduction of the Elements

Chapters 8 and 9 of *Le Monde* present the core of Descartes' vortex celestial mechanics, which in turn is the key to the entire system of natural philosophy taught in *Le Monde*. These and subsequent chapters will be studied more synthetically next in our Chap. 10. Here we adduce only a few of Descartes' opening points in Chapter 8, which serve as a base line for our close analysis of the vortex celestial mechanics to come.

In Chapter 8 in what amounts to a second cosmogonical passage, added to that in Chapter 6, Descartes first notes that the particles created by God's injection of motion into the world settle into a number of huge vortical motions:

> ... to consider this matter in the state in which it could have been *before God began to move it*, one should imagine it as the hardest and most solid body in the world. And, since one could not push any part of such a body without pushing or pulling all the other parts by the same means, so one must imagine that the action or the force of moving or dividing, which had first been placed in some of the parts of matter, spread out and distributed itself in all the others in the same instant, as equally as it could.
>
> It is true that this equality could not be totally perfect. First, because there is no void at all in the new world, it was impossible for all the parts of matter to move in a straight line; rather, all of them being just about equal and as easily divertible, they all had to come together in some circular motions. And yet, because we suppose that God first moved them diversely, we should not imagine that they all come together to turn about a single centre, but about many different ones, which we may imagine as diversely situated with respect to one another.[59] (emphasis added)

In this quick, daring and perhaps unconvincing piece of cosmogony one paramount point stands out: the ubiquity of circular vortical motion in the cosmos. Of course, Descartes is not saying that circular motion is natural in some Aristotelian sense of being simple and irreducible. He is suggesting that the laws of nature and properties of matter are such that huge circulating vortices cannot but be formed. To grasp this point we need only recall his doctrine of local circular displacement in a plenum medium and interpret it in the light of the laws of nature. Any body to which motion is imparted will tend to move in a straight line (third law), but the omnipresence of resisting, deflecting particles constrains all motion into curves. Curved motion, and ultimately, circular replacements, must occur without fail, but circular motion is not for that reason natural in the sense of the Schools.

As each vortex continues to rotate, the particles begin to sort themselves out into a definite distribution; those 'naturally less agitated or smaller, or both, toward the places nearest to the centers than toward those farthest away.'[60] This distribution is based on an important condition for the stability of the vortex—that no ring of corpuscles has more centrifugal inclination than the next outer ring:[61]

> For all of them having an inclination to continue their motion in a straight line, it is certain that the strongest (i.e.: the largest among those equally agitated and the most agitated among those equally large) had to describe the greatest circles, i.e. the circles most approaching a straight line.[62]

---

[59] AT XI 49; SG 32–33; MSM 79–81. Notice that the first paragraph of this passage, contrasted to the one cited above in Note 52, seems to presume that there is some time interval between God's creation of matter extension and his injection into it of particle-producing motion. Alternatively, to preserve a unified and total creation by God, one might suggest that the gap between creation of matter-extension and insertion of motion to shatter it is merely logical, there being no temporality in God's creative act. The consequences for the matter-theoretical cosmogonical narrative, as considered by us here, are irrelevant; but the consequences for articulating Descartes' natural philosophy to one theological position or another might be considerable.

[60] Ibid., p. 49; SG 33; MSM 81.

[61] Gaukroger (2002), p. 152, Note 19 citing Aiton (1972), p. 63 Note 78.

[62] AT XI 49–50; SG 33; MSM 81.

The implication is that the 'inclination to continue in a straight line' is measured by the agitation and size conjointly, and that as one moves away from the centre of the vortex this inclination, which gives rise to centrifugal tendency to motion, will increase, or at least not decrease.

To this point Descartes has only set out a condition for the variation in force of motion of the particles with distance from the centre of the vortex. Next, he specifies in detail the relative sizes and speeds of the particles making up successive rings from the centre out. Again invoking the continual impact of the particles among themselves, he describes a kind of steady state in which size varies in some inverse ratio with speed, such that while the size of the particles decreases with radial distance from the centre, their increased speed more than compensates. In this way the condition on force of motion can be maintained.[63]

> Thus, in a short time all the parts were arranged in order, so that each was more or less distant from the center about which it had taken its course, according as it was more or less large and agitated in comparison with the others. Indeed in as much as size always resists speed of motion, one must imagine that the parts more distant from each center were those which, being a bit smaller than the ones nearer the center were thereby much more agitated.[64]

Note Descartes conception whereby the acquisition of speed is inhibited in proportion to the quantity of matter. The idea seems to be that, given the increasing difficulty of imparting velocity to large particles, it will be the relatively smaller particles which will first assume the higher levels of force due to an overcompensating acquisition of speed. Thus the smaller particles will take their places in the outer regions of the vortex.

It is important to realize, however, that although, for the sake of exposition, we have called this a 'kind of steady state', this distribution of particles is not the final state of any vortex, nor the end point of this cosmogonical narrative. As the subsequent passages make clear, Descartes has been considering the system of particles before the production of the three permanent forms of particle, the elements.[65] As the particles circulate they collide, breaking off each other's rough edges and protuberances, with the smallest of these cosmic scrapings forming the first matter. A portion of these first matter particles is forced to the center of their vortex, forming a sun or central star, while the rest of the first matter fills the interstices left between the particles of the vortex. The particles smoothed by this process become the spherical *boules* of the second element, constituting the bulk of the rotating 'heavens'.[66]

---

[63] Obviously, the dynamical conceptions in play here are precisely those whose origins and use in *Le Monde* we have traced through our findings earlier in this study: most notably Sects. 4.2; 4.8.1; 9.5.2; 3.3.2, 3.3.3, 3.4 and 8.2.2.2. Force of motion is a function of size (quantity of matter) and speed (or instantaneous tendency to motion), so, as the size of particles in a vortex decreases, their speed must increase in order for the 'stability condition' to be maintained.

[64] AT XI 50–51; SG 33; MSM 81–83.

[65] Remembering that Descartes has already introduced his element theory in Chapter 5 in a 'non-cosmogonical' context, shaped by his didactic strategy at that point.

Note that in this cosmogony the first and second elements have evolved out of the original 'ur-particles', established when the block of matter-extension was shattered by the injection of motion. Particles of this type did not exist amongst the variety of originally created particles. But what of the particles of third matter? It turns out that they are assumed to have existed ever since that first creation of particles. Not every particle of the originally created matter changed into first or second element. There were some larger and more irregular parts in the beginning, and these retain the form of the third element which makes up the bulk of planets (including the Earth), planetary satellites and comets. Some of the original particles of this third element were so large and cumbersome that whenever they met they easily joined up. There were others, a second sub-category of third element, even larger ones that were instrumental in reducing the size of the other particles when they collided, whilst they themselves remained intact.[67] Nowhere in *Le Monde* does the third element change into either of the other forms. Such earthy, that is planetary, matter can never change into the matter of the 'heavens' that is the second matter of vortices or the first matter of stars. This bar in *Le Monde* on the transmutation of elements will be lifted in the *Principia,* as we shall see in Chap. 12, when we compare *Le Monde* and the *Principia*, identifying the much improved systematization of the latter, and its daring pro-Copernican stance, beyond anything offered in *Le Monde*.

Finally, returning to the constitution of stars in the centers of vortices, Descartes tells us that the highly agitated first matter occupying center of the vortex forms into 'perfectly liquid and subtle round bodies'. He also insists that a central star can agitate the surrounding particles of second matter of its vortex:

> [Stars] incessantly turning much faster than, and in the same direction as, the parts of the second element surrounding them, have the force to increase the agitation of those parts to which they are closest and even (in moving from the center toward the circumference) to push the parts in all directions, just as they push one another.[68]

On the one hand, as we shall discuss in the next chapter, this agitation of the first element of the star sets up additional lines of tendency to motion in the second element. These lines of tendency will later be identified with the propagation of light and thus will help explain the appearance of the full disk of the sun to an observer anywhere

---

[66] AT XI 52–53; SG 34; MSM 85.

[67] AT XI 56–57; SG 37; MSM 93: 'In order for me to begin to tell you about the planets and comets, consider that, given the diversity in the parts of matter that I have supposed [at the creation] even though most of them have—through breaking up and dividing as a result of collision with one another—taken the form of the first and second element, there nevertheless remains to be found among them two kinds [as described in the text above] that had to retain the form of the third element.' And, two pages later (AT XI 60; SG 39; MSM 99), describing the formation of comets and planets out of third matter, he opens with '... no matter where the parts of matter *that could not take the form of the second or the first element* may have been initially ...' (emphasis added) Thus Descartes reiterates the existence of third matter particles before the initial formation of the first and second element.

[68] AT XI 53; SG 34–35; MSM 85.

in the vortex. On the other hand, as we shall also see, the solar agitation is essential to this second and definitive set–up of Descartes' vortex celestial mechanics, because the central star provides extra agitation to the heavenly particles of the second element, thereby disturbing their initial distribution and making possible the stability of planetary orbits. Hence, Descartes' second distribution, involving the size and speed of the particles of the second element, differs from the size/speed distribution attributed in the first 'model' to the 'pre-elemental ur-particles' making up the, so to speak, 'primordial' vortices. This second distribution sets the framework for the actual details of his vortex celestial mechanics, and it will be taken up in the next chapter. There we shall see just how complex, serious and interesting Descartes' vortex celestial mechanics became.

By this point in Chapter 8 of *Le Monde*, the fable of the world has started to transform into a detailed and challenging cosmological and natural philosophical construct. We therefore need to depart from tracing the fable as such, with its increasingly misleading surface style, if we are to understand *Le Monde* as more than a narrative fancy, indeed also as a system of natural philosophy. It remains true, of course, that the rest of the vortex mechanics and cosmological optics, and all other matters in Chapters 8–13 are part of the fable,[69] but we must leave our blow by blow account of it and look for a synthesizing and generalizing perspective that will bring out the complementary nature of *Le Monde* as a system of natural philosophy.

References

Works of Descartes and Their Abbreviations

AT = *Oeuvres de Descartes* (revised edition, 12 vols.), edited by C. Adam and P. Tannery (Paris, 1964–76). References are by volume number (in roman) and page number (in Arabic).

CSM(K) = *The Philosophical Writings Of Descartes*, 3 vols., translated by John Cottingham, Robert Stoothoff, and Dugald Murdoch, and (for vol. 3) Anthony Kenny, (Cambridge, 1988) References are by volume number (in roman) and page number (in arabic).

HR = The Philosophical Works of Descartes, vol I translated by E.S. Haldane and G.R.T. Ross (Cambridge, 1968 [1st ed. 1911]).

MM = *René Descartes, The Principles of Philosophy*, translated by V. R. Miller and R. P. Miller (Dordrecht, 1991).

MSM = *Rene Descartes, Le Monde, ou Traité de la lumière*, translated by Michael S. Mahoney (New York, 1979).

SG = *The World and Other Writings*, edited and translated by Stephen Gaukroger (Cambridge, 1998).

[69] Chapter 14, on the properties of light, might be taken to begin a shift to the 'matching of appearances', which is then formally commenced in the final extant chapter, Chapter 15 where the appearance of the fabular world to its inhabitants completely matches the appearance of our world to us.

Other

Aiton, E.J. 1972. *The vortex theory of planetary motion*. New York: Neale Watson Academic Publications.

Alquié, F. 1950. *La découverte métaphysique de l'homme chez Descartes*. Paris: Presses universitaires de France.

Alquié, F. ed. 1963. *Oeuvres philosophiques de Descartes*, t.1. Paris: Garnier Frères.

Boschiero, Luciano. 2007. *Experiment and Natural Philosophy in Seventeenth Century Tuscany: The History of the Accademia del Cimento*. Dordrecht, Springer.

Buchdahl, Gerd. (1969) *Metaphysics and the Philosophy of Science*. Cambridge, MA: MIT Press.

Cohen, I.B. 1964. Quantum in se est", Newton's Concept of Inertia in Relation of Descartes and Lucretius. *Notes and Records of the Royal Society of London* 19: 147–8.

Collins, James. 1971. *Descartes' philosophy of nature*. Oxford: Blackwell.

Gabbey, A. 1980. Force and inertia in the seventeenth century: Descartes and Newton. In *Descartes: Philosophy, mathematics and physics*, ed. S. Gaukroger, 230–320. Sussex: Harvester.

Gaukroger, S.W. 2002. *Descartes' system of natural philosophy*. Cambridge: CUP.

Gaukroger, S.W., J.A. Schuster, and J. Sutton (eds.). 2000. *Descartes' natural philosophy*. London: Routledge.

Mouy, Paul. (1934) *Le développement de la physique Cartésienne*. Paris: Vrin.

Ranea, G. 2000. A "Science" for *honnêtes hommes*": *La recherche de la Vérité*. In *Descartes' natural philosophy*, ed. S. Gaukroger, J.A. Schuster, and J. Sutton, 330–346. London: Routledge.

Roger, Jacques. 1973. La Théorie de la Terre au XVII Siècle. *Revue d'Histoire des Sciences* 26: 23–48.

Verbeek, Theo. 2000. *The invention of nature: Descartes and Regius*. In *Descartes' natural philosophy*, ed. S. Gaukroger, J.A. Schuster, and J. Sutton, 149–167. London: Routledge.

Westfall, Richard. 1972. Circular motion in seventeenth century mechanics. *Isis* 63: 184–9.

Chapter 10
'Waterworld': Descartes' Vortical Celestial Mechanics and Cosmological Optics in *Le Monde*

10.1 Introduction: Uncommon Vortices

It is now time to switch gears as it were, approaching the remainder of *Le Monde* in a more synthetic way, by exposing its character as a systematic natural philosophy. We focus at first on the conceptual core and most difficult interpretative challenge in the text—Descartes' vortex celestial mechanics. If we can accomplish this, we can much more easily grasp the other pillar of the corpuscular-mechanics natural philosophy of *Le Monde*, the theory of light in its cosmological setting, as well as the other canonical natural philosophical topics whose treatment in *Le Monde* flow from these sources: weight, local fall, the tides, the motion of planetary satellites and comets.

Over sixty years ago, Thomas Kuhn, in his best selling and often reprinted, *The Copernican Revolution*, said this of Descartes' vortex universe: the 'vision was inspired'; the 'scope tremendous'; but 'the amount of critical thinking devoted to any of its parts was negligibly small'.[1] Typically more pointedly and poetically, Gaston Bachelard had in 1938 condemned Descartes' plenist universe, including the vortex mechanics, as the 'metaphysics of the sponge', an exemplary 'pre-scientific' monstrosity, in other words, the sub-scientific progeny of cancerous metaphor and Baroque ego projection.[2] Other more mundane brush offs could also be cited. Certainly, Descartes' vortices do not possess for us the straight, presentist scientificity of Newtonian mechanics, but, as this chapter will show, they do have a coherent conceptual structure, which makes a great deal of sense, given the prior evolution of Descartes' physico-mathematics and the manner of its flowing into a

[1] Kuhn (1959) pp. 240, 242.

[2] Bachelard (1965) p.79, 'La métaphysique de l'espace chez Descartes est la métaphysique de l'éponge.'

J. Schuster, *Descartes-Agonistes: Physico-mathematics, Method & Corpuscular-Mechanism 1618-33*, Studies in History and Philosophy of Science 27,
DOI 10.1007/978-94-007-4746-3_10, © Springer Science+Business Media Dordrecht 2013

systematic corpuscular-mechanical natural philosophy in *Le Monde*.[3] We can, in short display how and why the vortices were intellectually constructed and extend that understanding to the text of *Le Monde* as a systematic gambit in the natural philosophical contestation of Descartes' time.[4]

This chapter therefore brings together two lines of investigation and their corresponding results: First, as just noted, this chapter focuses on showing the natural philosophical seriousness of the vortex celestial mechanics as an intellectually constructed object and as a strategic gambit. It was Descartes' specific and technical way of addressing the natural philosophical challenge posed by realist Copernicanism. But, interesting as that may be, it is in the final analysis only a step toward a further aim, which is to show the ways in which *Le Monde* was simultaneously the climax of Descartes' trajectory in physico-mathematics and the first iteration of a systematic natural philosophizing, emergent from that carapace. The vortex celestial mechanics instantiates what that trajectory had come to be about. For, as we shall see, the celestial mechanics at the heart of *Le Monde* is a hybrid entity. On the one hand it depends upon the genealogy of his physico-mathematics and carries some of its conceptual DNA, especially in the form of operationalizing Descartes' principles of dynamics, themselves, of course, of physico-mathematical provenance. But, on the other hand, the vortex mechanics is clearly a piece of generic natural philosophical discourse, understandable as such by any member of the educated culture of natural philosophizing, and playing the central role in this new corpuscular-mechanical system of natural philosophy.[5]

The argument of this chapter will therefore unfold as follows: Sect. 10.2 contains the fulcrum of the argument, an extended intellectual reconstruction of the inner toils of the vortex mechanics of *Le Monde*. Section 10.3 steps back in time to pick up the thread of the genealogy linking the early physico-mathematics to the project of *Le Monde*. We already know a great deal about how Descartes' physico-mathematics—the

[3] And, as Aiton (1972) has shown, the vortex celestial mechanics was taken seriously and articulated much further down into the first half of the eighteenth century by committed Cartesians and anti- Newtonian mechanists.

[4] In saying this I in no way wish to imply that I introduced Bachelard and Kuhn above as mere straw men. These two historian/philosophers of science initially most influenced my understanding of the dynamics of seventeenth and eighteenth century natural philosophy. I have argued elsewhere that Kuhn and Bachelard indeed misunderstood the nature of that natural philosophy and the contestations over it—taking it as the necessary but pre-scientific backcloth to the temporally splayed crystallization of a heterogeneous set of new 'real' sciences. However, as I have also claimed, that is less important than the fact that their speculations prompted more positive modeling by historians of early modern natural philosophy, its nature, dynamics and trajectory. Schuster and Watchirs (1990), Schuster and Taylor (1996), and Schuster (2002) all set the groundwork for the model of natural philosophy presented above in Chap. 2.

[5] Of course, if *Le Monde* marked a node and climax, in Descartes' career, it was obviously a particularly transient and occluded one, rather internal to Descartes' development, not a public marker. Yet, to understand the later *Discours*, *Meditations* and *Principia* we need to understand how and why he arrived at this text, its genealogy, and its systematic character as a natural philosophy.

early hydrostatics and work on fall, as well as the optical triumph of the later 1620s—contributed along the way to the shaping of his commitment to some form of (non-systemic) corpuscular-mechanism, and provided its potential causal register, his dynamics of corpuscles. In Sect. 10.3 we shall see how Descartes' renewed acquaintance with Isaac Beeckman in 1628, after a nine year hiatus, presented Descartes with a challenge to speculate about the corpuscular-mechanical explanation of planetary motions in a Copernican system, and a rough conceptual exemplar for doing so. This last piece of the genealogy of the vortices will help make sense of their anatomy exposed in Sect. 10.2, and vice versa. Certain conceptual consequences of this insight will be discussed in Sect. 10.4. Then Sect. 10.5 will show how a charitable reading of the vortex mechanics makes prima facie sense of Descartes' further elaboration in *Le Monde* of theories of weight, local fall, the tides and motion of the moon. We shall, however, find certain interpretive challenges, and textual difficulties, in this connection, so in Sect. 10.6 we shall re-interpret the latter topics in *Le Monde*, finding that our principles of charitable interpretation need to be expanded a bit, chiefly by considering the ways in which Descartes' 'prior record' in physico-mathematics was shaping *Le Monde*, for better and worse. In Sects. 10.7 and 10.8 we will continue in this enhanced mode of charitable interpretation to deal respectively with the cosmological optics of *Le Monde* and the attempt to match appearances in our own world to those of the fabular corpuscular-mechanical world, chiefly in the theory of the appearances of comets. Finally, in Sect. 10.9 we will foreshadow our study of the systematic character of *Le Monde* and its status as a bold competitive gambit in the field of natural philosophy, to be explored in Chap. 11.

10.2 Descartes' Vortical Celestial Mechanics in *Le Monde*

10.2.1 Charitable Hermeneutics—Principles and Aims

Before we turn to a synthetic recounting of the vortex mechanics of *Le Monde*, the details of which will be examined in following sub-sections, it is important to note precisely what my interpretive strategy is, and what it is not, as well as how that strategy subserves the aims of this book, rather than some aims that might erroneously be attributed to my efforts here. The first point to make is that this reading depends on our already having established the nature of Descartes' dynamics, and its genealogy in physico-mathematics, running from the hydrostatics of 1619, through the climactic optical work of the 1620s, leading in the process of construction of *Le Monde* to the formalization of laws or rules of nature, for which Descartes was willing to advance voluntarist theological legitimations. In *Le Monde* that dynamics appears as the doctrine of causation of his natural philosophy, dealing with motions and tendencies to motion, through which he intended to 'run' the machinery of his vortex world. The genealogy of the dynamics has formed a central thread in our argument from Chap. 3, through Chap. 4 to the account of the creation of *Le Monde* in Chap. 8.

Turning directly, then, to our hermeneutical task, it has to be frankly said that Descartes does not communicate well to the reader in the sections of *Le Monde* dedicated to the theory of vortices. The entire work, of course, is incomplete, unpolished and remained unpublished in his lifetime, and the sections on the vortex mechanics suffer from these circumstances along with the rest of the text. Moreover, as we shall note in a couple of instances below, Descartes hardly helped his cause in the articulation of the vortex theory by his adoption of that commonsensical, *honnête homme* style which, we have seen, marked his treatment of the opening chapters of *Le Monde*. His appeal to commonly experienced analogies and observations—without explicating their limitations or precise modes of linkage to his underlying concepts and theories—tends to swamp and confuse his message about the vortex mechanics. But, and this is the key point, Descartes arguably did possess a coherent and well thought out theory of vortices, of which the surviving text of *Le Monde* is a rather poor representation. It is, however, a representation that can lead the hermeneut to that underlying theory, provided three conditions of reading and analysis are fulfilled: [1] As we have done and continue to do in this book, one must attend constructively to the likely trajectory of Descartes' work and struggle in natural philosophy and physico-mathematics in the decade or so leading up to the composition of *Le Monde*; [2] One must probe behind the breezy style of presentation in *Le Monde*, with its appeal to easy if somewhat misleading analogies. One must interpret the text charitably in the search for deep and coherent theorizing, consistent with and evolving out of the material studied in [1]; Finally [3] one must be willing to use the much more systematically and coherently developed explication of vortex mechanics in the *Principles of Philosophy* as an heuristic guide to what Descartes might possibly have been entertaining in *Le Monde* (without falling into a vulgar retrospective Whiggism). In many ways, therefore, this reading of *Le Monde* for a strong and complex underlying theorization rebounds upon our sense of the text itself, perhaps lending it the coloration of a more private, even solipsistic, document, a bit akin to those sets of working notes and drafts that, without ever seeing the light of day, scaffold our own public utterances.

Now, the aim of such a reading is not to conclude that *Le Monde* 'really' teaches such a coherent theory of vortices which later seventeenth century readers, and modern historians of science have, through some cognitive shortcoming, 'failed' to see. Nor is the aim to blame Descartes for failing lucidly to express what he had so systematically conceptualized. Such points are irrelevant in regard to my goals here. I aim, rather, to try to capture, via such a reconstruction, what arguably was the *state of theorizing* that Descartes had reached about vortices at the end of almost fifteen years of work—as his physico-mathematics developed and debouched in a systematic corpuscular mechanics with a vortex mechanics at its heart. *That* theory of vortex celestial mechanics lurks below the surface of *Le Monde*, but is recoverable from it. We shall see that Descartes' underlying theory was subtle and complex, reflecting upon and exploiting a sequence of technical achievements in physico-mathematics, as well as his own lived experience as an increasingly mature, and competitive, player in the struggle to forge a new natural philosophy embodying Copernican realism. In the latter sense, Descartes' underlying conceptualization was one instance of a more widely pursued problematic, worth studying as part of a mapping of other natural philosophical initiatives and aspirations of similar kind and intended scope—as in the work of Beeckman, Kepler, Gassendi, and Mersenne.

To all this one further and more pragmatic condition has to be added, as far as the account of the vortex mechanics offered here is concerned. My presentation will be synthetic and declarative.[6] There is no space here to offer the more analytical and textual critical account of how Descartes' theory has been teased out of the text; and exactly how textual juxtapositions and interpretations, as well as judicious appeals to the *Principles*, can be used to clarify his analogies, reorganize his diffuse and confusing order of presentation, and explicate certain half articulated points and claims. Appendix 2 is devoted to such a more detailed elucidation of these matters. It recapitulates some of the flow of my own initial interpretative engagement with this section of *Le Monde*, and provides backing for the systematic reading offered here. I shall, however, at various points indicate in footnotes the degree and type of interpretative work and reconstruction involved in presenting particular concepts and representations, signaling their place in Appendix 2.[7]

Amongst the concepts and representations I shall use in the following account: [1] Some arguably derive quite literally from the text of *Le Monde*. [2] Some arguably express Descartes' theoretical intentions in ways he did not quite accomplish in the text. [3] Some systematize or clarify concepts confusedly presented in *Le Monde* (but often better expressed later in the *Principia*) in a charitable attempt to elicit a coherent theory. [4] Some are novel, my own interpretive inventions, advanced again in a charitable attempt to elicit a coherent theory from Descartes' text. Arguably, they could have been constructed by Descartes himself or a

---

[6] Indeed in oral presentations of this paper at seminars and conferences I have used, not unsuccessfully, the following conceit in synthetically presenting the vortex theory: that this is a pro-Cartesian university lecture in Cartesian natural philosophy circa 1660, assuming fairly widespread consensual acceptance of the vortex mechanics. This allows the further conceit that the new diagrams and concepts I use below to explicate the vortex mechanics have actually become recognized parts of a Cartesian Scholastic tradition within a generation of his death. Perhaps if the remainder of this section is read in that spirit, the key points about the theory will come through, provided one remembers above all that I am not suggesting this was for anybody the explicit, publicly acknowledged version of the vortex celestial mechanics, but rather that this is very close, on a charitable reading, to Descartes' own best understanding of his vortex theory, as it related to his course of work and context of natural philosophical struggle up to the early 1630s. Cf. Sect. 10.5.4 below where in a similar conceit 'Descartes' himself speaks posthumously of the coherence of his vortex celestial mechanics.

[7] The more textual critical approach to teasing the underlying theory out the literal sense of *Le Monde* offered in Appendix 2 was begun in Schuster (1977). Amongst Descartes' inadequately or misleadingly expressed analogies and claims that—revised, criticized and explicated—will find their place the synthetic presentation of the theory below are [1] the appeal to the behavior of a large heavy boat compared to random flotsam in the confluence of two parallel rivers; [2] Descartes' mode of setting out the notion of a 'balance' of forces holding a planet in its orbit; and [3] the articulation of the key concept of 'massiveness' or 'solidity' of an orbiting body.

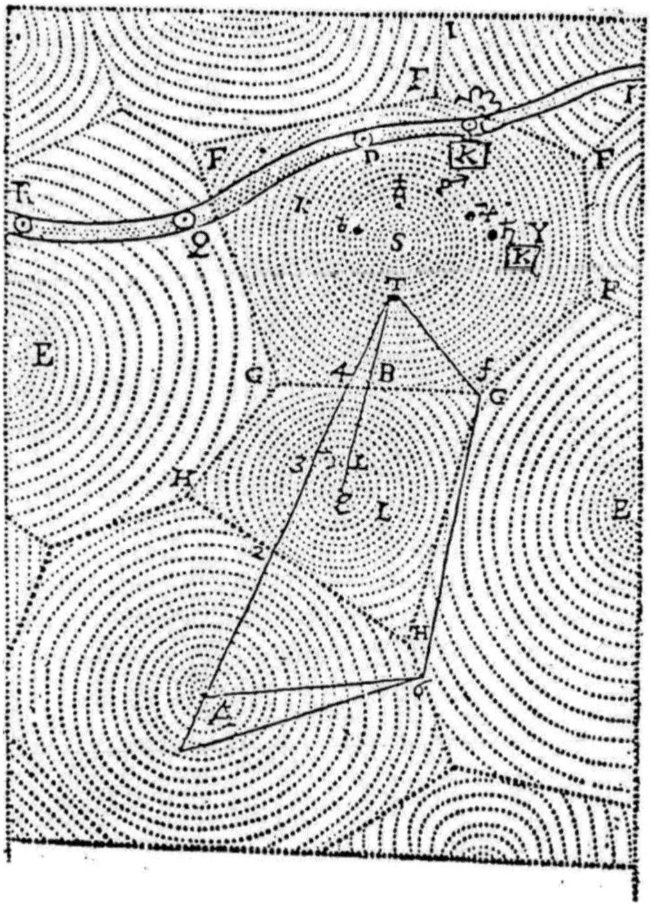

Fig. 10.1 The Vortex Cosmos, Descartes, *Le Monde,* AT XI p.55

contemporary, but were not to my knowledge. [5] Some representations and concepts correct misleading implications of some of Descartes' analogies in the interest of charitably supporting our vision of his underlying theory, and separating off misleading but understandable implications that have been or could be read into his surface analogies. With all these caveats, let us now begin the explication.

10.2.2 Size and Speed Distributions of Vortex Corpuscles and the Role of Central Stars

People often take the vortex celestial mechanics on its most superficial level, as if it were just an historical holding action waiting for Newton (Fig. 10.1): Descartes imagined whirlpools or vortices of second element, rotating around their respective central stars to sweep along their planets like boats in a strong current.[8] In fact the swishing along of the planets in the vortex was the least of his concerns. He thought that the mere existence of a whirlpool of second element accounted for the orbital movement. What interested him more deeply, and what shaped the conceptual architecture of the vortex celestial mechanics, were such non–trivial questions as: *Why does each of the planets maintain its own characteristic and yet relatively stable cosmic distance from the sun*; and, *Why do comets* (as he believed) *continually oscillate between vortices, spiraling in toward the central star of one vortex, up to a specific, theoretically given radial distance, and then spiraling out again into a neighboring vortex, down to a similar theoretically given minimum radial distance from its central star, and so on.*[9]

Now, the first step in unpacking the actual inner toils of the vortex celestial mechanics is to recall what we already noticed in our earlier examination of Chapter 8 of *Le Monde*: The overall condition for stability of the vortex is that there be a uniform and continuous increase in the centrifugal tendency of the particles making up the vortex as one goes away from the centre.[10] Let us think this through with Descartes, according to his dynamics—the causal register of his corpuscular mechanism, which we examined in Chap. 4 in relation to its emergence out of his physico-mathematical work in optics. We immediately see that centrifugal tendency is proportional to the force of motion in the tangent direction, and force of motion is measured by quantity of matter times speed, or more technically, quantity of matter and the instantaneously exercised 'principal' tendency to motion (as we learned to term it in Chap. 4). Descartes wants to specify how size and speed of the particles of the vortex vary with distance from the centre. We have already seen (Sect. 9.5.3 above) he does this twice in Chapter 8 of *Le Monde*—before and after the crystallization out of the elements—but now we shall need to attend closely to both moments in his exposition with a view to a deep dissection of the vortex celestial mechanics.

First Descartes describes the speed/size distribution of the particles making up the vortex in the earliest stages of vortex formation, prior to the production of his

[8] In fact in the key analogy used by Descartes, in a strong river current boats behave like comets, and it is light flotsam that behaves on analogy to planets. Thus, untutored intuition misleads as to Descartes' own preferred analogy (and hence misses the theoretical points he will be elucidating through the analogy).

[9] Additionally, as we shall see, he was also interested in relating a theory of local terrestrial gravity to his vortex celestial mechanics—a nice trick, since on Earth bodies of third element subjected to the local vortex fall down; but in the heavens, bodies of third element, subjected to the stellar vortex, find specific and stable orbital distances. Descartes thought there was a unified conceptual explication of these indubitable phenomena and he prided himself on designing it.

[10] Let us now call this the 'force-stability principle'. Strictly speaking, however, more is involved in Descartes' full conception of the orbital stability of the particles, or planets, orbiting at a given radial distance. Descartes' articulated version of the force-stability principle will be developed below, Sect. 10.2.3.

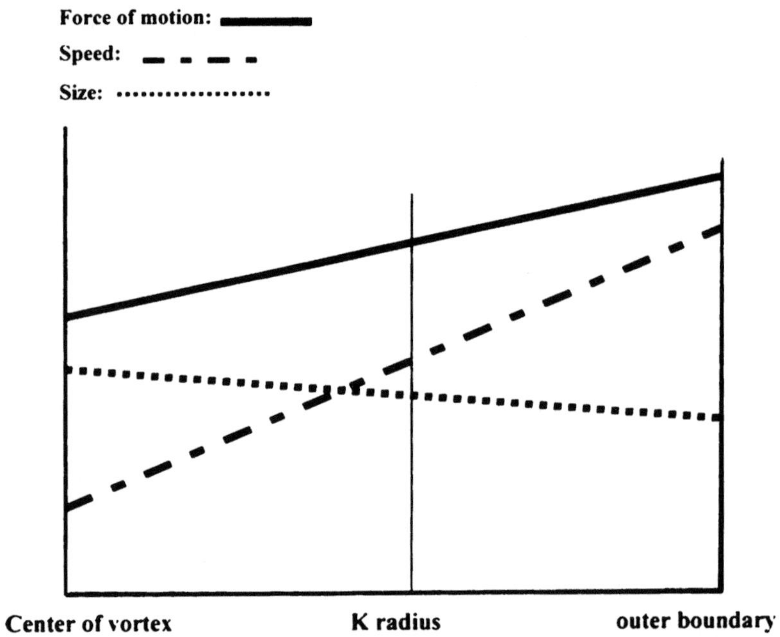

Fig. 10.2 Size, speed and force of motion distribution of vortex particles, prior to existence of central star

three types of stable particle, or elements, and hence prior to the formation of the sun, which, of course, is made up entirely of the highly agitated particles of the 'first element'—a critical moment in the theory as we shall shortly see. So, Descartes tells us that in this first, very early stage, as the vortex settled out of the original chaos, the larger corpuscles were, of course, harder to move, so there was a tendency for the smaller ones more easily to acquire higher speeds. Accordingly, in these early stages, the size of particles decreased and their speed increased from the centre out. But the speed of the particles increased proportionately faster, so that force of motion increased continuously.[11] In Fig. 10.2 we see Descartes' first declared distribution of size and speed of the particles making up the vortex in the period before the formation of the three elements and the emergence of a star in the

[11] AT xi. 50–51; SG 33; MSM 81–83, 'Thus, in a short time all the parts were arranged in order, so that each was more or less distant from the center about which it had taken its course, according as it was more or less large and agitated in comparison with the others. Indeed in as much as size always resists speed of motion, one must imagine that the parts more distant from each center were those which, being a bit smaller than the ones nearer the center were thereby much more agitated.'

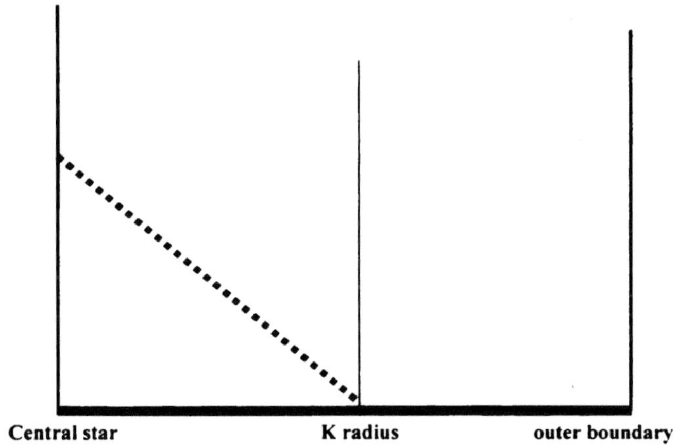

Fig. 10.3 Agitation due to existence of central star

centre of the vortex: Force of motion constantly rises, as does speed, while size decreases proportionately less than speed increases.[12]

Descartes' second description of the speed/size distribution of the particles making up the vortex applies to the period after the formation of the three elements.[13] Descartes explains that as the vortex rotates in its first stage, the particles collide with one another, breaking off their rough angles and points. These cosmic scrapings form the first matter, much of which is forced to the centre of the vortex, while the remainder fills the interstices left between the particles of the vortex. The latter

[12] Note in relation to this figure, as well as Figs. 10.3 and 10.4 below that they of course do not exist in *Le Monde* and are interpretative tools of my own design, used to picture the relationships Descartes sets out verbally. Additionally, it should be remembered that Descartes had no way of assigning empirically meaningful dimensions to the sizes and speeds of the *boules*. Nor would it have occurred to him to insist on any specific relationship for the variation of size and speed with distance. He limited his discussion to notions of proportionately greater or lesser increase or decrease of variables, which the figures then represent. *Had Descartes sketched figures like these, we might more easily recognize the traces of his style of physico-mathematics in the vortex mechanics.*

[13] Descartes adduces the elements at this stage in *Le Monde* in Chapter 8 (AT xi 51–55), but, as we have seen (Sect. 9.3 above) he has already adumbrated their properties at the end of Chapter 4, and described them in detail in Chapter 5 (AT xi 24–6; SG 17–18; MSM 37–39), writing there that, 'I conceive of the first, which one can call the element of fire, as the most subtle and penetrating fluid there is in the world... I imagine its parts to be much smaller and to move much faster than any of those other bodies. Or rather, in order not to be forced to imagine any void in nature, I do not attribute to this first element parts having any determinate size or shape; but I am persuaded that the impetuosity of their motion is sufficient to cause it to be divided, in every way and in every sense, by collision with other bodies, and that its parts change shape at every moment to accommodate themselves to the shape of the places they enter... As for the second, which one can take to be the element of air, I conceive of it also as a very subtle fluid in comparison with the third; but in comparison with the first there is need to attribute some size and shape to each of its parts and to image

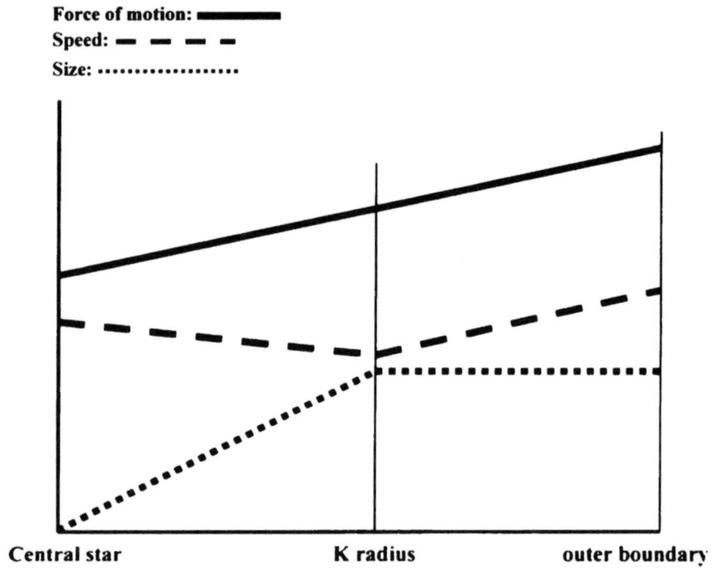

Fig. 10.4 Size, speed and force of motion distribution of particles of second element, in a stellar vortex

particles, smoothed and polished by this process, become the spherical *boules* of the second element. Grosser particles of third matter, assumed to have existed all along—as we found in Sect. 9.5.3—now form planets, their satellites and comets. The first matter at the centre of the vortex is highly agitated and forms 'perfectly liquid and subtle round bodies', that is, stars, including the sun at the centre of our vortex.[14] It is the presence of a star in the centre of a vortex that alters the first distribution of size and speed of particles in that vortex. This is absolutely crucial to the final theory, for the star's disturbing effect on the original size/speed distribution produces a second, quite different stable distribution of size and speed of the vortex particles, and it is this second distribution that both allows the planets to maintain stable orbits and explains the existence and orbital behavior of comets.

them as just about all round and joined together like gains of sand or dust. Thus, they cannot arrange themselves so well, nor press against one another, that there do not always remain around them many small intervals, into which it is much easier for the first element to slide in order to fill them. And so I am persuaded that this second element cannot be so pure anywhere in the world that there is not always some little matter of the first with it. Beyond these two elements, I accept only a third, to wit, that of earth. Its parts I judge to be as much larger and to move as much less swiftly in comparison with those of the second as those of the second in comparison with those of the third. Indeed, I believe it is enough to conceive of it as one or more large masses, of which the parts have very little or no motion that might cause them to change position with respect to one another.'

[14] AT xi 53, MSM 85; SG 34–35.

The sun is made up of the most agitated particles of first element; their agitation communicates extra motion to parts of the vortex near the surface of the sun; that is to those spheres of second element in the vortex lying near the sun. This increment of agitation decreases with distance from the sun's surface and vanishes to nothing at a certain radius, labeled by Descartes in Fig. 10.1 as K.[15] In Fig. 10.3 we represent Descartes' conception of the solar disturbance and its decrease with distance up to radius K. The solar effect alters the original size and speed distribution of the spheres of second element in the vortex, below the K layer.[16]

We now have greater corpuscular speeds close to the sun than in the pre-sun situation. But the force-stability principle, of course, still holds, so the overall size/speed distribution must change, below the K layer.[17] Descartes description of this situation may be represented in Fig. 10.4.

In the solar vortex as one moves away from the sun the agitation (speed) of the *boules* decreases, reaching a minimum at the distance K (where Descartes will locate the planet Saturn in our particular star–planets system). From K outward to the boundary of the vortex the agitation increases again. The size of the *boules* increases from a minimum near the sun to K; and from K outward the size remains constant or perhaps diminishes a little. From the sun to K the size of the *boules* of second element increases proportionately more than their speed decreases; from K outward the speed increases proportionately more than the size decreases. Thus we can draw a line of positive slope representing the force of motion of the *boules* (agitation × size) at each distance from the sun.

---

[15] Descartes insists that a central star can agitate the surrounding particles of second matter of its vortex: 'These (spherical bodies) incessantly turning much faster than, and in the same direction as, the parts of the second element surrounding them, have the force to increase the agitation of those parts to which they are closest and even (in moving from the center toward the circumference) to push the parts in all directions, just as they push one another.' (AT XI 53 MSM 85; SG 34–35) Note that in this exposition we often speak of our central star, the sun, as does Descartes. The theory, however, is quite general and applies to each and every central star and its respective vortex. No reader of *Le Monde* can be in any doubt about this fundamental point.

[16] The special radial locus at distance K is present in Descartes' own discussion. Here for expository purposes I introduce the term 'K layer' not used by Descartes. Note as well that the existence and location of the K layer are caused by the existence and action of the sun.

[17] Descartes' final distribution of the size and speed of the particles of the second element is as follows: AT XI 54–6; MSM 87–91; SG 35–37. (Fig. 10.1): 'Imagine… that the parts of the second element toward F, or toward G, are more agitated than those toward K, or toward L, so that their speed decreases little by little [as one goes] from the outside circumference of each heaven [vortex] to a certain place (such as, for example, to the sphere KK about the sun, and to the sphere LL about the star ε) and then increases little by little from there to the centers of the heavens because of the agitation of the stars that are found there....As for the size of each of the parts of the second element, one can imagine that it is equal among all those between the outside circumference FGGF of the heaven and the circle KK, or even that the highest among them are a bit smaller than the lowest (provided that one does not suppose the difference of their sizes to be proportionately greater than that of their speeds). By contrast, however, one must imagine that, from circle K to the sun, it is the lowest parts that are the smallest, and even that the difference of their sizes is proportionately greater than (or at least proportionately as great as) that of their speeds. Otherwise, since those lowest parts are the strongest (due to their agitation), they would go out to occupy the place of the highest.'

Three points are crucial about Descartes' model, and they are particularly clear in our representations in Figs. 10.2 and 10.4:

[1] It is the action of the sun that transforms the distribution of Fig. 10.2 into that of Fig. 10.4. This is absolutely crucial. The presence of the sun not only shifts the distribution of agitation, but it also as a consequence induces a change in the relative size distribution of the particles. This is due to the theoretical requirement that when the speed curve shifts, the size distribution must change accordingly, so that the force condition on the stability of the vortex is maintained.

[2] The K radius is the critical distance. It marks the locus beyond which the sun's added effect vanishes. Beyond K we have the *old*, stable pattern of size/speed distribution (of Fig. 10.2); below K we have a *new*, stable pattern of size/speed distribution—we still have force of motion increasing continuously with radius, but that comes about because size increases more quickly than speed decreases. As we shall see this new distribution permits the observed celestial motions to occur. In effect it turns the vortex into a special kind of machine—a machine that *locks* planets into their appropriate orbits below K and that *extrudes* them from inappropriate orbital distances.

[3] The existence of celestial vortices behaving as locking and extruding machines depends upon the fact that a star, made of first element, happens to inhabit the centre of each vortex, transforming the mechanical parameters and performance of that vortex. This is Descartes' version of the Keplerian emphasis (compared to Copernicus himself) on the physical-causal (or natural philosophical!) role of the sun in orbital mechanics. Interestingly, and crucially, the central location, and physical behavior of each vortex's star, are also essential to Descartes' theory of light in the cosmic setting—again it is the central star that completes the theoretical picture explaining the phenomena of light in the vortex universe, as we shall see below in Sect. 10.7.

10.2.3 *Locking and Extruding—Unpacking the Technical Core of the Vortex Celestial Mechanics*

We turn now to the technical details of how Cartesian vortices lock planets into orbits at definite radial distances from their central star below the K level, and why, according to Descartes comets continually oscillate between vortices, never penetrating below the K layer of any of them. Given what has already been established, all this occurs in what, to Descartes, seems a straightforward mechanical fashion.[18] Descartes' approach focuses on the centrifugal tendency of planets and of surrounding particles of second element in the vortex. Remember that, according to Descartes'

[18] Let us reiterate that the reconstruction that follows here skims over all the complexities of textual interpretation mooted above at the beginning of this section, including some presumably non-Whiggish appeals to clarifications in the utterances of the *Principles* eleven years later. For a recounting of a more analytical initial unfolding of these textual findings, see Appendix 2.

dynamics and his sling exemplar (Sect. 4.2 and Fig. 4.1), as a body or corpuscle moves on a curve, it has a certain force of motion along the tangent at any moment in its translation. Because it is constrained to move along a curved path, part of its tangential tendency manifests itself as a centrifugal tendency to recede along the normal to the path at that point on the curve. So, all bodies moving along a curve generate a centrifugal tendency to motion proportional to their size, quantity of matter, and instantaneously manifested tangential force of motion.

The key question in Cartesian celestial mechanics now becomes this: when and why is centrifugal tendency actualized as centrifugal motion, and when and why does that not happen? In the vortex, what plays the role of the sling, constraining the planet into a curved path and thus generating centrifugal tendency on its part? Well, it is of course the neighboring, superjacent particles of second element that do this job—they surround and penetrate the pores of every piece of third matter making up a planet.

Why then do planets maintain orbits and why at different distances—all within radius K—from their central star? This depends on the amount of resistance the superjacent second element can put up, and that is dependent upon how much second element can surround and *envelop* the parts of the planet, as what we may term a 'surface envelope'—a term of hermeneutical art that greatly helps our explication of *Le Monde* and the *Principles*.[19] The more matter of second element in this envelope, the more resistance the envelope will present to its being shoved aside by the planet's tendency to recede from the central star. Figure 10.5 is a schematic representation of this notion: a simple ball of third element in circular motion is surrounded by a smaller and a larger envelope of corpuscles of second element.

What determines how large a surface envelope is relative to a given planet? Obviously, it is the size distribution of the second element with distance from the star. Descartes recognized that the size of a surface envelop is dependent upon the volume to surface ratio of the spheres of second element. That ratio is function of their radii. The greater the radius of a sphere, the greater the V/S ratio.[20] To simplify the matter, as in Fig. 10.5, imagine a solid ball of third element in circular motion surrounded by an envelope of second element. As long as the spheres of second element are so small compared to the piece of third element that we do not reach the point at which only a few spheres of second element suffice to 'cover' its surface, we can get a great variation in overall, aggregate size of the envelop, hence its quantity of matter, and hence its resistance to being moved out of place by the centrifugal tendency of the piece of third matter.

[19] My notion of 'surface envelope' is a good example of a term of interpretative art belonging to my hermeneutical categories 2, 3 and 5, discussed earlier in Sect. 10.2.1. To see the reasons for its introduction, refer to Appendix 2.

[20] The second element, recall, is quite small compared to the pieces of third element, something Descartes goes out of his way to claim, in first describing the elements, as we saw above in Note 13: 'Its parts (third element) I judge to be as much larger and to move as much less swiftly in comparison with those of the second as those of the second in comparison with those of the third.' We are about to see one important reason why he has done this.

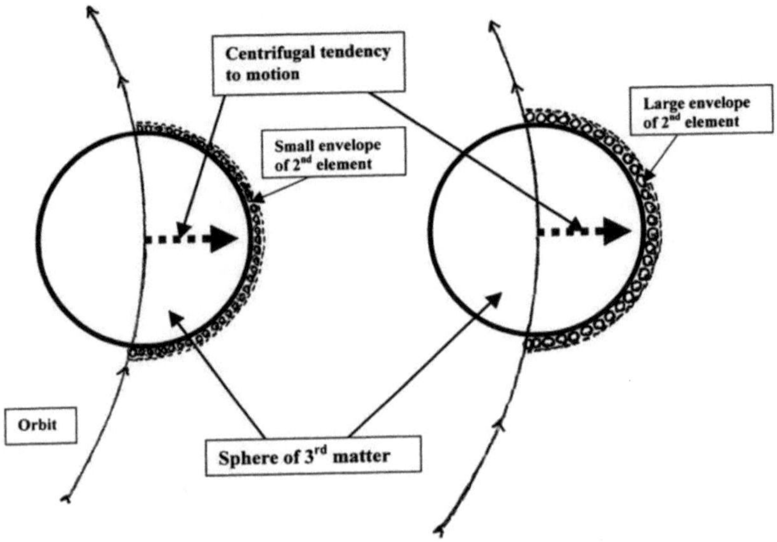

Fig. 10.5 'Surface envelopes' of second element. Small envelope *left*. Large envelope *right*. Envelope sizes are function of volume to surface ratios of spheres of second element

Next recall the size distribution of the second element (Fig. 10.4) We can turn this into a curve of volume to surface ratios, which in turn indicates the magnitude of the surface envelopes made out of the second element at different distances as related to a given piece of third matter (Fig. 10.6). The K layer marks an inflection point. From there outward, the spheres of second element get smaller not larger, and hence, surface envelops made out of them are progressively less capable of resisting a centrifugally tending piece of third matter.

The bottom line is this: planets will always be locked into the vortex at a radius below the K layer. If you like—and Descartes speaks this way obscurely in *Le Monde*, more clearly in *Principles*—a planet will drift outward due to actualized centrifugal tendency, until it reaches a layer of the vortex where the spheres of second element have a V/S ratio sufficient to make the surface envelope they form resist any further centrifugal translation by the planet. The planet is locked in somewhere along the V/S curve of the spheres of second element. In his discussion of this part of the theory Descartes spoke of the 'massiveness' or 'solidity' of a planet, meaning its aggregate volume to surface ratio.[21] This locking occurs at a radial distance from

[21] In *Le Monde* Descartes did this somewhat confusedly, improving his explication of massiveness and its role considerably in the *Principles*. On the emergence of massiveness or solidity as a key concept in de-coding *Le Monde*, based in part on its more clear deployment in the corresponding sections of the *Principles*, see Appendix 2. In short, I am reconstructing the underlying model in *Le Monde*, using a crisp hermeneutics of 'solidity' as aggregate volume to surface ratio and meshing that concept with my analysis of the size/speed distribution of the *boules* in the vortex. By using the graphical representations of these ideas, mediated by my interpretive construct of 'surface envelopes', the resulting decoding of the underlying model emerges. Note that in this process

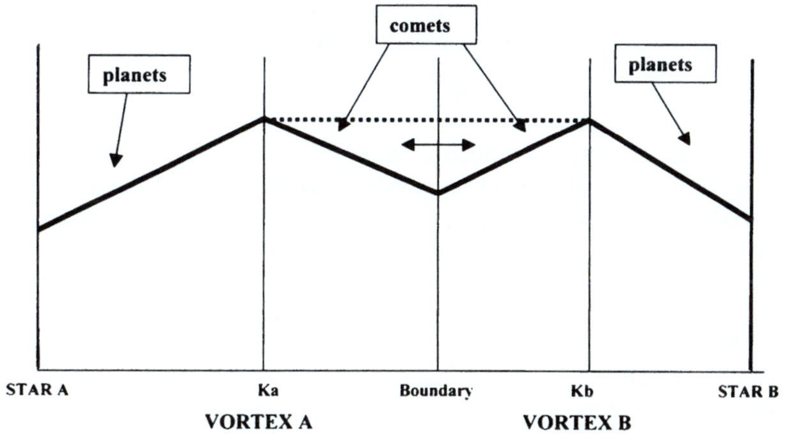

Fig. 10.6 'Resistance curve': derived from volume/surface ratios of spheres of second element

its central star at which the centrifugal tendency of the planet, a function of its massiveness, is exactly balanced by the resistance to centrifugal translation offered by the surface envelop in play at that location in the vortex. The greater a planet's V/S ratio or massiveness, the more distant that planet's orbit will from the central star.[22]

Imagine a planet, hypothetically finding itself not in its proper orbital place, literally too high up in the vortex given its degree of massiveness. It will not be able to develop sufficient centrifugal tendency and will be extruded downward by subjacent spheres of second element. It will stop 'falling' when a balance is realized on the one hand between the centrifugal force of the subjacent second element at that radius in the vortex and the resistance offered by the planet (owing to its degree of massiveness), and, on the other hand, the planet's own centrifugal tendency—conferred by its massiveness—balanced by the resistance of the superjacent surface envelope at that layer in the vortex.[23]

of reading, the verbal descriptions of the size/speed ratios come directly from the text, as does the concept of solidity—more clearly expressed in the *Principles* than in *Le Monde*, to be sure. The verbal descriptions are clarified and amplified graphically. The 'least Cartesian' notion used in this interpretation is that of 'surface envelopes', but even it has textual warrant in the overall direction of the theory, and in Descartes' various descriptions of the centrifugal tendency of planets (and comets) and the resistances they encounter at various levels of the vortex.

[22] This articulates the simple notion of centrifugal tendency as a function of size (quantity of matter) and force of motion only. In this mature application of the dynamics to a 'real' fluid vortex, it is clear that centrifugal tendency is a function of size, force of motion and 'solidity' (or massiveness), the latter taken in relation to the solidity of the relevant, resisting surface envelope.

[23] It must be reiterated that the systematic conclusions reached here constitute a charitable reading of the relevant passages in *Le Monde*, supplemented carefully by the somewhat more clear and cogent presentation in the *Principles*. As Appendix 2 shows, this reading depends to a large degree on using the discussion in the *Principles* to clarify and complete a *reductio ad absurdum* argument (concerning these various force relations) which is poorly and incompletely presented in *Le Monde*, but quite clear in the *Principles*.

We can therefore express the situation as follows: The condition for a piece of third matter to be in stable orbit in the vortex can expressed as

$$\mathbf{F^m{}_b} = \mathbf{R_{mu}} \textbf{ and } \mathbf{F^m{}_{ml}} = \mathbf{R_b}$$

Where $\mathbf{F^m{}_b}$ means Force of motion of the orbiting body; $\mathbf{R_{mu}}$ means resistance of superjacent layer of *boules* (upper medium) to being extruded downward by the orbiting body; $\mathbf{F^m{}_{ml}}$ means Force of motion of subjacent layer of *boules* (lower medium) and $\mathbf{R_b}$ means resistance of orbiting body to being extruded downward by subjacent layer of *boules*. All these terms need to be taken in their full explication, including the concepts of massiveness, surface envelopes, and the size/speed distribution of *boules* in the vortex. Note that the formula also expresses the conditions for a ball of second element to be in stable orbital motion as part of the total vortex, if we take $\mathbf{F^m{}_b}$ to mean the force of motion of the orbiting sphere of second element, and $\mathbf{R_b}$ to denote the resistance of that orbiting sphere of second element to being extruded downward by the subjacent layer of spheres. This, then, would conduce to a fuller understanding of what we above termed the 'force-stability' principle for constitution of the vortex.

Let us also stop for a moment here and reiterate that Descartes has constructed his mechanical heavens in such a way that mechanically efficacious stars are absolutely essential to the functioning of the celestial machine. If stars were inert, or if the second element filled the centers of the vortices, then two sets of consequences would follow: first, light would be propagated only along radial lines from the axis of revolution of the vortex (a matter we pursue below in Sect. 10.7); and, more importantly for our present concern, the resultant distribution of size and shape of the *boules* would not be proper for the existence of planets in stable orbits.

Descartes' theory of comets now follows with a kind of mechanistic inevitability: We already know that according to this theory of celestial mechanics, the more distant a planet is from the sun, the greater its V/S ratio or massiveness. Now, what if a planet is very massive, and it has the centrifugal tendency sufficient to overcome even the most highly resistant surface envelopes formed by second element at or near the K level? Well then, the object will pass by actualized centrifugal tendency beyond the K level, beyond the hump in the resistance curve in Fig. 10.6. Beyond K it will meet second element with decreasing V/S ratios, and less resistance, so that this object will move right on out of the vortex and stream into a neighboring one. The locking mechanism fails for these extremely 'solid' or 'massive' 'planets'.

When such an object of great 'solidity' is flung toward the center of a neighboring vortex, it meets increasing resistance to its centripetal trajectory—as we can see by looking at the curve in Fig. 10.6. The object picks up increments of orbital speed, until it starts to generate centrifugal tendency again, and again overcomes all obstacles—reading the curve in Fig. 10.6 backward—and gets flung back out of that second vortex. These, of course, are Cartesian comets, planets of high massiveness that oscillate between vortices, never penetrating any lower than the K level of any vortex—trapped on our representation in Fig. 10.6 in the resistance depression between K levels of adjoining vortices.[24]

To summarize, then, each vortex is a locking and extrusion device. Its corpuscular make-up, size and speed distribution, given Descartes' theory of planet/comet make up, entails that planets are locked into orbits of differing radii. Comets are objects extruded from vortex to vortex, first 'falling' into a vortex and being extruded out.[25] The existence and make up and mechanical behavior of the central stars are crucial, not to the existence of vortices, but to the existence of planet locking/comet extruding vortices. Otherwise extrusion would be the universal rule. Multiple vortices are conceptually necessary, as each vortex is set in a container made of contiguous vortices, exerting a kind of centripetal backwash at its boundary.

10.2.4 Genealogical and Systematic Dimensions of the Vortex Celestial Mechanics

We can now take stock of what we have learned, and what remains to be explored, about the genealogy of the vortex celestial mechanics and about its role in the natural philosophy of *Le Monde*, construed as a *system*. As to the latter, we can reap the benefits of getting serious about Descartes' own serious concern with vortices. We have seen that despite generations of simplistic glossing and easy dismissals, the vortices are the lynch pin, the engine room, of his corpuscular-mechanical universe of an infinite number of Copernican star/planet systems. As we shall see, the vortex mechanics are also present behind his intended explanations of other key, canonical natural philosophical issues, such as local fall, the tides, and motion of the moon.

We have also seen that the vortex celestial mechanics have much more internal conceptual density and delicate structure than usually thought. Therefore, we can

---

[24] There is of course much more to say about this theory of comets. First of all, it makes some concrete empirical predictions, which could have stood unrefuted for at least a generation after 1633; to wit, comets do not come closer to stars than a layer K; they are 'more massive' than planets, they move in spiral paths oscillating out of and into solar systems. In addition, in dealing with the phenomena of comets' tails, Descartes had to attribute odd optical properties to the K layer as part of his overall theory of cosmological optics—raising thereby issues quite telling about the origin and import of his theorizing, as we shall see in Sect. 10.8.

[25] The term 'falling' is chosen quite deliberately. In the previous argument about the placement of planetary orbits, a planet 'too high up' in the vortex for its particular solidity is extruded sun-ward, falling (and spiraling) down in the vortex to find its proper orbital distance. Below in Sect. 10.5 we shall see that Descartes' theory of local fall, and theory of the orbital motion of the moon, when taken in their simplest and most charitable acceptations, both also make use of this notion of falling in a vortex until a proper orbital level is found (assuming no other circumstances prevent completion of the process, as they do in local fall of heavy terrestrial bodies near the surface of the Earth). Ultimately, however, the interpretation of these two theories becomes more fraught, requiring additional interpretative attention, as we shall learn in Sect. 10.6.

supplement our analytical take on the vortex celestial mechanics in this section by realizing that they were the product of work and struggle, and that their construction was intimately related to the course of Descartes' physico-mathematics. This is what I mean by the genealogy of the vortex celestial mechanics. I see this genealogy as consisting in three main moments, two of which we have already studied in detail, whilst the third and final of which will be sketched in the next section.

The first moment in this genealogy was the physico-mathematics of the hydrostatics manuscript of 1619. As we recall from Chap. 3, the young Descartes' hyper-radical program had been to try to reduce Stevin's formal hydrostatics to an embryonic corpuscular mechanism, featuring a dynamical discourse concerning causes or 'forces', which in turn would provide the presumptive basis for unifying the mixed mathematical sciences under the explanatory umbrella of this sort of corpuscular-mechanically slanted physico-mathematics. The dynamic of research and concept formation this unleashed played itself out in his optical work in the 1620s—the second moment of the genealogy. We saw in Chap. 4 that these optical endeavors, in physico-mathematical mode, climaxed with his discovery of the law of refraction of light around 1627, whereupon Descartes immediately began to think about possible mechanical rationales or explanations for the newly discovered law. These attempts were intimately connected with a process by which he further crystallized his emerging concepts of dynamics directly out of a 'physico-mathematical' 'reading' of his geometrical optical results. In short, his optical researches marked the high point of his work as a physico-mathematician transforming the 'old' mixed mathematical sciences and had two main consequences: On the one hand his results confirmed his 1619 agenda of developing a (still unsystematized) corpuscular ontology and a causal discourse, or dynamics, involving concepts of force and directional 'determination' of motions or tendencies to motion. On the other hand, his results concretely advanced and shaped his concept of light as an instantaneously transmitted mechanical tendency to motion, as well as the precise principles of his dynamics.

Central to all these developments was the fact that Descartes had focused on results and phenomena in which, paradoxically no motion of bodies took place at all—in hydrostatics, and in the exemplary refracting of instantaneously transmitted light rays. In these 'statical' exemplars, or phénoméno-techniques' Descartes found crisp, clean messages about the underlying dynamics of the corpuscular world and indeed about its laws: he was involved in physico–mathematical attempts to 'see the natural philosophical causes' in and through well formed mixed mathematical results. Consequently it is easy to see how the prior developments in physico-mathematics made possible the construction of the vortex mechanics, with its reliance on the principles of Cartesian dynamics and its elaboration of concepts suited to explaining the orbital equilibrium—the stable orbital distances—of planets, or, given the breakdown of the stipulated conditions, the orbital disequilibrium, and hence 'rise' or 'fall', of planets (and comets). The vortex celestial mechanics has a conceptual and explanatory richness that can only be understood as having been drawn in large part from the genealogy of the physico-mathematics, especially as it involved the emergence of Descartes' dynamics.

There is, however, more to the genealogy—that third as yet unexplored moment which brings the vortex celestial mechanics of *Le Monde* into intimate contact with Descartes' mature physico-mathematics of the late 1620s. Again, as in 1618–1619, it was an encounter with Isaac Beeckman that crystallized and shaped the key ensuing events, and it is to these that we now turn.

10.3 Beeckman's Cosmic Balancing Acts—The Last Genealogical Step to the Vortex Mechanics

In late 1628 after a gap of nine years, Descartes re-established contact with Beeckman.[26] He found Beeckman ploughing through the astronomical works of Kepler, seeking to evaluate instances in which Kepler had invoked immaterial celestial forces or powers. In each case Beeckman sought to re-write the 'mechanisms' into corpuscular-mechanical terminology. As far as Beeckman was concerned, the key issues in astronomy did not involve the traditional activities of observation or even Kepler's work on elliptical orbits. Rather, Beeckman saw in Copernican astronomy, especially as transformed by Kepler, a broad, hitherto neglected field for natural philosophical explication, in particular corpuscular–mechanical explanation. Beeckman specifically identified his celestial mechanical speculations as desiderata for a *restitutio astronomiae*.[27]

We have seen that similar concerns would lay behind Descartes' celestial mechanics in *Le Monde*. Descartes, like Beeckman, avoided technical issues in observational astronomy, concentrating on plausible mechanical accounts of the causes of the motions of the planets in the Copernican system. Descartes and Beeckman were engrossed by the radical attempt to indicate how the latest conceptions in their 'physico-mathematics' might be brought to bear in explaining in a general way the causes of the motions of the planets in the Copernican system—allowing of course for the differences in content and trajectory in their respective versions of physico-mathematics, evident since 1619 and certainly quite further developed in Descartes' case by the late 1620s, as we have seen. In addition both Descartes and Beeckman sought to support their respective celestial physics by trading upon the suggestion that their celestial physics also explained the nature of light and thus was partially confirmed by its broad explanatory sweep.

Indeed, Beeckman's review of Kepler starts with a penetrating mechanistic critique of Kepler's theory of light: light is corporeal, consisting in a type of heat particle emitted by stars. Kepler's law of illumination is explained by the way

[26] See above Sects. 4.5.2, 4.7.4 and 8.3.3. Beeckman (1939–1953) iii p.114 note 3; Mersenne (1932–1988) ii p.222, 217–8, 233–44; AT x 341–3; Beeckman (1939–1953) iii p.103.

[27] Beeckman (1939–1953) iii p.103. In the period July 1628 to June 1629 roughly 21 out of 59 pages of Beeckman's journal deal with celestial mechanical and related matters. Material in this section was treated in more detail in Schuster (1977), pp. 507–520.

streams of light corpuscles spatially diverge from each other with distance from a source—an outcome impossible and unintelligible, he claims, on Kepler's own theory of light as an immaterial emanation.[28] This is crucial, because Beeckman's varied celestial mechanical speculations all play upon the idea of opposed, corporeally mediated forces that vary in strength with distance from source, hence constituting particular loci of equilibrium for the orbital placement of objects.

Beeckman then addresses a theory of lunar orbital placement: The moon is held in its orbit by a balance of attractive and repulsive actions delivered respectively, by rays of the sun reflected by the Earth, and rays of the Earth itself. The efficacy of the Earth rays decreases with distance more quickly than that of the solar rays. The solar rays are presumably Beeckmanian light rays—streams of corpuscles; the Earth rays are rays of Beeckman's version of Earth magnetism.[29] The 'attraction' and

---

[28] In early August 1628 Beeckman obtained a copy of Kepler's *Astronomia nova*. It is typical of Beeckman that his initial notes relate to the broader issue of mechanical explanation raised by the general tenor of Kepler's approach. In Chapter 36 of the *Astronomia nova* Kepler remarked that he erred in his *Ad Vitellionem Paralipomena* in postulating the weakening of the force of light with increasing distance from the source in order to account for the decrease in illumination over distance. He now saw that nothing is lost by the light. As much light moves from the source to a distant sphere of illumination as to a nearby one; but, since the larger, more distant sphere has more parts, the illumination offered by an equal quantity of light decreases. Beeckman leapt at the opportunity provided by this confession. He noted that Kepler would have been better advised to embrace a corpuscular view of light. In that case he could more easily have understood that the force of light does not decrease, but that as the light moves from the source equal quantities of light corpuscles must illumine spheres of increasing surface area. Beeckman chided Kepler for not seeing what 'obviously' must be granted—that light is corporeal: 'Truly Kepler was not able to know these things by a first intention and only when driven by necessity, because he falsely thinks that there is no distance between the particles, but merely what (scholastic philosophers) term *simple extension,* although they do not understand it themselves and foolishly avoid the assertion, which however must be made sometime or other, that light is a body [*...stulte vitantes dicere (quod tamen aliquando faciendum erit) lumen esse corpus*.]' (Beeckman (1939–1953) iii p.74).

Beeckman immediately drove home his point by applying it to the issue of the causes of celestial motion. For Beeckman this was the key problem raised by Kepler's enunciation of the elliptical orbit of Mars. He himself, Beeckman ironically suggested, had often pondered the celestial motions in a manner little different from Kepler, 'If I obtain some leisure and sometime or other free myself from this most burdensome office which is most unsuited to meditations, I shall discuss these things more accurately than Kepler, not only on account of the principle mentioned above which he refused to understand; namely that light is corporeal; but also because he did not know what is very true; *that all things once moved, always move, unless they are impeded.*' (*Ibid.*) Certainly, Beeckman seems to be saying, Kepler saw the problem of explaining the celestial motions, and he invoked some immaterial celestial powers, forces or emanations to that end. But, despite his great astronomical acumen, Kepler did not realize that light and other celestial emanations must be corporeal in order to be able to affect material bodies, and, moreover, he did not know that circular motion needs no explanation because it falls directly under Beeckman's general law of inertia.

[29] Beeckman (1939–1953) iii pp.74–5. 'Let the rays of the sun which are reflected by the earth have a force of attracting the moon and let the earth itself have a force of repelling the moon. . . I say that the sun's rays reflected by the earth retain their force much longer than the rays of the earth itself, because the sun's rays come to the moon from a more remote place and the distance between the earth and the moon is very small compared to the distance between the sun and the earth. Therefore, the solar rays have just as much force near the moon where it now is as they would have

'repulsion' attributed to these corporeal rays is unexplicated at the corpuscular level. (It is worth noting here in passing just how much better Descartes would later have judged his own constructions to be. In *Le Monde* he will have a plausible locking and extrusion mechanism deeply embedded in findings about hydrostatics, optics and a general dynamics of corpuscles.)

There were of course problems with the lunar theory, which Beeckman detected (perhaps aided by his French friend),[30] before he rushed onto a grander vision of the entire celestial mechanism: By substituting the fixed stars for the sun, and the sun for the reflecting Earth, his moon theory could perhaps be applied to the entire solar system. Sometime between October 1628 and late January 1629 Beeckman boldly writes:

[the same] thing can be said about all the planets (among which I also number the earth)... the light or corporeal virtue of the eighth sphere reflected by the sun draws the planets to the sun and the sun [itself] repels them. And thus each planet will be affected by each of the virtues according to its magnitude or rarity and therefore they will be located at different distances from the sun.[31]

This is indeed a striking speculation: the heavens are crisscrossed with the direct and reflected corporeal emanations of the fixed stars and the sun, the planets being located in the network of differential forces according to their 'magnitude' and 'rarity'.[32] However, Beeckman then noticed that the sun's own emanations were now repulsive in nature, and so he quickly reverted to a simpler picture of paired sun-planet interactions, based, as before, on a balance of forces—in this case planetary magnetic attraction, corporeally mediated, working against the mechanical repulsion arising from impact of solar heat and light corpuscles.[33]

---

if the moon were near the earth; but the earth has much more force near itself than near the moon, because the distance between the earth and the moon is very much greater than the distance from the earth's surface to the tops of its mountains. Thus, the earth strongly repels the moon when it is near the earth. The repelling force vanishes little by little as the earth-moon distance increases, and the repelling force of the earth is overcome at some point by the attractive force of the solar rays reflected by the earth. In this manner, the moon can never move further away from the earth nor approach it more closely.'

[30] One problem is that Beeckman realized that the unreflected rays of the sun would attract the moon to it. Beeckman (1939–1953) iii 75. Soon after Descartes' visit in October 1628 Beeckman returned to the problem and offered a solution based on a 'reduplication' of rays trapped between the Earth and the moon (ibid. p.100).

[31] Beeckman (1939–1953) iii. p.100.

[32] *ibid*. Two interesting issues arise in relation to Beeckman's model here: [1] Did Beeckman imagine this extended to a multi solar system universe of Cartesian type, or was he thinking only of a unique solar system and a chorus of fixed stars? We do not know for certain, but it is indeed hard to see how any given star can both be in the attracting chorus and be a local repellor of its own planets. [2] Note Beeckman's emphasis on the magnitude and rarity (density) of a planet. Beeckman was always acutely interested in how the volume to surface ratios of bodies, especially corpuscles, affected their mechanical interactions. The similarity in this respect to Descartes' later vortex celestial mechanics is obvious.

[33] *ibid*. p.101. 'Or, if it seems more plausible to avoid using an external agency in removing and attracting the planets to the sun or the moon to the earth, let us imagine that all magnetic virtues

Continuing to jot in his *Journal* as his speculations wandered, Beeckman shifted his ground again. He reverted to the fixed stars sending a flow of effluvia through the solar system. There are always more solar emanations immediately within the orbit of a given planet than immediately beyond it, thus fewer celestial emanations can make their way within the orbit and exert a back-pressure on the sunward side of the planet. Hence each planet suffers a pressure toward the sun arising from the incoming stellar rays which is to be balanced by the light/heat repulsion of solar emanations.[34] This indeed was a mechanical picture of orbital equilibria of causes which was to be supplied in a much more elegant fashion by Descartes' vortices.

By mid-1629 Beeckman had not achieved a settled view and in typical fashion he unceremoniously dropped the matter, although (as we learned in Chap. 8) later in the early 1630s he did plan a systematic treatise of natural philosophy.[35] Beeckman's work just pre-dates and overlaps the period of renewed contact with Descartes in late 1628. Arguably, the interest Descartes evidenced after 1628 in the problem of celestial mechanics, as well as his mode of approach to it, grew from his acquaintance with Beeckman's speculation. Descartes would have been all the more confident in his association in *Le Monde* of a corpuscular mechanical theory of light with his corpuscular-mechanical vortex celestial mechanics, if he recognized the advance in comprehensiveness, coherence and mechanical rigor achieved in his work as compared with these wranglings of Beeckman.

---

attract, but that there are many [sorts of particles], such as heat, light, etc. simultaneously flowing out [of the sun and earth] which repel. Moreover, let us conceive that the attractive force extends to a greater distance, so that the force of the heat particles taken at an equal distance is less. Thus the moon is driven away from the earth as long as the heat and other bodies flowing from the earth overcome the magnetic virtue; but, when they grow weak, the magnetic virtue still remains. Therefore, the moon is dragged to that place in which the forces are equal.' It is clear that Beeckman entertains a corporeal theory of magnetism, cf *ibid.* p.102. For Beeckman's corpuscular-mechanical theory of magnetism see also Beeckman (1939–1953) i 36, 101–2, 309; ii 119–20, 229, 339; iii 17, 76.

[34] Beeckman (1939–1953) iii p.103. Perhaps because he was displeased with the rather vague discussion of the magnetism as essentially involving attraction, implicated in the previous model, Beeckman was now in effect recurring to an account of magnetism he had used as early as 1615 in order to explicate the magnetic action of the fixed stars upon the solar system: Back in 1615, and more usually throughout his work, Beeckman had reduced apparent magnetic attraction to a differential repulsive action arising when magnetic effluvia drive air or aether out from between the lodestone and the piece of iron, thus allowing the pressure of air or aether on the remaining surfaces of the lodestone and iron to push the two together. Here in this final account of 'celestial' mechanics, there is a partial mobilization of a similar style of explanation. When the fixed stars send a flow of (magnetic) effluvia through the solar system, there will be more solar emanations immediately within the orbit of a given planet than immediately beyond it, and more celestial 'magnetic' effluvia outside the planet's orbit than within it. Solar heat/light effluvia push out, but celestial magnetic effluvia push in, and orbital equilibrium is achieved. As Beeckman continued toward the end of this sequence of speculations, he also speculated about countervailing forces arising from impact of corpuscular emanations to explain, amongst other things, the eccentricity of orbits and precession of the equinoxes. *ibid.* pp.102, 108.

[35] van Berkel (2000).

Had Descartes been quite a bit more charitable and magnanimous to his erstwhile mentor, he might just have acknowledged what is clear to us—the underlying spirit and structure of the argument of the celestial mechanics of *Le Monde* harks back to the notions behind Beeckman's shifting speculations of 1628–1629. That was not Descartes' style, as is well known.[36] One can imagine him much more readily agreeing with an uncompromising technical judgment which we can now offer, recalling our discussion of the vortex celestial mechanics in Sect. 10.2: Having been spurred by Beeckman highly interesting but inconclusive foray into a unified corpuscular–mechanical theory of light and celestial optics, Descartes was in a position to try to succeed where Beeckman was floundering, and he approached this by in effect cashing in the intellectual profit of his physico-mathematical endeavors since 1619. Instead of Beeckman's wandering and inconclusive jottings, Descartes elaborated his model of a celestial vortical locking and extrusion machine. He based himself on his principles of dynamics (the emergence of which was initiated in his hydrostatics of 1619 and articulated in his optical work of the 1620s); his theory of centrifugal tendency to motion; a theory of the make up of the stars and their surrounding vortices; and his notion of the 'massiveness' of planets and comets. A mechanistic theory of light as instantaneously transmitted tendency to motion could be fitted to this cosmic setting, providing the ultimate basis for the optical work and discoveries, and fulfilling the de facto challenge issued by Beeckman to render in corpuscular-mechanical and properly physico-mathematical terms the problematic of Kepler. *Le Monde* challenges Beeckman back, by saying in effect: '*Here is a physico-mathematical explanation of light in cosmic setting and of celestial 'physics'; causes are not multiplied; the same concepts of dynamics, applied to the nature of stars and vortices, explain everything!*'

10.4 Descartes' Celestial Vortex Mechanics as a 'Science of Equilibrium'

Now that we have a more adequate understanding of the conceptual structure and genealogy of the vortex celestial mechanics, and before moving on to Descartes' articulation of further natural philosophical issues in *Le Monde*, we should comment briefly on the nature of the vortex mechanics as a science or discourse of equilibrium, thus underscoring its sources in statical and hydrostatical exemplars, heavily refracted through the *sui generis* style of Descartes' physico-mathematics.

If our analysis of the vortex celestial mechanics is correct, it is fair to say that the forces at work upon a planet can only be fully specified when orbital equilibrium/stability has been attained. By this we mean specified conceptually, as in the formulae and accompanying explanation at the end of Sect. 10.2.3 above, not, of course, in

[36] van Berkel (2000) and Sect. 8.3.3 above.

terms of any actual measurements. The rise or fall of a planet (or comet) represents the effect of a breakdown (or non-establishment) of such equilibrium. If orbital equilibrium is modeled on classical statical and hydrostatical equilibrium in the sense of Archimedes or Stevin, so planetary/cometary rise or fall is the analogue of slippage or movement in an Archimedean/Stevinian machine or hydrostatic system. Slippages, celestial rises or falls, are not well defined mathematically, but their occurrence can be loosely explained as due to failure to attain, or breakdown of, equilibrium conditions. In all these respects celestial equilibrium and disequilibrium are analogous respectively to equilibrium and slippage in systems rigorously treated in classical statics. Moreover, in a curious way Descartes' celestial mechanics accords with Stevin's famous take on mechanics in general: Only equilibrium can be studied and explained rigorously, whilst slippage is understood (explained in a loose sense) as eluding these conditions.[37]

Of course, Descartes' vortex celestial mechanics is not a direct conceptual extension of classical statics and hydrostatics. This is because it deals with equilibrium conditions described and analyzed in terms of Cartesian dynamical concepts—solidity, centrifugal and 'principal' determinations, 'surface envelopes', and the like—all viewed against the background of stipulations about the size and speed distributions of the corpuscles of second element making up the vortices. The vortex celestial mechanics is therefore not in essence a science of volumes, densities and specific weights. Rather, it ultimately depends on the concept of relative solidity, or massiveness, a property defined by volume to surface relations aggregated over the constituent corpuscles of a planet or comet made of third element, or over relevant volumes of vortical *boules*. (This point is very important and shortly will help us to understand certain difficulties Descartes met in trying to deal with the local fall of 'heavy' bodies—a phenomenon he tended to want to describe using the vocabulary of classical hydrostatics, which, however, simply will not mesh with his vortex mechanics, a science, at bottom, of 'solidity' or volume to surface relations.)

In characterizing the vortex mechanics as a science of equilibrium, it is further useful to note a related and highly pertinent insight of Stephen Gaukroger, to the

[37] Stevin, had of course preferred pure Archimedean statics and so had rejected the dynamical approach to statics and mechanics characteristic of the pseudo-Aristotelian tradition of the *Mechanical Problems*. In Stevin's view systems in static equilibrium cannot be explained by considering the arcs through which bodies would move if they ceased to be in equilibrium, as in virtual or real displacements. You cannot deduce equilibrium conditions from the supposition that motion has or would occur—that is absurd, since if motion occurs the forces are not in equilibrium. This led Stevin to deny that the study of motion, that is, natural philosophy, could be pursued in a rigorous mathematical manner. (Stevin, 'Appendix to the *Art of Weighing*' in Stevin (1955–1966) vol. 1, 507–9; and '*The Practice of Weighing*, "'To the Reader'", *ibid.* vol. 1,. 297.) How ironic, then, that Descartes *sought natural philosophical capital*, by recourse not to the *Mechanical Problems* but to the purely statical, purely mathematical, equilibrium science of Stevin. He did this first in the hydrostatics manuscript of 1619 (see above Chap. 2, and Gaukroger and Schuster 2002 540, 545–9) and now in his mature work, his vortex mechanics was also at its core a science of equilibrium, not a science of motion and displacement.

effect that Descartes' vortex fluid offers no resistance whatsoever to the motion of a planet or comet.[38] We have seen that this is quite true. The business of the rotational motion of the vortex is to move planets and comets in a rotary manner, leading to the generation of centrifugal tendencies that are the proper objects of analysis on the central question of orbital stability and placement. Gaukroger points this out as part of an extended and enlightening comparison of Cartesian and Newtonian approaches to fluid mechanics, and hence mechanics in general. We have seen, as it were, the obverse of Gaukroger's point, having noted the off–hand manner in which Descartes deals with the orbital or rotational force, speed or 'agitation' of vortex *boules*, planets and comets. He never unequivocally declares a position on the relations of these parameters amongst themselves, and in so far as they are possessed by a planet or comet carried along with a portion of a vortex. His focus is on specifying orbital equilibrium or its breakdown, not on solving what seem to us as the exemplary fluid mechanical problems of relating the speed, force (or 'agitation') of medium and planet under certain conditions of rotary translation. Section 10.2.3 showed this, and Appendix 2 offers detailed textual exploration to back up this claim. Moreover, we shall see further consequences of this focus on orbital placement in preference to the mechanics of rotary translation of and by a fluid when we turn below to Descartes' treatment of local fall and the motion of the moon.

Finally, everything claimed in this section tends to support the idea that the vortex celestial mechanics has a strongly physico-mathematical character, in Descartes' sense of the term. Descartes is saying that at the descriptive level, in terms of appearances, orbital establishment and placement are 'statical phenomena', easily and well recognized. But, just as in the hydrostatics manuscript of 1619—where he insisted that a macroscopic, statical regularity be reduced to corpuscular-mechanical terms by means of a physico-mathematical redescription of the observed regularity—so in *Le Monde* he is insisting that beneath the observable radial equilibrium of orbits there reside: (1) a corpuscular-mechanical reality, 'an underlying machinery' (Sect. 3.4), the behavior of which, (2), is caused, and explained, by new concepts of dynamics of Descartes' own coinage. As the corpuscular-mechanics and dynamical reduction of the hydrostatic paradox was an exercise in physico-mathematics, so too is the vortex mechanics of 1633.[39] We shall see that the fact that the vortex celestial mechanics, the virtual pivot of the vision purveyed in *Le Monde*, is a discourse on equilibrium would affect much about the manner, and the results, of Descartes' further natural philosophizing in *Le Monde*.

[38] Gaukroger (2000)

[39] In fact, one might go a step further and suggest the following: Our interpretative diagrams, Figs. 10.2, 10.3, 10.4, display the type of 'figuring up' that characterized Descartes' physico–mathematics, and which, had he supplied them himself, might have aided comprehension of his physics for the last 350 years. Cf. above note 12, and Chap. 3, notes 51 and 105.

10.5 Applying the Vortex Mechanics to Local Fall, the Moon and the Tides: An Exercise in Charitable Interpretation

10.5.1 Aims of This Section and the Strategy of Reading

We can now return to the vortex mechanics of *Le Monde*, having looked at the last of the chief genealogical moments in its genesis, and at its conceptual structure, including its character as a discourse concerning the precise explanation of orbital equilibrium and the loose explanation of disequilibrium, or motion of rise or fall in a vortex. In the remaining sections of this chapter we shall survey the full range of issues that slot into place, given this genealogy and structural analysis. They fall into two categories: First there are those that depend quite directly on articulations of the vortex mechanics, as we have now come to understand it. These include the fall of heavy (third matter) bodies near the surfaces of planets; the motion of planetary satellites, and, for planets possessing both oceans and a moon, the nature and causes of the resulting tides. We treat these matters in the present section. The second category concerns Descartes' theory of light in its cosmological setting, that is, in the context of his universe of vortices, each centered on a light giving star—a topic also grounded in the genealogy and most enlightening about the status and aims of the vortex theory and the dynamics it expresses. We deal with the cosmological optics in the Sect. 10.7 below.

Again, a particular strategy of reading and exposition guides our discussion: First in the present section, we extend our charitable reading of the vortex mechanics, as proposed in Sect. 10.2.1 and applied in Sects. 10.2.2, 10.2.3 and Appendix 2, leading to a first order, simple yet coherent understanding of how 'weight', local fall, the tides and motion of the moon are all part of a unified theory of vortex mechanics. However, there are complications and difficulties in the portions of *Le Monde* devoted to these topics. These will be presented in Sect. 10.6, and whilst they necessarily complicate our first, simple reading, they continue to betray the marks of Descartes' physico-mathematical trajectory—the style, protocols, limits, and, if you will, illusions of that program. We shall therefore reinterpret these parts of *Le Monde* in a deeper yet still charitable way—by avoiding hasty criticism of Descartes for *ad hoc* or even idiosyncratic adjustments and tactics, and instead stressing how even some his oddest looking arguments and maneuvers have some sort of rationale and genealogical basis in at least one part or another of the trajectory of his physico-mathematics. So, our first, straightforward interpretation of how weight, local fall, motion of the moon and causes of the tides can be directly linked to the vortex mechanics as explicated above, we shall call the charitable[1] reading. Our second order charitable interpretation of further complexities in the account of weight and the motion of the moon, in Sect. 10.6, will be termed a charitable[2] reading. When we come to the cosmological optics and theory of comets' tails in Sects. 10.7 and 10.8 respectively, we shall also have to exercise the style of charitable[2] reading.[40]

10.5.2 Local Fall as Downward Extrusion (In Ultimate Search for an Orbit—Like the Moon's)

We begin with Descartes' vortex theory of the fall of heavy bodies near the surface of the Earth (or any planet). This theory, like the other articulations of the vortex mechanics listed above, contains quite a few conceptual and empirical problems, which are often emphasized in a kind of Whiggish perspective eager to move on to discuss Newtonian theory. Of course, every theory, indeed even Newton's, has its limits, its strengths, weaknesses, pointed difficulties, lacunae and, perhaps, contradictions. Descartes' bold, internally complex, and strategically thought out vortex mechanics is hardly an exception to this. But, in the spirit of a charitable[1] reading, I choose here to analyze the little appreciated coherences and strength of Descartes' account of local fall, as a token of my larger attempt to wring as much systemic cogency, and contextual and developmental 'reason' out of the vortex theory and its several articulations.

Descartes articulated his vortex theory of celestial mechanics, paving the way for these further developments, by claiming that smaller, local vortices form around the planets orbiting stars. First, he asserts that since each planet moves somewhat more slowly than the surrounding second matter, a small vortex will form around it, circulating in the direction ABCD in Fig. 10.7:

> ...since the parts of the heaven that are, say, at A move faster than the planet marked T, which they push toward Z, it is evident that they must be diverted by it and constrained to take their course toward B. I say toward B rather than toward D; for, having inclination to continue their motion in a straight line, they must go toward the outside of the circle ACZN they are describing, rather than toward the center S.[41]

The circulation of the *boules* about the planet imparts to it an axial rotation. Conversely the rotation of the planet helps determine the complete circuit of the *boules*.[42] The appearance of secondary vortices and axial rotation of the planets is thus a general consequence of Descartes' theory and presumably he would have expected all planets to manifest such behavior. Using the case of the Earth, Descartes attributes to its local vortex the explanation of the motion of the moon, the diurnal

[40] In sum, therefore: Charitable [1] reading deals with our interpretation of the vortex mechanics, as well as the theory of fall, lunar motion and tides, taken in their first order, most simple acceptations. Charitable[2] reading is open to the deeper complexities in the accounts of local fall, the motion of the moon, cosmological optics and comets' tails, but it is still charitable in looking for the physico-mathematical genealogy and conceptual keys to Descartes' texts.

[41] AT.XI.70; SG 45; MSM. 119.

[42] AT.XI.71–2; SG 45–46; MSM.119–121.

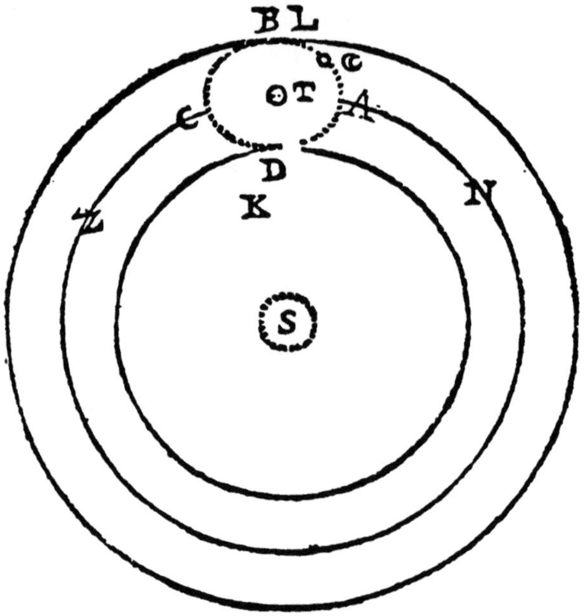

Fig. 10.7 The earth's local vortex with moon, AT XI, p.70

motion of the Earth, the local fall of 'heavy' bodies, and, conjointly with the moon's motion, and the existence of oceans on the Earth, the tides.[43]

In very general terms the local fall of heavy bodies follows on this theory as a case of extrusion downward in the planetary vortex of bodies possessing less centrifugal tendency than the surrounding matter of the local vortex.

> ...I want you to consider what the weight of this earth is; that is to say, what the force is that unites all its parts and makes them all tend toward the centre, each more or less according to the extent of its size and solidity. This force is nothing other than, and consists in nothing other than, the fact that, since the parts of the small heaven surrounding it turn much faster than its parts about its centre, they also tend to move away with more force from its centre and consequently to push the parts of the earth back toward its centre.[44]

In good mechanistic terms there is to be no question here of attraction or occult tendencies.[45] The tendency of terrestrial bodies toward the center of the Earth is due to their relative lack of centrifugal tendency compared to the *boules* of the Earth's vortex. It is true that particles of the third element making up the earth, water and air

[43] AT xi pp.64–83

[44] AT xi pp.72–3, SG 47; MSM 123.

[45] Aiton (1959) 27 links Descartes' ideas on the corpuscular nature of gravity to Gilbert's *De magnete*. The role of Beeckman as a channel for these ideas and source of new ones of a related nature should be kept in mind, especially in light of what we now know about Beeckman's recent 'celestial mechanical' speculations, based on Sect. 10.3 above.

all partake of the diurnal rotation of the Earth as a whole. If the universe were not a plenum, these particles would be thrown off the Earth by their centrifugal tendency. But, because the world is full and the Earth is surrounded by a rapidly rotating vortex, terrestrial particles cannot exercise their tendency, and, in fact, are extruded downward by the greater centrifugal action of subjacent *boules*.

Descartes' account, as is well known, meets an immediate large objection, as obvious to contemporaries as to us: Because the local vortex spins on an axis coincident with that of the Earth, fall on or near the Earth should be in direction normal to the axis of rotation, not radially toward the center of the Earth. Moreover, why do not falling bodies sweep laterally across the surface of the Earth, spiraling downward, rather than apparently falling in straight lines normal to the local surface of the Earth? Finally, to add to these commonly adduced puzzles, we are in a position to add a third one, grounded in our genealogy of the vortex theory: How in detail could Descartes consistently explain or reduce Stevin and Archimedes' rigorous, geometrical, and macro-descriptive statics and hydrostatics to a full vortex theory of fall and weight? In 1619, let us recall, he had been inspired by such an hydrostatics, but had only asserted a piecemeal corpuscular-mechanical explanation limited to claims about particles in the water, rather than a full vortex theory of weight, specific weight, the behavior of air, water and circulating vortex particles of second element. Descartes had things to say about all three issues, and we shall touch upon them in our 'charitable[2]' reading Sect. 10.6 below. In this section, however, we are emphasizing within our 'charitable[1]' reading the virtues of the theory, from Descartes' point of view, rather than exploring further its deficiencies and its (arguable) anomalies. Interestingly, we do not have to look far into Descartes' discussion of local fall in *Le Monde* to find strong hints as to how the charitable[1] interpretation should proceed.

In *Le Monde* Descartes furnishes an initial clue about what he thought was most striking about his theory of fall. It is precisely the fact that the theory of local fall, on his view, *is completely consistent with his vortex theory of planets and comets at the level of basic explanatory machinery*. In fact, what seems to guide the articulation of the theory of local fall is a direct analogy to what we now understand as the technical details of the vortex celestial mechanics: A terrestrial body undergoing local fall toward the Earth is in the same vortical mechanical situation as a planet in the solar vortex which, located at the 'wrong' orbital distance, 'falls' (indeed spirals) downward toward the star until it picks up enough centrifugal tendency to stabilize in an orbit—at a distance determined by the 'solidity' of the planet, and the speed and size distributions of balls of second element in the vortex, as we have explored in Sect. 10.2.

Indeed, the very first issue Descartes discusses after his explanation of fall is the following likely misunderstanding on the part of the reader:

> You may find some difficulty in this, in light of my just saying that the most massive and most solid bodies (such as I have supposed those of the comets to be) tend to move outward toward the circumferences of the heavens, and that only those that are less massive and solid

are pushed back toward their centers. For it should follow that only the less solid parts of the earth could be pushed back toward its center and that the others should move away from it.[46]

Descartes is directing us to his key concept of 'solidity' and to the fundamental theory of speed/size distribution of balls of second element in the star-centered vortex. He continues,

> But note that, when I said that the most solid and most massive bodies tended to move away from the center of any heaven, I supposed that they were already previously moving with the same agitation as the matter of that heaven. For it is certain that, if they have not yet begun to move, or if they are moving less fast than is required to follow the course of this matter, they must at first be pushed by it toward the center about which it is turning. Indeed it is certain that, to the extent that they are larger and more solid, they will be pushed with more force and speed. Nevertheless, if they are solid and massive enough to compose comets, this does not hinder them from tending to move shortly thereafter toward the exterior circumferences of the heavens, in as much as the agitation they have acquired in descending toward any one of the heavens' centers will most certainly give them the force to pass beyond and to ascend again toward its circumference.[47]

Using precisely the conceptual terms of the larger vortex theory, Descartes is distinguishing between: (a) the potential orbital distance of a body in a vortex, as determined by its overall volume-to-surface ratio or solidity, and (b) the amount of force of motion the body possesses at any given time and radial place in the vortex. Now, although he is more precise in his expression about this later in the *Principles*, even in *Le Monde,* it is clear that comets or pieces of terrestrial matter have definite solidities, which ultimately determine their placement in a vortex, but *only on condition* that they have gradually acquired circulatory motion and have begun to translate centrifugally. Hence, applying these concepts to the local planetary vortex, we may reason that a stone initially sharing only in the Earth's diurnal rotation will be forced down toward the center of the Earth, and that is what we habitually observe. But, Descartes is also saying that if the Earth's vortex were large enough and if the stone were released from a sufficiently great distance from the centre, it might acquire sufficient circulatory speed in the course of its descent to begin to rise as a result of the centrifugal tendency thus gained. It would rise through the vortex until it reached a level at which the resistance to centrifugal motion of the *boules* balanced its own centrifugal tendency. This hypothetical behavior of a stone would therefore correspond to that of a comet which first descends into a vortex and then gradually acquires enough speed of rotation and consequent centrifugal tendency to begin translating out once again.

The greater the solidity of a body, whether in a solar or planetary vortex, the more difficult it will be for the surrounding second matter to impart motion to it and the less first and second matter will be dispersed in its pores to agitate it in directions other than downward. Thus, upon being released, a body of great solidity will yield more readily to centripetal extrusion than one of lesser solidity. In local fall near the

[46] AT xi 73; MSM 125; SG 47

[47] AT xi 73–4; MSM 125; SG 47

Earth, heavy bodies generally do not attain sufficient centrifugal tendency to begin their ascent. Therefore the phenomena of weight have to do only with the extruding function of the celestial machine and do not pertain to the locking mechanisms of the vortices.[48]

Viewing stellar and planetary vortices under a unified theory of vortex mechanics, we can formulate the following theorem, reflecting the essentials of Cartesian celestial mechanics—'*comets, planets in the wrong orbits and 'heavy' bodies (bodies consisting of third element) released near any planet's surface are all doing the same thing for the same reasons grounded in the vortex mechanics*'. That is, a comet is continually either sinking toward the K layer of a vortex or rising out from such a K layer, because it is 'too solid' to find orbital equilibrium in any vortex at any time; a planet in the 'wrong' orbit, according to Descartes, is either extruded downward, falling until it finds orbital equilibrium, or is pushed out, rising until it finds such an equilibrium as determined by its degree of solidity; and, as we have just seen, terrestrial bodies in local fall basically are falling in a vortex and hitting the Earth's surface before they can acquire sufficient force of motion, and centrifugal tendency, such that they can rise in the vortex to find an orbital placement determined by their particular degree of solidity. Finally, all this leads us to see that there is at least one such example of an object of terrestrial matter 'in orbit' around the Earth—it is, of course, the moon.[49]

---

[48] This finding has very important consequences indeed for how one thinks about the status within the vortex mechanics of the phenomena of local fall (as well as planetary and cometary fall and rise). As we have seen, vortical mechanics focuses on explaining orbital equilibrium, as we have learned to define it above in Sect. 10.2.3. And, as we have argued in Sect. 10.4, the orbital spiraling down and up are 'predicted' and loosely explained in general terms by the theory, but no clear physico-mathematical description can be given of such a process, compared to the clear definition of the state of orbital equilibrium. This difference is reflective of deep tectonic alignments in the aims and evolution of Descartes' physico-mathematics 'turning-toward-corpuscular-mechanism'. We now see that local fall is also this sort of 'slippage' phenomenon, holding just when no orbital equilibrium has been effected—the moon, as we shall shortly see, would be an example of a 'terrestrial' body which has found orbital equilibrium in the vortex of the Earth! Therefore, on Cartesian vortex mechanical principles, local fall will not become an object of exact physico-mathematical study—a conceptual result that would hardly surprise the Descartes of 1633 and which would have finally made sense to him about his abortive physico-mathematical assault on local fall in 1619. To summarize in aphoristic terms: bodies in local fall are trivial, sub-scientific instances of the operations of the vortex mechanics; the moon is a planet– or comet–esque object illustrating the fine points of the vortex 'equilibrium mechanics'.

[49] We place scare quotes around the term 'in orbit' for the following reason: As we shall find out when we examine the theory of the moon in more detail in Sect. 10.6, Descartes probably thought of the moon as a 'comet-like' object, in that it was so 'solid' that it was indeed 'too solid to find a stable orbit inside the Earth's vortex'. Hence, in all probability, he meant that it was only held in orbit at the outer boundary of the Earth's vortex, by the strong resistance to centrifugal translation found at the boundary between the Earth's vortex and the encompassing solar vortex In any case, the moon is an example of a very 'solid' terrestrial body that, rather than continuing to fall toward the Earth, has 'risen', that is, centrifugally translated upward to the limits of the local vortex.

10.5.3 *Vortex, Earth, Sea and Moon: Descartes' Theory of the Tides*

Descartes managed to construct a general theory of the cause of the tides on the basis of the Earth-moon vortex already introduced. Evidently, he hoped to reduce this perennial problem in natural philosophy to the terms of his vortex celestial mechanics.[50] When the matter of the vortex of the Earth streams between the Earth and moon, it must, on Descartes' characteristic view, move more quickly than it does elsewhere so that the same total quantity of matter can pass the more narrow channel in the same period of time. According to Descartes, this increase in vortical speed deflects the Earth as a whole and thus entails that the center of the Earth no longer coincides with the center of the vortex, but rather is off-center toward the side of the vortex diametrically opposite the moon[51] (Fig. 10.8). The air and water are also deflected, but being fluid they are merely depressed under the moon and flow toward the wider portions of the vortex:

> ...it is evident that the same force that presses the earth in this way must also make them sink toward T, not only from the side 6,2, but also from its opposite 8,4, and in recompense cause them to rise in the places 5,1 and 7,3. Thus, the surface EFGH of the earth remaining round (because it is hard), that of the water 1 2 3 4 and that of the air 5 6 7 8 (which are liquids) must form an oval.[52]

The reason for the depression on the opposite side of the Earth from the moon is that the displacement of the Earth narrows the vortex on the opposite side, leading to the same increased vortical speed in the region above H.

The diurnal rotation of the Earth will convey each portion of the surface under the two bulges and depressions each day, thus producing two high and two low tides. Since the moon orbits the Earth in the same direction once a month, the tides will not occur at precise six hour intervals but lag behind about twelve minutes each day.[53] Attempting to confirm this theory, Descartes derives the consequence that there will be an endless flow in the seas from east to west as the Earth turns under the bulges. He alludes to the 'report of our pilots' that this flow causes voyages from the orient to be shorter than those from the occident.[54] Descartes seems to be drawing here upon one form of a common contemporary belief that the winds and/or currents favored voyages from the East Indies or even the Levant.[55] This piece of evidence gets pride of

[50] Hooper (2004), Biro (2006) which was revised and published as Biro (2009)

[51] AT.XI.81; SG 52; MSM.141.

[52] *Ibid.*

[53] AT.XI. 82–3; SG 52–3; MSM 141–143

[54] AT.XI. 82; SG 53; MSM 143.

[55] Cf. Galileo (1953) 441. Galileo alludes to shorter voyages in the Mediterranean from east to west than from west to east. He attributes this to a prevailing east wind caused by the lag of the air behind the diurnal rotation of the Earth. Burton, in the 'Digression of Air' in the *Anatomy of Melancholy* (Burton 1628 [1927], 409) asks why '...from Moabar to Madagascar in that Indian Ocean the merchants come in three weeks as Scaliger discusseth, they return scarce in three months,

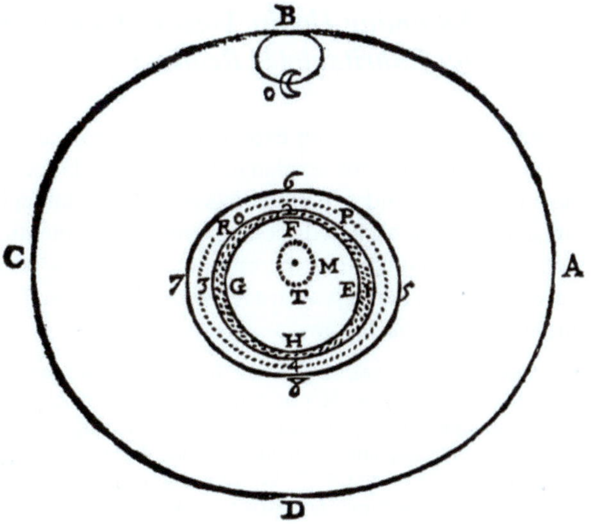

Fig. 10.8 Explanation of the tides in *Le Monde,* AT XI, p.74

place in his discussion over the derivation of spring and neap tides. These are explained by the elongation of the vortex which causes the tides to be higher when the moon is at D (new) or B (full) than when it is at A or C (quarters).[56]

The peculiarities of this theory should be noted. First, although it differs sharply from that of Galileo in attributing the central causal role to the moon, it differs from other lunar theories, such as Kepler's, in placing the low water directly under and opposite the moon.[57] Secondly, it seems to be a most obvious implication of this theory that the weight of bodies, which arises from the centrifugal tendency of the *boules* of the vortex, should increase markedly under the moon. After all, if the increased speed of the *boules* can displace the Earth and cause the tides, it should have some effect on the phenomena of weight. Whether by omission or design, Descartes ignores this issue.

---

with the same or like winds: the continual current is from East to West.' Bacon, in *De Fluxu et defluxu Maris* (1857–1874, vol v. 449) attributes the tides to the continental disruption of a permanent westward current derived from the diurnal motion of the heavens. Descartes' model *in Le Monde* makes no allowance for such continental disruption and treats the oceans in idealized terms (much like the Aristotelian 'sphere of water') as continuous, for the purpose of allowing the 'tidal bulges' to translate around the globe, as the model requires. As Biro (2009) has shown, in the *Principles of Philosophy* Descartes later addresses the tensions between his idealized ocean model and the state of geographical knowledge in his time. See below, Chap. 11 on this and other adjustments made in the *Principles* to difficulties in *Le Monde*.

[56] AT.XI. 83; MSM. 143; SG 53.

[57] Kepler attributes the tides to the moon's attraction of waters directly under it (Kepler 1938ff, III, 26)

10.5.4 Summing Up the Vortex Mechanics and Its Consequences on Charitable[1] Reading

Finally, let us consider an explanatory conceit, aimed at dramatizing the relative coherence of Descartes' explanation of celestial mechanics, local fall, the motion of the moon and the tides, on the basis of our charitable[1] reading. Imagine Descartes himself, brought back to discourse with us, commenting upon this unified theory, as well as other competing theories, including his post-mortem acquaintance with Newton's work. Perhaps such a revived, typically self-regarding and feisty Descartes might lecture us as follows:

> I know all of you are, so to speak, in love with Newton—he's like you, or so you think. Well, for me, he is like Kepler, brilliant but ontologically unsound. Here is Newton's leading question—orderly procedure starts with the right question. 'What single immaterial causal agency explains the motions of the planets, comets, satellites, the fall of bodies on Earth, as well as the tides?' Very elegant, is it not? And to be sure, nobody ever posed that precise question: not Aristotle obviously; certainly not Copernicus, not even Kepler—he multiplied such unintelligible immaterial causal agencies, rather than search for one elegant one.
>
> Very well, my question, the methodologically appropriate one, was: 'What unique and certain set of dynamical principles, applied to the vortex motion of corpuscles, explains the motions of the planets, comets, satellites, the fall of bodies on Earth, as well as the tides?'. You have seen my dynamics and general vortex theory and can work out for yourselves why they constitute a unified general theory of the key phenomena in question. Newton pursued the same 'problematic', as my modern countrymen would say. He had the benefit of my example. He grasped the aim of the problematic, but faltered badly on the issues of causation and ontology.[58]

10.6 Some Intricacies of the Theories of Local Fall and the Moon—Bearing the Imprint of the Genealogy of Physico-Mathematics

The last section ended with our supremely confident albeit fictional 'Descartes' speaking within the confines of a charitable[1] reading of the vortex mechanics and its extension into the account of local fall, motion of the moon, and theory of the tides. However, the latter matters are not in the end quite so simple, as we shall now see by examining the full complexity of the historical Descartes' dealing with them

[58] I am not advocating here history as mere literature or entertainment. Rather I believe that Descartes had intentions and conceptual structures reconstructable on the basis of textual and contextual evidence. My conceit is meant to motivate and focus proper historical scholarship on *Le Monde* and related texts, not to displace those texts or dissolve disciplined historical inquiry into more or less amusing creative writing. What 'Descartes' says here is also arguably a good heuristic guide to what to look for in post-Newtonian Cartesians. This conceit derives from the same tactics of role play mentioned above at Note 6.

in *Le Monde*, starting with the theory of local fall in the terrestrial vortex. These will involve what we have above dubbed a charitable[2] reading of these portions of *Le Monde*.

10.6.1 The Intricacies of the Theory of Local Fall and Weight

It is when Descartes attempts to give more detail to his extrusion mechanism that this interesting, and illuminating, complexity arises. Descartes tries to articulate the mechanism of local fall by insisting that it depends upon the differential centrifugal tendency of a terrestrial body and an *equal volume* of subjacent air and second element, the relevant parameter being the density or specific weight of these equal volumes. Hence he seems to be convinced that classical statics and hydrostatics are relevant here, in that they provide the macroscopic, mixed mathematical level generalizations to be reduced to corpuscular-mechanical terms. Let us consider this lengthy, curious and ultimately very telling passage, **(Fig. 10.8 now illustrating theory of fall)**

> ...the parts of the earth also cannot move away any farther than they do from center T, unless there descend in their place just as many parts of the heaven or other terrestrial parts as are needed to fill it. Nor, in turn, can they move closer to the center unless just as many others rise in their place. Thus they are all opposed to one another, each to those that must enter in its place in the case that it should rise and similarly to those that must enter therein in the case that it should descend, just as the two sides of a balance are opposed to one another. That is to say, just as one side of a balance can be raised or lowered only if the other side does exactly the contrary at the same instant, and, just as the heavier always raises the lighter so too the stone R, for example, is so opposed to the quantity (exactly equal in size) of air above it....that that air would necessarily have to descend to the extent that the stone rose. And, in the same way, it is also so opposed to another, like quantity of air below it,... that the stone must descend when this air rises.
>
> Now, it is evident that, since this stone contains in it much more of the matter of the earth than a quantity of air of equal extent (and in recompense contains that much less of the matter of the heaven), and since also its terrestrial parts are less agitated by the matter of the heaven than those of this air, the stone should not have the force to rise above that quantity of air, but, on the contrary, the quantity of air should have the force to make the stone fall downward. Thus, that quantity of air is light when compared with the stone, but is heavy when instead it is compared with the wholly pure matter of the heaven. And so you see that each part of terrestrial bodies is pressed toward T, not indifferently by the whole matter surrounding it, but only by a quantity of this matter exactly equal to the size of the part; that quantity, being underneath the part, can take its place in the case that the part falls.[59]

There are three substantial dimensions to this passage.

[1] First of all, the entire enterprise rests on an appeal to classical statics and hydrostatics in the rigorous style of Archimedes and more recently Stevin: rise and

[59] AT.XI. 76–77; MSM 129–133; SG 49–50.

fall have to do with relative weight, specific gravity. This is what Archimedes, or Stevin, would invoke to explain the behavior of water in air, or lead and wood in water. The only difference here is that Descartes does not judge density by ratio of matter to void in a given volume but by ratio of third matter to interstitial aether (*boules* of second element and shreds of tiny first element within them) in a given volume.

[2] Additionally, however, there a strong appeal to a *theory of fall* (and upward extrusion) based on a *dynamical* rendering of hydrostatical concepts. This move is partly explicit and partly tacit, but in any case totally unprecedented in Descartes' work to this point. He is boldly asserting fall to be an hydrostatical phenomenon, viewed dynamically (as had the young Galileo—unsuccessfully).[60] This is apparent in his analogy to the balance and his overall insistence that equal gross volumes of medium and body act against one another.[61] In other words he is attempting an hydrostatically based dynamics of fall. Moving down or up (and presumably speed or 'strength' of such fall or rise) are also determined by basic hydrostatical parameters.

[3] Hence, hovering behind and above [1] and [2] is a third dimension, a determined attempt to subsume and explain the hydrostatically based account of fall by means of the new corpuscular-mechanical theory of vortices! We see the slide between the first two dimensions and this ultimate, explanatory one, in Descartes, on the one hand, comparing the relative lack of centrifugal tendency of *equal volumes* of terrestrial body and air (with their respective loads of interstitial second element), whilst, on the other hand, still strongly implying that what really counts are the interstitial *boules* of second element that act against the *surfaces* of the terrestrial (third element) particles making up the body. A stone falls in air because it is less agitated than air by the interstitial second element. In other words, the classical hydrostatical study of relative weights, densities and volumes, with its traditional conclusions about the rise, fall or floating of various substances in various media, is wheeled in, on a dynamical basis, to supply a mixed mathematical account of fall. In turn, this nice mixed mathematical result is *really* explained by corpuscular impact, understood in the light of the vortex mechanics and the concept of 'solidity'. This move is exactly what one would expect from the main thrust of the vortex mechanics in *Le Monde*, and it also is generally reminiscent of Descartes' very first physico–mathematical attempts at the corpuscular-mechanical reduction of hydrostatics in 1619.

However, do dimensions [2] and [3] really fit together in a way that would allow the latter to be a serious candidate for explaining the former? Descartes has given the reader two understandings of weight: It arises from relative density (quantity of

---

[60] On the young Galileo's attempt to found a theory of fall on dynamical reading of hydrostatics, mediated by exploitation of the dynamical approach to statics found in the pseudo-Aristotelian *Mechanical Problems*, see Gaukroger (2000) and Gaukroger and Schuster (2002).

[61] Note, of course, that in this connection the medium is the terrestrial substance 'air' not the interstitial *boules* of second element.

third matter in a given volume) but also from relative lack of agitation derived from aetherial impacts. The former is a 'volume' phenomenon, grasped through classical hydrostatics; the latter a 'surface to volume ratio' phenomenon, grasped through the new Cartesian vortex mechanics. And despite Descartes' clear indication that the latter approach will subsume and explain the former, there is no simple or necessary reduction of the one to the other. Indeed, if we think back to the core concepts of the vortex mechanics, the relevant corpuscular level surface to volume ratio is not a matter of density in the classical statical sense at all. Rather, it is a matter of what Descartes' vortex mechanics properly calls 'solidity'. The stone, given its overall (third matter) corpuscular make up and texture, is more 'solid' than those aggregations of terrestrial corpuscles making up 'air' and is harder to 'agitate'. Descartes' difficulties here are indicated by slippages and occlusions involved in several passages where he attempts to smooth the articulation from classical statics/hydrostatics to his own corpuscular vortex mechanics.

Note first how Descartes compares the *relative lack of centrifugal tendency of equal volumes of body and medium.*[62] The talk of equal volumes, of course, derives from the traditional framework of hydrostatics, as in the work of Archimedes and Stevin. But, on the other hand, in accordance with his ambition for a corpuscular-mechanical natural philosophical explanation, Descartes tries to account for those very centrifugal tendencies by introducing the action of an omnipresent vortex of second element. And in the end, both he, and we (as charitable and informed students of his vortex mechanics), know that it is the action of the *boules* of second element moving between and against the particles of third matter that cause the relative centrifugal tendencies and hence the phenomena of fall or rise by vortical extrusion, seeking orbital equilibrium. He seems to be juxtaposing the language and conceptual grammar of the two approaches in hopes that they mesh and that the classical and mathematically rigorous hydrostatics can be coherently and convincingly subsumed under the corpuscular vortex mechanics.

The verbal and conceptual tension involved here is also neatly encapsulated by his overall insistence that gross volumes of medium and body act against one another, in some fashion, as on a balance.[63] This idea is expressed in the language of

---

[62] Recalling his words: 'Now, it is evident that, since this stone contains in it much more of the matter of the earth than a quantity of air of equal extent (and in recompense contains that much less of the matter of the heaven), and since also its terrestrial parts are less agitated by the matter of the heaven than those of this air, the stone should not have the force to rise above that quantity of air, but, on the contrary, the quantity of air should have the force to make the stone fall downward.' AT XI 76–7.

[63] Cf. Descartes' remarks cited earlier: 'That is to say, just as one side of a balance can be raised or lowered only if the other side does exactly the contrary at the same instant, and, just as the heavier always raises the lighter so too the stone R, for example, is so opposed to the quantity (exactly equal in size) of air above it....that that air would necessarily have to descend to the extent that the stone rose. And, in the same way, it is also so opposed to another, like quantity of air below it,... that the stone must descend when this air rises.' (AT XI 76) And, '...so you see that each part of terrestrial bodies is pressed toward T, not indifferently by the whole matter surrounding it, but only by a quantity of this matter exactly equal to the size of the part; that quantity, being underneath the part, can take its place in the case that the part falls.' (AT XI 77)

classical statics/hydrostatics in a way that seems be speaking about causation of the phenomena of rise and fall of gross terrestrial bodies. But, in Descartes' actual physico–mathematical cum natural philosophical agenda, it is always the case that reliable mixed mathematical descriptions require deeper corpuscular-mechanical explanations: there is no causal purchase in talking about 'volumes' of this or that 'density' opposing and displacing each other. The statics/hydrostatics 'talk' seems causal, but of course itself requires explanation, and moreover actually offers no help in formulating one. This is because there is no clear sense in which the vortex mechanics, as we have unpacked it above in Sects. 10.2 and 10.4 depends upon or involves such balances and conflicts of 'equal volumes' of body and medium (particularly the 'traditional' media of hydrostatics, such as **terrestrial** air or water, rather than the vortex celestial mechanical medium of *boules* of second element!). Indeed, in Sect. 10.4 we specifically concluded that classical statics is only a partial rather than exact model for the vortex mechanics. The latter certainly is a 'science of equilibrium' and bears important genealogical relations to statics, but a crucial difference exists in that specific gravity is a volume property, whilst solidity is a volume to surface property. In terms of persuasive rhetoric, Descartes' manner of expression *seems* to establish a relation, a relevance, between the causal-seeming talk of classical statics and his own natural philosophical explanatory procedures. But, the relation is illusory for the reader, and either mistaken or wish fulfilling for Descartes himself. There is no way that appeal to counteracting equal volumes of differing specific weight in statics, maps onto or is reducible to the vortex mechanics of corpuscular surface impact and pressure, mediated by the concepts of surface envelopes of vortex *boules*, and relative 'solidities' of terrestrial bodies and envelopes of *boules*.

So, whilst it is possible to give a simple, charitable[1] first-order reading of Descartes' theory of local fall, as we did in Sect. 10.5, deeper examination of wider swathes of the text reveals that Descartes has somehow wandered into some irreconcilable puzzles and tensions. This wandering may in fact have been a necessary consequence of how his evolving conceptions and practices intersected with his enlarged natural philosophical agenda. We can see some of this by reflecting on the similarities and differences between these endeavors in *Le Monde*, and Descartes' earliest recorded physico-mathematical *cum* corpuscular-mechanical work on hydrostatics, and local fall, in 1619.[64]

At first sight it may seem that the passages in *Le Monde* that we have been studying are simply offering the sort of subsumption of classical hydrostatics under a corpuscular mechanical physico-mathematics with which the young Descartes began his physico-mathematical career in 1619. It is quite true that an identical formal pattern of physico-mathematical *cum* natural philosophical utterance was in play, whether in the 1619 hydrostatics manuscript or the elaborated account of fall in 1633—or, let us recall, even in the dynamical rationalization of the law of refraction in the later 1620s. This general pattern, familiar to us since we studied the early physico-mathematics of 1619 in Chap. 3, is represented in Fig. 10.9.

---

[64] See Sects. 3.3 and 3.5.

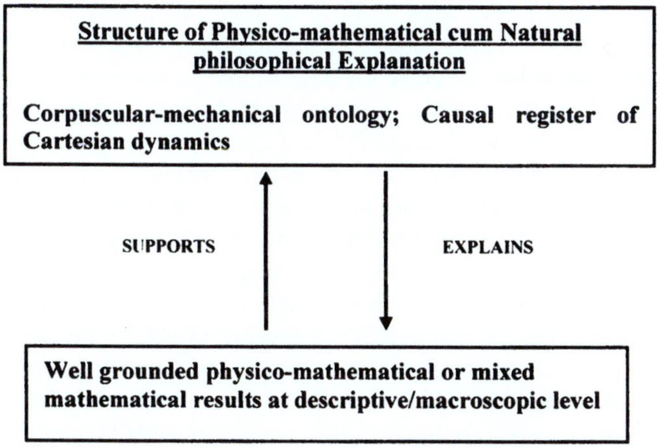

Fig. 10.9 Generic structure of physico-mathematics: hydrostatics Ms. 1619, optical work 1626–8

Nevertheless, there are crucial differences, between the 1619 hydrostatics manuscript and the detailed account of local fall in *Le Monde*, because Descartes in 1633 was no longer a struggling, embryonic physico-mathematician with a leaning toward 'physico-mathematicizing' natural philosophy, where the preferred species of natural philosophy is some sort of unsystematized corpuscular-mechanism. Rather, he and his work had evolved through physico-mathematical successes and toward a focus on a system of corpuscular-mechanical natural philosophy, expressing and shaped by his physico-mathematical style and practices. We can say that in the 1619 hydrostatics manuscript Stevin's key result, the hydrostatic paradox, was taken as a lone exemplar and was reduced/explained by corpuscular-mechanism, by Descartes literally inventing some bits of a physico-mathematical protocol and some embryonic concepts of a 'dynamics of corpuscles'. However, at that time there was no detailed attention to the full range of volume/density considerations established in classical hydrostatics, that is, to the very terms of Stevin's own proofs in the classic Archimedean style. Descartes did not offer any guidance as to how corpuscular-mechanical reduction of the hydrostatic paradox might be redirected to subsume in general the volume and density conceptual framework of traditional statics and hydrostatics. Moreover, in 1619 this work in no way was related to his contemporary study of fall. Indeed, we have seen the frustrating and unfruitful initiatives of Descartes and Beeckman about a physico-mathematics of fall at that time. Many gambits and hunches were in play, abortively as it turned out. But none of them included the tactic of formulating an account of fall (for subsequent corpuscular-mechanical reduction) based on a dynamical reading of hydrostatics, through the frame of the dynamical approach to the simple machines in the pseudo-Aristotelian *Mechanical Problems.*

Now, here in *Le Monde*, circa 1633, Descartes has a very much elaborated, putatively 'systematic' corpuscular-mechanical natural philosophy, and he now tries, at least to some extent, to grasp the body of hydrostatical volume/density considerations. This is because of the very important difference that he is now boldly using classical hydrostatics, interpreted in a dynamical manner, to provide the 'target' 'empirical' account of local fall which is to be the object of natural philosophical explanation using the vortex mechanics.[65] Hence he tries to explicate his extrusion mechanism with respect to considerations of volume and density, so that the vortex mechanics can capture the (sound, 'empirical') target conclusions of the hydrostatical theory of fall. *This procedure, however, ultimately will not accord with the underlying drive for a micro-mechanical account.* It is this situation and the demands it places on Descartes' account, that produces both the tensions we have discerned, and the odd ways in which Descartes' discussion misleadingly seems to resolve them.

These contrasts between the hydrostatics manuscript and the account of fall in *Le Monde* are represented in Fig. 10.10 We note that one of the most important features of the account of fall in *Le Monde*, as indicated in the figure, is that the vortex mechanics on the explanatory plane is explicated in terms of volume and density relations which are alien to it, but which supposedly allow the postulated hydrostatic model of fall both to 'speak to' or 'suggest' the form of explanation, and in turn to yield to such an explanation. Therein, as we have been arguing, resides Descartes most fundamental difficulty in *Le Monde's* treatment of fall, and yet, in accord with a 'charitable[2]' reading, his strategy broadly flows from his prior experience and results in a corpuscular-mechanized physico-mathematics!

We see another instance of these tensions emerging in the remainder of Descartes' discussion of local fall. The vortex mechanics theory of fall raises the problem of why bodies are not swept along circularly across the surface of the Earth as they fall. Descartes first notes that bodies share in the diurnal rotation of the Earth.[66] In addition he maintains that, (Fig. 10.8)

> ...the little heaven ABCD turns much faster than this earth, but that those of its parts that are caught in the pores of terrestrial bodies cannot turn notably faster than those bodies about the center T, even though those parts move much faster in diverse other directions, according to the disposition of these pores.[67]

But if this is true, why does it not argue against the circulation of planets in stellar vortices? The difficulty of reconciling centripetal extrusion with lack of rotary

---

[65] An additional pressure contributing to Descartes' perseveration on volume relations, and hence inclining him toward a implied hydrostatical account of fall comes, of course, from his doctrine of matter-extension, since given volumes of matter-extension (terrestrial bodies) cannot fall or move at all, unless equal volumes of matter-extension (under various elemental forms) also move to replace them. It was his aero-statics, first outlined to Reneri whilst *Le Monde* was being composed, that first betrayed this fault line of conceptual articulation. Cf. Sect. 8.2.3.3.

[66] In addition this forestalls objections made against the rotation of the Earth on the basis of the failure of bodies to fall to the west of the spot vertically under their point of release. It is standard cosmological Copernican fare.

[67] AT.XI.78; SG 50; MSM.135.

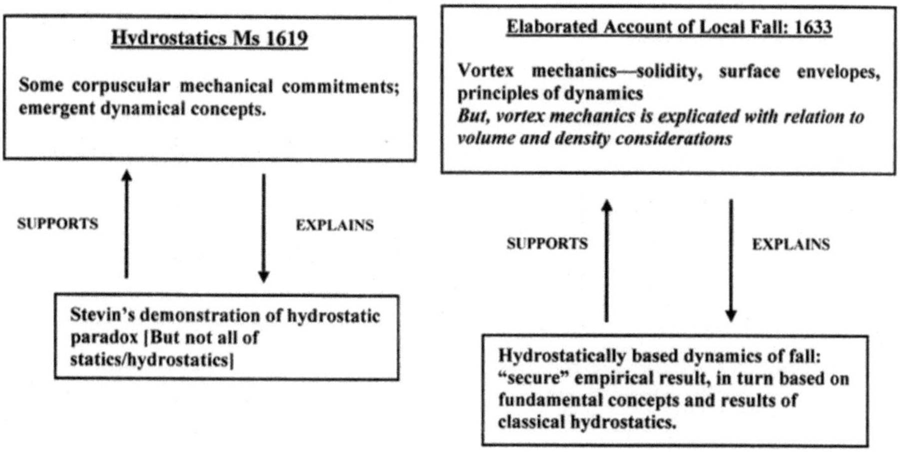

Fig. 10.10 Generic structure of physico-mathematics vs. account of fall in *Le Monde*

translation is even more apparent when he asks why a stone is not pushed more quickly across the surface of the Earth than an equal volume of air;

> You should also know, however, that, even though this matter of the heaven has more force to cause this stone R to descend toward T then to cause the air surrounding the stone to descend there, it should nevertheless not have more force to push the stone before it from the Occident toward the Orient, nor consequently to cause the stone to move faster in that direction than the air. To see this, consider that there is exactly as much of this matter of the heaven acting on the stone to cause it to fall toward T (and using its full force to that end) as there is matter of the earth in the composition of the stone's body, and that, in as much as there is much more matter of the earth in the stone than in a quantity of air of equal extent, the stone must be pressed much more strongly toward T than is that air. By contrast, to cause the stone to turn toward the orient, all the matter of the heaven contained in the circle R acts on it and conjointly on all the terrestrial parts of the air that is contained in that same circle. Thus, there being no more acting on the stone than on this air, the stone should not turn faster in that direction than the air.[68]

Descartes again attempts to ground the distinction upon the relative volumes of 'matter of the heaven' (*boules* of second element and interstitial scrapings of first element) acting in each case. In lateral translation just as much second matter supposedly acts against the air as against the stone; but, in fall, *hydrostatical considerations* suggest that the denser, less porous stone is pushed by more second matter; in fact, it is extruded by an amount of second matter equal to the volume of its parts. Again this will not do if we take seriously the aims of the vortex mechanics, as we have analyzed it: In micro-mechanical terms, both lateral and centripetal (or centrifugal) translations arise from corpuscular impact and must both be evaluated in these terms. If the stone is pushed more strongly in fall than the relevant volume of air, then why is it not also pushed more strongly by the lateral thrust of the medium?

[68] AT XI 79; SG 51; MSM 135–137.

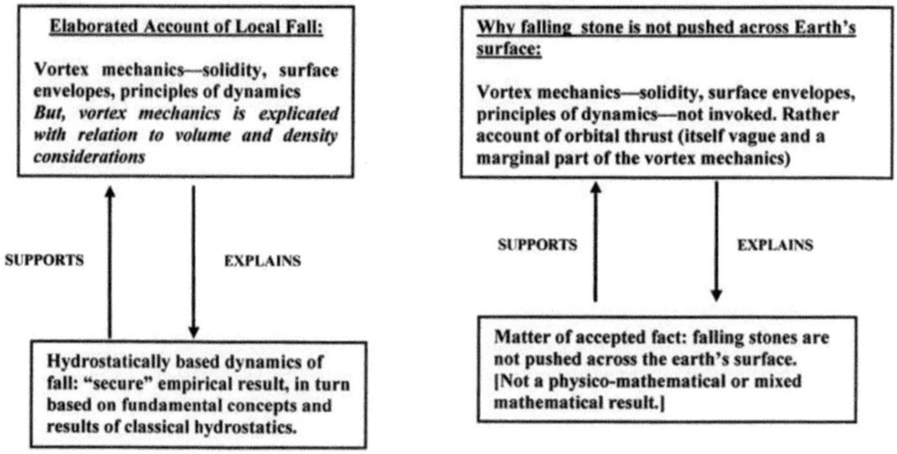

Fig. 10.11 Account of fall vs. account of why falling body is not pushed laterally, *Le Monde*

Descartes' difficulties here compound those in the case of his attempted account of fall, and are summarized in Fig. 10.11. First of all, the explanatory machinery in the theory of fall is the core of his well worked out vortex celestial mechanics (albeit 'augmented' with considerations of volume and density, for reasons we canvassed just earlier). In the case of explaining why falling heavy bodies are not swept across the surface of the Earth by the rotary thrust of the terrestrial vortex, the explanatory level is not occupied by the conceptual core of that vortex celestial mechanics—the account of locking and extruding. Rather it is occupied by an account of orbital circulation and thrust that, as we saw in Sect. 10.4, resides unelaborated, or ambiguously elaborated, on the periphery of the vortex mechanics.[69] Furthermore, in order to make his case for why there isn't a circulatory push on a falling body beyond that suffered by the circumambient terrestrial air, both sharing in the (local vortex caused) diurnal motion of the Earth, Descartes must further explicate this opaque and marginal part of the discourse of vortex mechanics by starting to make all kinds of claims about why falling 'heavy bodies' (which readily fall) and *equal volumes* of terrestrial 'air' (which do not fall) suffer the same degree of orbital thrust. Thus, once again, when his explanatory needs outran the resources of the core vortical mechanical theory of orbital stability, Descartes reached for statically based comparisons based on different micro-mechanical responses by differently constituted *equal volumes* of air or terrestrial objects. Finally, in the 'elaborated account of local fall', the move toward an explanatory discourse invoking statical categories of volume and density certainly undermined the conceptual grammar of his account of local fall. But, at least it had the virtue of looking as though it could be linked to his

[69] As in the problem about whether a planet moves with same agitation or force of motion as the surrounding *boules*. Cf. Sects. 10.2.3 and 10.4 above.

implied descriptive hydrostatical theory of fall. And that, in turn had some hope of counting as a nice, sound mixed mathematical finding to be further explained. By contrast, in the account of why falling bodies are not dragged along the surface of the Earth, that proposition itself, taken as a matter of fact, serves as the explanandum, whilst a peripheral and unfinished part of the vortex mechanics serves as the explanans. So Descartes' explicatory entanglement is even more serious in this case, but, again, we note in charitable[2] terms that the underlying frame of his analysis, and its presentation, derive from the experience and results of his physico-mathematics.

To invoke the charitable[1] interpretation again, one wonders whether Descartes would not have been better advised to say less, sticking with the central themes of the vortex mechanics: He might have then insisted that in the case of local fall, orbital thrust does occur across the surface of the Earth, but is vanishingly small because the time of fall is short, and the falling body develops negligible orbital agitation (beyond the diurnal rotation it shares with the surrounding air, and the main body of the Earth itself). He did not take that option, proceeding instead along the problematical lines we have been analysing. But, in accord with our notion of charitable[2] reading of these texts, we have at least been able to discern behind Descartes' presentation, the continuing shaping power of his experience and results in that activity he termed physico-mathematics, and some particular frames of discourse and explanatory protocols he had invented in the course of pursuing it.

10.6.2 Some Intricacies of Descartes' Moon Theory

In Sect. 10.5.2 we hinted at a simple explanation of the orbital motion of the moon: Taking our charitable[1] reading of the vortex celestial mechanics, we were able to suggest that the moon is 'THE' example of a terrestrial body of very high 'solidity' which has acquired sufficient centrifugal tendency to motion to drift up to an orbital equilibrium distance in the Earth's vortex.[70] Unfortunately our suggested reading is not cashed out in the detailed exposition of the motion of the moon which Descartes offers in Chapter 10 of *Le Monde*. Complications and entanglements ensue, again, consequent upon Descartes' attempting to articulate his account using resources and results embedded in his experience as a physico-mathematician *cum* emergent systematic natural philosopher. Hence we shall require a charitable[2] reading in order to unpack Descartes' discussion, even to a moderate degree.

Descartes first introduces the moon in *Le Monde*, just after his argument, cited earlier in Sect. 10.5.2, for the formation of planetary vortices, including one around

[70] Or alternatively it is an object of comet-like high solidity, and would translate out of the Earth's orbit if not constantly kept in check by the outer boundary of the latter with the encompassing solar vortex. We will learn more about the solidity of the moon and Earth in the present section.

the Earth. What if another planet were disposed to circulate at the same orbital distance from the sun as the Earth?

> ...if there should meet two planets unequal in size, but disposed to take their course in the heaven at the same distance from the sun, and the planets are such that the one is exactly as much more massive as the other is larger, then the smaller of the two, having a faster motion than that of the larger, will have to link itself to the small heaven around that larger planet and turn continually about it.[71]

That is, since the smaller planet is faster, sooner or later it will approach the larger planet, be swept into vortex, and continue to orbit it in the secondary vortex as a satellite.[72]

Passing lightly over the phrase 'two planets...disposed to take their course...at the same distance from the sun', we see that Descartes' lunar theory demands that: (1) the orbital velocity of the moon be greater than that of the Earth; (2) the Earth be larger than the moon, that is, have a larger gross volume including interstices filled by first and second matter; and (3) the moon be 'exactly' more massive than the Earth in the same proportion as the Earth is larger, where 'massive' as usual denotes the overall solidity or (v/s) depending upon the size and shape of the constituent particles of third matter.

If we take Descartes' account literally, several problems result. One might argue against point (3) that 'massive' here means 'dense' in the Cartesian matter-theoretical sense of ratio of volume of third matter to total volume (including interstitial second and first element). But then the Earth and moon would have the same total quantity of matter, and this result, combined with the greater density of the moon, would require that its velocity be less than that of the Earth in order to give rise to an equal force of motion.[73] (Descartes alludes twice to the notion that in principle the Earth and moon have equal agitation or force of motion.)[74] Whatever one might think of this line of reasoning, the smaller resulting velocity for the moon rules it out. However, one might still object to point (3) on the ground that solidity (aggregate v/s) is determinative of orbital distance and so must be equal for both bodies. The equivalence of solidities should have been the prime factor in this phenomenon for Descartes. Although he may have taken note of this in planning his argument, he soon slips into asserting an equivalence of 'force of motion', as noted. This slip parallels the careless locutions of the opening passages of the planetary theory of *Le Monde* discussed in Appendix 2. There, as compared with his later discussion in *Le Monde* and the passages from the *Principia* cited in Appendix 2, he suppressed

[71] AT.XI. 69–70; SG 45; MSM 119.

[72] AT.XI. 71–2; SG 45–46; MSM 119–121.

[73] What is meant is that in this option the total force of motion would be taken to be some function of the quantity of matter, speed and density (ease of passage through the medium).

[74] AT.XI. 68–69, 71; SG 44–46 MSM 117–121 Both loci imply that the moon and Earth should have equal centrifugal tendencies, and hence orbital placement in the solar vortex (leaving the complication of the moon's motion in the terrestrial vortex aside for the moment). They do this by alluding to the orbiting objects having as much 'agitation' or 'force' as the matter of the heavens by which they are being moved.

consideration of solidity (v/s) in favor of a focus upon equal 'forces of motion'. The lunar theory thus suffers from two related points of obscurity: The supposedly exact inverse ratio of solidity and size, and the attribution to the Earth and moon of differential solidities.

Some, but not all, of these difficulties can be resolved by viewing this passage more directly in the light of the planetary theory and comparing it to the more carefully worded argument of the *Principia.* We can address the first obscurity by taking the greater speed of the moon as a primitive given, to be explicated by an analysis of the other mechanical quantities involved. If in addition we take 'larger' to mean 'having more third matter', then the quantity of matter of the Earth will exceed that of the moon. Finally, in accordance with Descartes' mechanics we assume that force of motion depends on quantity of matter, speed and (v/s) conjointly, or,

$$F = mse,$$

where f is force of motion; m quantity of matter; s is speed and e is solidity or (v/s). Invoking Descartes' "careless" condition for equality of force of motion of the Earth and the moon, we can write the following equations (subscript e denoting Earth and subscript m denoting Moon).

1. $F_e = F_m$
2. $F_e = m_e s_e e_e = m_m s_m e_m = F_m$
3. $\frac{s_e}{s_m} = \text{const} = \frac{m_m e_m}{m_e e_e}$

That is, taking the difference of speed as given, the quantities of matter and solidities must vary in some exact proportion such that the equivalence is maintained. When Descartes says the moon is "exactly" as much more massive than the Earth as the Earth is larger, we might interpret this to mean that the moon is exactly more massive than the Earth is larger (has more third matter) such that the compound ratio on the right of (3) maintains the given value of the ratio of speeds on the left.

Of course, one error remains on this account, for in introducing differential solidities we violate the basic principle of the celestial mechanics as determined by our charitable[1] reading —that orbital distances are determined by the value (v/s). Again, if Descartes meant anything like the above, he could have lost sight of the equality of solidity as he slid into considerations of equality of force of motion.[75] In working out the force ratio he might have started playing with differential solidity upon which the force can depend. In addition the smaller size of the moon might have lead him unthinkingly to the consideration of relative solidities.

Some evidence that it was this type of conceptual slip which complicated the lunar theory of *Le Monde* is provided by the more lucid account given of the same theory in *Principia*:

[75] As he had in the celestial mechanics of *Le Monde* proper, see Appendix 2.

...because the Moon has as much force of agitation as the Earth; thus it must be situated in the same orbit around the sun: and since its bulk is less and it has the same force of agitation, it must move more rapidly.[76]

This account leaves out the curious inverse ratio of size and massiveness cited in *Le Monde*. It does indeed rest upon an equivalence of force or 'agitation', but since no condition of differential solidity is mentioned and since the Earth and moon do share the same orbit, we can safely conjecture that Descartes intends their solidities to be equal. He can therefore easily assert that the speed of the moon is greater, since its quantity of matter is less, for if we equate the solidities of the Earth and moon in the above equations, then:

$$\frac{s_e}{s_m} = \text{const} = \frac{m_m}{m_e}$$

The simplicity of this solution suggests that to clarify matters in the *Principia*, Descartes went through something like the back of the envelope calculations we have adduced here.

In sum we can say that Descartes' difficulties with the theory of the moon certainly have something to do with a degree of hasty conceptualization and lack of editing, but that, in accord with a charitable[2] reading, he was also struggling in a serious manner with the fact that unreconciled, opposing lines of articulation of his basic vortex mechanics position appeared when he started discoursing in detail about the moon. He had the same difficulty that he displayed with the opening of his solar vortex account, in hastily assigning equivalence of agitation, or force of motion, to bodies at a given radius from the centre—whether *boules* and planets in the earlier discussion, or moon and Earth in the solar vortex in the latter discussion. This caused difficulty in working out, and expressing, the actually foundational notion of orbital distance being determined by solidity and surface envelope relations (given the acquisition of sufficient orbital velocity to stop initial fall toward the center of a vortex). It is noteworthy, however, that in the *Principia* the lunar theory flows from the celestial mechanics with an ease reminiscent of our earlier charitable[1] reading. This argues that Descartes' vortex celestial mechanics does have considerable coherence on a charitable[1] reading, and that even when Descartes wandered into difficulties, they resulted from the as yet unclarified, but potentially resolvable, machinery of that theory, and not from some generally silly or fantastic nature of the conceit of vortices. When Descartes struggled with the moon in *Le Monde*, he struggled as a physico-mathematician with his own vortex celestial mechanics; that is, he behaved throughout as though the basic vortex mechanics, keyed to the underlying dynamics, were posing the terms of the problems and the terms of solution. And, in the *Principia* he reached a reasonable solution to his puzzles, redolent of our original charitable[1] reading.

[76] *Principia Philosophiae* part III para CXLIX. .....cum (Luna) non minorem habeat vim agitationis quam Terra, in eadem sphaera circa Solem debeat versari; et cum mole sit minor, aequalem habens vim agitationis, celerius debeat ferri.

10.7 Cosmological Theory of Light in Relation to Celestial Mechanics and the Style of Cartesian Physico-Mathematics in Corpuscular-Mechanical Mode

We have dealt with the natural philosophical topics that depend directly upon articulations of the vortex celestial mechanics. Accordingly, we can now turn to the two remaining key topics in *Le Monde*, the theory of light in its cosmological setting, in this section, and the explanation of the appearances of comets, in Sect. 10.8 to follow. Whilst these theories do not derive directly from the manipulation of the vortex mechanics, they do rely generally on the postulated existence of Cartesian vortices, the element theory and the principles of Cartesian dynamics. Additionally, when viewed in a charitable[2] manner, they display traces of Descartes' prior experience in physico-mathematics, particularly in regard to style of analysis and exposition. That is, the interpretation of Descartes' work on these topics benefits greatly from looking below the surface complexities of his texts for hints and clues about their underlying rationales from his physico-mathematical experiences.

10.7.1 Reprising the Dynamics and Laws of Nature

Having sketched his celestial mechanics, Descartes is now prepared in Chapter 13 of *Le Monde* to present his theory of light in its full cosmological setting. This in turn paves the way for Chapter 15 in which the optical phenomena of comets will be explained through an articulation of the theory of light annexed to the account of the motions of the comets. In this manner Descartes will be able to initiate a comparison of observed phenomena with the observational implications of his theory of the world.

Descartes opens his examination of light with a review of the dynamics of circular motion, articulated through his analogy to a stone flung by a sling. Earlier, in Chapter 7, he had analyzed circular motion in order to explicate the third law of nature and lay the foundation for the celestial mechanics. Now, in Chapter 13, he again stresses that the tendency to motion of a body is nothing like a desire or thought urging it in a certain direction. It is only a mechanically conditioned disposition to move in a certain direction, considered without regard to whether the body truly moves or is prevented from doing so.[77] Thus, tendency to motion in a certain direction is a directed magnitude, a determination, attributed to bodies at discrete instants of their motion or tendency thereto.[78] His focus is upon the stone itself and the tendencies

[77] AT.XI. 84; SG 54; MSM 147.

[78] Cf. Sect. 4.2 and Fig. 4.1 therein—Descartes' analysis of the stone rotating in the sling.

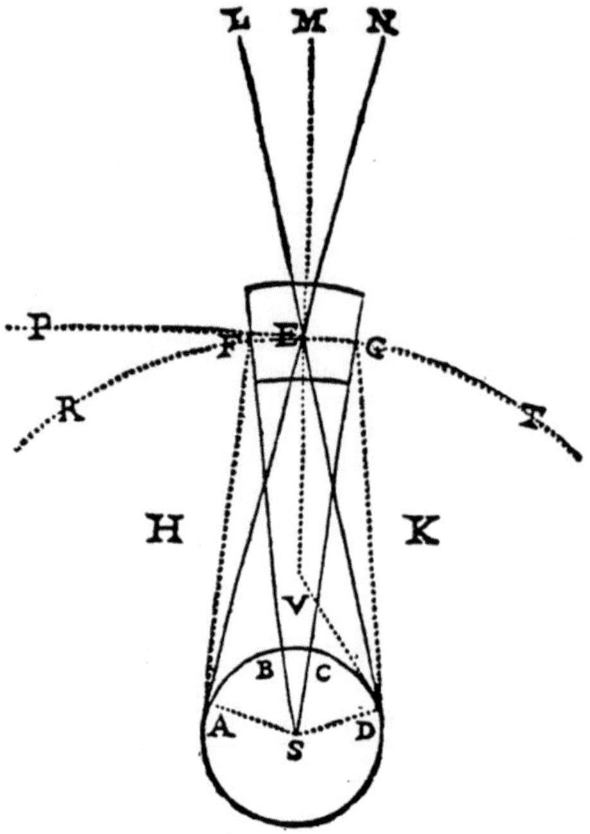

Fig. 10.12 *Le Monde:* production of light, visible disk of the sun, AT XI, p.87

internal to it in a given framework of constraints. He is only peripherally interested in what constraints must be applied to the body to force it along a given path. Hence, he recognizes the centripetal constraint of the sling, but does not build his analysis around it. Of course, a body in circular motion is being constrained; the interesting thing, however, for celestial mechanics and optics is how the body behaves under those conditions. We have already studied the details of this material in connection with its origin in Descartes' optical work of the late 1620s, and his voluntarist theological legitimation of the laws of motion developed in the course of writing *Le Monde*. Indeed, the genealogy of the dynamics has formed a central thread in our argument from Chap. 3, through Chap. 4 to the account of the creation of *Le Monde* in Chap. 8, so we need not revisit his discussion in any further detail here.

Turning to the proposed mechanism of production of light in the celestial vortices, Descartes can explain, on the analogy of the stone in the sling, the radial tendency of each *boule* away from the center of rotation (Fig. 10.12).

> ...you should think of each of the parts of the second element that compose the heavens in the same way as you think of this stone, to wit, that those which are, say, at E tend of their own inclination only toward P, but that the resistance of the other parts of the heaven which

> are above them cause them to tend (that is, dispose them to move) along the circle ER. In turn, this resistance, opposed to the inclination they have to continue their motion in a straight line, causes them to tend....toward M. And thus, judging all the others in the same way, you see in what sense one can say that they tend toward the places that are directly opposite the center of the heaven they compose.[79]

If the tendencies are identified with the action of light, one can understand how each individual *boule* conveys a ray of light which seems to issue from the axis of rotation of the vortex. This account does not explain how the *boules* act together to transmit lines of tendency to motion throughout the vortex. Nor does it explain how the entire disk of the sun is visible at each point of the vortex. In some sense, therefore, each *boule* contiguous with the visible disk of the sun must contribute a line or component of tendency toward the eye of an observer.[80] As we shall see, to meet these difficulties Descartes must develop his cosmological optics with a series of articulations and restrictions, which, on the one hand, seem push the entire theory to the brink of physical unintelligibility—even when viewed in its own terms—and, on the other hand, again reflect the characteristic style and practices of Descartes' physico-mathematical approach to forging a systematic corpuscular mechanism.

10.7.2 The Cosmological Theory of Light as Tendency to Motion Transmitted Instantaneously Through the Boules of the Vortex

To explicate the cosmological optics proper, Descartes must explain how it is that from each point on the surface of the sun lines of tendency to motion (rays of light) are propagated in all directions into the hemisphere imagined to lie above the plane tangent to the sun at that point (Fig. 10.12).

> But there is still more to consider in the parts of the heaven than in a stone turning in a sling: the parts are continually pushed, both by all those like them between them and the star that occupies the center of their heaven and by the matter of the star; and they are not pushed at all by those at M, or at T, or at R, or at K or at H, but only by all those that are between the two lines AF and DG together with the matter of the sun. This is why they tend, not only toward M, but also toward L and toward N, and generally toward all the points which the rays or straight lines, coming from some part of the sun and passing through the place where the parts are, can reach.[81]

---

[79] AT.XI. 86; SG 55; MSM.151.

[80] Or, at least seem to contribute; later Descartes argues that even in the absence of the sun, the vortex would produce the same cone of light. AT.XI.88, which should also be compared with AT. XI.109–10.

[81] AT.XI. 87–8; SG 55–6; MSM.151–3.

Leaving aside the role of the lines AF, DG for the moment, we can attend to the underlying mode of explanation involving the visual cone formed by AE, DE tangent to the sun at A and D respectively. Descartes seems to imply that the radial tendencies of the *boules* lying next to the surface of the sun can be considered to give rise to oblique lines of tendency which pass through the vortex in all directions above the tangent planes to each point of the surface. Once constituted, these lines of oblique tendency continue out into the vortex unaltered and without giving rise to any secondary diffusion of oblique tendencies at each intervening *boule*. Hence there is a curious and decisive difference between the behavior of the *boules* at the surface of the sun and all others in respect to their ability to transmit light-tendencies. The empirical situation has forced Descartes to articulate his mechanical-corpuscular theory of light in such a way that while he accounts for the appearance of the visible disk of the sun, he can deny to particles outside the cone AED any ability to contribute to the light at E.[82]

To clarify this account Descartes distinguishes two causes contributing to the production of light in the vortex. Even if the space occupied by the sun were void, an observer at E would supposedly receive lines of tendency within the visible cone AED. The rotation of the *boules* and the special property of the layer of *boules* at ABCD would be sufficient to cause the phenomenon.[83] The rotation of the sun, however, causes a centrifugal tendency among the particles of first element of which it consists. Descartes suggests that the radial tendency of these particles at the surface of the sun also contribute to the hemispherical diffusion of tendency from the *boules* of the boundary layer.[84]

In attempting to interpret this account, we can reintroduce the problem of the lines AF, DG which was bracketed above. In the course of uncovering the role of these lines we will discover more about the underlying rationale of the account just summarized. First, for purposes of comparison, it will prove useful to examine Descartes' explanation of the appearance of the visible disk of the sun in the *Principia* (Fig. 10.13).

> In addition, it must be noticed that not only do all the globules situated on the straight line SE push one another toward E, but also each is pushed by all the others situated between the straight lines drawn from one of these globules tangent to the circumference BCD. Thus, for example, the globule F is pushed by all those situated between the lines BF and DF, or, in other words, inside the triangle BFD, but is not pushed by any of those outside this triangle. Therefore, if the space marked F were empty, all those which are in the space BFD, and no others, would simultaneously advance as much as possible in order to fill it. And, moreover, just as we see that the weight which carries a rock in a straight line toward the center of the earth, when it is in the air, makes it roll sideways when it falls down a slope;

[82] Cf. passage cited in previous note, especially final sentence. The surface of the sun becomes per force, a privileged surface, in the same sense that these appeared in the hydrostatics manuscript of 1619, as discussed in Sect. 3.3.2, and again below. On this issue see the incisive article by Alan Shapiro (1974) especially pp.254–57, 265.

[83] AT.XI.88; SG 56; MSM 153–155

[84] Ibid. 96–7; SG 61–62; MSM. 169–171

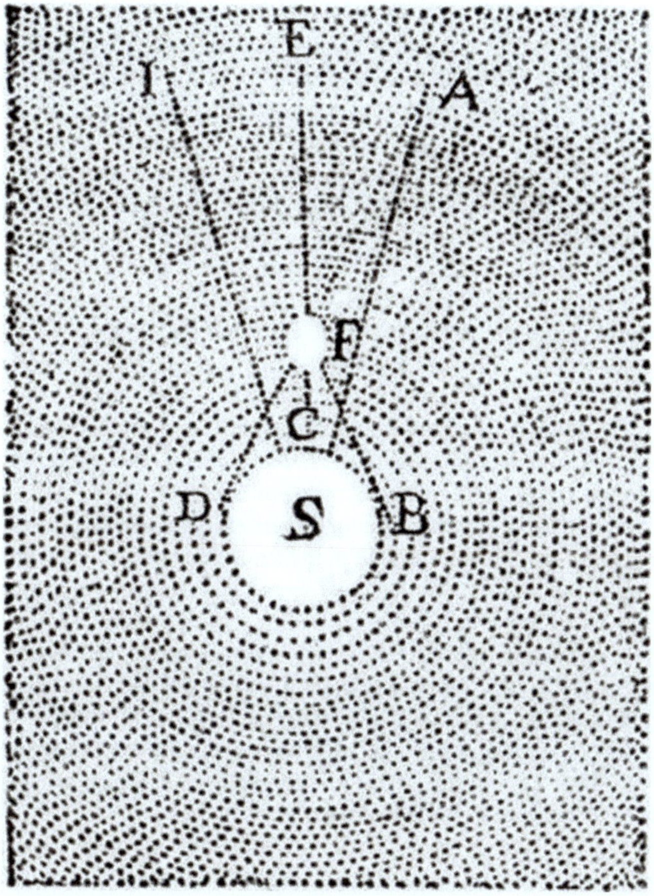

Fig. 10.13 Visible disk of the sun, *Principia Philosophiae* (1644), AT VIII-1, p. 113

> so, similarly, we must think that the force which makes the little globes in the space BFD tend to move away from the center S along straight lines drawn from that center can also make them move away from the center S along lines deviating [somewhat] from that center.[85]

This account is similar to that in *Le Monde*. Other than being a bit more specific about the origin of oblique tendencies to motion, it only differs in entirely omitting any reference to lines equivalent to AF, DG in Fig. 10.12.

[85] Principia part III para. LXII; Miller and Miller p.116 (the term in brackets of course enters the Miller and Miller translation from the 1647 French edition of the Principles). *Principia* part III para. LXII. 'Praeterea notandum est, non modo globulos omnes qui sunt in linea recta SE, se invicem premere versus E; sed etiam unumquemque ex ipsis, premi ab omnibus aliis, qui continentur inter lineas rectas ab illo ad circumferentiam BCD ductas, et ipsam tengentes. Ita exempli causa globulus F, premitur ab omnibus aliis, qui sunt intra lineas BF et DF, sive in spatio triangulari BFD;

In *Le Monde* AF and DG are not tangent to the sun. They touch the sun at A and D, the points of tangency of AE, DE respectively. The problem is: what role are they intended to play in the explanation, and what principles of construction, if any, account for their placement? These questions will be answered by offering a conjecture about the construction employed and then comparing the consequences of that conjecture to the remainder of Descartes' discussion of the appearance of the visible disk of the sun. First, however, an additional complexity must be dealt with. Let us compare Fig. 10.13, from the *Principia*, both with Fig. 10.12, which derives from Clerselier's 1664 edition of *Le Monde* (and which appears in all standard modern editions), and with Fig. 10.14, which derives from the first published edition (of that same year 1664). Note the absence of lines AF and DG in the first edition of *Le Monde* and the absence of analogous lines in the *Principia* diagram. However, unlike the case in Fig. 10.13, the points F and G do appear in Fig. 10.14, and the text, as cited, explicitly draws attention to the unrepresented lines AF and DG. This prompts the conclusion that whereas in the *Principia* Descartes dropped any consideration, textual or graphical, of points corresponding to F and G and lines AF and DG in Figs. 10.12 and 10.14, he did intend their existence, and argumentative use, in *Le Monde*, even considering the state of the Fig. 10.14 in the first edition. We can, therefore, continue to our reconstruction of what lines AF and DG were supposed to accomplish, and how they were constructed.[86]

I conjecture that construction may well have been as follows (Fig. 10.15). Descartes may have drawn AF and DG such that if straight lines FE and EG are drawn, two right angles, AFE and DGE, will result. It would then be the case that all the *boules* within right triangles AFE and DGE could have some tendency toward E arising solely from their respective radial tendencies. (The mechanical situation involved would be parallel to that Descartes invokes for *boules* on the surface of the sun both in *Le Monde* and in the *Principia*.) The construction would simply consist in describing semi-circles on the lines AE, DE and then inscribing right triangles, AFE, DGE, in these semi-circles by locating F and G at the intersections of the orbit of E and the semi-circles on AE, DE respectively. As the distance between E and sun increases, or as the radius of the sun decreases, AF and DG will approach AE and DE respectively, and thus leave vanishingly small sectors between the arcs FE, FG and chords FE, FG.

---

non autem sic a reliquis, adeo ut si locus F esset vacuus, uno et eodem temporis momento, globuli omnes in spatio BFD contenti, accederent quantum possent ad illum replendum, non autem ulli alii. Nam quemadmodum videmus eandem vim gravitatis, quae lapidem in libero aere cadentem recta ducit ad centrum terrae, illum etiam oblique eo deferre, cum impeditur eius motus rectus a plani alicuius declivitate; ita non dubium est quin eadem vis, qua globuli omnes in spatio BFD contenti, recedere conantur a centro S, secundum lineas rectas ab illo centro eductas, sufficiat ad ipsos etiam inde removendos, per lineas a centro isto declinantes.'

[86] Gaukroger, in his edition and translation of *Le Monde*, (SG, xxxvi) writes of the standard editions deriving from Clerselier's: 'The illustrations are not Descartes' own, although those in the *Treatise on Light* are undoubtedly based on sketches, no longer extant, by Descartes....Clerselier commissioned his own illustrations, which I have reproduced here, and these are slightly different from those of the first editions (of *Le Monde* and the *Traité de l'homme*).'

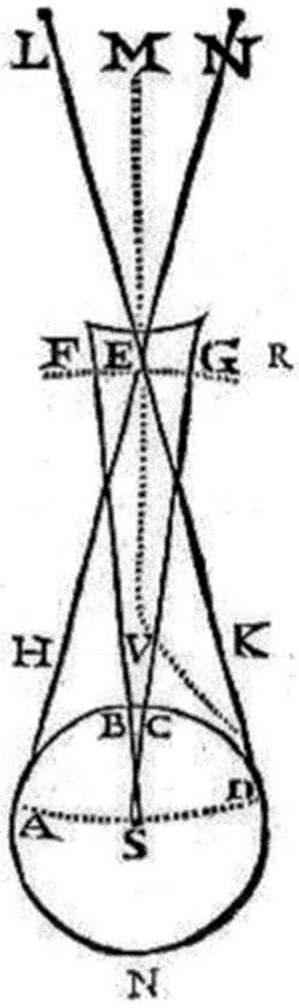

Fig. 10.14 Visible disk of the sun, first published edition of *Le Monde* 1664

So, in constructing the non-tangent lines AF, DG, Descartes apparently desired to come to grips with the implausibility of ruling out any effect on E by the particles outside lines AE, ED. On reflection, Descartes probably decided that, given the special property of ABCD in setting up the transmission of oblique tendencies throughout the cone AED, it was just too implausible to exclude all effect of outside particles upon E. We shall see that, on the one hand, Descartes' subsequent arguments reflect these considerations, but that, on the other hand, he had unwittingly introduced a distinction without a difference, because in fact no light can come from sectors AFE and DGE toward E lest the disk of the sun appear under an angle greater than <AED in Fig. 10.12 (=<LEN). In effect, the attempt to take cognizance of a plausible implication of the mechanical model was doomed from the outset, because that implication could not be taken seriously without violating the key empirical condition placed upon the entire explanation. In the *Principia* Descartes seems to have side-stepped this entire complex of issues (as he did so often in the *Principia* with difficult issues we have identified in *Le Monde*, involving, for example, the vortex mechanics itself, the theory of the moon, and now the cosmological optics).

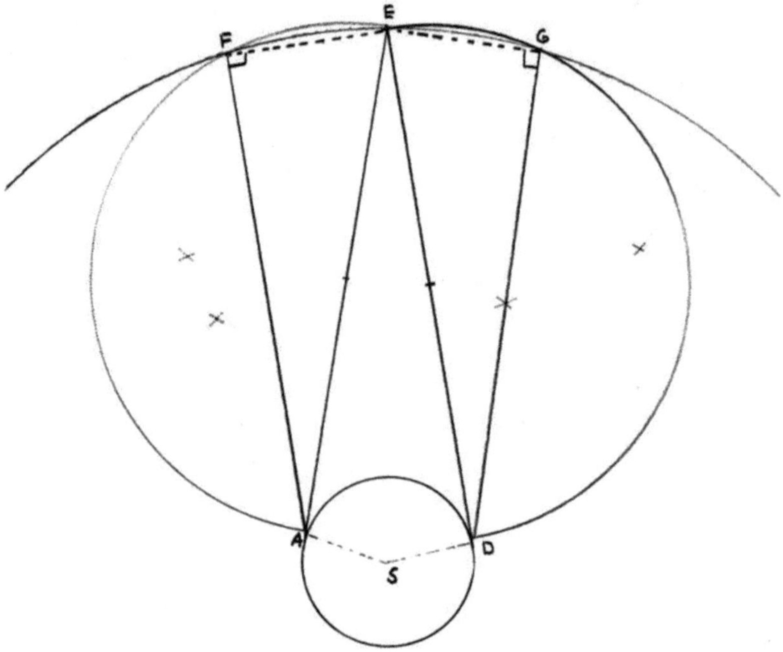

Fig. 10.15 How the operative 'sector' in Fig. 10.12 may have been constructed

Continuing with the text of *Le Monde*, Descartes attempts to rationalize his construction of sector AFGD by analyzing the behavior of the vortex when all first matter is removed from it and the space at E is left void. According to his now customary physico-mathematical protocols, he proposes to analyze the behavior of the *boules* upon the voiding of E and thus reveal the underlying pattern of tendencies to motion which exist even when E is full (Fig. 10.12).

Now the reason that impedes [*boules* from H and K] from tending toward that space [E] is that all motions continue, so far as is possible, in a straight line, and consequently, when nature has many ways of arriving at the same effect, she always *most certainly follows the shortest*. For, if the parts of the second element which are, say, at K advanced toward E, all those closer to the sun than they would also advance at the same instant toward the place they were leaving; hence the effect of their motion would be only that space E would be filled and there would be another of equal size in the circumference ABCD and that would become void at the same time. But it is manifest that the same effect can follow much better if those parts that are between the lines AF and DG *advance straightaway* toward E; and consequently, when there is nothing to impede the latter from doing so, the others do not tend at all toward E, no more than a stone ever tends to fall obliquely toward the center of the earth when it can fall in a straight line.[87]

---

[87] AT.XI. 89–90; SG 57; MSM.155–7 (emphases added).

In some sense the tendency toward E of all the particles in AFGD will be a shorter, easier, and better way of filling E than if other particles were to be involved. But is this criterion purely ad hoc? Is there some sort of rationale behind it, albeit not terribly well expressed in the text? Lurking implicitly behind Descartes' claim must be some sort of justification for the *definition and peculiar efficacy* of sectors AFE and DGE. After all, for the sake of argument, it may well be simpler and better for all particles in AFGD (and only those particles) to tend toward E; but, what determines the precise location of AF and GD? The answer, I believe, lies in connecting Descartes' last sentence to our reconstruction of how he constructed sectors AFE and DGE. On our reconstruction, Section AFDG is precisely that portion of the vortex all of whose parts vanishingly close to circumference TGEFR can tend toward E as a result of the decomposition of their own radial tendencies—that seems to be the underlying criterion, which licenses drawing the sectors, upon which the simplicity argument then depends.

So, in effect, Descartes wants us to imagine AFDG (for these mechanical purposes) as a 'unified' sector, all the parts of which 'straightaway' tend toward E on their own account upon the hypothetical voiding of E. Descartes is claiming that, given the immediate and inherent tendency of each portion of this sector toward E, it is mechanically cumbersome and conceptually implausible to assume some outlying particles can have any effect there. True, as we are about to see, he will concede there are particles in H and K which have the same sort of tendency toward E arising from their own radial tendencies. However, in tending toward E these particles will run up against the monolithic sector AFGD tending 'straightaway' to fill it.

10.7.3 The Physico-Mathematical Style and Protocols and the Celestial Optics

What Descartes has produced here, in other words, is yet another instance of his long standing style and protocols of physico-mathematical argument for identifying tendencies to motion, and the sectors and surfaces relevant and not relevant to their determination: If our reconstruction of how the sectors were drawn is correct, Descartes was trying to set down in geometrical terms some sort of condition determining the boundary between those sectors of the vortex that can and cannot contribute lines of tendency to motion (light rays) at E in the form of components of the radial, centrifugal tendency of the *boules* under consideration. In very physico-mathematical style, he thinks he has resolved the problem if he can set down a condition concerning which *boules* can contribute components of tendency to motion toward E, and then thinks of the sector so defined as a special, and specially efficacious chunk of vortex. That is, given the determination of the special sector, he can then launch further arguments about why a hypothetical voiding at E will be filled only by the immediate tendency to motion from *boules* within the sector. This is all quite reminiscent of the physico-mathematical hydrostatics of 1619 as explored in Sect. 3.3.2. Indeed the structures of the two arguments are virtually identical.

This line of interpretation is confirmed by Descartes' next argument, where he claims that the simultaneous and immediate tendency of every part of AFGD toward E is itself supported by considerations of simplicity (Fig. 10.12).

> ...consider that all the parts of the second element that are between the lines AF and DG must advance together toward that space E in order to fill it at the same instant it is void. For, even though it is only the inclination they have to move away from point S that carries them toward E, and this inclination causes those between the lines BF and CG to tend more directly toward E than those that remain between the lines AF and BF and DG and CG, you will nevertheless see that those latter parts do not fail to be as disposed as the others to go there, if you take note of the effect that should follow from their motion. That effect is none other than, as I have just now said, that space E is filled and that there is another of equal size in the circumference ABCD that becomes void at the same time...Now there is no shorter way of causing one part E of space to be filled while another, for example at D, becomes void, than if all the parts of matter on the straight line DG, or DE, advance together toward E. For, if it is only those between the lines BF and CG that were to advance first toward that space E, they would leave another space below them at V, into which those which are at D must come. Thus the same effect that can be produced by the motion of the matter in the straight line DG, or DE, can be made by the motion of that in the curved line DVE, and that is contrary to the laws of motion.[88]

If some parts of AFGD were privileged to move directly and immediately toward E, the remainder of the sector would only subsequently be able to exert its own proper tendency, and exert it in a roundabout manner. The argument from simultaneous and rectilinear tendency suggests inclusion of all of AFGD as *the* efficacious sector. Again, we may ask what determines this sector. The answer which seems to make the most sense is the construction conjectured earlier in Fig. 10.15.

As a final verification of the conjecture we should note that Descartes does not actually deny that there are particles outside AFGD which can tend toward E. Earlier in the discussion he remarked,

> ...those (*boules*) that are below that circle (TGEFR) but that are not contained between the lines AF and DG (such as those at H and at K) also do not tend in any way to advance toward that space E to fill it, even though the inclination they have to move away from point S disposes them in some way to do so. Thus, the weight of a stone disposes it, not only to descend along a straight line in the free air, but also to roll sideways on the slope of a mountain in the case that it cannot descend any other way.[89]

This remark is as curious as it is revealing. How, we may ask, can it be the case that the *boules* are 'disposed' to move toward E as a result of their radial tendencies, but do not 'tend' toward E to fill it? Is not a tendency to move a 'disposition to move' in Descartes' dynamics and by his own words? There is not enough logical space in that dynamics to make the kind distinction he seeks here, but on another level his intent is clear. What Descartes may very well mean is that there are parts

[88] AT. XI. 90–1; SG 57–58; MSM 157–61.

[89] AT XI 89; SG 57; MSM 155.

of H and K which tend to move toward E, but whose tendency has absolutely no role in filling E, because it is "straightaway" and immediately filled by particles from AFGD. The distinction is not between a 'disposition' and a 'tendency', but between a mere tendency and a tendency which in the nature of the case could be actualized as real translation, if the voiding at E were to occur. But *that* distinction must rest on something like the defining properties of the sector.

Some of the difficulties of Descartes' account should now be clear. A desire to take some account of the tendency of particles outside the visible cone has led him toward the edge of conceptual incoherence. On empirical grounds he really cannot grant that regions outside the cone AED tend toward E. Still, in mechanical terms there is something implausible about denying such an effect. The construction of sector AFGD was a compromise solution, assuaging some of the mechanical demands of the model, while radically cutting off all other particles from consideration. Unfortunately the compromise can accomplish nothing. It strains the overall plausibility of the mechanics of *boules*, and, depending upon whether one stresses empirical or conceptual difficulties, it either leads to a ludicrously wide apparent disk of the sun or to a total denial of the effect it is supposed to establish—the special character of surface ABCD in setting up oblique lines of tendency through the cone AED. Tellingly, the presentation in the *Principia*, as we have seen, side steps all of these complexities and quandaries (which is not to say that the underlying problem of production of light rays in the vortex had been solved). The more complex evidence from *Le Monde* has the compensating value that it reveals more about Descartes' difficulties, and his ways of trying to resolve them. Crucially, and in accord with our charitable[2] reading, his moves are deeply reminiscent of much that we have learnt in this study about his physico-mathematical style and practices, reaching back as far as 1619, and well before his work took its turn toward production of a system of corpuscular-mechanical natural philosophy.

10.7.4 The Physico-Mathematical Hydrostatics of 1619 as Precise Exemplar for the Celestial Optics

If we now dig a bit deeper into the celestial optics, we find not just echoes of, or random appropriations from, the physico-mathematical experience of Descartes, but instead the redeployment of precisely the technical and conceptual resources he first articulated in the hydrostatics manuscript of 1619. The cosmological theory of light bears a number of precise conceptual resemblances to the early physico-mathematical hydrostatics. It will be recalled that the strategy of Descartes was to rationalize selected empirically paradoxical yet theoretically reliable results by relating them to the putative mechanical behavior of the corpuscles constituting the fluid, that behavior being controlled by the principles of his emergent dynamics. The micro-mechanical theory of light in cosmological setting is, along with the vortex mechanics, the pinnacle of the process we have been studying in this

Fig. 10.16 Two of the Basins from the 1619 hydrostatics manuscript

chapter: The crystallization of an evolution from physico-mathematics, with un– or under–defined natural philosophical articulations, to physico mathematics expressed only in and through a systematic corpuscular-mechanism. As such, the cosmological theory of light draws heavily upon the earlier hydrostatics in both its overall style and detailed manner of argumentation.

For purposes of comparison let us recall two figures adapted from the hydrostatics manuscript of 1618 (**Fig. 10.16; Cf Fig. 3.3**). In arguing for the equality of pressure or total tendency to descend on the bottom of both basins (Stevin's rigorously demonstrated yet paradoxical conclusion), Descartes had tried to show that each point or part on the bottom of basin B is serviced by a line of tendency to descend from a point on the surface f.[90] In basin D there exists a one to one mapping of points on the surface and bottom by means of vertical lines of tendency. In basin B the surface area f is one third the area of the bottom. This precludes a one to one mapping. Descartes postulated a multiple mapping, such that each point on the bottom supports a line of tendency, even though the lines must be curved and several must issue from each point on the surface. The curvature of the lines is irrelevant, since only the vertical component of fall contributes to the effect. In order to justify the multiple mapping, Descartes invokes a hypothetical situation in which he imagines g, B, h of B and I, D, l of D to be simultaneously voided or opened. In basin D the points I, D and l will receive vertical lines of tendency to descent in the first instant they are opened. In basin B Descartes claims that in the first instant of descent point f on the surface will tend to move toward all three open points g, B and h. The hypothetical case reveals the underlying pattern of tendencies and shows that although the area of the top of B is only one-third the area of the top of D, the total pressure on the bottom is equal in the two basins.

[90] See Sect. 3.3.

The cosmological optics (Fig. 10.12) proceeds on a quite similar basis. In seeking to elucidate the structure of tendencies to motion in the vortex, Descartes invokes the hypothetical case of the voiding of E. Analysis of the tendencies of the *boules* in the first instant of tending toward the void E supposedly leads to claims about the behavior of light rays. In the optical theory there is also a privileged surface—the surface of the sun—from which the patterns of tendency are to be traced after the hypothetical voiding of E. There are even similar lacunae in both explanations. In the hydrostatic manuscript Descartes did not take account of the behavior of particles in regions such as X (Fig. 10.16). Surely those particles have weight and tend downward, but they simply do not enter into the mapping of lines of tendency from the privileged (top) surface to the void points. Similarly in the cosmological theory of light, Descartes has difficulty deflecting the plausible claim that particles in the regions H and K can have some tendency toward E. On empirical grounds this must be denied, and to that end Descartes employs the sector AFGD with its 'simpler' and more direct mode of mapping points on the surface of the sun to E.

These deep similarities to the style and protocols of the physico-mathematics of the 1619 hydrostatics manuscript reappear in the next stage of Descartes' discussion in *Le Monde*. Descartes considers how it happens that all the *boules* in a wide region, such as that just above ABCD in Fig. 10.12, can tend into a much smaller space and ultimately to the small void at E. The problem is compounded by the fact that each *boule* only tends into the narrow space by virtue of its own radial tendency. To illustrate this point Descartes produces Figs. 10.17, 10.18 and 10.19.[91] Each ball tends by its own weight to fall vertically (Fig. 10.17); yet, if we imagine the balls in a box in which an opening is made at 6 (Figs. 10.18 and 10.19), ball 50 will start to descend. Descartes claims it is clear that in the first instant of fall all balls marked 10, 20, 30 or 40 will tend obliquely toward the hole (Figs. 10.18 and 10.19).[92]

We can trace in all this the lineaments of the original physico-mathematics of fluids. Consider that what is shown happening in Figs. 10.18 and 10.19 is really only a geometrical and logical inversion of the paradoxical hydrostatic behavior of water in basin B in the hydrostatics manuscript of 1619. Let us invert Figs. 10.17 and 10.18 and make two sets of adjustments to them, to produce, respectively, Figs. 10.21 and 10.20. In Fig. 10.20 we have inverted Fig. 10.18, drawn in a bottom to the container, taken out the balls marked 5 and given the container a 'top' open in the region

---

[91] AT.XI. 93, 94; SG 59, 60; MSM 163, 165.

[92] Figure 10.19 has the balls marked '40' pressing laterally against each other and preventing further fall, so that Descartes can address and 'resolve' the seeming problem for his theory of light that this situation would present. He writes, 'At this point you will perhaps say to me that it appear from (Fig. 10.19) that the two balls numbered 40 and 40, after having descended a little, come to touch one another, which is why they stop without being able to descent further. In exactly the same way, the parts of the second element that must advance toward E (Fig. 10.12) will stop before having completely filled the whole space we have assumed to be there. But I reply to this that their being able to advance toward E at all is sufficient to establish perfectly what I have said, namely that since the whole space that is there is already filled by some body…the parts press continually on that body and strive against it as if to chase it out of its place.' (AT X 94–5. SG 60–61; MSM 165–7.)

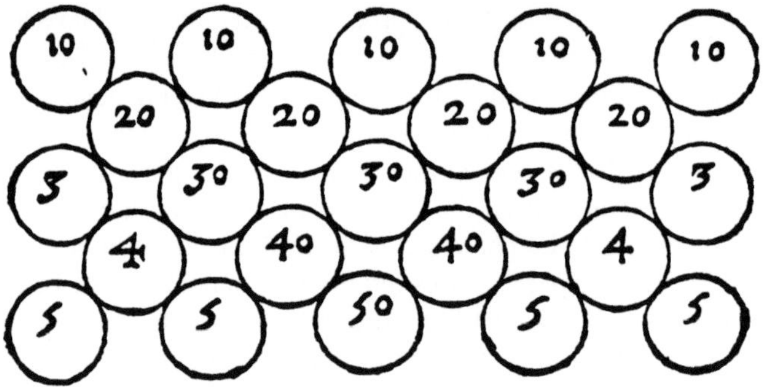

Fig. 10.17 From *Le Monde*, figure 8, AT XI, p.93

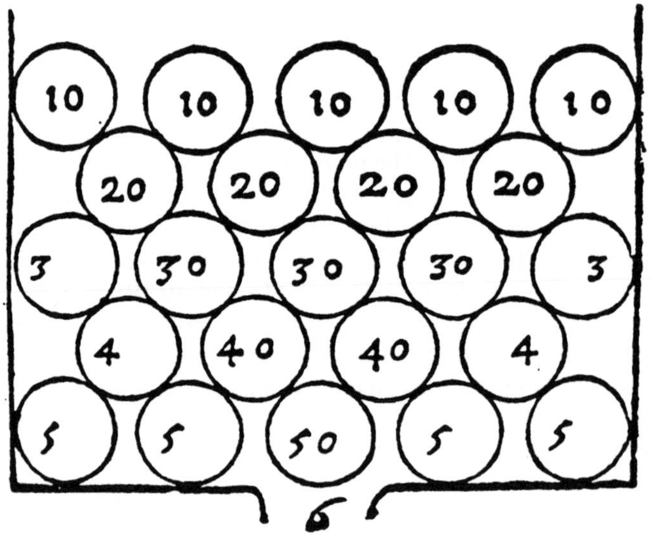

Fig. 10.18 From *Le Monde* figure 9, AT XI, p.93

of ball 50. In Fig. 10.21 we have inverted Fig. 10.17, and drawn in a bottom, leaving all other features the same.

Now, a good Cartesian physico-mathematician, following Descartes' work in 1619, could manipulate Figs. 10.20 and 10.21 in same way Descartes manipulated his lines of tendency to motion and special surfaces in basins B and D in Fig. 10.16. In particular it is clear that the total tendency to motion on the bottom of the container in Fig. 10.20 is the same as that on the bottom of the container in Fig. 10.21. If all the balls marked 10 in both containers started to fall at once, one could draw one to one lines of tendency in Fig. 10.21, and just as many lines of tendency in

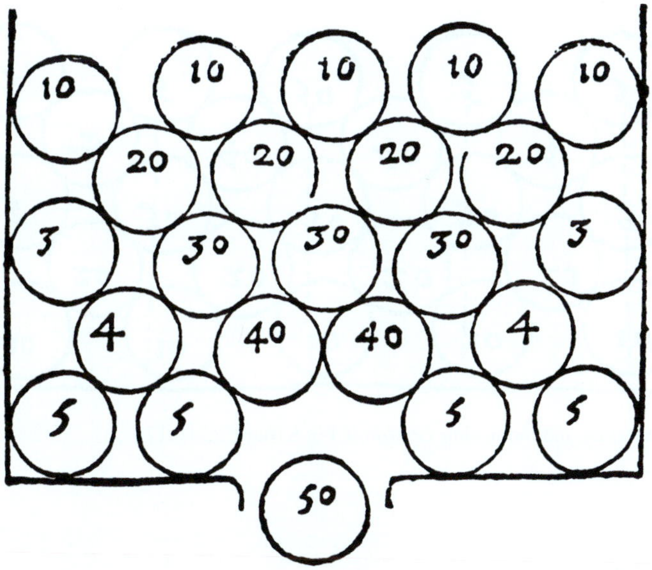

Fig. 10.19 From *Le Monde* figure 10, AT XI, p. 94

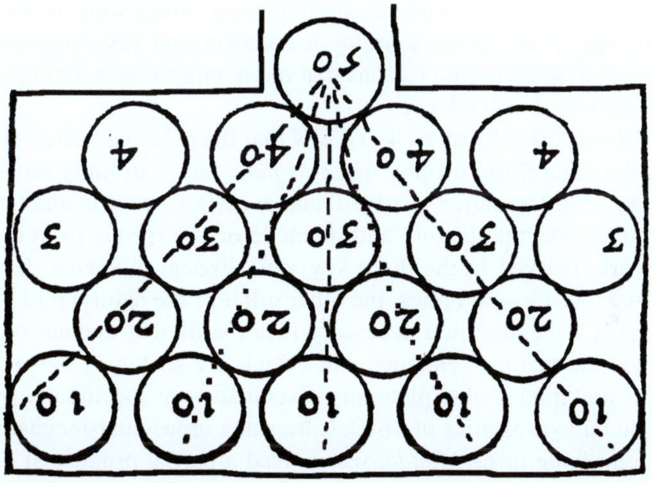

Fig. 10.20 Inverting and modifying *Le Monde* Fig. 9 (our Fig. 10.18)

Fig. 10.20, all ultimately connected to 50 which must be conceived to have several tendencies at once. Figures 10.20 and 10.21 present the physico-mathematical model for fluids, whether water or the fluid heavens.[93] The 'weighing down' appears as lines of tendency to motion and is evaluated by means of a hypothetical voiding

[93] In respect of the production of light, that is, not in relation to production of orbital rotation and settling to orbital equilibrium and establishment of orbital distance from the sun.

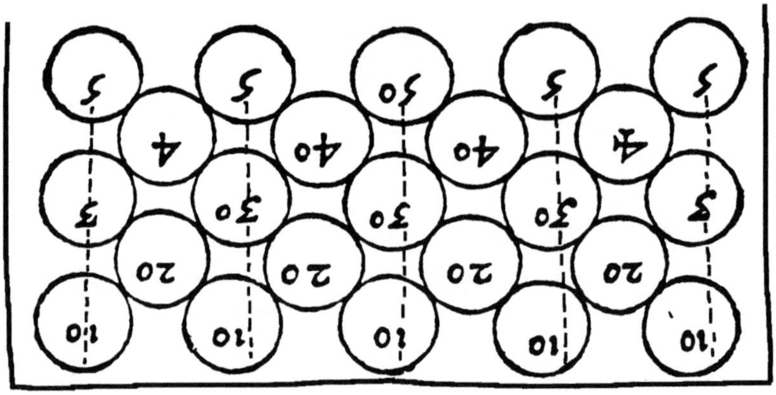

Fig. 10.21 Inverting and modifying *Le Monde* Fig 8 (our Fig. 10.17)

and analysis of consequent first tendencies to motion, whilst working the analysis in terms of privileged surfaces for sourcing tendencies, and simultaneously ignoring (virtually freezing) neighboring regions that might otherwise be thought capable of providing components of tendency.

Earlier in Sects. 10.2.4 and 10.4. we saw that the celestial vortex mechanics, a lynchpin of the natural philosophy of *Le Monde*, had a strongly physico-mathematical character in Descartes' evolved, and evolving, sense of the term, in relation both to its conceptual texture and its mode of emergence from the course of his earlier work, marked in the three key genealogical moments. Now we have discovered that the celestial optics, the other pillar of the natural philosophy of *Le Monde*, not only depended in a necessary if not sufficient manner on the vortex mechanics and the mature dynamics, but even more strikingly was shaped in its core concepts and modes of explanatory discourse very specifically by the physico-mathematical hydrostatics of 1619. Cartesian corpuscular-mechanical natural philosophy, emerging in *Le Monde*, was indeed a partial product of his physico-mathematical work, which had begun with no specifically systematic natural philosophizing aims at all, and need not necessarily have debouched in such a system. Only Descartes' trajectory of aspiration and struggle in mathematics, physico-mathematics, method and natural philosophy can explain this outcome, as we have been arguing all throughout this book. We shall say more in Chap. 11 about the genealogy of *Le Monde* and its intended standing in the natural philosophical agon. Additionally, we shall learn more there about its systematic character. For the moment we must turn, finally, to the last segments of the surviving text of *Le Monde*, the part that promised to match the fabular world of corpuscular-mechanism to the appearances of our own.

10.8 Matching of Evidence: The Appearance of Comets in Relation to the Celestial Mechanics, the Cosmological Optics and the Style of Cartesian Physico-Mathematics in Corpuscular-Mechanical Mode

The opening paragraph of Chapter 15, the last chapter of the extant version of *Le Monde*, makes plain the next level which Descartes' argument will assume,

> Having thus explained the nature and the properties of the action I have taken to be light, I must also explain how, by its means, the inhabitants of the planet I have supposed to be the earth can see the face of their heaven as wholly like that of ours.[94]

Having set out an account of the placement and motion of the celestial bodies and the cosmological theory of light, Descartes now proposes to attempt to match the major observable astronomical/cosmological phenomena of the real world against consequences drawn from his system of the world.[95] He first retails a few simple, but telling correspondences. Given all that has gone before, he can assert that a lighted solar disk, of appropriate relative size would be seen by an observer on a planet at T (Fig. 10.1); that numerous fixed points of light would be visible in all directions across the vault of the vortex; that infrequently a new light might appear or an old one vanish from the heavens; and that opaque comets and planets would appear in their proper places by means of light reflected from the central star.[96]

All the foregoing, however, is rather trivial and merely serves as an introduction to the chief subject matter of the chapter, the observable properties of comets. Descartes intends to address this timely problem, accounting for the appearances of comets by means of the celestial mechanics and theory of light.[97] The general theory of the orbits of comets already accounts for such gross patterns of appearances as

---

[94] AT.XI.104; SG 67; MSM.183.

[95] A slightly different interpretation is also possible: that the previous Chapter 14 may represent the beginning of the matching process. Chapter 14, with its corpuscular-mechanical explanations of the properties of light, adds little to what we already know as a result of our studies in this book. (See Gaukroger (1995, pp.258–60) for a good overview.) If it does begin the process of matching appearances, the stakes in that process are much raised in Chapter 15, with the turn to explaining fundamental astronomical/cosmological matters of fact.

[96] AT.XI. 104–9; SG 67–70; MSM 183–197. Thus Descartes makes brief allusion to the *novae* of 1572 and 1604, explaining them as due to the shifting and bending of intervortical boundaries, which can produce multiple images of a single star, or, so he claims, a star's sudden appearance or disappearance. In Chap. 12 we shall see that he presents a very different model for explaining *novae* (as well as the recently discovered phenomenon of variable stars) as part of a radical, new strategy of system building and system binding in the *Principles*.

[97] The problem is termed timely not because it was hotly debated at this stage in the early 1630s, but just because there was so little attention to comets in the period between the Galileo fomented controversy in 1618, and the work of the Accademia del Cimento and others in the 1660s. Descartes is boldly attacking an inviting, open and at present little studied issue.

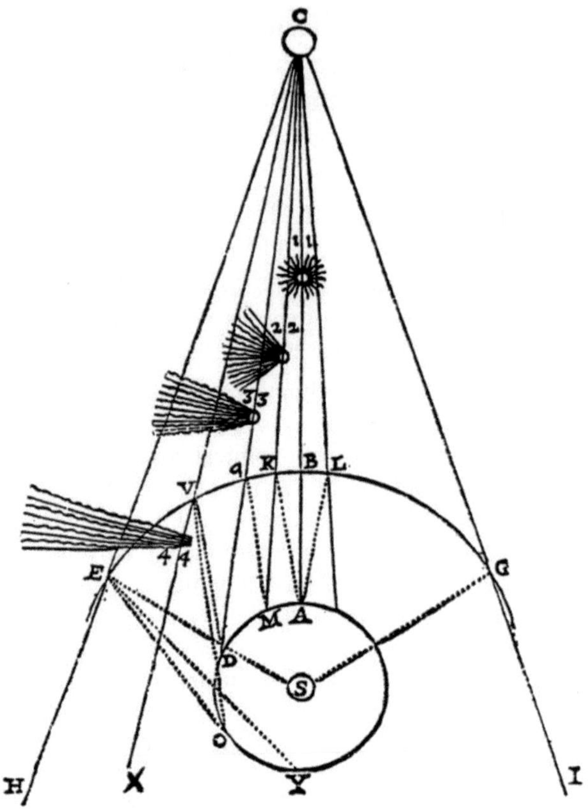

Fig. 10.22 Production of Comet's Tail, *Le Monde,* AT XI, p.113

their sudden visibility, their increasing brilliance as they sweep down near to the K layer and their subsequent dimming and disappearance as they are swept back out of the observer's vortex. The refraction of rays of sun light crossing vortical interfaces effectively prevents comets from being observed before they have entered one's own vortex.[98] But even more than the orbits of comets, Descartes is interested in their halos and tails. Characteristically Descartes considers them to be optical in origin and believes he can explain them by means of a peculiar celestial refraction of sun light reflected by comets back toward the planets.

This refraction takes place below the orbit of Saturn, where the *boules* begin to decrease in size as one approaches nearer to the sun. For purposes of exposition a rather sharp decrease in size is assumed to take place right at the K layer (Fig. 10.22).

Descartes claims that rays such as CV, Cq or CE, reflected by the comet, C, back toward the K layer (EKG) and orbit of the Earth (MAY), set up diverging lines of

[98] AT.XI.110; SG 70–1; MSM 199–203.

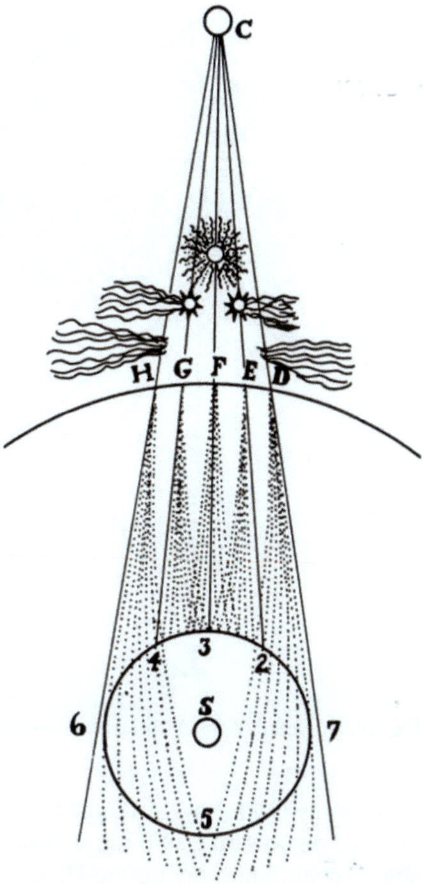

Fig. 10.23 Production of Comet's Tail, *Principia Philosophia,* AT VIII-1, p.189

tendency or secondary rays of light, upon reaching the K layer. The strongest resultant ray is the direct continuation of the original incident ray, such as EH, VX, qD, *etc.*[99] The other rays produced at the K layer diffuse in a fan-like pattern on the sunward side of the primary ray. To judge from the ray CES it seems that the innermost limit of the fan is marked by a ray near to the normal to the tangent to the K layer at the point of intersection of the original ray. However, the remainder of the diagram and Descartes' discussion make clear that he intends the fan to extend only some limited number of degrees to the sunward side of the primary ray. The pattern is somewhat clearer if we compare the diagram Descartes uses in the *Principia* (Fig. 10.23).

Returning to *Le Monde* and Fig. 10.22, we are told that for any given position of the Earth in its orbit, such as D, the head of the comet will appear by means of the

[99] Point X has been added to Clerselier's figure for ease of discussion

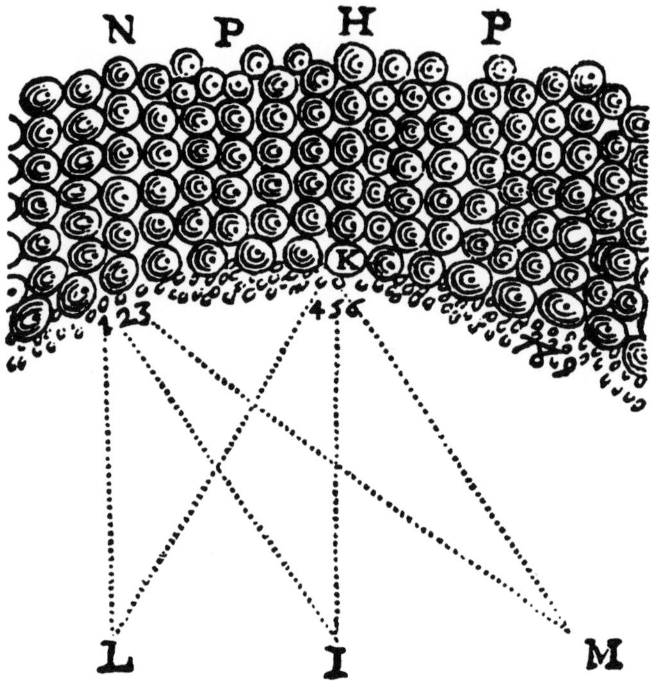

Fig. 10.24 Detail of *Boule* structure of vortex at and around K layer, AT XI, p.116

primary ray CqD. Rays striking the K layer between q and V will each contribute a diffused ray toward D. No ray striking the K layer on the sunward side of q will be able to contribute a ray toward D, and no ray striking the K layer beyond V toward E will be able to contribute a diffused ray back toward D. Thus the comet's head will appear as at 33, while its 'tail', formed by the rays refracted between q and V, will trail out to the left, and generally away from the sun. The appearance of the comet and tail can be constructed for other positions of the Earth and comet from the diagram.

A problem arises in respect to rays near CBA, when the comet is at or near opposition with the sun. When the Earth is at A the comet will appear to have a halo of rays formed by the contributions of rays refracted from the K layer between K and L. But since Descartes must claim that CB diffuses out symmetrically on all sides of B, it must be the case that near A part of the tail of the comet will appear on the sunward side of the comet. In addition it seems likely that some, if not all rays must diffuse into the region away from the sun as defined by their principle ray. If this were not the case, then only one ray, CBA, would diffuse in both directions, and it is implausible to assume that rays close to CBA do not behave in the same way.

The manner of the production of this refraction is of great interest. Descartes invites us to consider the *boules* around the K layer by means of a schematic diagram (Fig. 10.24) which posits a radical size differential at K,

> As for the refraction that is the cause of all this, I confess that it is of a nature very special and very different from all those commonly observed elsewhere. But you will not fail to see

> clearly that it should take place in the manner I have just described to you, if you consider that the ball H, being pushed toward I, also pushes toward I all those below it down to K, but that the latter, K, being surrounded by many other smaller balls, such as 4, 5 and 6, only pushes 5 toward I, and meanwhile pushes 4 toward L and 6 toward M, and so on. Nevertheless, it does so in such a way that it pushes the middle one, 5, much more strongly than the others, 4, 6 and those like them, which are on the sides. In the same way, the ball N, being pushed toward L, pushes the small balls 1, 2 and 3, one toward L, the other toward I, and the other toward M, but the difference is that it is 1, not the middle one, 2, that is pushed most strongly of all. Moreover, the small balls, 1, 2, 3, 4, etc., being thus all pushed at the same time by the other balls N, P, H, P, impede one another from being able to go in the directions L and M as easily as toward the middle I. Thus, if the whole space LIM were full of similar small balls, the rays of their action would be distributed there in the same manner as I have said are those of the comets within the sphere EBG (Fig. 10.22).[100]

At the inner surface of the K layer lines of tendency which have passed from N, P, H etc. without any secondary diffusion, initiate oblique lines of tendency which then pass unchanged further into the vortex. From one point of view the explanation depends on granting to the K layer the peculiar property of diffusing incident rays. Alternatively, one might say that whereas in his general theory of celestial optics Descartes, as we have seen, must deny that lines of tendency will diffuse in all oblique directions from each *boule* through which they travel, here he grants this mechanically plausible effect to one singular layer in the vortex.

This refraction is indeed as Descartes says 'very special and very different from those commonly observed elsewhere'. But it is not quite unique in the world of Descartes. We need only recall the curious behavior of light rays caused by the *boules* at the surface of the sun. That surface and the K layer are similarly privileged in that they produce oblique rays which are thereafter propagated undisturbed throughout the vortex. If anything. the peculiar refraction at the K layer is more implausible than that at the surface of the sun. At least in the latter case Descartes can ask the reader to grant a special effect at a well-defined, physically unique surface in order to meet an important piece of empirical data. On Descartes' own vortex mechanics model, however, there really is no such dramatically unique surface at K—with a discontinuity in *boule* size at that surface—and his attempt to meet this further difficulty involves him in serious difficulties which threaten to undermine the entire cosmological optics. Since his celestial mechanics demands that there be only a gradual decrease in size of the *boules* from K toward the sun, he retreats to the following position,

> If to this you object that the inequality between the balls N, P, H, P and 1, 2, 3, 4, etc. is much greater than that which I have supposed between the parts of the second element that compose the sphere EBG and those that are immediately below them toward the sun, I respond that one can draw no other consequence from this than that there should not take place as much refraction in the sphere EBG (Fig. 10.22) as in that composed by the balls 1, 2, 3, 4, etc. (Fig. 10.24); however, since there is in turn some inequality between the parts of the second element that are immediately below this sphere EBG and those that are still lower toward the sun, this refraction increases more and more as the rays penetrate farther.

[100] AT.XI. 116–7; SG 74–5; MSM 209–211.

> Thus, when the rays reach to the sphere of the earth DAF[Y],[101] the refraction can well be as great as, or even greater than that of the action by which the small balls 1, 2, 3, 4, etc. are pushed. For it is very likely that the parts of the second element toward this sphere of the earth DAF[Y] are no less small in comparison with those toward the sphere EBG than are those balls, 1, 2, 3, 4, etc. in comparison with the other balls N, P, H, P...[102]

This amounts to attributing the peculiar refraction to all *boules* below K and thus threatens to dissipate the unique ray-bearing property of these *boules* in a chaos of secondary rays diffusing from each particle. Although this behavior might be closer to our contemporary expectations regarding Descartes' fluid of spherical particles, it would undermine precisely the construction he must put on his fluid in order to make the general theory of celestial optics work.[103]

Two other sets of problems also follow from this passage. First, it seems that Descartes must maintain that the diffusion caused by all the layers below K will be the same as if one wide diffusion occurred at K. But there is no prima facie reason why this should be the case. If each succeeding layer contains *boules* only a little bit smaller than those of the layer above, and if diffusion of light along oblique components of tendency occurs at each level, then each layer is a source of oblique rays which will in turn contribute to the further widening of the diffusion fan at each succeeding lower *boule*. We might as plausibly imagine that a total diffusion will occur in all directions around the original line of tendency. Descartes cannot maintain that the width of the fan is set by the first diffusion at K, because that makes nonsense of his attribution of the refractive property to the lower layers.

Secondly, and more generally, the explanation of the tails of the comets, like others we have seen in *Le Monde*, such as that of the apparent disk of the sun, is characterized by the following tactic: To accommodate well grounded observational phenomena ('matters of fact') within the framework of corpuscular-mechanical cosmology, specific adjustments are made in the attribution of mechanical properties to the basic particles. However, to reiterate a key finding consequent upon our charitable[2] tactics of reading, these moves are not completely *ad hoc*, for, they depend upon Descartes mobilizing, albeit fluidly and perhaps confusedly, the style and even the protocols of his deeply embedded physico-mathematics. Below in Chap. 11 we shall return to this issue in the context of assessing the 'systematic' character of *Le Monde*, and the opportunities and tensions involved in Descartes' having moved from a physico-mathematics leaning toward some use of corpuscular-mechanical explanations, to a full blown, systematic corpuscular-mechanical natural philosophy which variously depends upon and borrows from that physico-mathematical experience.

---

[101] Descartes' text says DAF. The Gaukroger and Mahoney and translations therefore follow this. But Clerselier's figure lacks an 'F'. It is reasonable to assume that F is meant to be at the intersection of the Earth's orbit and line refracted ray GS.

[102] AT.XI. 117–8; SG 75; MSM 213.

[103] On this and related points see Shapiro (1970) Chap.II, 'Descartes' Cosmological Optics', especially. pp. 53–4, and Shapiro (1974).

Returning to the larger picture, we can say that as the first and only extant portion of the intended material on 'matching of appearances', the account of the appearances of comets represents a dramatic, high stakes bid for natural philosophical credibility and impact. There was to be relatively little study of comets in the period between the 1618 controversy on comets fomented by Galileo, and the revival of inquiry and debate in the 1660s, which swirled around and within the Royal Society and the Accademia del Cimento.[104] Descartes proceeds in *Le Monde* to treat the orbits and appearances of comets as challenging natural philosophical puzzles, which he proposes to resolve by use of his vortex mechanics and cosmological optics respectively, thus illustrating their separate and conjoint explanatory power. Whatever we may think of his explanations and their eventual standing in later debates about comets, we can still grasp, with Descartes, the boldness of the gambit to score natural philosophical credit on a major, dramatic public novelty of the time. This, to reprise a theme at the heart of our study, is competitive play in the natural philosophical contest with a vengeance: Descartes is claiming that comets pose both cosmological and optical problems; and that their existence, orbits and appearances only make sense if the world consists of an infinite array of Copernican vortices obeying the dynamics, corpuscular size and speed distributions, and the theory of light set down in *his* natural philosophical system.[105]

10.9 Looking Forward: *Le Monde* as Natural Philosophical System and Gambit

In 1619 Descartes had begun to develop his conceptions of force and tendency to motion in a hydrostatical context. By 1633, having been crystallized in his profound work in optics, they sat at the centre of the vortex celestial mechanics, which, as we have seen, was a veritable corpuscular-mechanical 'hydro-dynamics', that ran 'the world', or as my friend and occasional collaborator, Stephen Gaukroger, incisively dubbed the entire complex, Descartes' 'Waterworld'.[106] In very simple, but, I submit, accurate terms, the ambitious but embryonic physico-mathematical project of 1619 had borne some hefty dividends. Descartes, *physico-mathematicus*, had built a novel corpuscular-mechanical, pro-Copernican, natural philosophy that would entrain new, non-Aristotelian relations between natural philosophizing and the mathematically based physical disciplines.

---

[104] Boschiero (2007, pp.216–231); Shapin (1994, pp.266–91).

[105] In the *Principia*, as we shall see in Chap. 12, Descartes pursued a much improved version of this sort of strategy of empirical grip and systematic scope.

[106] The conceit arose out of Gaukroger's reflection on Gaukroger (2000), as well as issues arising in the composition of our joint study, Gaukroger and Schuster (2002). I have accordingly entitled the present chapter, as well as Schuster (2005) and previous conference and seminar presentations of this argument, 'Waterworld', in homage to Gaukroger's striking and amusing term.

We have dissected the details of that natural philosophy in this and the previous chapter. But that endeavor is only part of our task. We still need historically to evaluate this text in terms of the categories and historiographical premises and aims of this study. How, exactly, is that to be done? The threefold answer to that question has been inscribed in the approach taken in this volume all along: We must next examine *Le Monde* as a systematic statement in natural philosophy, whilst simultaneously understanding that decisions about its systematic form were in part shaped by considerations about contesting for hegemony in the field of natural philosophy, and *vice versa*. That is, the systematic character of *Le Monde*, such as we shall unpack it, interacted with its status as a daring competitive gambit in the field. But, beyond these two interpretive considerations there lurks a third: We know, and have signaled from the start, that *Le Monde* was the culmination of Descartes' early career in natural philosophy and physico-mathematics, but that it was not to be his mature, definitive natural philosophical statement. That we find in the *Principia philosophiae*. Fully to understand *Le Monde* (and the *Principia*), we need to investigate the latter in regard to its structuring as a system and the precise nature of the gambit for field hegemony it delivers. Accordingly, next in Chap. 11 we shall look at *Le Monde* as a natural philosophical system and as a competitive gambit in the field, and we shall also begin to open up the question of its comparison with the *Principles*. However, Chap. 11 will assay the latter task only in an initial way, because it turns out that the *Principia* bespeaks such a leap of systematizing vision—and such a raising of the stakes as a competitive gambit in the field—that an entire further chapter (Chap. 12) will be required fully to expound some new and quite surprising findings about these matters.

References

Works of Descartes and Their Abbreviations

AT=*Oeuvres de Descartes* (revised edition, 12 vols.), edited by C. Adam and P. Tannery (Paris, 1964–76). References are by volume number (in roman) and page number (in Arabic).

SG=*The World and Other Writings*, edited and translated by Stephen Gaukroger (Cambridge,1998).

MM=*René Descartes, The Principles of Philosophy*, translated by V. R. Miller and R. P. Miller (Dordrecht, 1991)

MSM=*Rene Descartes, Le Monde, ou Traité de la lumière*, translated by Michael S. Mahoney (New York, 1979).

CSM(K)=*The Philosophical Writings Of Descartes*, 3 vols., translated by John Cottingham, Robert Stoothoff, and Dugald Murdoch, and (for vol. 3) Anthony Kenny, (Cambridge, 1988) References are by volume number (in roman) and page number (in arabic).

HR=The Philosophical Works of Descartes, vol I translated by E.S. Haldane and G.R.T. Ross (Cambridge, 1968 [1st ed. 1911])

Other

Aiton, E.J. 1959. The Cartesian theory of gravity. *Annals of Science* 15: 27–49.

Aiton, E.J. 1972. *The vortex theory of planetary motion*. New York: Neale Watson Academic Publications.

Bachelard, G. (1965, 1st ed. 1938) *La formation de l'esprit scientifique*. 4th ed. Paris: Vrin.

Bacon, Francis. 1857–1974. *The works of Francis Bacon*, ed. James Spedding et al. 14 vols. London: Longmans.

Beeckman, I. 1939–1953. *Journal tenu par Isaac Beeckman de 1604 à 1634*, ed. C. de Waard. 4 vols. The Hague: Nijhoff.

Biro, Jacqueline. 2006. Heavens and earth in one frame: Cosmography and the form of the earth in the scientific revolution. Unpublished MA thesis, School of History and Philosophy of Science, University of New South Wales.

Biro, Jacqueline. 2009. *On earth as in heaven: Cosmography and the shape of the earth from Copernicus to Descartes*. Saarbrucken: VDM Verlag.

Boschiero, Luciano. 2007. *Experiment and natural philosophy in seventeenth century Tuscany: The history of the Accademia del Cimento*, Studies in History and Philosophy of Science, vol. 21. Dordrecht: Springer.

Burton, Robert. 1628 [1927]. In *The anatomy of melancholy*, ed. F. Dell and P.J. Smith. New York: Tudor.

Galilei, Galileo. 1953. *Dialogue concerning the two chief world systems*. Trans. S. Drake. Berkeley: University of California Press.

Gaukroger, S. 1995. *Descartes: An intellectual biography*. Oxford: Oxford Univeristy Press.

Gaukroger, S. 2000. The foundational role of hydrostatics and statics in Descartes' natural philosophy. In *Descartes' natural philosophy*, ed. S. Gaukroger, J.A. Schuster, and J. Sutton, 60–80. London: Routledge.

Gaukroger, S., and J.A. Schuster. 2002. The hydrostatic paradox and the origins of Cartesian dynamics. *Studies in History and Philosophy of Science* 33: 535–572.

Hooper, Wallace. 2004. Seventeenth-century theories of the tides as a gauge of scientific change. In *The reception of the Galilean science of motion in seventeenth–century Europe*, ed. Palmerino Carla and J.M.M. Thijssen, 199–242. Dordrecht: Kluwer.

Kepler, Johannes. (1938ff) *Gesammelte Werke*, ed. M. Caspar. Munich: Beck.

Kuhn T. S. 1959, 1st ed. 1957. *The Copernican revolution*. New York: Vintage.

Mersenne, M. 1932–1988. *Correspondence du P. Marin Mersenne*, ed. C. de Waard, R. Pintard, B. Rochot and A. Baelieu. 17 vols. Paris: Centre National de la Recherche Scientifique.

Schuster, J. A. 1977. *Descartes and the scientific revolution 1618–34: An interpretation.* 2 vols. unpublished Ph.D. dissertation, Princeton University.

Schuster, J.A. 2002. L'Aristotelismo e le sue Alternative. In *La Rivoluzione Scientifica*, ed. D. Garber, 337–357. Rome: Istituto della Enciclopedia Italiana.

Schuster, J.A. 2005. Waterworld: Descartes' vortical celestial mechanics: A gambit in the natural philosophical contest of the early seventeenth century. In *The science of nature in the seventeenth century: Changing patterns of early modern natural philosophy*, ed. Peter Anstey and John Schuster, 35–79. Dordrecht: Springer.

Schuster, J.A., and Graeme Watchirs. 1990. Natural philosophy, experiment and discourse: Beyond the Kuhn/Bachelard problematic. In *Experimental inquiries: Historical, philosophical and social studies of experimentation in science*, ed. H.E. Le Grande, 1–47. Dordrecht: Kluwer.

Schuster, John, and Alan B.H. Taylor. 1996. Seized by the spirit of modern science. *Metascience* 9: 9–26.

Shapin, Steven. 1994. *A social history of truth: Civility and science in seventeenth century England.* Chicago: University of Chicago Press.

Shapiro, Alan. 1970. *Rays and waves, a study in seventeenth century optics*. Unpublished Ph.D dissertation, Yale.

Shapiro, Alan. 1974. Light, pressure and rectilinear propagation: Descartes' celestial optics and Newton's hydrostatics. *Studies in History and Philosophy of Science* 5: 239–296.

Stevin, Simon. 1955–1966. *The principal works of Simon Stevin*, ed. Ernst Cronie et al. 5 vols. Amsterdam: Swets & Zeitlinger.

van Berkel, Klass. 2000. Descartes' debt to Beeckman: Inspiration, cooperation, conflict. In *Descartes' natural philosophy*, ed. S. Gaukroger, J.A. Schuster, and J. Sutton, 46–59. London: Routledge.

Chapter 11
Le Monde as a System of Natural Philosophy and Gambit in the Field

11.1 *Le Monde* as Competitive Gambit in the Natural Philosophical Field

In Chap. 2, we discussed the notion of a 'system' of natural philosophy and explored the ideal aim of achieving a good system. We were able there to hint at some of the usual techniques for constructing and solidifying a natural philosophy as systematic. It was also clear that all the modes and foci of contestation we discussed in that chapter regarding our model of natural philosophizing *ipso facto* supplied the typical avenues by which actors could advocate or contest the degree of systematicity claimed in any given case. Contesting in the field and building or attacking systematicity were two sides of the same disciplinary coin. This chapter looks at both issues in *Le Monde*. Note, however, that in so doing, we shall not revisit the more conventional avenues of 'system-binding' offered by metaphysical and theological elaboration. We have already surveyed Descartes' situation in relation to these and *Le Monde*: He had to hand a voluntarist theological legitimation of his dynamics and laws of nature, which does appear in the text, as well as the rudiments of a metaphysically argued ontological dualism, present between the lines, but not offered as a systematic legitimation for his claims, as it would be in its mature form in the *Principles*.

Instead, we concentrate here on more delicate issues of system and gambit inside the text of *Le Monde* as we have studied it in the preceding two chapters. First, in this Section we shall exploit our previous textual and genealogical analysis to assess *Le Monde* as an intended gambit in the natural philosophical contest of his time, viewing *Le Monde* in relation to key natural philosophical aspirations and strategies of similar contemporary actors, particularly Kepler and Beeckman. Like Descartes, they were attempting to displace Aristotelianism, install some version of realist Copernicanism, and create alternative hegemonic natural philosophical syntheses. Then, turning to the issue of systematicity in Sect. 11.2, we first reprise our general concepts of the 'core', 'vertical' and 'horizontal' dimensions of a natural philosophical system, and the 'system-binding' moves that are used to articulate the

J. Schuster, *Descartes-Agonistes: Physico-mathematics, Method & Corpuscular-Mechanism 1618-33*, Studies in History and Philosophy of Science 27,
DOI 10.1007/978-94-007-4746-3_11,

dimensions one to another, as outlined, Sect. 2.5.5. These are then applied to the case of *Le Monde*. Finally, Sect. 11.3 canvasses a few initial ways in which in the *Principia* Descartes repaired problems in *Le Monde* and articulated its strengths. This will lead to the conclusion that we require a more sustained analysis of system and gambit in the *Principia* itself, to which we turn in Chap. 12.

Chapters 9 and 10 have set out the underlying conceptual framework of *Le Monde*, while the chapters preceding them have explored the genealogy of that framework. One can now see that *Le Monde* was a work deeply symptomatic of a contemporary problematic in natural philosophy shared by certain bold, mathematically oriented anti-Aristotelian, pro-Copernican innovators, regardless of their own ontological differences, and ways of tacitly or explicitly understanding the general ideal of a physico-mathematics. The vortex celestial mechanics was not just a fanciful and amusing advertisement for Copernican realism in infinite universe mode, nor was it just a representation of Copernicanism inside a proffered, alternative system of natural philosophy. Descartes' 'Waterworld' was, in fact, a post-Keplerian play for hegemony in the field of natural philosophizing, in its particularly overheated and contested early seventeenth century state. (What we termed the 'critical period' or phase of 'civil war in natural philosophizing' in, Sect. 2.7.) That is, for all its problems of expression, incompleteness and lack of contemporary publication, *Le Monde* was potentially a major move in the natural philosophical *agon*, hinting at the high impact upon natural philosophical culture that would eventually be realized in the *Principia* of 1644.

Le Monde, as Descartes would have seen it, was built in part on the basis of a concatenation of achievements in natural philosophizing key chunks of the mixed mathematical sciences, a procedure he would have termed physico-mathematics, in his acceptation of the word. He had come to terms with, competed with, and, in his view, surpassed Beeckman's natural philosophical strivings, themselves partially shaped, as we have seen (Sect. 10.3), in the shadow of Kepler. In the mixed mathematical sciences, Kepler's own master strokes had been the elliptical orbit of Mars, and the laws of planetary motion in general. Descartes' competing jewels, in his view at least, were his corpuscular-mechanical reduction of hydrostatics, and his solution of the ancient and prestigious refraction problem. Moreover, he too had a celestial mechanics, which followed Beeckman's critique of what they took as Keplerian neo-Platonic ontological nonsense, but which out played Beeckman by being based on a coherent dynamics of corpuscles, itself the product of the same course of physico-mathematical research.

Consider this short list of the characteristics of Descartes' *Le Monde* program: Descartes was articulating Copernicus' claims to apply to a universe of innumerable solar systems; he was displaying what he thought was best dynamical practice—that is, best causal discourse practice—to explain planetary motion and the dynamical role of stars; he was associating in the same problematic local gravity, the behavior of satellites, orbital motions of planets, cometary motions, as well as the nature and causal role of the sun (or of any star) in all this, and in a theory of light in cosmic setting. Now, on each of these points, there are notable parallels to the enterprise of Kepler, allowing for complete difference of natural philosophical content (but not of aim). The problematic is the same in both cases: what Descartes unifies as *explananda* by virtue of his dynamics of vortices,[1] including the key role of stars within vortices, Kepler unifies by a theory of a set of hierarchically arranged causal forces, similar to each other in respect of their immaterial nature, and law-like, mathematical functioning. Both natural philosophers attempted a unified set of explanations under the aegis of a new, alternative natural philosophy prominently advertizing highly anti-Scholastic dynamical registers, or causal doctrines. In other words, as we have argued, the vortex mechanics were serious business, and, as Aiton brilliantly showed, they remained serious business amongst a small committed crew of serious celestial mechanicians, such as Huygens and Leibniz, well into the eighteenth century, in competition with the Newtonian view.[2]

Le Monde may well have been a daring gambit in the field of natural philosophizing. That, however, is not to say that as a system it did not display strengths and weakness of construction, upon some of which Descartes improved in his second attempt at a system in the *Principia philosophiae*. Therefore, we must next consider the systematic character of *Le Monde*, as well as some of its strengths and weaknesses, seen in comparative perspective with the *Principia*.

11.2 *Le Monde* as a System: 'Core', 'Horizontal' and 'Vertical' Dimensions and 'System-Binding' Moves

The first point about the systematic strength of *Le Monde* consists in our insight that the vortex celestial mechanics are the 'engine room' of the argument of *Le Monde*. We have seen that the vortex celestial mechanics, which have often been so simplistically glossed and so easily dismissed, are the lynch pin of the corpuscular-mechanistic system of *Le Monde*. A complicated and delicate conceptual construction in their

[1] Admittedly somewhat different types of vortices in detail—star centric and planet centric.

[2] The rigorously contextual approach of this book, in regard to understanding the vortex mechanics and its genesis should not be taken to signal a denial of larger, long term, diachronic relevances of this inquiry or its findings. One important diachronic dimension immediately presents itself to the technical and internalist historian of classical mechanics: The natural philosophical contestation carried out by Descartes and Kepler was pursued with special attention to the subsumption of astronomy; that is, realist Copernican astronomy, variously interpreted, and to its problem of celestial causation, in particular the function of stars. The nature of one's dynamics, the causal doctrine at the heart of one's system of natural philosophy, was thus focalized, and this drove both to contribute claims woven by later players in unintended and unforeseeable ways, into what we recognize as the process of emergence of classical mechanics. Similarly, we should note the role of optical inquiries, in natural philosophical contexts, in shaping the later crystallization of classical mechanics, a matter hinted at in this book and related work, and currently under serious study by Russell Smith (personal communication) and also Smith (2008, 2008a) It would seem, as Stephen Gaukroger has expressed to me in discussion of themes of this and related work, that the long term genealogy of classical mechanics should be written, at least in part, in terms of the concatenation of unintended conceptual windfalls bequeathed to the emerging discipline, by this and other nodes in the natural philosophical turbulence of the early and mid seventeenth century. I have also pursued this line of analysis in Sects. 2.1 and 2.3 of my article 'Cartesian Physics' forthcoming in the *Oxford Handbook of the History of Physics*, edited by Jed Buchwald and Robert Fox.

own right, they are also the result of Descartes' long and unusual trajectory of work and struggle, which stretched from the initial physico-mathematical gambits of 1619, climaxing in the optical work of the mid and later 1620s, and finally débouching, after 1628, in his growing commitment to the natural philosophical expression of that physico-mathematics. And, in turn, on our charitable[1] reading, the vortex mechanics take us a long way into the details of the corpuscular-mechanical world of *Le Monde*. At this level of interpretation the system projects considerable coherence and pleasing interrelation of parts—from the explanations of stars and stellar vortices, through planetary orbits, the behavior of satellites and comets, as well as local fall and tidal phenomena on planets. The same vortex mechanics, element theory and dynamics explain in broad terms the existence and behavior of light in the cosmological setting.

We might add that once one comprehends the structure of the vortex mechanics and its genesis in physico-mathematics, as well as the dynamics that physico-mathematics had also produced, one can supply an important new answer to the perennial question of 'Why did Descartes proclaim his natural philosophy to be mathematical?' The answer is, *'In part because at the core of that natural philosophy there resided a vortex celestial mechanics and dynamics grounded in his work and experience in physico-mathematics—even if the resulting natural philosophical claims were expressed discursively and not mathematically in the proper sense of the term'*. In the context of the competition to exploit mathematics in the service of anti-Aristotelian and pro-Copernican varieties of natural philosophy, this surely was—at least in Descartes' view—a major advantage of his system over others.

Of course, there were problems and tensions in the unfolding of the system of *Le Monde*, and we have had occasion to examine them in some detail. It was these issues that elicited our charitable[2] reading of the text, not only in regard to the problematical parts of the accounts of comets' tails, cosmological theory of light and the visible disk of the sun, but also in relation to the deeper toils of the accounts of local fall, satellite motion and the tides. Here we revisit some of this material under the aspect of the study of systematicity. To do that, we need to recall our frame of analysis set out in, Sect. 2.5.5 and its accompanying Figs. 2.3 and 2.4, which mobilize the following concepts: the explanatory 'core' of the natural philosophy; its 'horizontal' and 'vertical' dimensions of articulation into particular domains of explanation; and 'system-binding moves'. First, by the core of Descartes' system of physico-mathematical corpuscular mechanism in *Le Monde*, we mean the matter/element theory, the dynamics and the vortex celestial mechanics itself. By the horizontal dimension of articulation of the system, we denote the manners in which the core is explicated/developed/modified in order to launch explanations of results, 'matters of fact', 'solid findings' in various sub-disciplines and sub-domains of inquiry. Across the horizontal level one is asking how well the articulations of the core cohere over the spectrum of applications to differing domains. By the vertical dimension of articulation of the system, we mean how fully and coherently a given sub-discipline (such as a field of mixed mathematics) or domain of inquiry (such as local motion and fall, or magnetism) is grasped and explained by the (articulated) core of the system, and what sort of program of further inquiry, if any, is possible or proposed. By means of these two dimensions, therefore, we explore the arguable coherence of *extension* of the core to cover various sub-domains, and the arguable *depth* and *strength* of the core's explanatory grasp of those domains.[3]

11.2.1 Horizontal Analysis of the System in Le Monde

Let us begin with a simple example on the horizontal plane. Recalling Sect. 10.4, we know that the vortex celestial mechanics itself is a 'discourse of equilibrium'. The explanation of orbital equilibrium and the explanation of planetary rise and fall display two quite different orders of rigor, in that the condition of orbital equilibrium supposedly can be fully specified on the new Cartesian concepts (having postulated the nature and make up of the vortex as well), whilst fall or rise is 'explained' as due to over or under shooting of one or more of the relevant forces—no more, no less. The analogy is to rigorous Archimedean or Stevinite statics, where equilibrium can be fully specified and demonstrated, whilst movement eludes precise characterization and is explained as due to breakdown of one or more of the relevant conditions of stasis. In like manner, Descartes invokes the new concepts of his vortex mechanics to deal rigorously with the condition of stable orbital placement, whereas lack of stable orbital placement, extrusion up or down in a vortex, is a kind of slippage and is not capable of being fully explicated in terms of the concepts in play.

Similarly, we learned in Sect. 10.6.1 that in the case of the explanation of lack of rotary translation of falling bodies across Earth's surface, the core was present. But to some degree it was replaced at the explanatory level by 'alibi' concepts meant to make the particular explanation work more smoothly or plausibly. (See Chap. 10

[3] The reader is reminded, on the basis of the general account of the culture of natural philosophizing and its dynamics, offered in Chap. 2, that the criteria for assessing the goodness of a natural philosophy and the modes of applying such criteria to cases, were themselves objects of negotiation, part of the weave of the contestation in natural philosophy itself. What is proposed here is not meant as the only, best, or truest way of sizing up *Le Monde*, or any other system, and certainly not a set of criteria Descartes embraced. Rather, it is a self-consciously designed analytical tool for dissecting this, and hopefully other systems of the time, in the interest of building better accounts of the cultural process of natural philosophizing. We use it here further to elucidate the text of *Le Monde* and hence the way in which it was Descartes' first *system*, emergent from his particular physico-mathematical experience and practices, yet also shaped by the wider expectations and usages of the field of natural philosophy and the key nodes of debate at the time.

The careful reader should also note that when we first presented the general model of systematicity in Chap. 2, we had not yet attained our insights in Chap. 6 about Descartes' (or anybody else's) grand method discourse as mythic speech. It of course follows from our model of method discourse, and especially the rhetorical functions of method talk, discussed in Sect. 6.8, that a grand method doctrine can be employed in meta–level attempts to legitimate the goodness of horizontal or vertical systematizing moves in a natural philosophy. Our analysis here will not explicitly delve into such matters—they would have added too much complication to the exegesis, although they can be supplied by sympathetic readers of this book, to this point.

Figs. 10.9, 10.10, and 10.11) The explanation invoked to account for the rotary thrust of the terrestrial vortex, and why it does not propel falling heavy bodies, was not, in fact, the locking and extruding machinery provided by the vortex celestial mechanics. Rather, it was the account of orbital circulation and thrust that, as we have seen, sits unelaborated, or ambiguously elaborated, on the periphery of that vortex mechanics, involving as it does Descartes' quandaries about whether a planet moves with same 'agitation' or 'force of motion' as the surrounding *boules*.[4] So, to explain why there is not a circulatory push on a falling body beyond that suffered by the circumambient terrestrial air, both sharing in the diurnal motion of the Earth, Descartes had further to explicate this marginal part of the vortex mechanics with subsidiary claims about why falling 'heavy bodies' and *equal volumes* of (non falling) terrestrial 'air' suffer the same degree of orbital thrust.

In sum, we can form a continuum of degree of 'systematic quality' of explanation across four cases, where that quality is a function of the degree to which the core of the natural philosophy is deployed, and deployed in a relatively stable form: So, [a] orbital placement is a direct explanatory outcome of the core of the natural philosophy; [b] orbital fall or rise can be explained, less rigorously, by invoking failure of one or more of the factors determining stable placement according to the vortex mechanics; [c] orbital thrust, in turn, is a more loosely explained phenomenon, not unequivocally defined in terms of the concepts of the vortex mechanics; and finally, [d] the 'fact' that falling bodies are not thrust across the surface of the Earth uses the weakened terms of the previous case, with addition of appeals to comparisons of heavy bodies and equal volumes of air and vortical aether. The latter materials substitute or alibi for the proper deployment of the vortex mechanics.

Next, let us turn, still on the horizontal plane, from phenomena arguably of a celestial mechanical nature, to those concerned with light. In the cases of explaining the visible disk of the sun and the tails of comets, we have seen the following style of explanation, very much reminiscent of the hydrostatic manuscript of 1619: The reported/observed general phenomena are taken as given and assume priority in the shaping of how the core elements are articulated to accomplish the explanation, so that specific adjustments are made in the attribution of mechanical properties to vortical *boules* making up certain surfaces and loci: In the case of the visible disk of the sun, *boules* contiguous with the surface of the sun can propagate rays of light (lines of tendency to motion) in all directions, but *boules* within the vortex at large cannot. In the case of comets and their tails, it is the *boules* of the K layer that possess a special ability to propagate, across a fairly wide angular sweep, the light rays reflected by comets back toward the center of the vortex. Once established, these bands of refraction do not further widen as the rays, or tendencies to motion, pass back through the vortex *boules* toward the central star and planets.[5]

These two cases concerning light suggest a range of possible optimistic and pessimistic accountings of Descartes' systematic accomplishment, depending upon

[4] Cf. Sects. 10.2.3 and 10.4, as well as Fig. 10.11.

[5] For the moment, we leave aside the deeper difficulties we detected above in each of these two cases. They will enter the discussion shortly.

the viewer's standpoint and aim. Let us survey some of these possible accounts. In Chap. 10 we found in both cases that matters look rather *ad hoc* if viewed abstractly and, in a way, uninformed by a knowledge of Descartes' experience in physico-mathematics. But, our charitable[2] reading, thus informed, told us that what looks arbitrary actually follows closely the style, and, if you will, the protocols of physico-mathematics, evident as early as 1619: [1] The empirical demands of the situation shape the form of the explanation according to well worn patterns of construction of 'privileged surfaces', the behavior of whose particles is both unique in the system and causally central. [2] Particles in other parts of the system are denied these special properties and roles. The explanation then proceeds [3] by assuming an hypothetical voiding of certain loci—possibly accompanied by a graphical representation of the situation (what we have called a 'figuring up')—and then [4] arguing from analysis of the tendencies to motion supposedly brought into play consequent upon the voiding in the situation thus defined. Beyond that, the basic matter and element theory, vortex structure, and basic principles of dynamics invoked—the core in short—remain consistent.

So, on this view, the variance and fluidity of explanation across domains had less to do with arbitrary denial or modification of the core than with how, in Descartes' view, physico-mathematical explanation is accomplished. He himself doubtless thought he had considerable consistency of style and protocol here, whereas other observers, including ourselves, might see, as we did in the hydrostatics manuscript of 1619, a license for corpuscular mechanical storytelling, and picturing, uncontrolled by any plausible and consistent set of rules of procedure. This pessimistic line of evaluation is strengthened if we add to our considerations the fact that more was going on in Descartes' explanations than just (variously) applying physico-mathematical protocols to the core elements. In both the cases we are discussing, we saw that Descartes fell into further complexities: In the case of the visible disk of the sun, his difficulties involved first constructing and then rationalizing that odd 'effective wedge' of the vortex. And in the case of comets' tails, there were problems consequent upon Descartes' toying with illustrating his model in terms not of a special property of the K-layer, but of a K-volume of a 'depth' of a considerable number of ranks of *boules*.

Hence Descartes, or a favorable reader, focusing solely on the core of the system, and the use of the protocols of physico-mathematics, would see stronger systematicity than a reader who delves more deeply into the explanations, as we have done above. If one combines the somewhat different applications of the physico-mathematics protocol in the cases, with the further complexities of the respective explanations, one does not get the impression of smooth systematic extension of the core over various domains, but rather a suspicion that gaps, strains and inconsistencies are redounding against the credibility of the system as whole. Viewed this way, the entire system would show—both to us and to canny contemporary readers—a somewhat arbitrary and tendentious character, arising from the failure of these parts to congeal, combined with Descartes' strongly implied stance that, despite appearances, they have done so.

11.2.2 'System-binding' Moves on the Horizontal Plane

Nevertheless, before we leave the discussion of the horizontal dimension of systematicity, one further set of points needs to be made about Descartes' active concern with systemic goodness. These observations may boost the stock of anyone joining Descartes in applauding the systematic character of *Le Monde*. Within and across the horizontal dimension, Descartes makes a number of quite elegant and clever moves that arguably bind the system together and even lend extra theoretical credibility to some of his attributions of privileged surfaces, beyond mere postulation of their empirical necessity. For example, at the fundamental level of the matter theory and vortex mechanics, we have seen how Descartes sets up his exposition so that it is one thing to have a stable vortex *sans* central star; and quite another to have a star at the vortex's center, effecting the redistribution of types of *boules*, so that, reapplying the vortex mechanics, the existence of stable planetary orbits becomes explicable. Vortex mechanics, the 'dynamics' and the basic matter theory, on the one hand, and the theory of star structure and action, on the other hand, reinforce each other by making possible the resultant model that can meet the celestial mechanical challenge of realist Copernicanism. Vortex motion, and the dynamics, also explain light in its cosmic setting; but, turning defense into attack, only on condition that active stars of first element exist in the center of vortices, and (happily) condition the privileged nature of the layer of vortical *boules* contiguous with the surface of each star. The theoretical weavings here are tight, and they manage to include, that is, strongly suggest and rationalize, the presumed existence of the privileged surfaces of stars.

We can see the sturdy conceptual weave, and the correlative strong and plausible appeal for a privileged surface in another good example of system-binding: Consider, again, Descartes' privileged K-layer of *boules* in a stellar vortex: Far from this being taken as detrimental to a sense of strong system construction, Descartes could perhaps have turned that criticism on its head. After all, the K-layer at least has the virtue of arguably being well defined in terms of the distribution and properties of *boules* in the vortex, the vortex mechanics, principle of vortical stability, etc., once a central star is granted to be in existence, and, thence, as a consequence, the K-layer is capable of playing the key role with regard to the existence and orbits of comets, as well as their odd optical properties. This is a good example of dense and iterative use of the physico-mathematics protocol, because it is applied in structural layers of explanation. Without a central star, there is no distribution of *boules* such that the K-layer exists. Then, given that (on the prior explanation) the K-layer is not an arbitrary creation, it can explain the nature and orbits of comets, when taken in conjunction with the basic vortex mechanics. And finally, it can be appealed to for the needed refractive properties as well (all still in the style of physico-mathematics). What looks rather ad hoc when we begin from the direction of the model of comets' tails, looks highly systematic, indeed almost inevitable, if we follow out the system binding logic starting with the vortex mechanics and model of what stars are.

There is yet another system binding complex involving the Earth, moon and tides. Regardless of difficulties requiring our charitable [2] reading, it is true on the charitable [1] reading that the basic vortex mechanics and dynamics take account of the existence of an Earth-moon system, it being particularly attractive to view the moon as a heavy (i.e. third matter) object whose potential fall in the terrestrial vortex has been turned into orbital equilibrium. Now, nothing beyond the well known, relevant, qualitative facts of observational astronomy would follow from the existence and vortical motion of the moon were it not for the case that the Earth has quite a bit of fluid third matter, water, sitting in declivities on its surface. Given the moon-in-Earth's-vortex set up (including the diurnal motion of the Earth caused by the vortex) that fluid is going to be affected in precise ways, which observers have registered as the basic tidal phenomena. In other words, local fall (with some difficulty), diurnal rotation, the motion of the moon, and, triumphally, the tides, all follow from the vortex mechanics applied to the local vicinity of the Earth, plus the generalized empirical facts of existence and motion of the moon, and of oceans on Earth.

11.2.3 The Vertical Dimension of Systematicity

We now turn to the vertical dimension of systematicity, looking at the manner and depth to which the core, as articulated in any given case, grasped various subdomains, and the projects of explanation of those different domains that might be entailed, explicitly or implicitly. First of all, we need to consider the types of domain of inquiry that could come into play in building and extending the system. Descartes' physico-mathematics, by original intention and his own practice, focused on co-opting hard results in the mixed mathematical fields. The natural philosophy of *Le Monde* had inherited those genes and had a complex relation to several mixed mathematical disciplines—statics/hydrostatics, optics and astronomy.

Starting with the latter area, *Le Monde*, of course, was not a treatise about observational or geometrical astronomy. It was a text of natural philosophy, attempting the explanation and articulation of a certain sort of astronomy, or perhaps, as it were, the explanation of a region of novel and radical initiatives in astronomy, to wit, that new 'physical part of astronomy' summoned into existence by the demands of Copernican realism.[6] As we have seen, in the manner of Beeckman addressing the problematic of Kepler, Descartes too sought to explain planetary and cometary motions in star-centered systems, particularly orbital placement and stability (or lack thereof in the case of comets), as well as the physical role of central stars in shifting the corpuscular parameters of their vortices into the form of locking and extruding mechanisms. As we have seen, this was the daring, radical, 'Keplerian' side of the natural philosophical project of *Le Monde*, and it depended precisely on (and partially constituted) the core of that natural philosophy. Not only did

[6] Recall our discussion of this key 'hot-spot' of contention in the natural philosophical field, Sect. 2.5.4.

Le Monde buy into this new 'celestial mechanical' field of discourse, open to, and indeed required of realist Copernicans, but it also made potentially consequential predictions, for example, about the relative 'solidities' of the planets, and of comets compared to all planets, as well as about the orbital behavior of comets.

Statics and hydrostatics, at least in the form of the hydrostatic paradox, had been the target of explanation and reduction in Descartes' first physico-mathematical foray in the hydrostatics manuscript of 1619. *Le Monde*, however, sat uncomfortably in relation to classical statics, meaning that of Archimedes and Stevin. First of all, it might be noted that, even in 1619, the hydrostatics manuscript had not envisioned the explanation of all of classical statics. The physico-mathematical rationalization of the hydrostatic paradox had been the sole problem posed and marked the limit of Descartes' early aims. Now, in *Le Monde*, the situation *vis à vis* classical statics was even more occluded and unpromising, because what seemed a likely aim of complete cooptation into the new natural philosophy was very far from any hope of realization. We have seen that Descartes misleadingly waved his hands in the direction of importing into his treatment of local fall some sort of hydrostatically based theory of fall, which had he known it, would have been similar to the attempts of the young Galileo in his youthful *de Motu*. However, Descartes actually possessed no such theory. His earlier work on fall had foundered in physico-mathematical terms and certainly had not led to any such theory; and, most importantly of all, his new vortex mechanical approach to fall, as extrusion downward in a vortex due to lack of orbital equilibrium (broken or not yet established), left no space in which classical statics and hydrostatics could be deduced. As we have seen, at length, in examining the relevant texts on fall in *Le Monde*, classical statics and hydrostatics, sciences of volume and density relations (as traditionally defined by Archimedes) could not be mapped onto the shiny new vortex mechanics of surface-to-volume relations, and balances or imbalances of centrifugal tendencies and surface envelope resistances. At best, there remained in the text only the cognitively dissonant vanishing ghost of a hope of bringing statics under the core of the natural philosophy. In short, Descartes had indeed realized, in one version, the hope of late sixteenth and early seventeenth scholarly engineers and mathematically minded natural philosophers that resources from traditional mechanics, variously and widely construed, could be transformed into core causal doctrine in a natural philosophy. But, Descartes had done this not by transplanting into natural philosophy one or another conceptual nugget from either of the relevant mechanics traditions—Archimedean or pseudo-Aristotelian (in the *Mechanical Problems*). Instead, he had forged a new kind of mechanics, with *sui generis* dynamical principles and a vortex mechanics, a science of equilibrium using a newly minted set of concepts. This new dynamics and vortex mechanics, embedded, as intended, in a natural philosophy, was not, in fact, capable of grasping, explaining and reducing the very classical statics that had so inspired the younger Descartes, and other radical natural philosophical aspirants.

Our final area of mixed mathematics for consideration, geometrical optics, was, as usual, the realm of the most promising results for Descartes. His dynamics had emerged, to a large degree, from his optical work. Moreover, as our analysis of the

Dioptrique in Chap. 4 showed, at the time of writing *Le Monde*, he could muster cogent derivations for the laws of reflection and refraction using that dynamics: either applied overtly in the published proofs using the 'tennis ball model'; or, more convincingly, if read back to apply to the corpuscular-mechanical realm, where reflection and refraction of light (which consists in corpuscular tendencies to motion) are explained by instantaneous shifts of the instantaneously exerted force of motion, and/or its 'determinations' at optical interfaces. This achievement meant, in principle, that all of geometrical optics, in so far as it depended upon deployment of the laws of reflection and refraction, could come under the umbrella of this new physico-mathematical optics of corpuscular-mechanical natural philosophical temper. The same point applied to new work, such as the improved theory of lenses and explanation of the rainbow, which Descartes was able to produce, given his command of the law of refraction.

So, the physico-mathematical approach to optics under a corpuscular-mechanical variant of natural philosophy, presented to Descartes the ideal case of productive intercourse between a traditional mixed mathematical field and his emerging natural philosophical aspirations. The traffic between the traditional field, and the new physico-mathematical natural philosophizing was two-way, dense, and fruitful in ways that arguably stoked Descartes' hopes and aspirations, as they had grown and transformed from the initial physico-mathematics of 1619 down to the extended natural philosophical system building in 1629–1633. Yet even here, the potentials were ultimately limited and deeper puzzles resided, emergent, in part, due to Descartes' very success—as one would expect in any burgeoning frontier of 'research'. We may elect, with historiographical fastidiousness, to ignore the sorts of puzzles that would be posed later in the century to Cartesian-like theories of light, as wave or particle motion, by new phenomena, including instrumentally mediated phenomena, such as chromatic aberration, double refraction and diffraction.[7] Instead, we can find looming in Descartes' own work deep challenges to how his corpuscular-mechanism could further advance a physico-mathematical optics.

Consider for example the following problem, the initial appearance of which we noticed in Chap. 4. The issue concerns the contrast between Descartes' treatment of refractions in *Le Monde* and the *Dioptrique*, behind which we have seen reside the optical discoveries and dynamical theorizing of the late 1620s. In the latter work, Descartes shows how the deviation of a ray at a refracting interface arises from a change in its 'force of motion' combined with the conservation of its component of tendency to motion along the surface. We saw in Chap. 4 that ultimately the explanation is meant to attach to the mechanical properties of the particles constituting the medium, but, in fact, the 'demonstration' actually proceeds at what might be called the macroscopic level, where force of motion and 'determination' are rendered by means of geometrical figures—this, of course, being the splendid conceptual windfall of his path of discovery of the law of refraction. There is no consideration of the micro-structure of the interface, or how a tendency to motion is transmitted

[7] Dijksterhuis (2004).

through the corpuscular lattice of the medium by the interstitial aether. Indeed, we showed that had Descartes pursued such issues, his prized theory of refraction, in its existing conceptual and graphical form, would have collapsed into incoherence.[8] In *Le Monde*, however, in the theory of comets' tails, a form of refraction, a bending and diffusion of light rays, is held to result from the size differential of the transmitting corpuscles, and a number of problems were caused by taking that tack, as we have seen.[9] Moreover, the explanation in *Le Monde* is strong on visual imagery concerning what must occur at the corpuscular level, but it is weak in even potential physico-mathematical argument, in the style of the *Dioptrique* and the work we have uncovered leading up to it. At a gross level, both explanations derive from the corpuscular-mechanical approach to the study of light, but their fine structures are entirely different. It is difficult to imagine how either could be translated close to the conceptual and logical plane of the other. In the *Dioptrique*, Descartes had largely eschewed detailed corpuscular-mechanical story telling about optical interfaces, and the physical depth of optical media, choosing to skim along at the impressive level of his geometrical diagrams and arguments concerning force and determination shifts at mathematical interfaces. His problems, hinted at in the text of the *Dioptrique*, are on full display in the high stakes, headlined theory of comets' tails in *Le Monde*. Ultimately, the great physico-mathematical optics did resist being pushed under the umbrella of the new corpuscular-mechanical system.[10]

In sum, to analyze a natural philosophy in the vertical dimension amounts to a very searching examination, one that any system would have found difficult to pass with flying colours. But that is the point. Such analysis shows clearly what was and was not happening in the dynamic heart of a system, where reduction and co-optation of subordinate domains was expected, and often rhetorically claimed, but where,

[8] See, Sect. 4.4 .

[9] See Sect. 10.8 above. Additionally, at AT.XI.111-112 earlier in Chap. 15 of *Le Monde*, Descartes discusses the refraction of light crossing vortical interfaces. He does not attribute the refraction to different forces of light on either side of the surface, but rather to the curvature of the surface itself. Earlier he had insisted that the tendency to motion of the *boules* on either side of the surface must be equal, but of course they are also *opposite* and that might be expected to cause some deflection of light. But, how exactly does the characteristic force of light in traversing a given medium arise in a universe of omnipresent vortices, celestial and planetary? Descartes has left the ideal optical world of the *Dioptrique* where only the corpuscular make up of media determine optical properties, and not the underlying vortex mechanical circumstances as well. The situation here is structurally identical to his problems in relating the traditional ideal statics of volumes and densities to his 'real' world, where weight and fall are products of the vortex mechanics.

[10] We do not mention here the problems inherent in thinking through the simpler case of reflection of light at a detailed corpuscular-mechanical level. Descartes did not even so much as slip toward this problem in the *Dioptrique*, although he did in regard to refraction, because that was his main topic and he had to articulate issues concerning it to the brink of falling into the pitfall we have discussed. Newton, of course, would very early on raise issues about the Cartesian theory of reflection as coherent rebound of incoming particles off a (necessarily microscopically wildly uneven) reflective 'surface' of particles. On the larger history of the successes, pitfalls and failures of the seventeenth century study of optical phenomena in relation to the mechanics of collision, see the seminal papers of Russell Smith (2008, 2008a).

in the nature of the beasts under examination, smooth success is hard to find. It is never consensually granted by everyone, and is very much dependent upon the eye of the beholder. This is part of the field of possible contestations that made the game so inviting and difficult for the natural philosophical players. However, there is more than the delineation of a players' grammar involved here. I would contend that vertical analysis of a system of natural philosophy can have a value to the historian in ways less accessible to contemporary actors. This involves the long run tendency, from the mid seventeenth century onward, for natural philosophizing to be pursued less in terms of explicit and contending systems, and more in the form of fragments of 'experimental natural philosophy', ultimately leading, at least along some lines of descent, to crystallization of more specialized domains of inquiry, which over time, arguably began to look more like separate disciplines. This is the process discussed briefly in Sect. 2.7, regarding the long term dynamics of the field of natural philosophizing in the period 1600–1800, which amongst other things, amounts to a revision of the historiography of 'rise of experimental sciences' famously offered by Bachelard and Kuhn.[11] Recall, also, that we pointed out in relation to our model of systematicity (Sect. 2.5.5) that vertical analysis suggests the following rule of thumb: To the extent that in a given natural philosophy investigation within a subordinate domain is dictated by and co-opted toward the strengthening of horizontal systemic considerations, that natural philosophy is, or is intended to be, a system. To the extent that investigations within subordinate domains take on a life of their own—meaning amongst other things that horizontal systematic articulation is neglected, rather ad hoc, or merely rhetorically asserted as an afterthought—that natural philosophy is tending toward the genus 'experimental natural philosophy'. Moreover, as also noted in Chap. 2, to the extent that various natural philosophical inquiries tended to treat specific sub domains in that 'experimental' way, those sub domains took on *sui generis*, quasi disciplinary characters, and over time floated more and more free of any particular natural philosopher's systematizing ambitions.

So, taking a wider view of the trajectory of the field of natural philosophizing in the seventeenth century, one sees that Descartes was still a heavy systematizer, but, in tune with the rising competitive temper of the time, he was daring and aggressive in the ways he constructed, and presented, his system. His building of *Le Monde*, out from and in part upon physico-mathematics (as he understood it over time), shows this daring, as does his creative engagement with realist Copernicanism and his direct attack on the key 'articulation hot spot' of the causes of celestial motions in Copernican astronomy. He also played with and upon his strong successes in physico-mathematical optics, and was willing to gamble on daring explanations (and implied predictions) about such exotic, elusive, and famously contested objects of inquiry as comets. *Le Monde* was in all these respects, not only a system, but also a challenging and bold one at that.

[11] The revision of Kuhn's and Bachelard's manners of conceiving these processes is implicit in the discussion in Sect. 2.7, but is made explicit in Schuster and Watchirs (1990), and Schuster (2002) cited therein, as well as in Schuster and Taylor (1996, 1997).

11.3 Remediating Problems and Taking the Best Steps Forward in the *Principia*

The *Principia* displays distinct improvements in the realm of system-binding and the correcting, or evading, of some of the difficulties of *Le Monde*. What I have in mind here is quite beyond the obvious points that the *Principia* is a scholastic text book, originally published in Latin, and which includes a detailed explication of the legitimatory dualist metaphysics. All scholars of Cartesian natural philosophy are aware of these points, and they hardly need repetition in a chapter dealing with *Le Monde* and its genealogy. Here I want to comment briefly on some of the most salient system-binding features that relate intimately back to Descartes' earlier efforts in *Le Monde*. The last of these small examples will in turn trigger our inquiry in Chap. 12 into the much wider and deeper systematizing strategies which Descartes developed in the *Principia* and which have hitherto been little noticed by historians of science or philosophy.

We have already noted at least three areas in which problems of conceptualization or explication in *Le Monde* were remediated in the *Principia*: Most importantly, as noted throughout Sect. 10.2, and as Appendix 2 below demonstrates, the *Principia* contains a much clearer exposition of the core of the vortex mechanics. Not only is the key concept of solidity/massiveness better defined and deployed, but also, crucially, Descartes clarifies his key 'reductio' argument concerning the orbital placement and equilibrium of planets, and thus repairs omissions and confusions in the text of *Le Monde*.[12] Secondly, as we have had cause to mention, there are in the *Principia* deliberate attempts to improve (or evade) conceptual or graphical embarrassments in *Le Monde*. We saw this in relation to the odd 'unified vortical wedge' (Chap. 10, Fig. 10.12 sector AFEGD) in the theory of light in its cosmic setting, used to explicate the appearance of the visible disk of the sun. This completely disappears from the text and figures in the *Principia*. Thirdly, as shown in Sect. 10.6.2, in a similar manner, the statement of the theory of the moon is streamlined in the *Principia*, so that the range of possible glosses, and ambiguities, we displayed in the corresponding parts of *Le Monde* was curtailed.

Finally, as Jacqueline Biro has shown, there is in the tidal theory of the *Principia* an admirable recognition of the oversimplification of the theory presented in *Le Monde*, and a sophisticated attempt to address the issues of the plausibility and explanatory power of the models in play.[13] It certainly falls into the class of improvements over *Le Monde* in the *Principia*. Francis Bacon, we saw, had attributed the tides to the continental disruption of a permanent westward current derived from the diurnal movement of the heavens. Descartes' model in *Le Monde* makes no allowance for such continental disruption and treats the oceans in idealized terms (much like the Aristotelian 'sphere of water') as continuous, for the purpose of allowing the 'tidal bulges' to translate around the globe, as his model requires. Biro has shown that, in the *Principia*, Descartes was quite aware of the idealized nature of the *Le Monde* model

[12] See Appendix 2, and above Sect. 10.2.3.

[13] Biro (2009). This work derives from and articulates Biro (2006).

of the Earth and oceans and the ways it offended both basic geographical facts and his second, detailed model of the Earth offered in the *Principia*. For, whilst the *Principia* repeats the model of *Le Monde* in regard to the tides, it also teaches a very detailed second model of the Earth's structure, and its formation—the genesis and conformation of land forms, mountains, valleys, coastlines and ocean basins—consistent with the best geographical knowledge of the day.[14] Descartes knew what he was doing, and he moved to explicate and modulate the explanatory standing of the idealized tidal model. Biro reminds us that in *Le Monde's* theory of the tides (Fig. 10.8) the Earth is represented as completely surrounded by water, an idealized conception upon which Descartes did not remark at all.[15] Despite his failure to comment on the status of the model in *Le Monde*, Descartes explicitly refined the model in the *Principia*: 'let…1234 [be] the surface of the water, *which, for the sake of greater clarity, we are supposing* completely covers the Earth; and 5678, the surface of the air encompassing the ocean.'[16] Hence, here, Descartes was conceding that the model of the Earth encircled by water was not literally true but was 'a useful and appropriate conceptual device', as Biro puts it.[17]: About all this Biro concludes,

> In other words, Descartes maintained that, while oceans do not wholly cover the Earth, they are continuous to the extent that the westerly flow of water is unhindered as no stretch of land runs completely from north to south. The waters in the oceans can travel freely from east to west *as if* they covered the whole Earth.[18]

[14] The oceans formed, along with the other features, when an outer crust of the Earth collapsed forming protuberances and declivities, the latter filling with water. This process completes Descartes' narrative of the formation of Earth, and indeed any planet, from a star which dies when sun spots completely encrust its roiling outer surface of first matter, its light and activity are extinguished and it is eventually pulled into the vortex of a still active neighboring star. *Principia*, Part IV articles 1–44 deal with the formation, and resulting structure of the Earth; while the formation of sun spots on the surfaces of stars, which may ultimately form all encompassing crusts, transforming a star into a planet or comet, is dealt with in Part III articles 90–120. Note that the crust whose collapse leads to the creation of oceans and landforms is *not* the crust of sun spots that extinguished the original star. That crust sits deep inside the emerging planet, under all the third matter material of water, landforms and atmosphere, which derive instead from an original stellar 'aether', consisting mainly of third matter, which Descartes says surrounds all stars out to around the distance of their first orbiting planet. In the process of planet formation consequent upon sun spot encrustation of a star, this aether itself consolidates and then collapses in the latter stages of the process. These issues will be discussed in more detail below in Chap. 12, especially Sects. 12.9 and 12.10 when we examine their role in the daring system-binding strategies by which the *Principia* far surpasses *Le Monde*.

[15] Biro (2009) 106, Cf. Descartes, *Le Monde*: 'Because the air 5678 and water 1234 surrounding this earth are liquid bodies, it is evident that the same force that presses the earth in this way must also make them sink towards T, not only from the side 6, 2 but also from its opposite 8, 4, and in recompense cause them to rise in the places 5, 1 and 7, 3.' (AT X p.81; SG 52; MSM p.141)

[16] *Principles* Part IV art. 49; Miller and Miller p.206, emphasis added by Biro (2009) 107.

[17] Biro (2009) 107: Descartes writes '[I]t must be noted that the ocean does not in fact cover the whole Earth, as we assumed a little earlier; but because the Ocean extends around the Earth's entire periphery, as far as the general movement of the Ocean's water is concerned, it must be understood as if the Ocean did envelope the whole Earth.' (*Principles* Part IV art 55, MM.209):

[18] Biro, (2009) 106–7. Biro's Note 255 at this point reads: 'This interpretation is supported by the fact that Descartes chose to use a new term for oceans in the theory of the tides, as noted by Miller and Miller: '*The term used at this point (in respect to the tides) in the Latin is 'Oceanus' which*

Thus, the model in *Le Monde*, so reminiscent of the ideal Aristotelian picture of the element of water resting upon and completely enclosing the element earth, is relegated to heuristic status, and is replaced by a much more empirically plausible and geographically up to date model, which Descartes assures us, still has sufficient continuity of world oceans to accommodate the explanation of the tides.

This second model, however, was not introduced merely to put the simplistic conception of *Le Monde* into proper perspective. Descartes' new model of the Earth, with his narrative/explanation of its genesis from an extinguished star, was not an after thought or oddity, but rather one dimension of a complex and novel strategy of system-binding, wherein the *Principia* completely outgrew the vision of *Le Monde* and marked itself out as the mature statement of Descartes' systematic philosophy of nature. Recent investigations by Biro—of which so far we have only scratched the surface—as well as related research by Judit Brody, both independently and in collaboration with myself, have uncovered in the *Principia* a vast strategy of system-binding that dominates the structure and content of the latter portion of the work. This strategy required and was in part shaped by Descartes' recruitment and reframing of hosts of relatively recently discovered facts—about magnetism, sun spots, variable stars and *novae*. It turns out that unpacking his strategy and use of novel facts is both the key to understanding the most important systematizing improvements of the *Principia* over *Le Monde*, as well as the pathway to understanding the *Principia* as Descartes' mature statement in natural philosophy. Consequently, a separate, detailed consideration of these matters is required, not only to round off our assessment of *Le Monde*, but also to accomplish our other chief aim, as set out at the beginning of this volume, to wit, to delineate Descartes' mature position in natural philosophy, in contrast to the story we have traced of his youthful path to his first system.

References

Works of Descartes and Their Abbreviations

AT = *Oeuvres de Descartes* (revised edition, 12 vols.), edited by C. Adam and P. Tannery (Paris, 1964–76). References are by volume number (in roman) and page number (in Arabic).

SG = *The World and Other Writings*, edited and translated by Stephen Gaukroger (Cambridge,1998).

MM = *René Descartes, The Principles of Philosophy*, translated by V. R. Miller and R. P. Miller (Dordrecht, 1991)

refers to the collective total of the Earth's oceans. Previously, when referring to the ocean, the term 'mare' (uncapitalized) has been used. The French reflects this change as well.' In the English edition, they translate 'Oceanus' as 'Ocean' while 'mare' appears as 'ocean'. (Miller and Miller, editors' footnote 41, to Part IV, article 55, p. 209) Interestingly, when Descartes comes to explain the absence of tides in lakes and ponds he reasons that waters contained in lakes and ponds is not 'squeezed' and displaced by heavenly matter (like the water in oceans) as it is physically disconnected from the mass of water in the ocean and has only a small surface area. (*Principles*, Part IV article 55, MM 209).'

MSM=*Rene Descartes, Le Monde, ou Traité de la lumière*, translated by Michael S. Mahoney (New York, 1979).

CSM(K)=*The Philosophical Writings Of Descartes*, 3 vols., translated by John Cottingham, Robert Stoothoff, and Dugald Murdoch, and (for vol. 3) Anthony Kenny, (Cambridge, 1988) References are by volume number (in roman) and page number (in arabic).

HR=The Philosophical Works of Descartes, vol I translated by E.S. Haldane and G.R.T. Ross (Cambridge, 1968 [1st ed. 1911])

Other

Biro, Jacqueline. 2006. 'Heavens and earth in one frame: Cosmography and the form of the earth in the scientific revolution' Unpublished MA thesis, University of New South Wales, School of History and Philosophy of Science.

Biro, Jacqueline. 2009. *On earth as in heaven: Cosmography and the shape of the earth from Copernicus to Descartes*. Saarbrucken: VDM Verlag.

Dijksterhuis, F.J. 2004. Once Snell Breaks Down: From Geometrical to Physical Optics in the Seventeenth Century. *Annals of Science* 61:165–185.

Schuster, John. 2002. L'Aristotelismo e le sue Alternative. In *La Rivoluzione Scientifica*, ed. D. Garber, 337–357. Rome: Instituto della Enciclopedia Italiana.

Schuster, J.A., and Graeme Watchirs. 1990. Natural philosophy, experiment and discourse: Beyond the Kuhn/Bachelard problematic. In *Experimental inquiries: Historical, philosophical and social studies of experimentation in science*, ed. H.E. Le Grande, 1–47. Dordrecht: Kluwer.

Schuster, John, and Alan B.H. Taylor. 1996. Seized by the spirit of modern science. *Metascience* 9: 9–26.

Schuster, John, and Alan B.H. Taylor. 1997. Blind trust: The gentlemanly origins of experimental science. *Social Studies of Science* 27: 503–536.

Smith, Russell. 2008. Optical reflection and mechanical rebound: The shift from analogy to axiomatization in the seventeenth century, Part 1. *British Journal for the History of Science* 41: 1–18.

Smith, Russell. 2008a. 'Optical reflection and mechanical rebound: The shift from analogy to axiomatisation in the seventeenth century', Part 2. *British Journal for the History of Science* 41(2): 187–207.

Chapter 12
Cosmography, Realist Copernicanism and Systematising Strategy in the *Principia Philosophiae*

12.1 More Than Remediation: The *Principia* as a Triumph of Novel and Daring System-Binding

We said at the end of the previous chapter that we are now going to explore some differences between *Le Monde* and the *Principia* that go well beyond the small feats of remediation of systematicity that we noted there. What is in question here are not the well known differences between *Le Monde* and the *Principia philosophiae*; such as that the *Principles* is a textbook in the neo–Scholastic style; *Le Monde* an attempt at literary persuasion of *honnêtes hommes* in the vernacular; or, that the *Principia* offers a theory of the Earth, absent from *Le Monde*; or, that the *Principia* is much more elaborate in its presentation of laws of motion and numerous other natural philosophical topics; or even the most obvious difference, that the *Principia* sets out completely the metaphysical grounding of the natural philosophy, absent from *Le Monde*. Rather, what I have in mind are hitherto little noticed but daring systematizing strategies which more or less shape the content of the latter part of Book III and early part of Book IV of the *Principles*, where the most important parts of Descartes' radical realist Copernican vision of cosmology are presented, in ways completely unexpected and unprecedented on the basis of *Le Monde*. This is daring system building aimed at bidding for hegemony in natural philosophy.

We are going to see that these system binding strategies depend on changes in what Descartes has to say in the *Principia* about matter theory, the elements and cosmogony, but only on condition that we also appreciate the way these conceptual changes are entangled with his new sensitivity to, and interest in, ranges of novel fact concerning cosmic magnetism, sunspots, *novae* and variable stars.[1] In particular

[1] 'Novel' in this context does not necessarily mean newly adduced by the author in question. In the natural philosophical contest of the generation of Descartes, novel factual claims by others were routinely co-opted and reframed within one's own philosophy of nature. To be up to date in this style of work did not demand production of fresh claims about matters of fact. These rules of the game were to change considerably amongst the next generation of natural philosophers. (On this issue see also Sect. 2.5.2 above.) Descartes does not mention magnetism or sunspots in *Le Monde*. However, he alludes to *novae* ever so briefly. See above Sect. 10.8, note 96 and text thereto.

J. Schuster, *Descartes-Agonistes: Physico-mathematics, Method & Corpuscular-Mechanism 1618-33*, Studies in History and Philosophy of Science 27,
DOI 10.1007/978-94-007-4746-3_12, © Springer Science+Business Media Dordrecht 2013

I am going to suggest that far from being opposed intellectual practices, Descartes' moves in matter theory and his adoption, and re-framing, of wide swathes of novel and interesting matters of fact, were two sides of the same coin. And that coin I take to have been strategies for improving the systematic power, scope and consistency of the overall natural philosophy presented in the *Principia* compared to *Le Monde*. Moreover, the center of gravity of these strategies resides in his weaving ranges of novel matters of fact into explanatory and descriptive narratives with cosmic sweep and radical realist Copernican intent. We shall see that Descartes' dealings about such novelties as sunspots, variable stars and *novae* reveal how he strove to 'leverage' striking and novel matters of fact for systemic benefits, by which I mean the following: First, putatively reliable and agreed reports of such striking facts were taken up as *explananda*, things to be explained in the system. Then, secondly, such initially explained facts, now integrated into the explanatory machinery of the natural philosophy, were themselves leveraged into *explanans*, used to explain further, more complex or arcane phenomena. Indeed, I shall contend that the system itself may be viewed as a network of such moves.

What I have to say here has been shaped by my own work and that of two close colleagues, Jacqueline Biro and Judit Brody. I have worked collaboratively with each of these individuals, on separate occasions and in different time periods, whilst each has also accomplished significant independent work on matters relating to the contents of this chapter. In Biro's case, she published a monograph after we worked together as research supervisor and graduate student; in Brody's case, we collaborated on a paper which overlaps very significantly with this chapter, after she had published a monograph on key topics involved here.[2] The sources, collaborative and individual, for materials in this chapter will be carefully referenced as we go along. I am proud and delighted to mark the outcomes of this dense and productive sequence of encounters; but, of course, I take full responsibility for any mistakes or errors that may have crept into this, my present synthesis of what our various separate and conjoint efforts have revealed.

12.2 Cosmogony, Cosmology and Cosmography: Key Categories and Insights

As I explore Descartes' co-optation of facts regarding sunspots, *novae* and variable stars and his strategic exploitation of them in the system of the *Principles*, I shall be putting to work several explanatory insights which in turn depend upon understanding

[2] Biro (2009, 2006). Brody's monograph is *The Enigma of Sunspots: A Story of Discovery and Scientific Revolution* (2002). Brody then began a study of the *Principles,* interweaving its matter theoretical differences from *Le Monde* with her prior research on sun spots and variable stars. Consultations about this project led to our decision in mid 2009 to collaborate on what became by August 2011 a 14,000 word paper, Schuster and Brody, 'Descartes and Sunspots: Matters of Fact and Systematizing Strategies in the *Principia Philosophiae*', accepted for publication in early 2013 by *Annals of Science*, and published on line March 2012: DOI: 10.1080/00033790.2012.669703

three pursuits woven into Descartes' natural philosophizing: cosmogony, cosmology and cosmography. We are already acquainted in this volume with the first two categories, and I shall merely make explicit a few additionally needed points about them here. The latter category, cosmography, is about to become very important, and will be introduced in detail in this Section.

First of all I must clarify the distinction between the first two categories, which are often conflated in reading *Le Monde* or the *Principles*.[3] *Cosmogony* I take to consist solely in the short fabular narratives offered (in two different ways) in *Le Monde* and the *Principles*, dealing with how one gets from God's creation of matter to the point where the final, and continuing state of the cosmos has emerged, in regard to the number and type of elements, and the general fabric of innumerable, star centered vortices.[4] That final and continuing state of the cosmos—in which one can additionally count the nature and orbital behaviour of planets, comets and planetary satellites—we shall label Descartes' *cosmology*. This accords with the way the term was generally introduced in Chap. 2 as denoting that dimension of a natural philosophy dealing with matter, cause and structure in the universe.[5] Descartes' cosmogonies are short. They do not contain details about the final (quite elaborate) vortex mechanics. Moreover, although the cosmogonies are closely linked to claims about matter theory—the emergence of the final and continuing formats (types of element) in which all matter will be found—they omit some very important constituents of the Cartesian cosmos. For example, in the *Principles* the particles of the third element (terrestrial matter) are neither present in the cosmogony, nor produced by the cosmogonical process. They come into being from (some types of) first matter—and may also be transformed back into it—only during the business as usual cosmological patterns of activity on the surfaces of stars. Similarly, in the *Principles*, Descartes' theory of magnetism, in what I shall term its 'cosmic', rather than merely terrestrial applications, is crucial to how the final and continuing universe of vortices functions, but little of this elaborate model is even hinted at in the cosmogony. We shall also learn that Descartes' history of the Earth in the *Principles*, which actually stands in for the developmental history of any and all planets in his cosmos, belongs to his cosmology, and is not continuous with, or part of, the cosmogonical story in the *Principles*.

Secondly, we need to refine further our understanding of Cartesian cosmology. In the *Principles*, as in *Le Monde*, cosmology denotes the final, subsisting state of the cosmos. But, compared to the static picture of the cosmos in *Le Monde*, the *Principles* teach what we shall term a *'dynamic steady state'* cosmology. Although

---

[3] Note also that when we discussed cosmogony in *Le Monde* in Chap. 9, we were mainly concerned with understanding Descartes' fabular narrative in general, as part of unfolding the text in order. We were not concerned with cosmogony in detail, let alone with comparing how Descartes' cosmogony compares in the two treatises.

[4] As we shall see in Sect. 12.3, this statement is not quite correct in the case of the *Principles*, where the third element does not appear during the cosmogony, but only during the actual cosmological steady state.

[5] Cf. Schuster (2002) 337–338; (2009) 57–59, 64–65.

in the cosmology of the *Principles*, as well as that of *Le Monde*, the vortices and all the elements are present and accounted for, with planets and comets accomplishing their respective, appointed orbital duties, we are told in the *Principles* that some kinds of large cosmic changes routinely and rather randomly occur: Sunspots come and go—and they are the one and only place where the third (terrestrial) element is produced in the cosmos. Any star might become variable and even—completely encrusted with sunspots—die, leading to collapse of its vortex, the dead star becoming, depending upon circumstances, a planet or comet. Furthermore, such a planet, like the Earth, would then develop its final terraqueous structure, with its seas, continents, mountains, valleys, tidal phenomena, etc. as a result of a further sequence of natural events befalling the dead–star–turned–planet.[6] The cosmos of the *Principles* is dynamic. But, since there is no overall, macro level, directional or historical process involved in these kinds of changes, it is also steady state.[7]

This brings us to our third key category, '*cosmography*', and the interpretative insights about it used in this chapter. Following the recent work of Jacqueline Biro, I take cosmography to mean that part of a natural philosophy addressed to the relations between its matter and cause account of the heavens (its cosmology) and its theory of the Earth. This was an actor's category at the time and had emerged initially in the context of geo-centric natural philosophies, most notably Aristotelianism, in which the point of the 'relation' was certainly not identity or even similarity of matter and cause explanation.[8] However, for Descartes and other realist Copernicans, for whom the Earth was a heavenly body and the

[6] In Sect. 12.10 we shall see that the formation of planetary (Earth-like) structures is a necessary result of natural processes, given the contingent death of a star and its migration into/capture by a neighboring vortex. That the planet forming process is necessary has tended to lead commentators to conflate Descartes' Earth theory with his cosmogony. But his history of the Earth (or any planet) is not cosmogonical, rather a necessary process triggered by random events inside his dynamic, steady state cosmos. Indeed it may be said that Descartes' dynamic steady state cosmology resides entirely outside the purview, or implications, of his little cosmogonical story.

[7] In an unusually prescient comment McRae (1991, 159) noted that in Descartes' natural philosophy, 'If it is the relation of the fixed stars to one another which constitutes the form of the world, then...the universe does, according to Descartes, have a history of change from one world to another world as a result of the growth of sunspots and the death of stars'. This remark foreshadows the entire thrust of the argument in this chapter, although, as indicated in note 6, we do not quite attribute 'world-making and world-breaking' significance to the behavior of variable stars or births of planets as treated by Descartes in the *Principles*.

[8] Biro (2009) 8–9. Cosmography is defined by Biro, extrapolating from definitions by John Dee, Thomas Blundeville, Nathanial Carpenter and William Barlow, as 'that part of natural philosophy that provided within one explanatory framework the relationship between the heavens and earth', or as John Dee said, 'matcheth Heaven and the Earth in one frame'. Such early modern definitions usually say that cosmography requires the use of astronomy, geography and other disciplines. This demands some clarification. First of all, references to astronomy in this connection clearly are mistaken, if we are considering astronomy to be the mixed mathematical discipline devoted to construction of geometrical models of planetary motions. Cosmography was a domain within the field of natural philosophy, hence it is not astronomy that is being related to theorizing about the

traditional heavenly bodies were now arguably 'like' the Earth and closely 'related' to it, cosmography was a space of natural philosophical challenge and opportunity. The terms of argument shifted from the relation of 'the Earth' to everything else, that is, 'the heavens', to being about the relations, generally, of any and all planets, their structures and geneses, to any and all stars, their nature and developmental patterns. Most importantly, as Biro has shown, claims about the structure of the Earth could now be exploited for cosmographical ends, specifically realist Copernican ends: Arguably true claims about the structure and nature of the Earth were now endowed with the property of being *ipso facto* claims about a heavenly body, arguably therefore closely related to other heavenly bodies and processes.[9] We shall see that whereas *Le Monde* to some extent reflects this shift and form of strategy, large swathes of the *Principles* amount to one vast,

---

Earth but rather that dimension of natural philosophy dealing with structure, matter and cause in the cosmos, to wit, cosmology as we have termed it above. As to the other term in the relation, loosely called geography above, one has to recognize that geography had many acceptations in the period, mirrored today by historians of the field (Biro, *ibid.*, 12, note 19 thereto, discussing the views of Lesley Cormack and David Livingstone). The portion of geography considered to be part of cosmography might be taken to be mathematical geography. But there are difficulties here, as part of what was meant by mathematical geography was just that, a mixed or practical mathematical field with at best highly debatable relevances for natural philosophy and cosmology. In addition, the other parts of mathematical geography—such as the study of terrestrial gravity and magnetism, the study of exact locations, and deep articulations to cartography—constituted a diffuse and only partially natural philosophically relevant suite of concerns. Given all this, Biro adopted a contemporary term 'geognosy' in order to construct an historian's category of 'geognosic opinion' to serve as the 'Earthly' partner to cosmology in the cosmography pairing. Geognosic opinion would then be 'ideas and knowledge about the Earth's structure'; that is, geognosic knowledge claims concerned issues of *structure, matter and cause in regard to the Earth.* (Biro, *ibid.*, 16 and note 27 thereto) Within natural philosophical discourse this is to be paired, cosmographically, with cosmology as claims about *structure, matter and cause in the cosmos.* (In this chapter I simply denote the 'Earth' part of the heavens/Earth pairing as 'theory of the structure and nature of the earth'. Hence, for our purposes here, cosmography is that dimension of natural philosophizing in which cosmological and Earth theory claims were placed in relation to each other.)

[9] In other words, What is the nature of the Earth as a planet? What can be gathered about the Earth, for example, about its structure, its magnetism (Gilbert), its tides (Galileo and Descartes), the nature of local fall, that would support its construal as a planet amongst planets and allow for the motions realist Copernicanism required of it? For realist Copernicans the relation of 'the Earth' to everything else, that is, 'the heavens', changed, becoming the relation of any and all planets, their structures and geneses, to any and all stars, their nature and developmental patterns. Biro (note 8) has shown that claims about the structure of the Earth could now be exploited cosmographically, for realist Copernican ends: Early to mid sixteenth century technical developments in geography, consequent upon the re-discovery of Ptolemy's *Geography* and leavened by the findings of the voyages of discovery, were at first only grudgingly granted by the Scholastic Aristotelians, but were eagerly seized as a resource by natural philosophers advocating Copernican cosmology, with Galileo and Descartes offering late examples of such cosmographically focused tactics in a sequence of varied yet uniformly anti-Aristotelian natural philosophical gambits stretching from Copernicus himself, through Bruno, Gilbert and others. We further articulate Biro's initiative in our discussion below in Sect. 12.11 of the nature of Descartes' 'grand cosmographical gambit' in the *Principles*.

interrelated set of such radical, realist–Copernican cosmographical arguments.[10] Indeed, the *Principles* of Descartes offers a dynamic, steady state cosmography, from the genesis of third matter as sunspots on the surfaces of stars to the explanation of planets as collapsed and modified debris of dead stars, still internally structured (as were the parent stars) to accept incoming, oppositely axially directed left– and right–handed magnetic screw particles of first element. Descartes' games with sunspots sit squarely in the middle of this radical Copernican realist cosmographical nexus. This is what I mean by saying that the center of gravity of the system of the *Principia* will be revealed to reside in a place few have previously sought to locate it, in a network of systematically co-opted matters of fact about magnetism, sunspots and variable stars, reframed in Cartesian mechanistic and cosmographical terms, so that they can leverage further explanations in his realist Copernican cosmographic vision.

Given these clarifications of categories, the argument of this chapter will proceed as follows: First, the following section will compare in some detail the matter theories and cosmogonies in *Le Monde* and the *Principia.* Then, after a brief look in Sect. 12.4 at some points about Descartes' views about inter–vortical relations in the two treatises, we shall turn in Sect. 12.5 to his co-optation of the cosmographical tactics William Gilbert had deployed in his radical and influential 'magnetic' natural

[10] An example of the presence of a definite cosmographical orientation in *Le Monde* occurs when Descartes offers his first account of the elements, in Chap. 5 (AT XI 24–6; MSM 37–39; SG 17–18), a text we discussed in detail in Sect. 9.3, at note 42. In this passage, Descartes identifies his three elements with Aristotelian traditional ones: first element with fire; second element with air and third element with earth. It is a commentators' commonplace that Descartes was attempting here to preserve some continuity with (at least part of) traditional element theory. In *Le Monde*, as some suggest, he may have viewed his 'naming' his elements as yet another rhetorical ploy to keep the intended francophone *honnête homme* reader on side. But, his gambit would have arguably been quite unconvincing to just about any natural philosophically literate reader. Moreover, if that was part of Descartes' aim, it certainly seems he did not stick with it, dropping the pretense in the *Principles*. Not previously noticed, however, is a deeper motive, one grounded in systematizing tactics: This naming of the elements seems to have *cosmographical* significance in the sense we have given to the term. In this new system, neither air nor fire are elements found on and about a unique Earth. In the light of his radical Copernican realism, envisioning effectively an infinite number of star and planetary vortical systems, Descartes was saying to the aware reader that 'air' had been misconstrued by Aristotelians as the essential constituent of the local terrestrial atmosphere only. No, 'air' is ubiquitous in the cosmos, constituted of the spherical *boules* of second element that make up each and every stellar vortex. What natural philosophers have termed air is just a mixture of various kinds of earthy particles of third element, with the usual unavoidable interstitial 'filler' material of fugitive second and first element particles. Similarly 'fire' is not the Aristotelian element at home in some peculiar sense just below the Earth's moon. Again, no, for fire is the first element, the very stuff of every star, including our sun. Renaming the elements was less an unconvincing bow to traditional teaching than it was—as we have foreshadowed—a hint and sign of a new cosmography; that is, a new relation between all planets, in any vortex whatsoever, including our Earth, and all the stars and stellar vortices of the universe. If we are correct about this, we have here a nice example of Descartes' well known proclivities toward both elusiveness and allusiveness, in his simultaneous (and contradictory) appeal to the old element names and new cosmographical tactics. In any case, as this chapter argues, the *Principles* will display a much greater attention to cosmographical strategies and content.

philosophy. Sections 12.6 and 12.7 will deal respectively with claims about sunspots before Descartes and with his selection and theoretical reframing of those claims in the *Principia*. Then, after looking in Sect. 12.8 at the development of factual claims about variable stars in the period between Descartes' writing of *Le Monde* and the publication of the *Principia*, we shall examine in Sect. 12.9 how he leveraged his explanation of sunspots to account for both variable stars and *novae*. Section 12.10 will complete our tour of Descartes' cosmographical strategy in the *Principia* by looking at his account of planet formation anywhere in the cosmos, material usually treated merely as a 'theory of the Earth'. This will allow us in Sect. 12.11 to bring together the threads of our argument into a discussion of Descartes' 'grand cosmographical gambit' in the *Principia*, as the culmination of a tradition of cosmographically sensitive, anti-Aristotelian and realist Copernican natural philosophers. Finally, Sect. 12.12 will discuss how our findings about the *Principia* mark the end point of his natural philosophical trajectory.

12.3 Matter and Element Theory in *Le Monde* and the *Principia philosophiae*

The matter theory in *Le Monde* and the *Principia*, and the cosmogonical accounts related to them, are often seen as interchangeable. Such readings are defensible at a general level. After all, when most of the surrounding detail is stripped away, in both works we have in effect a divinely created infinite block of Cartesian matter-extension, precluding the existence anywhere and any time of even the smallest void space. Cartesian matter is the same incompressible, indestructible, homogenous substance in each and every particle, fragment, or corpuscle that might eventuate from the divine injection of motion into the block of matter-extension. Any and all differences that might exist amongst such pieces of matter arise solely from their size, shape, state of motion or rest. The three elements, once formed, are really three persistent formats, stipulating certain ranges of size, shape and distributions of degrees of motion, into which each and every corpuscle fits. No micro particle is not a member of one of those three classes or elements. In both works, sooner or later after a cosmogonical story, we have permanent differentiation amongst the three element formats: at any given moment in time thereafter matter appears only in one or another of the three guises.

Nevertheless, the differences between the two theories of matter are greater than usually acknowledged. This began to be recognized amongst historians of science with the ground breaking papers of Rosaleen Love and John Lynes, written a generation ago.[11] Most notably, as we have seen (Sect. 9.5.3), in *Le Monde* there is

[11] Love (1975) 127–37; Lynes (1982). At that time Lynes remarked (p.55) that explaining the development of Descartes' matter theory between *Le Monde* and the *Principia* had been a 'somewhat neglected task'. Love did not directly compare the matter theories of *Le Monde* and the *Principles*, but rather juxtaposed Descartes' implied matter theory in his *Essais* of 1637 to that of the *Principles*, as it were imputing the former to *Le Monde*, often in an erroneous sense it must be

no transmutation of elements, after their cosmogonical formation. Indeed, the third matter pre-exists the first and second produced by that cosmogony. In the *Principles*, as we shall see, again only the first and second elements emerge from the initial cosmogonical process, but in this case the third element is nowhere to be seen until the steady state cosmos has emerged, because in the *Principles* the third element only arises, under special and portentous circumstances, from the first element (and can be transformed back into it). We need, therefore, to look at matter theory and cosmogony in the *Principia* in a bit more detail and as compared to *Le Monde*. But we do this only as preparation for passing beyond such mere matter theoretical comparisons in search of bigger interpretive game. We shall find that in the *Principia* the third element, produced from certain types of first element, plays crucial roles in the dynamic steady state cosmographical processes which are central to the system-binding strategies of the *Principia*. Indeed, I shall argue that these processes constitute the heart of the *Principia* as a system of nature, and that their conceptualization depended upon Descartes' lively and concerted attention to, and co-optation of, significant ranges of matters of fact circulating in the natural philosophical culture.[12]

Recall that in Sect. 9.5.3, we concluded the following about Descartes' (second) cosmographical description of vortex and element formation: The first and second element evolve out of the original 'ur-particles' established when the block of matter-extension was shattered by the injection of motion, while it turns out that particles of third matter had existed ever since that first creation of particles. These large and irregular pieces of third matter, present from in the beginning, retain the form of the third element and go to make up the bulk of planets (including the

---

said· The particular problems raised by Love's manner of interpreting *Le Monde* are not the topic of the current chapter, but further comment on Love, and Lynes, appears below at Note 25. By 'matter theory' I mean Descartes' theories of the elements, or genres of micro-particles into which his matter-extension is taken to be divided in *Le Monde* and later in the *Principia Philosophiae*. Strictly, and most abstractly speaking, Descartes' theory of matter consists in his doctrine of matter-extension. However, that concept, taken in isolation, plays almost no role in the descriptions and explanations he offers in the working machinery of his natural philosophy, and it is these, rather than abstract doctrines on the metaphysical level with which we are concerned. Accordingly, throughout this chapter as we discuss Descartes' accounts of cosmology, cosmogony, magnetism, sun spots, variable stars, *novae* and the generation of planets, we indifferently label our object of study the 'matter theory' or 'element theory' of Descartes—or sometimes his 'matter and element theory'. It is worth recalling, in this regard, the sage words of T.S. Kuhn, discussing the inner workings of Cartesian natural philosophy in his *Copernican Revolution*: '...Descartes introduced a concept which since the seventeenth century has greatly obscured the corpuscular basis of his science and cosmology. He made the universe full. But the matter that filled Cartesian space was everywhere particulate in structure.' Kuhn (1959) 240.

[12] It has not always been the case that the matter theoretical contrasts between *Le Monde* and the *Principia* have been glossed over. Gabriel Daniel (1649–1728) for instance, who was a strong critic of Descartes, was not sure which of the two versions to accept: 'whether the third element be contemporary with the other two, as M. Descartes seems in some measure to suppose in his *Treatise of Light*: or, whether it be form'd by the Conjunction of several Parts of the first element hook'd to one another, as he seems to teach in the *Book of Principles*'. Daniel (1692) 261.

Earth), planetary satellites and comets. Nowhere in *Le Monde* does the third element change into either of the other forms. That is, although *Le Monde* takes a radical stance in cosmography—the Earth is just another planet in a realist Copernican universe of innumerably many star centered planetary systems—Descartes' Copernican unification of 'heavens and Earth' does not on this point go so far as matter and element theory. Once the cosmos is constituted, and stars and vortices have formed, Earthy, that is planetary, matter can never change into the matter of the 'heavens' that is vortices or stars. Nowhere in *Le Monde* does Descartes state the matter theoretical unity of heaven and Earth; that is stars and vortices and planets (plus comets and moons). The sun (and the other stars) differ from the Earth (and all other planets and comets). Descartes attributes to stars a nature 'totally contrary to that of the Earth because the action of their light is enough for me to recognize that their bodies are of a very subtle and very agitated matter.'[13] Here, again, we have an indication of the way the element theory in *Le Monde* is largely driven by the theory of light, as noted in Sect. 9.3. Hence the needs of Descartes' theory of light tend to run against the most radical implications of embracing an infinite universe realist Copernicanism, where such a strong 'bar' between 'planetary' and 'heavenly' types of matter would seem otiose and counterproductive. All this changes in the *Principia*, as we shall now discover.

In the *Principles* we also (eventually) find the same three elements; but their relations are quite different and their cosmogonical genealogies altered. Descartes steps away from the conceit of the simple cosmogonical cracking of the infinite block of matter-extension by God's injection of motion, thus producing a variety of micro particles, with the vortices evolving out of the chaotic state manifested at that initial corpuscle producing instant. In the *Principles* the ur-particles are now claimed to be equal in size and motion: being 'average' in these respects compared to the (first matter) particles that will later constitute stars, and the (third matter) particles that will later constitute the bulk of planets, comets and satellites. Descartes proclaims a type of principle of cosmic harmony or order, contrasting with the inchoate

[13] AT XI, pp.29–30; SG p. 20; MSM 45–7. It is, however, true that if by matter theory in Descartes, we were to mean solely the theory of matter-extension, then, of course, a unity of heavens and Earth was achieved from the start, and in principle Descartes *could* have gone on to assert in *Le Monde* the transmutability of the elements into which this matter-extension happened initially to be sorted. In fact, however, natural philosophizing was about producing detailed explanations of ranges of new and old facts, and 'systematisation' of the resulting suite of explanations. To 'do' natural philosophy, Descartes could not simply devote himself *ad infinitum* to 'analysis' of the doctrine of matter-extension and its possible implications. (Cf. note 11.) We see this already in the simple fact that the purpose of the cosmogonical story is to produce the elements and the types of structures—stars, vortices, planets—they constitute. In Cartesian natural philosophy, matter-extension as such lasts an instant (the instant of creation). While it exists in its pure state, no 'nature' or cosmos yet exists, so there is not yet any subject matter for natural philosophy. Similarly, although Descartes 'could' have had transmuting elements in *Le Monde*, based on his matter-extension doctrine, in articulating his natural philosophy in *Le Monde*, he specifically denied that possibility. Therefore, historians need to look to Descartes' aims and tactics in natural philosophizing for reasons for his insistence in 1633 on what became unnecessary to assert in 1644.

initial moments of the cosmos of *Le Monde*.[14] Additionally, we are informed that, 'All were moving with equal force in two different ways: each one separately around its own center but also several together around certain other centers'—a statement that strongly entails that the number and placement of (at least the initial set of vortices) is also inscribed in the cosmos at its moment of creation.

Leaving aside the new emphasis on pre-established harmony and pre-inscription of the vortex economy, the real puzzle here, not addressed by Descartes, but obvious to any contemporary or modern reader who understands his conception of completely full matter-extension is this: The original particles cannot have been all equal and all spinning around their own centers. John Heilbron has perspicaciously interpreted Descartes as speaking about equal perfectly cubic particles, completely space filling on that account, which begin to spin, each around its own centre, this immediately producing [1] spherical boules of second element, and [2] space filling debris of first element.[15] This is a nice and typically brilliant Heilbronian conceit. It convincingly decodes part of Descartes' text while obviously setting aside other parts of it. But it certainly has the benefit of capturing what turns out to be Descartes' clear intent in these *Principles* passages. The cosmogonical story issues only in second and first element. Third matter will come into being only later, for reasons we shall soon encounter, and only by virtue of the transformation of first matter. That is, in the *Principles*, regardless of the curious and tortured details of the opening of its cosmogony, it is clear that the original (supposedly equal) particles lose their initial shape[s] by constantly rubbing against each other, just as in *Le Monde*. Eventually they become spherical and are the building blocks of the second element. The debris, much smaller and therefore more agile, which fills the space between the globules (*boules*) of the second element is the first element. No third element particles were present at the creation, and none have been produced in the cosmogony described. How do they come into being?

When, in a given vortex, there are more first element particles created between the second element *boules* than necessary to fill in the space, then due to the revolution of the vortex the second element tends to recede toward the periphery and the first element flows into the centre thereby vacated, forming a star. From the manner first element particles are generated it follows that some move faster and some slower, some are larger and some are minute. Descartes tells us that the smaller and more agitated ones form the bodies of the stars.[16] (In essence, this is what happens

---

[14] 'Confusion seems less in accordance with the supreme perfection of God the creator of things than proportion or order' so he was 'supposing at this point that all the particles of matter were, initially equal in respect both of their size and their motion'. This point and the other textual references in this paragraph are located at: Principles III articles 46–47; AT VIII-1 102–3; CSM I 257; MM 106–107.

[15] Heilbron (1979) 31–33.

[16] Two versions of star formation are offered in the *Principles*, III, articles 54 and 72; AT VIII-1 107–8, 125; MM 111, 122–3. The former version corresponds to our text above; the latter gives an explanation more dependent on diametrically opposite axial inflows of first element from the equatorial areas of neighboring vortices toward the center of the vortex the creation of whose central star is being discussed. Alternatively, the second story might be interpreted as Descartes' detailed account of the movement of first element particles into and out of an already formed star. This latter account does map completely onto his explanation of the formation of oppositely handed, rimmed particles of first element which cause magnetic phenomena, given later in Book III Articles 87 through 93.

in *Le Monde* as well.) But in the *Principles* Descartes' focus shifts to the exact shape and nature of some of the remaining particles of first element, and to implications about their total range of variation. Considering that the spaces between the heavenly globules are roughly triangular, the particles of the first element remaining amongst them often have a triangular cross-section although they remain flexible enough to assume any shape. By constantly being forced in and out of the interstices of the second element, some of these particles become larger, more stable and acquire from the triangular interstices of the *boules* a more permanent channeled, grooved or rimmed surface with a distinctive right or left–handedness.[17] These particles are going to be used to explain magnetism, as we shall see later. For the moment I bracket those details, and, in the interest of our matter-theoretical inquiry, simply follow the cosmic pathways of some of these channeled, rimmed and handed particles of first element.

Firstly, there is a constant exchange of first element matter between neighboring vortices (Fig. 12.1). According to some implied principles of inter-vortical stability and spatial relations, the vortices arrange themselves in such a manner that they do not hinder each other's motion and so their poles touch as near as possible to the equators of the others.[18]

Due to the centrifugal tendency to motion generated by vortical rotation, some first element matter constantly leaves the equatorial part of one vortex and moves along the axis of the neighboring one. Some of these inter–vortex travelling particles of first element are those larger, interstitial ones just discussed, some of which can be channeled, rimmed and handed. In general, these larger first element particles move more slowly and adhere to each other more readily than the smaller ones. These are the ones most commonly found moving in straight lines from the poles towards the centers of the vortices because motion in straight lines requires less agitation. Thus, having entered the new vortex in diametrically opposite directions along the north and south directions of the axis of rotation of the vortex and its central star, the production of left and right handed channeled particles is completed or 'finished'.[19] These particles then penetrate into the polar regions of the central star where their progress is impeded by the first matter already in the star, (and the flow of oppositely handed particles coming from the

[17] The process of production of this sub-species of first element particles is related at *Principles* III articles 87–93; AT VIII-1 142–7; MM 132–6.

[18] Cf. Gaukroger (2002) 150. *Principles*, III, articles 65–67; AT VIII-1,116-119; MM 118–119.

[19] We put the matter this way because there is some ambiguity in Descartes' text on the issue of where and how the right and left handed rimmed particles are formed. There is no doubt he intended that the larger particles of first element, being pressed through the interstices of the spherical *boules*, can become rimmed and handed; but, on the other hand it is also clear that it is their passage along the axis of vortical rotation into the polar regions of a central star that gives the oppositely directed particles their opposite twists. We defer to the excellent hermeneutics of Gaukroger on this point, noting his reading at two places in his analysis of the *Principles*: [1] At Gaukroger (2002) 152 the production of the rimming is elided with the twisting into handedness during the axial transit. 'The larger parts of the first element have to pass around the tightly packed globules of the second element, and they become twisted into grooved threads, those coming from opposite poles being twisted in opposite directions, that is, having left- and right-handed screws (article. 91)'. [2] But, at pp.175–6 discussing Descartes' treatment of terrestrial magnetism in Book IV of the *Principles*, Gaukroger seems to interpret the twisting into handedness to be a generic result of

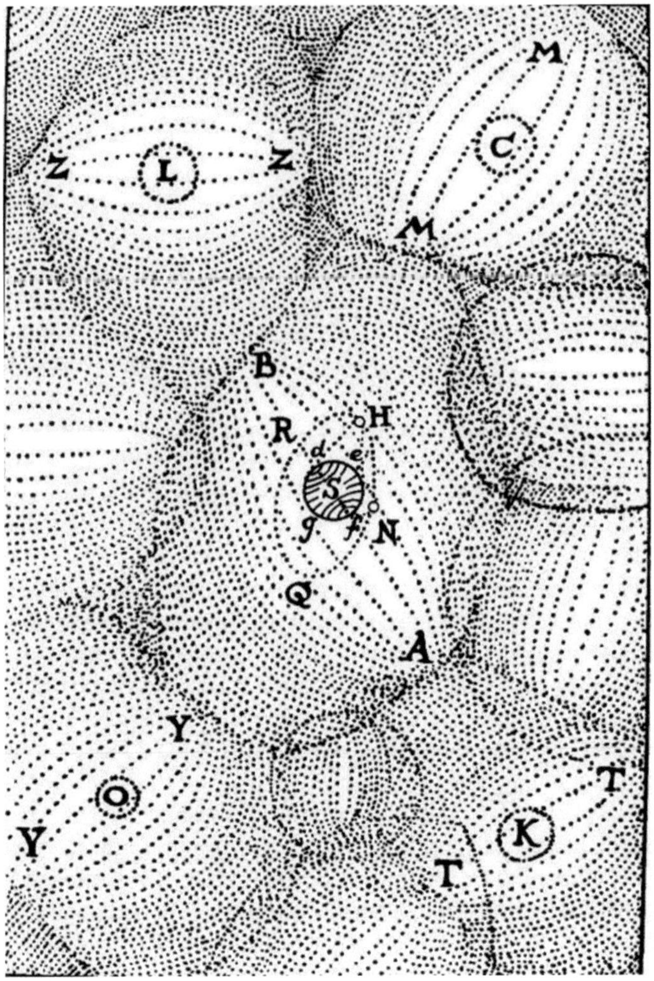

Fig. 12.1 *Principles*, AT VIII-1, p.141. Contiguous vortices tend to orient with axes of rotation as close to orthogonal to one another as possible

opposite axial direction). Since any backward flow is prevented by the particles continually flowing into the star behind them, these particles of first matter, including many of the newly finished, larger, left and right handed screw shaped particles, move

forcing through interstices of *boules*, and not necessarily (though perhaps sufficiently) a result of the cosmic transit along vortical axes of rotation: 'The generation of these grooved particles had been set out in Part III (articles. 87–93). Their grooves derive from the fact that they are squeezed through the interstices of contiguous spherical globules. As a result of this squeezing they end up as cylinders having three or four concave sides joined by rims….Moreover, because they rotate on being squeezed through these interstices, the channels or grooves are rotated, forming a stream of diagonally grooved, cylindrical fragments, some of which have a left-hand screw, some a right-hand screw, according to the direction of the twist'.

sideways and radially toward the star's surface, mainly in the polar regions, where they constantly 'bubble' out onto the stellar surface, there to begin a slow drift toward the star's equator.[20] This process occurs in all central stars, including of course our sun.

Descartes tells us that the particles of first element bubbling out onto a star's surface are sluggish since they have had no time to become purified and clarified by the heat, that is the high agitation and imparting of motion by the smaller, highly agitated particles of first element making up the body of the star. The added first element material floats like scum on a boiling liquid and sometimes forms,

> very large masses, which, being immediately contiguous to the surface of the heaven, are joined to the star from which they emerged. They resist that action in which ... the force of light consists; and are thus similar to those spots which are usually observed on the surface of the sun.[21]

In the next paragraph Descartes refers to these accretions not as being similar to sunspots but simply as sunspots.[22] This then is the origin of the third element. It comes into being by particles of the first element sticking together, and is manifested as the opaque, light–blocking material of sunspots—third matter in other words—which definitely lie on the surface of the sun. Sometimes such a body of third matter forms on the stellar surface only to be metaphorically 'boiled' away again by the roiling smaller first element particles which surround it. Hence, according to the *Principles*, third element originates from conglomerations of certain types of particles of first element on the surfaces of stars as sunspots, and it can also be changed back into first element again. Moreover, as we shall see in Descartes' further explanation of sunspots, variable stars and planet formation, stars can and actually do turn into planets, comets and satellites.[23] For the moment, at the level of matter theory, note that this certainly is not the case according to *Le Monde*, where he wrote: 'each part of matter tends always to one of their forms and, once it has been so reduced tends never to leave that form'. [24]

[20] *Principles* III articles 94–95; AT VIII-1 147–8; MM 136.

[21] *Principles* III article 94; AT VIII-1 147–8; MM 136. Gaukroger (2002) 153 comments: 'These grooved particles...move to the centre of the vortex. On account of their relatively small degree of agitation and their irregular surfaces, they easily lock together to form large masses at the surface of the star from which they emerge. Because of their size and small degree of agitation, they "resist that action in which we said earlier that the force of light consists" and as a result they appear as a spot on the surface of the Sun. Descartes compares the process by which they are formed to the boiling of water which contains some substance which resists motion more than the water: it rises to the surface on boiling to form a scum, which, by a process of agglutination, comes to acquire the character of the third element'.

[22] *Principles* III article 96; AT VIII-1 148. MM 136.

[23] *Principles* II article 23; AT VIII-1 52; CSM I 232. Descartes states explicitly 'celestial matter is no different from terrestrial matter'.

[24] AT XI 28; SG 19; MSM 43–5. But by January 1639 he must have begun to change his theory of matter, because in a letter to Mersenne Descartes says: 'some terrestrial particles continually take on the form of subtle matter when you crush them up; and some particles of this subtle matter attach themselves to terrestrial bodies, so there is no matter in the universe which could not take on all the forms'. (AT II 485; CSMK 133)

The foregoing comparison between *Le Monde* and the *Principles* operated mainly at the level of matter and element theory, although in order to explicate the novelties emergent in the *Principles* we perforce have had to touch lightly upon Descartes' theories of magnetism and sunspots. If we were to remain at this level of analysis, satisfied mainly with comparison of the respective matter theories treated in isolation from their systematic relations to other dimensions of the natural philosophy, we would miss exactly what we are seeking in this chapter. On the one hand, we would ignore Descartes' very interesting co-optation in the *Principles* of wide swathes of available matters of fact, and, on the other hand, his much more elaborate strategies of systematization in the *Principles* than in *Le Monde*. And, most importantly, we would not ask the key question, 'What is the strategic relation between Descartes' newly revealed thirst for hard, consensually agreed matters of fact and his breathtaking construction of improved systematicity in the *Principles?*'[25] However, we are not yet quite ready to open this wider inquiry, because there is one

[25] As we have noted, leading interpreters, such as Lynes (1982) and Love (1975), approached the problem of the differences between *Le Monde* and the *Principles* as centrally concerning matter and element theory. Additionally they looked for external triggers or motives for Descartes making the changes. For example Lynes p.72 placed emphasis on religious motivations, with Descartes striving to overcome the possibly heretical implications of his early supposedly atomistic-looking matter theory in *Le Monde* by means of his putatively better ability later to demonstrate the absence of any void in nature in the *Principles*. (In fact Descartes has a robust plenist account in both treatises.) Similarly Love's explanation for the changes in matter theory boils down to Descartes' increasing commitment to a plenist physics in the *Principles*: She maintained that Descartes must have revised his theory of matter between 1637 and 1644, basing her claim on the fact that in the *Discourse*, published in 1637, there is only one subtle element, while in the *Principles* there are two. Love suggested that the change from one subtle element to two could have been triggered by Morin's criticism of Descartes' theory of light, in particular the need of some matter to fill in the void between globules that transmit light. This for Love meant in all probability that the unpublished 1633 version of *Le Monde* only had one subtle element and thus is not identical to the one eventually published in 1664. Hence, Love p. 127, claimed that the differences between the two works 'follow from Descartes' well-known identification of substance with spatial extension, and his consequent rejection of the void'. We leave aside here the overwhelming evidence that a close analysis of the text of *Le Monde* and its course of construction undermine all this, since it is virtually certain that Descartes had the three elements in the original conception, and simply note that Love's explanation is based on a metaphysical driver, Lynes' on a theological one. In response to these and other guesses at circumstantial external drivers of Descartes' strategies and inscriptions, we suggest that the casting about for such putative causes is beside the point and actually rather ahistorical. When an actor is playing a competitive game in a field of contestation, the best initial explanation for the actor's moves resides in the best picture the historian can devise of the actor's assessment of the state of play, his resources and goals. (Cf. the seminal works on the socio-political dynamics of claim construction and negotiation in mature sciences by Pierre Bourdieu (1975); and Steven Shapin (1982), especially his discussion of actors' vested interests in their own field and discipline's state of play and likely directions of development, pp.164–69.) That is why this chapter, in accord with the basic premises of this book as a whole, stresses Descartes' systematizing goals inside the game of natural philosophizing. It is also why we have related those goals to Descartes' healthy respect for facts. Like any good, competitive natural philosopher (or later modern scientist) he knew facts need to be assessed, interpreted, selected for use, reframed in terms of the theory and claims under discussion, and argumentatively deployed for persuasion. His appetite for facts, their theoretical reframing and leveraging for further explanatory uses were intimately linked to his goals and strategies for building a winning system of natural philosophy, proclivities that will be display below, especially in Sects. 12.6, 12.7, 12.8, and 12.9.

more small but ultimately crucial textual difference that we need to canvass—some little noticed matters about inter-vortical behaviour in the *Principia*. These are almost completely absent in *Le Monde,* but they will be later shown to be functionally essential to the radical, realist Copernican cosmographical explanations/ narratives which power the *Principia*.

12.4 Inter-Vortical Phenomena

As Eric Aiton correctly observed in his classic study of the vortex theory of planetary motion in Descartes and his followers, there is little essential difference in the model between *Le Monde* and the *Principles*.[26] We have seen in Chap. 10 that Descartes' final cosmological model of the distribution of size, speed and force of motion of vortical spherical particles, and the dynamical role of central stars, are identical in the two treatises. As to the obvious differences, such as they are, we have discovered in Chap. 10 (and Appendix 2) that the exposition in the *Principles* is clearer, better ordered and argued than in *Le Monde*, so that it can be enlisted heuristically to help unpack Descartes' meaning in *Le Monde*. In Sect. 11.3, we counted these changes as amongst the small remediations of *Le Monde* accomplished in the *Principia*. Beyond that, we also have learned that the cosmogonical origins of the cosmological steady state, including the dynamics of pre-element vortices, are set out in more detail in *Le Monde*. In the *Principles*, as we have just seen, Descartes gives us his cosmogony of nearly identical Ur-particles, which from the moment of creation rotate around their own centers and move at 'average' speed around numerous proto–vortical centers. He explains how second and first element particles evolve in this situation, but makes no explicit statement about vortex dynamics and distributions of size, speed and force of vortical particles in relation to the cosmogony. These details are supplied only for the cosmological dynamical steady state of the *Principles* after the formation of first element, spherical second element, and most importantly, stars. None of these differences, therefore, suggest to us anything about the sort of large, strategic changes concerning system–binding and cosmographical argumentation that we seek in the text of the *Principia*. The more consequential and strategic changes regarding the universe of vortices in the *Principia* do not reside in the core of the celestial mechanics *per se*, but rather in some small corners of Descartes' exposition which are not usually taken as being of much consequence, especially in the area one might term 'inter–vortical' relations.

The relations amongst vortices are much simpler in *Le Monde* than in the *Principles*. In the former the main inter-vortical phenomena mentioned pertain to comets, their travel in the regions above the K-layers of vortices, and the fact that the light they reflect does not cross inter–vortical boundaries (Sects. 10.2.3 and 10.8 respectively). These phenomena are also described in the *Principles*. But in that text, as we have seen, there is painted a vast picture of the circulation of particles of

[26] Aiton (1972) 3.

first matter out of the equatorial regions of vortices and into neighboring vortices along the north and south directions of their axes of rotation. This is related to the implied conception of inter–vortical stability, mentioned above, which governs the arrangement of vortices so that their poles are as near as possible to the equators of their contiguous neighbours. Beyond this Descartes later in the *Principles* also inserts the idea that there is amongst neighbouring vortices a constant, dynamic jostling—pushing and shoving each other, thus causing slight deformations of vortical boundaries, hence vortical shape and size. Such movement of inter–vortical boundaries had been fleetingly mentioned in *Le Monde*, but in the *Principles*, as we shall explicate in detail below in Sect. 12.9, this becomes critically important in Descartes' treatment of variable stars and *novae* and is explicitly treated in relation to them. This is because the formation/destruction of crusts of sunspots on central stars is caused by these deformations (or vibrations) of vortical interfaces.[27] So, whilst these matters of inter–vortical behavior are not often commented upon, they are crucial to our reading of the strategies of the *Principia*.

This closes our comparisons of *Le Monde* and the *Principia* in terms of matter and element theory, cosmogony and inter-vortical behavior . These set the interpretative baseline from which we can move to the full exposition of the cosmographical strategies of the *Principia*. The starting point for that must be Descartes' theory of magnetism as a cosmic phenomenon, which we have so far simply touched upon as needed in the course of making our comparisons. The *Principia's* theory of cosmic magnetism underpins Descartes' entire account of the formation of sunspots on the surfaces of stars. The explanation of sunspots in turn becomes the veritable pivot of his vast cosmographical explanatory enterprise, ranging from *novae* and variable stars to the birth of planets and comets, and leading ultimately to the revelation of the generically 'earthly' structure of all planets—the closure of Descartes' cosmographical *tour de force*. Hence it is to Cartesian cosmic magnetism that we must first turn in our progressive dissection of the strategic core of the *Principia philosophiae*.

12.5 Co-opting and Re-framing Gilbert's 'Cosmic' Magnetism

Le Monde says nothing about magnetism, although the *Regulae* already show Descartes playing with the possibility of methodologically co-opting that field of experimental inquiry (See Sect. 6.4 for Descartes' abortive gambit in this regard.) Readings of the *Principia* usually emphasize Descartes' co-optation and reframing

---

[27] On the mention of the issue in *Le Monde*, see above Sect. 10.8, Note 96 and corresponding text. The large discussion in the *Principia* occurs in Book III, articles 111–116 and includes the key figure to which the entire discussion is referred [*Principia* Plate XII, Figure i which is introduced below as Fig. 12.2 in Sect. 12.9]. At one point (article 114) Descartes interestingly likens the movement back and forth of a vortical boundary and the accompanying formation/destruction of stellar crusts of sunspots to the behavior of a pendulum. Cf. note 79 below.

of Gilbert's 'lab' based experiments on magnetism, with Descartes re-writing Gilbert's manipulations in corpuscular-mechanical terms, using his left and right handed channeled magnetism corpuscles of first matter.[28] This focus ignores the kind of natural philosophical game and contest in which Descartes was involved, and misreads the nature of Gilbert's enterprise as well. Both Descartes and Gilbert had strategic cosmographical aims in mind for magnetism, which as a 'cosmic' cause was to play key roles in their respective systems of natural philosophy. Although writing a generation apart, they were both participants in that critical period of heightened natural philosophical contest, centrally, but not entirely focused on the meaning and destiny of realist Copernicanism. (Or, better put, at least Gilbert's very widely read *de Magnete* (1600) took part in that later contestation.) As has been argued throughout this volume, the unit of contest was systematic natural philosophy; competitors aimed at a scope of coverage of matter theory, cosmology and theory of causation (not to mention claims about method) similar to that offered by the neo-Scholastic Aristotelianism through which all players initially learned what a natural philosophical system was, and what the rules of formation of competing systems might be.[29] Moreover, as was observed earlier (Note 1), it was not yet incumbent on a contestant to adduce new matters of fact off his own bat. It sufficed to co-opt and reframe key facts from others, according to one's own systematizing strategies. Descartes did with Gilbert's lab facts precisely what he was to do with consensually accepted facts about sunspots and variable stars, as we shall see below.

Gilbert's *On the Magnet* (1600) was arguably the most influential and impressive new natural philosophical gambit of the turn of the seventeenth century. His program involved a new natural philosophical agenda and content, built on exploiting and metaphorically extending important experimental work he had done on the magnet and magnetic compass. Also indebted to a neo-Platonic view of ontology, Gilbert used a *cosmographical* strategy in the precise sense intended in this chapter, basing his new system of nature on a new theory of the Earth, according to which the Earth's magnetism, which he established as a fact, is a form of immaterial, spiritual power. The Earth's magnetic 'soul' is responsible for its spinning on its axis, and since other celestial objects similarly have magnetic 'souls', a host of celestial

[28] As Richard Westfall (1971, 36–37) describes the encounter over lab based manipulations: '…the mechanical philosophy had to explain away magnetic attraction by inventing some mechanism that would account for it without recourse to the occult. Descartes' was particularly ingenious. In considerable detail, he described how the turning of the vortex generates screw-shaped particles which fit similarly shaped pores in iron. Magnetic attraction is caused by the motion of the particles, which in passing through the pores in magnets and iron, drive the air from between the two and cause them to move together. What about the fact of two magnetic poles? Very simple, Descartes replied; there are left handed screws and there are right handed screws'.

[29] These points modeling the natural philosophical field in the critical phase of the scientific revolution c.1630–1660 derive from Sects. 2.5 and 2.7 above. See also Schuster (1990) 224–7, 232–8; (2002) 339–41, 344–8; and also Schuster (2012a).

motions could be explained. Gilbert worked in the first instance not on astronomical or cosmological questions, but on the structure and nature of the Earth. He co-opted and reinterpreted the craft knowledge and lore of miners and metallurgists, to argue that lodestone is the true elemental nature of the Earth; that the Earth is a gigantic spherical magnet; and that since magnetic force, even in a small magnet, is an immaterial, spiritual force, the magnetic nature of the entire Earth amounts to a cosmic soul or intelligence—capable of moving, or at least spinning the Earth. This natural philosophy, he claimed, showed the true nature of the Earth, as opposed to the superficial mutterings of Aristotelians about earth, air, fire and water.[30] All of this was in turn meant to 'leverage', in our terms, a cosmographical extrapolation by which Gilbert could, in the final book of *de Magnete*, hold forth about celestial causation and motion, attributed to the magnetic souls of the Earth and other heavenly bodies.

Now, it was this same 'cosmic' side of magnetism that Descartes chiefly sought to explain and systematize. Descartes borrowed from Gilbert (and from Kepler it must be said) the idea that magnetism is a cosmic force. But, he changed its ontology, of course, and also its functions, relieving it of its celestial mechanical role. Tellingly, Gilbert's cosmographical gambit had started with his ultimate laboratory artifact, the sphere of lodestone, or *terrella*, on which he modeled the magnetic properties of the Earth, using it to argue analogically, but with realist intent, to the essentially magnetic character of the Earth, which displayed the highest manifestation of magnetism, a magnetic soul. In contrast, Descartes' explanatory cosmographical tale ends with planets (including the Earth) which, born of sunspot encrusted stars, continue to display the causal imprint of their stellar origins, most notably in their retaining through their structure an ability to accommodate axial inflows of left and right handed magnetic particles.

On my reading of the *Principia* in terms of system–binding innovations, Descartes pursued a dual strategy of *co-option* of Gilbert's matters of fact and *displacement* of Gilbert's attempt to render magnetism 'the' key cosmic cause via a vast cosmographical gambit.[31] Descartes' response was also cosmographical, aimed at invoking magnetism in explaining how heavens and Earth are bound together. To this end the

[30] Similarly, Gilbert insisted that his knowledge was built on assiduous attention to experiments and to facts reported by craftsmen and artisans, and that it was productive of useful results, most notably improving the use of the magnetic compass in navigation.

[31] It might be asked whether I am maintaining that this strategy was deliberate on Descartes' part or whether it exists merely as an analyst's construct. I answer that it arguably was deliberate and part of his way of contesting for hegemony in natural philosophy. This is based on my reading the text of the *Principia* for its underlying goals and strategies, which I hold to be better than imputing motives based on circumstantial events or evidence. (Cf. above note 25 on Lynes and Love, and below Sect. 12.12, especially note 109, as well as the entire historiographical framework of Chap. 8 above, outlined in Sect. 8.1 of that chapter, used in explaining Descartes' career 'inflection' toward composing *Le Monde*) .

rewriting of Gilbert's experiments in corpuscular-mechanical terms was merely a necessary but hardly a sufficient move. Matter theory alone was not going to neutralize Gilbert's system and articulate a competing one. Descartes worked to insure that magnetism was not the principal cause guiding the planets in their orbits. That was the job of his vortex celestial mechanics which, considered in its narrow, technical senses, had no essential connection to his theory of magnetism (as the presentation in *Le Monde* proves). Nevertheless, in Descartes' mature natural philosophy magnetism retained, in three ways, something of the high cosmological status Gilbert had bestowed upon it: [1] There is a physical interweaving of each vortex and its central star with its neighboring vortex/star complexes, by means of axial input and equatorial output of magnetic particles; [2] The particles in question become fully capable of causing magnetic phenomena by being given right and left handed twists during their incoming journeys along the axes of rotation of vortices—vortical rotation is the final, necessary forge of magnetic particles;[32] [3] The ultimate possibility of formation of planets and comets has to do with these cosmic flows of magnetic particles, which can form sunspots which in turn can lead to star–death and planet/comet formation.

12.6 Claims About Sunspots from Galileo and Scheiner to Descartes

We now have in place all of the resources that Descartes adduced in the *Principles* to facilitate his explanation of sunspot formation, properties and behavior. Before we turn to Descartes' explanation, we need to look at the evolution of agreed matters of fact about sunspots in the larger natural philosophical community, as well as at Descartes' move from ignoring them in *Le Monde*, to featuring them in the *Principles*.

Galileo's claim to discovery of sunspots and consequent brilliant mixed mathematics style argument that they are on the surface of the sun or vanishingly close to it, established, for those who accepted his claims, on the one hand that the sun rotates, and on the other hand that changes could take place on a celestial body. Galileo was quite clear about his claims that

> …the solar spots are produced and dissolve upon the surface of the sun and are contiguous to it, while the sun, rotating upon its axis in about one lunar month, carries them along, perhaps bringing back some of those that are of longer duration than a month, but so changed in shape and pattern that it is not easy for us to recognize them.[33]

---

[32] See the comments on this point above at note 19.

[33] Galileo Galilei, *Letters on Sunspots*, in Drake (1957), 87–144, at p.102. Compare Galileo 20 years later in the *Dialogue Concerning the Two Chief World Systems* (Galileo 1953), 54, '[many spots] dissolve and vanish far from the edge of the sun, a necessary argument that they must be generated and dissolved'.

This showed that generation and corruption were taking place in the heavens, a notable argument on the cosmographical plane for the unity of heavens and Earth required by realist Copernican theory. But neither in 1613, nor over the next generation was there necessarily a consensus view, especially in the light of the masterful Jesuit astronomer Christoph Scheiner's competing claim (1612) that sunspots are small planets circling the sun.[34]

Although Descartes undoubtedly knew about sunspots at the time he wrote *Le Monde*, he did not even mention them in that book, while they are one of the cornerstones of the *Principles*. In October 1629 he wrote to Mersenne, asking him for information about recently observed phenomena around the sun without mentioning the name of the observer.[35] These were parhelia seen in March of that year by Scheiner.[36] Wishing to explain parhelia induced Descartes to drop other projects. His new direction at first extended to work on meteorology in general and later into a description of the whole world that eventually became *Le Monde*.[37] However, in the very same letter to Mersenne in which he asks for information about parhelia, Descartes alludes, without explicitly referring to, Apelles, the Greek painter who reputedly hid behind his board and listened to what people were saying about his painting. Apelles was the pseudonym Scheiner used in 1612 to announce his claim to discovery of sunspots. As mentioned above, in this publication sunspots were conceived of as small planets circling the sun. The connection between Descartes saying that he will be hiding to hear what others are thinking of his work and Scheiner's publication on sunspots has been pointed out by the editors of Descartes' collected works and is extremely unlikely to be a coincidence.[38] In other words, it may tentatively be suggested that, triggered by Scheiner's name, not only parhelia

[34] There are four contenders for the discovery of sunspots. Within about 18 months in 1611/2: Johann Fabricius (1611); Christopher Scheiner (1612) [under the pseudonym of Apelles]; and Galileo (1613), appeared and claimed discovery. Fabricius probably saw them as early as March 1611, Scheiner in spring 1611 and Galileo, who in 1613 responded to Scheiner's published claims of 1612, claimed observations 18 months earlier (this was in the published version of his first letter, to Welser, on sunspots, May 14, 1612, hence he was claiming observations as early as 1610. In the *Dialogue Concerning the Two Chief World Systems* (Galileo 1953, 345), he again claimed observations as early as 1610. Harriot, whose observations exist only in manuscript form, has notes on sunspots dating from December 1610, but began regular observations only about year later, following Fabricius' publication (Brody 2002, 68). It should also be noted that the painter and poet Raffael Gualterotti (1605) claimed to have followed for several days movements of spots on the sun. He explained them as resulting from a conjunction of Mars and Saturn which attracted exhalations and vapors which were drawn to the sun, purified and rarefied to become sunspots. Galileo knew Gualterotti and had corresponded with him (Brody 2002, 25–6, 55). Reeves et al. (2010) came to notice too late to be included in assessing these matters.

[35] Descartes to Mersenne, 8 October 1629, AT I 23; CSMK 6.

[36] Parhelia or mock suns or sun dogs are 'two concentrations of light on the small halo at the same altitude as the sun': Minnaert (1993) 214.

[37] On the process of emergence of the project of *Le Monde*, see above Sect. 8.4.

[38] AT I, 248 note referring back to p. 23 l.25–29. Judit Brody first pointed this out in drafts leading to our joint work on sunspots, *novae*, variable stars and the systematizing strategies of the *Principia* (Schuster and Brody, note 2)

but also sunspots were on Descartes' mind in October 1629.[39] In December of that year he asked Mersenne if sunspots have been more diligently observed 'de nouveau'.[40] He wrote to Mersenne for additional information about sunspots in January 1630 and again on March 4. He asked whether Gassendi had seen several at the same time and if so, how many; did they all move with equal speed and were they always round?[41] He also seems to express some scepticism about whether the spots can be small planets orbiting near the sun.[42]

So why did Descartes not mention sunspots at all in *Le Monde*? That is a question which we can only answer after evaluating some additional facts: First of all, even if Descartes had known Galileo's *Letters on Sunspots* prior to, or during, the drafting of *Le Monde*, it is clear from the resulting text that he had no inclination to co-opt Galileo's claims into his *Le Monde* cosmology. His letters to Mersenne certainly show that at the time of writing *Le Monde* he knew about sunspots, was interested in their nature; yet, he did not even mention them. The text of *Le Monde*, as we have reviewed it, specifically excluded Galileo's explanation, as well as others suggested at the time, involving the sun rather than nearby orbiting small planets.[43] Changes on the sun would have violated his matter theoretical bar

[39] Eventually he dealt with parhelia in the *Météores* and with sunspots in the *Principia*.

[40] Descartes to Mersenne, 18 December 1630, AT I, 102–103.

[41] Descartes to Mersenne, January 1630, AT I 112–113; CSMK 18; Descartes to Mersenne, 4 March 1630, AT I 125. Gassendi observed spots between 1618 and 1638. Descartes was seeking information by correspondence regarding as yet unpublished material. Gassendi's detailed reports on the 1626 observations and others only appeared in his *Opera Omnia* (1658) in the following locations: Vol.1 *Syntagmatis philosophici* pt 2 of pt 2 *De rebus caelestibus* pp.553–554 on spots; Vol.4 *Observationes Coelestes ab anno 1618 in annum 1655* (repr.1658). *Maculares solares* (observations in 1626 p. 99–100, in 1638 pp. 411–412); *Mercurius in Sole visus et Venus invisa...* 1631 (1632) pp. 499–505 (letters to W. Schickard: Mercury was so small that at first Gassendi thought it was a sunspot).

[42] To Mersenne, 4 March 1630, AT I 125, Descartes writes, 'Vous ne me dites pas de quel cofté font les pôles de cette bande, où fe remarquent les taches du Soleil , encore que ie ne doute point qu'ils ne correfpondent aucunement à ceux du monde, & leur ecliptique à la noftre'. This concerns the band to which sunspots seem confined, in particular, taking that band to be revolving around the sun, where the poles of its axis of rotation would be located. He doubts these poles correspond to the celestial poles and that the band's inclination to the celestial equator would equal that of our ecliptic. All of which seems to imply that at this time his view was that the sunspots are not planets, or at least are not like the known planets (and so might well be on the surface of the sun on this argument). Scheiner's original views had been supported by others, such as Jean Tarde (1620) and C. Malapertuis (1633), whilst Fortunius Licetus (1623) 124, held the interesting view, intermediate between theories of sunspots and orbiting planets, that spots cannot be solar exhalations because those would be more rarefied, not darker. He added that some falsely claim that there are craters on the sun. He thought they are parts of the aether condensing/rarefying in turn.

[43] For example: Leaving aside Gualterotti (1605) mentioned above note 34; Galileo likened 'sunspots to clouds or smoke' (Galileo 1957, 140); Kepler (1938ff, vol. 17, 36) in 1612 suggested to Simon Marius that spots might be like clouds originating from the fire of the sun and that perhaps cometary material also originates from the sun; J.R. Quietanus told Kepler, August 13, 1619, *ibid.* vol 17, 372, that he thought comets 'ex maculis solis colligitur et coacervatur' and Kepler told him in reply, August 31, 1619, *ibid.* vol 17, 376, that Marius agreed with this.

on the existence or generation/corruption of third matter in or on stars (or anywhere else). As he wrote in *Le Monde*, 'we have every reason to think that the Sun and the fixed stars have as their form nothing other than the first element'.[44] It seems that of the available explanations, only that of small planets orbiting close to the sun would have fitted with the matter theoretical scenario in *Le Monde*. But, to have adopted this view would have required some modifications to the vortex celestial mechanics, in the service of a factual claim Descartes seems, in March 1630, to have held to be dubious.[45]

However, in 1630 Scheiner published his *Rosa Ursina*, a huge volume containing his solar observations. Here Scheiner changed his mind and placed sunspots on the body of the sun.[46] Scheiner's careful observations are praised in the French

Marius (1619) himself argued that comets might come from the sun because for the last year and half (covering the period of the comet of 1618) there had been few spots on the sun. He also stated that he had seen spots on the sun with tails; and generally held that the surface of the sun is like molten gold, the spots being like slag; Willebrord Snell (1619) also discussing the comet of 1618 explained comets as 'maculae istae exhalationes...solis flagrantis atque ista ex recessu & interiore corpore per sua crateras eructantis quemadmodum in terris Aetna'.

[44] *Le Monde*, AT XI 29; SG 20; MSM 45. Also: 'we shall take one of those round bodies composed of nothing but the matter of the first element to be the sun, and the others to be the fixed stars', *Le Monde*, AT XI 53; SG 35; MSM 87. Cf. above note 13 and text to which it refers.

[45] Moreover in that case Descartes probably would have had to have taken some account of the strong claims for their appearance and disappearance, as mentioned above (note 33), often on the middle of the sun, a difficult challenge if they are planets (compared to their appearance and disappearance near the edges of the solar disk, which could be explained as visibility effects concerning continuously existing small planets). It should also be noted that when Descartes in the *Principles* accepts that the spots exist and form on the surface of the sun, there are celestial mechanical consequences with which he must deal: Observations of the spots indicate that the sun does not spin as quickly on its axis (in terms of linear velocity, not radial velocity) as the vortex theory would imply—that is, faster than any planet in its orbit. (Gaukroger 2002, 153 and *Principles*, III article 32, AT VIII-1 93; MM 97, where the rotational period for sunspots is given as 26 days.) For this and other reasons Descartes introduces the conception of stellar aether, an earthy atmosphere near a star, and extending out as far as its nearest planet, largely constituted by dissolved sunspots, which slows the rotational speed of the star (*Principles* III article 148, AT VIII-1, 196–7; MM 172). On other functions of the aether see below, note 59 and text thereto. Finally, the detection and description of transits of Venus or Mercury across the sun, posed many difficulties at the time, not to mention the complications introduced if one took sunspots actually to be conjunctions of small planets orbiting near the sun. For example, Scheiner had failed to observe a transit of Venus which he could have used early on to argue for the visibility of the other smaller planets whose conjunctions he claimed produced the appearances of sunspots (Brody 2002, 49). Gassendi in 1631 after hesitation, thinking he was observing a sunspot, claimed he had observed a transit of Mercury; while earlier, in 1607, Kepler had taken a sunspot for Mercury seen against the sun's disk (Brody 2002, 27). After Gassendi's observation there was more clarity about distinguishing a sunspot from a transiting planet. Hence by the time the transit of Venus was first observed in 1639 by Jeremiah Horrocks, as Brody (2002, 78) notes, 'the argument had already turned around. Previously the emphasis was on proving that the spots were not planets, now it had to be shown that a planet was not a spot'.

[46] Scheiner (1630), 537, 'maculae & faculae in ipso sole sunt'. Scheiner also stated that the spots grow, change, diminish, darken, lighten, disappear in the middle of the sun. *Ibid.* p.490.

edition of the *Principles*.[47] Descartes referred to the book in a letter to Mersenne in February 1634.[48] But, by that time he had already abandoned the plan of completing and publishing *Le Monde*. It is highly questionable that he saw *Rosa Ursina* any earlier, since his remarks to Mersenne in 1634 show a clear and seemingly fresh and recent grasp of the cosmographical implications of Scheiner's new view. He told Mersenne that he had heard that Jesuits had had a hand in Galileo's condemnation and that from the book he could see that Scheiner and Galileo were not on friendly terms. But, tellingly, he also asserted that since *Rosa Ursina* had furnished ample proof for it, Descartes could not believe that Scheiner did not 'share the Copernican view in his heart of hearts'.[49] Consequently, taking all these points into consideration, I conjecture that when he wrote *Le Monde* Descartes may well have been undecided between the two main theories and unhappy with the way each sat with key positions taken in *Le Monde*.[50] However, by 1634, possibly stimulated by his recent reflections on Scheiner's change of view, he was perhaps beginning to glimpse the cosmographical potential of a co-option of the now Galileo-Scheiner consensus on sunspots as entities subject to generation and corruption located on the surface of the sun.[51]

Whatever the dynamics of Descartes' views about sunspots over the next few years after 1634, we know for certain that in the *Principles* he was to take for granted the notable Galilean claims that the sunspots are generated and dissolved on the face of the sun and participate in its axial rotation. There is a sentence in the *Principles* to the effect that, 'spots which appear on the sun's surface also revolve around it in planes inclined to that of the Ecliptic', which could be interpreted as sunspots circling on the surface of a stationary sun.[52] However, there can be little

---

[47] *Principles,* III article 35; AT IX-2, 118; MM 98–99.

[48] Descartes to Mersenne, February 1634, AT I 281.

[49] *Ibid*, Mais d'ailleurs les obferuations qui font dans ce liure, fournissent tant de preuues, pour oster au Soleil les mouuemens qu'on luy attribuë, que ie ne sçaurois croire que le P. Scheiner mesme en fon ame ne croye l'opinion de Copernic; ce qui m'étonne de telle forte que ie n'en ose écrire mon fentiment... (Also see MM 99, note 29).

[50] Arguably neither theory was fully acceptable to Descartes at the time of composing *Le Monde*: To decide that sunspots are generated and destroyed on the surface of the sun would violate the matter theory of *Le Monde*; but, to accept sunspots as small planets orbiting very near the sun would require first overcoming the scepticism he had expressed to Mersenne in 1630 about this claim (see note 42), and second, significant further articulation of his vortex celestial mechanics.

[51] Additionally, let us also recall that, thanks to Beeckman, Descartes first saw Galileo's *Dialogo* in 1634 and so was potentially exposed to Galileo's persuasive deployment of his claims about sunspots, which in turn served as powerful arguments for the (Copernican) unity of heaven and Earth. Of course, Descartes saw the book for a short time only, for thirty hours, but he made some reasonable use of it for his own purposes, as in his later reported critique of the natural philosophical relevance of Galileo's abstract and idealized account of fall and projectile motion (To Mersenne, 11 October 1638, AT II 385).

[52] *Principles,* III article 35, AT VIII-1 95; MM 98.

ambiguity about his statement that, 'all the matter which forms the body of the Sun revolves' around a certain described axis.[53] Moreover, the overall force of his argument makes it clear that Descartes now took the spots completely seriously as matters of fact and accepted Galileo's proof that sunspots were on the body of the sun or at least so close as to make no difference, a claim that by 1630 even Scheiner famously now accepted. Descartes also now took for granted as matters of fact that most sunspots appear in a belt near the equator of the sun; that they have irregular shapes; and that they sometimes have a dark nucleus surrounded by lighter areas occasionally even giving rainbow effects; and that sometimes there are bright structures, called faculae, close to the spots.[54] As in the case of magnetism, the challenge was not to discover such new matters of fact, but rather first to co-opt them and then exploit them; that is, first to explain these properties and behaviors of sun spots within his natural philosophical system and then leverage the thus explained phenomena to aid in the explanation of additional facts and bind the system together.

12.7 Gaining Strategic Leverage: Sunspots as Explananda and Explanans in the *Principia Philosophiae*

We have seen in Sect. 12.3 how Descartes explains the circulation between vortices and through stars, and onto their surfaces, of particles of first matter, including that sub-set of them which are longer, channeled and left– or right–handed, having been, so to speak, finished and polished as magnetic particles on their trips from neighboring vortices, toward the north and south poles of stars, along their axes of rotation. Now we can examine how he uses that framework to address those matters of fact about sun spots largely accepted by the early 1640s. Recall that the sun as it were 'bubbles' near its poles with magnetic first matter particles (channeled and handed) and that this material on its surface moves constantly towards its equator, possibly forming sun spots of third matter under the conditions described earlier. Descartes now explains the observed properties of sunspots on the basis of his explanation of their generation within his system: We see most of them in a belt near the equator and not at the poles, because by the time they have managed to stick together into a mass big enough to be visible to our eyes they have covered a

[53] *Principles,* III article 74, AT VIII-1 129; MM 124.

[54] In addition, let us not forget that sunspots supplied observational evidence for the first time that the sun rotates. Although he does not say so, Descartes could not have wished for a better validation for his theory of vortices, notwithstanding the celestial mechanical issues requiring further adjustment, mentioned above at note 45. At the time of writing *Le Monde* he had passed up this advantage, which had been obvious to, and valued by Galileo and Kepler a generation earlier, when sunspots had first been observed.

considerable distance from the poles.[55] From the way they come into being, it naturally follows that they have irregular shapes. The spots, being on the sun's surface, are carried along by its rotation. The fact that the spots, sometimes have a dark nucleus surrounded by a lighter area is explained by Descartes by claiming that at the lighter parts the accumulation of third element is thinner and lets some light pass through, occasionally even giving rainbow effects.[56] Finally, the nearby especially bright areas or *faculae* are explained in the *Principles* by first element matter surging faster than the rest of the sun's substance out through the tight spaces around the spots. Sunspots cause a restriction in the movement of the sun's material which then tends to surge away at the edges of the spots, which thus become more luminous, while the mass of the spot itself prevents the tendency to motion being communicated through it, i.e. stops the light. [57]

Descartes writes that observations show some spots being destroyed 'in the same way as many liquids, by boiling longer, reabsorb and consume the same scum which they gave off in the beginning by bubbling up'.[58] His explanation of how they disappear is this: Sunspots, of third element material but originally generated from first element matter, get worn away by the rotating matter of the sun and disintegrate partially back into first element, partially into smaller but still relatively large and irregular (third element) stuff that then becomes the atmosphere around the sun slowing down its rotation (cf. note 45). This he terms aether. It surrounds the stars, consists mainly of third element and is inherited by planets resulting from the death of stars, becoming, as in the case of our own Earth, the ultimate source of their land masses, seas and atmospheres.[59]

[55] Judit Brody discovered that Descartes' thoughts were later echoed by the Swiss astronomer Rudolf Wolf (1816–1893). 'I compared the whole appearance of the sunspots to currents which proceed periodically from the two poles of the sun towards its equator.' (Wolf 1861, 27) (Brody manuscript research notes leading to composition of Schuster and Brody, 'Descartes on Sunspots' [forthcoming, note 2].)

[56] *Principles,* III article 97, AT VIII-1 149; MM 137. Descartes' explanation appeals to his explanation of prismatic colours in the *Météores* of 1637.

[57] *Principles,* III article 98, AT VIII-1, 149–50; MM 137–8; The explanation follows directly from Descartes' theory of light. The first matter surging around the edges of a spot not only contributes to a tendency to motion propagated out through the *boules* of the vortex, but also produces a more than normal intensity of that tendency, a set of stronger than normal rays. (It is crucial to understand that in Descartes' theory of light the propagation of the tendency to motion through the *boules* that constitutes light is always instantaneous, but the intensity or force of that tendency can vary. There can be weak or strong rays, albeit always instantaneously propagated. [This point was made clear in Chap. 4, and applied to reconstructing the development of Descartes' physical optics.] Returning to Descartes' explanation of *faculae*, strictly speaking he claims that a *facula* can form following the existence of a spot, and, by extension of the process described, a spot can turn into a *facula*; and vice versa, meaning that he claims that dark spots can turn into bright regions and vice versa.

[58] *Principles,* III article 96, AT VIII-1 148 MM 137.

[59] *Principles,* III article 100, AT VIII-1 150; MM 138–39. The central thread of Descartes' narrative of the formation of the Earth in Part IV of the *Principles* involves the formation of all the third matter on Earth that exists above the inner, unreachable, crust that suffocated the original star. This new planetary third matter is formed largely from material derived from the aether of the dead star (*Principles,* IV articles 1–7, AT VIII-1 203–6; MM 181–4). Cf Note 87 below, and Chap. 11 Note 14 above.

Coming back to our original matter theoretical concerns with the *Principles*, we see that according to the theory of stars, magnetism and sunspots in the *Principles*, third element originates on the surfaces of stars from conglomerations of first element particles, and it can also change back and become first element again. Moreover, we see that sunspots are theoretically constituted, their accepted properties re-derived from theory, and they can now be leveraged to be used (with rest of the machinery) as themselves *explicantes*—and this occurs in two dimensions [1] natural history of stars, as one might say—why there are *novae* and variable stars— and [2] the origin and nature of planets.

12.8 Claimed Matters of Fact About Novae and Variable Stars Before Descartes

I have already mentioned twice (in Sects. 12.2 and 12.5) that in Descartes' explanation of sun spots a fully encrusted star leads to further phenomena in stellar life patterns. Now we take these up, looking first at the matters of fact concerning *novae* and variable stars that Descartes was going to try to co-opt and systematically exploit. New stars (*novae*) had already famously been observed in 1572 and in 1604.[60] Many problems surrounded their explanation and indeed their characterization at the level of fact, even if a natural philosopher or astronomer intended to remain in the realm of natural causation, eschewing miraculous or supernatural causation.[61] Was it the case, for example, that all fixed stars were already in the catalogues? A faint star simply might not have been seen previously. Or, could it be suggested that only if a putatively new star was extremely bright, it was obviously new? Even with telescopes, parallax measurements were not easy and putatively new stars were difficult to tell apart from comets. Naturalistic explanations, such as causation by a conjunction of planets might even be made consistent with an Aristotelian perspective, but not other seemingly naturalistic explanations, such as the star had always existed and moved

[60] By modern definitions these of course were supernovae. The contemporary search for other *novae* included Johann Fabricius' claim regarding Mira Ceti in 1596 (which we discuss immediately below in the context of the later claims that it is in fact a variable); and Kepler and others' identification of a supposed *nova* in 1600 (Kepler acknowledged that it was first seen by W. J. Blaeu who put it on his celestial globe.) Cf. Hoskin (1977). The star of 1600 is now regarded as a LBV (luminous blue variable), hence it is neither a nova nor a supernova.

[61] Explanations invoking divine action could include the following: the star has been around since the creation but it was hidden and brought to the fore by God as a sign of his omnipotence; or, it had actually been newly created by God. A miracle could be carried out directly by God or through natural causes at the fiat of God. The latter might well violate the sense of 'natural' that previously held in a given natural philosophy. For example a Christian Aristotelian could take a new star as the result of God's decision to use (hitherto unknown but) natural causes in the heavens to generate a new star. Problems would be created for the natural philosophy as previously expounded.

towards the Earth in a straight line from infinity and back again.[62] It can also be said that in a general sense *novae* offered a prime opportunity to realist Copernicans to score points against the strict Aristotelian doctrine of incorruptibility and lack of change in the heavens. But everything depended upon the contents of one's natural philosophy and its cosmographical strategies. As we shall soon see, Descartes' dealings about *novae* fall into this category of maneuver.

Descartes would have been aware of *novae* as matters of fact his entire adult life, and he had briefly alluded to them in *Le Monde,* as we have seen. In contrast, the first well publicized claims bearing on the possible existence of variable stars fall into the interval between his writing *Le Monde* and the *Principles.* In 1596 David Fabricius (1564–1617) had made the first recorded observation of the star *Mira Ceti*, which was eventually subject of the first claim that a star could be variable, and later, periodically variable. It was considered, by Kepler and by others, a new star, similar to the one seen in 1572. When Fabricius saw it again in 1609, he still did not take note of any variability, nor, perforce, any periodicity.[63]

The story becomes much more interesting in the late 1630s. Putting the matter rather simply, *Mira Ceti* was recognized by J. P. Holwarda (1618–1651) as a 'new star' or 'phenomenon' that can appear, disappear and reappear. However, by 'new

[62] The latter possibility was discussed by Tycho Brahe in his *Astronomiae instauratae progymnasmatum pars tertia* (1916, vol. III, 204). This reports the opinions of John Dee and Gemma Cornelius that the new star moves away in a straight line. However there is also evidence that both Gemma (1573), and Michael Maestlin had thought the 1572 nova was newly created. Maestlin thought there were not enough exhalations and that the star was newly created by God. This was published in his *Demonstratio astronomica loci stellae novae, tum respectu centri mundi....* appearing pp.27–32 in Frischlin (1573). The key passage was recently cited by Granada (2007, 104). Maestlin's 'edificatory poem' (Granada 2007, 101) states that the star announces the second coming. Maestlin deals mainly with the location of the star, except for the key passage in question, which was also quoted by Tycho (1916, III, 60) as part of his reproduction of the entire document with commentary (1916, III, 58–62, with commentary, 62–67.) Tycho himself said that the new star was formed of matter from the Milky Way, but not of such perfection or solid composition as other stars, in the *Conclusio* to (1969 [1572]). Fortunius Licetus (1623), held that the phenomena are created and then annihilated. He also writes that there are also some people who think a *nova* is an old star, neglected, not observed by the ancients. Reisacher and Valesius (or Vallesius) thought an old faint star got brighter through sudden transformation of the air between it and us, so it was not a new creation (Dreyer 1890, 63–64). (Vallesius is quoted in Tacke 1653 and by Reisacher 1573.) Kepler, in his *De stella nova in pede Serpentarii* (Pragae 1606), Chapter 20 (Kepler 1938, I, 248–51) reports discussions with David Fabricius about where the material for the new star of 1596 (Mira Ceti) came from: whether the star had been around since the creation but hidden and then brought to the fore by God as a sign; or newly created either by God or by physical processes from existing material which must be all over the universe, since (*Ibid.*, Chapter 22, 259) the 'star in the whale', was not close to the Milky Way.

[63] David Fabricius (1612) wrote that *novae*, like comets, do not dissipate but can remain unseen, then reappear. Little note was taken of this claim, let alone any possible natural philosophical significances. Hence, in accord with modern understandings of the construction and attribution of discoveries in science, it would be quite wrong to credit Fabricius with the discovery of variable stars. See Arjen Dijkstra (2011) 77.

star' he meant that the phenomenon was not an actual star, but a solar emanation. Indeed, his claims gained notoriety, in part, because he couched them in natural philosophical terms, framed by a clearly stated anti-Aristotelian and pro-Copernican stance.[64] The young professor in Franeker first saw this 'phenomenon' in December 1638 while watching a lunar eclipse. At first he did not trust his own eyes but a fellow professor, Bernard Fullenius (1602–1657) saw it too. Holwarda kept watching until the phenomenon disappeared from view, to be seen again the following year. These observations were published in 1640 in a cleverly designed small volume, aimed at wide and easy distribution.[65] Jeremiah Horrocks (1619–1641) observed the star in January 1640, and we generally know that news of it was widespread, although it was only in the 1660s that Bouilliau (1605–1694) established the fact that the appearances of *Mira Ceti* are cyclical and provided an accurate calculation of its period.[66]

All this fits in chronologically with Descartes dramatically rearticulating his natural philosophy when he came to write the *Principles*. We do not know whether Descartes, who was in the Netherlands at the time, knew personally Holwarda or Fullenius.[67] However, he deals extensively with variable stars in the *Principles*. Hence it may safely be concluded that between late 1639 and sometime during the

---

[64] Dijkstra (2011) 86–87.

[65] Holwardus (1640), *pars secunda de novis phaenomenis, sive stellis*, 185–288. The star disappeared after he first observed it, and Holwarda failed to observe it all through the summer of 1639 ('frustra omnia', p.285). But, Holwarda saw it again about 11 months later, on Nov 7, 1639. By that time his book was being been printed, so he added an appendix (pp.277–88) about the reappearance. Here he pointed out that he had already suggested the phenomenon might disappear and reappear, and now identified the observations with a star in Cetus (Dijkstra 2011, 86–87, see also 89ff on the design and aim of Holwarda's book). A slightly different account of the timing of Holwarda's observations, making use of the work of Michael Hoskin (1977), is offered by Donahue (2006), 590–91, according to which Holwarda re-observed Mira Ceti in 1640 while his book reporting the initial discovery was in press, the appendix being added to report that reappearance. Note that, given Mira Ceti's 11 month cycle the 1640 observation by Holwarda must have been no earlier than October of that year.

[66] Bullialdus [Bouilliau] (1667) established Mira Ceti's period as about 333 days, allowing him successfully to predict future appearances. He proposed that the star rotates, periodically showing a more luminous region to earthly observers. So, as Dijkstra (2011, 92–97) convincingly shows, and as we might expect based on modern studies of the negotiation and attribution of discovery, the historical process of recognizing that a periodically disappearing and reappearing star had been found was long and hotly contested.

[67] Vermij (2002) says Descartes was in contact with many Dutch scholars (as is well known in any case), but offers no evidence concerning Holwarda. Terpstra (1981) says there is no proof that Descartes knew Holwarda, but also claims, p.67 that there is no doubt of Descartes' influence on natural philosophy in Franeker; that Descartes certainly influenced Holwarda; but, that there is no proof they met in person. This question is not definitively resolved. Judit Brody is currently exploring it further. Mersenne was quickly made well aware of Holwarda's work and the ensuing debate (Dijkstra 2011, 94–95), and so he may have been Descartes' main or initial informant on the matter.

composition of that text between 1640 and 1643,[68] he became acquainted with the possibility that variable stars might exist, without of course having any sense that they are strictly and characteristically periodic, since Bouilliau's work appeared much later. As we are about to see, it is clear that for his *Principia*, Descartes decided to re-frame and articulate Holwarda's claims into something like the following form: '*the disappearing and reappearing 'phenomenon' [Holwarda's 'new star'] might indeed be a star in the normal, cosmological sense, and further natural philosophical significances (explanations and systematic relations) might be attributed to this type of object. In particular novae and variables might be intimately related.* Descartes' bold working out of this strategy, indeed its deep cosmographical exploitation, is our next topic.

12.9 Extending the Strategy: Seizing upon Novae and Variable Stars in the *Principia Philosophiae*

Descartes explains the disappearance and appearance of certain stars and their change of apparent brightness using sunspots as explanatory devices; that is, using sunspots as already explained and framed within his corpuscular-mechanical matter theory, vortex celestial mechanics, inter–vortical dynamics and theories of star formation and magnetism. Hence dramatic *explananda* dating from the debates about Galileo's and Scheiner's claims, became in turn—in the total context of the system—*explicantes*. According to Descartes in the *Principles*, when first formed spots are soft and rarefied and easily trap other particles, but eventually their inner surface, the surface contiguous with the star, becomes hard and polished. Subsequently these spots are more stable and less easily reabsorbed. So, after a while it can happen that a spot gradually extends over the whole surface of the star and blocks its light.[69] This does not necessarily mean that there is absolutely no light coming from the direction of the star, since *boules* of second element constituting the vortex surrounding the star still exert tendency to motion away from the centre, but the light emitted may not be strong enough to cause sensation in our eyes.[70]

[68] Clarke (2006) Appendix 1 on 'Descartes' Principal Works'. Descartes was working on the *Principles* all during his controversy with Voetius and the University of Utrecht, the publication of the *Meditations* in 1641 and various entanglements with some Jesuits. It was only in January 1643 that he told Constantijn Huygens that he was currently working on the sections about magnetism. *Ibid.* 233. Clarke (note 30 to page 233) assumes this applies to the explanation of Gilbert's lab manipulations in Book IV of the *Principles*, but it might just as well apply to the cosmic magnetism prominent in Book III.

[69] *Principles,* III articles 102, 104; AT VIII-1 151–2; MM 139–40, 140–41.

[70] *Principles,* III article 111; AT VIII-1, 158–60; MM 144–5.

Moreover, such a previously disappeared star can also reappear again.[71] This reappearance is intimately connected with his conceptions in the *Principles* about the stability of vortices, which was briefly discussed earlier in Sect. 12.4: There is a constant interplay between vortices depending on their size, strength and situation.[72] Vortices are contained by neighboring vortices, but they can weaken and they can even collapse. In general, a vortex whose central star is covered in spots is weakened, because the first element in the body of the star is prevented from pushing away on the globules of the surrounding heaven. At the same time Descartes points out that spots have a great number of pores through which first element material can pass, but in one direction only, because, forcing their passage through a pore, the particles bristle up the material which then prevents their return.[73] It can happen that while the vortex of a star covered in spots is weakened, it is still stronger than some neighboring vortices and extends into their space. By this Descartes means the globules of the second element getting further away from each other, with first element particles filling in the space between them.[74] A star completely covered in spots cannot expand; but, as a result of the constant alterations of the shape and radial extent of the boundaries of jostling vortices (Fig. 12.2), the surrounding material—vortical *boules* of second element and interstitial first element particles—might move out further from its surface, allowing additional first element particles to pass through the pores from inside and cover the spots, with the result that a new star is born. This star now has a core (the old original star) then a crustal layer of spots and finally a new outer shell of roiling, agitated first element particles.[75] This shell building can continue and several layers can accumulate.[76]

If the star in question has never been observed during its occluded phase, but then comes into view for the first time, as far as humans are concerned, it is a *nova*.[77] If the star in question had been observed, then disappeared and now reappears, it is a variable, as very recently attested in European astronomy.[78] It is also

[71] Descartes refers explicitly only to *novae*, but here the reappearance at the same place is an important feature, as we shall see. *Principles*, III article 104; AT VIII-1 152; MM 140–1

[72] *Principles,* III article 111; AT VIII-1 158–60; MM 144–5.

[73] *Principles,* III articles 105–108; AT VIII-1 153–56; MM 141–143.

[74] One should recall that first element particles are constantly flowing into the central star from the north and south along its axis of rotation.

[75] *Principles,* III article 111; AT VIII-1 158–60; MM 144–5.

[76] *Principles,* III articles 112, 114; AT VIII-1 160–2; MM 145, 146–7.

[77] *Principles,* III article 104; AT VIII-1 152; MM 140–1. Descartes cites the 1572 *nova* in Cassiopeia, 'a star not previously seen'. He also mentions, more controversially: [1] the possibility of the disappearance of one of the Pleiades in ancient times, seven stars being mentioned in myth but only six reported by later Greek writers (MM 140 note 105)—such a star, if it once was visible, has obviously been occluded for over 2,000 years; and [2] the presumed fact that, 'We also notice other [more enduring] stars in the sky which formerly were unknown [to the ancients]', a claim which MM otherwise explain in their note 107 to p.141.

[78] *Principles,* III articles 112, 114, AT VIII-1 160–2; MM 145, 146–7. In contrast to the 1572 *nova* which he does report, Descartes does not name *Mira Ceti*, Fabricius, Fullenius or Holwarda. It is almost as though he is happier to offer the explanation in principle for a phenomenon of which he surely is aware in general, but without giving any firm citation of dates, discoverers or objects, thus revealing a still neo-Scholastic approach to the description and explanation of phenomena as 'generally well known'. Cf. Dear (1995). See also Sect. 12.12 below, point [3] and note 108.

clear that all these processes can occur suddenly or gradually: a known star can quickly or slowly disappear; a previously known star which has disappeared may reappear (hence now recognized as a variable) suddenly or gradually; a previously unobserved star, presumably long occluded from human observation, may suddenly or gradually come into view (a *nova* according to the model Descartes is expounding).[79]

Thus, overall, Descartes seems concerned to assert a set of general causes involving sunspot formation and dissipation—along with varieties of contingent outcomes amongst interacting sunspots, vortices, the surfaces of stars and stellar 'aethers'—that allow for a wide spectrum of in principle explanations of possible appearances.[80] Simultaneously, he is also implying it must be granted that human history and frailty have conditioned the appearances actually recorded, which perforce are the only ones we have that we can juxtapose to the explanatory resources he provides. Residing deeper in the tissue of his natural philosophical explanation are a number of key principles: all the processes are natural; no totally new balls of first element materialize inside vortices; there is no *ex nihilo* emergence or creation of a star where there has never before been one; any star may quickly or slowly disappear, and quickly or slowly [re-]appear;[81] but, the original sphere of first element is still there, possibly under additional alternating layers of third element crust and first element star stuff. This fully naturalizes *novae*, and renders them in explanatory terms a sub-class of variables, whose categorization is contingent upon the history of human observation of the star in question. Descartes thereby naturalizes, unifies and rationalizes the known empirical domains of *novae* and variables, subordinating to his natural philosophical strategy all the matters of fact he has chosen and framed as relevant. His next move, expressing and completing the cosmographical intentions of his system, involves relating the Earth, and indeed every single planet, comet and planetary satellite in the universe, to a certain pattern of possible stellar development.

[79] See for example: *Principles,* III article 104, AT VIII-1 152; MM 140. Speaking of *novae*, in particular the 1572 *nova*, Descartes says that such a star 'may continue to show this brilliant light for a long time afterwards, or may lose it gradually'. Cf. *Principles*, III article 111, AT VIII-1 159; MM 145: the 'almost instantaneous' appearance of a star; *Principles*, III article 112, AT VIII-1 160–1; MM 145: a star 'slowly disappearing'; and, *Principles*, III article 114, AT VIII-1 162; MM 146–7, the same star can alternately appear and disappear, which phenomenon Descartes elucidates with the analogy of pendulum motion (see note 27 above). An excellent exposition of Descartes' theories of comets, variable stars and *novae* (as a sub-species thereof) may be found in Heidarzadeh (2008), 67–81. Very helpful and well conceived diagrams accompany the discussion of the key points.

[80] *Principles* III article 101, AT VIII-1 151; MM 139: 'That the production and disintegration of spots depend upon causes which are very uncertain', a remark to be taken in conjunction with his explanations offered in the next 20 or so articles of the *Principles*, dealing with *novae*, variables and sunspots.

[81] The 're' is in brackets, because causally the star may be reappearing, but humans may only be noticing a star in that position for the first time; it is what European natural philosophers and astronomers had since 1572 called a new star.

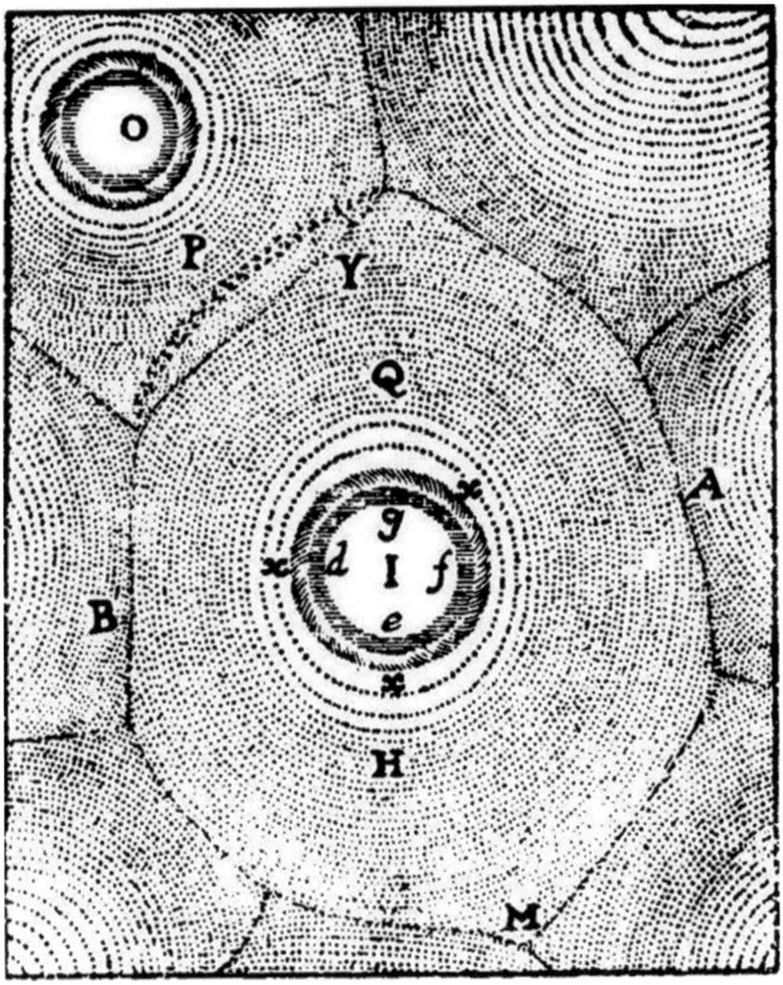

Fig. 12.2 *Principles,* AT IX , p.667, Planche IX, Figure 1, used for discussion of nova and variable formation, *Principles* Book III Articles 105–114, and showing changed position for inter-vortical boundary [P/Y] and possible shell formation around star. This plate in the French version of the *Principles* is clearer than the one used in the Latin edition; AT VIII-1, p.157 and several other times

12.10 Raising the Cosmographical Stakes: Genealogy of the Earth and All Other Planets in All Other Systems

We have seen that 'system-framed', sunspots (or starspots) occupy a central role in Descartes' system as presented in the *Principles*. They serve as explanations for the genesis of the third element and for variable stars and *novae*. But they have an additional, equally dramatic explanatory role. Occasionally a vortex collapses and the sunspot–encrusted defunct star in its centre is captured by another vortex, becoming a comet or a planet, entities that are composed mostly of the third element.[82] Here we encounter, on the systematic level, the material in the *Principles* most often treated as Descartes' 'theory of the Earth'. Indeed it is that, and it had significant impact on subsequent readers as a gambit in that domain, with huge theological and natural philosophical implications. However, properly understood in terms of systematizing strategies and cosmographical plays, the intended scope of Descartes' treatment is much wider.[83]

The dynamic of star spots encrusting and eventually destroying stars is what accounts in matter theoretical and structural terms for each and every planet and comet to be found in the universe. When the sun spots have completely encrusted the surface of a star, it is unable to help to maintain the overall centrifugal tendency of its vortex, and rather than a variable star eventuating, as just described, the entire vortex might instead collapse, with the dead, encrusted star itself being sucked into a neighboring vortex, there to become a planet or comet, according to its degree of 'solidity' or 'massiveness', and the usual workings of the vortex mechanics.[84] So, on this breathtaking vision, every planetary and cometary object in the cosmos traces its genealogy to the pattern of events that in principle might befall any

---

[82] *Principia* III, arts, 118–119; AT VIII-1, 166–168; MM 149–50.

[83] The narration/explanation of Earth formation and structure occurs at *Principia*, IV, arts 1–44, AT VIII-1, 203–231; MM181–203. Most of the attention paid to this material has been devoted to seeing Descartes as a founder of the early modern and enlightenment tradition of speculative theorizing about the Earth. (Cf. Roger 1973) The unfolding of this tradition, particularly in its English Protestant context, has been most perspicaciously analyzed by Peter Harrison (2000) who correctly suggested that the issue was not the substitution of a natural philosophical cosmogony for the account in Genesis, but rather the nuanced issue of which natural philosophical account best explicated or shed light on Genesis, a matter about which Descartes' account arguably had already displayed some sensitivity. Harrison argued that many historians mistakenly think that late seventeenth century English natural theologians and natural philosophers read Descartes' cosmogony and cosmology as a history, which therefore would have to agree with or contradict a history in Genesis. Against this Harrison pointed out that the central issue for the players was not whether Cartesian philosophy provided a parallel creation narrative. It was, rather, whether 'Cartesian or Aristotelian Philosophy would shed more light on the biblical account of creation.' It was not Descartes versus Moses on history, but which natural philosophy—Cartesian or Aristotelian—better explicated what Henry More, for example, had called 'The Physiological part of Mosaical Philosophy'. Clearly, once we understand the structure and dynamics of the field of natural philosophizing, as sketched and applied in this volume, we can see that similar points apply to Descartes himself. The issue for natural philosophical actors, including Descartes, was correctness in the natural philosophical field (Descartes vs. Aristotle), *and* the mode of articulation of natural philosophical utterances onto *Genesis*. Additionally, Harrison's ideas can be extended even further, in accord with the themes of the present volume, granted that we grasp the natural philosophical (rather than primarily theological) intention of the *Principia* cosmogony and earth history. As we are seeing in this chapter, the cosmogony and earth history play a brilliantly contrived, and controlled, role in the 'system-binding' of the natural philosophy taught in the *Principia*, well beyond the system which we discerned in the pages of *Le Monde*.

[84] As analysed in detail above in Sect. 10.2.3.

'star-in-a-vortex'.[85] Presumably, all planets, as opposed to comets, undergo the same further process of planetary shaping which is then described for the case of the Earth—the formation of land masses, with mountains and declivities, the latter filled with water to form oceans and seas subject to the phenomena of tides, which are a key cosmographical case for Descartes.[86] The account of the process turns most importantly on the results of the collapse of a crust, eventually formed from aetherial material of third, second and first element particles inherited from its dead, parental star.[87] Hence, all this material on 'Earth history' should not be treated in a piecemeal manner and as of marginal importance for the system of the *Principia*. Rather, the account of how planets and comets arise from stars, and the detailed theory of the process of formation of planet structure, arguably should be looked at in detail in relation to one another, as part of Descartes' strategy for securing the *Principia* as a coherent, extensive and novel—because so essentially cosmographically focused—system of natural philosophy.

12.11 Radical Realist Copernicanism and the Grand Cosmographical Gambit

We have reached the climactic point in our analysis, where it is appropriate to reflect upon the totality of what I shall term Descartes' 'grand cosmographical gambit' in the *Principles*. The gambit may be defined as follows: It begins with Descartes' theories of vortices and star structure and his corpuscular-mechanical co-opting of

[85] Satellites are also planetary in nature as we know from our detailed study of *Le Monde*'s celestial mechanics and theory of the moon. See also *Le Monde* AT X 69–70; SG 45; where the moon is termed a planet: '...if two planets meet that are unequal in size but disposed to take their course in the heavens at the same distance from the sun...'. In the *Principles*, of course, Descartes can rely on his genealogy of planets from encrusted stars— for example, at Book III article 146; AT VIII-1 195–96; MM 171: 'Concerning the creation of all the Planets' where it is clear that the planets of our solar system, along with the Earth's moon, the four satellites of Jupiter and the two Descartes attributes to Saturn all derive from encrusted stars in now defunct vortices, and are 'planetary' in nature.

[86] For Galileo and Descartes the tides provide a prime example of a phenomenon on Earth which, if well theorized, provides strong evidence for the motion of the Earth. Biro's analysis, which we followed in Sect. 11.3, devotes two chapters to their cosmographical use of theories of the tides (Biro 2009, 72–110). For Descartes in the *Principles*, tides are implied to be a feature of all planets, just as their magnetism is. Both sets of phenomena would be present on any and every planet, since their genealogies are identical to that of our Earth: Every planet carries with it the axial orientation of pores to accept the two species of screw shaped particles of first matter which it had as a star. Exactly how this is retained in the now third matter crustal layer[s] of the planet is detailed in Descartes' story of the Earth in Part IV of the *Principles*. Similarly the process of formation of oceans, mountains, valleys and atmosphere would be the same for all planets evolved from dead stars.

[87] The crust in question is not the primordial crust formed of sunspots which initially strangled the star. That crust remains deep in the planet, untouched by this process of creation of oceans, seas, landforms and atmosphere. Cf. note 59, and Chap. 11, Note 14.

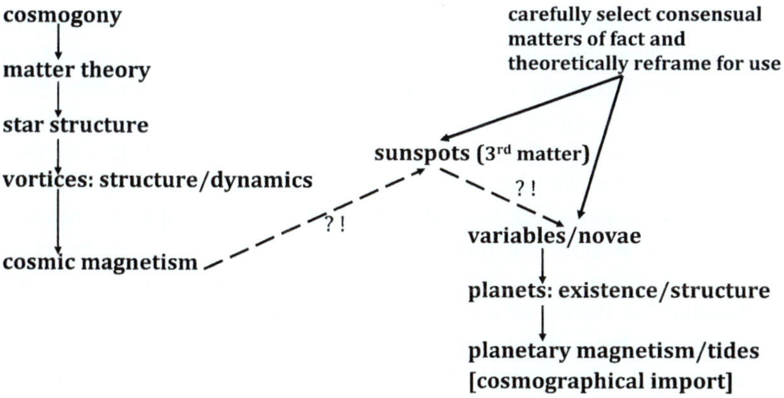

Fig. 12.3 The theory of sunspots is pivotal to binding together the sequences of cosmographical claims in the *Principia*

Gilbert's gambit of making magnetism a phenomenon of 'cosmic' significance. That Cartesian 'cosmic magnetism' is then the starting point of the rest of Descartes' cosmographical narrative/explanation, whilst his account of the formation of third matter sunspots out of first element magnetic grooved particles on stellar surfaces is its pivot, as, on that basis, the *Principia* goes on to explain *novae*, variable stars, the origins of planets and comets, and—cosmographically taking the Earth as its exemplary case of a 'known planet'—not only the structure of the planets, but also the common process of formation of their common structure.

Figure 12.3 illustrates the content of the gambit and where its most bold strategic moves were placed. Consider two sequences of natural philosophical claims which we now know were offered in the *Principia*: On the left we move from cosmogony, through matter theory to star structure and Descartes' vortical celestial mechanics and inter–vortical dynamics. On the right we move from claims about the nature of *novae* and variable stars through the genesis of planets (and comets) and via the 'theory of the Earth' to an account of the formation and structure of any planet, including the nature and cosmographical import of the tidal phenomena it will display. *Le Monde* had only offered an early version of the sequence on the left, plus the early version of his theory of the tides. The *Principia* offers both sequences, tied together by means of Descartes' theories of magnetism and of sunspots. His accounts of sunspots, *novae* and variable stars make use of judicious selection of available matters of fact and their framing for systematic natural philosophical use. The entire structure of cosmographical argument as presented in the *Principles* depends upon the way Descartes has elected to construct and place his theory of sunspots as generated by magnetic particles. The figure represents this point by linking the two sequences of claims through the claims about sunspots and by the dotted rather than full lines linking cosmic magnetism to sunspots, and then sunspots to variable stars and *novae*. The question marks and exclamation points attached to the dotted lines signal the strategic, novel and daring nature of the argumentative linkages flowing into and from the theory of sunspots.

The point to be noted is that Descartes did not necessarily have to do anything as daring or elaborate as this, even if he wanted to extend and improve upon *Le Monde* and take account of recently consensually agreed facts about sunspots. Descartes could have played it simpler and safer by just adding a theory of sunspots to his natural philosophy as a marginal extra, probably requiring the changes to his matter theory and cosmogony we have noted in the *Principles*, but nothing else. In other words Descartes could have put into the *Principles* a theory of third matter formation and sunspots without the further articulations 'back' to a theory of cosmic magnetism or 'forward' to *novae*, variables and planet formation, etc. Or, he could have elaborated his theory of cosmic magnetism and still used it to ground his theory of sunspots, but without going on from sunspots to *novae*, variables, planets, their structure and tides. Either of these smaller gambits would have involved changes only in matter and element theory and cosmogony, rather than the 'huge cosmographical gambit' we are discussing.

In fact Descartes took just about the most daring and radical path one could imagine in the circumstances. He brought the entire right hand sequence of claims into his system, that is, novelties about *novae* and variables linked further to planet formation, structure and the emergence of tidal phenomena, and he did this on the basis of his theory of sunspots, which he had developed as an elaboration of the sequence of claims on the left, which are articulations of material in *Le Monde*, *plus* the theory of magnetism in cosmic setting. The resulting structure, the grand cosmographical gambit, is hardly some careless or unintended outcome; nor is it lacking systematic natural philosophical coherence, a coherence extending over a range of claims far beyond that contained in *Le Monde*; nor do the key new claims lack an empirical basis, constituted as they are by timely appeals to novel but consensually received matters of fact of the day.

In saying that Descartes had important recourse to matters of fact, and hence that his natural philosophy is more factually grounded than perhaps is usually granted, we are not thereby falling into the tired topos that he was 'influenced' by certain facts to design and execute his gambit. Descartes actively selected, interpreted and reframed for systematic natural philosophical use empirical claims from the available set of relevant matters of fact.[88] As I have said, he selected relevant sunspot matters of fact as *explananda*, framed them in his own elaborate explanations—of element theory, magnetism, vortex and star structure—and then strategically leveraged them into *explanans* for the creation of third matter and the existence and structure of planets and comets (by way of variable stars and *novae*, about which he also selected recently announced matters of fact and treated them first as *explananda* and then as *explanans*). He *appropriated* the Galileo/Scheiner 'facts' about sunspots, but only on condition that he could frame them with an elaborate explanation linking back to his magnetic particles as *sources* for sunspots, and forward to variable stars and planets as *outputs* of their now framed properties

[88] The historiographical view point behind this remark was set down in Chap. 1, note 25 and has been adhered to throughout, but with particular reference to the issue of explaining Descartes' career inflection in Chap. 8 and whenever the agonistic dynamics of the field of natural philosophy have been in view, as here.

and modes of behavior. Descartes was trying to extend his natural philosophy, and systematically bind it together much better than he had in *Le Monde,* by scoring heavily in the realist Copernican cosmographical game of intimately relating the heavens and the Earth. And he did this, as we have found, by constituting the *Principles* as a set of radical, realist-Copernican cosmographical threads of narrative/explanation, tightly woven into a vast natural philosophical cloth.

To grasp fully the daring and scope of Descartes' cosmographical gambit, we need to follow Jacqueline Biro a bit further—beyond our use of her findings about the theory of the tides in Sect. 11.3, and her use of the category 'cosmography' in Sect. 12.2 above. This will help us appreciate that if Descartes was not the first ambitious realist Copernican natural philosopher to seize the cosmographical nettle, he may well have been the most daring and systematic to that point. Biro started out from a little noticed set of papers by Edward Grant, Thomas Goldstein and W.G.L. Randles (hereafter GGR).[89] These dealt with Medieval Scholastic quandaries over Aristotelian doctrine concerning the shape of the Earth, the placement of 'land' (the element earth) and relative amounts of earth and water. Prominent in these debates was a conceit wherein the land mass of the known world protruded—like a bobbing apple—out of a much larger and encompassing spherical mass of water, thus spoiling the perfectly spherical shape of the Earth, and earning this model the epithet, 'bobbing apple' theory of the Earth.[90] GGR variously show that these debates, including the rather widely known bobbing apple theory, were from the late fifteenth century overridden from outside the universities due to recovery of Ptolemy's *Geography* and the voyages of discovery, leading to the [re]emergence of Ptolemy's concept of the 'terraqueous' globe, consisting of a very nearly perfectly spherical mass of 'earth', marked by relatively small protuberances—mountains—and declivities, or relatively shallow hollows, containing water; that is, the seas and oceans. This reborn Ptolemaic terraqueous globe, enriched with the geographical findings of the voyages of discovery, was therefore very much a sixteenth century construction, taking place at first outside the universities, in the work of humanists, elite navigators, practical mathematicians and intellectually adventurous

[89] Thomas Goldstein (1972), Edward Grant (1984) and W. G. L. Randles (2000). Grant cites an article in French by Randles dated 1980. This suggests that the concepts in the English version of the work by Randles appeared in the earlier French article and therefore Randles' work predates that of Grant.

[90] In the thirteenth century, Aristotelians such as Sacrobosco and Michael Scot tried to reconcile the ideal picture of concentric spheres of the elements with the indubitable existence of dry land by proposing that the earth emerged slightly from the sphere of water. In the fourteenth century, Jean Buridan and Albert of Saxony articulated the 'floating apple' model of the Earth to square theory of the Earth with the additional belief, ascribed to Aristotle in some circles, that the sphere of water is ten times larger than that of earth. Biro (2009), 17–21, 23–25, following GGR. The Scholastic debates examined by GGR about the shape of the Earth and the distribution of water and earth certainly were cosmographical, having to do with fitting together the cosmological and Earth theoretical dimensions of the natural philosophy. This is especially true, given the fact that the issues studied by GGR concerned outright systemic tensions between the cosmologically dictated shape of the elemental realms—of aether, fire, air, water and earth—and considerations driven by need to define on Earth the place and extent of dry land above water.

non-Aristotelian natural philosophers.[91] The sharp end of GGR's findings focused on Nicolas Copernicus, with their contention that Copernicus was a relatively early convert to the newly re-minted terraqueous globe, and that his chapter on the shape of the Earth in Book I of *De Revolutionibus* reflects this, and is specifically used to advance the idea that only a truly spherical Earth (that is the terraqueous model as opposed, say, to the Scholastic bobbing apple model) was fit and able to rotate.[92]

Biro's fruitful insight was to extend the intellectual trajectory started by GGR, emphasizing cosmographical moves by a series of combative anti-Aristotelian natural philosophers—Bruno, Gilbert, Galileo and Descartes—and analyzing their gambits and tactics within the sort of view of natural philosophical contestation that has informed my work in this book.[93] These alternative natural philosophers of realist Copernican leanings found in the terraqueous globe a tool and a topic of natural philosophizing whereby increasingly articulated knowledge or speculation about the structure and make up of the Earth, could link to, support or ground realist Copernican cosmological arguments—the natural philosophical tactics and discourses evolving as one moved from Copernicus through the later cases. For pro-Copernican natural philosophers, the novel, terraqueous Earth, offered the possibility of articulating claims about that Earth that could lead to, support, and blend with their radical view of the heavens. Since the Earth is a planet, it must resemble the heavens, and the latter must resemble the Earth. In natural philosophical terms, this means that issues of resemblance, indeed identity of matter and cause were at stake, and that *cosmography*, as we defined it above in Sect. 12.2, following Biro, became for such players a preferred battlefield.

Opportunities might be available to argue from structure, matter and cause on Earth, near at hand and open to investigation, to the heavens. The terraqueous globe of the Earth had already played a small role for Copernicus himself in this regard, but more ambitious arguments could be built from further articulations of the nature of the terraqueous globe of the Earth, out to the heavens. All of Biro's cases show arguments of that form and direction. And so all can be grasped under the category of cosmography, in a radical pro-Copernican form, wherein new claims about earth structure form the theory of the Earth side of the Earth theory-cosmology pairing that constitutes cosmography.[94] Gilbert's natural philosophy and cosmology were

[91] In the late fifteenth and sixteenth century, controversy erupted with thinkers like Vadianus, Fernal, Nunes and Peucer rejecting the floating apple model of the Earth on the basis of knowledge gained from the voyages of discovery, and campaigning for the notion of a spherical, terraqueous globe derived from Ptolemy's *Geography*. It appears that the terraqueous globe entered university curricula only in the late sixteenth century through the efforts of Clavius. Biro (2009), 17–21, 30–36.

[92] Biro (2009), 28–30, 36–39, synthesizing the important claims by GGR on this little appreciated point.

[93] Reflecting of course her initial training in the School of HPS at the University of New South Wales, via her MA thesis, supervised by the present author.

[94] Biro is able to offer a most interesting historiographical observation in this connection: Despite all the excellent scholarly work recently expended upon Renaissance and early modern geography, the numerous attempts in that literature to answer the question, 'What did geography have to do with the Scientific Revolution?', have all been stuck on idea that geography provided models of

built almost entirely on the basis of moving out to the heavens after having established the structure, and essentially magnetic character of the Earth. Where Copernicus had exploited in this regard simply the newly reaffirmed spherical shape of the terraqueous Earth, Gilbert was focusing on its structure and characteristics. In addition to all his straightforwardly astronomical and cosmological work, Galileo, too entered this cosmographical competition, amongst Copernicans. By this stage the terraqueous nature of the Earth was not in doubt. Rather, Galileo took pains to try to refute Gilbert's magnetic Earth, moving tactically to replace that form of earth theory–to–cosmology argument with one of his own, according to which only the phenomena of the tides, explicated according to his theory, could provide terrestrial based evidence for the Copernican system.[95]

It was then left to Descartes to offer the most radical version of this sort of pro-Copernican cosmography, embedded in an anti-Aristotelian natural philosophy and articulated with extensive new claims about the structure, genesis and stellar heritage of the Earth, and indeed all planets in any vortex whatsoever. This is because in the *Principia*, his explanation *cum* narration of the heavenly origins of planets and their make up, drawing upon the vortex mechanics and theory of stars, cosmic magnetism and sunspots, and debouching in planet structure ripe for undergoing tidal phenomena, is not tangential to the system, but rather is the very core of its content, and its system-binding strategy.[96]

---

empirical method, or of appropriately utilitarian aims and values for 'the new science' (Biro 2009, pp.15–6). Biro's argument directs us not to the supposed methodological or normative contributions of geography, but directly to the issue of how part of its substantive content was played upon, and played into, the most dynamic and turbulent part of the process of natural philosophical contention that marked the period. In addition, going beyond Biro's point, there is the consideration that the usual arguments for methodological/normative 'influence' from geography to 'science' are simply redundant reiterations of arguments variously made, since the 1930s, for many of the domains of the practical arts and practical mathematics. What is needed is attention to the way active natural philosophical players adopted and adapted claims and hardwares from practical mathematics into their natural philosophical agendas and strategies, in accord with the sort of cultural process model of natural philosophy advanced in this book. See also on the practical mathematics and the Scientific Revolution question, including, on the historiographical issues involved, J.A. Schuster, 'Consuming and appropriating the mixed mathematical fields, or, being 'influenced' by them: the case of the young Descartes' available on my website: http://descartes-agonistes.com. Some points related to this study were made above, Sect. 8.4.2, concerning Descartes dealings in practical optics with Ferrier.

[95] Biro (2009) on Gilbert, 57–64; on Galileo, 73–94, on Descartes, 95–110.

[96] As for the long term strategic tendency of realist Copernican natural philosophers to pursue cosmography with novel Earth theory claims and extrapolations, we see an evolution from Copernicus' own concentration on the shape of planet Earth, through Gilbert's detailed natural philosophizing about the inner structure and make up of the Earth, down to Descartes' invocation of a process of heavenly generation to cement his cosmography and provide a developmental story for his claims about Earth's structure and formation. As Biro argues, for realist Copernicans the exploitation of strategic space in cosmography was a continuing theme in their corners of the natural philosophical field, and so Descartes' 'theory of the earth' is not so much the stark novelty that some historians of geology sometimes make it out to be, but a radical turn embedded in a longer running strategic campaign by the supporters of realist Copernicanism. This approach also allows Biro to compare and contrast the cosmographical strategies of various actors. For example, she is

12.12 Conclusion: Cosmographical System and Strategy in the *Principia*, the Culmination of Descartes' Natural philosophical Trajectory

The natural philosophical system in the *Principia*, unlike that in *Le Monde*, is cosmographical in essence[97]: (Some) stars are destined to be planets, products of processes involving cosmic magnetism and the now surprisingly cosmically significant sunspots; planets are transformed stars, and all planets necessarily are terraqueous, because dead, encrusted stars of less than 'cometary solidity' will undergo the further formative structural dynamics, ending in the production of a planetary crust, which collapses to produce uplifted mountains, and water filled declivities. Cosmography in this new dynamic steady state register becomes an essential component of the revised, mature system of natural philosophy. In Descartes' *Principles* the usually accepted keys to the system, shared with *Le Monde*—matter-extension, his laws of motion and vortex mechanics—are fused and entangled with his daring cosmography into the new style, theory-driven narrative of star/planet life. What was tactical or strategic for some Copernican natural philosophers had become, for Descartes, hyper-strategic and essential; that is, directly constitutive of the systematic natural philosophical utterance itself. His mature natural philosophy *is* (rather than rests upon) the dynamic steady state cosmography—there are not simply 'relations' or 'consistency' between Earth and heavens; rather, each Earth, each planet that is, was once a member of the highest class of macroscopic heavenly bodies—a star—and each star can in principle become a planet; and every planet must be terraqueous, magnetic and subject in principle to tides, and all this depends at its core upon how cosmic magnetism and cosmically indispensible sunspots are taken to work.[98]

Before concluding this chapter two objections and one qualification to the foregoing claims need to be addressed, if only briefly. They concern: [1] the status of Descartes' belief in the motion of the Earth and hence the possibility of his having been the kind of radical realist Copernican bespoken by the cosmographical contents

able to point out the interesting differences in modeling of oceans in Galileo's and Descartes' theories of the tides: For Galileo it is the *containment* of particular seas and oceans in their basins that allows the combined orbital movement and diurnal spin of the Earth mechanically to cause the tides. For Descartes, as we have seen in Sect. 11.3, the theory of tides depends on stressing *the fluid continuity* of all the Earth's sea and oceans.

97 We have of course seen important cosmographical elements in *Le Monde*: for example, the fundamental assertion that the Earth is just another planet, in a realist Copernican framework of infinitely many stellar systems; the overtones of the new element theory, discussed above in note 10, and the theory of the tides, as we have mentioned.

98 I gratefully acknowledge that number of the foregoing points in this paragraph emerged in course of my supervision of Biro's research toward her MA thesis (2006), which was then transformed into Biro (2009). I was then able to articulate these insights in my succeeding collaboration with Judit Brody on Schuster and Brody (Note 2).

and structure of most of the latter portions of the *Principles*; [2] the problem of the lack of expert reception of his putative grand cosmographical gambit; and [3] the precise degree of Descartes' openness to novel facts within his new tactic of forming large explanatory/descriptive cosmographical narratives.

[1] As is well known, Descartes was at great pains earlier in the *Principles* to establish a 'philosophical' (as opposed to vulgar) definition of motion. In such philosophically conceived motion, a body must translate from the vicinity of the layer of matter immediately contiguous to it at its initial position.[99] According to Descartes in the relevant early articles of Book III of the *Principles*, the Earth does not accomplish such motion.[100] But what is the status of this doctrine? Some quite excellent scholars take Descartes perfectly seriously on these points and accept that this was Descartes' default and fundamental position on motion, and hence motion of the Earth.[101] This can be argued by staying close to the relevant passages, but seems to raise problems when the totality of the *Principia* is read, particularly as we have now read it, stressing its deeply pro-Copernican and cosmographically oriented content and strategy. I therefore tend to agree with other, equally adept scholars, who would argue that what we have here is an elaborate smoke screen set down before the fact of possible theological objections (or worse) to the *Principia*, from either Catholic or Dutch Reformed forces.[102] Hauled before any university debate, or worse an inquisition or other ecclesiastical inquiry, Descartes could have sworn up and down the anti-realist Copernican tenor in the text based on his reasoned, philosophical denial of the motion of the

[99] Early in Book II of the *Principia*, at article 25, Descartes defines motion as 'the transfer of one piece of matter or of one body, from the neighborhood of those bodies immediately contiguous to it and considered at rest, into the neighborhood of [some] others' (AT VIII-1 53–54; MM 51). This is the philosophical definition of motion contrasted with vulgar or common understandings (Cf. Book II, article 24 'What movement is in the ordinary sense').

[100] *Principia*, III article 28, AT VIII-1 90; MM 94–95: '...no movement, in the strict sense, is found in the Earth or even in the other Planets; because they are not transported from the vicinity of the parts of the heaven immediately contiguous to them, inasmuch as we consider these parts of the heaven to be at rest. For, to be thus transported, they would have to be simultaneously separated from all the contiguous parts of the heaven, which does not happen'.

[101] Daniel Garber (1992), 181–88, discusses the matter with his usual care and perspicacity. In the end, p.188, Garber rejects the view that Descartes' theory of motion and its laws is an 'elaborate mask', a 'contrived stratagem' to allow him to deny motion to the Earth.

[102] Peter Dear (2001), 96, 'Descartes was not worried about the potential heresy inherent in his ideas about the extent of the universe or the nature of the stars. He major concern...was the unorthodoxy (as defined by Galileo's trial) of holding that the earth is in motion. Descartes published the *Principles*, with its more elaborate version of the same world–picture as that of *Le Monde*, only once he had thought of a way to deny the movement of the earth without compromising any of his cosmology. The trick (and that is what is really was) involved emphasizing the relativity of motion'. And, p.98, 'The subtlety of Descartes' theology was matched by the subtlety of his physics. As far as he could help it, no one would be able to accuse him of teaching that the earth moves'.

Earth.[103] Only a decade after the trial of Galileo, to thus prepare for the worst was the least any sensible, and very smart, Catholic natural philosopher and realist Copernican should have done, and he did it. We also know that very little in Descartes' writings or public behavior that touched on his person, his persona or his career was presented in a straightforward way by him, ever.[104] Hence there is no reason to believe that his sublimely radical realist, infinite universe Copernicanism would come into the world without some clever masking upon which he could rely if necessary. For excited seekers of natural philosophical novelty and forceful explication of realist Copernicanism, the message of the concluding two Books of the *Principia* would, however, be clear.

This brings us to [2] because it must be granted that no reader in his or the next two generations seems to have responded to the totality of what I have identified as Descartes' cosmographical gambit.[105] He certainly was taken as a Copernican. However, as with his system as a whole, so with his cosmographical weave in the *Principia*: it was taken to pieces by critics and by proponents focussed on one or another facet of the complete edifice. For example, his theory of the Earth was eagerly taken on board to be criticized, reformulated or surpassed, but by a new generation of Earth theorists, not cosmographical warriors, as the fight for Copernicanism was well and truly over.[106] Similarly, there were both vulgar recountings, expert articulations, as Eric Aiton showed, and withering criticisms of his vortex theory. Arguably, only Descartes ever adhered on a full technical level to the Cartesian system of natural philosophy. However, none of this impugns a reading of the text of the *Principles* itself, in the context of Descartes' career and proclivities, as a grand Copernican cosmographical *tour de force*, the culmination of a series of such attempts by innovative realist Copernican natural philosophers, starting with Copernicus himself.

[3] An important qualification needs to be added to what has been said about Descartes' openness to and use of novel matters of fact in his mature system. Inside the toils of his radical realist Copernican cosmographical explanations *cum* narratives, Descartes did not and could not aim at linear, deductive explanations of each and every particular state of affairs he recognized as a reliably reported matter of fact. Descartes' laws of nature do not function as premises of deductive explanations. Rather, his laws of nature in these parts of the *Principia* function as human laws do

---

[103] Readers familiar with legal proceedings, then or now, would recognize the strength of Descartes' position, if threatened in a legal context. He could have quoted, verbatim, extensive and connected published passages about the true, 'philosophical' definition of motion and the non-motion of the Earth, and read those passages with pointed literalness.

[104] Innumerable instances of Descartes' habitually secretive, reclusive, publicly masked and overtly tricky persona are captured with great panache in Desmond Clarke (2006).

[105] Although we might make an exception for Christiaan Huygens, who mocks exactly the interweaving of cosmographical claims into what we termed the explanatory and descriptive narrative in the *Principles*. Huygens wondered how Descartes. 'an ingenious man, could spend all that pains in making such fancies hang together' [*Cosmotheoros* (The Hague 1698), cited in Brody (2002), 84]. This mirrors a change in natural philosophical temper and rules in the next generation, leading to exactly the dissipation of the Cartesian system and piecemeal use and criticism of it that we discuss immediately below. However, Huygens (no modern historian!) misses the point about what the game of natural philosophizing was about in the preceding Baroque age, and how well Descartes had played.

[106] Cf. note 83.

in the making of legal arguments. The laws are woven, along with carefully selected matters of fact, into flows of argument, narrative lines of description-explanation, of the sort we have just canvassed.[107] Descartes proceeds by asserting a network of basic explanatory concepts involving matter theory, magnetism, vortices and sunspot formation/dissipation that in principle can explain, via *discursive causal story telling*, a spectrum of possible empirical outcomes. The causal stories are filled out according to the varieties of observed outcomes by appealing, loosely, to a variety of possible interactions amongst sunspots, vortices, the surfaces of stars, and the 'aether' of old dissipated sun spot material that floats in each stellar vortex near each star.[108] So, although when compared to *Le Monde*, Descartes' mature natural philosophy in the *Principia* values novel matters of fact, the system remained relatively closed to novel, deep discoveries at the theoretical level, because unexpected observational outcomes were accounted for at the level of contingent narrative formation, rather than by considering modification to the structure of deep concepts.

To return, finally, to our starting point, we have seen that commentators like Love and Lynes were on the right track in pointing out the consequential differences in matter theory between *Le Monde* and the *Principia* as part of the attempt to determine the relation between Descartes' prentice and mature systems. But Lynes and Love did not grasp the sort of game of competitive natural philosophical systematizing Descartes was playing, let alone realize that it was a game that necessitated the selection, reframing and deployment of available, more or less agreed novel matters of fact.[109] The differences between *Le Monde* and the *Principles* are not simply, or mainly about matter theory and presence of metaphysical grounding. It is the vast system-binding cosmographical gambit of Descartes, entraining the use and reframing of key, available matters of fact—in turn leveraged into explanatory resources—that best characterize both the difference between *Le Monde* and the *Principles* as well as the novelty and daring of the latter text, thus expressing and grounding a case for a realist, infinite universe Copernicanism of the most radical type. Moreover, by

[107] In the telling remark that ends Book III (AT VIII-1 202; MM 177), Descartes asserts that all inequalities of planetary motion can be sufficiently explained using the framework he has provided. Clearly he in no way intends that explanations will proceed by deductions from laws of motion, plus boundary conditions, leading to the exposure and study of various levels and types of perturbations. So, for example, it is not elliptical orbits, and their deviations that he wishes to study, leading to refinement of the relevant laws. Rather, he offers a 'sufficient' (verbal and qualitative) explanation of orbital phenomena and the general facts that no orbit is perfectly circular, and that all orbits display variations over time.

[108] Descartes introduces the section of Book III of the *Principles*, dealing with sunspots, *novae* and variable stars at Article 101 by stating: (AT VIII1, 151; M 139) 'That the production and disintegration of spots depend upon causes which are very uncertain.' Cf. above note 78 and text thereto.

[109] Cf. above notes 25, 31. By this juncture it is perhaps appropriate to point out that there was nothing defensive or reactive about Descartes' novel moves in the *Principles* which I have discussed in this chapter. Love (1975) and Lynes (1982) might each be read as depicting Descartes as motivated, even forced, to make matter theoretical changes by defensive consideration of real or possible theological or metaphysical criticism. But merely defensive gambits arguably would have taken quite different shapes, as we have hinted. Natural philosophical contestation may be decoded in part as like a game; its rules of utterance are in part determinable; and, as in other games, when master players make well considered, complex attacking moves, that is obvious to attentive spectators. I modestly offer this as a final, parting example of the historiographical policy and practice

looking at the *Principles* in this way and having appreciated the strategic aims and gambits Descartes employed, we see that these in themselves provide the 'reasons' behind not only his choice of changes in matter and element theory, but indeed the underlying design of his mature system of natural philosophy as a whole.

To conclude, therefore, we have reached the goal set for this chapter: to cap off the argument of this volume—in a way consistent with its historiographical principles and concerns—by capturing the structure, rationale and strategies underlying Descartes' mature system of corpuscular-mechanism in the *Principia*, in comparison to what we had already determined about those issues in regard to *Le Monde*, by virtue of our long excursion through his physico-mathematical, methodological and natural philosophical trajectory from 1618 to 1633 in Chaps. 3, 4, 5, 6, 7, 8, 9, 10, and 11. There remain only two more brief tasks for our next, and concluding chapter—first a 'Coda' rounds off the themes of this volume, and second an 'Epilogue' takes a few peeks at the career of the mature Descartes and his relation to the next phase of the Scientific Revolution.

References

Works of Descartes and Their Abbreviations:

AT=*Oeuvres de Descartes* (revised edition, 12 vols.), edited by C. Adam and P. Tannery (Paris, 1964–1976). References are by volume number (in roman) and page number (in Arabic).

SG=*The World and Other Writings*, edited and translated by Stephen Gaukroger (Cambridge,1998).

MM=*René Descartes, The Principles of Philosophy*, translated by V. R. Miller and R. P. Miller (Dordrecht, 1991).

MSM=*Rene Descartes, Le Monde, ou Traité de la lumière*, translated by Michael S. Mahoney (New York, 1979).

CSM(K)=*The Philosophical Writings Of Descartes*, 3 vols., translated by John Cottingham, Robert Stoothoff, and Dugald Murdoch, and (for vol. 3) Anthony Kenny, (Cambridge, 1988) References are by volume number (in roman) and page number (in arabic).

HR=The Philosophical Works of Descartes, vol I translated by E.S. Haldane and G.R.T. Ross (Cambridge, 1968 [1st ed. 1911]).

Other

Aiton, E.J. 1972. *The vortex theory of planetary motion*. New York: Neale Watson Academic Publications.

pursued throughout this volume: Historians' of nature-knowledge games should pose questions and seek answers that are framed by theorizing and testing of 'iceberg' type categories and models of those games, their mutual articulations and their contexts. (On iceberg categories, such as that of natural philosophizing as developed *passim* in this volume, cf. Sects. 1.3; 2.3; 2.5.0; 2.5.5.) This, I submit, on present evidence is preferable to simply accepting the often unreflected upon suite of problems (and forms of answers) that routine training in a given historical specialty might confer—in this case 'history of matter theory' or 'history of ideas' about same.

Biro, Jacqueline. 2006. Heavens and earth in one frame: Cosmography and the form of the earth in the scientific revolution. Unpublished MA thesis, School of History and Philosophy of Science, University of New South Wales.

Biro, Jacqueline. 2009. *On earth as in heaven: Cosmography and the shape of the earth from Copernicus to Descartes*. Saarbrucken: VDM Verlag.

Bourdieu, Pierre. 1975. The specificity of the scientific field and the social conditions of the progress of reason. *Social Science Information* 14: 19–47.

Brody, Judit. 2002. *The enigma of sunspots: A story of discovery and scientific revolution*. Edinburgh: Floris Books.

Bullialdus [Bouillau], Ismael. 1667. *Ad Astronomos Monita Duo*. Paris.

Clarke, Desmond. 2006. *Descartes, a biography*. Cambridge: Cambridge University Press.

Daniel, Gabriel. 1692. *A voyage to the world of Cartesius*. London: Thomas Bennet.

Dear, Peter. 1995. *Discipline and experience: The mathematical way in the scientific revolution*. Chicago: Chicago University Press.

Dear, Peter. 2001. *Revolutionizing the sciences: European knowledge and its ambitions, 1500–1700*. Princeton: Princeton University Press.

Dijkstra, Arjen. 2011. A wonderful little book. *The Dissertatio Astronomica* by Johannes Phocylides Holwarda (1618–1651). In *Centres and cycles of accumulation in and around the Netherlands during the early modern period*, ed. Roberts Lissa, 73–100. London: LIT Verlag.

Donahue, William. 2006. 'Astronomy' in the Cambridge history of science. In *Early modern science*, vol. III, ed. Park Katherine and Daston Lorraine. Cambridge: Cambridge University Press.

Drake, Stillman. 1957. *Discoveries and opinions of Galileo*, Trans. and ed. Garden City, NY: Doubleday.

Dreyer, J.L.E. 1890. *Tycho Brahe*. Edinburgh: Adam and Charles Black.

Fabricius, David. 1612. *Prognosticon astrologicum auff das Jahr 1615*. Nürnberg, J. Lauer: Landtrachtinger.

Fabricius, Johann. 1611. *De Maculis in sole observatis, et apparente earum cum sole conversione, narratio*. Witebergae.

Frischlin, N. 1573. *Consideratio nouae stellae*. Tubingen.

Galileo Galilei. 1613. *Istoria e dimostrazioni intorno alle macchie solari e loro accidenti.* Roma.

Galileo, Galilei. 1953. *Dialogue concerning the two chief world systems*. Trans. S. Drake. Berkeley: University of California Press.

Galileo, Galilei. 1957. *Discoveries and Opinions of Galileo*, Trans and ed. Stillman Drake. Garden City, NY: Doubleday.

Garber, Daniel. 1992. *Descartes' metaphysical physics*. Chicago: University of Chicago Press.

Gassendi, Pierre. 1658. *Opera Omnia*. Repr., 1964, Stuggart. 6 vols. Lyon.

Gaukroger, S.W. 2002. *Descartes' system of natural philosophy*. Cambridge: Cambridge University Press.

Gemma, Cornelius. 1573. *De peregrina stella.* Antwerp.

Goldstein, T. 1972. The renaissance concept of the earth in its influence upon Copernicus. *Terrae Incognitae* 4: 19–51.

Granada, M.A. 2007. Michael Maestlin and the new star of 1572. *Journal for the History of Astronomy* 38: 99–124.

Grant, E. 1984. In defense of the earth's centrality and immobility: Scholastic reaction to Copernicanism in the seventeenth century. *Transactions of the American Philosophical Society* 74: 20–32.

Gualterotti, Raffael. 1605. *Discorso sopra l'apparizione de la nuova stella.* Firenze.

Harrison, Peter. 2000. The influence of Cartesian cosmology in England' in Descartes. In *Natural philosophy*, ed. S. Gaukroger, J.A. Schuster, and J. Sutton, 168–92. London: Routledge.

Heidarzadeh, Tofigh. 2008. *A history of physical theories of comets from Aristotle to Whipple*. Dordrecht: Springer.

Heilbron, John. 1979. *Electricity in the 17th and 18th centuries: A study of early modern physics*. Berkeley: University of California Press.

Holwardus, Johannes Phocylides. 1640. *Panselenos, …id est dissertatio astronomica.* Franekerae.

Hoskin, Michael A. 1977. Novae and variables from Tycho to Bullialdus. *Sudhoffs Archiv für Geschichte der Medizin und der Naturwissenschaften* 61: 195–204.

Kepler, Johannes. 1938ff. *Gesammelte Werke*, ed. M. Caspar. Munich: Beck.

Kuhn, T.S. 1959, 1st ed. 1957. *The Copernican revolution*. New York: Vintage.

Licetus, Fortunius. 1623. *De novis astris et cometis.*

Love, Rosaleen. 1975. Revisions of Descartes' matter theory in *Le Monde'*, *British Journal for the History of Science* 8: 127–37. Repr., in *René Descartes, critical assessments.* vol. 4. Ed. Georges J.D. Moyal. London: Routledge, 163–174.

Lynes, John W. 1982. Descartes' theory of elements from *Le Monde* to the *Principles*. *Journal of the History of Ideas* 43: 55–72.

McRae, R.F. 1991. Cartesian matter and the concept of a world. In *René Descartes, critical assessments*, vol. IV, ed. J.D. Georges, 153–162. New York: Moyal.

Malapertuis, C. 1633. *Austriaca sidera.* Duaci.

Marius, Simon. 1619. *Astronomische und astrologische Beschreibung des Cometen…1618.* Nürnberg.

Minnaert, Marcel. 1993. *Light and color in the outdoors*. Trans. and Rev. L Seymour. New York: Springer-Verlag.

Randles, W.G.L. 2000. 'Classical models of world geography and their transformation following the discovery of America,' in *Geography. Cartography and Nautical Science in the Renaissance. The Impact of the Great Discoveries*. Aldershot:5–76.

Reeves, Eileen and Albert van Helden, Trans. and ed. 2010. *Galileo Galilei and Christoph Scheiner: On sunspots.* Chicago: University of Chicago.

Reisacher, B. 1573. *De mirabili novae ac splendidissimae stellae*. Vienna.

Roger, Jacques. 1973. La Théorie de la Terre au XVII Siècle. *Revue d'Histoire des Sciences* 26: 23–48.

Scheiner, C [under the pseudonym of Apelles]. 1612. *Tres epistolae de maculis solaribus.* Augustae Vindelicorum, Marcus Welser.

Scheiner, C. 1630. *Rosa Ursina, sive, sol ex admirando facularum et macularum suarum phaenomeno*. Bracciani.

Schuster, John. 2002. L'Aristotelismo e le sue Alternative. In *La Rivoluzione Scientifica*, ed. D. Garber, 337–357. Rome: Instituto della Enciclopedia Italiana.

Schuster, J.A. 2009. Descartes—Philosopher of the scientific revolution; or natural philosopher in the scientific revolution. *Journal of Historical Biography* 5: 48–83.

Schuster, J.A. 2012a. What was the relation of Baroque culture to the trajectory of Early Modern Natural Philosophy? In O.Gal and R. Chen-Morris eds. *Science in the Age of the Baroque, Archives internationales d'histoire des ideés* 209: 1–35.

Schuster, J.A., and Graeme Watchirs. 1990. Natural philosophy, experiment and discourse: Beyond the Kuhn/Bachelard problematic. In *Experimental inquiries: Historical, philosophical and social studies of experimentation in science*, ed. H.E. Le Grande, 1–47. Dordrecht: Kluwer.

Shapin, Steven. 1982. History of science and its sociological reconstructions. *History of Science* 20: 157–211.

Snel, Willebrord. 1619. *Descriptione cometae anni… 1618*. Lugduni Batavorum.

Tacke, J. 1653. *Coeli anomalon.* Gissae Hassorum.

Tarde, Jean. 1620. *Borbonia Sydera.* French trans. (1623). Paris.

Terpstra, H. 1981. *Friesche Sterrekonst.* Franeker.

Tycho Brahe. (1916) *Tychonis Brahe Dani Opera omnia.* vol.3 Ed. J.L.E. Dreyer. Hauniae

Tycho Brahe. 1969. *Tycho Brahe his Astronomicall coniectur of the new and much admired [star] which appeared in the year 1572.* Amsterdam: Theatrum Orbis Terrarum. Repr., New York: Da Capo Press.

Vermij, R. 2002. *The Calvinist Copernicans. The reception of the new astronomy in the Dutch Republic, 1575–1750.* Amsterdam Koninklijke Nederlandse Akademie van Wetenschappen.

Westfall, Richard S. 1971. *The construction of modern science: Mechanisms and mechanics*. New York: Wiley.

Wolf, Rudolf. 1861. *Die Sonne und ihre Flecken.* Zürich.

Chapter 13
Conclusion: The Young and the Mature Descartes *Agonistes*

13.1 Coda: Descartes' 'Youthful' Struggles Reconsidered

"Even revolutionaries like to be in a suitable tradition"—Eric Hobsbawm (*Interesting Times: A Twentieth-Century Life* [London, 2002])

From the beginning with Beeckman in 1618–1619, the young Descartes, as a budding *physico-mathematician*, pushed against the declared Scholastic Aristotelian 'rules' about the scope and application of mathematics within natural philosophy—not that he had much in the way of volume or density of results in those early days. Beeckman and Descartes were rebels of a sort, thumbing their noses at scholastic natural philosophical rules on the status and role of mixed mathematics, and even the ideal of systematization. Correlatively, they were willing to take on board a vague, but trendy concept such as physico–mathematics. As we have seen, Descartes added to that his home cooked version of the already circulating idea of a universal mathematics. He inflated both ideas with his own brand of aspiration and bravado. And Descartes soon went even further, to a putatively world–beating new analytical method. At each stage of these early adventures, Descartes was well pleased. To fancy himself a physico-mathematician, then a universal mathematician, and finally an all conquering methodologist, gave him some placement in a cultural debate, and provided a sense of who he was intellectually (and particularly as some sort of special specimen of mathematician). After ten years of these endeavors, and self-inflations; that is, after several notable but limited technical successes and a sequence of ever more grandiose fantasy agendas: 'physico-mathematics', 'universal mathematics', and, 'the method', it all blew up with the unfinished later portions of the *Regulae* in 1628–1629.

After the failure of the *Regulae*, and of universal mathematics and method with it, Descartes had to return to the two real but largely separate cultural games in town available to his talents. He retreated to a more isolated and independent, high level analytical mathematics; and he (re)turned, separately, to the field of natural philosophizing, both aping and hoping to surpass the sort of systematized natural philosophizing he had first imbibed under the Jesuits of *La Flèche*. But there was a catch,

J. Schuster, *Descartes-Agonistes: Physico-mathematics, Method & Corpuscular-Mechanism 1618-33*, Studies in History and Philosophy of Science 27,
DOI 10.1007/978-94-007-4746-3_13, © Springer Science+Business Media Dordrecht 2013

a lingering hankering after a grander, more legitimated, more unified and hence more culturally triumphant vision, arguably the sublimated remnant of the fantasies of universal mathematics and method. This was Descartes' remaining lifelong attempt to provide, through his dualist metaphysics, a grounding of certainty for both mathematics and, in the limited way discussed above, in his natural philosophy.[1] He still yearned to be the hegemon of both, not just through brilliant and novel work, but by immunizing both from scepticism, and the natural philosophy in particular from religious and politico-cultural radicalism. In other words, his early work showed little or no concern with the Aristotelian ideal of *Scientia*, in terms of a unified, true system of natural philosophy. He did hanker after *Scientia*, but under the aegis of his ultimately abortive universal mathematics and method. However, in the longer run, by the late 1620s, he was to return to the ideal of *Scientia* in natural philosophy, becoming precisely a systematizing natural philosopher, aiming to displace in its entirety the neo-Scholastic Aristotelianism he had been taught, including its relation to the subordinate mathematical fields. In the end, Descartes remained in the game, but wanted to radically transform it, whilst deposing the currently hegemonic version of how it should be played.

So, what we have learned of the young Descartes *agonistes* generates many useful hints and frames of reference for dealing with the more public, and published, mature Descartes.[2] One obvious example would be the demystification of Cartesian method and the understanding of its practical sterility yet power to convince, along with the likely path of Descartes' disillusionment with it (while continuing to exploit its persuasive packaging and selling power). A no less important result is our improved sense of the challenges faced by Descartes (and anybody else) in building and maintaining a system of natural philosophy. And indeed, we have made some surprising discoveries about what made his mature system of natural philosophy in the *Principles* tick, compared to *Le Monde*. A perhaps even more nuanced result requires a bit of spelling out, and concerns what we might make of Descartes' self-image as a mathematician, especially in later life when maintaining, in public, that his natural philosophy was mathematical, or geometrical.

We now know, of course, that Descartes' ambit claim that his natural philosophy is mathematical makes little sense. The best that can be said is that although his natural philosophy is discursive like all others, it tries to limit speech to in-principle quantitative properties. But, that really will not do, because that is not what any mathematician of the time would have seriously called mathematical. Part of the answer, I now suggest on the basis of our study of the young Descartes, is in his reflection upon his own prior trajectory in physico-mathematics *cum* natural philosophy. Descartes' physico-mathematics as presented in this book, with all its complexity and ambiguity at any point and over time, was still more mathematical than any natural philosophical discourse could be. And, we have seen that Descartes' corpuscular-mechanical natural philosophy had deep, complex, albeit

[1] Cf. above Chap. 6, note 19.

[2] For extended argument along these lines see Schuster (2009).

not entirely coherent and consistent roots and sources in his physico-mathematical trajectory. So, when he later called the natural philosophy mathematical, he was, in my view, hopefully, and wistfully, and perhaps disappointedly, alluding to this hidden iceberg of personal experience. 'Mathematical' was his short hand sign in public for all he had hoped for and tried to do, first in physico-mathematics with a corpuscular-mechanical flavor, and later in a corpuscular mechanical system of physico-mathematical source and 'type'. There was nothing settled or clear or truly mathematical in the final system; but he knew that its genealogy had all this mathematics–centered hope and activity laced into it. Not wishing in public to admit systemic defeat about the pure meaning of 'mathematical', and similarly not wishing to display his true 'history', he simply took a deep breath and said to the public, *'Take if from me, it's mathematical.'*[3]

Descartes' aspirations, and failures, in physico-mathematics, universal mathematics and method, as well as the wistful echoes of those dreams in his later work, just mentioned, invite us to sound one final, somewhat ironic chord at the end of this coda to our argument: We know that Descartes' fantasy projects peaked at two moments—first in 1619–1620 when he hit on universal mathematics, leading quickly to the first gleams of the method; and then late in the 1620s, after the optical breakthroughs, with the composition of most of the *Rules*. At these moments, Descartes presents not as a natural philosopher, but as a mathematician bent on displacing, or marginalizing, the entire field of natural philosophizing, with its systems in competition and densely institutionalised structure. The young Descartes, in effect, was saying, *'I personally shall displace the game of natural philosophizing with one or another, or all, of these other games, whose natures are essentially mathematical.'* But, these projects and dreams collapsed. Descartes realigned his identity and his agenda as a systematic philosopher of nature, with special interests in (and claimed special benefits derived from) physico-mathematical approaches to some of the traditional mixed mathematical sciences.

The historiographical irony here arises from the observation that, at present, many younger scholars seem to believe that such a wholesale displacement and destruction of the field of natural philosophers by mathematics and mathematicians is the key to understanding the Scientific Revolution. This view arises from taking a quick, headline gloss from the crucially important work of senior scholars such as Robert Westman, Peter Dear, Jim Bennet and Mario Biagioli, who have variously pioneered new approaches to the study of mathematics and mathematical traditions

[3] Descartes habitually displayed a secretive, reclusive, publicly masked and overtly tricky persona. This was a man who lived by the mottos 'he lives well who is well hidden' and 'masked I go forth.' Descartes announced 'Bene vixit, bene qui latuit' as his motto to Mersenne in April 1634 (AT I, p. 286). (Descartes proclaimed 'larvatus prodeo' (masked I go forth) in the middle of some fragmentary youthful ruminations preserved in AT X, p.213.) There is excellent evidence on these points for his mature career, and hints and clues about it for the earlier period. (Clarke 2006) Part of this, no doubt, was cultural, conditioned, in Descartes and others, by the superheated political and religious tensions of the Baroque age, which also elicited intense and elaborate courtesy as a defence against incipient social breakdown and chaos Schuster (2012a).

in the period, as well as of the contrasting roles and identities of natural philosophers and mathematicians (of various stripe).[4] The revealing and important literature they and others have produced invites inevitable simplifications and glossings, as the meaning of their findings is translated into graduate instruction and the chatter of a new conventional wisdom. Today, it is not difficult to meet history of science graduate students who apparently believe that what happened in the Scientific Revolution was a circulation of elites: Mathematicians as an intellectual tribe, and social group (or group of groups), displaced natural philosophers, and 'modern science' emerged as a mathematical enterprise. These claims are misguided and are mistakenly attributed to the founding scholars mentioned above. As we have learned in this volume, many of the moves to mathematicize one or another part of natural philosophizing were physico–mathematical in character and so involved *physicalization* of the mixed mathematical fields, not their *mathematization*. Moreover such moves were made within the natural philosophical field, not from outside to destroy or displace it. This was not an invasion of natural philosophy by mathematicians intent upon destroying or displacing it. The relevant players, such as Kepler and Descartes were mathematically adept natural philosophers/slash/natural philosophically literate, and aggressive mathematicians. Such people constituted one, small, intersectional sub-set of all European mathematicians and natural philosophers. No circulation or displacement of elites took place.[5]

It is against this background that the young Descartes, at the height of his fantasies of universal mathematics and method, ironically presents the very model of what the simplified thesis of 'mathematicians displacing natural philosophers' would, and should, look like. Descartes, in his most radical and in a sense anti-natural philosophical moments, was indeed the sort of player envisioned in these historiographical tales—the skilled and strategically minded mathematician, intent on displacing the traditional field of natural philosophizing in the name of a completely new and essentially mathematical dispensation. The problem is that the key historians mentioned above are being misread and no such larger process occurred. As for the young René Descartes, in particular, he eventually discovered the impossibility of his dreams that universal mathematics and method would replace discursive natural philosophizing as the core enterprise for gaining knowledge of nature.

---

[4] Westman (1980), Biagioli (1989), Bennett (1991), Dear (1995), Johnston (1996).

[5] The reader should recall our delineation of several species of physico-mathematician above, Sect. 2.5.3. Notable exceptions actually prove the point. Consider those master practical mathematicians, such as Simon Stevin, who played upon mixed mathematical fields from outside the realm of natural philosophizing, aiming not to make natural philosophical capital, but to expand and systematize the realm of practical mathematics. Indeed the mixed mathematical disciplines can be seen as a contested borderland in play between certain practical mathematicians and, as we have seen mathematically adept, ambitious natural philosophers. On this topic, see also J.A. Schuster, 'Consuming and appropriating the mixed mathematical fields, or, being 'influenced' by them: the case of the young Descartes', available on my website http://descartes-agonistes.com.

13.2 Epilogue: The Mature, Public and Published Descartes *Agonistes*

In 1637, Descartes finally produced his first publication, the *Discourse on Method* and three supporting *Essays*—the *Geometry*, *Dioptrique* and *Meteorology*. The *Discourse* introduced his method as well as an initial version of the metaphysical construction of ontological dualism. Descartes hoped that these abbreviated versions of his doctrines would lead the savants of Europe to his door, raising the public's appetite for the full system. The *Discourse* and *Essays* triggered much correspondence and debate, along with some recruitment to the Cartesian program in the United Provinces, scene of the first spread of Cartesianism into university teaching. But the overall reception was disappointing and Descartes could not move directly to the intended triumphal unfolding of his full system. In 1641, he took a strategic detour, publishing his *Meditations*, the fullest elaboration of the metaphysical arguments for dualism. Since the *Meditations* contain virtually no natural philosophical detail, they are often studied anachronistically, in isolation, as Descartes' inauguration of modern philosophical debate. However, as Gaukroger has convincingly argued, they should be seen in context as Descartes' attempt to set in place the metaphysical foundations of a mechanistic natural philosophy, without having to offer up debatable details.[6] Finally, in the *Principles of Philosophy* (1644, 1647) Descartes presented, in the form of a textbook, his full, mature system of mechanical natural philosophy and its explicit metaphysical legitimation—along with the vast cosmographical weave of explanation and co-optation of striking facts that we have discovered within it, by virtue of our analysis in Chap. 12. None of this activity, however, achieved what Descartes so much desired—the winning over of the learned world to his system of mechanism, his form of metaphysical legitimation, and his particular achievements and techniques in optics, mechanics and physiology. Descartes' later career was engulfed in controversy, debate and a constant struggle to defend and explicate his system.

One can illustrate the tenor of his later struggles, and the ways his concerns both deepened and shifted, by considering the evolution of his aims in medicine. Given the focus of this volume, we have not been able to trace the early development of Descartes' mechanistic physiology, emergent in the *Traité de L'homme*, composed in parallel with and meant to be part of *Le Monde*. We have, however, seen the importance of his early schematic approach to a mechanistic physiology of vision and perception in the design, and collapse, of the later *Regulae*. In any case, down through the mid 1630s, Descartes had thought that medical theory and therapy could follow directly from his mechanistic physiology. But, unsurprisingly, his medical program stumbled on the very complexity of the human condition as conceived in Cartesian ontological dualism. In humans the intimate and 'substantial' union of a reasoning, immortal and immaterial mind with a machine body entailed the existence

---

[6] Gaukroger (1995) 338, 345, 352, 362; Schuster (1995) 135–36.

of a subjective realm of emotions and internal sensations. From the late 1630s onward, Descartes increasingly recognised these difficulties and, by the 1640s, his medical theory had become focused on psychosomatic aspects of ethical and therapeutic issues. Accordingly, his last work, the *Passions of the Soul* (1649), explores the passions, emotions and internal sensations arising from and characteristic of the human mind-body union. Descartes' mechanistic medicine and his dreams of its therapeutic consequences were, in short, derailed by the dualism which he otherwise had to supply to shore up his overall natural philosophy. In his resulting engagement with ethics, the passions and the human condition, he tried to exploit the implications of that dualism for humans, whilst engaging the criticisms that his astringent dualism had elicited.

In the end, Descartes bequeathed to the next generation of the Scientific Revolution a powerful but particular version of the mechanistic philosophy—one but only one of the sources for the consensual experimental corpuscular-mechanism of what in Chap. 2 we labeled the 'CMF' period[7]—as well as startling, if often hotly debated, achievements in the traditional sciences and mathematics. He did not succeed in imposing upon his successors his personal program of corpuscular–mechanism, method and dualist metaphysics. Parts of his personal vision were variously altered, revised, adopted and rejected by the next generation of mechanists, who, unlike Descartes, had the luxury of being relaxed heirs to, rather than tortured inventors of, the mechanistic world vision and its program for the scientific domination of nature.

All the foregoing points reinforce the idea that Descartes was a fine example of the general patterns of natural philosophical struggle in the critical phase of the scientific revolution—his personal traits and patterns of behavior being not simply idiosyncratic, but to some considerable degree also exactly what one would expect from an engaged natural philosophical combatant of this age of Baroque struggle over systematics, foundations and significant, rebellious novelties in the field. Accordingly, one might be tempted simply to think of him as quite averse in style, goals and accomplishment to the natural philosophers of the coming CMF phase. This would not be entirely correct. Instead, by a kind of dialectical logic, one can move from an image of Descartes the exemplary warrior of the critical phase of the Scientific Revolution, and ask how, if at all, did his work, habits, and aspirations relate to the then just emerging patterns of the next 'CMF' phase of the Scientific Revolution. Recall our modeling of the CMF phase in Chap. 2, where we saw it characterized by a relative natural philosophical consensus on an experimentally oriented corpuscular-mechanism; a muting of public conflict about natural philosophical theory and systematization, especially within the new institutions; the beginning of the death of the ideal of *Scientia* and the incipient fragmentation of the field of natural philosophizing into successor scientific traditions, whilst the entire field enjoyed a moment of relatively increased autonomy from other branches of philosophy. There are several ways in which the mature Descartes was already

[7] Above, Sect. 2.7.

displaying hints, or even more than hints of these developments. We can even imagine possible ways in which, had he lived past 1650, he might have related to the emerging dominant tendencies in the field of natural philosophy, its organization, and the ongoing development of the entourage of subordinate fields.

First of all, there is the issue of Descartes' concern, as a natural philosopher and physico-mathematician with solid evidence and well established matters of fact. While we have little evidence on these scores from his early natural philosophical career, we have seen enough in the present study to know, at least, that his physico-mathematical and natural philosophical projects were grounded in concrete reference to experience and, in his view, established matters of fact.[8] The *Principia*, as we have learned, constitute his mature system of natural philosophy in part just because of how skeins of facts about magnetism, sunspots, variable stars and *novae* are co-opted into, and then leveraged as explanatory categories through, its system-binding novelties.[9] Even more to the point of the mature Descartes' general receptivity to novel matters of fact, any acquaintance at all with his later career reveals his overwhelming interest in experiments and gathering of hands-on experience in all fields, from lens grinding to animal anatomy; from medicinal botany to aerostatics; from medical diagnostics to pendulum motion. The mature Descartes read little, and often ignored books sent to him or dismissed them on cursory examination. But, this was not so that he could spend his late mornings in bed habitually meditating on metaphysics—something he told followers they should indulge in only once in a lifetime. Rather, it was so that, in his various Dutch hideaways, whose addresses were disclosed only to a tiny and select group at any time, he could experiment, peruse reports from others by voluminous correspondence, or, until he got into his fifties, work late into the night on his books and correspondence. Indeed, in the almost compulsive concern for evidence and experience that he certainly displayed in later life, Descartes was not so much different from those supposedly much more experimentally oriented corpuscular-mechanical savants one finds half a generation later at the Royal Society of London, or in and around the salons and lecture rooms of late-seventeenth-century Paris, where a veritable school or sect of Cartesian experimentalists thrived.[10]

What Descartes lacked, compared to the later so-called 'experimental philosophers', was a genuine commitment to collaborative research among recognized peers, and an imperative toward new organizational modes to facilitate the same. That was not Descartes' style or personality, as we have just mentioned, but then

---

[8] For example: hard results in mixed mathematics were to be used in physico-mathematics; his respect for the way the sheer opacity of raw experience of fall nullified his initiatives on a physico-mathematics of fall; his concern for what would count as empirically grounded and workable measures of dimensions in later *Regulae*; and the multiple ways *Le Monde* and the path to it were strewn with interest in and respect shown to putatively agreed experiences or matters of fact, as we have seen in Chap. 8, *passim*.

[9] However, see above, Sect. 12.12, point [3] on how Descartes' mature style of cosmographical explanatory/descriptive narrative created a situation in which factual detail tended to shape narrative detail, rather than prompt alterations in the structure of basic concepts.

[10] Clarke (1989), McClaughlin (1996, 2000).

again, as we have also reiterated above, he lived in an age of clashing systems of natural philosophy. He just missed out on the emerging more normalized, sedate and clubby world of later seventeenth-century natural philosophizing, where theoretical contestation certainly continued to exist, but was endemically somewhat muted, and ritually hidden from the public, at least, by the great new 'scientific' institutions.[11] Yet, Descartes in one behavioral dimension was almost there, for he was a canny tactician and negotiator of experimental work and its results, as for example in his dealings in the late 1640s, in person and by direct and indirect correspondence, with the young Blaise Pascal over barometric experiments and the supervening theoretical question of the existence and properties of vacuums in nature.[12]

Hence, one might hypothetically ask, 'How would Descartes have performed had he been more firmly entrenched in a public culture and dense immediate network of gentlemanly savants, bound by the demands of an etiquette of muted public disputation about theory?' So far as we can tell from the evidence of his mature career, Descartes could, to some degree, strike a balance between public as opposed to more back channel domains of intellectual engagement. For example, he habitually praised critics in public whilst rejecting and reviling them behind their backs. Moreover, Descartes could indeed turn on the elaborate Baroque etiquette and diplomacy. However, this was usually only up to a point, relatively quickly reached, when his resentment at personal or intellectual slights boiled over, or his creeping paranoia (sometimes well justified!) at intellectual or institutional cabals against him overtook his well-educated, rather neo-Stoical, attempts to control his passions rationally. In short Descartes' pragmatic, and sometimes even diplomatic, proceedings over experiments and experience were always overlaid with his prickly, devious and even paranoiac dealings and responses. Still, one should remember that Descartes was never actually placed in an immediate institutional environment such as the Royal Society or Parisian Academy, where the strong, immediate and unremitting demands of the new culture of public decorum over experiment and theory might have smoothed some of the edges of his 'Baroque age' personality, as arguably happened with other savants who bridged the two phases.[13]

[11] Shapin (1994) famously announced the advent at the early Royal Society of London, from the early 1660s onward, of an atheoretical 'experimental science' of 'matters of fact' exchanged in a culture of gentlemanly trust. He ignored the continuation of heated natural philosophical contestation, in muted circumstances and under a public rhetoric of 'matters of fact only'. Moreover, no fact in natural philosophy or any science is not 'theory-loaded', as almost any first year student in History and Philosophy of Science or Science and Technology Studies anywhere in the world can tell you. On the myth of merely matter of fact science at the Royal Society, see Schuster and Taylor (1997). For a clear demonstration that the contemporary Florentine Accademia del Cimento was also rife with natural philosophical agendas and conflicts, hidden below a public façade, see the important work of Boschiero (2007).

[12] Well canvassed in Clarke (2006), pp.535–60

[13] In his two somewhat contrasting accounts of the terrain traditionally covered by theses concerning a 'crisis' of the seventeenth century, Theodore K. Rabb (1975, 2006) has drawn attention to the contrasts— psychological, attitudinal and cultural—across varied intellectual and artistic pursuits

If we look at some of the content and techniques of Descartes' natural philosophizing we also see links, as well as significant differences, to the leading trends of the CMF phase. Consider, first, Descartes' pioneering of physico-mathematics both as an aspiration and as a set of results. The physico–mathematization of the older mixed mathematical fields was a long term and crucial development in the Scientific Revolution. Although Descartes never publicized his early programmatics and aspirations for physico–mathematics, his work in optics, at the very least, was on a main line of development. Additionally, his contribution to the eventual crystallization of classical mechanics through his forging of dynamical concepts for his natural philosophy, clear since the early studies of Gabbey,[14] also bespoke physico–mathematical practices, although this aspect of the development remained hidden from contemporaries (and later scholarly view). On the other hand, Descartes' physico–mathematics, like that of Kepler, had been premised on the peculiar conceit of directly 'seeing the causes' in well formed mixed mathematical results.[15] That was not an approach that was going to survive in natural philosophy generally or in any of the increasingly physicalized descendants of the mixed mathematical fields.

Much more entirely consistent with later developments, and contributing to them in a substantial way, was the manner in which Descartes conceived of explanation in corpuscular-mechanism in his mature and considered statements on the issue. We know from our study of the young Descartes how and why he passed out of the fantasy of a deductivist method spanning natural philosophy and physico-mathematics, amongst other disciplines. *Le Monde*, with its tacit metaphysical grounding, begins a process of more self-conscious reflection on the demands of corpuscular-mechanical explanation, which is fully expressed at the end of the *Principles*. As we have noted earlier, it is a mistake to take seriously the mature Descartes' occasional claims to have been able to deduce—as if according to a mathematical ideal of 'demonstration'—his

(including what he calls 'science') between the tense, anguished and deeply contested generations of the earlier seventeenth century and the modulated, controlled and more confident generations of the later seventeenth century. These map onto our Scientific Revolution phases of 'civil war in natural philosophy' and its contrasted sequel, the 'CMF' phase. As Descartes might have bridged these phases in a more mellow old age, so in fact did figures who had lived on into the Restoration following upon the more immediately manifested English political and religious crisis, such as Walter Charleton (Booth 2004), John Wilkins, Henry More, Seth Ward, or even the considerably younger Boyle, who inflected from a youthful van Helmontian with Puritan leanings into an impeccably establishment Restoration Anglican and carefully hedged corpuscular-mechanical natural philosopher. The Royal Society was founded by, and for a long time included in its membership, individuals who had survived the generation long English crisis, not those too young to have experienced it.

[14] Gabbey (1980). More generally on Descartes' unintended contributions to later developments in physical optics and the emerging 'classical mechanics', see Schuster, 'Cartesian Physics' in J.Z. Buchwald and R. Fox, *The Oxford Handbook of the History of Physics* (OUP forthcoming). Also cf. above, Chap. 11, note 2.

[15] On 'seeing the natural philosophical causes' directly in well grounded mixed mathematical results, see above Chap. 3, and papers on Descartes' and Kepler's physico-mathematical optics by Schuster, Raz Chen-Morris, Ofer Gal and Sven Dupré in *Synthèse* 185 (2012), in particular, Schuster (2012).

entire system of natural philosophy from absolutely certain metaphysical principles.[16] This folklore arose from the strictly deductivist tone of Descartes' method in both his formal and more offhand statements about it. It is clear that in his mature work, Descartes increasingly came to see that neither the details of particular explanatory models, nor the facts to be explained, could be deduced from metaphysics. Rather, he held that we may know with certainty, from metaphysical deduction, that the essence of matter is extension, but we cannot deduce from this truth more detailed explanatory models (concerning corpuscular sizes, shapes, arrangements and motions) that can explain various phenomena. The best one can say is that such models *should not contradict metaphysically derived certainties* (they must be 'mechanical' in some sense). Hence, corpuscular-mechanical explanatory models have a necessarily hypothetical character. Available evidence, and in particular the facts to be explained, also have an important bearing on the formulation of such detailed explanatory models and in the assessment of their "goodness" in terms of their explanatory power and scope of application. These views would not have been out of place at the Royal Society, and indeed, Hooke and Boyle could have calmly and to some degree acceptingly discussed it there with a visiting Descartes. All three would have had differences with the younger Newton's attempt to install a strangely authoritarian rhetoric of inductive certainty.[17]

Indeed, if we are willing to imagine the mature Descartes placed in the surrounds of the Royal Society, and hence in the general context of later seventeenth century natural philosophy, still one more hypothetical possibility emerges for reflection: It is fashionable, and correct, to maintain that later seventeenth mechanists, at least in England, made considerable capital out of what often was a necessity—political, religious or technical—of acknowledging some forms and types of 'spirit' and active agency in nature.[18] Nevertheless, it is not correct then to infer that the 'mechanical' philosophy is a misplaced or non-existent historical category. One simply needs to distinguish within the textual body of any particular mechanist's system a level or dimension of declaratory or self-glossing discourse, where the system purports to gloss itself-—sum itself up for presentation to audiences. Such glossing need not have been consistent with, or even an 'adequate' representation of the nuts and bolts of the system of which it was a part. At this level, most mechanical philosophies declared themselves opposed to magic, alchemy and the claims of 'spirit', quite independently of where and when some concessions needed to be made to non-mechanical agencies and processes elsewhere in the inner coils of the discourse.

---

[16] Chapter 6 Note 19.

[17] On Newton, Feyerabend (1970) and discussion of same above, Sect. 6.3. In this counter-factual conceit about a 'Restoration Descartes', it is certainly the case that in committing to a piecemeal, hypothetical corpuscular-mechanism, he would have had to give up not only metaphysical grounding, as mentioned, but also his mature practice in the *Principles*, of weaving large cosmographical explanatory narratives, just what Huygens criticized him for (Chap. 12, note 105. On how laws of nature and generalized grasp of facts fed into this mode of explanation in the *Principles* cf. also Sect. 12.12, point [3], and Sect. 12.9, note 78) .

[18] John Henry (1986) was perhaps the most important early advocate of this important insight.

Now, let us consider Descartes' corpuscular-mechanism in this light. Descartes, of course, insisted on the declaratory level that the system was fully mechanistic and had banned all spirit and immaterial substances and agency from non-human nature. There is a discursive and negotiational catch or loophole, however, which can teach us something about Cartesian natural philosophy, its fluidity and potential context dependence. Consider that Descartes' own description of the comportment of his first matter violated the putatively strictly mechanistic conditions placed on the system.[19] For example, the first matter can and must instantaneously readjust itself to accommodate the changing shape of a conserved total volume of interstitial 'space' amongst a collection of corpuscles of second and third matter. But, the question then arises: How can that occur, without equally instantaneous shifts of density of the first matter—although that is something clearly beyond its 'extension constituting' nature? After all, where there is first matter, space is already as full as it can be.

This need not be merely Whiggish snipping, as the following hypothetical scenario may demonstrate. Imagine René Descartes surviving the Swedish winter of 1649–1650, and then later moving to another, more tolerant Protestant land, Restoration England, where, as an elderly and honored guest, he flits around on the margins of the Royal Society. Here Descartes, like other English based mechanists, might well have considered that the appropriate declaratory gloss on his corpuscular-mechanism should stress the non-mechanical, hence spiritual nature and capabilities of his first matter. This 'spiritual' substance exhibits non-mechanical shifts of shape and density in order to maintain the universal plenum, and, in general, guarantee the overall law-like behavior and order of the system, including the other two purely mechanically acting elements. Thus, our hypothetical Descartes could have presented himself in England as the very model of a 'modern' mechanist, one who discerns exactly where and how a precisely defined spiritual entity fits into the total 'mechanical' system—just the strategy followed by his younger English friends and competitors.[20]

This certainly illustrates something, in general, about the negotiability of natural philosophical claims, the ways systems could be accounted for, and the likely rules of engagement in post-Restoration natural philosophizing. But more to the point, this scenario suggests that something so dramatic as Descartes' own personal adaptation

[19] I believe that Thomas Kuhn first made the sorts of points which follow, although I have not been able to find a statement of these points in his published work. He may well have argued them informally, in graduate seminars, or in personal discussion, during the time I studied in the Princeton HPS Program he headed (1969–1974).

[20] Alternatively, we can drop the hypothetical conceit and state these points as the results of a deconstruction of Descartes' natural philosophy: *Completely mechanist and anti-spirit at the declaratory level, Cartesianism harbored the curious 'first matter' which Descartes insisted behaves fully mechanically, but which just as easily might on occasion have been argued to have non-mechanical capabilities.* This is because consistency between declaratory glosses of the entire system and the meaning and function of particular parts thereof, was, as usual, up for negotiation amongst actors, including the author of the discourse. No rule, logic or imperative controlling the construction and negotiation of natural philosophical discourse could or did impose a single 'correct conclusion' about such matters.

to the demands of the CMF phase of the Scientific Revolution was not beyond the bounds of objective possibility, and therefore that the mature Descartes, product of the young Descartes we have studied here, was not in principle a figure totally averse to the oncoming conditions and opportunities of the CMF phase of the Scientific Revolution.

References

Works of Descartes and Their Abbreviations

AT=*Oeuvres de Descartes* (revised edition, 12 vols.), edited by C. Adam and P. Tannery (Paris, 1964–1976). References are by volume number (in roman) and page number (in Arabic).

SG=*The World and Other Writings*, edited and translated by Stephen Gaukroger (Cambridge,1998).

MM=*René Descartes, The Principles of Philosophy*, translated by V. R. Miller and R. P. Miller (Dordrecht, 1991)

MSM=*Rene Descartes, Le Monde, ou Traité de la lumière*, translated by Michael S. Mahoney (New York, 1979).

CSM(K)=*The Philosophical Writings Of Descartes*, 3 vols., translated by John Cottingham, Robert Stoothoff, and Dugald Murdoch, and (for vol. 3) Anthony Kenny, (Cambridge, 1988) References are by volume number (in roman) and page number (in arabic).

HR=The Philosophical Works of Descartes, vol I translated by E.S. Haldane and G.R.T. Ross (Cambridge, 1968 [1st ed. 1911])

Other

Bennett, Jim. 1991. The challenge of practical mathematics. In *Science, culture and popular belief in renaissance Europe*, ed. Stephen Pumfrey, Paolo L. Rossi, and Maurice Slawinski, 176–190. Manchester: Manchester University Press.

Biagioli, Mario. 1989. The social status of Italian mathematicians, 1450–1600. *History of Science* 27: 41–95.

Booth, Emily. 2004. *A subtle and myterious machine': The medical world of Walter Charleton (1619–1707)*, Studies in History and Philosophy of Science, vol. 18. Dordrecht: Springer.

Boschiero, Luciano. 2007. *Experiment and natural philosophy in seventeenth century Tuscany: The history of the Accademia del Cimento*, Studies in History and Philosophy of Science, vol. 21. Dordrecht: Springer.

Clarke, Desmond. 1989. *Occult powers and hypotheses: Cartesian natural philosophy under Louis XIV*. Oxford: Clarendon.

Clarke, Desmond. 2006. *Descartes, a biography*. Cambridge: CUP.

Dear, Peter. (1995) Discipline and Experience: *The Mathematical Way in the Scientific Revolution*. Chicago, Chicago University Press.

Dear, Peter. 2001. *Revolutionizing the Sciences: European Knowledge and its Ambitions, 1500–1700*. Princeton: Princeton University Press.

Gabbey, A. 1980. Force and inertia in the seventeenth century: Descartes and Newton. In *Descartes: Philosophy, mathematics and physics*, ed. S. Gaukroger, 230–320. Sussex: Harvester.

Gaukroger, S. 1995. *Descartes: An Intellectual Biography*. Oxford:OUP.

Feyerabend, P.K. 1970. Classical empiricism. In *The Newtonian heritage*, ed. R.E. Butts and J.W. Davis, 150–170. London: Blackwell.

Henry, John. 1986. Occult qualities and the experimental philosophy: Active principles in pre–Newtonian matter theory. *History of Science* 24: 335–381.

Johnston, Stephen. 1996. 'The identity of the mathematical practitioner in 16th century England', in *Der `mathematicus': Zur Entwicklung und Bedeutung einer neun Berufsgruppe in der Zeit Gerhard Mercators*. Ed. Irmgard Hantsche. Bochum, 93–120.

McClaughlin, Trevor. 1996. Was there an empirical movement in mid-seventeenth century France? Experiments in Jacques Rohault's *Traité de physique*. *Revue d'Histoire des Sciences* 49: 459–481.

McClaughlin, Trevor. 2000. Descartes, experiments, and a first generation Cartesian, Jacques Rohault. In *Descartes' natural philosophy*, ed. S. Gaukroger, J.A. Schuster, and J. Sutton, 330–346. London: Routledge.

Rabb, Theodore K. 1975. *The struggle for stability in early modern Europe*. New York: Oxford University Press.

Rabb, Theordore K. 2006. *The last days of the renaissance and the march to modernity*. New York: Basic Books.

Schuster, John. 1995. Descartes Agonistes: New Tales of Cartesian Mechanism. *Perspectives on Science* 3: 99–145.

Schuster, J.A. 2009. Descartes—Philosopher of the Scientific Revolution; Or Natural Philosopher in the Scientific Revolution. *Journal of Historical Biography* 5: 48–83.

Schuster, J.A. 2012. 'Physico-mathematics and the search for causes in Descartes' optics—1619–1637', *Synthèse* 185: 467–499. [published online Dec. 2011 DOI 10.1007/s11229-011-9979-4].

Schuster, J.A. 2012a. What was the relation of Baroque culture to the trajectory of Early Modern Natural Philosophy? In O.Gal and R. Chen-Morris eds. *Science in the Age of the Baroque, Archives internationales d'histoire des ideés* 209: 1–35.

Schuster, John, and Alan B.H. Taylor. 1997. Blind trust: The gentlemanly origins of experimental science. *Social Studies of Science* 27: 503–536.

Shapin, Steven. 1994. *A social history of truth: Civility and science in seventeenth century England*. Chicago: University of Chicago Press.

Westman, Robert. 1980. The Astronomer's role in the sixteenth century: A preliminary survey. *History of Science* 18: 105–147.

Appendix 1: Descartes, Mydorge and Beeckman: The Evolution of Cartesian Lens Theory 1627–1637

A.1 Introduction

In Sect. 4.5.2 it was foreshadowed that the analysis of the trajectory of Descartes' lens theory can provide crucial supporting evidence for my reconstruction of Descartes' discovery of the law and his attempts at its physico-mathematical rationalization. In particular, it can do this by allowing us quite firmly to date the material in the Mydorge letter to the period 1626/1627 shortly after the law of refraction was discovered, in cosecant form, by Descartes and Mydorge.

In this Appendix we are going to canvass the following points: [1] In constructing his lens theory Mydorge begins with the cosecant form of the law and only finds a sine formulation in the course of elaborating the theory. [2] His synthetic proofs of the anaclastic properties of plano-hyperbolic and sphero-elliptical lenses are similar to, but clearly pre-date those offered by Descartes later in the *Dioptrique* of 1637. Moreover, [3] Descartes' own synthetic lens theory demonstrations in the *Dioptrique* differ from those of Mydorge in another historically revealing way, the matter turning on a technical and aesthetic issue which Descartes seems to have learned from Beeckman in October 1628. In other words Descartes' lens theory developed during three moments between 1626/1627 and the publication of the *Dioptrique*: First, we have the earliest lens theory of Descartes and Mydorge in the Mydorge letter, whose content dates from 1626/1627; second, we shall see some consequential shifts and articulations in Descartes' theory as a result of consultations and negotiations with Isaac Beeckman in 1628; and, finally, we have the synthetic lens theory of the *Dioptrique* of 1637.

All these facts will therefore suggest that the Mydorge letter contains Mydorge and Descartes' *earliest lens theory*, and arguably *their first form of the law*, the cosecant form. The *material in the letter,* if not the artifact itself, pre-dates October 1628, certainly predates composition of the *Dioptrique* and very plausibly is as early as 1626/1627. So, this dating points to the cosecant form of the law as the first form Mydorge and Descartes possessed. This, as we have seen, is the key to reconstructing how they obtained it, because the other independent discovers first obtained

J. Schuster, *Descartes-Agonistes: Physico-mathematics, Method & Corpuscular-Mechanism 1618-33*, Studies in History and Philosophy of Science 27,
DOI 10.1007/978-94-007-4746-3, © Springer Science+Business Media Dordrecht 2013

it in the same *unequal radius form*.[1] We start, therefore, by returning to the Mydorge letter, intending to analyze all those parts of it not examined in our earlier discussion in Chap. 4. Having already looked only at Mydorge's Proposition 1, his statement and geometrical illustration of the cosecant law of refraction, we begin with his second proposition.

A.2 Mydorge's Refractive Index Instrument: Cosecants Not Sines

In Proposition 2 of his letter, Mydorge explains a device used to determine the refractive index of a given medium, in this case the glass Descartes and Mydorge apparently intended to use in the fabrication of lenses (Fig. A.1.1). Mydorge sends a ray, FG, through the triangular prism of glass ABC. The ray enters the prism normal to AB and is refracted at AC to E. DIH is the normal to AC at I.[2] The geometry of the device is elegant. The angle of incidence FID is equal to the angle BAC and hence is known in advance. The angle of refraction HIE is equal to the sum of angles FID and IEC.

Only one measurement, that of angle IEC need be made in order to determine the refractive index (RI) for

$$\frac{\sin \mathrm{FID}}{\sin(\mathrm{FID}+\mathrm{IEC})} = \mathrm{RI}$$

Curiously, however, Mydorge does not exploit the device in this manner, by taking the sines of the angles of incidence and refraction. Instead, he relies upon the radius form of the law taught in Proposition 1 (Fig. A.1.2). Around I he draws the arc of the circle of radius FI. He constructs FK parallel to AC cutting the arc FK at K; then from K he drops a line parallel to DIH cutting the refracted ray IE at L. The ratio IL:FI is the index sought.[3]

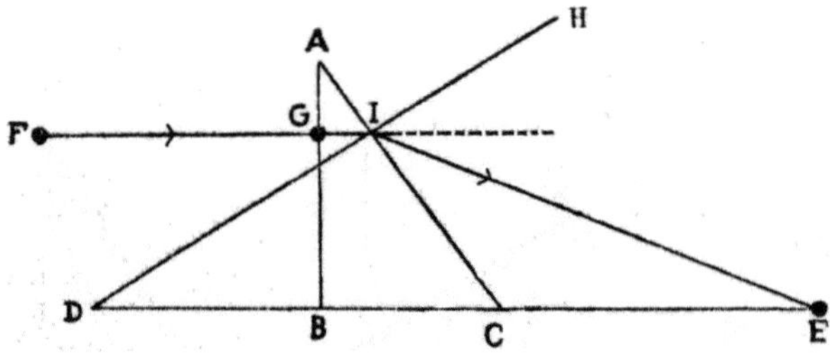

Fig. A.1.1 Simplified version of Mydorge, illustration to Proposition II, Mersenne (1938–1988) I, p.406

[1] Lohne (1959), (1963), Vollgraff (1913), (1936), deWaard (1935–36); Buchdahl (1972).

[2] Mersenne (1932–88) I. p.405

[3] *Ibid*. pp. 406–7. That is, the constant ratio IL:FI and the construction technique used in Proposition 1 will yield the paths of all other refracted rays. Cf. Above Sect. 4.5.1 and Fig. 4.5.

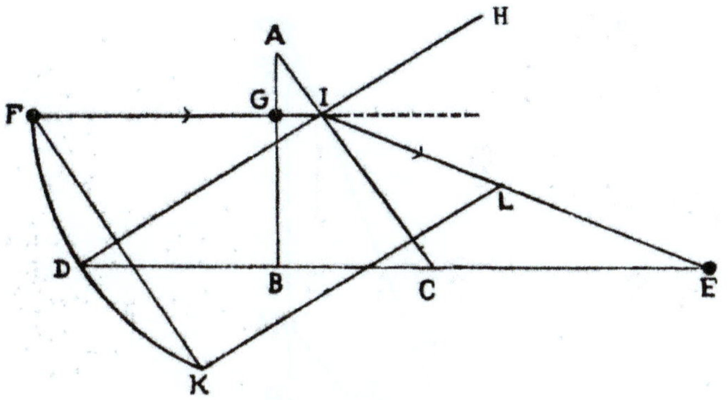

Fig. A.1.2 Mydorge, illustration to Proposition II, Mersenne (1938–1988) I, p.406

Now, had Mydorge discovered the law in what we have called sine form, he could have greatly simplified the whole discussion and diagram. A likely interpretation is that the radius form of the law was prior to the sine form. In seeking to exploit the law in an experimental situation, Mydorge apparently reached for the only form of it with which he was acquainted, the radius form. There is further evidence for this interpretation in the remainder of the letter.

A.3 Mydorge's Synthetic Propositions 3 and 4 on Anaclastic Surfaces: An 'Antique' Version of the Sine Law

Propositions 3 and 4 of Mydorge's letter are devoted to lens theory proper.[4] Applying the law of refraction to an hyperbola (prop. 3) and to an ellipse (prop. 4), Mydorge shows that if an incident ray, parallel to the transverse axis (t.a) of either of these conics, is refracted at its point of incidence with the section to the appropriate focus, then,

$$\frac{\sin \mathrm{i}}{\sin \mathrm{r}} = \frac{\mathrm{t.a.}}{\mathrm{f.d}}$$

where f.d. denotes the focal distance of either of these conic sections.

The case of the hyperbola is illustrated in Fig. A.1.3. CBA is the left branch of the hyperbola, D and E its foci, BF the transverse axis, CH the tangent to the section at C, and CI the normal to the section at C. The incident ray GC is refracted at C to the distant focus D. Mydorge uses the sine form of the law of refraction, representing the sine of the angle of incidence GCI by GL and the sine of the angle of refraction ICK by KM. We should note especially here for later reference the odd, or as

[4] *Ibid.* pp. 408–9

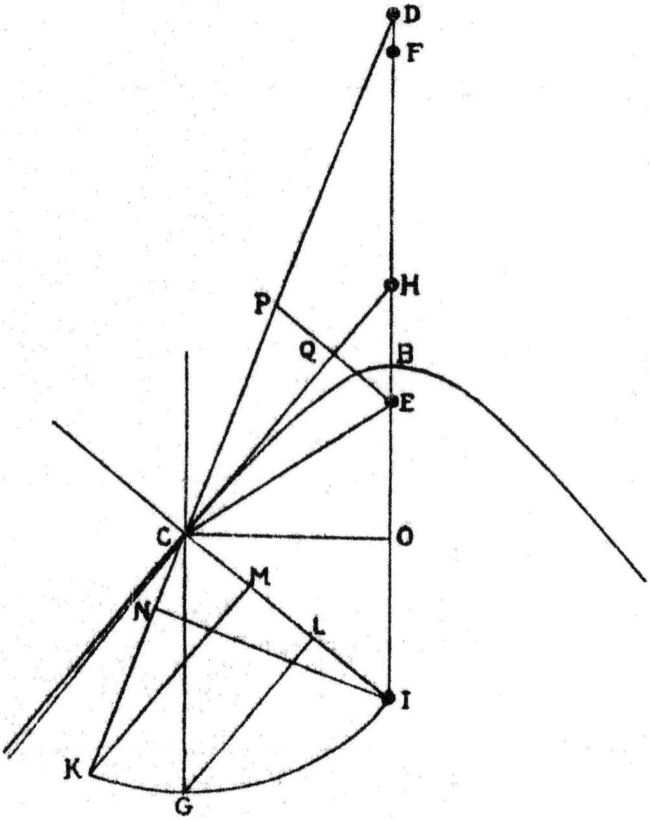

Fig. A.1.3 Mydorge, Proposition III, hyperbola as anaclastic curve, Mersenne (1938–1988) I, p.408

I shall call it 'antique', representation of the sine law, with the sines of incidence and refraction inscribed on the same side of the refracting surface.

Mydorge easily demonstrates that (GL:KM)=(BF:DE), the proportion sought between the ratio of the sine of the angle of incidence to the sine of the angle of refraction, and the ratio of the transverse axis to focal distance of the hyperbola.[5] Since ray GC and point C were selected at random, the demonstration applies to any such ray parallel to the transverse axis and refracted by the section to the distant focus. The relevance to lens theory is clear, although it is not spelled out by Mydorge in these propositions: If an hyperbola, defined by the ratio of transverse axis to focal distance of BF:DE, were embodied in a convex plano-hyperbolic lens made of a

[5] The proof proceeds easily and in routine fashion based on well known properties of the conics, chiefly by means of deduction through a sequence of equal and similar triangles inscribed in the figure. The proof tactics in these routine *concluding stages* are identical to those Mydorge uses in his Proposition 5 discussed below, and in Descartes' corresponding proof for the plano-hyperbolic lens in the *Dioptrique*. Here we are concentrating on *opening stages* of these proofs, where various representations of the law of refraction are adduced and further manipulated.

transparent material the index of refraction from which into air were equal to BF:DE, that lens would focus all rays incident parallel to its transverse axis to its distant focus—it would embody an anaclastic surface.

Mydorge's use of a sine form of the law of refraction, albeit in the unusual 'antique' form, might seem to undermine the claim that he and Descartes first discovered the law in radius form. This, however, is not the case, as we can see by placing Propositions 3 and 4 in the two relevant contexts which facilitate their accurate interpretation. The first of these contexts is the subsequent history of Propositions 3 and 4 down to their publication by Descartes in the *Dioptrique* of 1637. We shall see in the next Section that Mydorge's proofs sit at the very beginning of this history, during which the 'antique' version of the sine law was transformed into our familiar, let us say 'natural' form, in which the sines of incidence and refraction are assigned to their respective sides of the refracting interface. All this will strongly reinforce our earlier conjecture that the material in the Mydorge letter dates from 1626/1627, the very period of the initial discovery of the law of refraction. With that conclusion in hand, we will then turn in Sect. A.5 to the second context of Propositions 3 and 4, which is the surrounding text of Mydorge's letter itself; that is, Propositions 1 and 2, which we have discussed, and Proposition 5, his final proposition, which we will have to examine with great care. Proposition 5 shows how Mydorge connected the putatively original, radius or cosecant form of the law to a sine form of the law, but only in its first or 'antique' version. Additionally, because this material is quite early, we will be able to detect in Proposition 5 echoes of Mydorge's (and Descartes') earliest analysis of the anaclastic problem, the very beginning of their research on lens theory with a law of refraction in hand. *We will conclude in Sect. A.6 that the 'antique' sine form was evolved out of the radius form of the law during the course of this analysis.* In other words, we shall see that the sine law in its initial, 'antique' form was discovered during the course of an analysis of the anaclastic problem initially launched on the basis of the newly discovered radius form of the law. The 'antique' sine form, after having been uncovered in this way, was then deployed in the more synthetic Propositions 3 and 4, with Descartes' more 'natural' representation of the law—only unveiled in the *Dioptrique*—nowhere in sight (until Beeckman suggested it in October 1628).

A.4 Relating Mydorge's Propositions 3 and 4 to Descartes' Analogues in the *Dioptrique*: From 'Antique' to 'Natural' Representation of the Sines, Thanks to Isaac Beeckman in October 1628

First let us consider the place of Mydorge's propositions 3 and 4 in the development of Cartesian lens theory between 1627 and 1637. In the *Dioptrique* Descartes proves propositions identical to those of Mydorge, but they differ in one historically revealing way. Instead of setting up the sines of the angles of incidence and refraction by reference to a semi-circle on one side of the interface, as Mydorge had done, Descartes

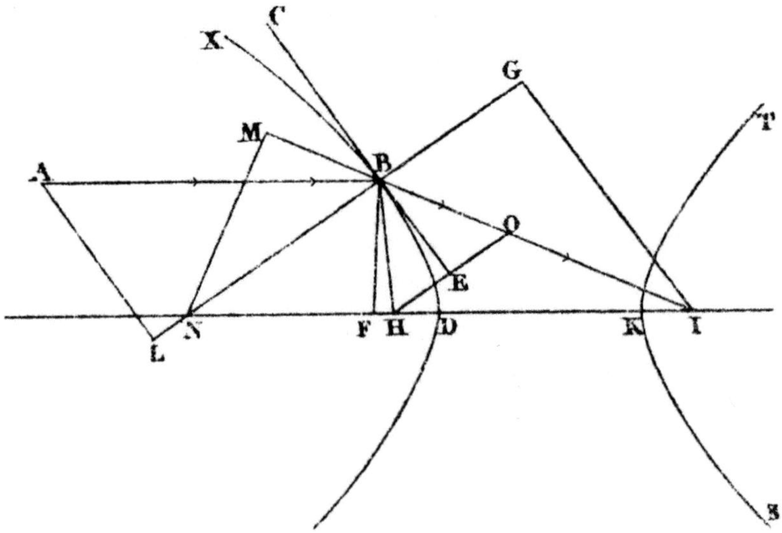

Fig. A.1.4 Descartes' hyperbola proof in the *Dioptrique,* AT VI, p.179

directly relates the sines to their respective rays. Consider Figs. A.1.3 and A.1.4 where this point is illustrated using Mydorge's and Descartes' figures for the case of the hyperbola. (Similar considerations would apply in the cases of their figures for the ellipse).

In Descartes' diagram (Fig. A.1.4) H and I are the foci of the hyperbola, D and K its vertices; ray AB is refracted at B to I; CBE is tangent to the left branch at B; LNBG is normal to CBE at B; BA is taken equal to BI so that AL and GI represent respectively the sine of the angle of incidence and the sine of the angle of refraction in the 'natural' representation of the law, familiar since that time. The proof that the ratio of the sines of the angles of incidence and refraction [AL:GI] equals the ratio of the transverse axis to the focal distance [DK:HI] again follows easily on routine knowledge of the conics, utilizing a sequence of relations amongst equal and similar triangles in the figure.[6] In Mydorge's diagram (Fig. A.1.3), we recall GL and KM are the sines of the angle of incidence and the angle of refraction respectively, given in 'antique' form as we have already seen.

Now, we can actually pin down the likely source of Descartes' later 'natural' representation. Descartes' friend Isaac Beeckman seems to have been the author of Descartes' mature representation of the sines in the context of lens theory. In 1628 Descartes asked Beeckman to provide a proof of the refractive properties Descartes had claimed for the hyperbola. Beeckman's proof omits several steps and does not fully specify the construction. But geometrically it is identical to Fig. A.1.4 and was 'approved' by Descartes.[7] At the same time, in 1628, Descartes showed to Beeckman

[6] *Dioptrique*, AT VI p. 178.

[7] AT x. 341-2; Beeckman (1939–53) fol. 338r.

an elegant proof for the case of the ellipse.[8] However, he did not subsequently use that proof in the *Dioptrique*, probably because the lines representing the sines of incidence and refraction are not related to their respective rays in the intuitively obvious way displayed in Fig. A.1.4. One can conclude that Descartes elected to use Beeckman's more 'natural' representation of the sines in both cases, ellipse and hyperbola, in the synthetic proofs in the *Dioptrique*, thus superseding his own elegant ellipse proof and Mydorge's early 'one sided' representation of the sines in Propositions 3 and 4. This episode with Beeckman, which we may imagine to have been in the nature of a negotiation (and set of mutual challenges, as befitted their previous interactions in 1618–1619) marks the second moment in the evolution of Descartes' lens theory (see Chap. 3).

To sum up so far: The development of the lens theory proofs places Mydorge's Propositions 3 and 4 very early in his and Descartes' researches. In terms of proof content and diagrammatic representation, Propositions 3 and 4 are the earliest proofs in their lens theory of which we have any record; and they are clearly the starting point for Beeckman's and Descartes' later improvements. Mydorge's demonstrations obviously pre-date Descartes' and Beeckman's discussions of lens theory in 1628, and hence they arguably date from the very period of the discovery of the law of refraction. This, accordingly, aids in our dating of all of the material in the Mydorge letter from 1626/1627. The dating becomes even more likely when one considers that by 1632 the Cartesian sine form of the law was well known to several of Descartes' associates, including Golius and Mersenne, in addition to Beeckman. In informing Golius about his optical work Descartes mentioned only the sine from of the law.[9] But in his letter Mydorge, Descartes' closest associate, does not initially use the sine form, and when he does introduce it, in his lens theory, he produces an early 'one-sided' version soon superseded in Beeckman's and Descartes' proofs. It is therefore most unlikely that the material in the letter was initially composed in 1631 or later, the possibility left open by De Waard when he tried to date the letter. All the evidence points toward the conclusion that the material in the Mydorge letter was an early and rather unsystematic and undigested report on his and Descartes' researches of 1626/1627.

[8] *Ibid.*

[9] Descartes to Golius, 2 February 1632, AT I. p.239ff. When Descartes met Beeckman in October 1628 he offered him a striking and very important mechanical analogy for the law-like refraction of light, appealing to a bent arm balance supporting identical weights, whose arms are immersed in media (upper and lower) of differing specific gravities. Section 4.7.4 above and Schuster (2000), 290–295, show how this analogy directly bespeaks Descartes' dynamical thinking about the absolute force of light and its determinations, before and after refraction. The bent arm balance, however, is presented to Beeckman using representations of the sines of the incidence and refraction of the arms, on their respective sides of the interface—the 'natural' representation of the sine law we are talking about. The issue is that whilst Descartes had by 1628 worked out this model for his dynamics of light, the representation of the sine law embodied in it was not applied back into lens theory until Beeckman suggested it. Presumably, Descartes still had to hand proofs resembling those of Mydorge from 1626/1627.

A.5 Decoding Mydorge's Proposition 5: The Cosecant Form Leads to 'Discovery' of the 'Antique'Sine Form Then Used Synthetically in Propositions 3 and 4

With these conclusions about the early date of the Mydorge letter in mind, we can now proceed to the second context of Propositions 3 and 4, the surrounding text of the letter, and in particular Proposition 5. What we are after is an explanation of Mydorge's use of the sine form of the law in Proposition 3 and 4, an explanation grounded in an understanding of the surrounding portions of Mydorge's text and framed by our now strong conviction that the material in the letter is indeed of very early provenance, dating back to the period of the discovery of the law of refraction.

Proposition 5 deals with the specification of hyperbolic and elliptical anaclastic curves in actual empirical cases. It amounts to a linking of Proposition 2 with Propositions 3 and 4. Mydorge starts by showing how to measure the index of refraction for rays passing into the air from the glass out of which the lenses are to be fashioned. Exactly as in Proposition 2, the index is determined by passing one ray through a triangular glass prism, and the index is expressed as a ratio of radii (not as a ratio of sines) by applying the radius form of the law of refraction to the given ray. That is, in Fig. A.1.5, which shows the first few steps in Mydorge's fifth proposition, the glass:air index is given as IL:FI.

Next, the empirically determined index is used to set the ratio of transverse axis: focal distance for the hyperbolas and ellipses in question. This construction is effected by exactly repeating a construction Mydorge had already given as a Corollary to Proposition 2.[10] Taking the case of the hyperbola only, Fig. A.1.6 shows how this construction is added to the material previously assembled in Fig. A.1.5.

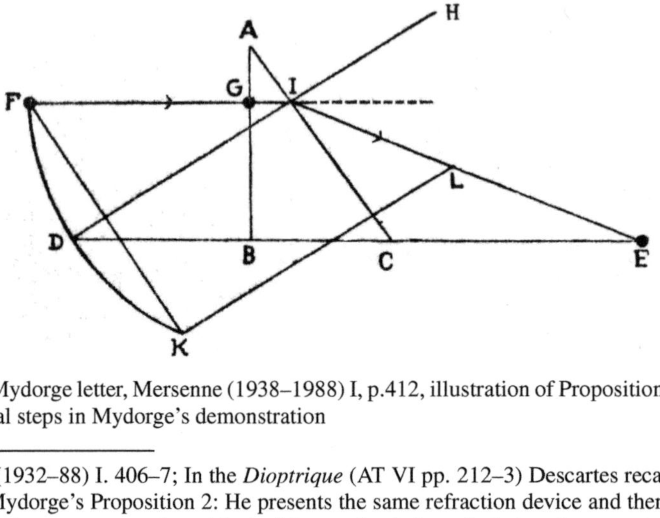

Fig. A.1.5 Mydorge letter, Mersenne (1938–1988) I, p.412, illustration of Proposition V, modified to show initial steps in Mydorge's demonstration

[10] Mersenne (1932–88) I. 406–7; In the *Dioptrique* (AT VI pp. 212–3) Descartes recapitulates the material in Mydorge's Proposition 2: He presents the same refraction device and then shows how to interpolate the transverse axis and foci of the anaclastic hyperbola into its geometry.

AC, representing the refracting surface of the prism, is considered to be tangent to the left branch of the hyperbola at I, the point of incidence. IE, the refracted ray, is taken to have passed through the distant focus at E. The near focus is then located by using the property of the hyperbola that a tangent to a point in the section bisects the angle between the lines drawn from that point to the two foci.[11] Thus AC bisects angle EIM and M is the left focus. IN (=IM) is marked off along IE. Then, according to the basic property of the hyperbola, the difference between IE and IN gives the length of the transverse axis, PQ, which can easily be inserted between the two foci. Finally, Mydorge must show that as IL:FI so PQ (or NE):ME. With the exception of its first step this proof is identical to that given in Proposition 3. The only difference is that Proposition 3 begins with a sine form of the law, whilst here we must start the proof with the radius form of the law, the product of Mydorge's introduction of the index measuring technique of Proposition 2.

The crucial step linking the radius form, IL:FI, to the sine form consists in a simple construction, which we add to Fig. A.1.6 in Fig. A.1.7.

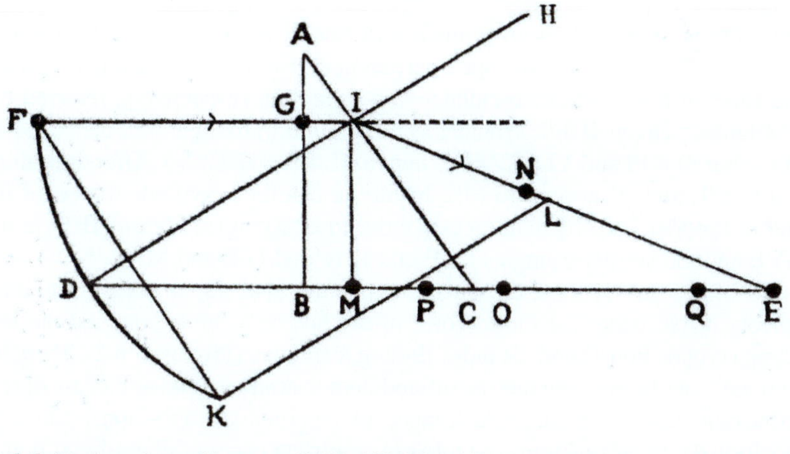

Fig. A.1.6 Mydorge letter, Mersenne (1938–1988) I, p.412, illustration of Proposition V, modified to show intermediate steps in Mydorge's demonstration

[11] Mydorge refers the reader to the relevant proposition in his own work on *Conics* first published in 1631; but he may of course be referring to draft material prior to that date. We can insist on dating the content of the Mydorge letter from 1631 or later, based on the publication history of his *Conics* and his use of routine material from it; or we can look closely at the absolutely novel aspects of the letter—the adducing of the cosecant form of the law and the working through from it to the 'antique' version of the sine law— placing all of that material in the context of whatever else we can reconstruct about Descartes' work on explaining the law of refraction and on lens theory. To reiterate, the latter considerations, argued here and in Schuster (2000), conduce to the hypothesis that the *material* in the letter dates from the very period of discovery of the law of refraction, 1626/1627, and not from the early 1630s when mature versions of all this material, well beyond the toing and froing of Mydorge's letter were known in the Descartes/Mersenne network, and Mydorge's presentation would have seen oddly out of date and out of touch.

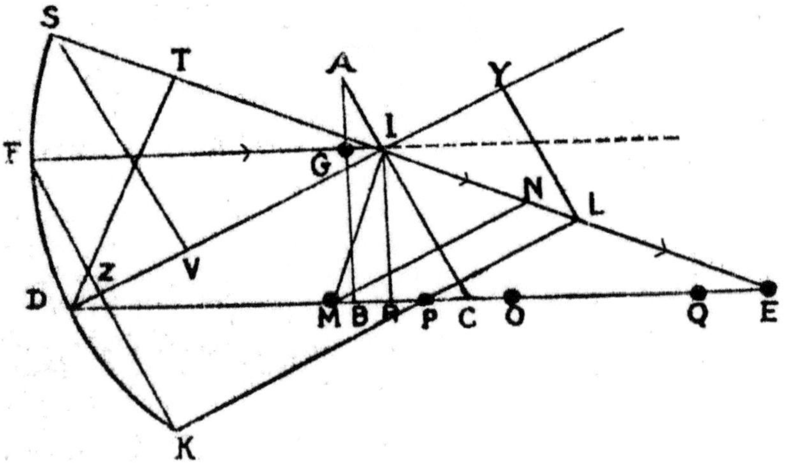

Fig. A.1.7 Mydorge letter, Mersenne (1938–1988) I, p.412, full illustration of Proposition V

Arc KDF is extended about point I, and ray IE is extended back to intersect the arc at S. SV and LY are dropped perpendicular to DI. Then FZ (=LY) and SV are the sines of the angles of incidence and refraction respectively, referred to the single reference circle KDFS. Triangle ISV is similar to triangle ILY, hence (SI:IL) = (SV:YL); but SI = FI and YL=KZ=FZ, hence (IL:FI) = (FZ:SV). After dropping DT normal to SIE and IR normal to ME, Mydorge can then continue the proof in the manner of Proposition 3 by making use of the equal triangles FIZ and IRD, and DTI and SVI; and the similar triangles DTE and IRE, and DIE and MNE.[12]

Proposition 5, as Mydorge presents it, marks the final step in his unfolding of his lens theory. First, using the radius form of the law only, he expounded the law of refraction (Proposition 1) and his index finding technique (Proposition 2). Then, using his 'antique' sine form of the law, he offered demonstrations linking the law of refraction to the ratio 'transverse axis:focal distance' of hyperbolas (Proposition 3) and ellipses (Proposition 4). In proposition 5 he turns to empirical cases of plano-hyperbolic and sphero-elliptical lenses, which means he must connect Proposition 2 to Propositions 3 and 4. This explains, in the context of the letter, Mydorge's construction of the 'antique' sine form of the law out of the radius form, and his redundant repetition of the proof structure of Propositions 3 and 4 in Proposition 5.

Of course, Mydorge's order of presentation of lens theory need not have corresponded to his and Descartes' order of research and discovery in this domain. If, as seems very likely, he initially had only the radius form of the law, he could not have pursued his synthetic lens theory through Propositions 3 and 4 without first having uncovered, by analysis, a way of linking the radius form to the fundamental properties of the conic sections. This way of linking was his construction of the 'antique'

[12] Cf above Note 5

sine form of the law out of the radius form. If one asks what Mydorge's (and Descartes') path of analysis might have looked like, a very plausible candidate springs to view—Proposition 5 itself. In the context of the letter, the bulk of the proof of Proposition 5 is redundant and repetitive; but, if Proposition 5 is read, as it were, backwards, as a remnant of an analysis, we obtain a story about Mydorge and Descartes' possible original analysis of the anaclastic problem which, given everything which has gone before, seems very plausible indeed.

A.6 Reconstruction of Descartes and Mydorge's First Analysis of the Anaclastic Problem, with the Cosecant Law of Refraction to Hand

Let us therefore try to reconstruct the analysis of which Proposition 5 seems to contain the remnants, and let us do this on the basis of the relevant facts we have already more or less established about Mydorge and Descartes' early optical work and intentions. First of all, as established in Chap. 4, we must imagine that Mydorge (and Descartes) obtained the radius form of the law by means of an image mapping technique similar to that used by Harriot. Next, we must hypothesize that with the radius form in hand, they moved to explore the possibility, hinted at by Kepler in his *Dioptrice* of 1611 (Proposition 59), that the hyperbola might be the anaclastic curve. Drawing upon their combined knowledge of the conic sections, they would have designed their elegant experimental device (if only on paper at first!) in such a way that they could easily interpolate an hyperbola whose defining property would be entailed by the geometry of the prism and the behavior of the empirically given ray, incident parallel to the intended transverse axis of the hyperbola. Then they would have had to attempt to prove the relation of the cosecant regularity to some expression of the defining property of the interpolated hyperbola, thus showing that the refraction to the distant focus holds for any parallel incident ray, and hence that the left branch of the hyperbola is an anaclastic surface. In the manner of classical geometry, and in accord with Descartes' explicit views on mathematical method, the analysis could then be reversed in so far as possible to guide the production of synthetic propositions such as Mydorge's Propositions 3 and 4.

With this background, the anaclastic problem would have taken the following form: Assume an incident ray parallel to the transverse axis of the hyperbola is refracted by the section to the distant focus. Can the law of refraction (reflecting the index of refraction) be related to the ratio 'transverse axis:focal distance' characterizing the hyperbola in question? So, let us imagine in Fig. A.1.8 what Mydorge and Descartes' analysis diagram might have looked like: Assume we are given hyperbola IPW, with foci M and E, and ray FI refracted at I to E. IC is tangent to the hyperbola at the point of incidence I. As accomplished geometers and experts on the conic sections, we also know that angle MIC=angle CIE; that IM=IN and NE=PQ = transverse axis. Next we construct the index of refraction in the only form we know, in radius form, as the ratio IL:FI.

Before proceeding any further, we need to make two observations. First, point D has not been located on the extension of ME, as in Proposition 5 (Figs. A.1.5, A.1.6 and A.1.7), because the exact location of D was probably discovered in the course of the analysis. Note also that Fig. A.1.8 has been constructed solely on the basis of (1) knowledge of the radius form of the law, (2) the suspicion that the hyperbola is the anaclastic curve, and (3) elementary knowledge of the properties of hyperbolas, which was second nature to Mydorge and Descartes.

Given Fig. A.1.8 the analytical problem is to show that (IL:FI) = (PQ:ME). Mydorge's synthetic demonstration of this relation in Proposition 5 relied on the establishment of the sine form of the law (FZ:SV) and on the relating of (FZ:SV) to (PQ:ME) via the set of equal and similar triangles pointed out earlier in Fig. A.1.7. Clearly, the successful analysis of the problem set in Fig. A.1.8, by Mydorge, Descartes or any one else, demands two crucial steps: One has to discover that point D must be located on the extension of EM, otherwise the series of interrelated equal or similar triangles does not materialize; and, one must also transform IL:FI into some ratio of lines relatable to the limbs of those triangles. *This is the very trick Mydorge accomplishes by constructing the 'one-sided' sine form of the law, FZ:SV in* Fig. A.1.7. Precisely how Mydorge and Descartes made those moves and in what order, we cannot know. That they performed an analysis of this general type is highly likely, given the close relation between Mydorge's Fig. A.1.7 and our Fig. A.1.8, constructed in light of what we can determine about the direction, background and tools of their early researches. This sort of analysis invites the construction of Mydorge's peculiar 'one-sided' version of the sine form, and such a route to

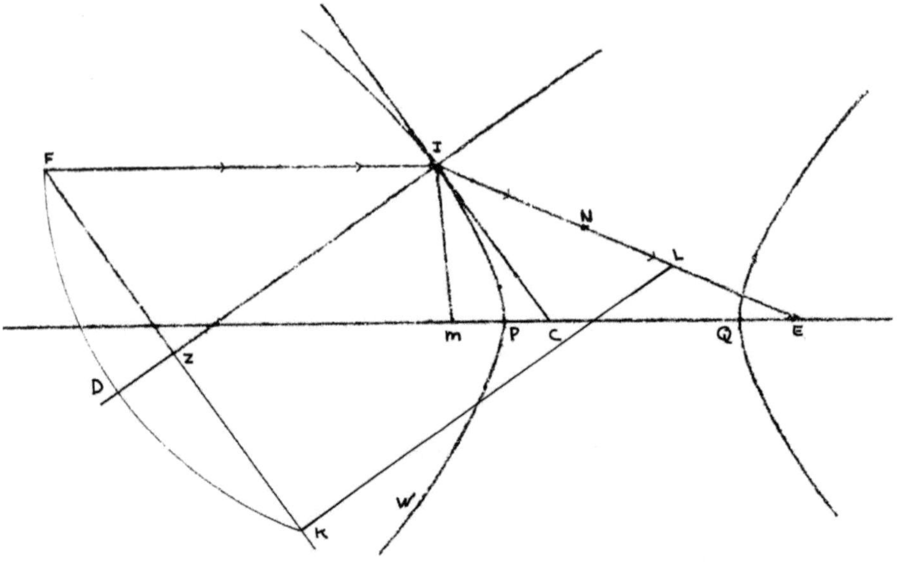

Fig. A.1.8 What Descartes and Mydorge's earliest analysis of the analclastic problem might have looked like, given initial possession of the 'radius' form of the law of refraction of light

Proposition 5 would explain Propositions 3 and 4 as later, synthetic versions of this material, launched, for simplicity's sake, on the basis of the 'antique' one–sided sine form. To put the matter quite generally, if you initially have only the radius form of the law of refraction and are analysing the anaclastic properties of hyperbolas, using the resources of classical and renaissance geometry, then you are very likely to construct the 'antique' sine form of the law in order to consummate the analysis. In this situation the 'one sided' form of the sine law is particularly useful and likely to turn up.

This necessarily technical section can be brought to a close by summarizing in 'synthetic' fashion the main conclusions we have reached through our complicated 'analysis' of the Mydorge letter, stage one of Descartes' lens theory. Below in drawing larger conclusions, we shall work in observations involving Stage two, the interaction with Beeckman in 1628, and Stage 3, the form of the theory published in the *Dioptrique* of 1637.

(1) The evolution of Mydorge and Descartes' lens theory shows that the content of the Mydorge letter dates from before 1628 and therefore approximates to the date of the discovery of the law of refraction in 1626/1627.
(2) Given (1), Mydorge's initial reliance upon the radius form of the law in his propositions 1 and 2 most likely indicates that this was the first form of the law with which he was acquainted, presumably because he and Descartes discovered the law through the 'Harriot-like' procedure of mapping image places using the traditional image placement rule, and deploying data which need have been no better than those supplied by Witelo.
(3) Given (1) and (2), Mydorge's Proposition 5 can be read as containing remnants of Mydorge and Descartes' initial analytical investigations of lens theory, using the radius form of the law as a tool. This path of analysis turned up the 'antique' sine form of the law, which was then used in devising the proofs of Propositions 3 and 4.

A.7 The Kramer-Milhaud Thesis: Discovering the Law of Refraction by Analysis of the Anaclastic Problem

Our results to this point permit us to evaluate a conjecture concerning the genesis of Descartes' law of refraction which has commanded a fair degree of credence for over a century. P. Kramer in 1882, followed by Gaston Milhaud in 1907, suggested that Descartes discovered the law of refraction as a result of posing and analysing the anaclastic problem in this form:

> Given an ellipse or hyperbola, on which a ray falls parallel to the focal axis, according to what geometrical condition will the ray be refracted to one of the foci?[13]

[13] Kramer (1882); Milhaud (1907).

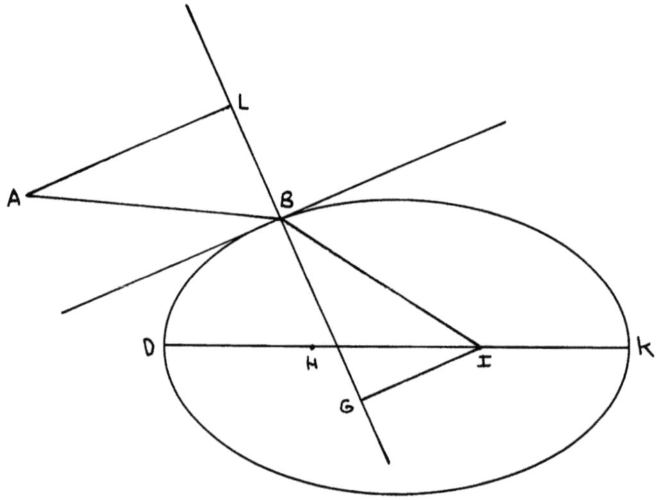

Fig. A.1.9 The Kramer–Milhaud conjecture: discovering the sine law of refraction by analysis of an ellipse assumed to behave as an anaclastic curve

Taking the case of the ellipse, Milhaud sketched an analysis beginning in the following fashion (Fig. A.1.9):

Ray AB enters the ellipse at B parallel to axis DK and is refracted to focus I. Lay off BA = BI and drop a normal LB to the tangent to the ellipse at point B. Then to this normal drop the sines of the angles of incidence and refraction, AL and IG respectively. Milhaud correctly stated that it can easily be shown that:

$$\frac{\text{AL}}{\text{IG}} = \frac{\text{DK}}{\text{HI}}$$

where DK is the transverse axis (t.a.) and HI the focal distance (f.a); hence that,

$$\frac{\sin \text{i}}{\sin \text{r}} = \frac{\text{t.a.}}{\text{f.d}}$$

and that the young Descartes could easily have done this.[14]

Milhaud did not give the rest of the analysis; but it would chiefly consist in a reversal of the steps found in Descartes' demonstration in the *Dioptrique* of the equivalent of Mydorge's Proposition 4. For Kramer and Milhaud this route of discovery had the virtue of exploiting Descartes' mathematical expertise whilst eliminating any appeal to experiment. (In addition one can point to the existence of a synthesis of the problem published by Descartes himself.)

Kramer and Milhaud were perfectly correct to believe the law could have been discovered in this fashion. Indeed, their conjecture could have been made even more

[14] Milhaud (1907) 226.

plausible had they defined less tendentiously the analytical problem they attribute to Descartes. There was no need to specify Descartes' (really Beeckman's) version of the sine form of the law as the fruit of the analysis. Mydorge's one sided version would have served equally well, as would any equivalent construction, for example, the neat construction used in Descartes' elegant but suppressed 1628 proof for the case of the ellipse, mentioned above. Strictly speaking, moreover, Kramer and Milhaud need not have specified any sine form of the law (or indeed any form of the law at all) in the data of the problem. It would have been more historically plausible simply to posit Descartes beginning with an ellipse or hyperbola and a parallel incident ray refracted to the distant focus. In such conditions, anyone with a knowledge of the conic sections could have easily discovered the relation between the ratio of the sines of incidence and refraction and the ratio of transverse axis to focal distance, by identifying the angles equal to the angles of incidence and refraction (or to their complements or supplements) and by applying the trigonometric law of sines.

Unfortunately, however, with or without such improvements, the Kramer-Milhaud conjecture suffers from one serious weakness: there is no positive evidence for it, and the evidence which can be teased out of the Mydorge letter runs directly counter to it. There is no evidence in Mydorge's letter of the law of refraction having been discovered by a straightforward analysis of the anaclastic problem. Mydorge's proofs are loaded with the one sided sine form and/or the radius form of the law; they are hardly the results of the elegant analysis envisioned in the Kramer-Milhaud thesis. Significantly, neither Harriot nor Snel give any evidence of having performed an analysis of that type. In addition, the evidence in Proposition 5 of Mydorge and Descartes' early analytical work in lens theory suggests that their analysis began with the radius form to hand. Their problem was to relate the radius form to the defining properties of an hyperbola or ellipse, operating on the not entirely wild suggestion of the authoritative Kepler that these conics could provide anaclastic surfaces. Descartes and Mydorge had the law, in cosecant form, already to hand, and needed to explore whether it could be related to the defining properties of the conics. Kramer and Milhaud require an initially entirely theoretical and mathematical procedure, producing a *candidate* law of refraction (in some trigonometric form or other as noted above), which then would have had to have been explored from an empirical point of view. However, it should be obvious from the total content of the Mydorge letter, properly interpreted, that the probability of the Kramer-Milhaud thesis being historically accurate is virtually nil.

A.8 Conclusions

[1] The reconstruction of the evolution of Descartes' lens theory confirms my claim in Chap. 4 that Mydorge and Descartes first stumbled on the law of refraction in cosecant rather than sine form, because it establishes that the sine form of the law only emerged during the course of analysis of the anaclastic problem, once the cosecant form was in hand.

[2] The reconstruction shows that the sine form provided more elegant lens theory propositions than the cosecant form, hence motivating its *overall* use in the *Dioptrique*, *even* in the problematical and confusing 'proof' of the law of refraction.

[3] This in turn further explains my finding in Chap. 4 that the 'natural' version of the sine law, when used in relation to the 'tennis ball model of light' proof of the law of refraction in the *Dioptrique,* created problems of exposition and understanding that would not have occurred had Descartes used the cosecant form, and a more explicit version of his 'dynamics of light (which on my reconstruction was derived from a physico-mathematical reading of the cosecant form).

[4] Similarly, the reconstruction shows that Beeckman introduced Descartes to the 'natural' form of the sine law for use in lens theory, whilst Descartes showed him an even more elegant representation for lens theory proof purposes. Because of its utility both in lens theory and in setting out the tennis ball model for the action of light in the attempted derivations of the optical laws, Descartes ultimately opted for the former over the latter. However, the original cosecant form of the law, which, as we have discovered, more accurately modeled the dynamical concepts underlying the optical proofs—having itself inspired their formulation—was never used by Descartes in geometrical representations of the law of refraction or in its supposed proof, although some of his verbal formulations in answers to critics of the *Dioptrique*, betray just that underlying conceptualization.[15]

[5] My argument shows that the Kramer–Milhaud reconstruction of Descartes' path of discovery of the law, which invoked a process of *de novo* and completely mathematically abstract analysis of the anaclastic problem, cannot be correct, given the documentary evidence available. However, it is fair to say that the Kramer/Milhaud conjecture was, as far as it goes, consistent with my claim that *the sine law did indeed emerge in the course of an analysis of the anaclastic problem*, provided, however, *Descartes and Mydorge already possessed and deployed in that analysis the cosecant form of the law*, itself having been obtained through other, quite different, and quite traditional mixed mathematical maneuvers in geometrical optics.

[6] In general, then, the sine form of the law emerged within, and became elegantly functional to, the development of lens theory, given the prior existence of the cosecant form of the law. By contrast, as we learned in Chap. 4, the original cosecant version of the law was intimately connected with Descartes' attempt to derive physico-mathematical capital from geometrical optics, first by reading out dynamical principles governing the behavior of light, and thence by promoting those principles to the level of a general dynamics of corpuscles.

[15] See for example Descartes' remarks to Mydorge for Fermat in March 1638, Chap. 4 Note 25 and Schuster (2000) Note 24.

References

Works of Descartes and Their Abbreviations

AT =*Oeuvres de Descartes* (revised edition, 12 vols.), edited by C. Adam and P. Tannery (Paris, 1964–76). References are by volume number (in roman) and page number (in Arabic).
SG = *The World and Other Writings*, edited and translated by Stephen Gaukroger (Cambridge,1998).
MM = *René Descartes, The Principles of Philosophy*, translated by V. R. Miller and R. P. Miller (Dordrecht, 1991).
MSM = *Rene Descartes, Le Monde, ou Traité de la lumière*, translated by Michael S. Mahoney (New York, 1979).
CSM(K) = *The Philosophical Writings Of Descartes*, 3 vols., translated by John Cottingham, Robert Stoothoff, and Dugald Murdoch, and (for vol. 3) Anthony Kenny, (Cambridge, 1988) References are by volume number (in roman) and page number (in arabic).
HR = The Philosophical Works of Descartes, vol I translated by E.S. Haldane and G.R.T. Ross (Cambridge, 1968 [1st ed. 1911]).

Other

Beeckman, I. 1939–1953. *Journal tenu par Isaac Beeckman de 1604 à 1634*, 4 vols, ed. C. de Waard. The Hague: Nijhoff.
Buchdahl, G. 1972. Methodological aspects of Kepler's theory of refraction. *Studies in History and Philosophy of Science* 3: 265–298.
de Waard, C. 1935–6. Le manuscrit perdu de Snellius sur la refraction. *Janus* 9–40: 51–73.
Kramer, P. 1882. Descartes und das Brechungsgesetz des Lichtes. *Abhandlungen zur Geschichte der Mathematischer (Natur) Wissenschaften* 4: 235–278.
Lohne, J. 1959. Thomas Harriot (1560–1621) The Tycho Brahe of optics. *Centaurus* 6: 113–121.
Lohne, J. 1963. Zur Geschichte des Brechungsgesetzes. *Sudhoffs Archiv* 47: 152–172.
Mersenne, M. 1932–1988. *Correspondence du P. Marin Mersenne*, 17 vols, ed. C. de Waard, R. Pintard, B. Rochot, and A. Baelieu. Paris: Centre National de la Recherche Scientifique.
Milhaud, G. 1907. Descartes et la lois des sinus. *Révue générale des sciences* 18: 223–228.
Schuster, J.A. 2000. Descartes *Opticien*: The construction of the law of refraction and the manufacture of its physical rationales, 1618–29. In *Descartes' natural philosophy*, ed. S. Gaukroger, J.A. Schuster, and J. Sutton, 258–312. London: Routledge.
Vollgraff, J.A. 1913. Pierre de la Ramée (1515–1572) et Willebrord Snel van Royen (1580–1626). *Janus* 18: 595–625.
Vollgraff, J.A. 1936. Snellius notes on the reflection and refraction of rays. *Osiris* 1: 718–725.

Appendix 2: Decoding Descartes' Vortex Celestial Mechanics in the Text of *Le Monde*

This appendix unfolds some of the grounds for the synthetic reading of the technical details of the vortex celestial mechanics in *Le Monde* offered in Chap. 10, particularly Sects. 10.2.3, 10.2.4, and 10.4. It reflects some of the process by which I arrived at the reading given in the text. Additionally, in the spirit of the 'charitable hermeneutics of *Le Monde*, discussed in Sect. 10.2.1 of that same chapter, it provides some of the justification for the concepts, terms and diagrams I have used in the reconstruction of the theory of vortex celestial mechanics. It is assumed that the reader has examined the synthetic interpretation offered in Chap. 10, before assaying this appendix.

Descartes' discussion in *Le Monde*, of the motion and placement of the planets and comets, is quite extensive, covering 13 pages of text in the Adam-Tannery edition. Moreover, for Descartes the explanation is remarkably repetitive, vague and back-handed. It will, therefore, require a good deal of explication and interpretation of the text to bring out the underlying pattern of explanation. In addition, we shall have to use some passages from the *Principia Philosophiae* (1644) to confirm or clarify parts of the interpretation. This procedure presents obvious pitfalls. One surely does not want to attribute some later theory of 1644 to 1633. My approach will be to limit such appeal to the later work to passages in which it is extremely likely that nothing new has been added or that the sense has not been altered. It should be recognized, however, that a clarification of meaning is probably not strictly distinguishable for a change of meaning, and that any anachronistic appeal to later work is open to challenge. The value and validity of this procedure will have to rest with one's judgment of the overall interpretation which emerges.

Descartes first notes that planets made up of the third element must eventually move with the same agitation (*de même branle*) as the matter of the heavens surrounding them in which they float.

> For, if at first they were moving more quickly than that matter, then, not having been able to avoid pushing it upon colliding with it in their path; in a short time they had to transfer to it a part of their agitation. And if, on the contrary, they had in themselves no inclination to move, nevertheless, being surrounded on all sides by that matter of the heaven, they neces-

J. Schuster, *Descartes-Agonistes: Physico-mathematics, Method & Corpuscular-Mechanism 1618-33*, Studies in History and Philosophy of Science 27,
DOI 10.1007/978-94-007-4746-3,

> sarily had to follow its course, just as we see all the time that boats and diverse other bodies floating on water (both the largest and most massive and those that are less so) follow the course of the water they are in when there is nothing else to impede them from doing so.[1]

The sense of *branle* here seems to be speed of motion; for if the planet initially moved faster than the surrounding *boules*, it would be retarded; whereas, if it moved more slowly, it would gradually acquire speed. This entails that the force of motion of the planet and its surrounding *boules*, and consequently their respective centrifugal tendencies to motion, would be functions of their respective quantities of matter. Nevertheless, later in the discussion, Descartes will contend that although planets have the same agitation as the surrounding medium, they need not move as quickly as it does.[2] Thus, in the latter case, agitation conveys more the connotation of total force of motion. All that can initially be gathered from the two passages is that Descartes' terms have a conceptual looseness which seems to preclude a unique mechanical interpretation.

Some of the conceptual fuzziness is explained by the fact that, for the moment, Descartes is interested in pursuing a somewhat different point by exploiting his analogy of bodies floating in a river. As the subsequent passages make clear, the real thrust of his analogy is to establish that bodies pushed along in a current can be classified as belonging to one of two types: those so 'massive' and 'solid' that the centrifugal tendency arising from their inertial force of motion will induce a real centrifugal translation; and those 'less solid and composed of less massive parts',[3] which will not have sufficient centrifugal tendency to translate across the direction of flow;

> And note that, among the diverse bodies that thus float on water, those that are rather solid (*assez dur*) and rather massive (as, ordinarily, boats are, principally the largest and most heavily laden boats) always have much more force than the water to continue their motion, even though it is from the water alone that they have received their motion. By contrast, those floating bodies that are very light like those lumps of white scum that one sees floating long the shores during storms, have less force to continue moving. Thus, if you imagine two rivers that join with one another at some point and then separate again thereafter before their waters...have a chance to mix, then boats or other rather massive and heavy bodies that are borne by the course of the one river will be easily able to pass into the other river, while the lightest bodies will turn away from it and will be thrown back by the force of the water toward the places where it is least rapid.[4]

---

[1] AT.XI. 57-58; SG 37-38; MSM 93-5.

[2] AT.XI. 68-9; SG 44; MSM 117.

[3] 3 AT.XI. 60; SG 39; MSM 97-99. '...*selon que chacune est plus ou moins solide, et composeé de parties plus ou moins grosse et massives.*' Recall that in Sect. 4.2, where Descartes' 'dynamics' was first discussed, we defined, in accordance with Descartes' laws of motion in *Le Monde*, the 'principal determination' of a body in motion or tending to motion as, *the directional quantity of force of motion directed along the tangent to the path of motion at a given instant.* Here we are using the term 'inertial force of motion' to denote the scalar quantity of the force of motion involved in the 'principal determination' at any instant of the motion—that is, the sheer amount of force of motion in play, which, in fact, is directed along the tangent to the trajectory at that point.

[4] 4 AT.XI. 58; SG 38; MSM 95. Mahoney renders 'assez durs' as 'rather solid'; Gaukroger translates it as 'rather big'. Larousse defines *dur* as *'ferme, solide, difficile à entamer...'*

Descartes can draw an analogy to the behavior of bodies floating in a celestial vortex. Comets are identified with the more solid and massive bodies, planets with those less so:

> By this example, it is easy to understand that, wherever the parts of matter that could not take the form of the second or of the first element may have been at the beginning, all the larger and more massive among them shortly had to take their course toward the outside circumference of the heavens that contained them and thereafter pass continually from one of these heavens into another without ever stopping for a very long period of time in the same heaven. By contrast, all the less massive had to be pushed, each toward the center of the heaven containing it, by the course of the matter of heaven. And (given the shapes that I have attributed to them) upon colliding with one another, they had to join together severally and compose large balls which, turning in the heavens, have there a motion tempered by all the motions their separate parts could have if they were in fact separate. Thus some tend to move toward the circumferences of those heavens, and others toward their centers.
>
> Know also that we should take those that thus tend to range toward the center of any heaven to be the planets, and we should take those that pass across different heavens to be comets.[5]

All this may very well be a suggestive, if mechanically vague, analogy; but it must be admitted that, to this point, Descartes has raised more questions than he has answered. The problem of whether a planet moves at the same speed as the medium still remains. In addition, Descartes has introduced the terms 'solid' and 'massive' without explaining what they mean, and whether, for example, they are conjointly reducible to density.[6] Finally, the analogy to rivers does not even begin to explain how it is that planets assume uniquely determined orbital distances from the center of a vortex. We must, therefore, pursue Descartes' exposition further. However, we can anticipate our conclusions for the sake of clarity: the problem of planetary placement will be resolved by a clearer understanding of the nature and role of 'solidity' and 'massiveness'; furthermore, as a result, the problem about planetary speed will lose its focal importance, although it will not be entirely resolved.

The next stage of Descartes' argument apparently represents a serious attempt to give a mechanical explication of his analogy. He advances a kind of *reductio ad absurdum* intended to show why a planet cannot but have the same 'force to continue in a straight line' as the second matter surrounding it. Considering first a planet ♄ (Saturn) following an orbit at radius K in Fig. A.2.1, he writes,

> But, in order to make you understand distinctly in what places the planets should stop, look for example at the one marked ♄, which I suppose to follow the course of the matter of the heaven toward the circle K, and consider that, if this planet had the slightest bit more force to continue its motion in a straight line than do the parts of the second element surrounding it, then, instead of always following that circle K, it would go toward Y and thus it would be more distant than it is from center S. Then, in as much as the parts of the second element that would surround it at Y move faster and even are a bit smaller (or at least are not larger) then those at K, they would give it still more force to pass beyond toward F, so that it would

[5] AT.XI. 60-1; SG 39-40; MSM 99-101.

[6] As we have seen, Descartes used '*solide*' at AT XI p.60; and '*dur*' at AT. XI. 58. It will be important to note where the notion of solidity appears, and does not appear, in the remainder of his presentation in *Le Monde*.

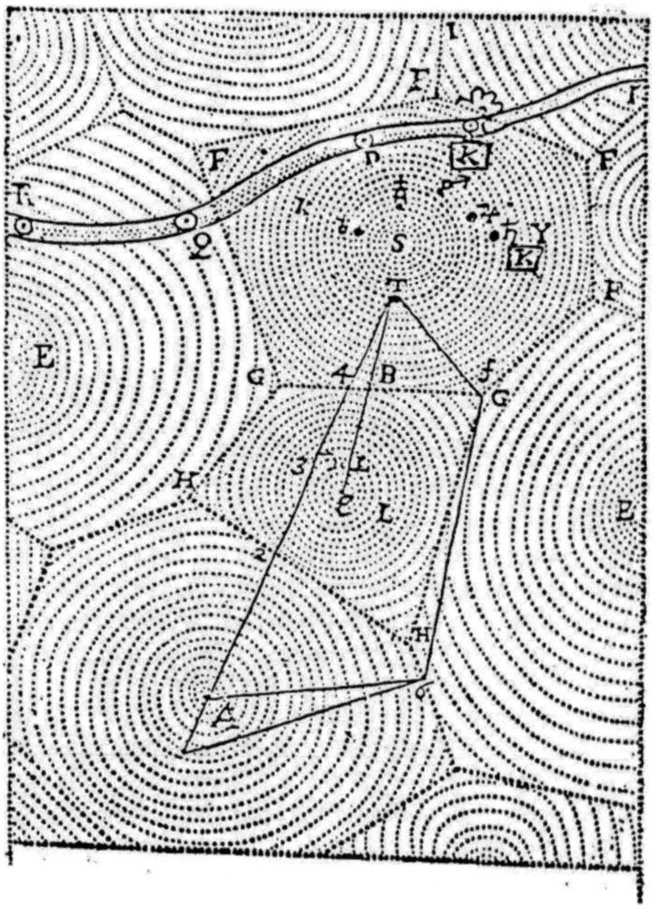

Fig. A.2.1 The vortex cosmos, Descartes, *Le Monde,* AT XI p.55

> go out to the circumference of that heaven, without being able to stop anywhere in between; then from there it would easily pass into another heaven, and thus, instead of being a planet, would become a comet.[7]

Thus, no planet can have more centrifugal tendency arising from its quantity of force of motion than does the second matter of the K layer; for, if it once started translating to the region beyond K, where the *boules* become swifter and smaller, more and more force of motion would be conveyed to it until it drifted right out of the vortex. Presumably, after penetrating the neighboring vortex to some distance, it would translate out again in the same manner, deriving increasing quantities of motion from each successive layer it passed. What is missing in this first branch of the reductio argument is an explicit statement about why such continuous centrifugal translation is caused beyond the K layer, but not below it.

[7] AT.XI. 64; SG 41-42,; MSM 109-111.

A similar lacuna appears in the second side of the reductio argument. Descartes takes up the case of the planet ♄ when it has 'less force than the parts of the second element surrounding it'. Here it happens that,

> those parts that follow it and that are placed a bit lower than it can divert it with the result that, instead of following circle K, it descends toward the planet marked 4 (Jupiter). The planet ♄ being there, it can happen that it is exactly as strong (*justement aussi forte*) as the parts of the second element that will then surround it. The reason for this is that, these parts of the second element being more agitated than those at K, they will also agitate the planet more; being in addition smaller, they will not be able to resist it as much. In this case, the planet will remain perfectly balanc ed in the middle of them and will there take its course in the same direction as they about the sun, without being at one time or another more or less distant from the sun, except insofar as they can also be more or less distant from it.[8]

In translating toward the sun, the plant meets *boules* which are increasingly more agitated and smaller than those farther out. By gaining more agitation, the planet will eventually attain sufficient centrifugal tendency to maintain a fixed orbit. Here we are helped by Fig. A.2.2 which we developed in Chap. 10 (where it was Fig. 10.4) to show the size, speed and hence force of motion distribution of the *boules* of the vortex. A glance at Fig. A.2.2 shows that this case is somewhat similar to the previous case of centrifugal translation, except for the fact that here the size of the *boules* must decrease in a greater proportion than their agitation increases, because their force of motion must decrease as one approaches the center.

The account is still rather vague. The planet drifts down to a level where the *boules* possess less force of motion but greater agitation than at the K-layer. The planet acquires some additional agitation from these lower layers of *boules* and eventually settles into a position, lower than its original one, in which a 'balance' has been struck. But in what does that balance consist? The passage cited implies that there is a balance between the agitation conveyed to the planet by the *boules* and the resistance they make to the motion of the planet. The agitation derived from the *boules* is so adjusted to the resistance to motion they offer that the planet now moves with exactly the same 'strength' (force, agitation, speed?) as the surrounding layer. This may be the case, but Descartes' very next remark only further clouds the issue, by mentioning an equality between the force of motion of the planet and that of the surrounding *boules*:

> But if this planet ♄ being at 4 still has less force to continue its motion in a straight line than has the matter of the heaven found there, it will again be pushed lower by the matter, toward the planet marked ♂ (Mars) and so on, until finally it is surrounded by a matter that has neither more nor less force than it.[9]

Thus, there appears to be a problem in reconciling the two interpretations of 'balance': The one between the planet's agitation and the medium's resistance; the other between their respective forces of motion. Furthermore, it is in this connection that we can note how this part of the reductio suffers from a lacuna analogous to the

[8] Ibid. 65.

[9] AT.XI.65-66; SG 42; MSM 113.

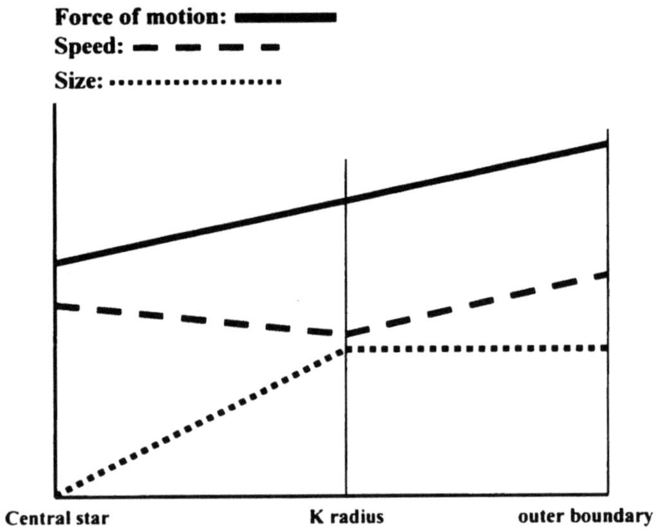

Fig. A.2.2 Size, speed and force of motion distribution of particles of second element, in a solar vortex

one in the first part; that is, 'Why is it that no continuous fall occurs here whilst in the first case continuous centrifugal rise occurs'?

So, let us ask in a pointed manner: Why is it the case that as soon as a planet well below K finds a layer in which its force of motion equals that of the surrounding matter, it does not start to translate outward again? After all, we know from Descartes' own discussion of the laws of nature that a centrifugal tendency arises from the constraint of what we called the 'principal' tendency to motion along a curved path.[10] Why does not that centrifugal tendency lead to a centrifugal translation out to layers of increasing force of motion, so that, as in the first stage of the argument from the K layer outward, the planet would continue right out of the vortex? In other words, the problem is why Descartes takes seriously the continuous centrifugal translation of planets once they pass the K layer, while he ignores this possibility for planets whose force is less than that of the *boules* of the K layer? For the region below K, all he does is invoke the notion of a balance of agitation and resistance, associated with the assertion that the planet will remain in a level of equal force. In summary, the reductio argument seems terribly inadequate; on the one hand, it suffers from a serious conceptual lacuna, while, on the other, it introduces a pair of unreconciled conditions for the placement of planets.

There is, however, a way out of this morass. Unfortunately it requires a brief mention of still one more peculiarity of the reductio argument. The discerning reader of *Le Monde* will probably notice that the reductio argument entirely omits any mention of the terms 'solidity' and 'massiveness'. This is odd, for we saw that

[10] See above Note 3.

in some way the analogy to bodies floating in a river involved differential placement according to their relative solidities or massivenesses. The key to Descartes' celestial mechanics lies in the correction of this omission. If we have been rather backhanded in arriving at it, that is because Descartes himself only falls into the solution in the passage following the reductio, and it is revealing to capture some of the confusion and disorder of the text of *Le Monde*.

As if taking note of the inadequacy of the reductio argument, Descartes writes,

> But, if I still have not made you understand well enough why it can happen that the parts of the heaven beyond the circle K, being incomparably smaller than the planets, do not cease to have more force than they to continue their motion in a straight line, consider that this force does not depend solely on the quantity of the matter that is in each body, but also on the extent of its surface. For even though when two bodies move equally fast it is correct to say that, if one contains twice as much matter as the other, it also has twice the agitation, that is not to say thereby that it has twice as much force to continue to move in a straight line; rather, it will have exactly twice as much if, in addition, its surface is exactly twice as extended, because it will always meet twice as many other bodies resisting it, and it will have much less force to continue if its surface is extended much more than twice.[11]

Thus, in considering the force of motion of bodies, or the resistance they encounter, one must take into account the role of the ratio of surface area to volume. In the *Principia* Descartes will explicitly call this quantity the 'solidity' of a body, a complex magnitude arising from the summed contributions of the volumes (quantities of matter) and surface areas of its constituent particles:

> What I understand here by the solidity of this star, is the quantity of the matter of the third element...in proportion to its volume and surface area.

and,

> That solidity does not depend on matter alone, but also on size and shape...[12]

Since shape contributes to surface area, we will subsequently express solidity as a ratio of volume to surface area (v/s), where solidity (and force of motion) increases as (v/s).

Given this passage from *Le Monde* and the definitions of solidity from the *Principia*, we can now clarify Descartes' entire line of argument. Let us recall the size and speed distribution of second element in the vortex (Fig. A.2.2). The resistance to being put in motion of the *boules* is a function of their solidity, their ratio of

[11] AT.XI. 66-7; SG 43; MSM 113-115.

[12] *Principia Philosophiae* part III. para. CXXI, 'Per soliditatem hic intelligo, quantitatem materiae tertii elementi....cum eius mole et superificie comparatam.' And CXXII. 'Soliditatem non a sola materia, sed etiam a magnitudine ac figure pendere.' Miller and Miller pp.151–153 (The reader will also be interested in the claims made by Miller and Miller in their notes 120 and 121 thereto.) Descartes mentions a 'star' in the first passage because he is, here, in the process of describing how a dead star, its surface encrusted with third matter, becomes a planet and assumes an orbit in the vortex of another still viable star. A.J. Aiton, who wrote extensively on this and other aspects of Cartesian science, curiously came to the contrary judgment that Descartes conceived solidity to depend on the 'proportion of third matter contained in the body' See Aiton (1957) 261, Note 42, where he cites these very passages from the *Principia*.

surface to volume. The larger the spherical particle the larger (v/s) becomes and the more resistance it offers to being moved aside. Furthermore, as developed in Chap. 10, we may think of the surface layer of *boules* completely surrounding the parts of a planet as a kind of surface body or envelope (Fig. 10.5 above) whose total quantity of matter is proportional to the (v/s) of the *boules* making it up. That is, as long as the *boules* do not grow so large that just a few of them can cover the surfaces of a part of the planet, they will form an envelope around all the parts of the planet of greater or lesser total quantity of matter, depending on their volumes. We can now take Descartes' earlier statement about a balance between force of motion and resistance to motion to mean that planets settle into orbits where their centrifugal tendency to motion is just counter-balanced by the resistance to being put in motion of the surrounding surface layer of *boules*. A planet in the K layer with sufficient centrifugal tendency to start to move away from the center will encounter surface layers made up of progressively smaller *boules*. The resistance to centrifugal tendency offered by these layers will, therefore, progressively decrease, never being as great as it was at K, where the *boules* are largest. Thus, no resistance arising from surface envelopes will be able to prevent the planet from translating right out of the vortex. Hence, in Chap. 10 we were able to depict the resistance of the *boules* to being set in motion in Fig. 10.6 which plots (v/s) with distance. It is as if there exists a hump in the resistance curve at K. If a planet can surmount that hump, it will move as a comet between the levels Ka and Kb of these neighboring vortices.

We can now supply the rationale for the placement of the planets. A planet does not translate toward and beyond layer K because it is not sufficiently solid. As it moves out from near the sun toward K, it meets surface layers of *boules* of increasing resistance to being set in motion. A planet will be located at that layer where the *boules* (and hence the surface envelope) are of such magnitude that they just counteract the centrifugal tendency to motion generated by the quantity of matter, speed and (s/v) of the planet. Since planets themselves vary in their overall solidity they will be located at different distances from the sun. Descartes clearly states in the *Principia* that:

> Thus, when we now see the principal Planets, Mercury, Venus, the Earth, Mars, Jupiter and Saturn being transported around the Sun at different distances, we shall judge that this occurs because {they are not all equally solid, and that} those which are closer to the Sun are less solid than those further away. And we have no reason to think it strange that Mars, although smaller than the Earth, is further from the sun, because size is not the only factor which determines the solidity of bodies, so that Mars, {though smaller}, can be more solid than the Earth.[13]

In effect, contrary to the rhetorical thrust of Descartes' initial discussion in *Le Monde*, planets translate out until they meet countervailing resistances to their

[13] Miller and Miller 172, material in brackets appears first in the French edition of 1644. The original Latin passage at *Principia* Part III para. CXLVII reads as follows: 'Sicque iam videntes primarios Planetos, Mercurium, Venerem, Terram, Martem Novem et Saturnum, ad diversas distantias circa Solem deferri, judicabimus id ex eo contingere, quod eorum qui Soli viciniores sunt, soliditas sit minor quam remotiorum; Nec mirabimur Martem terra minorem, ipsa tamen magis a Sole distare, quia solidior nihilominus esse potest; cum soliditas a sola magnitudine non peneat'.

tendency to centrifugal motion. They are locked into orbits by a balance of their own centrifugal tendency and the resistance of the surface envelope to giving way to that tendency.

This interpretation can be confirmed by comparing the reductio argument of *Le Monde* with the one Descartes provides in the *Principia.* The latter argument is really more properly called a reductio, because it attempts to show that the planet will always return to its place, if it is posited to be out of position. It is presumed that the planet has already reached a layer below K in which the *boules* possess sufficient solidity to resist any further centrifugal translation.

> For if it descended closer to the Sun, it would there find itself surrounded by slightly smaller heavenly globules which it would exceed in force to recede from the center around which it revolves. These parts would also be more rapidly moved, which thus would increase its own agitation along with its force, causing it to ascend. If, on the other hand, it receded further from the Sun, it would encounter there heavenly globules which were somewhat less rapidly moved and would thus decrease its agitation, and which were slightly larger and would thus have the force to drive it back toward the Sun.[14]

By comparison, note that in *Le Monde* Descartes does not consider the second branch of this argument—the ascent of the planet from the sun and its return to its original place. In addition, the first branch of this argument also differs from that in *Le Monde*, for, in this case, the planet returns to its original level, not to some level below its starting point. Third and most importantly, in the *Principia* Descartes does not say that the planet descends or ascends because it has more or less force of motion than the surrounding medium, as he did in *Le Monde*. Rather, the argument proceeds by asking what follows, if we assume that a planet of given (v/s) (and therefore of determinate orbital distance) is not in its proper orbit, but higher or lower. How or why it ascended or descended is not important. The entire problem is to show that it must of necessity return to its proper place. If it is lower than it should be, it will move up, because it will circulate with *boules* which lack sufficient solidity to resist the centrifugal tendency which the planet acquires in moving. The planet will drift up, propelled by the centrifugal tendency, until it is locked in at its appropriate level; that is, the level which can offer countervailing resistance to its centrifugal drift. If the planet is higher than it should be, it will be slowed by impact with ever slower-moving *boules*, and, having less solidity than they, will be displaced or extruded downward by the centrifugal tendency of the *boules* lying just below it. Eventually it will reach its proper level, where the underlying boules will not be solid enough to extrude it, and where its own centrifugal tendency is just resisted by the surface envelope. A balance is achieved, on the one hand, between the centrifugal tendency of the planet and the resistance of the surface layer, and, on the other,

---

[14] *Principia* pt. III para. CXL; Miller and Miller p.169. 'Quippe si proprius accederet versus Solem, ibi versaretur inter globulos coelestes paulo minores, ac proinde quos superaret vi ad recedendum a centro circa quod gyrat; et celerius motos, ac proinde a quibus ista eius vis simul cum agitatione augeretur, sicque inde rursus regredi deberet. Si vero a Sole magis recederet ei occurrerent globuli coelestes aliquanto minus celeriter moti, ac proinde qui eius agitationem minuerent; et paulo majores, ac proinde qui vim haberent, ipsum versus Solem repellendi.'

between the centrifugal tendency of the immediately subjacent *boules* and the resistance to downward extrusion offered by the body's solidity.[15] This, by the way, clarifies the two kinds of balance discussed in *Le Monde*. 'Resistance' is determined by (v/s) and has to do with the locking mechanism provided by the surface envelopes. Consideration of 'force of motion' refers to the centrifugal tendency of the subjacent *boules* which can extrude a planet from an orbit of too large a radius; that is, an orbit in which the great solidity of the *boules* allows them to generate greater centrifugal tendency than the misplaced planet.

In general, the reductio argument of *Le Monde* is marred by Descartes' insufficient distinction between these two orders of consideration. In *Le Monde*, the planet descends *because it has less force of motion than the* boules *of its original layer.* Solidity does not enter the reductio argument.[16] In the *Principia,* the planet is posited lower than it should be (it has hypothetically descended) because for some reason it has *less force of motion than its solidity potentially allows it to assume.* Because it is more solid than the *boules* in its new surroundings, it will translate up to its proper level as determined by its solidity. There it is locked into an orbit, and, as a matter of fact, it will circulate with a certain force of motion. Thus it is clear that solidity is determinative of orbital distance, whereas, in fact, the force of motion is peripheral to the argument.[17] We do not have to specify what we mean by 'the same agitation' or 'the same force of motion' as the surrounding heaven; we need only grant that circulating in a heaven entails speed, force of motion and centrifugal tendency in the body. Wherever a planet is located, it will derive force of motion from the medium in which it floats. The real problem is whether the surrounding medium is sufficiently 'solid' to resist the centrifugal tendency to motion, which must be generated in the planet by the mere act of circulating.[18]

---

[15] This is the source of the 'formula' for orbital equilibrium, the 'locking' of a planet into its orbital distance, given in Sect. 10.2.3.

[16] In *Le Monde*, the role of solidity does become slightly more clear later, when Descartes continues his discussion beyond the reductio argument, explicating his points about the distribution of surface to volume ratios amongst the *boules* at various distances from the central star, and extending his analogical remarks about types of bundles of third matter flowing in rivers. It is here that he comes closest to the interpretation we have developed. Continuing three paragraphs beyond the material cited above at Note 11, he writes, 'Whence you see how diverse planets can be suspended within circle K at diverse distances from the sun, and how it is not simply those that outwardly appear the largest, but those that are the most solid and massive in their interior, that should be the most distant.' [AT.X. 68; SG 44; MSM 117].

[17] As noted just above, Descartes finally achieves a clear statement of at least this point (solidity is determinative of orbital distance) somewhat late in his discussion, following the confused and confusing reductio passages we have been exploring.

[18] Recall Descartes' analogy of ships and flotsam in two confluent rivers to the behavior of comets and planets respectively. We can now see that when a planet gets too far out in the vortex for its solidity, it becomes 'flotsam' and is pushed down toward the central star. But, when a planet sinks closer to the star than its solidity, in principle, warrants, it becomes a 'ship' and drifts out. Since the planet's solidity is fixed, but the vortex corpuscles continuously vary in solidity and force of motion, as per the diagrams deployed above and in Chap. 10, the vortex is revealed as a locking and extruding mechanism.

In summarizing this tortuous and already overlong textual excursion, we can come to the following conclusions already embodied in our findings in Chap. 10. The Cartesian vortex is a kind of dual locking and extruding machine. There is a locking mechanism built into the (v/s) distribution of the *boules*. Planets of a solidity below a certain threshold value will be locked into the heavens below K at distances dependent on their solidities. Beyond K, the (v/s) distribution is such that planets cannot be locked in at any distance. Very solid objects which translate beyond K will become comets, which on Descartes' view oscillate between vortices above the K levels (as they orbit the centre of whichever vortex they presently occupy). All layers of the vortex are capable of extruding sun-ward a planet with too little centrifugal tendency. This is due to the fact that the force of motion of the *boules* constantly increases with distance from the sun. A planet of insufficient solidity, and consequently, too little centrifugal tendency will be displaced downward by the ascent of *boules* with greater centrifugal tendency.

References

Works of Descartes and Their Abbreviations

AT =*Oeuvres de Descartes* (revised edition, 12 vols.), edited by C. Adam and P. Tannery (Paris, 1964–76). References are by volume number (in roman) and page number (in Arabic).

SG = *The World and Other Writings*, edited and translated by Stephen Gaukroger (Cambridge, 1998).

MM = *René Descartes, The Principles of Philosophy*, translated by V. R. Miller and R. P. Miller (Dordrecht, 1991).

MSM = *Rene Descartes, Le Monde, ou Traité de la lumière*, translated by Michael S. Mahoney (New York, 1979).

CSM(K) = *The Philosophical Writings Of Descartes*, 3 vols., translated by John Cottingham, Robert Stoothoff, and Dugald Murdoch, and (for vol. 3) Anthony Kenny, (Cambridge, 1988) References are by volume number (in roman) and page number (in arabic).

HR = The Philosophical Works of Descartes, vol I translated by E.S. Haldane and G.R.T. Ross (Cambridge, 1968 [1st ed. 1911]).

Other

Aiton, E.J. 1957. The vortex theory of planetary motions 1. *Annals of Science* 13: 249–264.

MIX
Papier aus verantwortungsvollen Quellen
Paper from responsible sources
FSC® C105338

If you have any concerns about our products,
you can contact us on
ProductSafety@springernature.com

In case Publisher is established outside the EU,
the EU authorized representative is:
Springer Nature Customer Service Center GmbH
Europaplatz 3, 69115 Heidelberg, Germany

Printed by Libri Plureos GmbH
in Hamburg, Germany